Animal Homosexuality

Homosexuality is an evolutionary paradox in search of a resolution, not a medical condition in search of a cure. Homosexual behaviour is common among social animals, and is mainly expressed within the context of a bisexual sexual orientation. Exclusive homosexuality is less common, but not unique to humans. The author invites the reader to embark on a journey through the evolutionary, biological, psychological and sociological aspects of homosexuality, seeking an understanding of both the proximate and evolutionary causes of homosexual behaviour and orientation in humans, other mammals and in birds. The book also provides a synthesis of what we know about homosexuality into a biosocial model that links recent advances in reproductive skew theory and various selection mechanisms to produce a comprehensive framework that will be useful for anyone teaching or planning future research in this field.

DAMAGED

Animal Homosexuality

A Biosocial Perspective

Aldo Poiani
Monash University, Victoria

with a chapter by Alan Dixson

CAMBRIDGE UNIVERSITY PRESS
Cambridge, New York, Melbourne, Madrid, Cape Town, Singapore,
São Paulo, Delhi, Dubai, Tokyo, Mexico City

Cambridge University Press
The Edinburgh Building, Cambridge CB2 8RU, UK

Published in the United States of America by Cambridge University Press, New York

www.cambridge.org
Information on this title: www.cambridge.org/9780521196758

© A. Poiani 2010

This publication is in copyright. Subject to statutory exception
and to the provisions of relevant collective licensing agreements,
no reproduction of any part may take place without the written
permission of Cambridge University Press.

First published 2010

Printed in the United Kingdom at the University Press, Cambridge

A catalog record for this publication is available from the British Library

Library of Congress Cataloging-in-Publication Data

Poiani, Aldo.
Animal homosexuality : a biosocial perspective / Aldo Poiani ; with a
chapter on primates by Alan Dixson.
    p. cm.
Includes bibliographical references and index.
ISBN 978-0-521-19675-8 (hardback)
1. Homosexuality in animals. I. Dixson, A. F. II. Title.

QL761.P65 2010
591.56′2–dc22   2010019223

ISBN 978-0-521-19675-8 Hardback
ISBN 978-0-521-14514-5 Paperback

Cambridge University Press has no responsibility for the persistence
or accuracy of URLs for external or third-party internet websites referred
to in this publication, and does not guarantee that any content
on such websites is, or will remain, accurate or appropriate.

For Marisa, with love

Contents

Contributors	*page* xi
Acknowledgements	xii
Preface	xv

1 Animal homosexuality in evolutionary perspective — 1
  Historical overview of studies on homosexual behaviour — 3
    A list of hypotheses — 28

2 The comparative study of homosexual behaviour — 33
  Epistemological caveats — 33
  Definitions — 35
  Sources of the datasets and potential biases — 38
  Comparative analyses — 39
    Phylogenies — 41
  Path analysis — 44
  Meta-analytical methods — 45
  Is same-sex mounting affected by conditions of captivity? — 46

3 Genetics of homosexuality — 55
  Some genetic models of homosexuality — 58
    Recurrent mutation as a genetic mechanism underlying homosexuality — 64
  Candidate genes for homosexual behaviour — 65
    The *fruitless (fru)* gene in *Drosophila* — 66
    The *genderblind* and *sphinx* genes in *Drosophila* — 67

Testicular feminisation mutation
(*tfm*) in mice and rats ... 68
Apoptosis and the *Bax* gene ... 69
The *c-fos* proto-oncogene ... 70
The *trp2* and olfactory receptors
in mice ... 71
Genes studied in humans ... 71
Genes on the X-chromosome and
homosexuality ... 72
Genes on the Y-chromosome and
homosexuality ... 74
Other possible genetic markers of
homosexuality ... 76
    Sex-biased mutation rates ... 77
    A study across the human genome ... 78
DNA methylation and
X-chromosome inactivation ... 79
    DNA methylation and genomic
    imprinting ... 80
Human sex chromosome
aneuploidies and homosexuality ... 81
Family analysis ... 82
Twin studies ... 85
Evidence from selection studies ... 89
    Does homosexuality have any
    link with increased female
    fecundity? ... 91
    Kin selection and homosexuality ... 91
Summary of main conclusions ... 95

4 Ontogenetic processes ... 97
Hormonal effects during early
development ... 100
    The 2nd:4th digit ratio ... 106
    Handedness ... 111
    Dermatoglyphics ... 112
    Otoacoustic emissions and auditory
    evoked potentials ... 113
    Tooth crown size and waist-to-hip
    ratio ... 113
    Prenatal exposure to steroid
    hormones and sex-typed toy
    preferences ... 114
Birth order and family size effects ... 116
    Intrauterine positioning ... 117

Stress effects ... 118
    Stress effects and human
    homosexuality ... 124
    Adrenarche and the development of
    homosexuality ... 128
    Fluctuating asymmetry ... 130
Developmental effects of endocrine-
disrupting environmental chemicals ... 131
    Diethylstilboestrol (DES) ... 132
Early social experience and learning ... 134
Juvenile sex play and development of
sexual orientation ... 134
    Juvenile sex play and homosexuality
    in humans ... 138
    Do 'sissy' boys and 'tomboy'
    girls become adult homosexuals? ... 140
What is the role of learning in the
early ontogeny of human
homosexual orientation? ... 145
    Fetishism and homosexuality ... 147
    Contributions of social learning ... 148
Other-sex mimicry: an alternative
route to animal homosexuality? ... 149
Evolution and ontogeny: is
homosexuality a neotenic trait? ... 152
Summary of main conclusions ... 157

**5 The endocrine and nervous systems: a
network of causality for homosexual
behaviour** ... 159
Human endocrine disorders and
homosexuality ... 159
    Altered testosterone biosynthesis ... 160
    5α-Reductase deficiency ... 162
    Congenital aromatase deficiency ... 162
    Complete androgen insensitivity
    syndrome ... 163
    Congenital adrenal hyperplasia ... 164
Circulating hormones in the
adult and homosexual behaviour ... 166
    Hormones and homosexual
    behaviour in primates ... 166
    Hormones and homosexual
    behaviour in other mammals and in
    birds ... 170

Non-steroid hormones	171
Sex hormones and the *Lock-in Model* for homosexual behaviour in macaques	172
Bos taurus: hormones, sexual behaviour and the buller steer syndrome	178
Ovis aries: the case of homosexual rams	179
Crocuta crocuta: selection of masculinised females	185
Circulating hormones and sexual orientation in adult humans	191
Does the use of anabolic androgenic steroids by adults affect sexual orientation?	191
Does sexual orientation in men and women correlate with levels of circulating hormones?	193
Do orchiectomy and/or oophorectomy lead to a homosexual orientation?	195
Effects of endocrine-disrupting chemicals on the expression of sexual behaviour in adult mammals and birds	198
The neuroendocrine system and homosexual behaviour	199
Sexual behaviour and brain plasticity	201
Sexual dimorphisms in the adult brain	206
Homosexual behaviour and the central nervous system	206
The hypothalamus and homosexuality	208
Beyond the hypothalamus	213
Central nervous system, hypersexuality and homosexual behaviour	220
Orgasm, sexual arousal and homosexuality	224
Masturbation and homosexuality	232
Learning processes in adults and their role in homosexual behaviour	236

	Femininity, masculinity and homosexuality	239
	Behavioural plasticity and sexual orientation	243
	Brain, emotions and homosexuality	246
	Aggressiveness, empathy and homosexuality	248
	Androgyny and emotional coping	249
	Homosexuality and cognitive processes	250
	Influences of sexual selection on homosexuality	255
	Brain, complex social behaviour and homosexuality	259
	Summary of main conclusions	263
6	**Immunology and homosexuality**	**265**
	Immunity and Fraternal Birth Order effect in the development of a homosexual orientation in males	266
	Fraternal and Sororal Birth Order effect in the development of a homosexual orientation in females	270
	Parental age effects	271
	Can immunology explain the development of homosexuality in males and/or females?	272
	Immunity and the parental age effect	274
	Non-immunological explanations for birth order and parental age effects	275
	Family dynamics	276
	Summary of main conclusions	281
7	**Sexual segregation effects**	**283**
	List of hypotheses proposed to explain sexual segregation	298
	A review of sexual segregation studies	301
	Caprinae	302
	Other bovids	304
	Cervids	305
	Giraffids and equids	307
	Bears, polecats, marsupials and rodents	308

Bats, elephants and pinnipeds	309
Cetaceans	310
Birds	311
Sexual segregation in primates	312
Evolutionary models for the potential link between sexual segregation and same-sex mounting	316
Birds	317
Mammals	319
Summary of main conclusions	321

8 The social, life-history and ecological theatres of animal homosexual behaviour — 323

Reproductive Skew Theory	323
A historical overview of *Reproductive Skew* Theory	324
Queuing to breed	328
Early-developmental effects	330
Mate choice and incest avoidance	330
Tests of *Reproductive Skew* Theory	331
The *Synthetic Reproductive Skew Model of Homosexuality*	331
A comparative test of the *Synthetic Reproductive Skew Model of Homosexuality*	336
Birds and the *Synthetic Reproductive Skew Model of Homosexuality*	336
Is joint nesting an indication of homosexuality?	347
Mammals and the *Synthetic Reproductive Skew Model of Homosexuality*	348
Path analyses	361
Full correlation matrix	363
Direct comparisons between birds and mammals	367
Homosexuality in humans and interest in having children	373
Same-sex sexual behaviour and STD transmission	375
Summary of main conclusions	379

9 Homosexual behaviour in primates — 381
Alan Dixson

Primate diversity and evolution	381
Primate homosexual patterns: phylogenetic considerations	384
Functions and derivations of homosexual behaviour	388
Hedonic aspects of sexual behaviour	392
Human homosexual behaviour and homosexuality	395
Summary of main conclusions	399

10 *A Biosocial Model* for the evolution and maintenance of homosexual behaviour in birds and mammals — 401

The long and winding road to the *Biosocial Model of Homosexuality*	403
The *Synthetic Reproductive Skew Model of Homosexuality*	411
The *Biosocial Model of Homosexuality*	413
Bisexuality vs. exclusive homosexuality	416
Homosexuality in men and women	417
What does not seem to work	419
Some interesting areas for future research	420
Is homosexuality a pathology?	421
One last thought	424

Appendix 1: Glossary	427
Appendix 2: Predictions of the *Synthetic Reproductive Skew Model of Homosexuality* and results obtained in the comparative tests of the model carried out in birds and mammals	435
Appendix 3: Comments on further results of comparative analyses of independent contrasts reported in the full correlation matrices of birds and mammals	441
References	443
Index	535
Colour plates can be found between pages 240 and 241	

Contributors

Aldo Poiani
School of Biological Sciences
Monash University
Victoria 3800
Australia

Aldo Poiani is an evolutionary ecologist at Monash University, Australia. He has studied social behaviour, sexual behaviour and parasitism in birds, having published widely in the fields of avian cooperative breeding and host–parasite interactions, including aspects related to behaviour, endocrinology and comparative biology. He has carried out field research in Australia, Europe and South America and is editor of the book *Floods in an Arid Continent*.

Alan Dixson (Chapter 9)
School of Biological Sciences
Victoria University of Wellington
Wellington 6140
New Zealand

Alan Dixson is an internationally renowned primatologist at Victoria University of Wellington, New Zealand. He has made substantial contributions to our understanding of primate reproductive biology and behaviour, having held research and academic positions in four continents. He is the author of the classic book *Primate Sexuality: Comparative Studies of the Prosimians, Monkeys, Apes and Man*. His most recent book is *Sexual Selection and the Origins of Human Mating Systems*, published in 2009.

Acknowledgements

First and foremost I thank Alan Dixson for accepting my invitation to contribute Chapter 9. He was also kind enough to find time, in between the writing of a couple of his own books, to provide constructive feedback on the other chapters. Alan's shrewd comments substantially improved the quality of the text and confronted my ideas with critical challenges that, I believe, finally strengthened the various arguments. Responsibility for what I wrote, however, falls only on my shoulders.

Many thanks to the following colleagues for also providing very valuable and constructive feedback: Prof. Ary Hoffmann (Chapter 3); Prof. Harold Gouzoules (neoteny section of Chapter 4), Dr Andrew Iwaniuk (Chapter 5), Prof. Charles Roselli and Assoc. Prof. Eric Vilain (*Ovis aries* section of Chapter 5); Prof. Steve Glickman for providing information about F–F mounting in *Crocuta crocuta* (Chapter 5). Dr Javier Pérez-Barbería (Chapter 7); Assoc. Prof. Mariella Herberstein (Chapter 8), Dr Sonia Altizer (for suggesting examples for the STD section of Chapter 8); Dr Joseph Soltis (for providing information about F–F mounting in *Macaca fuscata* on Yakushima Island, Chapter 8). A very special thank you to Prof. Douglas Futuyma for commenting on Chapters 1, 2 and 3.

I am especially indebted to Martin Griffiths of Cambridge University Press for believing in this project and to Rachel Eley, Dawn Preston and Lynn Davy for their invaluable assistance during the production process. The project also benefited from the insightful comments and encouragement of three anonymous referees, to whom I am most grateful.

Acknowledgements

I am thankful to Assoc. Prof. Ian Jamieson for providing the pukeko photograph for the cover that was also used in Figure 2.1, the photograph was taken by Nic Bishop. I thank Prof. Luděk Bartoš for Figure 2.5 of male–male mounting in *Dama dama*. Figure 3.2 was published in *Cell*, vol. **121**, Demir, E. and Dickson, B.J., *fruitless* Splicing specifies male courtship behaviour in *Drosophila*. pp. 785–94, Copyright Elsevier 2005. I am grateful to Elsevier for granting permission to reproduce this figure. Figure 3.3 is reproduced with permission from the National Academy of Sciences, USA from Forger, N.G., Rosen, G.J., Waters, E.M. *et al.* 2004. Deletion of *Bax* eliminates sex differences in the mouse forebrain. *Proceedings of the National Academy of Sciences, USA* **101**: 13666–71, copyright (2004) National Academy of Sciences, USA. Figure 3.4 is in the public domain. Figure 4.1 is reproduced from Gahr, M. 2003. Male japanese quails with female brains do not show male sexual behaviors. *Proceedings of the National Academy of Sciences, USA* **100**(13): 7959–64, copyright 2003 National Academy of Sciences, USA, I am grateful to the National Academy of Sciences, USA for granting such permission. Dr Richard Bowen kindly granted permission to reproduce Figure 5.1. Figure 5.3 is published with kind permission from Springer Science + Business Media: *Archives of Sexual Behavior*, Sexual reward via vulvar, perineal, and anal stimulation: A proximate mechanism for female homosexual mounting in Japanese macaques, vol. **35**, 2006, 523–32, Vasey, P. and Duckworth, N., Figure 2, photo by S. Kovacovsky. I am also grateful to Prof. Charles Roselli for Figure 5.4, Dr Christine Drea for Figure 5.5, Prof. Dick Swaab for providing Figure 5.10 and Ms Vanessa Woods for making available the bonobo G–G-rubbing photos used in Figures 2.2 and 5.2; the photos were taken in the *Lola ya* bonobo sanctuary, Democratic Republic of Congo. I am grateful to the Nature Publishing Group for granting permission to reproduce Figures 5.6 (Figure 1 of Fenstemaker *et al.* 1999), 5.9 (Figure 6 of Lösel and Wehling 2003) and 5.17 (Figure 1 of Thompson *et al.* 2001). Figure 5.11 is reproduced from Savic, I., Berglund, H. and Linström, P. 2005 Brain response to putative pheromones in homosexual men. *Proceedings of the National Academy of Sciences, USA* **102**: 7356–61, I am grateful to the National Academy of Sciences, USA for granting such permission. Figures 5.12 and 5.13 are published with kind permission of Springer Science + Business Media: full credit *Psychopharmacology*, The effect of hetero- and homosexual experience and long-term treatment with fluoxetine on homosexual behaviour in male rats, vol. **189**, 206, pp. 269–275, Habr-Alencar, S.F., Dias, R.G., Teodorov, E. Bernardi, M.M., Figures 3 and 4. Figure 5.14 is reprinted from *Brain Research*, Vol. **1024**, Komisaruk, B.R., Whipple, B., Crawford, A. *et al*. Brain activation during vaginocervical self-stimulation and orgasm in women with complete spinal cord injury: fMRI evidence of mediation by the vagus nerves, pp. 77–88, Copyright 2004, with permission from Elsevier. Figures 5.15 and 5.16 are, respectively, Figures 3 and 6 of Holstege *et al.* (2003) and are reproduced with permission from the Society for Neuroscience. Dr Kathreen Ruckstuhl provided the photos of *Capra ibex* used in Figure 7.1 A,B, the photos were taken by Dr Peter Neuhaus. Figures included in Chapter 9 have been reproduced courtesy of Oxford University Press (Figures 9.1, 9.4, 9.5, 9.8, 9.9, 9.10 and 9.11); Elmtree Books, Hamish Hamilton Ltd and Edinburgh University Press (Figure 9.2); Edinburgh University Press (Figure 9.3); Zoological Society of London (Figures 9.6 and 9.7); W.B. Saunders and Co. (Figure 9.12) and Harvard University Press (Figure 9.13).

I am also thankful to J. Michael Bailey, Albert Bandura, Michael J. Baum, Sandra L. Bem, Ray Blanchard, Anthony Bogaert, William Byne, Anne Innis Dagg, Lisa Diamond, Milton Diamond, Anke Ehrhardt (photo by Sandra Elkin), Steve Emlen, Richard C. Friedman (photo by Marite Jones), Albert Galaburda, Richard Green, Dean Hamer, Simon LeVay, Eleanor Maccoby, Heino Meyer-Bahlburg (photo by Eve Vagg, from the Audiovisual Department of the New York State Psychiatric Institute), Letitia A. Peplau, Qazi Rahman, Charles Roselli, Joan Roughgarden, Frank Sulloway, Dick Swaab, Robert Trivers, Paul Vasey, Sandra Vehrencamp,

Martin Weinberg, George C. Williams, Edward O. Wilson and Kenneth Zucker for allowing their photographs to be published as part of Figure 1.1. The images of: Richard von Krafft-Ebing and his wife Marie Louise, Karl-Maria Benkert, Karl-Heinrich Ulrichs, Sigmund Freud and Carl Westphal are all in the public domain. The photos of Vern Bullough, Michel Foucault and Günter Dörner are also published in the understanding that they are in the public domain. Havelock Ellis's photograph is taken from his autobiography *My Life* (1940, Heinemann, London) and is no longer subject to copyright restrictions. Alan Bell's photograph is published with kind permission from Springer Science + Business Media: *Archives of Sexual Behavior*, In Memoriam: Alan Bell, 31(5), 2002, 389, Martin S. Weinberg. I am grateful to The Kinsey Institute for Research in Sex, Gender, and Reproduction, Inc. for granting permission to reproduce the images of Alfred Kinsey (photograph by William Dellenback) and John Money (photo graph by José Villarrubia), and to Shawn Wilson in particular for following up on my request. We often forget that the name at the beginning of an article or on the cover of a book is (or was) a real person, deserving of personal homage even though she or he might not always have got it right. After all, knowledge progresses thanks to the collective effort of all who contribute with serious and honest research and we learn from both our successes and our mistakes. I apologise to those whom I did not contact and who, with hindsight, should have been contacted.

The manuscript was completed while I was adjunct researcher in the School of Biological Sciences at Monash University. I thank Prof. Gordon Sanson, Head of School when I joined the University, and Prof. John Beardall, who subsequently replaced Prof. Sanson at the helm of the School, for their hospitality. A big thank you also to Assoc. Prof. Martin Burd in particular for generously sharing his space and computer facilities, not to mention a good friendship and inspiring conversations about evolution, and to Prof. Ralph MacNally for use of his computer facilities.

And in the often last but never least thank you, I also express my gratitude to my wife Marisa and daughter Catiray, for their heroic patience during this intellectual journey of learning and discovery into the depth and breadth of animal homosexuality.

Preface

Sexually reproducing animals are faced with various challenges in order to find, approach and copulate with a mate in reproduction, achieve fertilisation and finally ensure that the offspring survive and in turn reproduce. In social species, this whole process includes additional dimensions involving sometimes intricate relationships of competition and cooperation; further contributing to the overall complexity. Reproduction is central to the survival of the lineage, its suppression leads to extinction. As a consequence of this basic tenet of evolutionary biology we are obviously bound to be puzzled by the emergence of behaviours such as same-sex sexual intercourse, and by the occurrence of exclusively homosexual individuals. Homosexuality is indeed an evolutionary paradox, but one that can be resolved within the broad framework of the theory of evolution itself, after we take into account the many variables and scenarios that make homosexuality likely to be expressed in the first place and then maintained across generations.

In this book the reader will discover that we humans are not the only mammal species that expresses exclusive homosexuality and that some of the evolutionary processes that may explain the emergence of homosexual attraction in humans may be common to other sexually reproducing species as well. More importantly, it will be also shown that across taxa same-sex sexual behaviour is pervasive in the context of bisexuality and its expression takes modalities that can blend in the same individual, whether male or female, characteristics that are both feminine and masculine.

It is very unlikely that the evolutionary paradox of homosexuality will be resolved by appealing to a single cause or mechanism. On the contrary, this book provides both empirical evidence and theoretical arguments to support a scenario of multicausality for homosexual behaviour. Based on the available comparative evidence, the conclusion is that homosexual behaviour can be broadly understood in the context of adaptive evolution and therefore it is not a malfunction of sexuality. The various dynamics of human homosexuality can be also explained, to a great extent, adaptively, whether as mediator of social cooperative or competitive interactions or as a side effect of selection for traits such as high reproductive rates, high mutation rates and others.

A complex brain produces sexual behaviour that is especially plastic in our species, but we are not the only vertebrate exhibiting plastic sexuality. Other primates, but also rodents, various social mammals and some bird species express intimate sexual contacts between members of the same sex on a regular basis and they do so following complex and dynamic patterns, as part of their normal and evolutionarily adaptive behavioural repertoire.

The aim in this book is to convey not only a sense of wonder when faced with diversity and complexity, but also a sense of reassurance that such complexity can be integrated in an evolutionary synthesis. The challenge is not only to explore as many as possible of the various dimensions that define the expanse of homosexuality and its levels of causation, but also to cross the boundaries between various disciplines, each contributing unique information on the many specific aspects of sexual orientation. In the process, the limits and barriers imposed by political ideologies, barriers that tend to hinder rather than foster our understanding of sexual orientation, have been intentionally ignored.

This book was written in a somewhat asymmetrical partnership, with Alan Dixson mainly contributing one chapter and many insightful suggestions for the improvement of the other chapters, which were written by Aldo Poiani. The author of each chapter takes personal responsibility for what he writes by using the first-person singular. Finally, during the production of this book I tried to achieve the rather ambitious objective of both ranging wide across the many aspects of same-sex sexuality in various taxa and also digging as deeply as possible into each one of those aspects. I can only hope that the results of this effort may both inform and inspire the reader.

Animal homosexuality in evolutionary perspective

More often than not, great scientific journeys start with a paradox. For instance, Charles Darwin was intrigued by the non-reproductive castes of some insect species, to the extent of considering them a major obstacle to his theory of evolution by natural selection. In the sixth edition of *The Origin of Species* he wrote:

> I will not enter on these several cases, but will confine myself to one special difficulty, which at first appeared to me insuperable, and actually fatal to the whole theory. I allude to the neuters or sterile females in insect communities: for these neuters often differ widely in instinct and in structure from both the males and fertile females, and yet, from being sterile, they cannot propagate their kind
>
> (Darwin 1872a).

The search for a resolution to Darwin's sterile castes paradox produced a wealth of research that mainly took off in the 1960s and 1970s and that from the elegant William D. Hamilton's formula (also known as *Hamilton's Rule*) and John Maynard Smith's concept of kin selection, through Edward O. Wilson's impressive *Sociobiology: The New Synthesis*, has led us nowadays to *Reproductive Skew Theory*.

More than 80 years after Darwin's formulation of his non-reproductive castes paradox, the ecologist George Evelyn Hutchinson (1959) was able to turn the apparently most trivial of questions: 'Why are there so many kinds of animals?' into a major evolutionary ecological paradox leading to the study and better understanding of the ecological processes, such as interspecific competition, that determine the structure and dynamics of natural communities. Over evolutionary time, such ecological processes can eventually produce character displacement and, ultimately, speciation.

In the early 1970s it was the turn of sex to become paradoxical. When John Maynard Smith (1971) enquired about the origin of sex in his *The Origin and Maintenance of Sex*, what he meant was the origin of sexual reproduction: Why should some species reproduce sexually, a process that implies a 'waste' in reproductive capacity in the form of males, rather than asexually as parthenogenetic species do? In other words, Maynard Smith was able to transform heterosexual sex, and here I want to stress the word *heterosexual*, into an evolutionarily paradoxical phenomenon. In the same way as even heterosexual sex, i.e. the *prima facie* most unparadoxical of biological phenomena because it can lead to reproduction, can be seen as a paradox, homosexual sex has also gained in recent times the status of evolutionary paradox.

Homosexual behaviour can be defined as an interaction that is **sexual or of sexual origin** and that is performed between two or more individuals of the same sex. Homosexuality, in turn, is a sexual orientation that is characterised by sexual attraction to individuals of the same sex. Leaving internal mental states aside for a moment, from the perspective of manifest behaviour, the *attraction* involved in the definition of homosexuality may imply exclusive, sometimes life-long, preference for engaging oneself in sexual behaviours with members of the same sex, or shorter-term experiences. In other words, an individual who usually mates heterosexually but who does mate homosexually if, for instance, only individuals of the same sex are available is behaving in a homosexual fashion in the latter circumstance. If the behaviour is freely expressed, then some degree of attraction may also be involved. Whether it is exclusive or occasional, in all those cases individuals are described as behaving homosexually, but the differences between one and the other should obviously be acknowledged and studied.

In studies of homosexuality, *sexual attraction* and *sexual behaviour* are two distinct phenomena that should be the target of specific empirical research. For instance, in humans consensual occasional homosexual behaviour may also involve temporary same-sex sexual attraction, but if the occasional homosexual behaviour is performed under coercion (e.g. in a prison), it may not involve sexual attraction by one of the two partners; on the other hand, individuals exclusively attracted to members of the same sex may engage in sexual behaviour with members of the other sex under specific circumstances (e.g. social pressure to marry, but also desire to have a child). Whereas homosexual behaviour is a major focus of this book, sexual attraction aspects of homosexuality are explicitly identified here in the context of empirical studies carried out across various species, including humans, that involve the opportunity to choose at the same time between male and female sexual partners. All of these complexities are encapsulated in the definition of homosexuality. An interesting historical study of the definition of homosexuality can be found in Sell (1997).

At this point I also have to mention the concepts of bisexuality and bisexual behaviour. Bisexuality involves sexual attraction towards members of both sexes, attraction that will be expressed in sexual behaviour performed with members of both the same and the other sex. Such sexual attraction for and sexual behaviour performed with members of both sexes may be concurrent (*simultaneous bisexuality*; Weinberg *et al.* 1994), i.e. the individual may engage in sexual activities with either males or females, or homosexuality and heterosexuality may represent non-overlapping phases in the individual's lifespan (*sequential bisexuality*; Weinberg *et al.* 1994). In the latter case we have a situation where an individual is either homosexual or heterosexual during specific periods of his/her life, but bisexual over the lifetime. The example of the 'occasional homosexual' that I mentioned above should be more appropriately described as a case of sequential bisexuality.

Manifest behaviours are also associated with mental states that are more accessible in some species (e.g. in humans through language) than others, with gender identity, or self-definition of own gender, being especially dependent on language in humans. In this book the focus will be on manifest behaviours and on mechanisms that can cause, proximately and ultimately, the maintenance of, but also changes in, those behaviours. Whether similarities in behaviours across species are associated with similarities in internal mental states such as gender identity or not is an issue of no easy resolution. Given the current difficulties of studying sexual identity in a cross-species comparative perspective, the emphasis in the various chapters will be more on sexual behaviour, sexual orientation (homosexuality, heterosexuality, bisexuality) and gender role (masculinity, femininity, androgyny) than on identity.

The **evolutionary paradox** of homosexuality can then be formulated in this manner: *If sexual behaviours such as mounting or genito-genital contact have originally evolved in the context of reproduction, why is it that they occur between members of the same sex where those behaviours cannot obviously lead to immediate fertilisation?*

The aim in this book is to try to resolve the evolutionary paradox of homosexuality. As suggested by Richard Dawkins (1982) the paradox is especially intriguing whenever homosexual behaviour is found to be heritable. The behaviour is also especially intriguing from an evolutionary perspective when the individual is an exclusive homosexual and actively prefers to engage sexually with conspecifics of the same sex given a choice of sexual partners. The assumption, implicit in the definition of the evolutionary paradox of homosexuality, that *mounting or genito-genital contact have originally evolved in the context of reproduction* is amply supported by the widespread direct association of mounting behaviour with fertilisation, whereas mounting performed in social interactions not associated or only indirectly associated with the act of fertilisation, although not uncommon, is nevertheless relatively less widespread in animals.

Throughout this book homosexual behaviour, same-sex sexual behaviour, and isosexual behaviour will be regarded as synonyms. Alternative terms to *homosexuality*, such as androphilia and gynephilia or gynecophilia (see, for example, Vasey *et al.* 2007; Diamond 2009), that refer to the orientation of individuals who are sexually attracted to males or females respectively independent of their own sex, and alternative terms to *homosexual* such as 'women who have sex with women' and 'men who have sex with men' that are increasingly common in the medical and psychological literature to emphasise the behaviour more than the gender identity, are also acknowledged and sometimes used in this book. Some of the authors that are quoted have also used the terms 'gay man' and/or 'lesbian' in their works and their choice of lexicon has been maintained in the text; however, in all cases the terms are used in this book simply to mean male and female homosexual, independently of any additional cultural, political or philosophical meaning that the terms may have. Interestingly, the term *queer* (see, for example, Jagose 1996), which took on a connotation of 'homosexual' in the late nineteenth century, although it has been more or less consistently in use since the 1990s (Sell 1997), nevertheless appears infrequently in the current scientific literature on homosexuality. Finally, I decided to spare the reader any lengthy discussion of the pros and cons of the use of one or the other term; they all have some positive and some negative aspects with regard to their usefulness in the study of homosexuality. The terms used in this book are defined in the text and in the Glossary (see Appendix 1) and the usage strictly adheres to such definition.

Although homosexual behaviour in our own species will obviously feature prominently in this book, it is our intention to approach the study of human homosexuality in a comparative manner. The particular focus is on mammals and birds, the two behaviourally most complex classes of vertebrates, which, however, are separated by a period of about 250 million years of independent evolution. What will be shown in this book is that same-sex sexual behaviour is more common among mammals than among birds and that although those two taxa display some similarities in the modalities of same-sex sexual behaviour, they also show many differences.

In the following sections I provide a historical overview of the study of homosexual behaviour, then list the major hypotheses that have been proposed to explain homosexual behaviour at different levels of analysis and causation. Finally, I briefly describe the plan of the book and the contents of the remaining chapters.

Historical overview of studies on homosexual behaviour

The scientific study of homosexual behaviour and orientation has come a long way since its initial steps in the nineteenth century. Today regular scientific meetings are organised and journals are published that report and discuss the findings of state-of-the-art research (see, for example, Zucker 2008a; Patterson 2008). At the broader community level, scientific initiatives such as the exposition about homosexual behaviour in animals organised by the Oslo Natural History Museum in 2006 (British Broadcasting Corporation 2006) and press releases regarding specific cases of same-sex sexual behaviour in animals (e.g. Smith 2004), in spite of their shortcomings, do help in closing the gap between scholarly research on sexual behaviour and orientation across the animal kingdom and the broader community. Undoubtedly, such initiatives help to dissipate the mist of taboo that still engulfs sections of our society, including sections of the professional scientific community, regarding evolutionary research on homosexuality.

The term *homosexuality* was coined by Karl-Maria Benkert, an activist for the cause of the civil rights of homosexuals, in 1869. Benkert wrote under the pseudonym, later to become his legal surname, of Karl-Maria Kertbény (Herzer 1985; Byne & Parsons 1993), a choice that is usually interpreted as an attempt by Benkert to make his surname sound more Hungarian, although it is an interesting coincidence that he chose such a surname at a time when homosexuals were referred to as 'inverts', a concept first introduced in sex research by Carl Westphal (1869). Other terms that were in common use in the nineteenth century and that referred to same-sex sexual behaviour included: Casper's *paederasty*, Ulrichs' *uranianism* (Ulrich referred to homosexuals as 'urnings'), Westphal's *contrary sexual feeling*, Hössli's *man-love*, Heismoth's *homophily* and Römer's *homoiophily* (Herzer 1985). Vern L. Bullough (2004), in his biographical article on Alfred Kinsey published in the *Archives of Sexual Behaviour*, made the interesting point that many of the sex researchers who were more or less Benkert's contemporaries in the period between the end of the nineteenth and beginning of the twentieth centuries were German-speaking Jewish physicians. His explanation for this is worth a full quote and it should be a matter for some meditation:

> One reason for this Jewish predominance was not so much their more positive attitudes about sexuality than the Christians of their time, but the fact that in Germany Jewish physicians were discriminated against in many of the developing specialities and denied university appointments although they could attend university. Investigation into sexuality gave them an opportunity to explore new fields and to gain new insights into patient well-being, independent of the university
>
> (Bullough 2004: 278).

What I want to highlight with this quote is that a novel, controversial, and yet socially important field of inquiry was, at least in part, relegated to somehow marginalised communities of researchers, according to Bullough, perhaps because most of the academic establishment was too afraid to take risks? One is only left to wonder how much things have really changed since then in universities around the world (see Kempner 2008).

According to Hubert Kennedy (1997) it was Karl Heinrich Ulrichs who first proposed a scientific theory of human homosexuality between 1864 and 1865. Ulrichs viewed homosexuality not as a pathological condition, but as a 'riddle of nature' (Herzer 1985). Ulrichs' theory emphasised other-sex sex role and other-sex sexual identity aspects of homosexuality: *anima muliebris virili corpore inclusa*, 'a woman's soul trapped in a man's body' in the case of males and, for females, *anima virilis muliebri corpore inclusa*, 'a man's soul trapped in a woman's body', an unfortunate bias that still permeates many of the current studies of homosexual behaviour in humans and other animals. Although Ulrichs ultimately preferred to see homosexuals as a 'third sex', a view shared with Magnus Hirschfeld (1914; see also Crozier 2000), later in life he also recognised the occurrence of both masculine and feminine male homosexuals (Kennedy 1980/81).

The inversion model of homosexuality was also supported by other early students of sexual orientation such as Sigmund Freud (1905), Richard von Krafft-Ebing (1886) and Henry Havelock Ellis (1928, see also Crozier 2000, 2008). Masculine and feminine gender roles vary, however, and they may be affected by interactions with the external social milieu and other environmental components during development. They can also vary more or less independently of sex and sexual orientation. Masculinisation and de-feminisation, feminisation and de-masculinisation are distinct processes of development (Money 1987). The gender role inversion view of homosexuality has been rejected by various authors (e.g. Kinsey *et al.* 1948; Veniegas & Conley 2000; Peplau 2001), as the combinatorial capacity of gender roles can explain an almost continuous gradation of masculine, feminine and androgynous characteristics among homosexuals, heterosexuals and bisexuals within and across cultures (Kinsey *et al.* 1948; Bem 1974; Shively & De Cecco 1977; Ross 1981; Storms 1981; Whitam 1983; Klein *et al.* 1985; Bem 1987; Deaux 1987; Maccoby 1987; Weinrich 1987; McCabe 1989; Cramer *et al.* 1993; Peters &

Cantrell 1993; Stein 1999; Kauth 2000; Lippa & Tan 2001; Kauth 2002; Dewar 2003; Roughgarden 2004).

Another recurrent idea that is still entertained by some even today is that the development of a homosexual sexual orientation is favoured by masturbation at early ages (see Martin 1993 for a review), an alleged causal link that took hold only at the beginning of the twentieth century with the rise of psychoanalysis (Martin 1993). We will see in this book that there is no comparative evidence for a direct causal link.

The distribution of sexual orientations within the human population has also been the battleground of significant intellectual clashes that drag on even today (Crozier 2000). Krafft-Ebing (1886) favoured a bimodal distribution with peaks formed by homosexuals at one end and heterosexuals at the other, a view subsequently also supported by Rado (1940). More recently, a bimodal view has been embraced by LeVay (1996), Pillard & Bailey (1998), Rahman & Wilson (2003a) and Rahman (2005a). Other authors see the distribution of sexual orientations as continuous and multimodal, with bisexuality being a proper sexual orientation that can be significantly represented in the adult population, rather than an unstable transient state in the progression from heterosexuality to homosexuality (Fliess 1897 (cited in Sulloway 1979); Moll 1897; Freud 1905, 1931; Hirschfeld 1914; Kinsey 1941; Bell & Weinberg 1978; Wilson 1978; McConaghy 1987; Ruse 1988; Money 1990; Gorman 1994; Van Wyk & Geist 1995; Kirk et al. 2000; Kirkpatrick 2000; Peplau & Garnets 2000; Peplau 2001; Dewar 2003; Kangasvuo 2003; Adriaens and DeBlock 2006; Diamond 2008b; Worthington & Reynolds 2009). The multimodal distribution of sexual orientations among humans has been especially highlighted by the recent empirical works of Weinrich & Klein (2003) and Worthington & Reynolds (2009), whereas Stokes et al. (1997) provided empirical evidence that the majority of male bisexuals in their sample remained bisexuals (49%), whereas only some of those bisexuals made the transition to full homosexuality (34%) and an even smaller proportion (17%) made the transition to full heterosexuality.

One of the major criticisms of human sexuality studies carried out in the late nineteenth and the early twentieth centuries is that they were based, with only a handful of exceptions (Gathorne-Hardy 1998: 152), on the in-depth analysis of a small number of case studies, and from those limited cases psychologists derived some very impressive theoretical constructs. In principle, there is nothing wrong with considering single-case studies, as they may be seen as the 'first line of evidence' (Levin 2007) in the testing of a theory. However, the theory that was inspired by the limited sample of cases subsequently needs to be subject to stringent empirical tests that usually require larger sample sizes. It was not until the middle of the twentieth century that such a necessary step in the testing of theories of human sexuality was taken at a scale that allowed proper statistical analyses to be carried out. It is generally acknowledged that the *Kinsey Reports* (Kinsey et al. 1948, 1953) are a real turning point in the study of homosexuality (Bullough 2004). Interestingly, Alfred Charles Kinsey was a graduate in both biology and psychology who, after starting his scientific career as a zoologist, finally ended up heading the most ambitious study of the various aspects of human sexuality undertaken until then. This was based upon the collection of face-to-face, in-depth interview data (Gathorne-Hardy 1998; Bullough 2004). His dataset included more than 18 000 case studies collected in the USA, a very impressive number by anyone's standards. Kinsey's works, including the *Kinsey Scale*, will be mentioned in various chapters of this book. Incidentally, I should point out that a conceptualisation of sexual orientation as a continuum of states had also been expressed in a scale fashion by Magnus Hirschfeld 40 years before Kinsey did the same (Brennan & Hegarty 2007).

Kinsey founded what is now known as the Kinsey Institute for Research in Sex, Gender, and Reproduction in 1947 at Indiana University (USA); extremely valuable summary statistics on various aspects of human sexuality collected from diverse sources can be freely accessed from the Institute at www.kinseyinstitute.org/resources/FAQ.html. Among many other contributions, Kinsey identified and quantified a variegated distribution of sexual orientations ranging from strict heterosexuality to various degrees of bisexuality to strict homosexuality. The importance of Kinsey's work in highlighting the relevance of human bisexuality – it is possible that he might have been bisexual himself (Gathorne-Hardy 1998) – has recently been acknowledged by the *Journal of Bisexuality*, which in 2008 devoted an issue to the commemoration of the sixtieth anniversary of the publication in 1948 of Kinsey, Pomeroy and Martin's *Sexual Behavior in the Human Male* (Suresha 2008).

Bisexual sexual behaviour is far from uncommon in human populations, with estimates ranging from 1.8%–33% in men and 2.8%–65.4% in women (Kinsey *et al.* 1948, 1953; Hunt 1974; Janus & Janus 1993; Mosher *et al.* 2005; Francis 2008; Santtila *et al.* 2008; see Blumstein & Schwartz 1977; Morrow 1989; Pryor *et al.* 1995 for reviews). This suggests that the distribution of sexual orientations in most human populations is not bimodal (see also Weinberg *et al.* 1994). A bimodal distribution may be trivially obtained through straightforward sampling biases, for instance, if one is looking for exclusive gay/lesbians and exclusive heterosexuals for strict methodological purposes associated with the aims of the study (see, for example, Hamer & Copeland 1994) or if participants in the study are recruited from specific sources where bisexuals are hard to find. More importantly, a bimodal distribution of sexual orientations, when it occurs, may be a potential result of social factors such as the political polarisation of heterosexuals and homosexuals (Weeks 1985; Jagose 1996) that may squeeze bisexuals out of the public arena (see, for example, Balsam & Mohr 2007; Diamond 2008b). In fact, Shively and De Cecco (1977, cited in Feldman 1984) noted a decrease in the reported frequency of bisexuals between the early (1940s–1950s) Kinsey reports and the later report of 1978 (Bell & Weinberg 1978). Although this shift could be interpreted as a greater tendency for homosexuals to 'come out' during the 1970s, the alternative that the admission by an individual of being bisexually oriented might have been affected (e.g. suppressed) in the 1970s by political polarisation should also be explored (Weinberg *et al.* 1994 make a similar point in a study that analysed data collected in the 1980s).

More recently, Balsam & Mohr (2007) have suggested that, although bisexuality can be a stable sexual orientation, its stability may be jeopardised by the level of prejudice and stigma against bisexuals coming from both heterosexuals and homosexuals in the community. In the case of women, Lisa Diamond (2008a) has recently reported a trend for an increased level of heterosexual–homosexual bimodalism in the distribution of sexual behaviour over time (i.e. as the respondents aged) in a group of women interviewed regularly between 1995 and 2005. She interprets such a trend as a result of the subjects increasing the level of stable, long-term partnerships with age. The distributions of sexual attraction, rather than sexual behaviour, however, were more even (i.e. less bimodal) and did not change dramatically during the 10-year period of the study. As suggested above, this discrepancy between actual behaviour (bimodal) and attraction (more evenly distributed) may well be a result of increased political definition in the public sphere as women mature. However, Diamond's (2008a) alternative explanation, that behavioural bimodality is a necessary outcome of increased long-term monogamous partnerships across sexual orientations as individuals age, is also plausible (see also Weinberg *et al.* 1994b). Overall, Diamond too supports a variegated distribution of sexual orientations among women, with bisexuality being a proper sexual orientation rather than a transient state (see also Sanders *et al.* 2008).

Among men, a shift against an otherwise desired state of bisexuality has also been explained by respondents as a strategic decision motivated by various factors, including lack of social support – from family, gay groups, heterosexual members of society at large – for a bisexually oriented person, and the need to establish a monogamous relationship (Weinberg *et al.* 1994b; Matteson 1997 make similar points). In an excellent recent work authored by Eric Anderson (2008) some respondents engaged in rather interesting semantic juggling in order to reconcile heterosexuality and same-sex sexual behaviour without mentioning the word 'bisexual':

> 'And when I asked if he thought men who have sex with men are gay he said, 'Not really, no. They can be, but don't have to be. *And gay men can have sex with women too*. It doesn't mean they are straight'
>
> (Anderson 2008: 111, italics mine).

Anderson obviously does realise that there is an issue here:

> 'Interestingly, none of the men in either group used the label of bisexuality to describe their sexual identities either. I suggest this reflects either a defensive maneuver to protect themselves from higher rates of bi/homophobia outside of cheerleading culture or a growing polarization of sexual categorization among men in this age cohort more broadly'
>
> (Anderson 2008: 112).

In this context, it is also worth mentioning the work of Carrier (1985) on Mexican male bisexuality. Although in Mexico masculine and feminine gender roles are quite well defined, when it comes to sexual intercourse, same-sex sexuality is well accepted and widespread among men, with the inserter individual retaining his full masculinity in the prevailing cultural stereotype. Most men who have sex with other men do also have sex with women, a pattern that defines them as bisexuals. According to Carrier (1985) such activities are, at least in part, a result of the cultural practice of female chastity before marriage, and their social acceptance, at least in the case of the inserter individual, removes an important barrier to their widespread occurrence. An equivalent system is found in the Brazilian state of Santa Catarina, where feminised male homosexuals known as *paneleiros* engage in sexual intercourse with masculinised *paneleiros lovers*, who are functionally bisexual men (Cardoso 2005), and also in Independent Samoa where mostly feminised males known as *fa'afafine* have sex with masculinised men especially when women are not easily available (Vasey *et al.* 2007).

Although Cardoso (2009) suggests that masculinised men participating in sexual intercourse in an 'active' (mounter) role with feminised men – who assume a more 'passive' (mountee) role – do not self-identify as homosexuals across many cultures, in my opinion they none the less do qualify as bisexuals. Moreover, it still remains uncertain whether they would change their opinion about their own sexual orientation and identity were social pressures and prejudice against homosexuals and bisexuals eliminated. We will see in Chapter 8, however, that a heterosexual self-identification of males adopting a mounter role in occasional homosexual sexual encounters is also consistent with an ancient socio-sexual role of same-sex sexual behaviour, a role that may be adaptive in the context of dominance and/or cooperation across many taxa, including many primates. Purely hedonistic aspects of the sexual act may also play a role, of course. Such an evolutionary scenario, however, is also fully consistent with a bisexual self-identification. For instance, the practices of ritualised homosexuality that will be mentioned below (Herdt 1984a) would not be possible unless human sexual orientation included an important component of bisexuality.

Rieger *et al.* (2005), in a recent work, also accept the occurrence of male bisexuality in terms of behaviour and identity, but question its reality in terms of sexual attraction and arousal towards both males and females. They carried out an empirical study in the USA where 30 heterosexual, 33 bisexual and 38 homosexual men viewed a neutral film (e.g. a landscape) for 11 minutes, then the participants were exposed to various sexual films for two minutes, followed by a neutral film again. The sexual films showed either two males or two females engaged in sexual intercourse. The authors measured sexual arousal through both the degree of penile erection recorded during the screening of the various films and also the level of subjective sexual arousal expressed by the participants. Subjective arousal did show an increase for the least preferred sex in participants falling into the bisexual sexual orientation region, as measured by the *Kinsey Scale*. Heterosexuals and homosexuals showed low subjective sexual arousal when shown films involving the least preferred sex (i.e. males for heterosexuals and females for homosexuals). However, the same result was not repeated for penile responses (measured using a mercury-in-rubber gauge). They also plotted the values of 'male sex film' minus 'female sex film' contrasts for sexual arousal and found a trend for an increase, with diminishing returns, in the value of the contrast with increasing degree of homosexual preference. Overall, they interpret their results in terms of 'bisexuals' consisting, in fact, of individuals with either a homosexual or a heterosexual type of attraction and arousal, and they conclude that 'with respect to sexual arousal and attraction, it remains to be shown that male bisexuality exists' (p. 582). I offer an alternative interpretation of their results.

First, levels of arousal that can cause penile erection may be associated with a single sex at any given time in bisexuals, but the preference may shift from one sex to another through time at different time scales (weeks, months, years). Thus studies of arousal should be also carried out with the temporal dimension in mind, and not only by assuming that bisexuality is just a specific blend of strong homosexual and strong heterosexual inclinations that manifest themselves constantly and to a fixed degree over time (see also Weinberg *et al.* 1994: 44–5 for arguments in favour of bisexuality as a continuum rather than a discrete state). Moreover, their result showing that the level of genital arousal caused by the less arousing sex is nevertheless greater than that caused by the neutral stimulus, also suggests that bisexuality should be studied as a continuum between the two extremes of exclusive homosexuality and exclusive heterosexuality, not as a specific and narrowly defined discrete state. Their values of the male–female sexual arousal contrasts also show a consistent trend to increase with the level of homosexual preference in the participants, again indicating that bisexuality is a continuum of states. They also observed that some individuals who were categorized as homosexuals had male–female sexual arousal contrasts around the 0 region, which to me suggests that, in fact, they were probably bisexuals. The work of Rieger *et al.* (2005) has also recently been criticised by Worthington & Reynolds (2009).

Finally, a recent Finnish study published by Santtila *et al.* (2008) in the journal *Biological Psychology* strongly suggests that the potential for homosexual behaviour, and thus the level of underlying bisexuality, could be quite high in the general population (32.8% for men and 65.4% for women) and it could be genetically heritable. We will see in this book how functional bisexuality is in fact the norm, not the exception, among mammals and birds that engage in same-sex sexual behaviours. We will also see that in humans the level of shift in the individual preference for partners of one sex or another is higher in women than in men. Moreover, I will also argue that the conditions for the evolution and maintenance of bisexuality are far less restrictive than those concerning exclusive homosexuality.

Cross-cultural studies of homosexuality have also produced some interesting generalisations. First, it is clear that homosexuality is found across many human cultures (see, for example, Whitam 1983). Second, focusing on male homosexuality, an early review by Crapo (1995) suggested that three major groupings of male homosexuals exist:

(a) those displaying *intragenerational* homosexuality, where individuals who are engaged in same-sex sexual behaviours are of similar age,
(b) those of the *intergenerational* kind, where the age of the individuals is different and the older participant often takes on a mentorship role towards the younger, and
(c) all the rest.

More recently, Cardoso & Werner (2003, cited in Cardoso 2005) have suggested a different, and in my view better, cross-cultural classification:

(a) *pathic* or *gender-stratified* systems of homosexuality, where males displaying culture-specific traits (usually feminised) have sex with masculinised males, the latter being functionally bisexual in practice, although they tend to self-identify as heterosexuals; this is the most common type of system around the world;
(b) *age-stratified systems*, akin to the *intergenerational* systems described by Crapo (1995); and

(c) *egalitarian* systems where two individuals of the same sex, who self-identify as homosexuals, engage in sexual behaviours with each other. Egalitarian systems are less frequent than the other systems and they seem to have originated mainly among Europeans and people of European descent (Cardoso 2005). Egalitarian systems clearly indicate the independence between sexual orientation and gender role: homosexuals may be feminised, masculinised, androgynous or 'third gender'. Early mentions of a kind of *egalitarian* system are already available in Plato's *Symposium* (Halperin 1990: 268–9).

Gilbert Herdt (1984a) also recognised a specific type of same-sex sexual behaviour called *Ritualised Homosexuality*, which involves 'culturally conventionalized same-sex erotic practices' (Herdt 1984a: ix). Because of the symbolic nature of *Ritualised Homosexuality*, age-stratified systems are referred to in this context as *boy-inseminating rites*. *Ritualised homosexuality* is strongly dependent on cultural transmission for its maintenance. This has been dramatically illustrated by Bruce Knauft (2003: 148) in his study of the Gebusi of Papua New Guinea: 'During the course of 16 years, cohorts of young Gebusi men had gone from actively and proudly practicing sex with other men to apparently not even knowing about it'. From a sexual orientation perspective, *Ritualised Homosexuality* is fully consistent with a bisexual capacity in humans.

With regard to women's homosexuality, it is becoming clear from various cross-cultural studies that women are more plastic than men in terms of their ability to undergo changes in sexual orientation throughout their lifetime (see Diamond 2008a for a recent review). Diamond (2008b) describes women's sexual orientation as more 'fluid' than that of men. I believe that generalisations about any kind of mechanisms purported to explain the maintenance of homosexuality in female and male humans can only be achieved after a sufficient number of cross-cultural studies is carried out. Therefore we will draw from what cross-cultural knowledge is available in order to achieve a better understanding of those mechanisms in the various chapters of this book.

The view that regards homosexuality as 'pathology' was established early on by psychologists such as Krafft-Ebing (1886, see also Hartwich 1951), a view that at the time was opposed by Ulrichs and that I also vehemently reject. At the beginning of the twentieth century Magnus Hirschfeld also rejected the notion of homosexuality being a 'biological degeneration', a notion that in those years would soon give way to very dramatic developments through the application of 'racial hygiene' policies in Germany (Amidon 2008). According to Amidon (2008: 68; see also Brennan & Hegarty 2007) Hirschfeld's books provided a considerable part of the fuel at 'the public Nazi book-burnings of 10 May 1933'. On the other hand, Ellis' book *Sexual Inversion*, published in 1898, was regarded as plainly 'filthy' by the judge in a trial in London against George Bedborough, the secretary of the *Legitimation League*, a group of activists who fought for the legalisation of de facto marriages and the legal recognition of 'illegitimate' children, who had the book on sale. At the trial the judge addressed Bedborough so:

> you might at the outset perhaps have been gulled into the belief that somebody might say that this was a scientific book. But it is impossible for anybody with a head on his shoulders to open the book without seeing that it is a pretence and a sham, and that it is merely entered into for the purpose of selling this filthy publication
>
> (Ellis 1967: 309).

Krafft-Ebing, however, did temper his position with regard to the pathological character of homosexuality later in life, according to Havelock Ellis (1946: 193), and he also produced some remarkable insights in his *Psychopathia Sexualis* that I agree with and further develop in this book (Chapter 4). For instance, he was one of the first to suggest that homosexual behaviour may be what we would now call a neotenic trait. Neoteny is an evolutionary process that involves the slowing down of the rate of development of somatic tissues compared with reproductive tissues.

This process produces descendant species with adult individuals that look like juvenile stages of their ancestor. Havelock Ellis (1946) also mentioned the association of homosexuality with both behavioural and morphological traits reminiscent of 'infantilism', a situation expected from the action of evolutionary neotenic processes.

Specific evolutionary theories aimed at explaining homosexuality were already available in the 1950s, such as George Evelyn Hutchinson's suggestion that recessive alleles determining homosexual behaviour could be maintained by heterozygote advantage (Hutchinson 1959). George C. Williams (1966), on the other hand, suggested that homosexuality is plainly maladaptive. The great sociobiology debate of the 1970s brought an emphasis on kin selection (Trivers 1974; Wilson 1975, 1978) socio-sexual functions of same-sex sexual behaviour (Wilson 1975, 1978; see also Dewar 2003), parental manipulation (Trivers 1974), reciprocal altruism (Trivers 1971), mutualism (Trivers 1971) and sibling rivalry (Trivers 1971). More recently, models based on sexual selection, e.g. female choice for feminised males (Cramer *et al.* 1993; Sprecher *et al.* 1994; Miller 2000; Dewar 2003), reproductive skew (see, for example, Dickemann 1993), sexually antagonistic selection (Gavrilets & Rice 2006; Camperio-Ciani *et al.* 2008a) and sexual segregation (see, for example, Dickemann 1993) have attracted great interest, and even selectively neutralistic views have been also put forward (Vasey 2006a).

As biologists were developing their own ideas about the evolution of homosexuality, sociologists, some psychologists and others were developing their *social constructionist* (also known as *social constructivist*) theories in parallel. Social constructionism had one of its greatest advocates in the French philosopher Michel Foucault (Foucault 1976). Sexuality in general is viewed by Foucault mainly in its socio-sexual context, where language, through discourse and therefore communication, is both a means of exerting power and control over other individuals, what an ethologist would call 'dominance', and also a medium to achieve cooperation. If sexual behaviour is the mean of expression of this process of communication involving relationships of power between two individuals, what it is proximately achieved with it is pleasure. That is, pleasure can be manipulated to achieve control over others, in Foucault's view; conversely, sex may also acquire a function to mediate control of power for the attainment of pleasure. In the words of Eric Anderson (2008: 104): 'Social constructionism attributes the creation of gendered identities to a complex process of cultural, institutional and organizational influences ... alongside individual agency ... with the "power of discourse" ... serving a system of exchange between these systems' (see also De Block & Du Laing 2007 for a concise introduction to social constructionism). DeLamater & Hyde (1998) define what they refer to as the *constructionist paradigm* in the following terms, that I reproduce verbatim except for some slight modifications to adapt them to the specific case of sexuality: (1) Our sexual experience is ordered, (2) language provides the basis on which we make sense of that sexual experience, (3) the reality of sexual life is shared, (4) shared typifications of sexual life become institutionalised, and finally (5) sexual knowledge may be institutionalised at the level of society, or within groups. The so called *Ecological Theory* is a recent derivation of social constructionism that seeks explanations of sexual behaviour at three different levels: the individual level (microsystem), the level of immediate interindividual relationships (mesosystem) and that of broader relationships within the community at large (exosystem) (see, for example, Henderson *et al.* 2008). Applications of social constructionist views to the understanding of homosexual behaviour, that rely on the action of learning mechanisms, can also be found in Storms (1981), Tyler (1984), Hogben & Byrne (1998), Ågmo & Ellingsen (2003) and Anderson (2008). We will see how social interactions and (to a variable extent) learning are very important, but by no means the only, causative explanations of homosexual behaviours.

Psychoanalytical theories that are inspired by Sigmund Freud's works (e.g. 1905, 1919, 1931) are the object of some of my criticisms in various chapters (see also Bell *et al.* 1981). Two points that I salvage from Freud's views on human homosexuality, however, are: (a) the relevance of social stress during early postnatal ontogeny on development of a homosexual orientation, (b) the importance of bisexuality (see also De Block & Adriaens 2004). With regard to subsequent developments of psychoanalysis, I reject the views such as those of Sandor Rado (1940) and Charles Socarides (1978) who regarded homosexuality as 'pathology' (see Friedman & Downey 1998 for a historical review). Homosexuality is seen in this book as one of the various sexual orientations that can develop in humans as a result of the action of one or more internal and external factors, a sexual orientation that is not necessarily maladaptive. Homosexuality, therefore, is just as susceptible to taking a 'pathological form' as heterosexuality. Moreover, we will see in Chapters 3 and 4 that homosexuality can be a more or less developmentally canalised (i.e. difficult to modify) trait across taxa, including humans. The more canalised the trait is, the more unlikely it is that psychoanalytical (or any other) therapies will be able to achieve any change in sexual orientation. Unless the level of developmental canalisation of homosexuality is understood, strategies such as the so called 'reparative therapies' used by some psychoanalysts to try to change homosexuals into heterosexuals (see, for example, Nicolosi 1997), may be nothing more than a good recipe to cause pain and suffering, with very little to show for it.

The scientific study of non-human homosexual behaviour is much more recent than that concerning humans (see, for example, Zuckerman 1932) and those initial works prompted Havelock Ellis (1946: 188) to write that

> The fundamental and what may be called 'natural' basis of homosexuality is manifested by its prevalence among animals. It is common among various mammals, and, as we should expect, is especially found among the Primates most nearly below Man.

The same view also pervades the work of Alfred Kinsey, whose professional background in entomology and evolutionary biology clearly shows in his choice of the titles for his two major books on human sexuality: *Sexual Behavior in the Human Male* and *Sexual Behavior in the Human Female*, where he used the expressions 'human male' and 'human female' rather than 'man' and 'woman'. In the volume on the human male he wrote (p. 666): 'The homosexual has been a significant part of human sexual activity ever since the dawn of history, primarily because it is an expression of capacities that are basic in the human animal'.

Interestingly, the initial reviews of homosexual behaviour in animals other than humans focused on female–female (F–F) rather than male–male (M–M) mounting (Beach 1968; Parker & Pearson 1976). The first extensive cross-species survey of both male and female same-sex sexual behaviour was carried out by Anne Innis Dagg (1984; see also Gadpaille 1980) who focused on mammals. Her early work reached some major conclusions that have stood the test of time: same-sex mounting in mammals occurs in the context of social play, aggression, sexual excitement and non-playful physical contact. It took another 15 years, however, for homosexuality in non-human animals to take centre stage with the publication of Bruce Bagemihl's book *Biological Exuberance: Animal Homosexuality and Natural Diversity* (Bagemihl 1999).

Unfortunately, I have to disagree with Bagemihl's (1999) interpretation of same-sex sexual behaviour being a result of an 'excess of energy' available to animals to be used in activities that have no adaptive purpose (or just for fun), a view that reminds me of Herbert Spencer's (1890: 452–3) theory of the origin of singing in birds: '... like the whistling and humming of tunes by boys and men, the singing of birds results from overflow of energy'; see my criticisms of Bagemihl's ideas in Table 1.1 and also criticisms by Harvey (1999) and Roughgarden (2004). His facile dismissal of all the major evolutionary mechanisms that have been put forward to explain homosexual behaviour is not very impressive either, and I also disagree with his lumping together of all sorts of social activities that cannot be uncritically interpreted as sexual, let alone

Table 1.1. A taxonomy of the hypotheses proposed to explain same-sex sexual behaviour in male and female birds and mammals. Some of the hypotheses may overlap to various degrees

Hypothesis	Mechanism	Sex[1]	Taxon[2]	Reference
Proximate mechanistic hypotheses				
Genetic				
(1) *High Mutation Rate*	Alleles determining the development of homosexual orientation (e.g. those affecting neuroendocrine processes during early ontogeny) periodically recur in the population as a result of mutation	M + F	B + M	Hamer & Copeland 1994
(1a) *Male Mutation Bias*	Chromosomes found in males are more likely to accumulate mutations due to higher rates of germ cell divisions in males than females. With regard to sex chromosomes, if mutations accumulate in the Y chromosome, then they will be necessarily transmitted to males through the paternal line of descent. If they accumulate in the X chromosome, then they will be paternally transmitted to daughters, but be more likely expressed in grandsons, especially if mutations are recessive	M	B + M	e.g. Montell *et al.* 2001
(1b) *Muller's Ratchet*	The male Y chromosome in mammals mostly does not recombine, being therefore more susceptible to accumulate mutations over time, some of which may affect the development of a homosexual orientation. This mechanism may act in synergism with the male mutation bias (see hypothesis 1a)	M	M	e.g. Gabriel *et al.* 1993
(1c) *X-chromosome*	Many loci on the X chromosome do not have a homologue on the Y chromosome. Therefore males in mammals are more likely to express recessive alleles located in their single X chromosome than females, as the latter may be heterozygous for X-chromosome loci. Females may be also affected by recessive alleles at X-chromosome loci; e.g. if they are homozygous recessive, but, for any specific locus, the relative frequency of female homozygous recessive phenotypes in the female population should be lower than that of recessive males in the male population (roughly 1:2 if sex ratio is even).	M + f	M	Craig *et al.* 2004
(2) *DNA Methylation* or *Genomic Imprinting*	Homosexual sons are more likely to be born of mothers who have specific patterns of X-chromosome methylation, leading	M	M	Macke *et al.* 1993; Bocklandt & Hamer

	to the silencing of some relevant loci. Effects of specific maternal or paternal methylation patterns on autosomes are also possible		2003; Bocklandt *et al.* 2006
(3) *Heterozygote Advantage*	Alleles that encode for homosexual traits are advantageous (e.g. they may increase reproductive success) when occurring in a heterozygous genotype, but they decrease reproductive success when they occur in a homozygous combination	M + F / B + M	Hutchinson 1959; MacIntyre & Estep 1993; Gavrilets & Rice 2006
(4) *Sexually Antagonistc Genes*	Genes controlling the development of homosexual orientation in males, and that are associated with decreased male fitness, may also (a) control increased fertility in females,[a] or (b) control the provision of better parental care by females.[b] In principle, the same argument could be applied to female homosexuality. More generally, the hypothesis is known as the *Sexually Antagonistic Selection* hypothesis, which could also involve two or more linked loci	M + F / B + M	[a]Trivers 1972; Rice 1992; Gavrilets & Rice 2006; [b]Getz 1993
(5) *Pleiotropic Effects*	Genes that control the expression of homosexuality also have other phenotypic effects on the same individual that compensate for any loss of fitness due to decreased fertility rate. One important compensatory phenotypic effect is increased parental care that improves survival rate of the fewer offspring produced (see also hypothesis 4). Another compensatory effect is increased cooperation directed to kin (see also hypothesis 39)	M + F / B + M	Miller 2000
Endocrinological			
(6) *Sexual Hormones*	Redirected copulations to same-sex partners should be more likely in individuals that are endocrinologically primed to mate, whether in the mating season for seasonally breeding species or at any time of the year for non-seasonally breeding species. This includes F–F mounting in pregnant mammals	M + F / B + M	e.g. Lincoln *et al.* 1970; Guinness *et al.* 1971; Eisenberg *et al.* 1971; Berger 1985; Yeager 1990; Niven 1993; Davis *et al.* 1998
(6a) *Length of Female Oestrus*	There is an intermediate level of female oestrus length where frequency of same-sex mounting among males is maximum. A very short female oestrus should decrease the probability of M–M mounts as males will be constantly moving around in search for oestrous females,[a] whereas a very long oestrous period, especially when associated with female multiple matings, will give many males plenty of opportunities to	M	[a]e.g. Koprowski 1992

Table 1.1. (Cont.)

Hypothesis	Mechanism	Sex[1]	Taxon[2]	Reference
	copulate heterosexually, thus decreasing the frequency of same-sex mounting. This effect may be counteracted by female mate choice and socio-sexual interactions among males. From the point of view of F-F mounting, a very large or very short oestrous period should increase the probability of homosexual mounting among females as, in those cases, the chances of not accessing a sexually motivated male increase (i.e. trivially in the case of very short oestrous periods, and in the case of very long oestrous periods a male may not be available at many points in time when the female is motivated to copulate)			
Neurobiological				
(7) *Large Brain*	Same-sex sexual behaviour should be more prevalent in species possessing larger brains. This is expected especially in social species, where such behaviour may have socio-sexual functions	M + F	B + M	Wrangham 1993
(7a) *Other-sex Brain Organisation*	Relevant aspects of brain organisation in homosexuals follow other-sex patterns.[a] This hypothesis is in the psychoanalytical tradition of seeing homosexuals as sexually 'inverted' individuals[b]	M + F	B + M	[a]Dörner *et al.* 1975; [b]e.g. Krafft-Ebing 1886
(7b) *Third-sex Brain Organisation*	Relevant aspects of brain organisation in homosexuals do not follow either heterosexual male or female-typical patterns	M + F	B + M	Swaab 1995; Swaab & Hofman 1995
(8) *Sensory Processing*	Same-sex sexual partner preference is due to a mechanism operating at the sensory level (e.g. peripheral or central processing of olfactory, visual or auditory stimuli)	M + F	B + M	Roselli *et al.* 2004b
(9) *Pleasure*	Same-sex mounting is a way of achieving psychologically pleasurable rewards, especially in those species exhibiting orgasm. With regards to humans, the concept that homosexual intercourse is a 'play-like' behaviour performed in order to achieve pleasure was already well developed in ancient Greece[a]	M + F	B + M	Rasa 1977; Vasey *et al.* 1998; Bagemihl 1999; Sommer & Vasey 2006; Fruth & Hohman 2006; [a]Carson 1990
Immunological				
(10) *Maternal Immune*		M + F	B + M	MacCulloch & Waddington 1981;

	Description	Sex	Reference
	Maternal immunity may alter offspring brain tissues development in a sex-specific manner, leading to the expression of homosexuality	M	Blanchard & Klassen 1997
Behavioural			
(11) *Mistaken Identity*	Males mount other males, mistakenly believing that they are females. This hypothesis does not assume that behavioural and/or morphological similarity between males (or between some males) and females is a trait that has been selected via mate deception (as hypothesis 32 does). The mistaken identity mechanism is more likely to apply to sexually monomorphic species or to sexually polymorphic species where younger males resemble females and also to colonial species with high levels of extra-pair copulations. Mounting that results from mistaken identity usually elicits fleeing behaviour from the part of the mountee	B + M	Sick 1967; Geist 1968; Craig 1974; Birkhead 1978; Thomas *et al.* 1979; Huber & Martys 1993; Brown & Brown 1996; Wagner 1996
(12) *Social Tradition*	Same-sex mounting may also spread in the population simply by copying.[a] This is akin to the *Behavioural Contagion* hypothesis[b]	M + F	[a]Vasey 2006a; [b]Jones & Jones 1994, 1995
(13) *By-product of Masturbation and Orgasm*	Masturbating to orgasm provides a positive reinforcement of the perception of individuals like self as sexual partners; thus the individual becomes conditioned to develop a homosexual orientation. This hypothesis is not supported by the available comparative evidence. Many more males and females masturbate than develop a homosexual orientation	M + F	b + M	Alexander 1971, 1974, 1975; Ruse 1981; Pfaus *et al.* 2003
(14) *Social Cognitive*	Gender role and sexual orientation result from postnatal social interactions and learning	M + F	B + M	Bandura 1989
(14a) *Constrained Social Cognitive*	Social effects on development of homosexuality are constrained by individual characteristics	M + F	B + M	Mischel 1973
(15) *Social Constructionist*	Homosexuality is a result of learning processes that are influenced by power struggles between members of a society	M + F	B + M	Foucault 1976
(16) *Bisexuality*	Homosexual behaviour is a result of a bisexual potential in the individual. Here bisexuality is an adaptive trait of a plastic central nervous system (see also MacIntyre & Estep 1993 for a genetic theoretical argument). A similar hypothesis can be applied to gender roles: femininity and masculinity are not exclusive to females and males, respectively, but they can be manifested in both sexes (androgyny).[a] Bisexuality and androgyny may co-occur or not in the same individual.	M + F	B + M	Freud 1905; Hirschfeld 1914; Kinsey 1941; Bell & Weinberg 1978; [a]Habibi 1987a,b

Table 1.1. (*Cont.*)

Hypothesis	Mechanism	Sex[1]	Taxon[2]	Reference
(17) *Hypersexuality*	Homosexual mounting may result from sex-specific elevated levels of sexual activity, leading to interactions between equally motivated individuals	M + F	B + M	Inspired by Klüver & Bucy's (1939) work
(18) *Mirroring of Heterosexual Mating*	This hypothesis suggests that homosexual mating expresses the same behavioural patterns as heterosexual mating but with the involvement of same-sex partners. Thus, homosexual mounting should manifest effects usually observed in heterosexual mounting such as inbreeding avoidance, and it should also express specific strategies of the sex involved that are manifested in heterosexual mounting such as mate choice	M + F	B + M	Symons 1979
(19) *Female Weaning Period*	In some primates (e.g. gibbons, *Hylobates lar*) females may disengage themselves from sexual activity during weaning of their young; this produces a simultaneous lack of sexual partners for adult males and also lack of maternal comfort for the young. In this case, same-sex sexual contacts between two males, one adult and one young, especially of the ventro-ventral type, may be a result of both of them seeking two different kinds of gratification: mainly sexual in the adult and mainly comfort in the young	M	**Primates**	Edwards & Todd 1991

Life-Histories

Hypothesis	Mechanism	Sex[1]	Taxon[2]	Reference
(20) *Sex Ratio Bias*	Same-sex mounting is the outcome of a biased adult sex ratio during mating periods. Unavailability of other-sex sexual partners and availability of same-sex sexual partners during a period of elevated mating activity may lead to same-sex mounting. In species with developed heterosexual pair bonds, a biased sex ratio, especially during mating periods, may lead not only to same-sex copulations but also to the establishment of same-sex consortships. However, socially bonded individuals will be regarded as homosexuals only if they engage in same-sex behaviours	M + F	B + M	Hunt & Hunt 1977; Hunt *et al.* 1980; Conover & Hunt 1984a,b; Wolfe 1986; Huber & Martys 1993; Bried *et al.* 1999; Frigerio *et al.* 2001
(21) *Sexual Segregation*	Sexual segregation due to ecological and social factors paves the way to the expression of same-sex sexuality in sexual or sociosexual contexts. However, sexual segregation per se does	M + F	B + M	Ruse 1981

not necessarily lead to exclusive homosexuality throughout the development of the individual

Ontogenetic (Developmental) Hypotheses

Endocrinological

(22) *Developmental (Organisational) Effects of Sexual Hormones*	Sexual hormones affect processes of development of central nervous system tissues (e.g. apoptosis) in the young. This in turn may affect the development of a homosexual orientation in the adult	M + F	B + M	Phoenix *et al.* 1959; Grady *et al.* 1965; Money & Ehrhardt 1971; Baum 1976; Götz *et al.* 1991
(23) *Endocrine-Disrupting Chemicals*	Environmental chemicals may affect neuroendocrine processes at any stage of development, including adults, thus contributing to changes in sexual orientation	M + F	B + M	Fratta *et al.* 1977; Ehrahrdt *et al.* 1985; Kerlin 2005
(24) *Effects of Stress During Ontogeny*	Release of corticosteroids, steroids, catecholamines and/or opioids under stress during pre-, peri- and/or postnatal life may affect the development of central nervous system tissues that control homosexual behaviour	M + F	B + M	Ward 1972
(24a) *Maternal Stress*	Maternal release of corticosteroids, steroids, catecholamines, opioids as a result of stress during gestation may affect the development of central nervous system tissues that control homosexual behaviour in the offspring.[a] Maternal effects in birds will include the alteration of the egg yolk chemical composition	M + F	B + M	[a]Ellis *et al.* 1988; Roper 1996

Neurobiological

(25) *Gestational Neurohormonal*	Mutations may alter the development of relevant brain regions by altering exposure of those tissues to sex hormones during specific periods of the early ontogeny, thus leading to the development of a homosexual orientation	M + F	B + M	Ellis & Ames 1987

Behavioural

(26) *Early Ontogenetic Effects*	Early same-sex sexual experience affects adult sexual orientation	M + F	B + M	Owens 1976; Tulloch 1979; Zenchak *et al.* 1981; Rothstein & Griswold 1991; Leonard *et al.* 1993
(26a) *Early Sexual Play Effects*	In particular, sexual play behaviour may have some early ontogenetic effects on development of a homosexual orientation in some individuals. In this book I argue that this hypothesis is unlikely to explain the development of exclusive homosexuality in adults	M + F	B + M	Koford 1957; Henry & Herrero 1974; Reinhardt 1985

Table 1.1. *(Cont.)*

Hypothesis	Mechanism	Sex[1]	Taxon[2]	Reference
(26b) *Parental Manipulation*	Parents control the development of offspring to enhance the probability of some of them developing a non-reproducing, cooperative phenotype. This trait may be also associated with a homosexual orientation whenever the mechanism of decrease of reproduction does not involve suppression of sexual behaviour. Benefits accruing to parents may also include the release of resources that offspring may have otherwise used for reproduction. This hypothesis is also linked to hypothesis 39 in the context of *Reproductive Skew Theory*	**M + F**	**B + M**	Trivers 1972; Ruse 1981
(26c) *Exotic Becomes Erotic*	Sexual attraction is developed towards the sex that is perceived as being 'different' from oneself. Therefore an individual identifying him/herself with members of the other sex since young ages will be more likely to develop a homosexual orientation in adulthood	**M + F**	**Humans**	Bem 1996
(26d) *Psychoanalytical*	Homosexuality develops as a result of postnatal experiences at young ages. In particular, male homosexuality results from a disdain of women due to a castration complex (i.e. women are seen as castrated individuals). In females, homosexuality is a result of a masculinity complex according to this theory	**M + F**	**Humans**	Freud 1931
(27) *Learning*	Same-sex mounting, especially at young ages, affords an opportunity to learn copulatory skills that will be useful later in a heterosexual context; play mounting may afford such a learning opportunity	**M + F**	**B + M**	Ploog *et al.* 1963; Barash 1973; Reinhardt *et al.* 1986; Trail & Koutnik 1986; Westcott & Smith 1994; McGraw & Hill 1999; Davis *et al.* 1998
(27a) *Imprinting*	Preference for a same-sex partner in copulation may develop during critical periods when the sex of preferred partners is established, in some species, through an imprinting learning process. Imprinting on same-sex sexual partners, if it occurs, is expected to be further enhanced by sexual segregation, i.e. distribution of young individuals in the population in monosexual groups (see also hypotheses 28 and 28a). In this book I argue that imprinting is unlikely to be a major	**M + F**	**B + M**	Kotrschal *et al.* 2006

Life Histories

(28) *Early Sexual Segregation Effects*	Early same-sex sexual experience effects may be enhanced by sexual segregation (but see 27a)	M + F	B + M	Leonard *et al.* 1993; Bertran & Margalida 2003
(28a) *Erotic Orientation*	Sexual segregation during specific periods in the development of the individual is conducive to homosexuality in humans. In this book I argue that sexual segregation in social birds and mammals is not conducive, as such, to the development of an obligate homosexual sexual orientation (see comments in 27a)	M + F	Humans	Storms 1981

Adaptive Functional Hypotheses
Endocrinological

(29) *Reproductive Suppression*	By mounting subordinates the dominant influences their reproductive status (e.g. reproductive suppression). Among females in some mammals, F–F mounting may cause pseudopregnancy in subordinates	M + F	B + M	Albonetti & Dessì-Fulgheri 1990
(30) *Activational Effects of Hormones in Adults*	Hormones (e.g. sex steroids) may activate same-sex sexual behaviour in a specific manner in adult individuals	M + F	B + M	O'Neill *et al.* 2004a
(31) *Stress Coping*	Affiliative same-sex mounting may dampen the negative neurophysiological effects of stressful situations	M + F	M	Hanby 1974; Sommer & Vasey 2006

Behavioural

(32) *Other Sex Mimicry*	Subordinate males may mimic female behaviour and become recipient of mounting attempts by dominant males to avoid dominant attack and potentially achieve heterosexual copulations inside the dominant's territory[a] or simply to be tolerated within a harem by a harem holder. In the latter case young males mimicking female behaviour may be expelled from the harem as they grow older, e.g. *Antilope cervicapra*[b]	m + F	B + M	[a]Sick 1967; Geist 1968; Foster 1987; Trail 1990; Slagsvold & Saetre 1991; Huber & Martys 1993; Stutchbury 1994; Westcott & Smith 1994; [b]Dubost & Feer 1981
(33) *Maintenance of Erection*	Homosexual copulations in males allow the maintenance of erection that can decrease the time-lag for successful engagement in heterosexual copulations when the opportunities arise	M	M	Nishida 1997

Table 1.1. (Cont.)

Hypothesis	Mechanism	Sex[1]	Taxon[2]	Reference
(34) *Defence Against Male Harassment*	By engaging in homosexual behaviours females avoid being molested by males	F	B + M	Martin *et al.* 1985
(35) *Sperm Transfer*	Transfer of sperm by proxy. Male birds that receive sperm from another male via a homosexual copulation may then transfer such sperm to a female. Obviously, this hypothesis can only apply to same-sex mounting with ejaculation, and it is also more likely to apply in species where copulation occurs by cloacal contact	M	B	Møller 1987; Lombardo *et al.* 1994; Davis *et al.* 1998; Werner 2006
(36) *Dominance*	Same-sex mounting functions as a way to establish or reinforce a dominance hierarchy.[a] This effect should be especially enhanced by hormonal influences during mating periods, especially if the same hormone/s (e.g. testosterone) may control both mating and aggressive behaviours[b]	M + F	B + M	[a]Innis 1958; Wickler 1967; Geist 1968; Guinness *et al.* 1971; Dixson 1977; Collias & Collias 1978; Fujioka & Yamagishi 1981; Jamieson & Craig 1987a; Yamagiwa 1987; Orgeur *et al.* 1990; McGraw & Hill 1999; [b]Katz 1949; Komers *et al.* 1994
(36a) *Reproductive Manipulation*	Subordinates, by behaving homosexually, may manipulate the dominant male into believing that they are not sexual competitors for access to females	M	B + M	Perry 1998
(36b) *Redirected Copulations*	Copulations are redirected to a same-sex partner when a selected heterosexual partner is not sexually motivated. This hypothesis may explain cases of a dominant mounting a subordinate and vice versa. The behavioural pattern may be explained by two mechanisms: (a) same-sex mounting is just an outcome of neuroendocrine mechanisms determining sexual motivation in the context of availability of same-sex sexual partners and unavailability of heterosexual partners (see **Exaptive Hypotheses** section below), or (b) same-sex mounting may have the function of increasing sexual motivation of potential other-sex sexual partners who are witnessing the homosexual interaction (see hypothesis 36c)	M + F	B + M	Hanby 1974; Jamieson & Craig 1987a

(36c) *Heterosexual Arousal (Signal Mounting)*	Same-sex mounting may increase motivation of potential heterosexual partners to engage in copulation	m? + F	M	Beach 1968; Parker & Pearson 1976; Pal *et al.* 1999
(36d) *Mating Diversion*	Dominant males may allow subordinate males to mount them in order to divert their sexual activity from females. This is different from hypothesis 36, where the dominant is the mounter	M	B + M	Jamieson & Craig 1997a
(36e) *Family Dynamics*	Competitive interactions within the family promote development of same-sex sexual behaviours. This is akin to *Compromise* models of *Reproductive Skew* theory (see Chapters 6 and 8)	M + F	B + M	Sulloway 1995
(37) *Affiliation*	Same-sex mounting signals and reinforces a cooperative bond between interactants.[a] The capacity to establish social bonds might have initially evolved outside a current context of same-sex sexual behaviour (e.g. heterosexual share of parental care, short-term heterosexual pair bonds as a mate-guarding strategy in sperm competition, parent–offspring bond, etc.)[b]	M + F	B + M	[a] Layne 1954; Bernstein 1975; Fairbanks *et al.* 1977; Rasa 1977; Dixson 1977; Würsig & Würsig 1979; Yamagiwa 1987; Smuts & Watanabe 1990; Moynihan 1990; Wrangham 1993; Heg *et al.* 1993; Silk 1994; Kempenaers 1994; Félix 1997; de Waal 1997; Bertran & Margalida 2003; Fítos Anestis 2004; [b]Conover *et al.* 1979; see also Lagrenade & Mousseau 1983
(37a) *Heterosexual Signalling of Same-sex Affiliative Bonding*	In polyandrous trios where males provide parental care, male same-sex mounting may signal the strength of the cooperative bond between males to the female sexual partner	M	B + M	Smuts 1987
(37b) *Reciprocal Altruism*	Same-sex mounting reinforces cooperative bonds that sustain interactions of reciprocal altruism between non-related individuals	M + F	B + M	Trivers 1971
(37c) *Confluence*	Cooperative interactions within the family promote development of same-sex sexual behaviours. This is akin to *Transactional Models* of *Reproductive Skew* theory (see Chapters 6 and 8)	M + F	B + M	Zajonc *et al.* 1979

Table 1.1. *(Cont.)*

Hypothesis	Mechanism	Sex[1]	Taxon[2]	Reference
(38) *Ejaculate Waste (Sperm Competition)*	Same-sex mounting is a mountee's tactic to make the mounter waste his ejaculate	M	B + M	Birkhead & Møller 1992
Life-Histories				
(39) *Kin Selection*	Presence of homosexuals in social units formed by close relatives increases the individuals' inclusive fitness. Moreover, affiliative, cooperative interactions mediated by same-sex mounting may be especially intense among kin. This is potentially counteracted by mechanisms of inbreeding avoidance that may be operational for heterosexual mounting and could be retained in the context of homosexual mounting. The combination of both effects leads to the prediction that same-sex mounting, when it occurs in groups of kin, should be more frequent between individuals of intermediate genetic relatedness (e.g. cousins)	M + F	B + M	Williams 1966; Wilson 1975
(40) *Interdemic Selection*	Homosexuality associated with sociality and cooperation may lead to the occupation of better habitats that will allow higher reproductive rates by heterosexual and bisexual relatives of the homosexuals, leading to (a) the production of large numbers of dispersers or (b) decreased probability of extinction and increased geometric mean of fitness of the population.[a] This hypothesis should not be confused with group selection à la Wynne-Edwards (1962) that implies self-sacrifice for the benefit of the species. Such group selection is unlikely to be a major force in evolution (Maynard Smith 1964) and the arguments for its potential relevance in the evolution of homosexuality have been rebuffed by Sommer & Vasey (2006). I prefer to stress the distinction between the two mechanisms by using the concept of *interdemic selection* (Wilson 1975; Wade 1985; see the review of Kirby's (2003) work in Chapter 3). The selection process, however, is frequency dependent as groups with an excessive number of exclusive homosexual individuals will probably produce fewer dispersers than groups with a relatively smaller number of homosexuals	M + F	B + M	[a]Cook 1996; Kirby 2003

(41) *Ecological Constraints*	Under conditions of ecological constraints for independent breeding, co-breeding may be selected among sympatric members of the same sex. Homosexual mounting may then occur between the social partners in either sexual or socio-sexual contexts	M + F	B + M	Gjershaug *et al.* 1998
(42) *Mating System*	This hypothesis predicts an association between same-sex mounting and mating system that may be variable: some taxa may have evolved more same-sex mounting among polygamous, sexually dimorphic species resulting from female mimicry by younger males for instance; other taxa may display more same-sex mounting among monogamous, sexually monomorphic species resulting from mistaken identity. However, due to the association of polygamy with a biased *operational* sex ratio, a stronger link between homosexuality and polygamy is expected	M + f	B + M	Wrangham 1993
(43) *Pathogenic*	Pathogens may affect host behaviour directly, through their action on the host's central nervous system for instance, so that their rate of transmission is increased; in this case homosexuality itself could be a result of parasitism	M + F	B + M	Cochran *et al.* 2000
Exaptive Hypotheses				
(44) *Best-of-a-Bad-Job*	Homosexual behaviour is the next best option for individuals that are neuroendocrinologically motivated to engage in sexual intercourse but lack other-sex sexual partners. Indeed, such individuals face the option to either express homosexual behaviour, masturbate or avoid sexual activities altogether (asexuality) (see also hypothesis 20)	M + F	B + M	e.g. Dawkins 1980
(45) *Quest for Sexual Novelty*	In mating systems where extra-pair copulations (EPCs) are common, same-sex mounting may follow from the tendency of males and females to engage in EPCs	M + F	b + M	Wolfe 1986
Adaptively Neutral Hypotheses				
(46) *Selectively Neutral Behaviour*	Same-sex sexual behaviour has no specific adaptive function	M + F	B + M	Vasey 1996; Sommer & Vasey 2006

Table 1.1. (Cont.)

Hypothesis	Mechanism	Sex[1]	Taxon[2]	Reference
(47) *Biological Exuberance*	This hypothesis was proposed by Bruce Bagemihl (1999) in his book *Biological Exuberance: Animal Homosexuality and Natural Diversity*. Sommer & Vasey (2006: 36) summarise the hypothesis in this way: 'Solar energy exists in excess of what is needed for organisms to grow and stay alive. The surplus energy is used for "extravagant activities" such as sexual reproduction and also for non-procreative behaviour including homosexual interactions'. I agree with both Paul Harvey (1999) and Joan Roughgarden (2004) that this hypothesis adds conceptually very little to our understanding of homosexuality or, for that matter, to our understanding of the evolution of sexual reproduction. The hypothesis' emphasis on the diversity of same-sex sexual behaviours in nature is something that is obviously welcome; what the hypothesis fails to do, however, is to explain such diversity. Why should this 'excess energy' be used for 'extravagant' sexual reproduction in some species and asexual reproduction in other? Do sexual species have more 'excess energy' than asexual species? Why should some individuals spend their 'excess energy' affiliative-mounting each other at one point in time and allogrooming at another? Does the hypothesis predict that the shift between affiliative-mounting and allogrooming is completely random? The hypothesis in its current state of development does not provide answers to those questions simply because it is too vague	M + F	B + M	Bagemihl 1999
Adaptively Disadvantageous Hypotheses				
(48) *Pathology*	Homosexual behaviour is selectively disadvantageous and therefore it represents a pathology of sexual behaviour	M + F	B + M	Krafft-Ebing 1886; Williams 1966
Phylogenetic Hypotheses				
(49) *Phylogenetic Constraint*	Individuals engage in homosexual behaviours in specific contexts (e.g. affiliation) rather than in alternative, equally effective behaviours (e.g. allogrooming) as a result of past evolutionary events that led to the adoption of one over the	M + F	B + M	e.g. Gould & Lewontin 1979

	other. The maintenance of same-sex sexual behaviour can be understood adaptively, but its current use instead of an alternative such as allogrooming is no indication of current adaptive superiority			
(49a) *Exaptation*	The current adaptive context of homosexuality does not correspond to the adaptive context that initially influenced the evolution of the trait	M + F	B + M	Vasey 1995
(50) *Genetic Accomodation*	Homosexual behaviour is an evolutionary outcome of selection for adaptive developmental plasticity	M + F	B + M	Baldwin 1896, 1902; Crispo 2007
(51) *Genetic Assimilation*	Development of homosexuality may become canalised under favourable adaptive circumstances (e.g. sociality)	M + F	B + M	e.g. Crispo 2007
(52) *Neoteny*	Same-sex mounting is a juvenile trait that is retained in adults as a result of neotenic evolutionary processes	M + F	b + M	Bromhall 2004
Methodological Hypotheses				
(53) *Observer Error*	The alleged homosexual behaviour may result from a mistaken identification from the part of the observer. This problem is of particular concern for single observations carried out on sexually monomorphic species in the wild	M + F	B + M	e.g. Goddard & Mathis 1997

[1]Sex concerned: M, male; F, female; a letter in lower case (f, m) means that the relative relevance of the hypothesis for that sex is comparatively low;
[2]Taxon relevance: M, mammals; B, birds; a letter in lower case (m, b) means that the relative relevance of the hypothesis for that taxon is comparatively low. The taxonomy of hypotheses broadly follows the conceptual structure of Niko Tinbergen's (1963) levels of analysis (the four areas of biology) and its modification by Randolph Nesse (2000, **www.personal.umich.edu/~nesse/fourquestions.pdf**), who also draws from Ernst Mayr (1982), and an additional extension to other dimensions such as adaptive neutrality and methodological factors. Note that, although this kind of taxonomy necessarily adopts an analytical approach by pulling potentially co-occurring and interacting mechanisms apart, the models that will be introduced in Chapters 7, 8 and 10 integrate many of those mechanisms in a functional manner. Most of the hypotheses will be critically evaluated in the various chapters.

homosexual, behaviour as he seems to have done; e.g. forming monosexual groups, co-nesting between two individuals of the same sex. I do, however, praise Bagemihl for his effort in stirring the interest in the topic. As we all know, or should know, it takes a great deal of courage to swim against the current.

The most recent reviews of animal homosexual behaviour are the book edited by Volker Sommer and Paul Vasey (2006; see also Vasey 2006a) who have been fair in giving space to a diversity of views about evolutionary factors that could explain the origin and maintenance of same-sex sexual behaviour in vertebrates, and the article published in the journal *Behavioral Ecology* by Geoff MacFarlane, Simon Blomberg, Gisela Kaplan and Lesley Rogers (MacFarlane *et al.* 2007), which is also the first study of non-human homosexual behaviour that has used modern phylogenetically based comparative analyses. The most recent comparative study of animal homosexuality of which I am aware is my own article published in *The Open Ornithology Journal* in 2008. Although the books by Bagemihl (1999) and Sommer & Vasey (2006) include works on both mammals and birds, the articles by MacFarlane *et al.* (2007) and Poiani (2008) are taxonomically more restricted, focusing on birds only. Bailey & Zuk (2009) provide a brief review of some of the evolutionary aspects of same-sex sexual behaviour across vertebrate and invertebrate taxa, but did not carry out comparative analyses.

One major historical divide in the fields of inquiry on homosexuality is that occurring in association with the dichotomy *Biology vs. Culture* (*Nature vs. Nurture*), which is somehow expressed in a split among professionals between the so called 'Biological Essentialists' and the 'Social Constructionists'. Following DeLamater & Hyde (1998), biological essentialism could be defined as 'a belief that certain phenomena are natural, inevitable, universal, and biologically determined' (p. 10), whereas social constructionism could be defined as a belief that 'reality is socially constructed', as I have also explained above. Even today 'Biological Essentialism' and 'Social Constructionism' are pitched as two competing views on homosexuality. Against the Essentialism/Constructionism dichotomy have emerged those who could be labelled 'Integrationists' (or 'Transactionalists'; Diamond 1965; Sameroff 1975; Ehrhardt 1987). In fact the integration of the 'biological' with the 'social constructionist' views has been gaining pace since the late 1980s, although very early integrationist ideas are already present in Hirschfeld's (1914: 432, 'Evolution provides the key to understanding same-sex love', cited in Amidon 2008) and Havelock Ellis' (1946) writings. For instance, Beach (1987) proposed a resolution of the evolutionary paradox of homosexuality through a biosocial integration in an ontogenetic perspective; a similar integration into a biosocial transactional model of gender differentiation was proposed by Ehrhardt (1987) in the same year. Albert Bandura also seemed to have been sensitive to the issue of a biosocial integration of human behaviour, even in his *Social Cognitive Theory* (Bandura 1989). Bandura's emphasis on the neurophysiological plasticity of the human brain, which leads to both cognitive and behavioural plasticity through language, is a position that I strongly adhere to, as will be seen in Chapter 5. In that chapter I will explain that I only reject extreme interpretations that regard brain plasticity as being exclusively constrained by external environmental inputs, with disregard to the constraints imposed by the internal structure and functions of the brain. The relevance of brain plasticity in the expression of diverse sexual orientations was also emphasised by Futuyma & Risch (1984), Seaborg (1984) and, more recently by Diamond (2008a).

Subsequent examples of biosocial integration include Gooren (1993), Gladue (1994), Ferguson (1995), Baldwin & Baldwin (1997), Diamond (1998), Horvath (2000), Kirkpatrick (2000), Ross & Wells (2000), Maccoby (2000), Muscarella (2000), Muscarella *et al.* (2001), Mustanski *et al.* (2002b), James (2005), and the recent work of Pieter Adriaens and Andreas de Block (2006) who see the socio-sexual aspects of homosexual behaviour, such as alliance formation, as one of the linchpins that can bridge the divide between the 'biological'/ 'individual' and the 'social'. I suggest that, more generally, a major linchpin is *social/pair bond* (see also Weinrich 1987; Ross & Wells 2000; Roughgarden 2004). Andrew Francis (2008) has also suggested the need to adopt a biosocial perspective in the understanding of homosexuality, but he did not propose a specific model. A basic integrationist model, however, was proposed by Philip Hammack in 2005.

Biosocial integration has also been advocated by some feminist authors (e.g. Tuana 1983; Birke 2000; Diamond 2008b) and psychoanalysts (e.g. Friedman 1992; Friedman & Downey 1993, 2008). David Sloan Wilson (2005; see also De Block and Du Laing 2007) coined the expression *evolutionary social constructionism* to highlight the integrationist view, whereas John DeLamater and Janet Hyde (1998) used the expression *conjoint approaches*. The evolutionary psychologist David Buss (1995: 13) expresses the issue in a very unambiguous manner: 'Evolutionary psychology advocates integration and consistency of different levels of analysis, not psychological or biological reductionism'. It looks somewhat paradoxical to me, then, that at a time when researchers of human sexuality from different backgrounds are advocating an integrationist view, Engle *et al.* (2006) report an apparent shift among sociologists in the USA towards 'essentialist' interpretations of homosexuality. I hope that this book will provide a balanced view that may help all concerned avoid the often unnecessary jumping from one extreme to the other of the spectrum of possible explanations of homosexuality. Of course, a predilection for an integrationist approach does not mean to uncritically accept any biosocial mechanism that may be proposed, as it will be abundantly illustrated in the various chapters. All specific mechanisms will be treated as hypotheses subject to potential empirical falsification (see also Byne & Parsons 1993; Byne 1994).

The plasticity of the central nervous system that underscores the plasticity of sexual orientation both throughout the early stages of development and also during adult life – more in females than males, however, as far as transitions from one sexual orientation to another are concerned – is a very important, although not the only aspect of the biosocial approach that is emphasised in this book (see also Moore 1991; Peplau *et al.* 1999; Singh *et al.* 1999; Baumeister 2000; Peplau & Garnets 2000; Diamond 2008a,b). The *Biosocial* integrationist approach will stress both the more plastic and the more rigid mechanisms that can produce manifest behaviours throughout development. On this regard, Chapter 4 will be especially important as it will show how developmental programmes can be plastic or canalised to various degrees in comparisons within and between species.

In an evolutionary perspective, it is the biosocial underpinning of behaviour that can lead not only to the evolution of anatomical and physiological traits, but also to the evolution of cultural traits in an interactive process that is known as *gene–culture evolution* (Pulliam & Dunford 1980; Cavalli-Sforza & Feldman 1981; Dawkins 1982; Boyd & Richerson 1985; Findlay 1992; Mesoudi *et al.* 2004; Stearns 2007).

The integrationist views behaviour as a result of proximate mechanisms involving the expression of genes throughout the lifespan of the individual, physiological, nervous system and endocrinological activities and the effects of other relevant constituents of the organism in interaction with its surroundings, including other members of the same species, and of both short-term (ecological) and long-term (evolutionary) processes that involve the population at large. In this context then, learning occurs when the appropriate structures and functions of the organism retain sufficient plasticity that allow them to, for instance, respond in a novel and consistent manner to changes in the external environment. Not all structures and functions, however, are expected to be equally plastic. Although the integrationist stance adopted will be clear throughout this book from the review of published works, I will also show through comparative analyses that an integrationist *Biosocial* approach delivers a far more successful understanding of homosexual behaviour in birds and mammals than any of the more restricted alternatives. The psychologist Eleanor Maccoby (2000: 405) expresses the need for a *Biosocial* perspective thus:

The fuller incorporation of the ethological and psychobiological perspectives on gender into our existing frameworks should enrich the research agenda of students of social development.

In the sharp and provocative words of John Money (1988: 50):

The postnatal determinants that enter the brain through the senses by way of social communication and learning also are biological, for there is a biology of learning and remembering. That which is not biological is occult, mystical, or, to coin a term, spookological. Homosexology ... is not a science of spooks;

and:

the converse of biology is not social learning and memory. Logically, the converse of biology should be spiritualism and the astral body

(Money 1987: 14).

The neurobiologist Simon LeVay fundamentally agrees with this integrationist view, although I cannot assure that he would necessarily agree with Money's blunt wording of it: '... even the most nebulous and socially determined states of mind are a matter of genes and brain chemistry too' (LeVay 1993: xii). In a nice demonstration of poetic skills, LeVay devotes the last two lines of his book to exactly this issue, writing: 'Like waterlilies, we swing to and fro with the currents of life, yet our roots moor us each to our own spot on the river's floor' (p. 138). I find it difficult to improve upon this description of the integrationist programme.

A list of hypotheses

Table 1.1 lists 76 hypotheses that have been proposed over the years to explain the evolution and maintenance of homosexual behaviour in humans and other vertebrates at different levels of analysis (see also Chapter 2). The organisation or taxonomy of the hypotheses is inspired by Niko Tinbergen's (1963) levels of analysis (i.e. the four areas of biology: proximate causation, adaptation, ontogeny and evolution) and its modification by Randolph Nesse (2000) – who also draws from Ernst Mayr (1982) – and I also add an extension to other dimensions such as adaptive neutrality and methodological factors. It should be noted that, by organising a list of hypotheses into a taxonomy, I do not imply that all those hypotheses are mutually exclusive; i.e. the taxonomy is not a list of alternative hypotheses. Some of them can indeed be complementary (e.g. *Heterozygote Advantage* and *Sensory Processing*; Table 1.1). Moreover, not all of those hypotheses are independent from each other; some may be subtle variations around a common theme and therefore may overlap to various degrees (e.g. *Kin Selection* and *Confluence*). Although some indications of my standing with regard to the plausibility of the various hypotheses are given in Table 1.1, the hypotheses will be further evaluated in the appropriate chapter/s throughout the book.

Figure 1.1 shows a partial list of past and current researchers of sexual behaviour, sexual orientation and evolution, who will be mentioned throughout this book along with many others. In the figure are also included some of the researchers who contributed knowledge in areas that in this book I integrate within various models of same-sex sexual behaviour. I decided to include pictures of some of the researchers whose works are quoted, not only as a way of honouring their intellectual contributions, but also as a reminder that behind a name there is always a real person and a life experience subject to the ups and downs of the human condition.

This chapter concludes with a brief description of the main contents of the rest of the book. Chapter 2, *The comparative study of homosexual behaviour*, is a methodological chapter in which some epistemological issues such as those concerning multicausality and the so-called principle of simplicity, which are important in order to understand my approach to the study of homosexuality, are introduced. The comparative, meta-analytical and path-analytical methods used in this book will be also explained and a detailed account of the sources of the phylogenies will be provided. Most of the technical terms to be used throughout the text are included in a Glossary (see Appendix 1), but the usage of some key terms is explained here in more detail. The chapter ends with a methodological

Historical overview of behavioural studies

Figure 1.1. Some of the past and present contributors to our knowledge of sexuality, but also psychology, development and relevant aspects of evolutionary biology, that we mention in this book. The sequence is in alphabetical order.
(1) J. Michael Bailey, (2) Albert Bandura, (3) Michael Baum, (4) Alan Bell, (5) Sandra L. Bem, (6) Karl-Maria Benkert (Kertbény), (7) Ray Blanchard, (8) Anthony Bogaert, (9) Vern Bullough, (10) William Byne, (11) Anne Innis Dagg, (12) Lisa Diamond, (13) Milton Diamond, (14) Günter Dörner, (15) Anke Ehrhardt,

Caption to Figure 1.1. (*cont.*) (16) Henry Havelock Ellis, (17) Steve Emlen, (18) Michel Foucault, (19) Sigmund Freud, (20) Richard C. Friedman, (21) Albert Galaburda, (22) Richard Green, (23) Dean Hamer, (24) William D. Hamilton, (25) Magnus Hirschfeld, (26) Alfred Kinsey, (27) Richard von Krafft-Ebing and his wife, (28) Simon LeVay, (29) Eleanor Maccoby, (30) Heino Meyer-Bahlburg, (31) John Money,

Caption to Figure 1.1. (*cont.*)   (32) Letitia A. Peplau, (33) Qazi Rahman, (34) Charles Roselli, (35) Joan Roughgarden, (36) Frank Sulloway, (37) Dick Swaab, (38) Robert Trivers, (39) Karl Heinrich Ulrichs, (40) Paul Vasey, (41) Sandra Vehrencamp, (42) Martin Weinberg, (43) Carl Westphal, (44) George C. Williams, (45) Edward O. Wilson, (46) Kenneth Zucker.

analysis of the contrast between studies carried out in the wild and in captivity.

In Chapter 3, *Genetics of homosexuality*, the major hypotheses and evidence suggesting a heritable genetic component for the expression of homosexual behaviour will be explained and critically discussed. Great emphasis will be given to candidate genes that affect same-sex sexual behaviour in various taxa and to twin and family studies in humans.

Chapter 4, *Ontogenetic processes*, will focus on developmental processes. The effects of early exposure to hormones, stress and endocrine-disrupting environmental chemicals, along with the role of early experience, including juvenile play, on the development of homosexual behaviour will be critically analysed in a comparative perspective. A major neoteny hypothesis that could explain the evolution of homosexual behaviour in adults via maintenance of juvenile-play-like sexual behaviours will be thoroughly discussed.

In Chapter 5, *Endocrine and nervous systems: a network of causality for homosexual behaviour*, the endocrinological and neurological mechanisms that could explain various aspects of animal homosexual behaviour will be evaluated. There is an emphasis on the adult stage of development.

Chapter 6, *Immunology and homosexuality*, will focus especially on the potential effects of immune mechanisms that could facilitate the development of a homosexual sexual orientation, especially as they relate to the fraternal birth order effect and the parental age effect. Current models will be critically reviewed and alternatives will be proposed.

Chapter 7, *Sexual segregation effects*, will focus on sexual segregation as a widespread behavioural pattern that occurs among many mammalian and some bird species. Here the potential evolutionary effects that sexual segregation may have on the manifestation of homosexual behaviour among adult individuals will be explored in detail.

In Chapter 8, *The social, life-history and ecological theatres of animal homosexual behaviour*, a detailed comparative analysis of the effects of group living, mating system and alternative mating strategies will be carried out, along with an analysis of the effects of sexual dimorphism, social structure and dynamics, sex ratio and others on the evolution of homosexual behaviour in birds and mammals. A specific *Synthetic Reproductive Skew Model of Homosexuality*, which is an extension of *Reproductive Skew Theory*, is tested in this chapter.

Chapter 9, *Homosexual behaviour in primates*, will specifically focus on primate same-sex sexual behaviour in a comparative perspective. The chapter will describe the diversity of behaviours that could be labelled homosexual in primates and their different functions. The prevalence of bisexual sexual behaviours will be especially emphasised.

In the final Chapter 10, *A Biosocial Model for the evolution and maintenance of homosexual behaviour in birds and mammals*, the major conclusions are drawn together and a novel evolutionary synthesis of animal homosexuality is proposed: the *Biosocial Model of Homosexual Behaviour*. This model extends the *Synthetic Reproductive Skew Model of Homosexuality* introduced in Chapter 8 to include additional endocrine, neurological and genetic mechanisms that have been discussed in previous chapters along with a series of major evolutionary processes: sexual selection, neoteny, kin selection, sexually antagonistic selection and others.

Each of Chapters 3–9 will also end with a summary of the main conclusions for the chapter. The various empirical aspects and theoretical models of same-sex sexual behaviour analysed in the various chapters will be addressed in some detail, and additional background information regarding the different mechanisms involved will be also provided. The objective is to give the reader a relatively deep insight into the issues at hand rather than a superficial brush-over. In so doing it is hoped that this book may provide an initial platform for future debates over the importance of this or that mechanism based on thorough knowledge of the available facts. The progression of the different chapters broadly follows a levels-of-analysis pattern, and all taxa (birds and mammals, including humans) are considered in each chapter, the only exceptions being Chapter 9, which specifically focuses on primates, and Chapter 6, which has a stronger human focus.

The comparative study of homosexual behaviour

Any attempt to understand the behaviour of a complex organism such as a bird or a mammal is no easy task; trying to do so across dozens of species in a comparative perspective is rather more difficult, but not impossible. Moreover, it is necessary if an understanding of the evolutionary aspects of the behaviour is to be achieved. This chapter introduces the comparative techniques, methods and sources of empirical data used in this book. The major techniques and sources of phylogenies will be described in some detail for the benefit of those who want to know precisely how the analyses were carried out. Non-specialists may skip those sections if they wish. Some especially important terms are also defined here, the chapter ending with an analysis of the effects of captivity and wild conditions upon the expression of same-sex sexual behaviour in birds and mammals. Let us start, however, with some epistemological caveats.

Epistemological caveats

> At present all too many scientists ... seem to think that theories based upon the notion of 'nothing-but' are somehow more scientific than theories consonant with actual experience, and based upon the principle of not-only-this-but-also-that.
> Aldous Huxley (1963) *Literature and Science*: pp. 77–8

There are very few guiding principles in modern science that can compete in degree of consensus regarding usefulness and importance, with the so called *Ockham's Razor*. This principle states that 'one should not multiply entities beyond necessity'. The *Ockham's Razor* principle was apparently never stated as such by its alleged author, the English medieval philosopher William of Ockham (sometimes also spelt 'Occam', a Latinized form), but it is certainly consistent with his philosophical views. *Ockham's Razor* reflects a widespread view – also encapsulated in similar principles such as that of *Simplicity*, or of *Parsimony* or of *Economy of Thought* – that when we are seeking a scientific explanation for any phenomenon our hypotheses should avoid unnecessary complications whenever simpler alternatives are available. The principle has been the subject of an extremely illuminating analysis published more than 50 years ago by Lewis S. Feuer (Feuer 1957). Feuer distinguished between the methodological and the metaphysical principles of simplicity, the latter being a belief that nature is governed by intrinsically simple laws. In Feuer's 1957: 110) words:

> The plain fact, however, is that the metascientific creed of simplicity has nothing to do with the scientific methodological principle of simplicity. Occam's Razor, in other words, doesn't owe its validity to any doctrine that Nature is simple.

That is, if in order to explain an evolutionary pattern, or a behaviour such as same-sex sexual behaviour, we have to recur to the interaction among a series of variables because the phenomenon is multicausal, by doing so *Ockham's Razor* is not being compromised at all. Nature is not necessarily simple; in fact most of the time it is rather complicated.

Ockham's Razor simply states that we should refrain from making nature even more complicated than it actually is. Thus, *Ockham's Razor* does not require any specific form of causality (whether mono- or multi-), it just precludes unnecessary complications during our enquiry on causation. By restricting the enquiry to a parsimonious level of causal complexity, *Ockham's Razor* prevents the drifting of hypotheses into the realm of unfalsifiability (Feuer 1957; Popper 1959).

The acceptance of the potential complexity of many phenomena in nature does not at all prevent the applicability of the scientific method as it is practised in the *strong inference* (Platt 1964) or *conjectures and refutations* (Popper 1969) methodologies. What it prevents, however, is missing the full causality of a phenomenon because we may be obsessed with simplicity or monocausality. Ecologists, psychologists and sociologists were among the first to advocate the use of specific statistical tools that could deal with multicausality whenever multicausal mechanisms are likely to provide the most parsimonious and effective of the available explanations (see, for example, Hilborn & Stearns 1982). Variables that co-occur to determine a multicausal phenomenon may have a strictly cooperative causal effect whereby if one or another is missing the phenomenon may not occur. Or they may be redundant, i.e. the phenomenon occurs equally if any of them occurs, or they may be non-redundant but also unnecessary: that is, if one of them is missing the phenomenon occurs but not in the same way as if a different variable is missing or as if they are all present (see Mackie 1965; Hilborn & Stearns 1982). Two consequences of the multicausality of a phenomenon are the production of a diversity of variants and the appearance of that phenomenon (e.g. same-sex sexual behaviour) under diverse circumstances (see also Baldwin & Baldwin 1997).

With multicausality in mind, the epistemological approach adopted in this book and that will be clearly exemplified by the models developed in Chapters 7, 8 and 10 is what Mitchell & Dietrich (2006: S76) call *integrative pluralism*: '... a view of the diversity of scientific explanations that endorses close study and modelling of different causes and different levels of organization but calls for integration of the multiple accounts in the explanation of concrete phenomena'. This is a view that springs straight from the issue of levels of analysis in the explanation of biological phenomena.

In a seminal article published in 1963 in the *Zeitschrift für Tierpsychologie* (now *Ethology*) the Dutch ethologist and Nobel Prize winner Niko Tinbergen laid down the four major aims of ethology, and in doing so he also explicitly unravelled the problem of multiple causation at diverse levels in the broad field of biology. Tinbergen's (1963) four aims of biology are the understanding of (a) proximate causation, (b) adaptation, (c) ontogeny and (d) evolution of any trait or set of traits (see also Alcock & Sherman 1994). In an article published two years earlier, Ernst Mayr (1961) had also stressed the need to focus on multicausation and to distinguish proximate and ultimate levels of causation. The whole organisation of this book follows this general framework, as will be apparent from the sequence of chapters, ranging from genetics to ontogeny and neuroimmunoendocrinology, to behavioural adaptations, ecology, life-histories and evolution, and from the organisation of the diverse hypotheses that have been proposed to explain same-sex sexual behaviour as they are summarised in Table 1.1. As will be shown in this book, across species, homosexual behaviour is a multicausal phenomenon that can be explained by processes occurring at different levels of analysis.

I conclude this epistemological section with some thoughts about the concept of adaptive neutrality. Ever since Stephen J. Gould and Richard Lewontin published their influential article titled *The spandrels of San Marco and the Panglossian paradigm: a critique of the adaptationist programme* (Gould & Lewontin 1979), adaptation-bashing has become a sort of light entertainment for many evolutionary biologists. I certainly take seriously Gould and Lewontin's warning that seeing and seeking adaptations everywhere may just be the byproduct of our fecund imagination. However,

I regard the practice of describing phenotypic traits as selectively neutral simply because an adaptive function is not forthcoming or, worse still, because the researcher does not even seek one for fear of being publicly 'spandrelled', as being equally pernicious. Between the imaginative just-so stories of adaptationism and the adaptive void of the lazy neutralistic researcher there must be some middle ground where the good old *strong inferences* and *conjectures and refutations* scientific method takes hold and any hypothesis, whether imaginative or not, complex or simple, adaptive or selectively neutral, can be empirically tested. My evolutionary view of animal homosexuality is **not** what we may call 'panadaptationist'. I regard it as a well accepted fact in evolutionary biology that evolutionary change can occur through a diversity of specific mechanisms that range from genetic mutations leading to phenotypic changes that can affect the relative fitness of organisms and that can therefore be subject to natural selection, thus leading to adaptation (and sometimes astonishingly fine-tuned adaptations); to adaptively neutral changes produced by mutations and genetic drift; and to adaptive changes that are far from being 'optimal' (e.g. from an engineering perspective) because they are constrained by conflicting aspects of the biology of the organism and the past evolution of the lineage. The latter process led the French biologist François Jacob to coin the very felicitous metaphor of evolution as a process of tinkering ('bricolage') (Jacob 1977). Likewise, exaptations (i.e. traits that have a current function that differs from that originally evolved through natural selection; e.g. flexibility of human hands currently used for writing) (Gould & Lewontin 1979) are also 'real-life' outcomes of evolution, being consistent with the basic 'tinkering' character of evolutionary processes. In my view, the complexity of the real world requires that our mind make use of the epistemological aid of integrative pluralism, but ultimately whatever specific explanation will be finally accepted for whatever specific aspect of homosexuality in the various taxa is, and can only be, a matter of empirical evidence.

Definitions

Although I refer the reader to the Glossary available at the end of this book for the definition of the various technical terms that are used, I provide here a more specific and detailed justification of the usage of some of the key concepts. The central theme of this book is homosexuality and by homosexuality I mean behavioural patterns of same-sex sexual behaviour and their associated mental states. Both manifest behaviours and internal mental states characterise a sexual orientation, be it homosexual, heterosexual or bisexual. Following this definition, I am reluctant to limit homosexuality to humans and to use other terms such as isosexuality to refer to same-sex sexual behaviours in non-human animals. The reason for this is that, although human sexuality may well be different from sexuality in other species, I have no reason to believe that sexuality in all the other vertebrates, or even among the other primates, represents a unitary phenomenon, with humans being the only exception. If I could put it in an Orwellian fashion, I do not subscribe to the idea that 'all animals are equal, but humans are more equal than others'. Moreover, in humans, as a consequence of language, we can identify internal mental states associated with sexuality (e.g. sexual identity) relatively easily, but we cannot assume that other vertebrates do not possess internal mental states of some kind that endow them with their own brand of sexual identity. Whether individuals of species other than *Homo sapiens* do or do not possess a sexual identity should not be a matter of superficial judgement. For this reason I prefer to use the terms homosexual, isosexual and same-sex sexual as equivalent. This choice allows us to put all species in equal terms from a **methodological** perspective, and study them in comparative analyses that will detect differences and similarities in the association of same-sex mounting with various other variables and contexts (Daly & Wilson 1999). That is, differences and similarities in patterns of sexual orientation are not established a priori, by definition, but I expect them to emerge from the various analyses. The specificities of human and other species' homosexuality will be studied as appropriate following the various themes of the chapters and sections within chapters. With this I obviously imply that, in spite of the various difficulties and caveats, results deriving from studies of different species can be compared and similarities and differences studied in a broad evolutionary framework (see also Baum 2006).

I regard the terms homosexuality, heterosexuality (i.e. patterns, and their associated mental states, of sexual behaviour performed with conspecifics of the other sex), and bisexuality (i.e. patterns, and their associated mental states, of sexual behaviour performed with conspecifics of both sexes), as categorisations of a continuum of sexual orientations. Following the approach of the *Kinsey Scale* (Kinsey *et al.* 1948), bisexuality will describe a very broad distribution of tendencies to engage in sexual behaviours with individuals of either sex. In addition, two or more conspecifics of the same sex who occur together or even establish a social bond are referred to as part of a 'monosexual' group or partnership; if the sexes of the partners are different, the relationship or group composition is referred to as 'disexual'. It is only when the same-sex partners engage in sexual interactions with each other that the term 'homosexual' is used. So, monosexual groups may be formed by heterosexual individuals, homosexual individuals, bisexual individuals or even asexual individuals or a mixture of the four. Note that in the study of sexual orientation, asexual individuals are those who do not engage in sexual behaviour; the term is not to be confused with 'asexual reproduction' (e.g. parthenogenesis).

Gender role is used here in a broad definition to mean behavioural characteristics and their associated mental states that are defined as being *masculine* (male-typical) or *feminine* (female-typical) and that are displayed by the individual. Gender roles, however, may vary in a continuous fashion from more to less masculine and from more to less feminine, with both dimensions co-occurring in the same individual to potentially produce a variegated combination of androgynous states. This means that some feminine characteristics may be expressed by males, including heterosexual males, and some masculine characteristics may be expressed by females across sexual orientations. That is, gender role and sexual orientation may vary independently. This dynamic view of gender role and its relationship to sexual orientation is central to the analysis of same-sex sexual behaviour across species as it is carried out here.

As I have already mentioned in Chapter 1, the issue of gender identity (i.e. self-definition of the own gender) in non-human animals cannot be addressed more specifically with our current tools to access their minds from the animals' perspective. It seems clear, however, that for an organism to possess a gender identity it should also be capable of displaying self-consciousness. This means that if individuals of any species other than humans display a gender identity, the number of those species is probably limited. The neurobiological evidence that will be reviewed in Chapter 5 provides a link between neuronal activity and either external sensory inputs and/or internal endocrine states and then with behaviour. Such a line of evidence could be used to identify 'mental states', but it will still remain dubious what such 'mental state' will mean to the animal, beyond what we can deduce from its manifest behaviour. This technical limitation does not deny animal sexual identity at all, it just complicates its scientific study. For this reason, in this book there is only a limited use of the current terminology associated with the various (and ever-increasing) sexual identities described in humans (see for example, Worthington & Reynolds 2009). If sexual identities are a manifestation of brain plasticity, what we can predict is that they may be also precisely described in the future in animals whose brain is also ontogenetically plastic. At the moment, however, I feel unable to go beyond that.

Although this book mainly focuses on manifest behaviour and within this category it emphasises mounting behaviour (e.g. ventro-dorsal, ventro-ventral), sexual behaviour does obviously comprise other categories as well, such as courtship, touching, kissing, embracing, caressing, etc. These additional sexual behaviours will be addressed as well, but in less detail. The focus on mounting is dictated by the need to use a behaviour that is widespread among both mammals and birds, and that is of clear sexual origin across taxa. By contrast, licking, caressing, touching, kissing, etc. cannot be automatically assumed to be of sexual origin, or to have a sexual function, in all species. This can be exemplified by considering a behaviour such as *touching*. Although it is usually true that two individuals engaged in sexual behaviour do touch each other, we cannot deduce from this that touching is necessarily a sexual behaviour. People across cultures touch each other more or less frequently in various social contexts (DiBiase & Gunnoe 2004), but only a small subsample of that touching could be regarded as sexual. I regard the lumping together of a disparate variety of social behaviours into a common label of homosexual or same-sex sexual behaviour simply because they are performed between two or more members of the same sex as ludicrous. Aggregating in monosexual groups, allogrooming among individuals of the same sex, or cooperating with members of the same sex in the acquisition of resources or the caring for offspring are all social behaviours that cannot be assumed to be homosexual unless unequivocal sexual behaviours such as mounting, genital rubbing and so forth are also expressed. Incidentally, this is one of the major drawbacks – although not the only one – of the extensive compilation carried out by Bruce Bagemihl (1999) in his book *Biological Exuberance*.

Given that the issue of gender identity will not be thoroughly examined in this book, I will set aside phenomena such as those involving asexual homosexuals (asexual gay men and lesbians) (Bell & Weinberg 1978). Individuals who define themselves as either gay or lesbian but simply do not participate in sexual intercourse with anybody are, from an evolutionary perspective, more likely to fall within the definition of homosocial (see Chapter 7, where I analyse the issue of sexual segregation). Homosociality is a tendency shown by individuals to preferentially interact socially with conspecifics of the same sex (Gagnon & Simon 1967; see also Lipman-Blumen 1976). The self-definition of those same individuals as gay men or lesbians, however, will be understood in this book as a property of the plasticity of the central nervous system. It is such plasticity that ultimately allows anybody to define him- or herself in the way or ways he or she may choose. Homosociality, same-sex affection and mutual support, companionship and sympathy are all traits that can be shared among members of the same sex in the context of either homosexual, heterosexual or bisexual sexual orientation depending on the culture, subculture and situation (see, for example, Tripp 1975). Asexuality could also be seen as a proper sexual orientation (see, for example, Green & Keverne 2000). In this case, the 'asexual homosexuals' of Bell & Weinberg (1978) could be redefined as 'asexual homosocials'.

Although cases of intersexuality in humans and other species are going to be mentioned in various chapters of this book, this will be done with the aim of unravelling the specific biological underpinnings of homosexual behaviour and orientation, in the context of a broader biosocial approach. A focus on intersexuality as such would require a book in itself (see, for instance, Cohen-Kettenis & Pfäfflin 2003; Preves 2003).

Same-sex mounting as it is understood here may involve genito-genital or genito-anal contact or not, and, in the case of male mammals, it may involve anal penetration or not. In addition, it may involve ejaculation, in the case of males, and/or orgasm in the case of both sexes, or not. All those possible variations of mounting will be considered in this book in the various chapters (see Figures 2.1 and 2.2 for examples). The proximate function of mounting may be also variable: from sexual intercourse to manifestation of dominance or affiliation. Those various functions and their interactions will be also specifically studied.

Finally, when referring to pairs of individuals (siblings, twins, partners) of different sexes I use the expression 'different sex' or 'other sex' rather than the often used 'opposite sex'. Males and females are not 'opposite'; they are just different.

Figure 2.1. Male–male mounting in the polygynandrous *Porphyrio porphyrio*. Photo courtesy of Ian Jamieson and Nic Bishop.

Figure 2.2. Female bonobos (*Pan paniscus*) mounting ventro-ventrally (G–G-rubbing). The photo was taken in the *Lola ya* bonobo sanctuary, Democratic Republic of Congo, by Vanessa Woods.

Sources of the datasets and potential biases

The major comparative analyses included in this book were carried out using datasets from 72 avian and 107 mammalian taxa. Initially, the list of species and subspecies came mainly, but not exclusively, from Bagemihl (1999); the original and other references were subsequently consulted to determine the actual presence or absence of same-sex mounting in the various taxa. The sample is therefore biased towards taxa that have been observed engaging in homosexual mounting. The emphasis on obtaining a large sample of these taxa is dictated by the relatively low frequency of homosexual behaviour recorded in animals compared with heterosexual behaviour, and therefore the need to obtain a sufficiently large sample size that will allow the study of the effect of various variables on the behaviour.

Obviously, proving the occurrence of a behaviour is comparatively easier than proving its absence; therefore the species that are classified here as not showing homosexual behaviour could be either real cases of absence of same-sex sexual behaviour or cases where the behaviour is expressed at low frequency. In either case, however, they will represent a valid source of comparison. I also rely on the assumption that researchers studying sexual behaviour in birds and mammals will report the occurrence of same-sex mounting if it is observed, suggesting that, if no such report is available in published studies of the species' sexual behaviour, chances are that same-sex mounting, at the very least, is infrequent in that species. The dataset may have been affected also by other factors impinging on the detectability of same-sex mounting such as the differences in the frequency of copulations among species, which may alter the likelihood of observing same-sex copulations; time in the day when individuals mount (for example, species that copulate at night are obviously difficult to observe); length of mating periods/season (shorter seasons obviously afford a smaller window of opportunity to observe same-sex mounting). If mounting occurs in secluded places it will be more difficult to observe. Mounting events themselves may be of brief duration, thus making them less detectable, and some species are more conspicuous during their mounting than others. Unfortunately, not all those factors can be controlled; however, I rely on: (a) the diversity of authors, (b) the diversity of species and (c) the large total number of taxa for both birds and mammals to diminish the systematic impact of those potential biases.

For a discussion of the limitations of various sampling methodologies used in studies of sexual orientation in humans, see Meyer & Wilson (2009).

One interesting potential source of bias in studies of homosexuality that was suggested by Feldman (1984; Bell 1975, Jackson 1997 and Stacey & Biblarz 2001 also make a similar point) is the actual sexual orientation of the researchers. I am not sure whether this is necessarily an issue of genuine concern, as, in a remark made by Havelock Ellis (1967: 179) in his autobiography that to some extent could apply to the case of heterosexual authors writing about homosexuality: 'after all, it is the spectator who sees most of the game ...'. None the less the idea was taken on board, as 'players' do obviously bring specific insights into the understanding of the 'game' that complement those of the 'spectators' (Williams 1993). Although I cannot control for the sexual orientation of the authors of all articles and books quoted here, an effort was made to 'control' for potential effects of the sexual orientation of the authors of this book. Consequently, draft versions of the chapters were sent to a very broad diversity of colleagues who are also experts in the various fields. Their suggestions were seriously considered and changes were made to the text where I found myself in agreement with their criticisms.

In a deceptively trivial aside, I would also like to state that during the production of this book the adoption of an integrationist view required a critical reading of all relevant sources of information, theoretical or empirical, that I could find, no matter who authored it and in which journal the article was published. I tried to rescue what I considered was the wheat of each work and pointed out the chaff as well. This means that I was happy to rescue the points that I found convincing and that were made by authors with whom I may not completely agree on every score.

Comparative analyses

As a result of evolutionary processes some taxa, or lineages, may show patterns of phenotypic *convergence* with other taxa or of phenotypic *divergence*, whereas still others may retain phenotypic states that were already present in their ancestor (*phylogenetic trait conservatism*). Phenotypic *convergence* between very different taxa is more likely to be explained by the similarity of adaptive processes. On the other hand, phenotypic *divergence* over evolutionary time may be a result of adaptation but also of non-adaptive drift, whereas *phylogenetic trait conservatism* may be a result of either the absence of specific factors (e.g. environmental, genetic) that select or favour phenotypic change or of ongoing, stabilising selection for the current trait. *Phylogenetic trait conservatism* therefore does not necessarily indicate the selective neutrality of the trait, although neutrality is always a possibility. An important point I want to make about *phylogenetic trait conservatism* is that even if the conserved trait is maintained by current selection across taxa, phylogenetic conservatism may still indicate that the past evolution of the lineage conditions the kind of adaptive traits that will evolve and be maintained within that lineage in response to specific environmental challenges. In other words, a similar ongoing environmental challenge may stabilise (conserve) trait M in one lineage, but trait N in another lineage, owing exclusively to the past evolutionary history of those lineages. For instance, a common environmental alteration such as increased habitat aridity may have very different evolutionary effects on different species that, however, may result in similarly successful adaptive responses to the change. One selective effect may be the increase in motility that will allow the finding of better habitats, whereas another may be the evolution of the ability to stay put and become dormant. Species that already possess the capacity to disperse over long distances (e.g. birds) may be more likely to evolve increased motility and nomadism and less likely to evolve dormancy, whereas species that have less ability to disperse in a situation of drought (e.g. freshwater fish, amphibians) may be more likely to evolve dormancy in one or more of their life stages (e.g. eggs, adults). That is, in this example broad similarities in the evolutionary response to aridification of the habitat *within taxa* and differences *between taxa* are the concomitant result of both current adaptive value and phylogenetic constraints. To put it simply, if nomadism is displayed by all the closely related bird species within a taxon, this may reflect a common adaptive response to drought, but that each one of those species responded to increased environmental aridity by evolving nomadism rather than other alternative adaptive traits, such as dormancy, is due to phylogenetic constraints. This means that whenever we want to test adaptive hypotheses through the analysis of multispecies datasets we have to control for the phylogenetic relationship between the species we are using, as the similarity of certain traits within that lineage may still be affected by phylogenetic constraints even though the trait is currently undergoing selection. The consequence of phylogenetic trait conservatism is to disproportionately increase the effect of data deriving from specious, monotypic taxa on the analysis. Modern comparative methods are designed to tackle that problem, while studying concomitance of evolutionary trends among the traits that are investigated in order to understand evolutionary processes, adaptation in particular (Harvey & Pagel 1991).

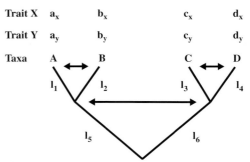

Figure 2.3. Schematic diagram of the comparative method of independent contrasts, l_i (i = 1, ..., 6) = branch length.

The chief method that will be used for the comparative tests of the evolutionary models of same-sex mounting developed in this book is the *Comparative Analysis by Independent Contrasts* (Felsenstein 1985). Figure 2.3 shows the basic logic of the method. Four hypothetical species A, B, C and D are related to each other according to the phylogenetic tree shown in Figure 2.3. Two traits, X and Y, are measured in these species. Although all four species share a common ancestor down the phylogeny, the differences in the values of traits X and Y for the A–B (and the C–D) pair of sister species arose after they differentiated from their most recent common ancestor. Before that differentiation, they just were the same taxonomic entity. From this it can be seen that although the values of traits X and Y across species are not independent, owing to their common ancestry, the differences in those values between sister taxa are. A phylogenetically independent contrast is therefore a difference in values for any trait between two tips of a phylogeny, two nodes, or a tip and a node (Felsenstein 1985). In this comparative method, independent contrasts are standardised in order to be used in parametric statistical analyses. Such standardisation is achieved by dividing the contrast by the square root of the sum of its branch lengths. For instance, for the A–B contrast and the X trait, $a_x - b_x$ is divided by $\sqrt{(l_1 + l_2)}$. The standardised independent contrasts can then be used for simple (or multiple) regression analyses through the origin (Garland et al. 1992).

The calculation of phylogenetically independent contrasts was performed with the PDAP program (Midford et al. 2005) that runs in *Mesquite* (Maddison & Maddison 2006). All analyses were performed twice: once after all branch lengths were set equal to unity, this approach approximates a punctuational mode of character evolution; followed by a second run that this time used branch lengths set according to Grafen's (1989) method. In this method the height of each node is measured as the number of species that are below the node minus one; from this the branch length is calculated by subtracting the height of the lower node from the height of the higher node (Grafen 1989, 1992). I used the phylogenies that are described in the next section. Phylogenies are a compound of the information taken from various sources that used diverse techniques, a situation that complicates the use of the original branch lengths. For this reason, although the topology of the phylogenetic trees directly derives from the published literature, branch lengths have been set by using the above criteria. Simulation studies indicate that, provided that standardised independent contrasts are used, as it was done here, the effect of errors in branch lengths is far less important (Díaz-Uriarte & Garland 1998) than the effect of errors in the topology of the tree (Symonds 2002).

Whenever contrasts for any of the variables used were significantly correlated with log-body mass contrasts, residuals from the regression were used. Residuals were calculated as: observed value minus the expected value from the regression. Before using the residuals, however, I checked that they were not correlated with log-body mass contrasts; they never were.

Qualitative traits (e.g. mating system) were codified in the manner explained in the text in order to semi-quantify them. For instance, in the case of mating system the values of the code increased from monogamy to polygamy, mirroring the general trend for increased number of sexual partners involved in the various mating systems.

Phylogenies

In the comparative analyses of independent contrasts reported in Chapters 7 and 8, one phylogeny was used for each one of the two classes of vertebrates included in this study: birds and mammals. Each phylogeny is a compound of various phylogenetic data taken from published works. Given that the various authors used different methods to reconstruct their phylogeny, only the topology of the phylogenetic tree will be set on the basis of the information taken from the literature. Branch lengths will be assigned following Grafen's (1989) method and the punctuational evolutionary method that were explained in the preceding section.

The bird phylogeny is shown in Figure 2.4. Higher-level nodes were assigned following Sibley & Ahlquist (1990), Mindell et al. (1997), Slack et al. (2006) and Livezey & Zusi (2007). Several Orders did not require the availability of a within-Order phylogeny as the number of species was either one or two (Apodiformes, Pelecaniformes, Columbiformes, Procellariiformes, Piciformes, Sphenisciformes, Gruiformes and Falconiformes); they only required

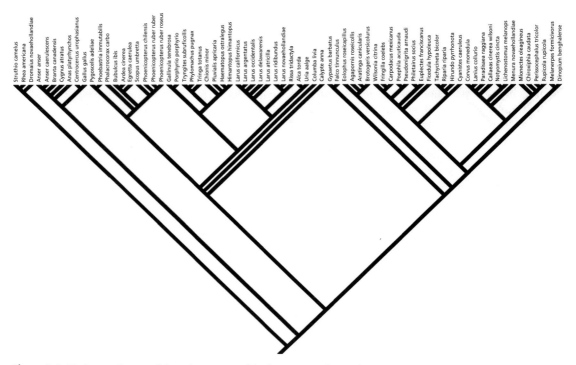

Figure 2.4. Phylogenetic tree of the avian taxa used in the comparative analyses.

a higher-order phylogeny for their placement in relation to the other Orders. For the Psittaciformes I used Sibley & Ahlquist's (1990) phylogeny based on DNA–DNA hybridisation. The Struthioniformes, in spite of contributing only three species to the sample, posed some phylogenetic challenges. Three possible fully resolved phylogenies can be reconstructed with *Struthio camelus australis*, *Dromaius novaehollandiae* and *Rhea americana*, the three species available in the sample, and all three phylogenies receive support from various authors: *Struthio* and *Rhea* as sister taxa (Livezey & Zusi 2007), *Rhea* and *Dromaius* as sister taxa (Sibley & Ahlquist 1990; Harrison *et al.* 2004; Slack *et al.* 2006; Gibb *et al.* 2007, who used either DNA–DNA hybridisation or mitochondrial DNA) and *Dromaius* and *Struthio* as sister taxa (van Tuinen *et al.* 2000; Haddrath & Baker 2001; Paton *et al.* 2002; Pereira & Baker 2006, using mitochondrial DNA sequences). I finally decided to use the recently published phylogeny of Livezey & Zusi (2007), which is based on a cladistic analysis of morphological traits, because it also includes fossil taxa. The phylogeny of the Anseriformes followed Livezey (1997), who compounded information from both DNA and morphology studies. For the Ciconiiformes I followed the DNA–DNA hybridisation phylogenies of Sibley & Ahlquist (1990) and Sheldon & Slikas (1997); Sibley & Ahlquist (1990) was also the source for the phylogeny of the Phoenicopteridae. The phylogeny of Charadriiformes was reconstructed by using Friesen *et al.* (1996), Crochet *et al.* (2000), Thomas *et al.* (2004), Baker *et al.* (2007) and Fain & Houde (2007), who used various techniques: mitochondrial and nuclear DNA sequences, DNA–DNA hybridisation and allozymes. Finally, the Passeriformes phylogeny was reconstructed following Christidis (1987), Sibley & Ahlquist (1990), Irestedt *et al.* (2001), Ericson & Johansson (2003), Barker *et al.* (2004), Spicer & Dunipace (2004), Sheldon

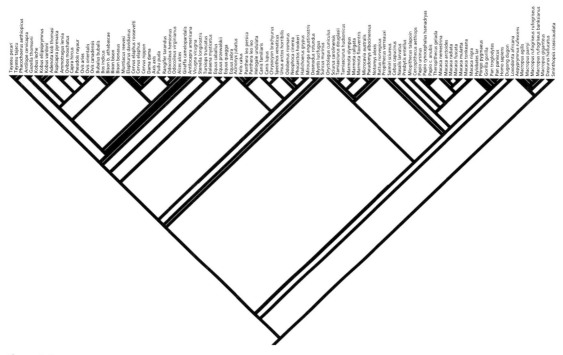

Figure 2.5. Phylogenetic tree of the mammalian taxa used in the comparative analyses.

et al. (2005) and Treplin (2006), who based their works on nuclear and mitochondrial DNA sequences, DNA–DNA hybridisation, allozyme electrophoresis and karyotype.

The phylogeny used for the mammalian sample is shown in Figure 2.5. The higher-level nodes follow the phylogenies of Novacek (1992) and Murphy *et al.* (2001). For the Sirenia I used Domning's (1994) phylogeny based on morphology. Purvis' (1995) composite phylogeny that compounds various sources of data provided the topology for the Primates. Cardillo *et al.* (2004) also reconstructed their phylogeny of the Marsupialia by compounding information from various sources and methods. The Carnivora phylogeny was reconstructed following Bininda-Emonds *et al.* (1999), who used various sources of data, and Flynn & Nedbal (1998), Flynn *et al.* (2005) and Wesley-Hunt & Flynn (2005), who used either DNA sequences or morphology. DNA sequences were also used by Milinkovitch *et al.* (2000) and Nikaido *et al.* (2001) for their phylogenies of the Cetacea. The phylogeny of Chiroptera was reconstructed following the DNA sequence-based phylogenies of Miyamoto *et al.* (2000) and Nikaido *et al.* (2001). The morphological phylogeny of Berta & Wyss (1994) was used for the Pinnipedia. The works of DeBry & Sagel (2001), Piaggio & Spicer (2001) and Ford (2006) that used analysis of DNA sequences (mitochondrial and/or nuclear) were the sources for the phylogeny of the Rodentia. The phylogeny of Perissodactyla was based on Norman & Ashley (2000), who used DNA sequences, and Groves & Bell (2004), who used morphology data. Finally for the Artiodactyla I used inputs from the morphological phylogeny of Geisler (2001), and the phylogenies of Pitra *et al.* (2004), who used mitochondrial DNA, Price *et al.* (2005), which is based on DNA sequences and morphology, and the variously sourced phylogeny of Stoner *et al.* (2003).

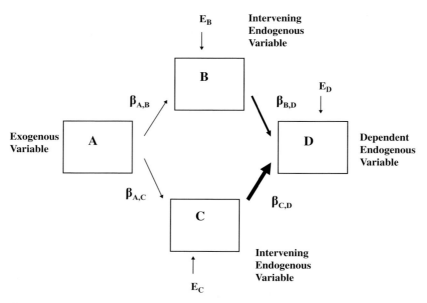

Figure 2.6. Schematic diagram of the path analysis method.

Path analysis

In Chapter 8 I will introduce the *Synthetic Reproductive Skew Model of Homosexuality* that applies classic *Reproductive Skew* theory to the understanding of same-sex sexual behaviour. The model is called 'synthetic' because, in addition to classic reproductive skew mechanisms, it also considers both the mainly sexual aspects of homosexuality (e.g. those associated with mating system and mate choice) and also the socio-sexual aspects (e.g. those associated with dominance and affiliation). A series of tests of the model will be carried out by using path analyses of standardised independent contrasts (Garland *et al.* 1992).

Path analysis is a statistical method devised by Sewall Wright in the first half of the twentieth century (Wright 1921, 1934) to analyse the conjectured paths of causation among a series of variables (see also Alwin & Hauser 1975). Figure 2.6 shows the basic components of a path analysis. Variables *A*, *B*, *C* and *D* are hypothesised to be linked by causal relations, whose directionality is depicted by an arrow. In the hypothetical model illustrated in Figure 2.6 a variable *A*, called exogenous because it is supposed to be independent from the value of any of the other variables in the model, affects variables *B* and *C*. The effect of one variable over another is measured through the standardised coefficient of regression β, also called path coefficient. The path coefficient is calculated as:

$$\beta_{ky} = (s_k / s_y) \times b_{ky}$$

where s_k is the standard deviation of the *k* independent variable, s_y is the standard deviation of the dependent variable and b_{ky} is the partial regression coefficient between the two variables (Abdi 2003). If the path depicts a situation of monocausality, then the path coefficient (β_{xy}) reduces to $(s_x/s_y) \times b_{xy}$. Variables *B* and *C* are called intervening endogenous variables because they act as both the dependent and the independent variable in successive regressions, whereas variable D is called the dependent endogenous variable, which is hypothesised to be affected by both B and C. Such effects are measured through a multiple regression between D, the dependent variable, and the independent variables B and C. In this case $\beta_{B,D}$ and $\beta_{C,D}$ are the standardised *partial* regression coefficients. The value of each dependent variable in a regression is predicted with an error (E in Figure 2.6), called the residual error term or disturbance term, which includes unexplained variance and measurement error effects. The disturbance term is calculated as $1-r^2$, where r^2 is the coefficient of determination that indicates the fraction of the variance in the dependent variable explained by its correlation with the independent variable/s. The causal effects in the model are divided into direct effects (e.g. the effect of B on D) and indirect effects (e.g. the effect of A on D via its effect on B). The total causal effect in the model is calculated by adding the total direct effects to the total indirect effects. As already mentioned, direct effects are measured through the standardised regression or partial regression coefficient as the case may be. Indirect effects, however, are calculated as the product of the standardised regression coefficients along each path leading to D and by the addition of those products across all alternative paths. In the case of Figure 2.6 there are only two paths leading from the exogenous variable to the dependent variable: $A \rightarrow B \rightarrow D$ and $A \rightarrow C \rightarrow D$. Therefore the indirect effect of A on D is calculated as $(\beta_{A,B} \times \beta_{B,D}) + (\beta_{A,C} \times \beta_{C,D})$. The arrows that indicate the direction of the hypothesised causation vary in thickness proportionally to the absolute value or modulus of the standardized coefficient of regression β. In this study the thinnest arrow will represent $0 \leq |\beta| \leq 0.2$, the medium arrow will represent $0.2 < |\beta| \leq 0.4$ and the thickest arrow will represent $0.4 < |\beta| \leq 1$.

Path analyses will be used to calculate total effects of three alternative models, as they apply to the avian or mammalian datasets. The first model is a simplified version of the *Synthetic Reproductive Skew Model of Homosexuality*, which, however, retains all of its major components, the second model is the *Synthetic Reproductive Skew Model of Homosexuality* restricted to the core *Reproductive Skew* components and the mainly sexual variables, and the third model also retains the core but restricts the other variables to those that are mainly socio-sexual. The core variables that affect *Reproductive Skew* are: Ecological Constraints, Mating System, Group Living and Genetic Relatedness. My hypothesis is that the total effect calculated in the path analysis should be higher in the case of the more complete *Synthetic Model* than in either of the two more restricted alternatives. Note that because negative and positive standardised partial regression coefficients, which take absolute values between 0 and 1, are multiplied and then added up across alternative paths, there is no guarantee a priori that a model with more paths (i.e. the complete *Synthetic* model in this case) would necessarily have a higher value of total effect. Because we want to understand the causation of same-sex mounting we are after models that have a high positive value of the total causal effect and, because regressions between standardised independent contrasts will be used, the total effect should be interpreted as: ***the evolutionary increase in same-sex mounting expected from evolutionary changes in the variables of the model as they relate to each other in the specific manner hypothesised by the model***. From this it should be clear that higher values of total effect in a model will mean that the specific pathways of evolutionary change specified by that model have a higher effect in increasing same-sex mounting in the taxon in question (birds or mammals) over evolutionary time than the alternatives.

Meta-analytical methods

Although we are used to envisaging the empirical testing of a hypothesis, model or theory as an exercise in implementing a crucial experiment (i.e. Bacon's *experimentum crucis*), rarely is an experiment sufficiently crucial that no repetition is needed by the same or other research teams. In fact, it is the accumulated evidence of multiple and independent tests of a hypothesis that finally makes it acceptable, even if temporarily, or dooms it to oblivion (Kuhn 1962). It is therefore legitimate to ask ourselves: what is the status of a hypothesis with regard to the works that claim to have tested it empirically? Has the hypothesis been sufficiently supported? Is the evidence available sufficient to reject (falsify) the hypothesis? If there is no unanimity in the results, should we suspend judgement or should we give more weight to the works that represent a more stringent test of the hypothesis and ignore the rest? Giving a satisfactory answer to these questions is no trivial matter and, in fact, there is a whole area of statistics that is devoted to studying methods for the evaluation of hypotheses on the basis of the analysis of results of the empirical tests of that same hypothesis. This is the field of *meta-analysis* (see, for example, Gurevitch & Hedges 2001). In this book I mainly used Type III meta-analytical methods that utilise the data (means, frequencies) from the original published works (Blettner *et al.* 1999) and I have checked, where appropriate, for potential biases due to sample size effects.

Is same-sex mounting affected by conditions of captivity?

Some of the studies that report same-sex sexual behaviours are carried out on animals held in captivity (see Table 2.1). Conditions of captivity could potentially facilitate same-sex sexual behaviours (e.g. same-sex mounting) if animals undergo extended periods of sexual activity due to, for instance, ad libitum access to food, if captive conditions modify socio-sexual interactions among individuals, or if a variety of behaviours, including same-sex mounting, are expressed out of sheer boredom (see Dagg 1984 for an early review of this issue). A biased sex ratio might also facilitate same-sex mounting under those conditions. It is therefore important to understand to what extent captivity may affect same-sex sexual behaviour in the sample of avian and mammal species. Note that Table 2.1 also includes our own species. In this case 'captivity' stands for a situation of relative or total spatial confinement that in the case of humans is found, for instance, in prisons and boarding schools.

In the avian sample there are 90 studies included, of which 40% were reports of same sex mounting carried out in the wild, 13.4% reported same-sex mounting in captivity, 36.6% were carried out in the wild but did not report same-sex mounting and 10.0% were studies that did not report same-sex mounting in captivity. The difference between these categories is that expected by chance ($\chi_1^2 = 0.022$, $p = 0.44$). The case of the mammalian sample is different. A total of 186 mammal studies were included and, of those, 43.0% reported cases of same-sex mounting in the wild, 40.3% reported same-sex mounting in captivity, 13.4% were carried out in the wild but did not report same-sex mounting and 3.2% were carried out in captivity and did not detect same-sex mounting. This distribution is not expected by chance ($\chi_1^2 = 7.716$, $p = 0.0027$).

It seems from these analyses that although in the avian sample captive conditions are not significantly associated with presence or absence of same-sex mounting, in the mammalian sample significantly more reports of same-sex mounting are available from studies carried out in captivity (92.5% of all captivity studies) than from studies carried out in the wild (76.1%). Although this result may well be explained by the easier observability of rare behaviours in captivity than in the usually more difficult conditions of the field, I would regard the exact understanding of how conditions in captivity may favour the display of same-sex sexual behaviour in mammals as a legitimate area of scientific enquiry (Figure 2.7). On the one hand, living in a confined environment may simply produce specific conditions that are found in the wild only under unusual circumstances. In this case, same-sex sexual behaviour in captivity will just be a normal manifestation of the behavioural repertoire of the species in response to specific environmental and/or social circumstances. The situation, however, may be somewhat worrying with regard to the correct interpretation of behaviour if the

Table 2.1. Same-sex mounting observed in birds and mammals in studies carried out in captivity and in the wild

Species	Sex involved	Captive (C) or Wild (W)	Mounting	Reference
Birds				
Columba livia	M?	W	Y	Brackbill 1941
Phoebastria immutabilis	M	W	Y	Fisher 1971
Philomachus pugnax		W	N	Dit Durrell 2000
Phoenicopterus chilensis	M > F	C	Y	King 2006
Phoenicopterus ruber ruber	M = F	C	Y	King 2006
Phoenicopterus ruber roseus	F	C	Y	King 2006
Anas platyrhynchos		W	N	Lebret 1961
Cygnus atratus	M	W	Y	Braithwaite 1981; Brugger & Taborsky 1994
Branta canadensis		C	N	Koplam 1962
Anser c. caerulescens		W	N	Quinn *et al.* 1989
Anser anser	M	W	Y	Huber & Martys 1993; Kotrschal *et al.* 2006
Bubulcus ibis	M	W	Y	Fujioka & Yamagishi 1981
Ardea cinerea	M	W	Y	Ramo 1993
Egretta caerulea	M	W	Y	Werschkul 1982
Gallinula tenebrosa	M	W	Y	Garnett 1978
Porphyrio porphyrio melanotis	F > M	W	Y	Jamieson & Craig 1987a, b
Alca torda	M	W	Y	Wagner 1996
Uria aalge	M	W	Y	Hatchwell 1988
Melanerpes formicivorus	(M = F)?	W	Y	MacRoberts & MacRoberts 1976
Dinopium benghalense	M?	W	Y	Neelakantan 1968; Hutchins *et al.* 2002
Calypte anna	M	W	Y	Stiles 1982
Gypaetus barbatus	M	W	Y	Bertran & Margalida 2003
Falco tinnunculus	M	W	Y	Olsen 1985
Chionis minor		W	N	Bried *et al.* 1999
Tryngites subruficollis	M	W	Y	Prevett & Barr 1976; Myers 1979
Pluvialis apricaria		W	N	Parr 1992
Haematopus ostralegus	F	W	Y	Heg & van Treuren 1998; Ens 1998
Himantopus h. himantopus	F	W	Y	Kitagawa 1988
Tringa totanus	M	W	Y	Hale & Ashcroft 1983
Struthio camelus australis		W	N	Sauer 1972
Dromaius novaehollandiae		W	N	Coddington & Cockburn 1995
Rhea americana		W	N	Codenotti & Alvarez 1997
Pygoscelis adeliae	M	W	Y	Hunter *et al.* 1995; Davis *et al.* 1998
Phalacrocorax carbo sinensis		C	N	Kortlandt 1995
Agapornis roseicollis	F	C	N	Dilger 1960
Aratinga canicularis	F	C	N	Hardy 1963
	M	W	Y	Buchanan 1966
Brotogeris versicolurus		C	N	Arrowood 1988
Eolophus roseicapillus		C	N	Rowley 1990
	F	C	Y	Rogers & McCullock 1981
Scopus umbretta		W	N	Cheke 1968
Larus ridibundus	M	C	Y	van Rhijn & Groothuis 1987

Table 2.1. (*Cont.*)

Species	Sex involved	Captive (C) or Wild (W)	Mounting	Reference
Birds				
	M	C	Y	van Rhijn & Groothuis 1985
	M	C	N	van Rhijn 1985
Larus atricilla		C	N	Hand 1985
	M	C	Y	Noble & Wurm 1943
Larus argentatus		W	N	Shugart *et al.* 1987
		W	N	Shugart *et al.* 1988
		W	N	Shugart 1980
Larus occidentalis	F	W	Y	Hunt *et al.* 1984
	F	W	Y	Hunt & Hunt 1977
		W	N	Pierotti 1981
		W	N	Hunt *et al.* 1980
Larus novaehollandiae	M	W	Y	Mills 1994
Larus delawarensis		W	N	Conover 1984
		W	N	Conover & Hunt 1984a,b
		W	N	Fox & Boersma 1983
		W	N	Kovacs & Ryder 1985
		W	N	Kovacs & Ryder 1981
		W	N	Conover 1989
		W	N	Lagrenade & Mousseau 1983
Larus californicus		W	N	Conover 1984
Rissa tridactyla		W	N	Coulson & Thomas 1985
Menura novaehollandiae	M	W	Y	Smith 1988
Centrocercus urophasianus	F	W	Y	Gibson & Bradbury 1986
Gallus gallus	F	C	Y	Guhl 1948; Banks 1956
Fringilla coelebs		W	N	Sheldon 1994; Browne 2004
Carpodacus mexicanus	M	C	Y	McGraw & Hill 1999
Ficedula hypoleuca		W	N	Slagsvold & Saetre 1991
Lanius collurio		W	N	Fornasari *et al.* 1994
Cyanistes caeruleus		W	N	Kempenaers 1994
		W	N	Kempenaers 1993
Corvus monedula		W	N	Roell 1979
Callaeas cinerea wilsoni		W	N	Flux *et al.* 2006
Euplectes franciscanus	M	C	Y	Craig 1980; Craig 1982
Poephila acuticauda	M	C	Y	Langmore & Bennett 1999
Wilsonia citrina		W	N	Niven 1993
Chiroxiphia caudata	M	W	Y	Sick 1967
Mionectes oleagineus		W	N	Westcott 1992
Perissocephalus tricolor		W	N	Trail 1990
	M	W	Y	Snow 1972
Rupicola rupicola	M	W	Y	Trail & Koutnik 1986
Pseudonigrita arnaudi	M	W	Y	Collias & Collias 1980; Perrins 2003
Philetairus socius	M	C	Y	Collias & Collias 1978
Tachycineta bicolor	M	W	Y	Lombardo *et al.* 1994

Table 2.1. (*Cont.*)

Species	Sex involved	Captive (C) or Wild (W)	Mounting	Reference
Birds				
Riparia riparia	M	W	Y	Carr 1968
Hirundo pyrrhonota	M	W	Y	Brown & Brown 1996
Paradisaea raggiana		C	N	Frith & Cooper 1996
Notiomystis cincta	M	W	Y	Ewen & Armstrong 2002
Lichenostomus melanops cassidix	M	W	Y	Franklin *et al.* 1995
Mammals				
Suncus murinus	M	C	Y	Matsuzaki 2004
Dugong dugon	M	C	Y	Anderson 1997
Loxodonta africana		W	N	Buss & Smith 1966; Laws 1969
Felis catus	M > F	W	Y	Yamane 2006
Panthera leo persica	F	W	Y	Chavan 1981
	M	W	Y	Pati 2001
Panthera leo leo	M	C	Y	Cooper 1942
	F	C	Y	Cooper 1942
	M	W	Y	Schaller 1972
Acinonyx jubatus		W	N	Caro & Collins 1987
Helogale undulata rufula		C	N	Rasa 1979
	M ≫ F	C	Y	Rasa 1977
Canis familiaris	M ≫ F	W	Y	Pal *et al.* 1999
Canis lupus		C	N	Derix *et al.* 1993
Chrysocyon brachyurus	M	C	Y	Kleiman 1972
Speothos venaticus	M	C	Y	Kleiman 1972
		C	N	Drüwa 1983
Ursus arctos horribilis		W	N	Craighead *et al.* 1969
Desmodus rotundus		W	N	Wilkinson 1985, 1984
Myotis lucifugus	M	W	Y	Thomas *et al.* 1979
Balaena mysticetus		W	N	Würsig *et al.* 1993
Stenella longirostris	M	W	Y	Silva *et al.* 2005
Tursiops truncatus	M > F	W	Y	Mann 2006
		W	N	Connor & Smolker 1995
	M	C	Y	Shane *et al.* 1986
	M	C	Y	Östman 1991
Odobenus rosmarus	M	W	Y	Sjare & Stirling 1996
	M	W	Y	Miller & Boness 1983
		W	N	Miller 1976
	M	W	Y	Miller 1975
Halichoerus grypus	M	W	Y	Backhouse 1960
Mirounga angustirostris		W	N	Le Boeuf 1974
Neophoca cinerea	M ≫ F	W	Y	Marlow 1975
Phocarctos hookeri	M	W	Y	Marlow 1975
Rattus norvegicus	(M = F)?	C	Y	Barnett 1958
Oryctolagus cuniculus	F	C	Y	Albonetti & Dessì-Fulgheri 1990
Notomys alexis	M	C	Y	Happold 1976

Table 2.1. (Cont.)

Species	Sex involved	Captive (C) or Wild (W)	Mounting	Reference
Mammals				
Pseudomys albocinereus	M	C	Y	Happold 1976
Microcavia australis		W	N	Rood 1970
Marmota flaviventris	F	W	Y	Oli & Armitage 2003
Marmota caligata	F	W	Y	Barash 1974
Marmota olympus	F > M	W	Y	Barash 1973
Tamiasciurus douglasii	M	W	Y	Smith 1968
Tamiasciurus hudsonicus	M	W	Y	Smith 1968
	F	W	Y	Layne 1954
Sciurus carolinensis		W	N	Koprowski 1993, 1992; Thompson 1977, 1978
Macropus giganteus		W and C	N	Poole 1973; Poole & Catling 1974
Macropus rufogriseus banksianus		W	N	Johnson 1989
Macropus agilis	M	C	Y	Strirrat & Fuller 1997
Macropus parryi	M	W	Y	Kaufmann 1974
Macropus rufogriseus rufogriseus	F	C	Y	La Follette 1971; Strahan 1983
Aepyprymnus rufescens	F	C	Y	Johnson 1980
		W	N	Frederick & Johnson 1996
Sminthopsis crassicaudata		W	N	Morton 1978
Dasyurus hallucatus		W	N	Schmitt *et al.* 1989
Kobus leche kafuensis		W	N	Nefdt 1995
Kobus vardoni		W	N	Rosser 1992
	F	W	Y	de Vos & Dowsett 1966
Kobus ellipsiprymnus		W	N	Wirtz 1983
	F ≫ M	W	Y	Spinage 1982
Gazella thomsoni	M	W	Y	Walther 1978a,b; Weckerly 1998
Adenota kob thomasi		W	N	Ledger & Smith 1964; Leuthold 1966
	F	W	Y	Buechner & Schloeth 1965
Antilope cervicapra	M ≫ F	W	Y	Dubost & Feer 1981
Antilocapra americana	M	W	Y	Gilbert 1973; Maher 1997; Carling *et al.* 2003
	M ≫ F	W	Y	Kitchen 1974
Odocoileus hemionus columbianus	F	C	Y	Wong & Parker 1988
Odocoileus virginianus	M	W	Y	Hirth 1977
Alces alces		W	N	Van Ballenberghe & Miquelle 1993
Cervus elaphus	M	W	Y	Lincoln *et al.* 1970
	F	W	Y	Guinness *et al.* 1971
Cervus elaphus roosevelti	M	C and W	Y	Harper *et al.* 1967
Cervus nippon	M	C	Y	Chapman *et al.* 1984 (cit. in Bartoš & Holečková 2006)
Dama dama	M	C	Y	Holečková *et al.* 2000 (cit. in Bartoš & Holečková 2006)
Elaphurus davidianus	(M = F)?	C	Y	Schaller & Hamer 1978; Wemmer *et al.* 1983 (both cit. in Bartoš & Holečková 2006)

Table 2.1. (*Cont.*)

Species	Sex involved	Captive (C) or Wild (W)	Mounting	Reference
Mammals				
Muntiacus reevesi	M	C	Y	Barrette 1977 (cit. in Bartoš & Holečková 2006)
Rangifer tarandus	(M = F)?	C	Y	Bergerud 1974 (cit. in Bartoš & Holečková 2006)
Axis axis	M	W	Y	Schaller 1967 (cit. in Bartoš & Holečková 2006)
Pudu puda	M	C	Y	Bartoš & Holečková 2006
Giraffa camelopardalis	M	W	Y	Coe 1967
	M > F	W	Y	Pratt & Anderson 1985
	M	W	Y	Innis 1958
	M	W	Y	Leuthold 1979
Vicugna vicugna	M?	W	Y	Koford 1957
Tayassu tajacu	M	Semi-C	Y	Dubost 1997
Tayassu pecari	(M = F)?	Semi-C	Y	Dubost 1997
Phacochoerus aethiopicus	F	W	Y	Somers *et al.* 1995; Silva Downing 1995
Bison bonasus	F	W	Y	Krasinski & Raczynski 1967
	M	W	Y	Cabón-Raczyńska *et al.* 1987
	F	W	Y	Jaczewski 1958
Bison bison	M	W	Y	Rothstein & Griswold 1991
	M	C	Y	Reinhardt 1985a
	M > F	Semi-C	Y	Vervaecke & Roden 2006
	M	Semi-C	Y	Reinhardt 1985b
Bison bison athabascae	M	C	Y	Komers *et al.* 1994
Bos indicus	M ≫ F	Semi-C	Y	Reinhardt 1983
Bubalus bubalis	F	W	Y	Tulloch 1979; Nowak 1999
Ovibos moschatus	M > F	C	Y	Reinhardt & Flood 1983
Equus caballus	M > F	W	Y	Feist & McCullough 1976
Equus przewalskii	F	C	Y	Boyd 1991
		W	N	Van Dierendonck *et al.* 1996
Equus quagga		C	N	Schilder & Boer 1987
	M	Semi-C	Y	Schilder 1988
Equus zebra zebra	M	W	Y	Penzhorn 1984
Rupicapra pyrenaica ornata		W	N	Lloyd & Rasa 1989; Rasa & Lloyd 1994
		W	N	Lovari & Locati 1993
Pseudois nayaur	M	W	Y	Wilson 1984; Lovari & Ale 2001
Ammotragus lervia	M ≫ F	W	Y	Habibi 1987a,b
	M	C	Y	Katz 1949
Ovis orientalis musimon	F > M	C	Y	McClelland 1991; Silva & Downing 1995
Ovis aries	M	W	Y	Orgeur *et al.* 1990
Ovis canadensis	M	W	Y	Hogg 1987; Hass & Jenni 1991
Capra hircus	M	W	Y	Orgeur *et al.* 1990
Cebus capucinus	M	W	Y	Robinson & Janson 1987; Perry 1998
Propithecus verreauxi	M	W	Y	Richard 1974

Table 2.1. (*Cont.*)

Species	Sex involved	Captive (C) or Wild (W)	Mounting	Reference
Mammals				
Saimiri sciureus		W	N	Mitchell 1994
	F	C (all-F group)	Y	Talmage-Riggs & Anschel 1973
	M	C	Y	Ploog *et al.* 1963
Nasalis larvatus	F = M	W	Y	Yeager 1990; Boonratana 2002
Miopithecus talapoin	M	W	Y	Rowell 1973; Blaffer Hrdy & Whitten 1987
Cercopithecus aethiops[a]	M > F	W	Y	Gartlan 1969; Whitten & Turner 2004
Hylobates lar	M	W	Y	Edwards & Todd 1991
Presbytis entellus[b]	F	W	Y	Sommer 1988
	F	W	Y	Srivastava *et al.* 1991
	M > F	W	Y	Sommer *et al.* 2006
Papio ursinus	M	W	Y	Hall 1962; Altmann & Alberts 2003
Papio cynocephalus hamadryas	(F = M)?	C	Y	Kummer 1995
Papio cynocephalus anubis	M	W	Y	Smuts & Watanabe 1990
	M	W	Y	Smuts 1987
	M ≫ F	W	Y	Owens 1976
Theropithecus gelada	M ≫ F	C	Y	Bernstein 1975; Kawai *et al.* 1983
Macaca fuscata	M ≫ F	C	Y	Hanby & Brown 1974
	F	W	Y	Vasey & Duckworth 2006
	F	C (free range)	Y	Vasey 2006a
	F	C	Y	Vasey 2002a
	F	C	Y	Vasey Gauthier 2000
	F	C	Y	Vasey *et al.* 1998
	F	C	Y	Vasey 1998
	F	C	Y	Vasey 1996
	M	C	Y	Hanby 1974
	F	C	Y	Wolfe 1986
	F	W	Y	Takahata 1982
	(F = M)?	W	Y	Enomoto 1974
	M	W	Y	Takenoshita 1998
	F	C	Y	Lunardini 1989
	F	C	Y	Chapais & Mignault 1991
	F	C	Y	Rendall & Taylor 1991
	F	C	Y	Corradino 1990
Macaca arctoides	F	C	Y	Slob *et al.* 1986; Blaffer Hrdy & Whitten 1987
	(M > F)?	C	Y	Chevalier-Skolnikoff 1976
	F > M	C	Y	Chevalier-Skolnikoff 1974
		C	N	Gouzoules 1974
Macaca mulatta	F	C	Y	Akers & Conaway 1979
	M	C	Y	Reinhardt *et al.* 1986
	F	C	Y	Fairbanks *et al.* 1977
	M	C	Y	Gordon & Bernstein 1973

Table 2.1. (*Cont.*)

Species	Sex involved	Captive (C) or Wild (W)	Mounting	Reference
Mammals				
	M > F	C	Y	Bernstein et al. 1993
	F	C	Y	Loy & Loy 1974
	F	C (free range)	Y	Kapsalis & Johnson 2006
	F	C	Y	Huynen 1997
Macaca nemestrina	F	C	Y	Giacoma & Messeri 1992
	(M = F)?	W	Y	Oi 1990
Macaca radiata	M	C	Y	Silk 1994
Macaca tonkeana	(M = F)?	Semi-C	Y	Thierry 1986; Lindefors 2002
Macaca nigra	M	W	Y	Silva & Downing 1995; Reed et al. 1997
	(F = M)?	C	Y	Dixson 1977
Pan paniscus	F ≫ M	W	Y	Hohmann & Fruth 2000
	F ≫ M	W	Y	Fruth & Hohmann 2006
Pan troglodytes	F	C	Y	Firos Anestis 2004
		W	N	Nishida 1997
	F	C	Y	Yerkes 1939
Gorilla gorilla	M	W	Y	Yamagiwa 2006
	M	W	Y	Yamagiwa 1987
	F	W	Y	Harcourt 1979
	M F	W	Y	Nadler 1986
	M	W	Y	Robbins 1996
	F	C	Y	Fischer & Nadler 1978
Pongo pygmaeus	M	W	Y	Fox 2001, 2002
Homo sapiens	M > F	W	Y	Kinsey et al. 1948; Brown et al. 2007
	M > F	W	Y	Kinsey et al. 1953
	M	C*	Y	Hensley 2000 (* Prison)
	F	C*	Y	Hensley 2000 (* Prison)
	M	C*	Y	Ashworth & Walker 1972 (* Boarding school)
	F	C*	Y	Ashworth & Walker 1972 (* Boarding school)

[a] *Cercopithecus aethiops* is currently known as *Chlorocebus aethiops*.
[b] *Presbytis entellus* is currently known as *Semnopithecus entellus*.

increased levels of same-sex mounting in the captive population of a mammal species are proven to be the result of, say, an industrial pollutant. In such a case, using those studies as models for the understanding of same-sex sexuality in the wild or in an evolutionary perspective may be problematic and much care should be taken about which kind of inference is drawn from the study. As far as this work is concerned, the issue of captivity vs. wild is unlikely to significantly affect the analyses of the mammalian dataset as only 24.2% (26/107) of the species were studied exclusively under strict captive

Figure 2.7. Male–male mounting in fallow deer (*Dama dama*) in captivity. Photo courtesy of Luděk Bartoš.

conditions. Moreover, a similar percentage of the avian species were studied exclusively in captivity 22.2% (16/72).

Other, more specific methodological issues will be addressed in the text as they arise. In the next chapter I review both the theoretical advances and the empirical findings regarding the specific genetic mechanisms that may contribute to the evolution and expression of same-sex sexual behaviour.

Genetics of homosexuality

> I never have seen any literature on homosexuality with which I agreed. I have my own theory that the tendency to homosexuality is basically inherited. It may be brought out through environment. The parents are usually not homosexual but the uncles and aunts may be. I have a suspicion that this might be worth investigating.
> (Michael D., homosexual participant in Henry's (1941) study, p. 144)

That homosexuality, as most other phenotypic traits, depends directly or indirectly, along a large or small causality network, on the expression of one or more genes should be regarded as a truism of little informative power. Studies on the genetics of homosexuality, if they are to be useful, should address more specific questions, such as the ones listed in Box 3.1.

Needless to say, addressing such issues is far from easy, but progress is being made as a result of research on non-human animals; Michael D. would probably have been thrilled to know that advances are also occurring in the study of the genetic mechanisms underpinning human homosexuality, as I will show in this chapter.

The various aspects of the genetic basis of homosexuality could be encapsulated in the following three key questions:

(a) Which genes are involved in the development of the trait, how and when are they expressed in the organism and what is that they actually do?
(b) How are those genes inherited?
(c) How did they arise in the population in the first place and how are they maintained over time?

That is, what we have to address are the issues of identity and mode of operation of genes, the modalities of inheritance and the evolutionary processes involved in their persistence.

In an early review, Michael Ruse (1981) speculated that new mutant genes directly controlling the expression of homosexual preference could be retained in the population, in spite of their apparent negative effect on the individual's lifetime reproductive success (see, for example, Bobrow & Bailey 2001; Rahman & Hull 2005), by well known mechanisms such as heterozygote advantage (Hutchinson 1959), kin selection (Williams 1966; Wilson 1975) or parental manipulation (Trivers 1974). To those, we should also add other potential mechanisms such as increased fertility of females in the case of sexually antagonistic genes (Trivers 1972; Rice 1992), pleiotropic effects (Miller 2000), linkage disequilibrium with positively selected loci such as parental care in the case of reproductively active homosexuals and bisexuals (Miller 2000) or reciprocal altruism or mutualism (Trivers 1971), interdemic selection (Wilson 1975) and high mutation rate (Hamer & Copeland 1994). Note that recurrent mutation could explain the maintenance of a trait at low frequency in a population even if it is highly detrimental to the mutant's survival and reproduction; moreover, given a fixed mutation rate, the larger the population the more likely it is that some low-frequency phenotypes may be maintained by recurrent mutation alone, with no involvement of selection or even in the face of selection against mutants.

> **Box 3.1. Some of the questions that need to be answered in order to understand the genetics of homosexuality.**
> **Many of these issues will be addressed in this chapter and throughout the rest of the book.**
>
> - Which loci are involved in the determination of which aspect of the trait and what is the mapping of those loci on chromosomes?
> - What is the specific role of gene products (polypeptides) and nucleic acids (e.g. RNA) on processes leading to expression of behaviours such as same-sex sexual partner preference (e.g. regulation of expression of genes affecting the development and differentiation of neural tissues that control choice of sexual partner)?
> - What are the heritability values and the modes of inheritance of the trait (paternal, maternal)?
> - Are there pleiotropic effects (multiple phenotypic effects of a single locus or several closely linked loci)?
> - Is the trait polygenic (whether more than one gene is involved)?
> - Are there epistatic effects (interactions between genes)?
> - What is the level of penetrance (the degree of phenotypic expression) of the genes involved?
>
> Is there any
> - Linkage (joint inheritance of alleles at different loci);
> - Dominance effects (dominant, recessive alleles);
> - Codominance (expression of both alleles at a locus) or overdominance (greater phenotypic expression in heterozygotes than homozygotes);
> - Heterozygote advantage (heterozygote fitter than either homozygote);
> - Polymorphism (multiple alleles present in the population);
> - Genomic imprinting (parental effects on probability of expression of specific alleles);
> - Way in which alternative splicing of the same gene (i.e. different ways of cutting the same primary RNA transcript producing different secondary transcripts that are then translated into different polypeptides) may produce different phenotypes including one or more homosexual variant?
>
> - Are the loci involved in homosexuality subject to sexually antagonistic selection whereby one or more of them may have reproductively detrimental effects if expressed in males but reproductively beneficial effects if expressed in females?

The idea that genetic variability can be maintained by high mutation rate in spite of natural selection has a long history and derives from the early work of Ronald A. Fisher (1922) and John B. S. Haldane (1928). It is encapsulated in the concept of *mutation–selection equilibrium,* which is a balance in the frequency of alleles over time resulting from the combined effect of selection against those alleles, which decreases their frequency, and mutation, which increases it. It has recently been shown that *mutation–selection equilibrium* is a plausible mechanism to explain the maintenance of some traits that decrease fitness (Zhang & Hill 2005). From the standpoint of the current exercise, the trait in question would be exclusive homosexuality.

LeVay & Hamer (1994) and Lim (1995) reviewed the early research on the genetic basis of human homosexuality, based on linkage analysis, family tree analysis and twin studies that were published at the time and that I will update here. More recently, Kirkpatrick (2000) has highlighted that in human societies there is a marked trend for homosexual behaviour to be more prevalent among men than women, with 52.3% (11/21) of the societies surveyed showing male-only homosexuality, 42.8% (9/21) showing both male and female homosexuality and 4.9% (1/21) female-only homosexuality. Although the potential causes of such a trend are likely to be many, and not necessarily mutually exclusive, a sex-chromosomal involvement in the inheritance of the trait is clearly consistent with such a male bias, whether genes contributing to the development of a homosexual orientation are located in the Y chromosome (only found in men), in the X chromosome, in which case they are more easily expressed in men (who are the heterogametic sex), or in both X and Y chromosomes.

In principle, therefore, any serious student of sexual orientation should attempt to formulate testable hypotheses that link the expression of homosexual behaviour in animals to specific loci. Emmons & Lipton (2003) reviewed the genetic mechanisms of sexual behaviour known in male vertebrates and invertebrates, listing several mouse (*Mus musculus*) genes that may be involved in homosexual behaviour, such as *trp2* that affects discrimination of the sexes, *αER* and *βER* that affect sexual behaviour and, in the prairie vole (*Microtus ochrogaster*) *V1ar* that affects affiliative sexual behaviour. Emmons & Lipton (2003) also reviewed work done in *Drosophila*, clearly indicating that loci controlling the development of adult sexual behaviour such as *tra-2*, *fruitless* (*fru*) and *dissatisfaction* (*dsf*) are involved in the display of same-sex sexual behaviour as well. In this chapter I review many of the loci that have been proposed as being responsible for various aspects of homosexual behaviour in several species.

Some of the major themes that will be developed in this chapter have also been reviewed by Craig *et al.* (2004) and Dawood *et al.* (2009). From a genetic point of view, mutations that affect the ability of steroid hormones to modify the development of the brain have been the focus of much research. Although the very early stages of sexual development in some mammals are not under hormonal, but under direct genetic control (Craig *et al.* 2004), hormones are central to the subsequent pre-, peri- and postnatal development of various areas of the brain. Specific endocrine effects of mutations in the development of human sexuality, such as 5α-reductase deficiency, androgen insensitivity syndrome, congenital adrenal hyperplasia, congenital aromatase deficiency and altered testosterone biosynthesis, will be thoroughly reviewed in Chapter 5.

Although we will see that many genes are likely to be involved in determining aspects of homosexual behaviour, it is not surprising that loci in the sex chromosomes have been a major focus of research in the search both for specific genes and also for the broader genetic mechanisms explaining male and female differences in same-sex sexual behaviour. Mammal males are hemizygous for most sex chromosome-linked loci (i.e. they have only one copy of the gene, rather than the usual two; genes in the PAR region of the Y chromosome are an exception) (Arnold 2004) and therefore are more likely to express recessive alleles carried by the X chromosome that may decrease reproductive success. In birds, the hemizygous sex is female (ZW) not male (ZZ).

Even though one could unequivocally prove that one or more loci in the X chromosome affect specific developmental pathways leading to homosexuality, not all carriers of the allele/s may in fact become homosexual. Just to remain within the genetic realm and assuming a sex-chromosome linked locus, females may be heterozygous for the hypothesised locus, penetrance may vary, or the gene may be inactivated in some crucial tissues. Thus carriers of the allele, even if dominant, may never develop homosexual preferences. Chromosomal inactivation, a process that will be reviewed in this chapter, could even explain marked phenotypic differences between 'identical' female twins (see, for example, Weksberg *et al.* 2002). Thus evidence showing that female twins are discordant for sexual orientation is not unequivocal proof against the existence of a 'gene for homosexuality' (Turner 1995; Craig *et al.* 2004).

That specific genetic mechanisms explaining some aspects of homosexuality are very likely to be found is one thing; to actually prove their mode of operation in empirical studies is quite another. The study of the genetics of same-sex sexual behaviour is fraught with potential methodological pitfalls. Just to mention a few, research on homosexuality in humans that relies on questionnaires to determine sexual orientation must also include molecular genetic evidence for the degree of relatedness between siblings (e.g. monozygotic or dizygotic twins) and other relatives. Samples should be as complete as possible with regard to relatives and as representative as possible with regard to the whole population (McGuire 1995). Heritability estimates in twin and family studies also require a sufficient sample size – to reach adequate statistical power – that usually runs into the hundreds of families and a few thousand twin pairs (McGuire 1995). Other problems include samples that may be non-random (e.g. when recruitment of participants is through specific gay societies in human studies) or when crucial tests to tease apart genetic and environmental effects are not conducted (e.g. studies of monozygotic twins reared apart *from birth* vs. monozygotic twins reared together vs. dizygotic twins reared together and apart) (McGuire 1995) thus jeopardising the validity of the conclusions of the study. Studies of non-human animals are also subject to equally specific methodological constraints but in that case the researchers are relatively less hampered by the ethical issues that necessarily constrain research on humans.

In what follows I will start by reviewing some of the major theoretical genetic models that have been proposed to explain the evolution of homosexuality, I will then continue with the role of specific genes that have been studied in both invertebrates and vertebrates and also the potential role of sex chromosomes and processes involving DNA-methylation. I will subsequently review studies that have used family analyses and then twin studies, concluding with a section focusing on selection studies. I will offer syntheses of the major findings regarding the diverse genetic aspects of homosexuality throughout, along with ideas for further research and a critique of the current evidence.

Some genetic models of homosexuality

My view of a scientist's ideal world is one where theoreticians and empirical scientists work in close collaboration to produce the best hypotheses, which are quickly and efficiently tested using data from controlled experiments. In reality, theoreticians and empiricists tend to live, more often than not, in island fortresses that sometimes stand many miles apart from each other. Although progress in science can, and indeed does, occur anyhow, it is when those intellectual islands are connected by communicating bridges that growth of knowledge may progress at a faster pace. I therefore start my review of the genetic basis of homosexuality with an exposition and constructive critique of some major evolutionary genetics models of homosexuality.

Models of homosexuality have been produced that address the issue of the genetic basis of same-sex sexual behaviour at different levels of analysis. Ellis & Ames' (1987) *Gestational Neurohormonal Theory* focuses on the potential role of mutations in determining the exposure of the developing brain to testosterone, oestradiol and other sex hormones. According to this hypothesis the changed conditions of such early developmental exposure to steroids result in the development of homosexual preferences. Although relying on mutations, this model puts more emphasis on endocrine-controlled developmental processes, which will be reviewed in Chapter 4 and at the beginning of Chapter 5.

Wayne Getz (1993) produced a selection model that involves an autosomal locus affecting both mating success (e.g. through sexual orientation whereby males behaving homosexually have less success at fertilising females compared with heterosexual males) and parental care. The main assumptions of the model are: infinite population size, non-overlapping generations and random segregation of alleles. Readers will recognise these as the usual assumptions required for the application of Hardy–Weinberg equilibrium in genetic modelling. Those conditions are not necessarily fulfilled by all or even most vertebrate populations in the wild, an issue that already poses a challenge to this kind of modelling. This approach, however, does provide a benchmark against which data can be compared, and if the model is falsified, then specific alternatives may be produced that relax some of the assumptions. The author considers the conditions for invasion of a population carrying a wild β allele by a mutant α allele. The α allele is assumed to encode for increased parental behaviour and, in males, also increased homosexual mate preference. The phenotype of the heterozygous αβ individuals is assumed to be intermediate between those of the homozygous αα and ββ individuals. Getz analyses two main scenarios: (*a*) that α increases parenting behaviour when it is expressed in both sexes (sexually mutualistic scenario), and (*b*) that α is most beneficial in terms of parenting when expressed in females (sexually antagonistic scenario) whereas β is most beneficial from a parental behaviour point of view when it is expressed in males. Scenarios for the stable polymorphism of the two alleles in the population are derived by Getz and he concludes that recessive autosomal alleles for male homosexuality are unlikely to invade the original population even though they may increase parenting in females (scenario *b*). However, if the α allele is dominant in both sexes then a stable polymorphism between α and β is possible in the population and homosexual males may coexist with heterosexual males if their mother (who is homozygous or heterozygous for the α allele) displays an increased level of parental care. The α allele could also be retained in the population in a stable polymorphism if it improves the reproductive success of male and female as a cooperating breeding pair (scenario *a*), in which case males may display a bisexual sexual orientation. In both cases, the α allele is more likely to be retained in the population as the parental care effect increases. Therefore in this model both sexually antagonistic and sexually mutualistic scenarios are considered.

Selection for increased parental care in both sexes and increased degree of homosexuality in males may involve the occurrence of breeding pairs where the male is actually bisexual (*sexually mutualistic selection*), as I have already mentioned; whereas if the increase in parental care is in females only, then the scenario is one of *sexually antagonistic selection*. However, note that an allele that increases 'parental care' but decreases heterosexuality in males is one that is expected to also produce an *alloparenting* male phenotype (i.e. a phenotype that provides care to the offspring of others, those of close relatives for instance) if the male, rather than being bisexual, is an exclusive homosexual, a scenario consistent with *kin selection*. That is, in my opinion, from Getz's model one may derive an evolutionary scenario whereby *sexually antagonistic selection*, *sexually mutualistic selection* and *kin selection* may affect the evolution of homosexuality in a concerted manner: alleles that increase fecundity in females and homosexuality in males may become associated with alleles that increase parental care in both males and females. This may subsequently lead to the production of some exclusively homosexual offspring that provide alloparental care to close relatives. This concept will be further developed in this chapter and also in Chapter 10.

Cook (1996) argued that populations with a greater percentage of homosexuals would have lowered population growth rates and dampened population fluctuations. Such population dynamics are usually associated with lower probability of extinction and increased geometric mean of fitness in a variable environment. This interdemic selection argument, however, is more relevant to the understanding of female homosexuality than male homosexuality, as it is usually females that are the limiting sex in the determination of population growth rate and fluctuations. Many males do not reproduce in most populations of wild vertebrates (see Chapters 7 and 8), without being homosexuals, and this may be inconsequential to population growth provided that all females are inseminated. Males can become more limiting as soon as they are needed for the provision of parental care, but in contrast with Getz (1993), in her model Cook (1996) assumes that males only contribute genes. Under this scenario and the additional absence of kin selection, as assumed by Cook (1996), recurrent mutation is needed for the maintenance of homosexual genotypes in the population at low frequency. The model assumes two loci: an A locus affecting viability and an H locus affecting

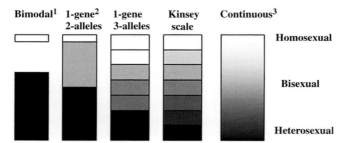

Figure 3.1. Some alternative models for the inheritance of homosexual behaviour. [1]This reflects a bimodal distribution of phenotypes expected from 1-gene, 2-alleles, one dominant and one recessive, mode of inheritance. [2]This reflects a co-dominance model with one locus and two alleles. [3]A polygenic, multiallelic trait that is also subject to environmental effects will produce a continuous distribution of phenotypes. The Kinsey scale is a recognition of the diversity of phenotypes ranging from heterosexual to bisexual and homosexual. Adapted from MacIntyre & Estep (1993).

homosexuality (the latter being subject to recurrent mutation). Cook (1996) models four scenarios: (a) the homosexuality locus is autosomal and the allele may be either dominant or (b) recessive and (c) the homosexuality locus is located on the X chromosome and the allele may be either dominant or (d) recessive. The author concludes that, under the above conditions, an allele for homosexuality will have its greatest fitness effect (in terms of increased geometric mean of fitness in the population under fluctuating population sizes) when it is located on the X chromosome in close linkage with the viability locus or, alternatively, both loci are located in an autosome and the homosexuality allele is recessive.

A more explicit interdemic selection model has been proposed by Kirby (2003). In this model an association between homosexuality and sociality and cooperation may lead local populations that have higher proportions of homosexual individuals to occupy the best and most productive habitats. This, in turn, would lead to increased local population growth within those habitats as heterosexual and bisexual carriers of the hypothesised homosexuality allele/s would reproduce, taking advantage of the abundant resources available. Habitats containing local populations that include homosexual individuals will therefore produce more dispersers which will spread the allele over broader areas provided that the allele is carried by reproducing members of the population (e.g. heterozygotes), leading to the maintenance of homosexuality over time. If homosexuality is a polygenic and multi-allelic trait or if it is determined by a single locus with alleles showing variable penetrance, then the trait will display a degree of phenotypic variability across individuals; consequently, even some homosexuals may be reproductive. This would further increase the probability of maintenance of the trait in the population. An interdemic selection argument for the evolution of human homosexuality based on the benefits of cooperation and decreased inter-male competition in reproduction was also put forward by Schuiling (2004).

MacIntyre & Estep (1993) provide a graphic depiction of the potential modes of inheritance of homosexuality and how a relatively simple genetic model of two loci with two alleles per locus can already provide enough phenotypic variability in the population to be a reasonable representation of the observed distribution of sexual orientations as measured, for instance, by the *Kinsey Scale* (Kinsey *et al.* 1948): from heterosexuality to homosexuality going through various gradations of bisexuality (Figure 3.1).

Although MacIntyre & Estep (1993) support, on theoretical grounds, the heterozygote advantage hypothesis first proposed by Hutchinson (1959), they also add an additional twist by highlighting the potential selective advantage of heterozygous males at the homosexuality locus if such heterozygotes display a bisexual behaviour (e.g. in the case that the alleged alleles are co-dominant). They suggest that bisexual males may be more successful at competing for access to females both through direct female choice and also through preventing other males, i.e. their partners in same-sex intercourse, from mating with females. This latter scenario makes sense only if what same-sex sexual behaviour is doing is to mediate dominance relationships for instance that may limit subsequent access to females by the mountee male, or it may decrease female choice for such males. In this model exclusive male homosexuality may derive from an initial selection for bisexuality.

More recently, Gavrilets & Rice (2006) have proposed a one locus, two allele model of homosexuality for a diploid population of infinite size, displaying random mating and non-overlapping generations. The dominant allele **A** is assumed to have little or no effect on sexual orientation, whereas the recessive allele **a** masculinises or feminises both sexes in such a way that it leads to homosexual behaviour in the discordant sex (e.g. an **aa** male will be feminised and therefore more likely to develop homosexuality). As I have already mentioned in Chapters 1 and 2, sexual orientation (homosexual, heterosexual, bisexual) and gender role (feminine, masculine) are independent as there is a great region of overlap in masculinity and femininity among homosexuals, bisexuals and heterosexuals in humans. Therefore this initial aspect of Gavrilets & Rice's (2006) model may cause some confusion. I will assume here that what they meant by 'masculinising' and 'feminising' was not gender role but sexual orientation: a masculinised female prefers other females as sexual partners, whereas a feminised male prefers other males, with no implications for gender role. In their initial approach Gavrilets & Rice (2006) assume autosomal inheritance of the gene and consider three alternative scenarios:

(a) that there is heterozygote advantage in both sexes,
(b) that there is heterozygote advantage in one sex and directional selection in the other, and
(c) that there is sexually antagonistic selection, i.e. that the allele that feminises males decreases their fitness but increases the fitness of females.

They then consider the alternative that the gene is not autosomal but X chromosome-linked. The different scenarios in this case are:

(a) heterozygote advantage in the homogametic sex (i.e. females in mammals but males in birds),
(b) sexually antagonistic selection in the feminising allele (allele **a** is beneficial to females but detrimental to males),
(c) sexually antagonistic selection in the masculinising allele (allele **a** is beneficial to males but detrimental to females).

They finally consider maternal effects and combined maternal and directional selection effects.

Their results suggest that if the gene is autosomal, heterozygote advantage is a plausible scenario in both sexes although they prefer the alternative of heterozygote advantage in one sex and directional selection in the other. Thus a feminising allele may be advantageous to males in the heterozygous combination but always advantageous to females, especially if it comes in the homozygous form. If there is sexually antagonistic selection on allele **a**, then their model predicts that the allele will be selected if the gain in fitness accrued by females offsets the loss in fitness of males due to their homosexual orientation. That is, homosexual males should be born of mothers who have higher reproductive rates. In the case that the gene is X-linked and there is heterozygote advantage in the homogametic sex, homosexuality can be maintained provided its fitness costs are not too high. In the case of sexually antagonistic selection for the feminising allele, the allele will be selected more easily if it is dominant and X-linked, and therefore more likely to also be expressed in female mammals (the homogametic sex) where it is more beneficial. If the allele is masculinising and there is sexually antagonistic selection (the allele confers higher fitness to males than females) then **a** can reach fixation if it is recessive. This is easily seen as a recessive X-linked masculinising allele will be less likely to be detrimental to females (only homozygous recessive females will be masculinised). Gavrilets & Rice's (2006) model, which simultaneously considers the potential contribution of specific types of chromosomal inheritance, the effect of overdominance (i.e. heterozygous advantage) and both directional and sexually antagonistic selection, has been highly praised by Savolainen & Lehman (2007) in a recent review. Indeed, one of the advantages of such a model is the ability to explore the consequences of many evolutionary scenarios within the same theoretical framework, an integrative approach that I strongly support.

One of the most recent genetic modelling for the evolution of human homosexuality that I am aware of is that of Camperio-Ciani *et al.* (2008a). The authors explore the performance of various genetic mechanisms in terms of which one is better able to explain the maintenance at low frequency of a male homosexual phenotype in the population, and, importantly, they also test their various alternative models with pedigree data from a human population. They refer to 'genetic factors influencing male homosexuality or bisexuality' or GFMH and study various modes of inheritance of those factors: single locus or two loci, assuming, as usual, non-overlapping generations and infinite population size. The GFMH-associated allele is assumed to be dominant or overdominant, in the latter case the expression of same-sex sexual behaviour will be higher in heterozygotes. The various genetic models that they test are summarised in Box 3.2.

In brief, the following major evolutionary mechanisms are considered in the work of Camperio-Ciani *et al.* (2008a) in various combinations: one/two-locus determination, autosomal/sex chromosome inheritance, overdominance (heterozygote advantage), maternal effects (including genomic imprinting), sexually antagonistic selection and directional selection. Their results suggest that homosexuality is unlikely to be a trait controlled by one locus, whereas two-locus models seem to be better able to explain the maintenance of the trait at low frequency in the population. In particular, the model that best fits their human pedigree data is a two-locus model, involving two X-linked loci and sexually antagonistic selection. The model not only predicts the maintenance of male homosexuality at low frequency in the population, but it also predicts that homosexual males should be more frequent along the maternal line of descent and that mothers of homosexual males should have higher fecundities than mothers of heterosexual males. Those are predictions that fit the available data. For instance, Iemmola & Camperio-Ciani (2009) show how average fecundity of mothers of homosexuals (2.73 offspring) is significantly higher than that of mothers of heterosexuals (2.07 offspring). A significant difference was also found for fecundity of maternal aunts, maternal grandparents and sons and daughters of maternal grandparents. The only significant difference in fecundity that they obtained along the paternal line was for paternal uncles: paternal uncles of heterosexuals had higher fecundity than those of homosexuals. Therefore this work tends to support *sexually antagonistic selection* as a major mechanism maintaining homosexuality in the population. Although their model points to an X-linked GFMH, they also leave the possibility of an autosomal contribution open.

A recent variation on the sexually antagonistic selection theme has been published by Rice *et al.* (2008). They provide a model of sexually antagonistic zygotic drive based on competition between siblings of different sexes that they suggest could explain the emergence of female homosexuality in humans. This particular model, however, is still in its infancy.

In sum, the theoretical work produced so far on the potential genetic mechanisms leading to the evolution and maintenance of homosexuality is almost as diverse as the contexts in which homosexuality is manifested. With regard to chromosomal determination of the trait, models produce scenarios that could bias determination towards the sex chromosomes or autosomes, but the possibility that

Box 3.2. Genetic models considered by Camperio-Ciani et al. (2008a).

One-locus Models

The various one-locus models that they test include, following their notation:

(1a) autosomal locus and heterozygote advantage,
(1b) autosomal locus with directional selection in females and overdominance in males,
(1c) autosomal locus and sexually antagonistic selection,
(2a) X-linked locus and overdominance in females,
(2b) X-linked locus and sexually antagonistic selection for an allele that favours females or
(2c) for an allele that favours males,
(3a) autosomal locus exerting maternal effects on males (maternal selection) and producing selectively positive effects on females,
(3b) X-linked locus having the same effects as in 3a.

Two-locus Models

They also test the following two-locus models:

(4) X-linked locus with either
(4a) one autosomal locus or
(4b) a second X-linked locus and sexually antagonistic selection,
(4c) two autosomal loci and sexually antagonistic selection,
(5) one autosomal locus with either
(5a) one X-linked locus or
(5b) a second autosomal locus with also overdominance in males and, in females, directional selection;
(6) two independent autosomal loci with one of them undergoing maternal genomic imprinting, and finally
(7) two autosomal loci undergoing maternal selection in males and, in females, direct selection.

---

loci in both kinds of chromosomes could be involved is not necessarily excluded. The theoretical emphasis has focused on the following major mechanisms: (a) selective value of homosexuality via cooperation enhanced by interdemic selection (Kirby 2003), interdemic selection mechanisms were also emphasised by Cook (1996) and Schuiling (2004); (b) association of homosexuality with increased parental care with emphasis on sexually antagonistic or sexually mutualistic selection (Getz 1993; Camperio-Ciani et al. 2008a); (c) heterozygote advantage in the context of sexual selection (MacIntyre & Estep 1993); (d) Gavrilets & Rice (2006) and Camperio-Ciani et al. (2008a) provide what are probably the most synthetic models available to date that take into account heterozygote advantage, sexually antagonistic selection, directional selection and maternal effects (parental manipulation), although tests of a model that includes those evolutionary mechanisms suggest that sexually antagonistic selection may be particularly important (Camperio-Ciani et al. 2008a;

Iemmola & Camperio-Ciani 2009; see also Cook 2008). Additional empirical support for sexually antagonistic selection in the evolution of homosexuality also comes from studies carried out by Zietsch *et al.* (2008) in humans and experiments carried out by Stellflug & Berardinelli (2002) on Rambouillet sheep (see the section '*Ovis aries*: the case of homosexual rams' in Chapter 5 for a more detailed account of this work). (e) Kin selection has not been strongly emphasised in these recent models, but, as I already indicated in my discussion of Getz's work, it remains a relevant hypothesis. In fact, I will rescue it in the final section of this chapter and integrate it into a biosocial model along with other selective mechanisms in Chapter 10. (f) A final group of models, however, emphasises what may be probably the most neglected mechanism that could explain the appearance and maintenance of homosexuality but one that has great chances to also be proven of importance: recurrent mutation (Hamer & Copeland 1994; Cook's 1996 model also relies on recurrent mutation).

Recurrent mutation as a genetic mechanism underlying homosexuality

We tend to underestimate recurrent mutation as an important factor in the persistence of a trait as mutation rates in most loci are usually very low in most species. For this reason evolutionary biologists tend to explain the maintenance of a trait in a population more in terms of genetic drift, natural selection or linkage with selected loci. However, in large populations or in species with particularly high mutation rates per locus, a trait could be potentially maintained at low frequencies by recurrent mutation alone, thus explaining the persistence through time of traits that may be detrimental to direct fitness (i.e. lifetime production of own offspring). In humans, for instance, estimated mutation rates per genome per generation are high for a vertebrate with a relatively low individual reproductive rate: about 175 new mutations per genome per generation (Nachman & Crowell 2000) and about 1.6 new deleterious mutations per genome per generation (Eyre-Walker & Keightley 1999). Moreover, if the probability that those mutations are expressed in tissues of the human brain is high, it will make their phenotypic effect on behaviour even more likely. In fact, more genes are expressed in the human brain than in the brain of other primates such as chimpanzees and rhesus macaques, suggesting that if mutations occur in those genes they may be expressed as well (Cáceres *et al.* 2003). Cáceres *et al.* (2003) described 169 genes that are differently expressed in the cortex of humans and chimpanzees, and about 90% of the genes studied are more highly expressed in the human brain. If exclusive homosexual behaviour is mainly maintained by recurrent mutation (i.e. the locus or loci are at *mutation–selection equilibrium*; see Zhang & Hill 2005) then the frequency expected in each generation is low and it may require selection for genome-wide increased mutation rates to persist. A mechanism relying mainly on recurrent mutation seems to run against the evidence of family effects in the distribution of homosexuality that will be mentioned below (see 'Family analysis' section in this chapter). However, to the best of my knowledge, nobody has complemented the information on family effects with studies of mutation rates between families. The high mutation rate hypothesis predicts that family groups that tend to produce more homosexual offspring should also show higher mutation rates than the rest of the population. This is a prediction that these days should be very easy to test. Some preliminary data from a Brazilian study do suggest that families characterised by higher levels of inbreeding may be also more likely to have a homosexual male offspring (Kerr & Freire Maia 1983), a result that is consistent with a mechanism of increased expression of recessive mutations.

Moreover, if homosexuality can result from a diversity of mechanisms where more than one locus is involved (see Camperio-Ciani *et al.* 2008a for a theoretical and empirical argument and Mustanski *et al.* 2005 for additional empirical evidence), then mutation rates per locus must be added up across all the loci involved, with those mutations being subsequently available for selection. This leads us to an important corollary: in populations with high mutation rates, additional mechanisms such as sexually antagonistic selection and kin selection will significantly increase the probability that the homosexual trait will be maintained over time. In other words, alongside specific arguments in favour or against individual evolutionary mechanisms, we also require arguments that consider an integration of the various mechanisms (some or all) into a more complex model (see Chapter 2).

Candidate genes for homosexual behaviour

As soon as we become convinced that there must be genes that are somehow contributing in a non-trivial manner to the development of homosexual phenotypes, the hunt for those genes is obviously on. How do we know that this quest is not just an unhealthy fixation likely to lead our research effort astray? For those who do not regard the genetic contribution to homosexuality as a truism, I offer two simple reasons that make such an enterprise worth pursuing, at least from a technical perspective: the first, and most important one, is that genes have already been found in both vertebrates and invertebrates that control sexual behaviour and aspects of sexual orientation. The second is that even in those cases where we can prove that homosexual behaviour can develop through early experiences in specific social environments, we know that not all individuals experiencing such environments will become homosexuals (see, for example, Peplau *et al.* 1998) and, conversely, not all homosexuals have experienced that kind of environment.

How can we explain such a variability of outcomes? Although I argue throughout this book that the importance of environmental effects (e.g. the social milieu, maternal environmental effects during gestation) on the biological processes determining homosexuality is undeniable, I can only hope that even the staunchest social constructionist would agree that a reasonable possibility is that different genetic make-ups may account for the different responses of individuals to similar environments. That the environment does not just mould passive individuals throughout their development even in humans is probably most dramatically exemplified by the phenomenon of resilient children (Cove *et al.* 2005 and references therein), i.e. children who are able to conduct a normal adult life in spite of having experienced dreadful social conditions during their development. Where do their phenomenal coping abilities come from? It is true that a social environment of hope and material support at some later stage in their life may help in the process, but what about those who did it pretty much on their own?

We will see in a later section how the issue of disentangling the genetic and environmental contributions to the expression of a homosexual phenotype can be addressed by the study of twins. Finally, as we already know from much quantitative genetic analysis (Falconer & Mackay 1996), traits can be similar between parents and offspring because of genetic, environmental or a combination of genetic and environmental contributions.

At this point most books of this kind also make an ethical comment regarding the use and potential misuse of genetic research into homosexuality. In this book, ethical issues will be addressed in the last section of Chapter 10.

Not surprisingly, important advances have been made in the study of genes controlling sexual behaviour in invertebrates, organisms that are simpler in structure and usually easier to keep and breed in a laboratory than mammals or birds. Although here I mainly focus on vertebrates, some of the studies done in invertebrates may be of great help to understand the genetics and evolution of vertebrate same-sex sexual behaviour. The fruit fly in particular (genus *Drosophila*) has provided one of the best biological models for the study of loci controlling same-sex sexual behaviour (see Emmons & Lipton 2003 for a recent review).

Table 3.1. Courtship behaviour of *fru Drosophila* mutants

Genotype	Male–male courtship CI(%)[a]	n	Male–female courtship CI(%)	n
wild-type	4	10	84	7
fru^1/fru^1	51	25	61	21
fru^3/fru^3	32	31	15	20
fru^4/fru^4	41	25	29	20
fru^3/fru^4	42	16	22	13

[a] CI, courtship index. Table adapted from Ryner *et al.* 1996.

The *fruitless (fru)* gene in *Drosophila*

Fruit flies have a male-specific elaborated courtship display (Hall 1994) which is determined by several genes; chief among them is the *fruitless (fru)* gene (Ryner *et al.* 1996). Some mutants at the *fru* locus court both males and females, and if only males are available they form courtship chains among themselves. The *fru* gene has been cloned by Ryner *et al.* (1996), who also carried out a series of experiments using *Drosophila* males differing in their specific mutations at the *fru* locus. The experiments were designed to determine the differences in courtship behaviour of males towards other males and towards females across various mutant genotypes as compared with the wild genotype. Table 3.1 is taken from Ryner *et al.* (1996) and it shows values for the courtship index of wild and four mutant genotypes of males in monosexual or disexual pairings. The courtship index is the percentage of time that a given test male spent courting another individual (either male or female) during periods of observation of between 5 and 8 minutes. From Table 3.1 it is clear that fru^1, fru^3 and fru^4 mutants court both males and females, whereas wild-type males show a marked preference for females. It is also interesting to note from Table 3.1 that some mutations seem to be more selective than others. For instance, although fru^1 males court males and females at the same rate, both fru^3 and fru^4 males court other males at almost double the rate they court females. This pattern of bisexual behaviour in the mutants may be explained, for instance, by their undergoing altered processes of development of their central nervous system. The *Drosophila* brain contains a cluster of interneurons called *fru*-mAI that are male-specific. In males, Fru^M, i.e. isoforms of the *fru* gene obtained by cutting the DNA in alternative manners, a process known as *splicing*, inhibit programmed cell death (apoptosis) of the *fru*-mAI neurons (see Chapter 5 for more details on the process of apoptosis), whereas in females apoptosis does occur in this region of the brain, producing the sexual dimorphism. Male flies bearing the fru^1 mutation do not have the *fru*-mAI cluster (Billeter *et al.* 2006).

Ryner *et al.* (1996) also studied the expression of the *fru* gene by cells of the central nervous system of males and females and found that, in fact, nerve cells of both sexes do express the gene. However, the primary RNA transcript of the gene is further spliced in a sex-specific manner in male and female neurons, which may explain the different effects of the gene on male and female behaviour. Such sex-specific splicing of *fru* is under the control of *tra* and *tra*-2 genes (Ryner *et al.* 1996). Moreover, a single locus such as *fru* can control different aspects of a complex behaviour such as courtship by producing diverse polypeptides, resulting from the alternative splicing of the gene as mentioned above (Billeter *et al.* 2006).

Demir & Dickson (2005) studied the role of male-specific splicing of the *fru* P1 transcripts in sexual behaviour of males. They produced four isoforms of the *fru* gene by homologous recombination: fru^F, which prevents male-specific splicing, fru^M and $fru^{\Delta tra}$, which produce male splicing, and fru^C, a control isoform that should leave splicing unchanged. fru^F Males court other males significantly more than any of the other three treatments, thus mirroring the results obtained for fru^1, fru^3 and fru^4 mutants by Ryner *et al.* (1996). However, Demir & Dickson (2005) also introduced the same mutations in females and found that fru^M and $fru^{\Delta tra}$ females that undergo male splicing of their *fruitless* primary transcript court other females with a 40% value of the Courtship Index (Figure 3.2). Moreover, their courtship behaviour of other females resembles normal male behaviour (Demir & Dickson 2005).

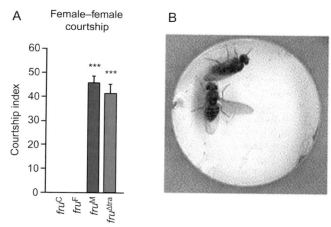

Figure 3.2. (A) Values of the courtship index for female fruit flies of various genotypes pairing with a wild-type virgin female. (B) A fru^M female, with one extended wing, courting a wild-type virgin female. From Demir & Dickson (2005).

An insect's perceived world is strongly affected by environmental chemicals. Crucial questions concern how those chemicals are detected by the sensory systems of the individual and how activated peripheral neurons connect with the brain. The same is valid for many mammalian species, but apparently less so in birds, which do not seem to rely as much as mammals and insects on chemical cues (but see Ball & Balthazart 2004 for arguments in support of a role for olfaction in avian sexual behaviour). In the chemical world of many insects, some volatile chemicals have a role in communication between conspecifics. In particular, some such pheromones may be used to attract a sexual partner.

In *Drosophila*, specialised cells called oenocytes have been suggested as the source of pheromones used in sexual partner attraction. Male and female *Drosophila* oenocytes produce sex-specific pheromones and if mutations occur that feminise male-produced pheromones then the mutant males will attract other males (Yamamoto & Nakano 1999). Moreover, Demir & Dickson (2005) produced masculinised females that courted in male-fashion males that had feminised oenocytes! Using transgenic technologies, it is also possible to produce male *Drosophila* with feminised sensory tissues (e.g. antennae lobe glomeruli) that will court other males rather than females (Yamamoto & Nakano 1999). Other modifications of the nervous system can be experimentally produced that alter the sexual behaviour of male fruit flies towards same-sex sexual attraction. For instance, feminisation of the mushroom body that is innervated by antennal lobe neurons results in bisexual courtship by males (O'Dell *et al.* 1995). That is, same-sex sexual behaviour can be elicited in male *Drosophila* after genetic modification of either central or peripheral neurons or sensory organs. If those genetic modifications occur as mutations in wild populations, then the resulting phenotype will be exposed to the vagaries of natural selection and persist in time, or not, according to the current environmental and social circumstances.

The *genderblind* and *sphinx* genes in *Drosophila*

Chemical compounds that are sex-specific can be perceived by *Drosophila melanogaster* through taste (e.g. 7-tricosene) or olfaction (e.g. *cis*-vaccenyl acetate). Flies that for whatever reason cannot produce the above pheromones become sexually attractive to males, and males that are unable to sense those pheromones may court other males. Grosjean *et al.*

(2008) have recently carried out a remarkable set of experiments showing that what they refer to as the *genderblind* (*gb*) locus, which encodes for a glial amino acid transporter, regulates the strength of glutaminergic synapses that in turn control male–male sexual behaviour. In fact the wild form of *gb* inhibits male–male homosexual behaviour, but if the gene is disrupted by, for instance, the insertion of a transposon, males display homosexual behaviours, including attempted copulations. Across various male genotypes, Grosjean *et al.* (2008) found that the frequency of male homosexual courtship was inversely proportional to the amount of genderblind protein produced. Those males, however, were also capable of courting females to various degrees. In this case, the *gb* mutant males seem to be sexually attracted to males in response to male-produced pheromones such as 7-tricosene and *cis*-vaccenyl acetate. Another *Drosophila* gene that when disrupted or knocked down releases male–male courtship is *sphinx* (Dai *et al.* 2008).

Testicular feminisation mutation (*tfm*) in mice and rats

In genetic research the house mouse (*Mus musculus*) and the rat (*Rattus norvegicus* in particular) could be easily regarded as the '*Drosophila*' of the laboratory mammals (students of rodents would probably see it the other way around). Several mutations have been studied in rodents that affect sexual behaviour. The testicular feminisation mutation (*tfm*) of the androgen receptor gene found in the mouse X chromosome (Migeon *et al.* 1981) makes neurons expressing the mutation insensitive to androgens by impairing the functionality of the androgen receptor in the cell membrane (Sato *et al.* 2004). The mouse *tfm* locus is homologous to the human androgen receptor (*AR*) locus (Migeon *et al.* 1981). Androgen insensitivity may affect the development of certain brain regions that are associated with sexual behaviour and that show male- or female-specific developmental pathways following differential exposure to androgens during pre- and/or postnatal life (Morris *et al.* 2005). As we will see in Chapters 4 and 5 in greater detail, sex-specific androgen production (usually by the male gonads) during development affects the size of specific brain regions (i.e. those with cells containing androgen receptors) by preventing programmed cell death (apoptosis). Very broadly speaking, this process ultimately leads to differentiation of some sexual behaviours between males and females. In fact, we will see in Chapters 4 and 5 that the mechanisms involved are far more complex, but this explanation will suffice for the current purposes. Therefore any gene able to control apoptosis is likely to be of primary importance in the feminisation/de-feminisation or masculinisation/de-masculinisation of brain areas that could affect specific aspects of sexual behaviour.

Following the initial work by Ohno *et al.* (1974) on *tfm*, recent research has used genetic engineering methodologies to produce mutant mouse males and females lacking functional androgen receptors: $AR^{L-/Y}$ males and $AR^{L-/L-}$ females (Sato *et al.* 2004). Although mutant males showed female-like external genitalia, when placed with wild-type males they were mounted but did not respond by adopting the female-typical lordosis posture, a reflex behaviour adopted by females inviting copulation (Sato *et al.* 2004). Lordosis was not induced in $AR^{L-/Y}$ males even after administration of oestradiol alone or oestradiol followed by administration of progesterone (Sato *et al.* 2004). Similarly, $AR^{L-/L-}$ females did not perform male-typical sexual behaviour. The work of Sato *et al.* (2004) therefore suggests that *tfm* causing impairment of the androgen receptor does not feminise males in their sexual behaviour, neither it does masculinise females (see also Matsumoto *et al.* 2005). However, *tfm* male mice prefer male-soiled bedding rather than the female-soiled bedding preferred by wild-type males, and both mouse and rat males show a reduction in masculine sexual behaviour compared with wild-type males (see Zuloaga *et al.* 2008 and references therein). Moreover, with regard to the size of various brain centres that are associated with control of sexual behaviour, *tfm* male rats tend to be either feminised (e.g. suprachiasmatic

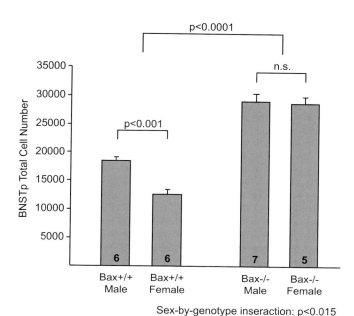

Figure 3.3. Total number of cells in the BNSTp of wild-type (*Bax+/+*) and *Bax* knockout (*Bax–/–*) mice. The BNSTp is sexually dimorphic, having more cells in wild-type males than in females. Deletion of the *Bax* gene not only increases total cell number but also eliminates the sex difference. From Forger et al. (2004).

nucleus, ventromedial hypothalamus, posteromedial nucleus of the bed nucleus of the stria terminalis) or intermediate between the sexes (e.g. posterodorsal amygdala) (see Table 1 of Zuloaga et al. 2008). Therefore, it seems that *tfm* is at the very least associated with behavioural and neuroanatomical de-masculinisation of male rodents.

Apoptosis and the *Bax* gene

The use of mouse models has also permitted detailed understanding of other aspects of the genetic regulation of developmental processes affecting sexual behaviour. During early development the brain architecture is literally sculpted by a process of differential cell death in various centres. These apoptotic processes are likely to be affected by proteins of the Bcl-2 family, with some of these polypeptides (e.g. Bcl-2 and Bcl-xL) inhibiting apoptosis, whereas others (e.g. Bax) promote apoptosis in a neural-area-specific manner (Forger et al. 2004). Because gonadal steroid hormones are able to regulate the expression of Bax proteins and also of proteins of the Bcl-2 family (Forger et al. 2004 and references therein), these proteins are primary candidates for the processes leading to sexual differentiation of specific areas of the central nervous system that control sexual behaviour and orientation. Forger et al. (2004) produced knockout C57BL/6 strain mice for the *Bax* gene (i.e. mice that have the gene shut down and therefore do not produce the Bax protein) by crossing male and female mice heterozygous for the *Bax* gene. This kind of crossing produces offspring that are able to synthesise the Bax protein (homozygous dominant *Bax+/+* and heterozygous *Bax+/–*), but also knockout mice that cannot produce the protein, i.e. homozygous recessive *Bax–/–*. As expected from the known effects of the Bax protein to promote apoptosis, *Bax* knockout mice (both males and females) have larger numbers of neurons in the principal nucleus of the bed nucleus of the stria terminalis (BNSTp) (a sexually dimorphic nucleus that is normally larger in males than in females) (see Figure 3.3). Interestingly, knockout *Bax* males have also a significantly larger number of neurons in the BNSTp than wild-type males, indicating that a degree of apoptosis is also occurring in wild-type *Bax* males, not just females.

In terms of volume of the BNSTp, however, knockout *Bax* females had a BNSTp 25% smaller than that of knockout males (Forger *et al.* 2004).

Forger *et al.* (2004) also compared the anteroventral periventricular nucleus (AVPV) of the hypothalamus, which is larger in females than in males in mice. Although the results for the AVPV generally mirror those found for the BNSTp, with $Bax-/-$ males and females having larger AVPV than wild-types and displaying no sex difference, *Bax* deletion did not affect dopaminergic nerve cells in the AVPV, suggesting that different mechanisms of apoptosis operate for different nerve cell types even within the same nucleus in regions of the central nervous system such as the hypothalamus (Forger *et al.* 2004).

Do *Bax* mutations affect sexual behaviour and orientation? Jyotika *et al.* (2007) have recently reported the results of experiments using male and female $Bax+/+$ (wild-type) and $Bax-/-$ mutant mice. They tested both sexes in trials with either a stud male, to detect feminine behaviour, or with a sexually experienced female, to detect masculine behaviour. In the feminine behaviour trial, gonadectomised experimental males and females of all genotypes were injected with oestradiol followed by progesterone to provide the hormonal trigger that usually elicits female-like behaviour in those animals whose brain is capable of producing feminine behaviours. In the masculine behaviour experiment, the gonadectomised mice were fitted with silastic capsules filled with testosterone to elicit masculine behaviour. Jyotika *et al.* (2007) found that hormonally feminised $Bax-/-$ males and females displayed lower levels of lordosis (a typical female reflex in mice exhibited in the presence of a male and inviting copulation) than $Bax+/+$ females, but similar levels of lordosis to $Bax+/+$ males. This result suggests that $Bax-/-$ de-feminises females. Does the mutation also masculinise both sexes? In the presence of a sexually receptive female both wild-type male and female mice copulated with her, in fact 70% of $Bax+/+$ females and a smaller 40% of $Bax+/+$ males copulated with the female. In contrast, 5.5% of $Bax-/-$ males and none of the $Bax-/-$ females copulated with the female. In this case, rather than the expected masculinising effect, both male and female $Bax-/-$ mice seem to be sexually unmotivated by a sexually receptive female. Therefore *Bax* seems to mainly have a de-feminising effect in female mice.

The *c-fos* proto-oncogene

Mounting of a female is associated in male mice with increased expression of the *c-fos* proto-oncogene (which encodes for the Fos protein, a nuclear transcription factor) in several limbic–hypothalamic regions of the brain, including the medial preoptic area (mPOA), the posterodorsal preoptic nucleus (PDPN), the medial amygdala (MeA) and the BNST (Baum *et al.* 1994). Through crossing heterozygous males and females, Baum *et al.* (1994) obtained mice that were genotypically wild-type (+/+), heterozygous (+/−) or homozygous mutant (−/−) for a *c-fos* mutation that incapacitated the individual to produce the Fos protein. Males of different genotype were then tested with an oestrous female for mounting behaviour. Homozygous mutant males displayed a lower number of mounts per test and per minute than heterozygous or wild-type males (Baum *et al.* 1994); moreover, the Fos protein was expressed in nerve cells of all the above nuclei plus the paraventricular nucleus (PVN) and the central tegmental field (CTF) in heterozygous males that copulated with an oestrous female, in comparison with heterozygous males that remained unmated (Baum *et al.* 1994). The experiments of Baum *et al.* (1994) also indicate that the lower copulation rate displayed by mutants was not due to a generalised lack of sexual arousal, as they tended to perform a higher number of ano-genital inspections of the oestrous female ($n = 18$) per test than heterozygous ($n = 10$) and wild-type ($n = 5$) males. Although this may be suggestive of rejection of females as sexual partners, equivalent experiments where male subjects are tested with a male potential sexual partner are needed before we can assess the relevance of these findings to the understanding of male homosexuality in mammals. More direct, although still preliminary, evidence of the role of the Fos protein in the regulation of homosexual behaviour comes from the work of Heimovics & Riters (2005) on European starlings (*Sturnus vulgaris*). The authors carried out Fos immunocytochemistry on 24 male starlings held in captivity that were in breeding condition and that were also implanted with silastic tubes filled with testosterone propionate. Two of those birds (i.e. 8.3%) displayed homosexual behaviour and, upon carrying out Fos immunohistochemistry, the authors found that they had higher levels of expression of the protein than any other bird in the study in all the brain regions studied: medial preoptic nucleus, ventral tegmental area, high vocal centre and arcopallium, all brain nuclei that are associated with control of sexual behaviour.

The *trp2* and olfactory receptors in mice

Chemicals are used in sexual attraction not only by insects, as we have already seen, but also by many mammals. Apart from the taste sensory system, *Mus musculus* and other mammals possess two other main organs that are sensitive to chemicals: the main olfactory epithelium, which mainly detects volatile chemicals, and the vomeronasal organ, which can detect both non-volatile and volatile chemicals (Trinh & Storm 2003). In the vomeronasal organ, olfactory receptors are activated through the trp2 cation channel (Leypold *et al.* 2002). Using the now familiar technique, Leypold *et al.* (2002) genetically engineered $trp2 +/-$ mice that they then crossed to produce $trp2 -/-$ mutants and $trp2 +/+$ and $trp2 +/-$ controls. Mutants do not produce the trp2 protein of the cation channel and are therefore expected to have a vomeronasal organ insensitive to chemicals. Sexual behaviour of mutant males was determined in tests with other males and also in independent tests with non-oestrous females and males plus females combined. Interestingly, $trp2 -/-$ males mount females at the same frequency as controls, but mount males at a significantly higher frequency than controls. Moreover, previous experience mounting females further increases the mounting rate of $trp2 -/-$ males with other males. In addition, if $trp2 -/-$ males are given the choice to mount either a female or a male, both being simultaneously available in the test cage, although they prefer to mount females as control males do, they continue to mount males at a higher frequency than controls (Leypold *et al.* 2002). Leypold *et al.* (2002) also demonstrated another striking characteristic of mutant male mice: their low level of aggressiveness. This remarkable set of experiments provides strong evidence that, at least in *Mus musculus*, specific mutations affecting the perception of olfactory stimuli normally used in sexual partner recognition can lead to same-sex sexual behaviour. Similar results were also obtained by Stowers *et al.* (2002). Although they interpreted male–male mounting of their trp2 protein-deficient males as evidence of lack of sexual discrimination due to their inactivation of the vomeronasal organ, their data (Figure 4 in Stowers *et al.* 2002) actually show a slight preference of those males to mount other males that had been swabbed with male urine rather than males not swabbed with urine. That is, in spite of their $trp2 -/-$ genotype male mutants retained some ability to discriminate – and prefer – males, perhaps using their olfactory epithelium.

Genes studied in humans

Although we all recognise the enormous phylogenetic differences between *Drosophila* and rodents, or between rodents and primates, including humans, the increased complexity of central nervous system and flexibility of behaviours found in vertebrates may render more difficult, but certainly not impede, the study of the genetics of homosexual behaviour in those taxa. The best research scientists are usually excited by the challenge of complexity, rather than turned off by it. These considerations notwithstanding, studies of genes potentially involved in the determination of homosexual behaviour in humans are fraught with problems. To begin with, the ethical issues faced by students of invertebrate or non-human vertebrate genetics pale into insignificance when compared with those faced by students of human genetics and sexuality. This constrains, and rightly so, the kinds of methodology and technology that can be used in this research area. For instance, genetically engineering humans for the purpose of studying the genetic basis of behaviour, as we routinely do for mice and fruit flies, is simply unthinkable. In addition, the genetics of human homosexuality is bound to produce much public and media interest that may hamper rather than foster reasoned and constructive debate. In spite of these issues, efforts have been made to advance our understanding of the genetic underpinning of homosexual behaviour in humans. On the positive side, complete pedigrees are available for family studies of the mode of inheritance of various human traits, and the detailed knowledge of the human genome that we have will certainly facilitate the location and study of specific genes.

In this section I focus on studies of an X-linked region (Xq28) and of microsatellites and other loci. Studies of loci of the Y chromosome will be also reviewed, along with studies of sex chromosome aneuploidies. I finally review the single published study that spans a broad area of the human genome (Mustanski *et al.* 2005).

Genes on the X-chromosome and homosexuality

Few studies can rival both the excitement and the controversy sparked by Dean Hamer, Stella Hu, Victoria Magnuson, Nan Hu and Angela Pattatucci's work on Xq28 (Hamer *et al.* 1993; the background story to this work is vividly recounted by Dean Hamer and Peter Copeland in their 1994 book *The Science of Desire*). When they are compacted (e.g. in metaphase), chromosomes show a typical structure with two sections: a 'p' and a 'q' arm separated by a centromere. Each arm can be characterised by regions that are numbered in increasing order from the centromere to the tip of the arm, the numbers being equivalent to the recombination frequency occurring among homologous chromosomes: regions closer to the centromere recombine less frequently than regions closer to the tip of the chromosome. Hamer *et al.* (1993) carried out a pedigree analysis and a study of DNA linkage in families that indicated an association between homosexuality in males and genetic markers at the Xq28 region (Figure 3.4). Their pedigree analysis (see Table 3.2) indicates that for two kinds of subject (known as probands in genetic studies), homosexual random and homosexual sib-pair (i.e. two homosexual brothers), homosexual relatives are significantly more common through the maternal than through the paternal line of descent. This result is strongly suggestive of an X-chromosome mode of inheritance of homosexuality in this pedigree. However, the pattern as such, even if statistically significant, is just suggestive of X-chromosome inheritance of homosexuality. This is because controls were not carried out for behavioural influences of maternal versus paternal relatives on the psychological development of probands that could affect sexual orientation. If maternal relatives are more influential on the development of a child than paternal relatives, then homosexual probands are more likely to have homosexual relatives through the maternal than through the paternal line of descent. However, Hamer *et al.* (1993) also performed a linkage analysis using the homosexual sibling pairs and estimated the probability that those brothers shared various X chromosome markers by descent. Most of their significant results suggest that homosexual siblings have specific similarities at the q28 region of the X chromosome.

Unfortunately, Hamer *et al.* (1993: 324) did not include heterosexual brothers of the homosexual

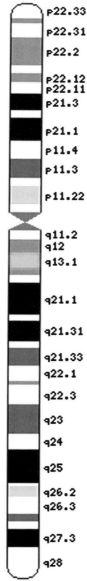

Figure 3.4. Human X-chromosome, with the q28 region visible towards the tip of the long arm of the chromosome.

Table 3.2. Relative frequency of individuals with a homosexual orientation among male relatives of homosexual males

Relationship	Homosexual/Total	Percentage
Random probands ($n = 76$)		
Father	0/76	0
Son	0/6	0
Brother	14/104	13.5**
Maternal uncle	7/96	7.3*
Paternal uncle	2/119	1.7
Maternal cousin, aunt's son	4/52	7.7*
Maternal cousin, uncle's son	2/52	3.9
Paternal cousin, aunt's son	3/84	3.6
Paternal cousin, uncle's son	3/56	5.4
Sib-pair probands ($n = 38$)		
Maternal uncle	6/58	10.3**
Paternal uncle	1/66	1.5
Maternal cousin, aunt's son	8/62	12.9**
Maternal cousin, uncle's son	0/43	0
Paternal cousin, aunt's son	0/69	0
Paternal cousin, uncle's son	5/93	5.4
Population frequency		
Uncles and cousins of female probands	14/717	2.0

Data taken from Table 1 of Hamer *et al.* (1993).
Significance: **$p < 0.01$, *$p < 0.05$

subjects in the analysis. This is a crucial control, as their hypothesis would predict that those markers shared by homosexual brothers are not shared with heterosexual brothers. Incidentally, Hamer & Copeland (1994) seemed to justify the absence of such a control on the grounds that chances of finding a locus linked to homosexuality increase if the research focuses on exclusively homosexual individuals. This is obviously correct; the problem is that the comparison with heterosexual brothers is still needed to prove that the gene found to be linked with homosexuality is a relevant one. Anyhow, they corrected the problem in a subsequent article published two years later (Hu *et al.* 1995). There they studied a new sample of families having gay male sib-pairs, but this time they also included heterosexual brothers of the homosexual sib-pairs in the analyses. In addition, they extended the study to lesbian and heterosexual women as well. They focused on three regions of the X chromosome: Xq28, Xq27 and the Xq/Yq homology region. The only statistically significant result obtained by Hu *et al.* was that homosexual brothers shared significantly more genes by descent in the Xq28 region than heterosexual brothers did, whereas all the other regions showed no difference. They could not detect any difference in any region in women. The results obtained by Hamer's team could not be replicated in an independent study carried out by Rice *et al.* (1999), although the latter work is not free from problems either (Wickelgren 1999).

The hypothesis that the Xq28 region may contain genes controlling the expression of homosexual behaviour was also supported by an independent pedigree analysis carried out by Turner (1995) on a total of 249 families; whereas Bailey *et al.* (1995) showed that 80% of sons of gay fathers were actually heterosexuals, a result that is supportive of an X-chromosome location for putative genes affecting the development of homosexual orientation: fathers do not transmit their X chromosome to their sons. A pedigree analysis by Camperio-Ciani *et al.* (2004) of 98 homosexual and 100 heterosexual men from northern Italy, based on questionnaires, also suggests a potential X chromosome transmission of homosexuality in men, although these authors also realised the potential for a cultural transmission of homosexuality through the maternal line of male relatives. In addition, Camperio-Ciani *et al.* (2004) report increased fecundity of maternal relatives of homosexuals than maternal relatives of heterosexuals; this result supports a sexually antagonistic selection model for homosexuality (Trivers 1972; Rice 1992).

The study by Camperio-Ciani *et al.* was replicated by Rahman *et al.* (2008) using a population sample of 'white' and 'non-white' (i.e. 'Black', 'South-Asian', 'East-Asian', 'Hispanic' and 'Other', the last including ethnic groups such as Turkish) homosexual and heterosexual men from London (University of East London and Soho district). Their results suggest that in general homosexual men have an excess of homosexual male relatives through the maternal line compared with heterosexual men, a result that is consistent with an X-chromosome-based mechanism of inheritance of homosexuality. Interestingly, however, they also reported that although fecundity is higher in maternal aunts of 'white' homosexual men than in aunts of 'white' heterosexual men, suggesting both an X-linked control and also a sexually antagonistic selection of homosexuality in 'white' men, in the case of 'non-white' men various kinds of both male and female relatives (i.e. paternal aunts, paternal uncles, maternal grandparents, mother, paternal grandparents) of heterosexual men had **higher** fecundities than those of homosexual men. Moreover, family size of 'non-white' heterosexual men was larger than that of homosexual men. Although the difference between ethnicities could well be explained by genetic differences, social and developmental mechanisms should be also considered in this case. If stress is an ontogenetic factor acting during pre-, peri- and/or postnatal periods, that could contribute to the development of a homosexual orientation (see Chapter 4), and if we reasonably assume that the immigration process implies a significant degree of stress, stress that could be felt differently by different ethnic groups, then we may expect that among the diverse groups of immigrants homosexuality may be more frequent in small family groups that may lack the support afforded by a large network of relatives than in large family groups. For better socially integrated ethnic groups the reverse may be expected, as in that case the large family group may be a greater source of stress than a small family group within an urban context (e.g. if financial resources are limited, leading to high levels of intrafamily conflict). These issues will be further explored in subsequent chapters.

Genes on the Y-chromosome and homosexuality

We know that X and Y chromosomes in males mostly do not recombine during meiosis (some recombination does occur in the PAR region, however) and that therefore the Y chromosome is likely to accumulate mutations over time (Craig et al. 2004). Females do not have such a problem, as their pair of X chromosomes undergo crossover and therefore genes reshuffle in different combinations from one generation to another. This may make males more susceptible to accumulate multiple mutations in their Y chromosome over the generations that may result in less reproductively active genotypes, a process known as *Muller's Ratchet* (Gabriel et al. 1993). The capacity to accumulate mutations, however, is not necessarily a bad thing as the apparently maladaptive genes may well manifest increased fitness in some of the mutants as the environmental conditions change; e.g. in social contexts where socio-sexual functions of homosexual behaviour can be very adaptive, as we will see in Chapter 8. On the other hand, from an X-chromosome perspective, because many loci on the X chromosome do not have a homologous locus on the Y chromosome, males are more likely to express extreme phenotypes (e.g. those encoded by recessive alleles in their single X chromosome) than females, which may be heterozygous. Therefore, in a sex-chromosomal context (X or Y) males are expected to manifest exclusive homosexual behaviours with more frequency than females (Diamond 1993; Craig et al. 2004).

Several recent studies have addressed the mechanisms by which genes on the Y-chromosome may control sexual behaviour. In a remarkable experimental paradigm, De Vries et al. (2002) were able to breed male *Mus musculus* that had a mutated Y chromosome (Y^{129}) deleted for the *Sry* gene (the testis-determining gene). *Sry* is a highly conserved Y chromosome gene in mammals (Tiersch et al. 1991) that affects brain development through the control of gonadal development. The gonads subsequently secrete the sex hormones affecting the brain ontogeny. This model, however, has been recently challenged, not only by findings that sex phenotypic differences can develop before differentiation of the gonads (see Arnold et al. 2004 and Craig et al. 2004 for reviews) but also by discoveries that *Sry* can affect male brain development directly (i.e. without mediation of testicular hormones) by altering the function of neurons in specific areas of the brain (Dewing et al. 2006). Therefore, directly or indirectly, changes in the *Sry* gene may affect sexual behaviour through its effects on central nervous system function and architecture. Males with the Y^{129} mutation (XY^- males) develop ovaries. De Vries et al. (2002) were able to reverse the effects of the Y^{129} mutation by producing XY^- transgenic mice that had the *Sry* gene inserted in an autosome (XY^- *Sry* males). Such males not only develop testes but are also fully fertile. Breeding of XY^- *Sry* males with wild-type XX females produces four kinds of genotype in the offspring: XX females, XY^- females, XY^- *Sry* males and XX *Sry* males. It can be seen that this set of four genotypes allows the independent study of the effects of gonadal sex (i.e. male and female) and of XX and XY sex chromosome complement on brain development and sexual behaviour (De Vries et al. 2002). Litters were of mixed genotypes, therefore the different genotypes were exposed to the same pre- and postnatal environment, including litter effects (De Vries et al. 2002: 9006). After reaching sexual maturity, mice were bilaterally gonadectomised and fitted with Silastic implants filled with testosterone, thus homogenising the effect of current circulating testosterone across genotypes. One week after this intervention, they were test-exposed to a stimulus female primed with oestradiol benzoate and progesterone to induce mating activity. Although all individuals were gonadectomised and fitted with testosterone implants, intact (XY) males tended to mount the test female more often than intact (XX) females. In addition, XY^- *Sry* and XX *Sry* males copulated significantly more frequently with the test female than XY^- and XX females did (Figure 3.5). This result confirms that it is the specific action of *Sry* on central nervous system development that, in this case, affects the expression of masculine behaviour in mice. However, De Vries et al. (2002) also compared the density of vasopressin immunoreactive fibres in the lateral septum, which are

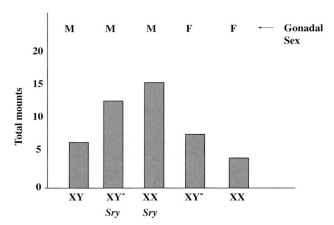

Figure 3.5. Total mounts recorded during sexual interactions with an oestrous female for chromosomally diverse male and female mice with or without the testis-determining gene *Sry*. Modified from De Vries *et al.* (2002).

usually more abundant in males than females, and found that XY⁻*Sry* males had more fibres than XX*Sry* males, thus suggesting that Y-chromosome loci other than *Sry* are also involved in masculinising central nervous system circuitry during development. These experiments clearly demonstrate the specific effect of a locus on masculine sexual behaviour, leading to predictions that mutations at the *Sry* locus (and other Y-chromosome loci) could potentially also alter male sexual orientation.

Other possible genetic markers of homosexuality

Other genetic markers have also been studied in humans. We will see in Chapter 6 that there is a current interest in the potential role of immune mechanisms in the development of homosexual phenotypes during gestation in mammals, humans in particular. Here I mention work published by Gangestad *et al.* (1996) that describes an association between left-handedness (a phenotypic trait shared by many homosexual men and that will be analysed in detail in Chapter 4) and human leukocyte antigen (HLA) genetic markers that are associated with autoimmune disease. Left-handed persons in the study by Gangestad *et al.* (1996) were more likely to carry the B8 and DR3 alleles and also the A1/B8 haplotype at the HLA locus, and possession of the B8 allele and A1/B8 haplotype was negatively associated with the number of offspring produced, perhaps owing to sexual orientation effects at least in some cases. Although the potential link suggested here between genes and homosexuality is tenuous: alleles affecting immune function of mother → producing developmental instability in the embryo → affecting handedness and sexual orientation, it is a theoretically plausible mechanism.

Another gene, the *CYP17* gene, can suffer alterations and produce single nucleotide polymorphisms (SNP) that are associated with high circulating levels of oestradiol, progesterone and testosterone. Bentz *et al.* (2008) compared the frequency of one such *CYP17* SNP allele among 102 male-to-female, 49 female-to-male Caucasian transsexuals and Caucasian controls (males and females) of unspecified sexual orientation. Allele frequencies and genotype distributions differed significantly between female-to-male transsexuals and female controls, but there was no significant difference between male-to-female transsexuals and controls (both males and females). This is consistent with results that will be reported in both Chapter 4 and Chapter 5 regarding the effect of elevated steroids on the development of a same-sex sexual orientation in females. A review of studies of the *CYP19* gene will be provided in the "Congenital aromatase deficiency" section of Chapter 5.

In general, genetic disorders afford an opportunity to study the genetic basis of sexual orientation, but not because we should consider homosexuality as a 'disorder', 'pathology' or 'defect'; in fact I will provide in Chapter 5 and then again in Chapter 10 an extensive argument in rejection of the 'pathology' view of homosexuality. In this book homosexuality is mainly seen as an evolutionary paradox in search of a resolution in the context of evolutionary theory (see Chapter 1). However, the study of well-known medical syndromes that are also associated with homosexual behaviour can help unravel not only the genetic, but also the neurobiological and developmental mechanisms determining same-sex sexual orientation outside the context of those syndromes. Here I will review work published by Comings (1994) on Tourette syndrome (TS), which is associated with attention deficit hyperactivity disorder (ADHD), because the author used a genetic approach.

TS individuals manifest a variety of traits that include obsessive behaviours, learning problems, anxiety and what in the psychiatric literature is variably described as inappropriate, compulsive or obsessive sexual behaviour. TS is genetically inherited, although the mode of inheritance seems to be complex (Comings 1994). Comings (1994) used questionnaires to evaluate compulsive and sexual behaviours in patients diagnosed with TS and compared the scores for patients with those of controls: mainly foster parents and step-parents of the patients and participants from a thyroid cancer clinic. Participants were categorised according to a progressive loading of the genes associated with the syndrome (*Gts* genes): controls presumably have the least loading and the TS probands were considered to have the highest loading (Comings 1994). It may be useful to recall that 'probands' are the subject of interest in genetic studies. The author found that 9.6% of TS probands defined themselves as either homosexuals or bisexuals whereas only 1.4% of controls did so.

Many other loci may potentially contribute to the development of same-sex sexual behaviour. Microsatellites in regulatory regions for the expression of hormone receptors in brain cells could control the expression of phenotypes that have more (or fewer) functional or non-functional receptors, thus potentially affecting sexual behaviour, including sexual orientation (Hammock & Young 2005). Genes controlling whole developmental programmes supporting morphogenesis of the vertebrate brain such as *Otx1* and *Otx2* could also be a potential source of alternative phenotypes that could differ in sexual orientation (Simeone 1998; Pilo Boyl *et al.* 2001). For instance, in a recent study, Klar (2004) compared hair-whorl direction (clockwise or counterclockwise) in men frequenting a beach in Delaware (USA) that was allegedly used by gay men, with hair-whorl direction of two kinds of control male population where the proportion of gay men was expected to be comparatively less. The sample from the gay beach had a higher percentage (29.8%) of counterclockwise individuals than either of the two control samples (8.2% and 10.7%, respectively). Klar (2003, 2004) suggests that this difference is due to the effect of the *RGHT1* gene that is hypothesised to control aspects of both brain and scalp development. However, a recent study by Rahman *et al.* (2009) has not found an association between the number of counterclockwise hair whorls and sexual orientation in men (14%, $n = 100$ in heterosexuals vs. 18%, $n = 100$ in homosexuals) sampled from the London area in the UK.

The more we understand about the complexity of the chemical, cellular, morphological, developmental and physiological processes leading to same-sex sexual behaviour and orientation, the more regulatory and structural genes we will be able to target for study.

Sex-biased mutation rates

Across mammal and bird species that exhibit homosexual behaviour it is males that are relatively more likely to engage in same-sex mounting than females (see Chapter 8). Although this pattern could be trivially explained at one level of analysis by the fact that mounter behaviour is a typically male behaviour in heterosexual sexual copulation, its use in less typical contexts such as same-sex mounting also coincides with a well-known male bias in mutation rates found across vertebrate taxa, mammals and birds in particular (Makova & Li 2002). Such a trend is consistent not with being homogametic or heterogametic, as in mammals males are heterogametic (XY) whereas in birds it is the females that are heterogametic (ZW). Instead, higher mutation rates in males of both vertebrate taxa are more likely to be due to germ cells undergoing more mitoses in the males, which produce large numbers of spermatozoa per ejaculation and can ejaculate frequently, than females, which produce a comparatively smaller number of ova. Therefore, the chromosomes more likely to be transmitted through the male line of descent are also more likely to carry mutations (Montell *et al.* 2001). In mammals this chromosome is Y and it is known that DNA sequences in the Y chromosome evolve faster than in the X chromosome. The same pattern occurs in the avian *CHD1* gene, which is found in both the Z and the non-recombining part of the W chromosome, but evolves faster in the Z chromosome (Montell *et al.* 2001). Moreover, although in mammals the X chromosome also spends time in males, and therefore mutations can also accumulate on this chromosome, the avian W chromosome is only transmitted through the maternal line. We can thus formulate a *Male Mutation Bias* hypothesis that predicts a male-biased frequency of homosexuality in both birds (due to mutations on the Z chromosome) and mammals (due to mutations on the Y chromosome) but a larger bias in birds (because the W chromosome is female-specific) than in mammals (because mutations on the X chromosome can also occur in males and be expressed in both their daughters and their daughters' male and female offspring). We will see in Chapters 7 and 8 that, in the sample used, 39 avian species display male same-sex mounting and only 11 species female–female mounting, indicating a male–male mounting-biased ratio of 39/11 = 3.54, whereas in mammals 77 species display male same-sex mounting and 54 female same-sex mounting for a male same-sex mounting-biased ratio of 77/54 = 1.42 (species that display both male–male and female–female mounting are counted twice). That is, in mammals same-sex mounting is a male-biased behaviour, as it is in birds, but the male bias is less accentuated in mammals than in birds, as predicted by the *Male Mutation Bias* hypothesis. Although I will carry out proper phylogenetically based comparative analyses of the effect of sex on same-sex mounting in Chapter 8, this preliminary species-level comparative test, in which most of the species that I included pertain to different genera (see Table 2.1), indicates that the difference between birds and mammals is statistically significant ($\chi^2_1 = 5.00$, $p = 0.012$).

A study across the human genome

Mustanski *et al.* (2005) carried out a multipoint linkage analysis, using 403 microsatellite markers, for human chromosomes 1–22 in 146 unrelated families, with 137 of those families including two gay brothers and 9 families including three gay brothers. They found associations of homosexuality with markers on chromosomes 7 and 8 of both maternal and paternal origin, and they also found a link with chromosome 10 of maternal origin. The markers they used, however, were not sufficiently specific to detect linkage with the X chromosome. Their statistically stronger association between male homosexuality and a chromosome region was with the q36 region on chromosome 7 (7q36) from both mother and father. They point out that there are various genes of interest that are found on chromosome 7 in humans. One of those, the vasoactive intestinal peptide receptor type 2 (*VIPR2*) locus encodes for the VIPR2 neurotransmitter that can also act as a hormone affecting the development of the suprachiasmatic nucleus in mice. As we will see in Chapter 5, homosexual men have been found to possess an enlarged suprachiasmatic nucleus (Swaab & Hofman 1990). The work of Mustanski *et al.* (2005) clearly indicates that homosexuality is very likely to have multigenic causes.

A recent work by Ellis *et al.* (2008) also points to a potential autosomal contribution towards the development of a homosexual orientation. In fact, they found a statistically significant difference in the frequency of blood type A between heterosexual Canadian college students (88% white): 31.9% in males and 33.1% in females, as compared with non-heterosexual students (18.5% males and 44.2% females). They also found a marginally non-significant difference in the distribution of Rhesus factor Rh^- : 17.4% and 19.4% in heterosexual males and females, respectively, and 29.2% and 31.6% in non-heterosexual males and females, respectively ($p = 0.06$). Both blood type and the Rh factor are genetically inherited traits, the former is controlled by alleles located on chromosome 9 and the latter by alleles on chromosome 1 (Ellis *et al.* 2008). Although this is not a study of mechanisms, but of descriptive association, it is nevertheless suggestive of a potential autosomal link with homosexuality.

DNA methylation and X-chromosome inactivation

Vertebrates are especially prone to manifest a phenomenon called DNA methylation, in which cytosine in specific areas of the genome, the CpG islands, acquires a methyl group, a reaction catalysed by the enzyme DNA methyltransferase 3-like (Dnmt3L) (Bird 2002; Bourc'his & Bestor 2004). An important adaptive function of DNA methylation is the silencing of parasitic DNA (e.g. retrotransposons) (see, for example, Falls *et al.* 1999; Wilkins 2005). Bourc'his & Bestor (2004) have shown that if the *Dnmt3L* gene is deleted methylation does not occur, retrotransposons can be expressed and spermatocytes that do not synthesise Dnmt3L undergo meiotic failure. The silencing of retrotransposons can occur because CpG islands are frequently found close to promoter regions (Bird 2002), which are specific DNA sequences that mark the initiation of DNA transcription into RNA, leading to the final synthesis of polypeptides. Methylation impedes the activity of RNA polymerase that is necessary for transcription (Molloy 1986). DNA methylation occurs *de novo* in germ cells or at the early embryonic stages (Bird 2002). Although DNA methylation has been extensively studied in the X chromosome, it is a phenomenon that occurs across the genome, also involving autosomal chromosomes (Singal & Ginder 1999; Shen *et al.* 2007).

DNA methylation is a mechanism that contributes to an explanation of another common phenomenon in mammals: X-chromosome inactivation. Female mammals have two X chromosomes in each diploid cell nucleus, compared with a single X chromosome in males. Such X-chromosome 'advantage' is compensated for during early embryonic development by inactivation of one of the female X chromosomes (Heard 2004). Whether it is the X chromosome of paternal or maternal origin that is inactivated may have profound implications for the development of the individual and/or the development of her offspring during gestation. Bocklandt *et al.* (2006; see also Bocklandt & Hamer 2003) carried out a study of 200 women: 97 mothers of one or more gay sons and 103 controls. Sexual orientation of gay sons was assessed by using the *Kinsey Scale* (Kinsey *et al.* 1948), with gay sons having an average score of 5.65. They compared both groups of women in terms of the bias of X-inactivation (whether the same or different X chromosomes are inactivated across cells) between them at both the androgen receptor (*AR*) and the Fragile X mental retardation (*FMR1*) loci. Although the *AR* gene has been implicated in the determination of human homosexuality, its exact role remains controversial (Quigley 2002). For instance, Macke *et al.* (1993) provide genetic evidence that the *AR* gene may not be involved in the determination of homosexual phenotypes in humans. The *FMR1* gene, on the other hand, is found in the Xq28 region, which has been associated with human homosexuality

(Hamer et al. 1993; Hu et al. 1995; Turner 1995). What Bocklandt et al. (2006) found was that mothers of gay sons had a higher propensity to inactivate the same chromosome across cells than mothers of heterosexual sons, whereas mothers with two or more gay sons had a more pronounced X-inactivation bias than mothers of only one gay son. Although the origin of the inactivated X chromosome (maternal or paternal) could not be ascertained by Bocklandt et al., and the exact cellular mechanisms explaining these findings remain unclear, the results are none the less consistent with the possibility that some genes may be more targeted for DNA methylation than others and that such methylation bias may be associated with a tendency to develop a homosexual orientation.

DNA methylation and genomic imprinting

DNA methylation has also been implicated in another phenomenon: genomic imprinting. In the mid-1980s genetic work on mice uncovered a phenomenon whereby chromosomes would function differently according to their origin; whether they come from the father or from the mother makes a difference. This was called by Surani et al. (1984) *Genomic Imprinting* (reviewed by Reik & Walter 2001). Normal embryonic development requires that genes of both maternal and paternal origin be imprinted. Imprinted genes are usually rich in CpG islands and the two parental alleles differ in their relative degree of DNA methylation (Reik & Walter 2001). Parental-specific methylation occurs during germ cell development, into either spermatozoa or ova. Within the new organism (i.e. the offspring) the original parental imprint is maintained, but the imprint is erased in the offspring's germ cells, which are imprinted again in an individual fashion. Following the antiparasitic theory of DNA methylation, the new imprinting in the offspring permits a control of new DNA parasitic infections acquired by the offspring (Reik & Walter 2001). Although other hypotheses have also been proposed to explain the evolution and current adaptive function of genomic imprinting (see review in Wilkins & Haig 2003), the antiparasitic hypothesis mentioned above is particularly appealing.

It has been suggested by Bocklandt & Hamer (2003) that the development of male homosexuality, if affected by loci on the X chromosome, could also be influenced by genomic imprinting (Green & Keverne 2000 also proposed a genomic imprinting mechanism involving the X chromosome that could explain familial distribution of male-to-female transsexuals). Bocklandt & Hamer based their hypothesis on studies carried out on Turner syndrome. Turner syndrome individuals have only one sex chromosome (i.e. X), rather than the normal two, and are therefore XO and girls. Because of this, Turner syndrome individuals could express either the paternally imprinted X or the maternally imprinted X chromosome (but not both). Turner syndrome girls with a paternally inherited X chromosome seem to have better communicative abilities than girls having a maternally inherited X chromosome; this observation suggested to Bocklandt & Hamer that there may be a gene that is maternally silenced, but paternally expressed, and that is 'feminising'. This suggests the possibility that variability in the kind of maternal imprinting may allow the expression of 'feminising' alleles or silencing of a 'masculinising' allele or both in one or more of her sons (Bocklandt & Hamer 2003). Although I disagree with the exclusive association of male homosexuality with 'femininity' and female homosexuality with 'masculinity' (see Chapter 1) the basic mechanism suggested by Bocklandt & Hamer (2003) is plausible if at least some loci associated with the expression of sexual orientation are located near CpG islands. If so, they could be silenced or not according to specific patterns of individual methylation. This opens up the possibility for adaptive variability of the patterns of genomic imprinting (Pollard 1996) associated with sexual orientation, providing new uses for a mechanism that, presumably, originally evolved as an antiparasitic adaptation. More specific research on this potential mechanism would clearly be welcome.

Human sex chromosome aneuploidies and homosexuality

Nuclear genes are found in chromosomes and somatic cells of birds and mammals usually have only two homologous versions of each chromosome: one inherited from the father and the other from the mother. Occasionally, during meiosis, a pair of homologous chromosomes may remain together rather than dividing up and becoming part of a haploid spermatozoon or ovum. Whenever such chromosomally abnormal reproductive cells are involved in the production of a zygote, the resulting individual will be an aneuploid, i.e. his or her cells will have an altered number of a specific chromosome. Aneuploidies such as trisomies have been described for various chromosomes in humans, including several autosomes (2, 7, 10, 13, 14, 15, 16, 18, 21, 22) and both X and Y sex chromosomes (Hassold & Hunt 2001). To date, there is no proven association between any of the autosomal trisomies, including trisomy 21 (i.e. Down Syndrome) and homosexuality (Kessler & Moos 1973; Shepperdson 1995). With regard to sex chromosome aneuploidies, Klinefelter syndrome (KS) has been the subject of especially intense study.

KS individuals are typically males who usually carry a 47,XXY trisomy, that is they have an excess X chromosome, although other karyotypes also occur at lower frequencies: 48,XXXY; 49,XXXXY; 48,XXYY; including 46,XY/47,XXY mosaics. Prevalence of KS is around 2.2 per thousand live births (Kessler & Moos 1973). Early reviews of sexual orientation in KS individuals did not indicate a high prevalence of homosexuality in this group, being less than 10% (Hoaken *et al.* 1964; Swanson & Stipes 1969; Orwin *et al.* 1974; Činovský *et al.* 1986) in sample sizes that were usually very small (small sample sizes tend to inflate percentages). The trend for a very low prevalence of homosexuality among KS individuals has been confirmed in more recent publications (e.g. Blanchard *et al.* 1987; Diamond & Watson 2004). Some authors have reported lower masculinity scores in KS teenagers than controls after administering the Bem Sex Role Inventory, but they did not show higher femininity and even their alleged lower masculinity might have been simply a result of a generalised lower level of self-esteem (Bancroft *et al.* 1982), a trait that may well develop in these individuals as a result of stressful experiences in childhood (see, for example, Orwin *et al.* 1974; Zastowny *et al.* 1987). Zastowny *et al.* (1987) have described a case of a homosexual KS subject who displayed a feminine gender identity, but the individual also was reported to have experienced a very stressful family environment during his childhood.

KS individuals are described as timid, introverted and passive but also cooperative and eager to please, all traits that could be interpreted as 'feminine' in a gendered society (Diamond & Watson 2004), thus giving the impression of 'gender dysphoria'. Some of those 'feminine' KS males do develop female gender roles as adults to the extent of requesting sex reassignment, but it remains to be determined to what extent this is the result of ontogenetic processes influenced by interactions within their social milieu or not.

The XYY trisomy (47,XYY) is also quite rare, showing a prevalence of about 1.5 per thousand. There is no evidence that prevalence of homosexual orientation is especially high in these men compared with XY men (Kessler & Moos 1973; Blanchard *et al.* 1987).

Other aneuploidies are associated with females. Females characterised by trisomy X are usually 47,XXX but they may also have other chromosomal configurations such as 48,XXXX and 49,XXXXX, although the latter two are rare. Turner syndrome women may be either XO, that is they lack one of the usually two X chromosomes characterising females, or XX but with one of the X chromosomes altered. Both kinds of aneuploidy occur at similar frequencies, with mosaics also having been reported (45,X/46,XX). There is no evidence to suggest that either trisomy X or Turner syndrome is conducive to an elevated development of homosexuality (Kessler & Moos 1973, but see the section above on DNA methylation).

Therefore, what the available evidence suggests is that it is not the gross imbalance between sex chromosome numbers per se that is relevant for the development of sexual orientation.

Family analysis

If homosexuality is heritable, family studies should reveal clusters of homosexuality within families (e.g. among siblings and other close relatives) rather than random distributions of homosexual individuals across families. They should also show parent–offspring similarity in homosexual tendencies in those cases where an adult with a homosexual orientation does reproduce. Of course even if such associations are found and are statistically significant, they may well be derived from a shared environment or social effects rather than genetic inheritance. For instance, if an elder brother or sister is homosexual (for whatever reason) the younger brother or sister may also become homosexual under the social influence of the elder sibling. However, if no family clusters or parent–offspring correlations or sibling–sibling correlations are detected, then the chance that the trait is genetically heritable decreases.

In the analysis of the evidence coming from family studies, we have to distinguish two concepts. The first one is *heritability* and the second one is genetic *determination*. In classic quantitative genetics, the realised heritability (h^2) is defined as the ratio between the difference in mean value of the trait between the offspring and the parental generation, called *response to selection* (R), and the difference in the mean value for the trait between the population and the parents being selected, called *selection differential* (S). The more the trait is changing from one generation to another under the effect of a specific selection regime and the smaller the difference in the trait in the parental generation, the more heritable the trait is expected to be. More formally:

$$h^2 = R/S$$

The value of h^2 can be directly measured as the slope of the curve for the offspring value of the trait (e.g. sexual orientation as measured through the *Kinsey Scale*) vs. the mid-parent value (Falconer & Mackay 1996). It can be seen that high values of h^2 for a trait such as same-sex sexual preference could be found as we study groups of siblings and their close relatives (e.g. parents) whether the trait has a relatively straightforward genetic determination contributed by several known loci with two or more alleles each in the population, or whether the phenotypic similarity is maintained by purely cultural mechanisms of learning. In the latter case the measured values of heritability will be high but the genetic determination of the differences in the values of the trait between individuals will be low. Determining whether heritability of homosexual orientation is mainly due to the effect of gene expression is easy in theory as the environment of genetically similar (or different) individuals can be controlled: changed or made constant. This issue will be also addressed in the next section on twin studies.

In this section I review studies carried out on families that contain one or more homosexuals (male or female) and I will compare the rate of homosexuality among sisters and brothers of those homosexual men and women with the rate expected from the sample population at large. If homosexuality is heritable, then the rate of homosexuality among siblings of homosexual men and women should be higher than the rate of homosexuality found in the sample population. The hypothesis of genetic heritability of the trait will be falsified if there is no familial effect upon the distribution of homosexuality.

Table 3.3 summarises the available information on rates of homosexuality among the brothers and sisters of homosexual men and women reported in the literature. All studies were carried out in the United States of America and therefore this is the control population for a comparison of rates of homosexuality expected among non-related individuals. The relative frequency of homosexuality

Table 3.3. Homosexuality rates among non-twin brothers and non-twin sisters of homosexual probands

Proband's sex and reference	Source country	Methods for assessing sexual orientation	Number of probands	Rate of homosexuality among	
				Brothers (%) (n/N)	Sisters (%) (n/N)
Males					
Pillard & Weinrich 1986	USA	Kinsey scale, questionnaire, interviews	51	18	n.a.[a]
Bailey et al. 1991	USA	Kinsey scale, questionnaire	116	10.5 (15/143)	1.5 (2/132)
Bailey & Bell 1993	USA	Kinsey scale, interviews	686	2.8 (14/500)	0.6 (3/476)
Bailey et al. 1995[b]	USA?	Kinsey scale and questionnaire?	n.a	9	4
Bailey et al. 1999					
Sample 1	USA	Kinsey scale questionnaire	350	7.3 (33/453)	3.8 (15/394)
Sample 2	USA	Kinsey scale questionnaire	167	9.7 (25/259)	2.8 (6/214)
Females					
Pillard 1990	USA	Kinsey scale, sibling report	n.a.	13	25
Bailey & Benishay 1993	USA	Kinsey scale, interviews, questionnaire, sibling report	61	7.2 (8/110)	12.1 (12/99)
Bailey & Bell 1993	USA	Kinsey scale, interviews	292	3.8 (8/210)	1.6 (3/188)
Pattatucci & Hamer 1995	USA	Kinsey scale, self report	177	5.0 (11/219)	5.5 (9/165)
Bailey et al. 1991	USA	Kinsey scale, questionnaire	72	3.3 (1/30)	4.1 (1/24)

[a]Although Pillard & Weinrich (1986) report percentages of brothers that are predominantly or exclusively homosexuals, they only have combined homosexual–bisexual figures for sisters. The value, however, (8.1%) is consistent with the general trend of larger prevalence of homosexuality among brothers than among sisters of homosexual men found in most studies reported here.
[b]Unpublished data reported in Table 1 of Pillard & Bailey (1998).

among men in the USA is estimated at 3.1% (Laumann et al. 1994; Kendler et al. 2000), whereas for women the estimated rate is between 1.5% (Laumann et al. 1994) and 2.5% (Kendler et al. 2000). For calculation purposes I will use the arithmetic mean of the Laumann et al. (1994) and Kendler et al. (2000) values for female homosexuality, i.e. 2.0%.

On average, 6.1% of sisters of homosexuals (men and women combined) are themselves lesbians (Table 3.3). This is a statistically significant departure from the expected proportion of women in the USA population at large who have a lesbian sexual orientation (2.0%) (Student's t-test after log-transformation of data: $t_9 = 2.02$, $p = 0.03$). A total of 8.14% of brothers of homosexuals are themselves homosexuals (Table 3.3), a figure that again is significantly higher than the expected percentage of male homosexuals in the USA population at large (3.1%) (Student's t-test after log-transformation of data: $t_{10} = 4.42$, $p = 0.0006$). Therefore, homosexual siblings are more likely to occur than expected given the percentage of homosexuality in the general population.

We may next ask whether the proportion of homosexuals among brothers and sisters varies according to the sex of the homosexual proband. Before we address this issue, however, we have to check for potential sources of bias among the data from the literature sources that I used. Although the approach is to use Type III meta-analytical methods that make use of the actual data from the original published works (Blettner et al. 1999; see also Chapter 2), such an approach does not guarantee complete freedom from biases that could produce spurious results. One obvious potential problem is that smaller sample sizes may produce larger percentages. If so, sample size and fraction of siblings who are homosexuals should be negatively correlated. This is in fact the case: Pearson product-moment correlation, $r_{14} = -0.26$, $p < 0.05$. Next I asked whether the four samples: homosexual brothers of homosexual male proband, homosexual sisters of homosexual male proband, homosexual brothers of homosexual female proband and homosexual sisters of homosexual female proband, actually differ in their mean value of sample sizes. If they do not, the above correlation would be unproblematic. Unfortunately, the answer is that they do (see Table 3.3), with samples for siblings of female probands being smaller than samples for siblings of male probands (Two-way Factorial ANOVA after log-transformation of data: $F_{1,12} = 7.08$, $p = 0.02$; the interaction is not significant). Therefore siblings (males and females) of female probands are expected to show higher percentages of homosexuality owing to sample-size effect alone. However, a two-way Factorial ANOVA for homosexuality data for siblings across sibling sex and proband sex categories shows that there is no significant effect of sibling sex ($F_{1,17} = 3.22$, $p = 0.09$), whereas there is a significant effect of the interaction ($F_{1,17} = 5.29$, $p = 0.03$). This means that the sample size effect is not enough to override other effects. The significant interaction between factors is caused by a greater percentage of homosexual brothers (9.55% ± 4.96, $n = 6$) than homosexual sisters (2.54% ± 1.46, $n = 5$) among male probands, but a greater percentage of homosexual sisters (9.66% ± 9.41, $n = 5$) than homosexual brothers (6.46% ± 3.95, $n = 5$) among female probands.

The results of the above meta-analysis are suggestive of a familial effect on human homosexuality that is sex-specific: females seem to be concordant with sisters' homosexuality, whereas males are concordant with brothers' homosexuality. The same pattern has recently been described by Schwartz et al. (2010). Of course, such a trend could well be explained by social influences exerted by same-sex siblings on each other's sexual orientation development. A common X-chromosome inheritance of homosexual orientation for both men and women, however, seems to be refuted by these results. If homosexuality is X-chromosomally inherited in both sexes in humans then brothers of lesbians should be especially prone to also show a homosexual orientation as they would be likely to express any allele for homosexuality, even if recessive. Homosexuality in both sexes cannot obviously be explained by genes on the Y chromosome. *The conclusion from the above results is that if there is a specific genetic determination of homosexuality in humans, it is different for women and men.* This is in accordance with the finding of Mustanski et al. (2005) that autosomal loci are also likely to be involved in the determination of same-sex sexual behaviour in humans, suggesting a polygenic determination of the trait that may well differ between the sexes. Although this is a plausible scenario that should encourage further research on the genetics of sexual orientation, the potential effect of social interactions between siblings and with parents and other relatives is equally plausible and should also be the focus of research. The overall evidence reviewed in this book strongly suggests that both specific social and equally specific genetic factors co-occur in the determination of homosexuality.

The fact that all the research included in Table 3.3 was carried out in the USA might be seen as an advantage, in terms of controlling a set of important socio-cultural variables. Equally, it could be seen as a disadvantage, as we cannot ascertain whether the results, if affected by culture, are culturally specific or not. On the other hand, even when similar data are gathered for diverse ethnic groups living in their own country of origin and different results are obtained, a purely socio-cultural explanation would be premature. Thus, ethnically diverse groups might differ not only in their culture but also in their genetic make-up (Cavalli-Sforza 2001). This means that broad analytical approaches, although useful and certainly interesting, leave open too many gaps for uncontrolled variables to squeeze in. Very specific and detailed experimental studies testing clear mechanisms (genetic or otherwise) will ultimately be needed. Another methodological issue arising from Table 3.3 is that the vast majority of works were done by a team (or teams) that share one common co-author. This may well be irrelevant, but in meta-analytical studies it would be ideal to analyse results derived from independent sources. Convergence of results among completely independent groups certainly provides much more powerful evidence to support or reject a hypothesis.

The result obtained in the meta-analysis follows a coincidence in proband sex-specific trends common to the vast majority of studies reported in Table 3.3 and previously recognised by students in the field (Bailey & Bell 1993). Although Bailey & Bell (1993) did mention that previous work indicated the presence of a sex-specific concordance of homosexuality in family contexts, in their empirical study they found a different pattern where homosexual females have more homosexual brothers than sisters and also more homosexual siblings of both sexes than heterosexual probands. It should be pointed out, however, that from the data summarised in Table 3.3 this trend is unique and it was not replicated by the other studies.

Twin studies

When it comes to researching the potential contribution of specific genes to the development of a homosexual orientation, few 'natural experiments' can improve upon the study of twins. Monozygotic (MZ) twins are genetically identical, whereas dizygotic twins (DZ) are not. DZ twins share the same prenatal environment in the womb but their genetic make-up is the same as that of any other pair of full siblings. Non-twin siblings have a degree of genetic relatedness similar to DZ twins but they have not shared a common pre-natal environment as DZ twins did. Moreover, DZ twins may be either same-sex or different-sex.

Genetic and environmental effects can, in theory, be perfectly controlled in these studies by comparing MZ twins reared together with MZ twins reared apart, and also with DZ twins, non-twin siblings and non-related (e.g. adoptive) siblings reared apart or together. Different pairs of MZ twins reared apart could have been separated at different ages and therefore the effect of a shared environment may be studied at different stages of the twins' development. This is an extremely important issue, as we will see in Chapter 4, because MZ twins separated at an older age may be more concordant in their sexual orientation than DZ twins separated at a younger age, not because of their greater genetic similarity, but because of the difference between the two twin categories in the degree of pre- and early postnatal ontogeny they shared in a similar environment. In addition, MZ twins who have lived together in the same environment from birth may also influence each other behaviourally and therefore become more concordant in their sexual orientation owing to this social factor. On the other hand, female MZ twins may have undergone different patterns of X-chromosomal inactivation during the early stages of development, leading to the expression of different phenotypes in spite of sharing the same genotype. In this case they would be discordant for genetic and ontogenetic, rather than purely environmental, reasons. All these complications can affect the results of twin studies (see McGuire 1995 for a thorough review of the methodological pitfalls plaguing most twin studies of sexual orientation), but their effects can be controlled to some extent by careful choice of participants and the use of large sample sizes.

Table 3.4 summarises the results of available studies on sexual orientation of male and female twins. The table only reports studies that published raw data (i.e. clearly identifiable frequencies) permitting the calculation of the prevalence of homosexuality among different kinds of twins and other related and non-related siblings. Studies that did not include original raw data identifiable in the manner described above will be also reviewed below. A meta-analysis of the data collated in Table 3.4 is not straightforward, for the following reasons.

1. The methodologies used to assess sexual orientation or zygosity differ across studies.
2. The sampling methods used to recruit participants vary: from population-wide twin registries to participants recruited through gay and lesbian organisations, publications or events.
3. Sample sizes are very different across most studies.
4. Some studies reported data specifically referring to homosexual subjects, others pooled homosexuals and bisexuals together.
5. It is unclear from most studies at which age the twins became separated: from birth, or at which other postnatal stage of development.

In spite of these shortcomings, a broad analysis is still possible comparing the prevalence of homosexual/bisexual (i.e. non-heterosexual) orientation between MZ and DZ twins. A simple Sign Test will allow the contrast of the two twin-categories within a study, thus eliminating the effects of the factors that are common to both. If the within-sample effect of other variables is controlled in this way, then a comparison is also possible for MZ–DZ twin-pair values across studies. This is the basic logic of all matched-pairs tests. Because the relationship of homosexual prevalence values between MZ and DZ twins seems to go in the same direction for females and males, data were pooled to increase the sample size. A Sign Test suggests that prevalence of non-heterosexuality among MZ twins is significantly higher than among DZ twins ($p = 0.039$); this is expected to be the result of a higher degree of concordance between MZ twin-pairs in their sexual orientation than among DZ twin-pairs. As such, the result is supportive of a specific genetic basis for sexual orientation. This meta-analysis, however, may still be subject to the vagaries of publication biases (Gurevitch & Hedges 2001), although the variability in research teams listed in Table 3.4 may provide a partial safeguard against those effects. A bias that cannot be controlled in the analysis, however, is one that could have affected the initial quality of the data if only specific kinds of people are keen to participate in sex research studies, whereas other categories are not (Dunne et al. 1997).

Although the initial report by Kallmann (1952) indicated a 100% homosexual concordance among monozygotic male twins, all current estimates are far less dramatic, indicating a co-contribution of both genetic and environmental factors in the development of homosexuality. Early studies of the sexual orientation of twins tended to focus on patients attending psychiatric hospitals (Heston & Shields 1968) and this, by itself, could have biased the level of concordance among twins. Apart from the odd report by Kallmann, the reported levels of concordance between MZ twins indicate that a value of about 50% for male twins is a more realistic figure (see, for example, Heston & Shields 1968; Bailey & Pillard 1991; King & McDonald 1992). Mechanisms that could explain sexual orientation concordance of less than 100% in MZ twins are varied and range from environmental to genetic and gene–environment interactions during ontogeny. Turner (1994 and references therein) lists causes such as skewed X-chromosome activation, post-conception DNA modifications such as methylation, inequalities in the supply of blood between twins during gestation, and diverse levels of penetrance of the gene/s involved in the trait. Heritability estimates for same-sex sexual preference vary dramatically between studies from 0.28–0.65 (Kendler et al. 2000) to 0–0.48 (Hershberger 1997), 0.34–0.39 (Långström et al. 2008) and 0.30–0.58 (Kirk et al. 2000), thus suggesting a complex modality of inheritance.

Comparing MZ twins with DZ twins is not enough to disentangle the effects of genetic and environmental components of heritability. Complete designs should allow the comparison of groups of individuals that show all combinations of genetic and environmental effects (Eckert et al. 1986): MZ twins reared together (i.e. same genes/same environment) and reared apart (i.e. same genes/different environment) vs. DZ twins reared

Table 3.4. Homosexuality rates among MZ and DZ twins and non-twin siblings

Proband's sex and reference[a]	Source country	Methods for assessing sexual orientation	Method of zygosity assessment	Rate of homosexuality among[b]				
				MZ % (n/N)	DZ-same sex % (n/N)	DZ-opposite sex % (n/N)	Non-twin brothers % (n/N)	Non-twin sisters % (n/N)
Males								
Heston & Shields 1968	UK	Interview	Appearance, blood groups, plasma proteins, fingerprints	66.6 (4/6)	—	—	—	—
Bailey & Pillard 1991	USA	Interview, *Kinsey scale*	Questionnaire	50 (25/50)	24 (11/46)	—	—	—
Eckert et al. 1986	USA	Interview	Blood typing	75 (3/4)[c]	—	—	—	—
Buhrich et al. 1991	Australia	Questionnaire, *Kinsey scale*	Questionnaire	4.7 (9/190)	0.8 (1/126)	—	—	—
King & McDonald 1992	UK	Questionnaire	Twins' self classification	55 (22/40)	54 (27/50)	—	—	—
Whitam et al. 1993	USA	Questionnaire, interview, *Kinsey scale*	Photo comparison	83.8 (57/68)	72.9 (27/37)	77.7 (14/18)	—	—
Bearman & Brückner 2002	USA	Interview	Self or mother's report	9.9 (26/262)	9.8 (27/276)	16.8 (31/185)	7.9 (47/596)	7.3 (31/427)
Females								
Eckert et al. 1986	USA	Interview	Blood typing	50 (4/8)	—	—	—	—
Whitam et al. 1993	USA	Questionnaire, interview, *Kinsey Scale*	Photo comparison	87.5 (7/8)	77.7 (14/18)	—	—	—
Bailey et al. 1993	USA	Interview	Questionnaire	48 (34/71)	16 (6/37)	14 (10/71)	—	—
Bearman & Brückner 2002	USA	Interview	Self or mother's report	7.6 (20/264)	6.6 (17/259)	5.3 (10/190)	8.3 (35/423)	7.5 (45/601)
Sexes pooled								
Kendler et al. 2000	USA	Phone interview, self-administered questionnaire	Self report	3 (19/648)	3.1 (15/480)	—	—	—

[a] Here we only include works that report raw data.
[b] Only two studies report rates of homosexuality among adopted siblings: Bailey & Pillard (1991), who reported a 19% (6/31) rate of homosexuality among adoptive brothers of homosexual or bisexual men; and Bailey et al. (1993), who reported a 6% (2/35) homosexuality among adoptive sisters of lesbians.
[c] MZ male twins reared apart.

together (i.e. different – but not too different – genes/same environment) and reared apart (i.e. different – but not too different – genes/ different environment) and also a comparison with non-related (e.g. adopted) siblings (i.e. very different genes/ same environment) and a random sample of the population at large (i.e. very different genes/different environment). Bailey & Pillard (1991) and Bailey *et al.* (1993) have produced two of the few studies that also compared the rate of homosexuality in adoptive siblings, males and females respectively. In both cases, the rate of homosexuality was lower in adoptive siblings than in either monozygotic or same-sex dizygotic twins. The same is true for non-twin sisters and brothers (Bailey *et al.* 1993; Bearman & Brückner 2002; Heston & Shields 1968): their rate of homosexuality tends to be lower than that found among twins. Results for different-sex DZ twins are conflicting. Some reports suggest that they tend to share a homosexual orientation at an even higher rate than either MZ or same sex-DZ twins (Bearman & Brückner 2002), whereas other works carried out on both male and female twins indicate a lower rate of concordance in homosexual orientation than MZ twins (see Table 3.4). Although twins reared apart sometimes are concordant with regard to their homosexual orientation, on other occasions they are discordant (Whitam *et al.* 1993). Buhrich *et al.* (1991) carried out a multivariate analysis of the data available in the Australian NHMRC Twin Registry to conclude that several variables measuring sexual identity and orientation in childhood and at adult ages are correlated with zygosity of twins in such a manner as to suggest combined environmental and additive genetic effects in the determination of homosexuality. Although twin registries are usually less biased sources of data for homosexuality studies than, for instance, newspaper advertisements, or recruitment of volunteers from gay/lesbian organisations, they are not completely free from biases either (Lykken *et al.* 2007).

Kirk *et al.* (2000) also carried out an analysis of the large Australian NHMRC Twin Registry and found greater correlation values for MZ than DZ twins in their homosexual feelings, partners and attitudes, as expected from the presence of significant additive genetic variance for those traits, although some evidence of environmental effects was also found in their results in the form of correlation coefficients between MZ twins being less than twice those for DZ twins. Their explanatory models are also consistent with combined additive genetic and environmental effects contributing to the variance in sexual orientation in both men and women.

There are reports of homosexual twins reared apart who express sexual attraction towards each other when they meet, although this is not always the case for twins reared together (Eckert *et al.* 1986; Whitam *et al.* 1993 and references therein). King & McDonald (1992) reported a prevalence of 21% of same-sex twins having had a homosexual relationship with each other, although their small sample size (n = 33 twins) is likely to overestimate percentages. Why some twins become sexual partners could be explained by several potential mechanisms that are not necessarily mutually exclusive. Of course, for those cases of twins reared together, if both twins have inherited a propensity to develop a homosexual orientation (e.g. if they are MZ), they have each other readily available at the time when they become sexually active, this tendency, however, may be constrained by inbreeding avoidance mechanisms (see Eckert *et al.* 1986; Whitam *et al.* 1993 and references therein, mentioned above) that may have been originally selected in a heterosexual context but may remain active in a homosexual context. King & McDonald (1992), however, provide a socio-sexual interpretation of sexual interactions between same-sex twins in terms of dominance relationships between them, and we will see in Chapter 8 that socio-sexual mechanisms associated with affiliative behaviour are an additional possibility.

With regard to sex differences, the work of Whitam *et al.* (1993) suggests that the level of concordance in homosexual orientation is higher in MZ female twins (75.0%) than in MZ male twins (64.7%), and that it is also higher in DZ female twins than in DZ male twins. Hershberger (1997), after analysing data from the Minnesota Twin Registry, concluded that concordance in same-sex sexual preference is more stable over the years (i.e. before and after 25 years old) in female MZ twins than in male MZ twins. Kendler *et al.* (2000) reached the same conclusion, also finding greater concordance between female MZ twins than between male MZ twins. This, however, stands in contrast with other studies (see Table 3.4).

Finally, twin studies have detected some important trends suggesting an early onset of homosexuality during ontogeny (see also Chapter 4), such as the positive correlations found between MZ twins in their level of 'Childhood Gender Nonconformity' (Bailey & Pillard 1991; Buhrich *et al.* 1991).

In sum, in spite of the many methodological caveats regarding the correct interpretation of results coming from twin studies of homosexuality, comparisons among twin and other sibling and non-sibling categories remain a major source of evidence for some degree of genetic control of homosexuality in humans. However, compelling as this evidence is, it also indicates that even 'identical' male twins do not necessarily both become homosexual; indeed in about 50% of cases one brother may be homosexual whereas his twin is heterosexual. Thus, the contribution of postnatal environment may be very important even in the case of twins reared in the same families. Or, alternatively, the assumption of the 'identical' status of monozygotic twins is an over simplification of what occurs during foetal development in such cases. Either way, these results should raise a red flag and make researchers cautious when considering explanations of homosexuality that over-emphasise the role of 'gay genes' in human beings. We will see throughout this book that a biosocial approach is better able to make sense of all the available empirical information.

Evidence from selection studies

If same-sex sexual preference has a heritable genetic component, then it could be easily selected in a population under the appropriate selective regime. Therefore selection studies are central to the whole issue of whether homosexuality is a trait that can undergo adaptive evolution. Of course, demonstrating that the trait has enough genetic heritability to respond to selection is one thing; understanding the adaptive value of homosexuality in wild populations of the organism is quite another. The latter requires the testing of specific models of the benefits of the trait (e.g. in enhancing inclusive fitness through cooperation; or enhancing direct fitness through the formation of alliances for cases of non-exclusive homosexuality and bisexuality, etc.). Because selection experiments are more easily carried out in invertebrates than in vertebrates, I will start with a review of some selection studies carried out in insects.

Same-sex mounting in male insects is not uncommon (Thornhill & Alcock 1983). It is usually assumed that such a behaviour in males is a result of lack of ability to recognise conspecifics of different sexes (Serrano *et al.* 1991). However, if such errors were costly (e.g. in terms of time and energy wasted), then we would expect discriminating mutants to have been selected rather quickly. Yet the behaviour seems to be persistent in natural populations of many insects, suggesting the possibility that mounting same-sex partners may have either negligible costs or be a trait under current selection, assuming that enough genetic variability for the trait is available in the population. José Serrano, Laureano Castro and collaborators carried out a study of the red flour beetle (*Tribolium castaneum*) to explore these issues. Although courtship behaviour between males and females has not been described in this species, it is known that males secrete a pheromone that is attractive to both sexes (Levinson & Mori 1983). In order to study the genetic underpinning of male–male mounting in *T. castaneum*, Serrano *et al.* (1991) carried out an experiment using six different inbred lines of the beetle in a set-up involving mate choice between two males and a variable number of virgin females (one, two or four). In general, they found that rates of male–male copulations were expected by a random association of individuals following a non-discriminatory choice of mating partner, except in additional cases where the number of males was large and the sex ratio heavily biased towards males. In such cases males tended to aggregate at about the centre of the Petri dish used for the experiment and therefore their copulation partners became male-biased. Serrano *et al.* (1991) also found that male sexual activity (mountings per unit of time) increased with the number of individuals available, irrespective of their sex. Interestingly, rates of homosexual copulations between males were much higher within inbred lines than within F_1 generations of crosses between those different inbred lines. This is strongly indicative of a very fine mechanism of sexual partner discrimination, presumably chemically based. In fact, Serrano *et al.* report that the rate of homosexual mounting between two inbred males of the same line was even higher than the rate of heterosexual mounting between an inbred male and a non-inbred female. These results were interpreted by the authors as an indication of homosexual mounting being an unselected outcome of mating activity triggered by a genotype-specific recognition system in males that does not discriminate between the sexes. However, the fact that males tend to cluster when in large numbers (a pattern found in other insects too and that has been associated with the presence of a pheromonal aggregation signal in males; Soares Leal *et al.* 1998), and that there is strong between-inbred-lines discrimination, may be also indicative of a potential selective process being in operation. For instance, ejaculatory products may be transferred between males (e.g. spermatozoa and other components of the seminal fluid) during homosexual copulation that could have a function in male–male competition for reproduction (see Poiani 2006 for a review), a possibility that is worth exploring experimentally. Levan *et al.* (2009) did carry out such a test and their results will be discussed below.

Castro *et al.* (1994) subsequently carried out a selection experiment for *T. castaneum* homosexual mounting behaviour. In an experimental set-up where two males were placed in a glass vial with two virgin females from an unselected base population, the authors maintained a divergent selection regime (four replicates per line) for the rate of homosexual copulations among males over three generations. The realised heritability value for homosexual copulation in the base population was estimated by Castro *et al.* at 0.11 (standard error corrected for drift). They observed a statistically significant divergence in homosexual copulations between the lines positively and the lines negatively selected for the trait after just three generations of selection. They calculated realised heritabilities for the increase in homosexual behaviour at 0.17, whereas for decrease and for overall divergence, realised heritabilities were lower (0.08 and 0.11, respectively). These results are consistent with the presence of genetic variance for homosexual behaviour in the population. What is maintaining this level of heritability? Several alternatives are possible. One, of course, is that the trait may not be under strong selection (directional or stabilising) in wild populations, therefore genetic diversity may be maintained by random mutations alone. However, there is also the potential for antagonistic selection between males if, for instance, homosexual copulations are a way to transfer spermatozoa or seminal fluid to other males, as mentioned above. In this case, high heritabilities may be maintained by divergent effects on reproduction produced by loci that favour male–male mounting and loci that favour male–male resistance to being mounted. Evidence for sperm transfer during male–male copulations in this species, however, is weak (Levan *et al.* 2009).

Levan *et al.* (2009) have carried out experiments on *T. castaneum* to test alternative selective models for male–male copulations in this species. Their results did not support the hypotheses that male homosexual mounting has a function in establishing a dominance hierarchy or as a practice to refine copulatory behaviour that can then be used in heterosexual mating. Some weak support for the hypothesis that male homosexual mating may serve the function of sperm transfer (sperm competition hypothesis) was obtained, but sperm translocation was observed in only 7% of the same-sex sexual pairs and the efficiency of this sperm transfer in fertilisation was less than 0.5% (percentage of offspring sired). Levan *et al.* (2009) suggest that, instead, male–male homosexual mating may serve the proximate adaptive function of sperm renewal (see also the 'Masturbation and homosexuality' section in Chapter 5).

Although selection experiments, such as the ones described above for insects, are not as easily carried out in large vertebrates, some studies of the fitness consequences of homosexuality are available for some human populations. In humans, it is known that homosexual men have, on average, a lower lifetime reproductive success than heterosexual men (see reviews in Bobrow & Bailey 2001 and Rahman & Hull 2005; see also Moran 1972; Bell & Weinberg 1978; Iemmola & Camperio-Ciani 2009; Santtila *et al.* 2009). This does not seem to result from sperm characteristics being different between homosexuals and heterosexuals, as similarly great variability has been detected in sperm traits among homosexuals and also among heterosexuals (Parr & Swyer 1960; Kolodny *et al.* 1971; Doerr *et al.* 1973).

If homosexuals have lower reproductive success than heterosexuals, from a natural selection point of view the question we need to answer is: How is the genetic variability underpinning the trait being maintained? Or, put in another way, why is it that any allele contributing in a non-trivial manner to the determination of a homosexual orientation has not become extinct?

Does homosexuality have any link with increased female fecundity?

We have already seen previously in this chapter that even simple *mutation–selection equilibrium* mechanisms could maintain a fitness-decreasing trait at low frequency in a population. *Sexually antagonistic selection* and *heterozygote advantage* are also major potential mechanisms. Sexually antagonistic selection, in particular, has been strongly emphasised by recent theoretical and empirical work that was reviewed earlier in this chapter (Gavrilets & Rice 2006; Iemmola & Camperio-Ciani 2009). Moreover, Camperio-Ciani *et al.* (2008b) have suggested that sexually antagonistic selection may also explain selection of bisexuality. In their work they provide evidence that although fecundity of heterosexual men (0.63 children on average) is higher than that of homosexual (0 children) and bisexual men (0.21 children) in their sample, fecundity of mothers of the subjects that they interviewed was higher for both homosexual and bisexual men compared with heterosexual men (Camperio-Ciani *et al.* 2008b). Such results are consistent, for instance, with the presence of an X-linked allele that confers increased sexual attraction towards males to both male and female carriers of the allele. This postulated common basis for sexual attraction towards males in both men and women is known as *androphilia* (see also Vasey *et al.* 2007). Schwartz *et al.* (2009) also provide evidence for a role of sexually antagonistic selection in the maintenance of homosexuality in the human population. In fact, in their sample from the USA and Canada, homosexual males had more relatives than heterosexual males.

Kin selection and homosexuality

However, another obvious mechanism that could explain the maintenance of a trait that decreases *direct fitness* of an individual is one in which the same trait increases the *indirect fitness* of that same individual; this is the *Kin Selection* hypothesis for the evolution and maintenance of homosexual behaviour (Williams 1966; Wilson 1975). The hypothesis has been criticised on several fronts; for example, if homosexuality has advantages in terms of helping kin, why is it that human male homosexuals spend a large amount of their time in same-sex sexual activities and comparatively little time actually helping relatives (Bobrow & Bailey 2001)? Also, when homosexuals in modern industrialised societies help somebody, close kin are not necessarily the preferred recipients (Bobrow &

Bailey 2001; Rahman & Hull 2005). Moreover, Robert Trivers (1985) makes the additional point that if receiving help is a consequence of having a homosexual offspring why is it that in modern, industrialised societies parents are so afraid of having a child who may develop a homosexual sexual orientation? Although these criticisms seem to be *prima facie* valid arguments against a current kin-selected adaptive value of homosexuality in modern societies, they are of very little relevance to the issue of the potential role of kin selection in the initial evolution of the trait and the current maintenance of the same trait in non-industrialised societies. How the trait was initially selected can be understood only in the context of the ancestral *Environment of Evolutionary Adaptedness* (EEA) (Bowlby 1969). In more recent ecological and demographic contexts, homosexuality might have been kin-selected not necessarily because homosexuals actively provided help to kin (e.g. alloparental care), although this is a possibility, but because by not reproducing, or reproducing less, they made resources available to kin for reproduction (see also Chapter 8). A similar argument has been put forward to explain other cases of indirect help to kin by withdrawal from reproduction. Just to mention one example, see, for instance, the tendency for Catholic priests born between the second half of the nineteenth and the first half of the twentieth centuries to come more often from families with a male-biased sex ratio among siblings, in a sample of Irish Catholic families (Deady *et al.* 2006). In this case withdrawal from reproduction might have decreased the pressure on distribution of family property, thus, presumably, indirectly benefiting the reproducing brothers and their offspring.

In addition, an offspring who is an exclusive homosexual is of no 'use' to the parents in terms of fitness if he is the only child, as in that case there are no brothers and/or sisters to help. The specific kin-selection argument for homosexuality makes more sense in the context of relatively large families. Moreover, from a kin selection point of view it is even expected that parents of very small families would feel (at least initially) 'disappointed' by the birth of a child who is not going to reproduce. Indeed, such an initial feeling in very small families is expected if the child is not going to reproduce whatever the circumstance, whether the offspring is homosexual or heterosexual.

Bobrow & Bailey (2001) recruited homosexual volunteers through gay publications in the USA and asked them to complete a questionnaire about the kinds of feelings they had for relatives and their willingness to give to, receive from or channel resources toward relatives. They did not find any evidence that homosexual men display any particular generosity towards relatives; the same result was obtained by Rahman & Hull (2005) in a study carried out in England. Both Bobrow & Bailey (2001) and Rahman & Hull (2005) did realise some obvious shortcomings of their work, such as that the current family and social environment in industrialised societies may well differ in some crucial aspect from the social environment where homosexuality might have been adaptive. However, their analysis suffers from an even more serious misunderstanding of the control of adaptive traits in higher vertebrates with a complex central nervous system, such as humans. Behaviourally plastic organisms such as humans are expected to respond quickly to changes in local conditions. For instance, Darwinian fitness is measured over the lifetime of the individual, therefore for a long-living species like *Homo sapiens* we would expect selection not only for genes that ensure reproduction but also for those that ensure an extended life. Personal wellbeing would be expected to be a proximate measure of the chances of a long lifetime. Under kin selection homosexuals would maximise their inclusive fitness not only if they help kin but also if they do so for many years. Therefore they may be expected to resist any situation that may jeopardise their wellbeing and their chances of survival. If relatives increase the costs of the trait for homosexuals through increased distress – as may occur under the influence of prejudice and discrimination – then a counter-reaction is expected from the homosexual. Although homosexuality may be expressed in any event (perhaps because of inherited propensities) it will lead to redirection of cooperation where the rewards in terms of survival are higher; e.g. towards a sexual partner willing to also provide assistance and companionship of some kind. *This argument implies that in the EEA homosexuals were better integrated in their social family milieu than they are in some societies today.* Such better social integration in the EEA would have produced both a longer life for the homosexual and consistent help to kin, thus maximising his lifetime inclusive fitness. Of course, we cannot rewind the film of our evolution to observe our ancestors behave in their environment, but we can certainly use information from archaeology and comparative anthropology and ethnology to understand both what the ecological and social conditions might have been in the past for our species and how humans behave today when living in similar conditions. I will comment on one such example in a moment.

Maintenance of homosexuality under a scenario of variable adaptive value of the trait is obviously facilitated if homosexuals reproduce at some stage, as in this case cooperation may not only benefit their indirect but also their direct fitness. The non-zero reproductive success of homosexuals reported in many studies (Bobrow & Bailey 2001) is an indication that homosexuality (functional bisexuality in this case) is probably less selectively costly than one may think. In this regard, the finding that homosexuals have lower reproductive success than heterosexuals does not necessarily embark them on a route to extinction; it just marks a route to phenotypic 'minority' (see, for example, Laumann *et al.* 1994; Kendler *et al.* 2000).

Trivers' criticism could be easily addressed by using analogous arguments that consider changes in the modern, urbanised socio-demographic context. If the size of the family is small, why should not parents be disappointed, to some extent at least, by the birth of a child who is not going to produce grandchildren? Moreover, if by having a homosexual child parents who do not need the labour of that child to help in the domestic economy are at risk of suffering a social stigma, then a homosexual offspring becomes a 'liability' with little to compensate for it. From this we can predict that whenever the help from offspring makes a difference to the economy of the family (e.g. in traditional rural settings where labour may be required from all members of the family), homosexuals should be better integrated into the social frame, and they should also be better integrated when the family is wealthy if there is less bigotry in the society at large. Interestingly, this view predicts that in small rural societies a homosexual son could be rejected by his family if he is the cause of withdrawal of support from the rest of the community (see, for example, D'Augelli & Hart 1987), but he would be accepted and his help gladly received if the community do not see having a homosexual son as a reason to ostracise the whole family. The very fine balance between acceptance and rejection of male homosexuals in a rural community in the USA is vividly described by Walter Boulden (2001; see also references therein), but more studies are needed from other ethnic groups as well.

The need to test the *Kin Selection* hypothesis in conditions that could be described as being relatively closer to those that probably prevailed in the EEA has been brilliantly exemplified by the recent work of Paul Vasey, David Pocock and Doug VanderLaan (2007) from the University of Lethbridge, Canada, on *fa'afafine* male homosexuals in Independent Samoa. The *fa'afafine* are mainly feminine, exclusive (98.5%; Vasey & VanderLaan 2009a) homosexual males who engage in sexual intercourse with masculinised men; from a sexual behaviour point of view, the latter should be described as bisexuals, although they tend to self-identify as 'straight men'. *Fa'afafine* do not reproduce and are not involved in sexual activities with each other (Vasey & VanderLaan 2009a). In Samoa, families tend to be large and relatives live somewhat close to each other. In addition, *fa'afafine* are fully integrated in both their family network and also the society at large. After interviewing both *fa'afafine* and 'straight men' from the islands of Upolu and Savai'i, Vasey *et al.* (2007, see also Vasey

& VanderLaan 2009a) reported significantly greater avuncular tendencies (i.e. tendencies to behave like an 'uncle' in terms of kindness, indulgence and material support) and marginally not significant tendencies ($p = 0.08$) to be more generous among *fa'afafine* than among 'straight men'. Among the statistically significant avuncular tendencies displayed by *fa'afafine* were: babysitting, buying toys for the children, helping to expose the children to the arts, and contributing money to the children's education. *Fa'afafine* also displayed higher avuncular tendencies than the materteral tendencies (i.e. tendencies of females to behave like an 'aunt') displayed by women with and without children (Vasey & VanderLaan 2009a). Vasey & VanderLaan (2009a, 2009b) clearly show that *fa'afafine* behave unlike heterosexual males and females (*third-gender* pattern), as the frequency of performance of a specific kind of kin-directed (avuncular) cooperative acts was higher than that observed in both heterosexual males and females in their society. Note that in order for a phenotype such as *fa'afafine*'s to be selected it is just required that conditions prevail where reproduction by all individuals, including potentially the *fa'afafine*, has the effect of decreasing the reproductive success (inclusive fitness, to be more precise) of all individuals in the extended family group. In such a situation, forfeiting reproduction and helping kin instead may be selected; this is more likely to occur when the family group is relatively large and intensity of competition for resources is also large. This suggests that more studies like the one carried out by Paul Vasey and his collaborators should be repeated for traditional societies living on islands (whether oceanic like those of the Samoan studies or ecological such as the fragmented habitats of human populations living in arid or semi-arid regions or in mountainous regions).

In sum, what Vasey and his collaborators' work demonstrates is that in circumstances where the help from homosexuals is needed by the family group, and the family group fully integrates the homosexual into the broader social network, homosexuals who do not reproduce tend to divert their resources and help towards kin, as predicted by *Kin Selection*. Moreover, their patterns of behaviour are specific (*third-gender*) rather than gender-inverted, suggesting a potentially selected specialisation. More work focusing on rural and traditional societies will obviously be welcome.

The above issues notwithstanding, helping relatives does not require homosexuality. Help can be received or given in a perfectly heterosexual social context. The issue, however, is whether kin selection had anything to do with making a homosexual mutant adaptive. In this context, a homosexual phenotype willing to cooperate (even if conditionally) may have a better chance of transmitting his/her genes than a selfish homosexual (see, for example, Salais & Fischer 1995; Kirby 2003). In the context of the kin selection hypothesis, alleles favouring homosexuality should be in linkage disequilibrium with alleles favouring cooperation, whether the cooperation is direct or indirect via release of resources, as explained above. An interesting result that may be indicative of the association between homosexuality and cooperation is provided by the study of dreams of homosexual and heterosexual men carried out in the 1970s by Winget & Fanell (1972). In their work they described 'trends in the dreams of homosexuals toward increased presence of characters, lowered aggressive-hostile content, and a higher incidence of friendly interactions' (p. 119).

The same linkage between homosexuality and cooperation is also favoured by other selective mechanisms such as reciprocal altruism and mutualism that do not require, but are enhanced by, close relatedness between interacting individuals. If the expression of cooperation in homosexuals retains a degree of plasticity, then cooperation can be shifted from close relatives in the context of a small traditional village to fellow gay men in the context of urbanised, modern societies according to the rewards received in terms of personal well-being.

A linkage between cooperation and homosexuality might have been selected during periods of population expansion, where homosexuality could be associated with larger families (Camperio-Ciani *et al.* 2004, 2008a): the more cooperative members a family has, the more it can afford to increase its numbers. That is, *kin selection* and *sexually antagonistic selection* may act in synergy. Per se, *sexually antagonistic selection* does not predict the expression of cooperative behaviours on the part of homosexuals; it just predicts increased probability of expression of homosexual phenotypes in larger families. In addition, the larger the family is, the less the relative cost of having a homosexual child is in terms of parental fitness. The latter possibility is akin to the 'compensatory fitness' interpretation proposed by Camperio-Ciani *et al.* (2004), where parents with homosexual children have more children in order to compensate for the loss of grandchildren expected by the presence of homosexuals in the family. We will see in Chapter 8 that although the probability of producing homosexual offspring is expected to increase with group size, it is also predicted to reach a maximum at some point and then to decrease as group size increases further.

In sum, from a selection perspective human homosexuality becomes less of an evolutionary paradox if: (a) homosexuals exact some direct reproductive success (e.g. if they actually are sequentially bisexual), (b) the trait is linked to cooperative behaviours that can be directed to kin under the appropriate circumstances (*kin selection*), (c) men and women prefer to mate with co-operative individuals (*sexual selection*), (d) female heterosexual relatives of male homosexuals (or male heterosexual relatives of female homosexuals) have higher reproductive success than heterosexual relatives of heterosexuals (*sexually antagonistc selection*).

Summary of main conclusions

- Various loci have been identified or hypothesised to control the expression of same-sex sexual behaviour in various animals, humans included. Although in humans there has been an emphasis on loci of the sex chromosomes, autosomal loci are also likely to be involved.
- Most genetic models of homosexuality and empirical data tend to emphasise sexually antagonistic and sexually mutualistic selection, linking homosexuality with cooperation and parental care; heterozygote advantage, especially in the context of sexual selection; and recurrent mutation. Kin selection has fallen out of favour in recent years, but results of recent research call for a reappraisal of its potential importance.
- Family analyses of homosexuality in humans tend to support a sexually antagonistic model for selection of homosexuality.
- Chromosome aneuploidies in humans are not associated with homosexuality.
- The trend for men to display exclusive homosexuality at higher frequency than women is associated with (a) higher mutation rates in male than female germ cells, with sons inheriting Y-chromosome mutations from their father; and (b) males being the heterogametic sex, which makes them more likely to express recessive X-chromosome mutations inherited from the mother.
- Potentially, DNA methylation could affect the expression of loci controlling some aspects of same-sex sexual orientation.
- The genetic contribution to homosexuality in humans seems to be different in men and women.
- Monozygotic twins are more concordant in terms of sexual orientation than dizygotic twins, supporting a genetic contribution to homosexuality. The concordance between monozygotic twins, however, is about 50%, in fact suggesting a variable contribution of genes and environment.
- Some evidence from human traditional societies suggests that kin selection may play a role in the evolution and maintenance of human homosexuality in synergy with other selective mechanisms, such as sexually antagonistic selection.

Although genes can be expressed in tissues of the adult organism, thus affecting sexual behaviour, the most dramatic effects of genes are those affecting the organism in its early stages of development. In the next chapter I review the current knowledge of the early ontogenetic mechanisms implicated in the development of same-sex sexual preference in males and females.

4

Ontogenetic processes

The events occurring at the very early stages of development are of key importance for the future of the individual, as they can have significant consequences later in life. As soon as an ovum is fertilised, the zygote begins a developmental journey known as ontogeny that will take the organism through various phases of growth and change. Understanding the genetic and also the non-genetic factors, both internal to the individual and environmental, that affect such ontogeny is critical to the understanding of the how, when and why homosexuality may be expressed. This chapter mainly focuses on the effects of events occurring in the early developmental stages of a mammal or a bird: from gestation (pre-laying and incubation in birds) to attainment of sexual maturity. However, additional analyses of early developmental effects, such as human syndromes specifically involving endocrine mechanisms, prenatal immune processes and others, will be carried out in Chapters 5 and 6, where the adult stage of the individual will be also considered.

The importance of early events affecting the development of sexual behaviour and sexual orientation has been recognised by students of both human psychology (e.g. Freud 1905) and animal behaviour (e.g. Tinbergen 1963; Lorenz 1970) since the late nineteenth and early twentieth centuries. As a result of the accumulated knowledge on the development of sexuality that occurred especially during the first half of the twentieth century, two major views confronted each other in the 1960s and 1970s that exemplify the two broad schools of thought on the ontogeny of sexuality that somehow remain in place even today: the 'psychosexual sexuality-at-birth' (PSAB) and the 'psychosexual neutrality-at-birth' (PNAB) theories (Diamond 1965). In the truly manicheistic or polar fashion typical of some scientific debates, the role of specific genetic influences on the development of sexual orientation (i.e. the 'nature' pole of the debate) was minimised by supporters of the PNAB theory (e.g. Money 1988; Money & Ehrhardt 1996, but see especially John Money and Anke Ehrhardt's earlier works: Money & Ehrhardt 1971), whereas proponents of the PSAB theory undervalued the role of learning processes (i.e. the 'nurture' pole) while emphasising genetically predisposed tendencies to develop a specific sexual orientation (Diamond 1965). Diamond (1965) did, however, acknowledge that the genotype just narrows down the limits of the available developmental pathways, while still leaving a degree of variability that could be expressed in the adult individual. What the early supporters of a mainly genetic control of homosexuality seem to have missed, however, is a clearer reconciliation between the obvious flexibility conferred by some processes, such as learning, especially in vertebrates possessing a complex brain such as primates, and the genetic makeup of their cells. As I will show below, recent work on developmental stability/instability, phenotypic plasticity and canalisation affords a better theoretical framework to understand sexual orientation in an ontogenetic and evolutionary perspective.

On the other hand, what the supporters of learning approaches to homosexuality seem to have missed in their early works is that the action of genes is inescapable. At one or more points in the web of causation of any trait there is a direct product of a gene that plays an important role. The challenge is therefore more to understand how genes are regulated and expressed during ontogeny and what is the exact role of their products in the causation of homosexual behaviour, rather than to establish whether a trait is genetically or environmentally determined (see Chapter 3).

To be fair to both John Money and Milton Diamond, but also Anke Ehrhardt, I should also clarify that in later years their views tended to converge towards a biosocial interpretation of human sexuality (see Ehrhardt 1987; Money 1990; Diamond 1998). This trend towards a more balanced biosocial approach is also in agreement with the views of evolutionary biologists, who see the relationship between genetic make-up – the genotype – and phenotype as the result of developmental processes, where both expression of genes and environmental influences determine the characteristics of an organism at any specific time in its life (Crews & Groothuis 2005).

As we have seen in Chapter 3, mutations occur all the time, they are a true fact of life, and some of those mutations may have effects on the development experienced by the organism. In situations where the population is under *stabilising selection*, however, resilience of developmental processes to perturbations such as mutations and random environmental fluctuations would be highly adaptive. Stabilising selection occurs whenever phenotypic variants produced by mutations are selected against (i.e. they die young or reproduce less) if they have values of the trait that are extreme (too high or too low) compared with the rest of the population. Under stabilising selection, unstable ontogenies are more likely to produce an organism whose phenotype departs from the mean value of the population, and under those circumstances such a phenotype will be more likely to be maladaptive. The concept of *canalisation* was first introduced by Conrad Waddington (1942) to describe the ability of an organism to achieve full development and functionality in spite of the many perturbations both internal (e.g. mutations) and external that the organism may experience. The more a population is under stabilising selection, the more canalisation the ontogeny of the members of that population is expected to experience. This situation may lead to an equilibrium in the expression of genes at different loci during development that has been termed *developmental stability* (Siegal & Bergman 2002). On the other hand, mutants that undergo processes of *developmental instability* (Polak 2003) – which is a variability of phenotype that individuals express within the same environment (see, for example, de Witt *et al.* 1998), as in the case of genetically similar organisms growing fur of different thickness even when they are all exposed to the same temperature – may be selected against if the population is under strong stabilising selection (for example, if the individual happens to grow fur that is too thick when the temperature is high it may overheat), but notice also that they may be retained in the population during periods of *directional selection* if the environment, including – or perhaps especially, in the case of same-sex sexual behaviour – the social environment, changes consistently over a sufficient period of time and the unstable developmental programme produces better adapted phenotypes.

Potentially, developmental instability expressed within a given environment could be also associated with phenotypic variability across environments; the latter is known as *phenotypic plasticity* (de Witt *et al.* 1998). An organism is phenotypically plastic if it can adaptively respond to changes to its surroundings (e.g. grow thicker fur when the temperature decreases in winter). Phenotypic plasticity is therefore expected to be selected for when the environment does not remain constant over time (de Witt *et al.* 1998; Crispo 2007; Pigliucci 2007). If we apply this concept to the evolution of homosexual behaviour, it can be easily seen how the complex and changeable environment resulting from living in social groups could produce the context required for selection of plastic sexual behaviour. This kind of plasticity could be associated, for instance, with various socio-sexual functions of same-sex mounting such as expression of dominance or affiliation.

Phenotypic plasticity may operate following two modalities at least, which just for illustrative purposes could be called *unidirectional plasticity* and *bidirectional plasticity*.

Unidirectional plasticity. For the sake of the argument let us imagine development as a pathway branching out into various alternative roads at a series of intersections along the way. The organism will follow one and only one of those various alternatives at any given point during its development. Once a developmental path has been taken it will condition the subsequent development, including which options will be available at the next 'developmental crossroads'. This kind of phenotypic plasticity allows each organism to potentially follow alternative developmental routes. However, different individuals, including clones, who experienced different ontogenies and therefore have developed more or less different phenotypes, remain different thereafter (e.g. exclusive homosexuals vs. exclusive heterosexuals), hence the term *unidirectional*.

Bidirectional plasticity. In this case the organism will still be travelling along a road that branches out at various points along the way, but now and then it can go back (phenotypically) and try an alternative route. That is, if plasticity is bidirectional the organism retains the ability to modify its phenotype (e.g. sexual orientation) throughout various stages of its development, including the adult stage, being able to move back and forward to alternative phenotypic states (e.g. heterosexuality, homosexuality). From a sexual orientation perspective, bidirectional plasticity may be expressed by sequential bisexuals (Weinberg *et al.* 1994b). In this metaphor, simultaneous bisexuality could be represented by a confluence of two roads (homosexuality and heterosexuality) rather than a branching out.

From the above it follows that homosexual behaviour could initially have arisen in the population owing to selection for adaptive plasticity in complex social environments, a process known as *genetic accommodation* (West-Eberhard 2003; Crispo 2007; Moczek 2007; see also Baldwin 1896, 1902). Subsequently, however, those same homosexual behavioural traits could be retained and even become canalised if the social conditions that made them adaptive initially became stable and predictable over time, a process known as *genetic assimilation* (Waddington 1953, 1957, 1961). The bonobo's (*Pan paniscus*) G–G-rubbing behaviour (de Waal 1997) that will be analysed in Chapters 5 and 9 is a trait that could have been selected through such a process.

The interplay of *selection for canalisation* and *selection for plasticity* in response to either a stable or a variable (social) environment is likely to provide a broad scenario for the evolution of homosexuality, both as a plastic and also as a more rigid behaviour across species and sexes within species, as follows:

(a) Sexual orientation such as heterosexuality is usually canalised under stabilising selection because heterosexual individuals reproduce more.
(b) Sociality introduces environmental complexity.
(c) Phenotypic plasticity (perhaps resulting from an initial developmental instability) may be selected under such environmental complexity,
(d) leading to the expression of new and diverse sexual behaviours. At this stage same-sex sexual behaviour could be expressed as a plastic trait in the context of either unidirectional or bidirectional plasticity. However,
(e) if same-sex sexual behaviour becomes very adaptive (e.g. if it mediates cooperative and/or competitive interactions), and
(f) the conditions for the adaptiveness of the new behaviour stabilise, then
(g) random mutations that canalise the new behaviour may be selected. This final canalisation is consistent with both widespread bisexual capabilities in the population and also the reliable emergence of some more extreme cases of exclusive homosexuality in the same population. How could such canalisation be selected? We

have seen in Chapter 3 how processes such as sexually antagonistic selection, kin selection, sexual selection, reciprocal altruism and others may be involved.

Novel behaviours produced by a plastic phenotype could be also more or less reliably transmitted inter-generationally through learning, and become widespread and consistently expressed through cultural traditions. Perhaps we may refer to this process as 'cultural canalisation'. In fact, *gene–culture evolution* may also occur (see Chapter 1), whereby cultural traditions produce the environment that selects the new mutations that were mentioned above. This evolutionary view clearly detaches homosexuality from its still common study in the context of 'pathology' of development.

From a more proximate perspective, mutations may affect genes that alter the hormonal milieu of the organism during ontogeny, and it is then the altered endocrine environment that leads to the development of diverse phenotypes. Alternatively, mutations may affect the central nervous system architecture and functions directly, leaving the hormone levels intact, which may then lead to changes in behaviour. However, mutations may also alter genes affecting whole developmental programmes (see also Chapter 3) that could in turn alter the timing of developmental processes affecting different parts of the organism, a process known as heterochrony (Gould 1977). In particular, when juvenile traits stop or slow down their development and are retained in the reproductive, adult stage those adults may express juvenile-like behaviours. This specific process, known as neoteny, will be thoroughly reviewed at the end of this chapter and it can be an important evolutionary process that may explain the appearance of same-sex sexual behaviour in adults in some species. As we will see in this chapter, same-sex mounting is very common among juveniles and sexually immature young of many mammals, especially in the social species.

Environmental factors acting on the endocrine system may also play an important role in constraining the development of sexual orientation. These include the social environment that may affect the stress level of the pregnant mother in mammals or female birds; in the latter case stress may alter the egg's hormonal content, for instance, thus potentially altering the development of the offspring. In addition, here I also review other environmental factors such as endocrine-disrupting chemicals that may have an effect on the developing organism.

Hormonal effects during early development

Hormones can have both an activational and an organisational effect on homosexual behaviour (Ehrhardt & Meyer-Bahlburg 1981; see Diamond 2009 for a recent review). Activational effects in which hormones modulate the expression of already well developed behaviours will be reviewed in Chapter 5. In the same chapter I will also review human syndromes such as congenital adrenal hyperplasia and others that involve potential changes in sexual orientation and gender role due to alteration of endocrine functions during development. Depending on the case, the endocrine mechanisms behind such syndromes may remain in operation throughout an extended period in the life of the individual, including adult ages. This is the reason why they have been consigned to Chapter 5.

Organisational effects of hormones refer to their action at the early stages of the development of the organism, which may have consequences on which kind of sexual behaviour and orientation are eventually expressed by the adult individual. Here I focus on some major mechanisms of the early organisational effects of hormones, and also on the association between homosexuality and phenotypic traits such as 2nd:4th digit length ratio, waist-to-hip ratio, otoacoustic emissions, auditory evoked potentials, tooth crown size, dermatoglyphics and handedness that are potentially sensitive to the prenatal hormonal environment.

The classic model for the organisational role of hormones in the development of sexual behaviour is that genes directly determine the development of the gonads, that then produce steroid hormones, which finally direct the development of the central nervous system in a female-typical or male-typical fashion (Phoenix et al. 1959; Grady et al. 1965; Money & Ehrhardt 1971; Baum 1976; Götz et al. 1991), with a variable degree of behavioural androgyny being also a common outcome (Goy & Goldfoot 1975). Such changes may also affect preference for sexual partners of a specific sex, in which case alterations of sexual orientation may occur.

In both birds and mammals the 'default' sex in the absence of steroid activation during ontogeny is the homogametic sex (female in mammals, male in birds). The development of sexual traits produces phenotypes that are *sex-specific*, i.e. those that are unique to each sex, but also traits that are *sex-typical*; the latter can be shared to a variable extent by individuals of both sexes. As we have already seen in Chapter 3, this classic model has been challenged by studies showing direct genetic control of some processes of central nervous system tissue differentiation (see Swaab 2007 for a review).

Suggestive evidence that both sexual hormones and direct genetic mechanisms are important in the early development of sexual behaviour comes from some remarkable experiments. For instance, in a study carried out on the Japanese quail *Coturnix japonica*, Gahr (2003) grafted a section of the brain primordium that eventually develops into the forebrain, including the hypothalamus, into males or females at embryonic day 2, i.e. before differentiation of the gonads occurs and therefore before circulating gonadal steroids can exert any action on development. The procedure involved transplants from male donor to female host (MF), female to male (FM), male to male (MM) and female to female (FF). The gonadal anlage, or undifferentiated gonadal tissue, was untouched in the chimaeras. Oestrogen is produced by the gonadal anlage in quails at about day 6 of embryonic life and directs the development of the undifferentiated gonad into an ovary. Inhibition of production of oestrogen in females masculinises them. In birds, therefore, oestrogen de-masculinises a default male-brain development in females (Gahr 2003 and references

therein). Gahr (2003) reasoned that if this model is correct, with the default development of the brain being *male* in birds and differentiation into female being under the direct control of the hormonal milieu, then female brain tissues transplanted into a male before the onset of gonadal endocrine activity should develop in a male fashion. He therefore proceeded to produce the above four types of chimaera and found that the chimaeras' anlage developed the kind of gonads and embryonic levels of circulating 17β-oestradiol and testosterone expected from the host's genetic sex. However, adult FM chimaeras had very small testes and low circulating levels of steroids (Figure 4.1A, B). Low levels of circulating testosterone in adults were associated with lack of mounting in FM males, whereas both control males and MM males did mount (Figure 4.1F). Moreover, FM males did not show female-typical sexual receptivity, in spite of their lack of mounting behaviour (Figure 4.1G). Although these behavioural results may be due to low circulating levels of both testosterone and oestradiol and therefore low levels of activation of sexual behaviour in FM males, Gahr (2003) also analysed the hypothalamic preoptic area (POM), a sexually dimorphic nucleus in quail (larger in males than in females) that is involved in control of sexual behaviour, and found that mounting was associated with a larger POM in general and that males with a female implant (FM) had a smaller POM (Figure 4.1C, D) and did not mount, as mentioned above. Therefore adult FM male behavioural de-masculinisation, but also lack of behavioural feminisation, was associated with (a) feminisation of the size of their POM – which, it will be remembered, was of female origin – and (b) normal male-typical levels of embryonic testosterone produced by their male gonads. In conclusion, Gahr's (2003) study suggests that both embryonic sexual hormones *and* genetic make-up of brain cells affect the development of brain regions and sexual behaviour.

To complicate the matter further, experiments carried out on zebra finches (*Poephila guttata*) also suggest a combined contribution of early hormonal exposure and early learning on development of

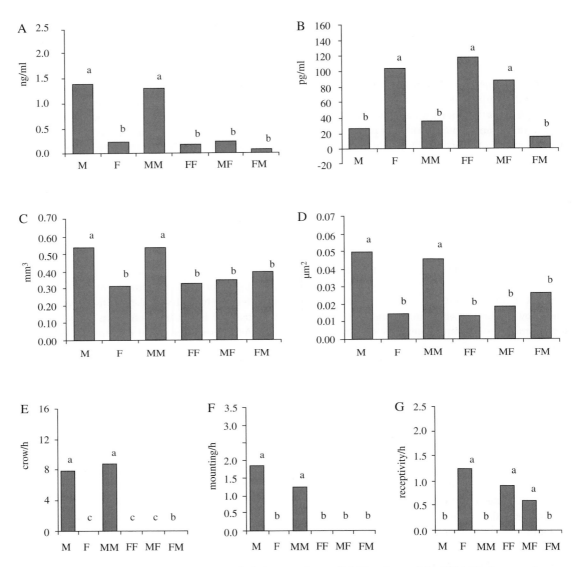

Figure 4.1. Levels of plasma testosterone (A) and of plasma 17β-estradiol (B), volume of the POM (C), the area of vasotocin-immunostained structures of the POM (D), and crowing activity (E), mounting attempts (F), and receptivity (G) of adult male Japanese quail controls (M), female controls (F), MM chimaeras, MF chimaeras, FF chimaeras, and FM chimaeras. Statistical similarity is indicated by similar letters. From Gahr (2003).

sexual behaviour, producing a more complex scenario for the ontogeny of sexuality that implicates genes, hormones and learning.

Mansukhani et al. (1996) treated female zebra finch nestlings with oestradiol benzoate (EB) and when they were 100 days old they were administered testosterone propionate in silastic tubes in order to activate sexual behaviour. Some of the female finches were reared in monosexual groups, whereas others were reared in the company of males. Some of the EB-treated females attempted mounting of other females (a result similar to that found in chickens by Sayag et al. 1989) irrespective of their social rearing environment; however, only those EB-treated females that were reared in the monosexual group also attempted to establish a same-sex pair bond. This suggests that both early organisational hormonal and learning effects contribute to the development of various aspects of sexuality.

These experiments illustrate a theme that will recur in this book, which is that early development of sexual behaviour is under the combined control of genes, hormones and central nervous system, the last affecting and being also affected to variable degrees by learning processes (see also Wallen 2005).

The specific effect that hormones may have in the development of gender roles and perhaps sexual orientation at the organisational level during the early stages of development may be also dose-dependent. This may explain Udry's (2003) observation regarding the increasing difficulty of socialisation to achieve behavioural feminisation of young girls as their prenatal exposure to androgens increases. Moreover, such effects may also be dependent on critical periods of central nervous system development.

For instance, early work by Grady et al. (1965) described the display of a female-typical sexual behaviour (lordosis) in adult male rats only if they were castrated before day 10 from birth; additional female-typical sexual behaviours could be elicited from males only if they were castrated before postnatal day 5 (see also Goldfoot et al. 1969). These results show that there are specific critical periods or 'windows' that the organism goes through during its development that will determine the specific effect of hormones on sexual behaviour. Similarly, female ferrets display male-like thrusting and mounting of another female only if they are testosterone-treated on postnatal days 0–15 (Baum et al. 1982). This level of hormonally induced masculinisation in female ferrets, however, did not de-feminise their sexual behaviour. This suggests that there are different, independent mechanisms controlling sexual behaviour that involve feminisation and masculinisation available initially in the developing individual that may be modulated in a sex-specific manner. Baum et al. (1990) described two critical periods of steroidal organisational effects on sexual behaviour in ferrets, one beginning at about embryonic day 25 and a second period beginning soon after birth and extending until postnatal day 20.

Various degrees of masculinisation/de-masculinisation and feminisation/de-feminisation occur during sensitive periods of the early ontogeny, and such sensitive periods have been described in various mammals during prenatal, perinatal and postnatal development (Grady et al. 1965; Gerall et al. 1967; Baum et al. 1982; Baum & Erskine 1984; Mann & Fraser 1996; see Cohen-Bendahan et al. 2005 for a review). In humans there is a prenatal surge of foetal testosterone in males between weeks 10 and 18, with a second peak occurring perinatally, circulating testosterone declining again after one or two weeks postnatally. A second postnatal peak occurs at week 8 (Cohen-Bendahan et al. 2005; McIntyre 2006). In birds, Ottinger et al. (2001) described a peak of plasma androgens at days 10–12 of embryonic life in *Coturnix japonica*, and then a second one at about one day post-hatching. Males become de-masculinised if treated with the steroids testosterone propionate or oestradiol benzoate at embryonic days 10–14; later treatment is ineffective.

The initial model of sexual behaviour differentiation described above involves the direct effect of gonadal androgens, testosterone in particular, on the development of specific brain areas. This is a mechanism common to both non-human primates and humans (see Wallen 2005 for a review). In humans, for instance, behaviourally masculinised girls are born of mothers with elevated circulating testosterone levels, whereas mothers with elevated sex-hormone binding globulins (SHBGs) in their blood have daughters showing more feminine behaviour (see review in Cohen-Bendahan et al. 2005). SHBGs prevent testosterone from interacting with androgen receptors in the cell membrane. Similar masculinisation and de-feminisation effects of testosterone have been also described in *Macaca mulatta* (Wallen 2005).

The testosterone-mediated mechanism was subsequently challenged by the discovery of aromatase activity in several brain regions involved in sexual behaviour (see reviews in Beyer 1999 and Roselli 2007). It is now clear that during development gonadal androgens can induce aromatase activity in certain regions of the brain that leads to production of oestrogens from aromatisation of testosterone in various mammals, birds and fish, and that in those central nervous system regions it is oestrogens that regulate development, leading to sexual behaviour differentiation.

The action of oestrogens on brain development is diverse and profound: they affect growth and development of neurons, they can modulate apoptosis (i.e. programmed cell death) of neurons, and through apoptosis the organisation of the brain can literally be sculpted, with variable results in terms of function and therefore behaviour. Oestrogens can also affect neuroblast migration and aggregation into specific regions of high cell numbers. Such regions are known as nuclei. They can even affect synapse formation among neurons within a specific area, interconnections among them through regulation of neurite growth and also production of specific neurotransmitters, along with contributing to neuroprotection (Beyer 1999; Roselli 2007). The mode of action of oestrogens can be either genomic (i.e. slow) or non-genomic (i.e. rapid) (Beyer 1999). The accumulated weight of the above evidence led MacLusky & Naftolin (1981) to postulate that the development of sexual behaviour is dependent on the aromatisation of testosterone into oestrogens (*Aromatisation Hypothesis*).

Of course oestrogens can only affect the activity of neurons that posses oestrogen receptors in their cell membrane, leaving open the possibility for the existence of mechanisms other than the one involving aromatisation of testosterone if other kinds of receptors, e.g. androgen receptors, are also present. That the oestrogen-dependent mechanism is not the only pathway towards regulation of development of brain regions involved in sexual orientation has become clear with new research in this area (Beyer 1999). In particular, there are neurons in brain areas traditionally associated with control of sexual behaviour that indeed possess androgen receptors; in humans, mechanisms involving the androgen receptor may be especially important in the development of sexual orientation (see Chapter 5 and the recent review by Zuloaga *et al.* 2008). In rodents, both androgen and oestrogen receptors play a role in masculinisation and de-feminisation of sexual behaviour (Zuloaga *et al.* 2008 and references therein).

Roles of the androgen receptor notwithstanding, disruption of oestrogen action during brain development may also potentially contribute to feminisation of specific male brain regions that affect some aspects of sexual orientation. Genes encoding for oestrogen receptors are found in chromosomes 14 and 6 in humans (Beyer 1999; Kruijver *et al.* 2003), which is in agreement with the suggestion by Mustanski *et al.* (2005) of an autosomal contribution to the genetic aspects of homosexuality (see Chapter 3).

Additional experimental evidence for the organisational effects exerted by androgens and oestrogens on sexual behaviour and orientation are provided by Vega Matuszczyk & Larsson (1995) in Wistar rats, Henley *et al.* (2009) in Long–Evans rats, Bakker *et al.* (2002b) in mice, and Adkins-Regan & Wade (2001) in the zebra finch (*Poephila guttata*).

In primates, brain organisational effects of oestrogens are consistent with results obtained in male infant *Macaca mulatta*, which express the oestrogen receptor in the pituitary, the hypothalamic preoptic area, hippocampus and various cortical areas (MacLusky *et al.* 1986). That is, brain areas that are involved in the expression and regulation of sexual behaviour. MacLusky *et al.* (1986) detected the highest level of oestrogen synthesis in the hypothalamic preoptic area, a region of the brain strongly associated with sexual and homosexual behaviour in humans and other mammals, as we will see in Chapter 5. Sholl & Kim (1989) have also shown that the oestrogen receptor is expressed in the medial basal hypothalamus, the level of expression increasing with developmental stages, in both male and female *M. mulatta* foetuses. In humans, cultured neurons from both the brain cortex and spinal cord of embryos 8–13 weeks old do express the oestrogen receptor-β (Fried *et al.* 2004). The oestrogen receptor and aromatase activity are also expressed in neurons of the adult male hypothalamus, lateral septum, thalamus and the amygdala of various primates, including humans (Blurton-Jones *et al.* 1999; Roselli *et al.* 2001; Kruijver *et al.* 2003). Davidson *et al.* (1983) described one case of a bilaterally castrated man whose sexual function was maintained by oestrogen–progestin treatment. This evidence suggests that in primates both organisation during early development and activation of sexual behaviours at adult ages can be modulated by mechanisms that involve oestrogen receptors, apart from those that involve androgen receptors. Other authors, however, have been critical of this view (e.g. Gooren 1985; Bagatell *et al.* 1994).

In particular, in a recent review Thornton *et al.* (2009) stress what is the chief evidence against an organisational role of oestrogens in the early ontogeny of primate sexual behaviour, i.e. that both the aromatisable testosterone and the non-aromatisable dihydrotestosterone have a similar organisational effect on sexual behaviour in *Macaca mulatta*. However, if both androgen and oestrogen receptors are expressed in specific areas of the brain, then experiments using testosterone and dihydrotestosterone are not conclusive. In fact, if the androgen receptor is blocked with flutamide, rhesus monkeys still display masculinisation of sexual behaviour (see Thornton *et al.* 2009 for a review), a result that could well be explained by endogenous oestrogens activating oestrogen receptors. Moreover, in a review of hormonal influences on development of sexual behaviour in primates, Kim Wallen (2005: 17) wrote, referring to flutamide effects on mounting behaviour in *M. mulatta*: 'Thus contrary to our hypothesis, that flutamide late in gestation would block juvenile masculinization, it paradoxically seems to have hypermasculinized these males'. Overall, these results could be better understood if **both** kinds of receptor (oestrogen and androgen) can be expressed in relevant areas of the primate brain during early development, so that if one type of receptor is inactivated the development of sexual function could be achieved by the activation of the other. This suggests that the role of steroids in the ontogeny of sexual behaviour, and presumably sexual orientation as well, is likely to be complex in primates.

It is traditionally assumed that feminisation of brain nuclei in mammals is the default ontogenetic pathway whereas masculinisation requires specific hormonal input (the opposite is suggested for birds). This view has been challenged by recent studies showing that feminisation in mammals is also a result of direct hormonal activity (Fitch & Denenberg 1998 and references therein; see also Collaer & Hines 1995; Bakker *et al.* 2002b for reviews). In particular, oestrogen exerts a feminising effect on the development of the female brain after the ontogenetic period when it could exert a masculinising effect (Fitch & Denenberg 1998). During the 'masculinising' period the female brain is protected from the effect of oestrogen by alpha-fetoproteins (AFP) that bind to oestrogen, inactivating it. AFP production decreases in the postnatal period in females (Fitch & Denenberg 1998).

In sum, the combined evidence available clearly shows that changes in the hormonal milieu during early ontogeny do affect sexual behaviour, gender role in particular, and they can also potentially affect the development of sexual preferences for conspecifics of the same or the other sex in non-human vertebrate models. Hormonal action produces these outcomes by means of feminising, de-feminising, masculinising or de-masculinising relevant brain areas that control various aspects of sexual behaviour. Does the action of prenatal hormones also affect the ontogeny of homosexuality in humans? Although I have already provided in the preceding paragraphs some initial evidence for the potential role of hormones on the development of same-sex sexual behaviour in humans, large-scale studies based on the direct measurement of foetal circulating hormones obviously face some technical, but above all ethical challenges when it comes to performing them in our species. This has led to the use of proxy variables that are correlated with foetal circulating steroids. Below I review some of the major studies carried out in this area that explore the association between prenatal steroids and the development of a homosexual orientation in humans.

The 2nd:4th digit ratio

The development of broad regions of an organism is under the control of specific regulatory genes. Genes of the *Hox* family control the development of both limbs and the genital tract in vertebrates. Within the *Hox* gene family, *Hoxa* and *Hoxd* in particular control the sexual differentiation of the genital anlage and also the development of the digits (Manning *et al.* 1998). It is therefore possible that development of sexuality may be correlated with development of some specific morphological traits seemingly unrelated to the reproductive system if both are affected by the same genes. In this case the genes may control the development of the gonads that will produce the sex hormones that, in turn, may affect the development of the brain to produce homosexual phenotypes, and those same genes may affect the development of more easily measurable digit characteristics, also through prenatal exposure of those individuals to gonadal steroids.

In a seminal work that has inspired considerable research activity, John Manning and collaborators (Manning *et al.* 1998) showed that the ratio of the second digit (the index finger) to the fourth digit (the ring finger) length (i.e. the 2D:4D ratio) is lower in males than in females and that the ratio is set early in development (at or before postconception weeks 13 or 14; Putz *et al.* 2004), to remain unchanged thereafter. They also found that adult males with smaller (i.e. more masculinised) 2D:4D ratio also had higher levels of circulating testosterone, whereas circulating oestrogen was positively correlated with the 2D:4D ratio especially in females, with a weaker pattern in the same direction also found for prolactin. Therefore hormonal effects on early sexual development could be studied by proxy through measurements of the hand digits! Needless to say, the finding produced considerable excitement among students interested in the ontogeny of sexual orientation and the technique has been used to study several aspects of homosexuality.

Manning *et al.* (1998), however, did not measure exposure of the foetus to androgens, but circulating hormones in adults. This problem was corrected by Manning and co-workers in a later publication (Lutchmaya *et al.* 2004) where they measured foetal testosterone and other steroids from amniocentesis samples and also measured digit length ratio when the child was two years old. This was a necessary technical step to validate the use of digit ratio as a proxy measurement of prenatal androgen exposure. They found that male foetuses were exposed to more elevated levels of testosterone than females and detected no significant difference in the levels of oestradiol. Lumping all sexes together, they found that foetal testosterone levels correlated negatively with the 2D:4D ratio, although the relationship was not statistically significant. Foetal oestradiol was positively, but non-significantly, correlated with 2D:4D ratio and the foetal testosterone/foetal oestradiol ratio was negatively and significantly correlated with the 2D:4D ratio (right hand only). The work of Lutchmaya *et al.* (2004) strongly suggests that digit ratio is not the best predictor of foetal testosterone but of the ratio between foetal testosterone and oestradiol. Moreover, Manning *et al.* (2002) also suggest a potential association between CAG microsatellite length and the 2D:4D ratio in humans. CAG repeats are associated with both the androgen (*AR*) and the oestrogen (*ER*) receptor genes and thus they may affect the sexual development of the child. This can be better understood in light of the proposed mechanism that links foetal sex steroids and 2D:4D ratio. Sex steroids can alter bone growth by acting on oestrogen receptors α and β in metaphyseal tissue and bone growth plates; the link with testosterone is through the latter's aromatisation into oestrogen, but direct testosterone effect on bone growth through its action mediated by the androgen receptor is also a possibility (McIntyre 2006) as also suggested by Manning *et al.* (2002). The combined contribution of both testosterone and oestrogens to the development of sexual orientation has been summarised by Collaer & Hines (1995) into their *Gradient Model* and its stronger version, the *Active Feminisation Model*, where testosterone is suggested to exert a masculinisation and de-feminisation role, whereas estrogens have a feminisation and de-masculinisation action in humans. The recent review by Zuloaga *et al.* (2008) challenges these models, however.

Williams *et al.* (2000) studied a sample of men and women from the San Francisco area in the USA. They not only confirmed the original finding by Manning *et al.* (1998) of lower values of the 2D:4D ratio in men than women, but they also described a greater sex difference in the right than in the left hand. They found a statistically significant difference in the ratio between homosexual and heterosexual women (right hand measurements), with the former having a more masculinised (i.e. lower value) ratio than the latter. The difference between homosexual and heterosexual men was not significant, but the value of the ratio for homosexual men was intermediate between that of heterosexual women and heterosexual men. That is, the study by Williams *et al.* (2000) supports the notion of ontogenetically masculinised homosexual women and somewhat feminised homosexual men. However, an analysis of the 2D:4D ratio among homosexual men with more than one older brother reveals a value of the ratio even smaller than that found in heterosexual men, suggesting that, in this context, younger homosexual brothers of men are 'hypermasculinised' (Williams *et al.* 2000). Same-sex attraction was also associated with a masculinised 2D:4D ratio in women in a study carried out in Switzerland by Kraemer *et al.* (2006), but these authors did not find any effect of sexual orientation on 2D:4D ratio in men.

Lippa's (2003b) study of a USA sample not only confirmed the sex-specific trends in 2D:4D ratio already described by previous works (Manning *et al.* 1998; Williams *et al.* 2000), along with the right hand bias in the magnitude of the ratio found by Williams *et al.* (2000), but they also described a significant effect of ethnicity (see also Manning *et al.* 2000; Loehlin *et al.* 2006), with 2D:4D right-hand values being larger (i.e. more 'feminised') among Whites than Hispanics, with the latter having a larger ratio than Asians. The left hand showed 2D:4D ratio values that were larger among Whites than Asians, and the latter had higher values than Hispanics. Overall, homosexual men had higher values of the 2D:4D ratio (both hands) than heterosexual men, especially for Whites and Hispanics. The trend for Asians is less clear-cut, but the sample is small. With regard to female participants in Lippa's (2003b) study, lesbians have larger 2D:4D ratios than heterosexual females, in support of a 'hyperfeminisation' model for lesbians, however this trend disappeared after controlling for ethnicity.

The results of five studies on 2D:4D ratios, none of which overlap in authorship, have been re-analysed in a single test by McFadden *et al.* (2005) using the original raw data. The analysis by McFadden *et al.* (2005) confirmed what is the strongest and most consistent result found in these studies, i.e. that heterosexual males have smaller 2D:4D ratios than heterosexual females. When it comes to analysing the trends for homosexual males and females, however, discrepancies are found among studies. One possible explanation for these diverse trends is, of course, that the studies may have differed in regard to the uncontrolled effect of additional variables. McFadden *et al.* (2005) in their analysis controlled some of those variables, chief among them: ethnicity, age, outliers and methods of measuring digit length. They pruned the original datasets to restrict the sample to White or Caucasian subjects only (see Table 4.1) and after a re-analysis of the data found that the 2D:4D ratio for females was consistently higher among heterosexuals than homosexuals (i.e. lesbians tended to have a masculinised ratio), whereas for males the effect of sexual orientation was not significant, although they detected a significant interaction between sexual orientation and study, with some studies (e.g. Lippa 2003b; McFadden & Shubel 2002) showing feminised 2D:4D ratios in homosexual males, whereas others (e.g. Robinson

Table 4.1. Mean 2D:4D ratios for left and right hand in five studies

Study	Group[a]	n	Left hand	Right hand
Lippa 2003b	FHt	285	0.974	0.961
	FHm	314	0.974	0.964
	MHt	154	0.955	0.940
	MHm	304	0.969	0.953
McFadden & Shubel 2002	FHt	37	0.969	0.980
	FHm	19	0.959	0.963
	MHt	38	0.943	0.951
	MHm	19	0.966	0.968
Rahman & Wilson 2003a	FHt	47	0.986	0.989
	FHm	50	0.965	0.960
	MHt	51	0.968	0.976
	MHm	54	0.962	0.960
Robinson & Manning 2000	MHt	88	0.983	0.976
	MHm	82	0.964	0.968
Williams et al. 2000	FHt	138	0.965	0.972
	FHm	148	0.964	0.961
	MHt	106	0.956	0.956
	MHm	271	0.959	0.953

Adapted from Table I of McFadden et al. (2005).
[a] FHt, female heterosexuals; FHm, female homosexuals; MHt, male heterosexuals; MHm, male homosexuals.

& Manning 2000; Rahman & Wilson 2003a) showed hypermasculinised 2D:4D ratios in homosexual males. This suggests that homosexuality in men may have diverse causes, even within the same ethnic group.

Putz et al. (2004) also found that homosexual females have a masculinised 2D:4D ratio, and male homosexuals have a hypermasculinised 2D:4D ratio. The authors also measured other psychological traits that they correlated with 2D:4D ratio and found an association between the hypermasculinised 2D:4D ratio for homosexual men and a tendency among those male homosexuals for uncommitted sexual intercourse. Within lesbians, Brown et al. (2002b) analysed 2D:4D ratios between 'butch' and 'femme' lesbians in a California sample, describing a statistically non-significant trend for 'butch' lesbians' ratios to be more masculinised than 'femme' lesbians' ratios.

Differences between males and females in their 2D:4D ratios that parallel the patterns found in humans were also found in mice (*Mus musculus*): females had a larger 2D:4D ratio than males (Brown et al. 2002a). In Guinea baboons *Papio papio,* measurements of the 2D:4D ratio in males reported larger values than in females (Roney et al. 2004), a trend different from that described in humans, although lower 2D:4D values were associated with higher circulating levels of testosterone in males, consistent with the trend observed in humans (Manning et al. 1998). Digit ratios have also been investigated in our closest living relatives, the chimpanzee and the bonobo. McIntyre et al. (2009) have shown that overall the 2D:4D values for both left and right hand are larger; i.e. more feminised, in bonobos than chimpanzees, with humans' 2D:4D ratios being even larger than those of bonobos. Therefore, regarding their 2D:4D finger ratios, the two species with the highest levels of same-sex sexual behaviour are both more feminised and more similar to each other than either of them is to chimpanzees.

Interestingly, 2D:4D ratio studies have also been carried out in birds, offering an opportunity for an important comparative contrast. The yolk of avian eggs may contain androgens of maternal origin that can affect the development of morphological, behavioural and life-history traits of the chick (Gil *et al.* 2007). If androgens (e.g. testosterone, androstenedione) are experimentally injected into an egg, male chicks may develop more conspicuous morphological masculine traits (e.g. sexually dimorphic plumage), but also an increased level of aggressiveness, affecting dominance relationships with conspecifics (Strasser & Schwabl 2004; Eising *et al.* 2006).

Burley & Foster (2004) carried out a study on the zebra finch (*Taeniopygia guttata castanotis*), a plumage sexually dimorphic passerine, and found that the 2D:4D ratio in the toes increased for both sexes with laying order of the egg – female zebra finches lay several eggs in a clutch in sequence. The digit ratio was lower for females than for males, a pattern similar to that found in humans (McFadden & Shubel 2002), as the zebra finch toes are homologous to the human toes, not the fingers, and in humans toes show a sex-dimorphic 2D:4D ratio that is reversed (females have a smaller ratio than males) to the one found in the fingers. Such a pattern in birds could be explained by differential deposition of testosterone in the egg before laying. This laying-order effect, however, could not be replicated by Forstmeier (2005). Moreover, Forstmeier (2005) did not detect any sexual dimorphism in digit ratio in zebra finches and suggested that the trait has a sizeable degree of genetic heritability ($h^2 = 0.80$), a value not too different from that calculated in humans for fingers (i.e. 0.71 for the left hand, Paul *et al.* 2006). However, he did find that females with a more feminised digit ratio (i.e. lower values of 2D:4D for toes) spent more time close to males than females with a more masculinised digit ratio did.

Twin studies have also been recently carried out focusing on the 2D:4D finger ratio. van Anders *et al.* (2006) studied finger ratios from photocopies in a sample of twins from Ontario and British Columbia (Canada), comparing same-sex twins (males and females) with different-sex twins. Their results follow the well-established pattern found previously for the two sexes: 2D:4D ratios are lower in males than females, a result that was significant for the left hand only. They did not find a significant difference between same-sex male twins and the male member of a different-sex twin pair, but the 2D:4D ratio was significantly lower (i.e. more masculinised) in the female member of a different-sex twin pair than in same-sex female twins. The trend was significant for the left hand only. This suggests that the prenatal hormonal environment experienced by a female twin growing in the company of a male twin may affect at least some features of her development.

Manning *et al.* (2000) carried out an intriguing study correlating the value of the 2D:4D ratio for the female and male member of a couple with the number of their children (a measure of fecundity) across several ethnic groups. More precisely, they calculated a wife *minus* husband ($f - m_r$) finger ratio (defined as: female 2D:4D – male 2D:4D) in each couple and correlated this value with the number of children in the family for each of several samples: from England, Spain, Germany, Hungary, Poland, Jamaica, Finland and South Africa. What they found in the English sample was a positive relationship between the value of $f - m_r$ and the number of children per couple. Clearly, larger values of $f - m_r$ occur when females are ontogenetically feminised and males are ontogenetically masculinised, thus suggesting that fecundity increases in this sample as a function of hormonal effects during development that are concordant with the chromosomal sex. From this result one would be tempted to conclude that there is a cost in terms of fitness in undergoing developmental processes leading to feminised males and/or masculinised females. However, such a conclusion is unwarranted as we also consider the results obtained from other ethnic groups. In fact, the analysis by Manning *et al.* (2000) for the other ethnic groups indicates that the situation is far more complex than we may think. In the Spanish sample, they found a negative correlation between 2D:4D and the number of children in males (i.e. more masculinised males have more children), but no correlation in the female sample. That is, to have more children, Spanish women do not seem to need to be ontogenetically feminised. The German sample follows the general trend observed in the English sample: ontogenetically masculinised males and feminised females tend to have more children. The same trend for females was maintained in the Hungarian sample, but the relationship between fecundity and 2D:4D ratio was not significant in males, although in this case a small sample size might have been a problem. The Polish sample provides a very interesting trend, although again subject to the caveat of a small sample size, with a positive association between 2D:4D values and number of children in males and a negative one in females. That is, in this sample ontogenetically feminised males and masculinised females have more children. What does all this mean? Manning *et al.* (2000) suggest that the above pattern could be explained by the selection of genes that influence prenatal hormonal levels, the expression of such genes being affected by the sex of the developing embryo. If such a mechanism of control of phenotypic plasticity is enacted, however, it has to be even more flexible than that, as the fitness of a masculinised or feminised male or female seems to be somewhat variable across ethnic groups. In my view, the results of Manning *et al.* (2000) suggest that variability exists in human populations with regard to the level and type of canalisation of hormonal influences on development; which end-phenotype is selected, and which is not, may depend on cultural aspects characterising different ethnic groups. For instance, ontogenetically masculinised females may have greater fitness in some social contexts than in others, and this may affect sexual selection for such females from the part of males who, in turn, could be ontogenetically masculinised or even feminised as a result of sexual selection exerted by female mates. If morphological masculinisation or feminisation is also associated with behavioural masculinisation and feminisation, in regard to gender roles for instance, then those results may suggest a degree of diversity in the ways in which gender role combinations can increase fitness in different ethnic groups. We may even push the argument a bit further and perhaps speculate that such ethnic diversity in gender roles, in turn, may also be related to variability of expression and social acceptance of homosexual behaviour across those same ethnic groups. For instance, everything else being equal, lesbians may be better accepted in societies where heterosexual masculine females are preferred sexual partners, whereas gay men may be better tolerated where feminine heterosexual males are preferred sexual partners. At this point in time these ideas are obviously just speculations, but they are perfectly testable. I will return to the issue of the relationship between masculine and feminine gender roles and homosexual behaviour in Chapter 5.

Interestingly, Saino *et al.* (2006) have recently shown that Italian males do not prefer female right hands digitally modified to have a masculinised 2D:4D ratio and, vice versa, Italian females do not prefer 2D:4D feminised right hands, suggesting that sexual selection may be also playing a role in the maintenance of the dimorphism. In contradiction with this view, however, Koehler *et al.* (2004) have indicated that the 2D:4D ratio does not correlate with any of their measures of masculinity involving body and face variables. However, when it comes to testing for sexual preferences of external traits, morphologically masculinised or feminised men could be preferred by women, depending on their ethnic and cultural background (see, for example, Manning *et al.* 2000).

Martin & Nguyen (2004) also carried out a study, in several locations in the USA, of morphological structure of the hands, but in their case the variable used was the hand width:length (W:L) ratio. What they found was that heterosexual males had a larger ratio than heterosexual females in both hands, whereas homosexual males and females had a ratio similar to each other and intermediate between both heterosexual sexes, that is homosexual males are somewhat W:L-feminised and homosexual females are somewhat W:L-masculinised (a pattern known as *gender shift*; Lippa 2008a). They found the same trend for the arms length:stature ratio as well (Martin & Nguyen 2004).

In sum, studies of digit ratios show a considerable degree of variability regarding the presumed role of prenatal hormones in the development of sexual orientation. Three studies suggest a feminising action of prenatal androgens on homosexual males (Williams *et al.* 2000; McFadden & Shubel 2002; Lippa 2003b), but four studies also suggest a hypermasculinisation effect of prenatal androgens on male homosexuals (Williams *et al.* 2000; Robinson & Manning 2000; Rahman & Wilson 2003a; Putz *et al.* 2004). With regard to the development of homosexual orientation in females, four studies suggest a prenatal masculinisation of female homosexuals (Williams *et al.* 2000; Putz *et al.* 2004; McFadden *et al.* 2005; Kraemer *et al.* 2006) and only one study suggested a hyperfeminisation effect of prenatal androgens on development of female homosexuality (Lippa 2003b). If anything, these variable results suggest that human homosexuality is a diverse phenomenon that is determined, at the very least, by the interaction of prenatal hormones with other variables (genetic, environmental) to lead to alternative ontogenies and therefore different phenotypes.

Handedness

Most humans (about 90%) use the right hand preferentially to perform any manual task (Zucker *et al.* 2001). This lateralisation or dextrality (i.e. right-handedness) is established during the early foetal life and it is believed to be affected by the level of foetal androgens. The *Geschwind–Galaburda model* (Geschwind & Galaburda 1987) posits that testosterone present in male and female foetuses slows down the development of the brain's left hemisphere, so that the right hemisphere develops relatively more rapidly. Because levels of circulating testosterone are higher in male foetuses than female foetuses, this lateralisation of the brain is more accentuated in males. Brain lateralisation produces a series of behavioural effects, according to the *Geschwind–Galaburda model* that include preferential right-handedness. If homosexual sexual orientation is determined during foetal life by the low brain exposure to androgens, then the *Geschwind–Galaburda model* predicts that homosexuals should be more likely to be left-handed, or at least less consistently right-handed, than heterosexuals.

Lalumière *et al.* (2000) carried out a meta-analysis of the studies of handedness and sexual orientation in men and women published until 1998 and concluded that most studies indicate an association between non-right-handedness (that includes exclusive left-handed people and also those who are inconsistent right-handers) and homosexuality in both men and women. Homosexual men in the study by Lalumière *et al.* (2000) had odds of 31% of being non-right handed, whereas the percentage for homosexual women was 91%. Preferential left-handedness in boys showing cross-gender identification (19.5% vs. 8.3%) was also found by Zucker *et al.* (2001) in a Canadian sample. Mustanski *et al.* (2002a) also found an association between handedness and sexual orientation, in a study carried out in the USA, but only among females, with a larger percentage of homosexual women being ambidextrous (6%) than heterosexual women (0%). More recently, Lippa (2003b) carried out a study of the association between handedness and homosexuality in California and confirmed the trend found in previous works that both homosexual males and homosexual females tend to be more non-right-handed than heterosexuals, but in this case the effect was stronger in males (homosexuals have 82% greater odds of non-right-handedness than heterosexuals) than females (homosexuals have 22% greater odds of non-right-handedness than heterosexuals) as predicted by the *Geschwind–Galaburda model*. Interestingly, Lippa also reports an association between greater tendency to be left-handed and self-ascribed level of femininity in heterosexual men, with the trend being in the same direction for homosexual men, but statistically not significant in the latter case. Conversely, self-ascribed masculinity is associated with greater levels of right-handedness in homosexual males. Right-handed women tended to be more feminine, whereas left-handed women were more masculine. Although the *Geschwind–Galaburda model* predicts the observed male bias in left-handedness among homosexuals, the percentages of non-right-handed individuals among homosexuals are not very high and are not dramatically different between men and women; e.g. 19% vs. 14.3% respectively in Lippa's (2003b) work. This suggests that other mechanisms may be at work. In particular, I will consider the effect of prenatal and early postnatal stress on the development of homosexual orientation later on in this chapter.

Dermatoglyphics

Dermatoglyphics, more popularly known as 'fingerprints', are characteristic patterns formed by the dermal ridges in the skin of fingertips, and other parts of the body, that develop in humans between weeks 8 and 16 of foetal life (Holt 1968). This timing, it will be remembered, coincides with a period of testosterone surge in the foetus (McIntyre 2006). Although dermatoglyphics have a significant genetic heritability, the foetal environment also affects their development, as indicated by differences in dermatoglyphic patterns between monozygotic twins (Sorenson Jamison *et al.* 1993). Sorenson Jamison *et al.* (1993) described a positive correlation between adult male salivary testosterone levels and left–right hand (i.e. directional) asymmetry in some dermatoglyphic variables, from this they inferred a potential effect of testosterone that occurred in foetal life, following the *Geschwind–Galaburda model*. However, their interpretation remains speculative as they only measured adult testosterone levels. Most men and women have a right-hand-biased number of ridges (80%–85%), whereas women prevail among those individuals with the unusual left-hand bias (23.9% of women vs. 12.9% of men) (Hall & Kimura 1994; Kimura & Carson 1995). This sexual asymmetry provides a proxy variable to infer ontogenetic masculinisation or feminisation during foetal life among men and women with varying sexual orientation.

The relationship between dermatoglyphics and sexual orientation in humans has been investigated by several authors with variable results. In a Canadian sample of 182 heterosexual and 66 homosexual men, Hall & Kimura (1994) found no statistically significant difference between the two groups in total ridge count, whereas the directional asymmetry was mainly rightward in both groups, with the less common leftward asymmetry being relatively more frequent among homosexual men. Therefore homosexual men seem to have both male-typical patterns in their dermatoglyphics (i.e. total number of ridges) and female-typical patterns (i.e. leftward asymmetry). In addition, among homosexual men, dermatoglyphic leftward asymmetry is associated with an incidence of adextrality (i.e. preferential use of the left hand), an association that is not found in either heterosexual men or women (Hall & Kimura 1994). Forastieri *et al.* (2002) analysed a sample of 60 homosexual men, 76 heterosexual men and 60 heterosexual women from Salvador, Brazil, comparing both the total ridge count and some dermatoglyphic patterns such as arches, ulnar loops, radial loops and whorls among the three groups. Their analysis shows that homosexual men have a (statistically insignificant) higher number of ridges than both heterosexual men and women, and that the degree of asymmetry in dermatoglyphic parameters was not significantly different between males of the two sexual orientations. Thus in this study, homosexual men seem to be only slightly hypermasculinised with regard to total ridge count. Mustanski *et al.* (2002a) carried out an analysis of a USA sample of 429 males and 457 females and found that total ridge count was higher in men than women, although the two sexes did not differ in the level of directional asymmetry. However, they did not find any statistically significant effect of sexual orientation on dermatoglyphics.

The above results clearly indicate a very significant level of variability among the different studies with regard to the specific association between dermatoglyphics and sexual orientation. Disentangling the specific relationship, if there is any, between sexual orientation and development of dermatoglyphics from the effect of other variables is a challenge for future studies in this area.

Otoacoustic emissions and auditory evoked potentials

In both male and female humans the inner ear produces weak sounds known as otoacoustic emissions (OAEs). We cannot hear the OAEs that we produce, but we can measure them by using an appropriate microphone connected to an amplifier. McFadden & Pasanen (1998) analysed two kinds of OAEs: click-evoked OAEs (CEOAEs) and spontaneous OAEs (SOAEs) that are sexually dimorphic. Females have stronger CEOAEs than males and also have more numerous SOAEs in their right ear than males have in their left ear. These sexual differences are established during prenatal development. McFadden & Pasanen (1998) proposed a role of prenatal androgens in the development of the sex-specific patterns of OAEs, predicting that higher levels of prenatal androgen exposure should be associated with the masculinised pattern of fewer or weaker OAEs. This has been recently demonstrated experimentally by McFadden et al. (2009) in sheep, where females administered with testosterone prenatally developed masculinised CEOAEs, whereas prenatally testosterone-treated males did not differ from control males in their CEOAEs patterns. In their study, McFadden & Pasanen (1998) found that homosexual and bisexual women had CEOAEs of smaller amplitude than heterosexual women, whereas among men there was no statistically significant difference among the different sexual orientations. In terms of CEOAE strength, heterosexual women had stronger CEOAEs than both homosexual and bisexual women, but again there was no difference between male categories. This suggests a trend for homosexual women to be masculinised, a conclusion that was confirmed by an analysis of questionnaire responses obtained from the same subjects (Loehlin & McFadden 2003).

Auditory evoked potentials (AEP) are potentials that are triggered by brief acoustic stimuli. AEPs also show sexually dimorphic patterns (McFadden & Champlin 2000). McFadden & Champlin (2000) carried out a study of AEPs that not only confirmed the already known sexual dimorphism in AEPs but also demonstrated a masculinisation shift in non-heterosexual (i.e. homosexual and bisexual) females and a hypermasculinisation shift in non-heterosexual males.

Tooth crown size and waist-to-hip ratio

Dental crown size is another sexually dimorphic trait that starts to develop during early foetal life from about four weeks until completion of tooth development at about eight years of age (Dempsey et al. 1999). Although some aspects of dentine and enamel formation in teeth are under direct genetic control, hormonal effects may also be implicated. In order to study the potential role of hormones in tooth crown size sexual dimorphism, Dempsey et al. (1999) carried out a study of monozygotic twins (males or females), dizygotic twins (males, females or mixed sex) and singletons (males or females) as a non-intrusive methodology to control for individual exposure to hormones in utero (e.g. the ones coming from a twin if present). Of course, all foetuses were equally exposed to maternal hormones. The authors found a clear trend for masculinisation of tooth size of the female twin in different-sex twin pairs, compared with female twins in same-sex twin pairs or singleton females, whereas no such trend was found among male twins. Although Dempsey et al. (1999) did not carry out their study in the context of the ontogeny of sexual orientation, they certainly highlighted the importance of developmental effects of hormone diffusion in utero among twins, different-sex twins in particular. The alleged prenatal effect of elevated androgen exposure, however, seems to affect more female than male development.

The value of waist-to-hip ratio is affected by the amount of abdominal adiposity present in the individual. In adult men, higher circulating testosterone levels tend to be associated with lower levels of abdominal adiposity. However, McIntyre *et al.* (2003) suggest that prenatal androgen exposure in men is associated with increased adult waist circumference due to adiposity deposition. If so, waist-to-hip ratio could also be used to infer foetal exposure to androgens. What McIntyre *et al.* (2003) found by studying a sample of gay men from Boston, USA, is that adult circulating testosterone (measured from saliva) was negatively associated with waist circumference, whereas prenatal androgen exposure, as inferred from 2D:4D ratio, was positively associated with waist circumference. Although they did not carry out a direct comparison with heterosexual men, their work suggests that the waist-to-hip ratio could also be used as an additional variable to study the role of foetal androgen exposure in the development of homosexuality. Johnson *et al.* (2007) did compare the waist-to-hip ratio of heterosexual and homosexual men and found that heterosexuality was associated with higher values of the waist-to-hip ratio.

Recent work carried out by Anthony Bogaert using British data from the National Survey of Sexual Attitudes and Lifestyles-2000 (NATSAL 2000) also indicate that although lesbians did not differ from heterosexual women in either height, weight or the values of a body mass index, gay/bisexual men were shorter, lighter and also had significantly smaller values of the body mass index than heterosexual males (Bogaert 2008). Such differences could be established during early development, especially through processes affecting bone growth.

In sum, the above studies of proxy variables that are correlated with prenatal exposure to androgens, but perhaps also other sex steroids, in humans indicate very diverse effects of prenatal steroid exposure on development of a homosexual orientation in men, but a comparatively clearer effect on development of homosexuality in women. More elevated prenatal exposure to androgens in human females tends to be broadly associated with developmental masculinisation and in some cases with a homosexual sexual orientation. The degree of variability in results across studies, however, is large.

Prenatal exposure to steroid hormones and sex-typed toy preferences

The interaction that a child has with a toy may be rather complex, with the specific choice of toys to play with, at least in Western societies, having potentially quite dramatic consequences for the child, leading, in some cases, to the diagnosis of Gender Identity Disorder if the child happens to amuse him- or herself with the 'wrong' kinds of toy. A toy is an object that possesses both basic and more complex characteristics. For instance, toys may vary in colour, shape, texture, smell, taste, thermal conductivity, ability to produce sound if shaken, or motility, just to mention a few. On the other hand toys may also vary in some more complex properties; for example, a truck, a doll, or a guitar may convey additional meanings in terms of function in specific social contexts. In particular, some toys may be regarded more apt for boys ('masculine') or girls ('feminine') or they may be regarded as gender-neutral by adults (see Alexander 2003 for a review).

Whenever we observe a child preferentially playing with some specific toys, questions arise whether she or he is doing so because of a preference for one or more basic properties of the toy or because of a preference for the more complex characteristics. In addition, whenever we see a consistent preference for certain kinds of toy based on the sex of the child, questions usually also arise over whether such sex-specific preference is affected by learning processes that may be guided by the child's carers or whether the preference is established independently from learning. In particular, prenatal and perinatal effects of hormones on central nervous system development could potentially affect the use of toys, based, for instance, on preferences for one or more basic properties, leading to the characterisation of the child as more feminine or more masculine.

A recent study carried out by van de Beek *et al.* (2009) at the University of Utrecht in The Netherlands compared use of specific sex-typed toys by 13-month-old boys ($n = 63$) and girls ($n = 63$) with levels of testosterone, oestradiol and progesterone measured in the mother. Hormones were measured in both the amniotic fluid and the maternal serum between weeks 15.3 and 18.0 of pregnancy. Children were subsequently observed playing with a set of nine different toys that were simultaneously available to the child and that had been classified by adults (parents and non-parents) into 'masculine', 'feminine' and 'neutral'. The mother was with the child during the trial, although she was requested not to interfere and when she did respond to requests for attention from the part of the child, she was asked not to guide the child towards any specific toy; whether we could consider the mother as being a 'neutral' component of this experimental set-up is something that may be debatable. Be that as it may, the authors found that boys were exposed to elevated levels of testosterone in the amniotic fluid, whereas girls were exposed to elevated levels of oestradiol. When they compared the patterns of use of toys, it emerged that girls used 'feminine' toys more frequently than boys, whereas boys used 'masculine' toys more frequently, no difference was observed for use of 'gender-neutral' toys. With regard to more specific dose-dependent effects, the authors only found one direct association between prenatal hormone levels and use of toys: amniotic fluid progesterone levels were positively correlated with use of 'masculine' toys in boys. However, boys with more older brothers had a tendency to use fewer 'masculine' toys, that is they were 'toy-usage-feminised', whereas boys born to parents with a lower educational level also were feminised in their use of toys. Although the last two patterns could be explained by the kind of postnatal stress effects that I will analyse in a section of this chapter below, van de Beek *et al.* (2009) suggest that as far as understanding children's choice of specific kinds of toy is concerned, a biosocial approach, where both pre- and postnatal neuroendocrinological mechanisms and also social learning mechanisms are both likely to play a role, is probably the approach that will prove most productive. I certainly agree with this conclusion.

In a somewhat amusing experimental work, Alexander & Hines (2002) tested 44 male and 44 female 2–185-month-old vervet monkeys (*Cercopithecus aethiops sabaeus*) in captivity for their preference to handle six different toys simultaneously offered to them: a ball, a police car, a soft doll, a red pan, a picture book and a stuffed dog. The toys had been categorised as either 'masculine', 'feminine' or 'neutral' on the basis of which ones are preferred by boys and girls. Contact with toys differed between the sexes, with female vervets having greater contact with 'feminine' toys (doll and red pan), whereas male vervets had more contact with 'masculine' toys (orange ball and police car). Although there was no significant difference between the sexes in the use of 'neutral' toys, males showed a slight tendency towards greater contact with the picture book and the furry dog than females. Of course, the authors did not argue for some kind of primate-wide preference for young males to play with police cars; instead they suggest that young males and females may possess specific perceptual biases towards objects that have some basic characteristics of shape, texture, colour, etc. In fact, this is a potentially very important mechanism that may explain gendered preferences to play with specific objects. Such bias in preference could be caused by sex-specific steroid-dependent neuronal changes occurring in peripheral sensory tissues (e.g. the retina) and/or specific brain areas (e.g. the visual cortex) during the early ontogeny of males and females. Potentially, the same mechanisms could also explain some of the preferences displayed by individuals of the same sex but different sexual orientation.

Gerianne Alexander (2003) has recently reviewed this model. The postnatal increase in circulating steroids that affect the development of various parts of the brain may also affect the visual cortex, thus leading to sex-specific ontogenies of the visual neuropathways. In particular, human males seem to develop specific competencies for object movement, whereas females develop particular competencies for form and colour (Alexander 2003).

Moreover, such competencies are further perfected through experience that can modify brain regions controlling specific cognitive and social functions. We will see in Chapter 5 (e.g. Table 5.5) how some such sex-specific cognitive abilities, which may develop early in life, also show different patterns between homosexuals and heterosexuals, especially men.

What is the relationship between use of specific gender-typed toys during early childhood and subsequent development of homosexuality in adults? Empirical evidence suggests that homosexual men tended to play with female gender-typed toys when they were children (Whitam 1977), whereas congenital adrenal hyperplasia (CAH) girls, who are more exposed prenatally to adrenal steroids, tended to be masculinised in their choice of toys and also somewhat predisposed to develop a same-sex sexual orientation (Berenbaum & Hines 1992; see also Chapter 5). From this, however, we should not conclude that all children who show a preference for other-gender-typed toys will necessarily develop a homosexual sexual orientation when adult! We will see in the section 'Do "sissy" boys and "tomboy" girls become adult homosexuals?' in this chapter that gender-discordant behaviours in childhood are no guarantee of development of homosexuality in adulthood.

A greater ability to predict adult homosexuality based on toy preference at young ages will be achieved once we better understand the specific perceptive nature of that toy preference, the developmental process that led to it, and the association of those processes with the development of homosexuality. There is a potential for a link with prenatal steroids (see, for example, Berenbaum & Hines 1992, van de Beek *et al.* 2009), but specific empirical tests of this link are still scanty and, in any event, other factors (genetic, learning) are also likely to play a role.

Birth order and family size effects

Having older brothers increases the probability of developing a homosexual sexual orientation (see Cantor *et al.* 2002 for a review); this is the so called

Birth Order Effect. Recent models have involved the action of the maternal immune system on the developing foetus as a major mechanism to explain birth order effects. I will review immune mechanisms in Chapter 6, here the focus is on some non-immune mechanisms.

In his review of birth order effects in homosexuality, Miller (2000) listed some patterns found in earlier works indicating that younger siblings display less 'sex-appropriate activities', more feminine personality traits and also being physically less active than their older siblings. Miller (2000) supports models of birth order effects of homosexuality that imply in utero mechanisms, rather than postnatal mechanisms involving interactions with siblings, parents, other relatives or unrelated individuals (see also Chapter 6). In addition, he leans more towards prenatal parental manipulation and kin selection evolutionary processes than, for instance, sibling competition to explain why younger brothers seem to be more likely to become homosexual. I agree with Miller that kin selection is an important mechanism to consider when analysing family effects on homosexuality, but we will see in Chapter 6 that intrafamilial interactions that may affect the development of homosexuality are likely to be far more complex than Miller seems to suggest, involving both cooperation and competition. In addition, we will see in this chapter and in Chapter 6 that dismissing postnatal mechanisms as an explanation of the birth order effect is premature (Cantor *et al.* 2002). Moreover, Cantor *et al.* (2002) estimate that about one in seven homosexual men in their sample is consistent with the older brother effect, suggesting that this is likely to be a significant, but by no means the only, proximate cause of homosexuality in human males.

The crucial issue of whether birth order effects in the context of homosexuality are more likely to be explained by prenatal maternal effects or postnatal social mechanisms has been tackled by Anthony Bogaert in a recent article (Bogaert 2006a). Bogaert's (2006a) study demonstrates that, after controlling for maternal age, homosexuality in an individual is associated with having older biological brothers, regardless of whether they had been reared together or not, and it is not associated with having non-biological older brothers, or with having been reared or not with older biological or non-biological brothers. Although Bogaert's (2006a) results seem to clearly support a prenatal mechanism for the older brother effect in the development of homosexuality, there is a major methodological issue in this and similar studies that I am not convinced has been properly tackled. When studying the effect of biological siblings reared apart or of non-biological siblings reared together, the timing of separation or of union is essential. We have seen in this chapter how factors such as hormones exert their effect on the development of behaviour during specific sensitive periods. If there is one or more postnatal sensitive period/s during which social interactions with siblings can exert an influence on the ontogeny of homosexual orientation, then cases where the siblings have been separated after such a sensitive period will obviously show an older sibling effect. However, such an effect will not necessarily be a result of prenatal mechanisms. Similarly, if an older stepsibling is integrated into a family after the postnatal sensitive period of his stepbrothers, then he will be obviously irrelevant for the development of homosexuality in the stepbrothers, but for reasons that do not exclusively relate to the prenatal determination of the older brother effect.

King et al. (2005) also carried out a study of the older brother effect and found that homosexual men not only had more older brothers, but also more older sisters than heterosexual men. Moreover, the prevalence of homosexuality was higher among adopted men than non-adopted men (King et al. 2005). In addition, King et al. (2005) showed a significant effect of family size on homosexuality, with the effect being stronger for the paternal than the maternal number of relatives. The results obtained by King et al. (2005), if confirmed, suggest that social factors acting postnatally can also play a role in the development of homosexuality, as the older sibling effect also included sisters, whereas stepsiblings also tended to develop a homosexual orientation.

I certainly agree in pointing to prenatal developmental mechanisms for an explanation of at least a percentage of the cases of birth order and family size effects in homosexuality, but I am also concerned about a premature dismissal of postnatal mechanisms, especially social effects operating in the early postnatal life (see, for example, James 2001). We are an altricial species and children continue to develop after birth, being subject to the influence of the social milieu. I will return to the issue of prenatal effects on the ontogeny of homosexuality in Chapter 6, where I will thoroughly review the older brother effect in the context of prenatal immunological models, but also consider in detail the potential effect of postnatal offspring–parent and sibling–sibling interactions especially in larger families.

Intrauterine positioning

Masculinisation of the female twin in a disexual twin pair was first described by Frank R. Lillie in an article in *Nature* published in 1916. The phenomenon was later named the *Freemartin Effect*, following the title of Lillie's article, but it is now more generally known as the *Intrauterine Positioning Effect* (see vom Saal 1981 for a review of the early work in this area).

Studies carried out on polytocous mammals, i.e. species that regularly produce more than one offspring at each pregnancy, indicate that the positioning of females with respect to their brothers in utero may affect feminisation or masculinisation of those females. Male prenatally produced testosterone may be released into the amniotic fluid, thus finding its way into adjacent siblings, sisters in particular, and perhaps affecting the development of steroid-sensitive brain regions. vom Saal (1981) reported higher levels of aggressiveness and lower degree of attractiveness to males in female mice that developed between two males (2M females) compared with females not developing in the immediate vicinity of males (0M females) in utero. Male mice were not equally affected by the kind of sibling they were in contact with in utero in vom Saal's (1981) study. This strongly suggests that the effect is a result of testosterone action, oestrogen in foetal blood is inactivated during these early stages of development in mice by α-fetoproteins (vom Saal 1981). Clark & Galef's (1998) study on the Mongolian gerbil (*Meriones unguiculatus*), however, does suggest that males can also be affected by their sibling environment in utero as sexual behaviour of 2F males was less competent than that of 2M males and they also had less circulating testosterone when adults than 2M males. Females prefer to copulate with 2M males. Clark & Galef (1998) suggest that less masculinised males may show greater parental care, thus compensating for their reduced fertilisation capability with an increased probability of offspring survival. The effect has been described in several mammals with variable results in males and females: e.g. behavioural masculinisation of 2M females in mice and decreased aggressiveness in 2F males in swine (see review in Ryan & Vandenbergh 2002).

In humans not all studies have produced supportive evidence for this effect. For instance, Gaist *et al.* (2000) could not detect a significant difference between females in monosexual or disexual twin pairs in either handgrip strength or several anthropometric variables; neither could Rose *et al.* (2002) detect a significant difference in age at puberty and attitudes regarding femininity and fertility. Bearman & Brückner (2002) also suggest that the female co-twin is not more likely to become homosexual or, as indicated by Henderson & Berenbaum (1998), to develop a preference to play with boys' toys. Medland *et al.* (2008) have recently shown that there are no differences in 2D:4D digit ratio between individuals who developed in a monosexual or disexual twin-pair environment in the womb. However, several studies do point to positive effects of masculinisation of the female twin in disexual twin pairs.

We have seen above how the study by van Anders *et al.* (2006) of different-sex twins indicated a masculinisation of the 2D:4D digit ratio in the female member of the pair (see also Voracek & Dressler 2007). Masculinisation of the auditory system was observed by McFadden (1993), whereas Cohen-Bendahan *et al.* (2005) described a masculinised pattern of females from disexual twin pairs to be more aggressive than females from monosexual twin pairs. This latter effect may also potentially explain some intriguing ethnological findings. Margaret Mead (1935), for instance, described how among the Mundugumor of Papua New Guinea both men and women were highly aggressive, with women also displaying low levels of body contact with their newborn. Infanticide was apparently common among the Mundugumor, except when twins were born. Mead described how the frequency of twinning was higher in this ethnic group than among other tribes in the region. It is possible that the freemartin effect could have contributed to the relatively high level of aggressiveness that was present in this group, especially in females. Such aggressiveness would have subsequently permeated through various aspects of culture. Cross-cultural studies of twinning and larger sample sizes will be able to shed more light on this issue and on the potential effects of co-twins on development of sexual orientation and gender role in humans. In addition, studies should make sure to compare not only disexual vs. monosexual twin pairs, but also twins vs. singletons.

Stress effects

Early studies of prenatal and neonatal effects of hormone treatment on the development of sexual behaviour in rats indicated that prenatal and neonatal androgen deprivation in males (e.g. achieved by treatment with the antiandrogen cyproterone acetate or by castration) resulted in display of lower levels of male-typical copulatory behaviour and higher frequency of female-typical lordosis compared with controls (see Ward 1972 for a review of research done in the 1960s). On the other hand, female rats that were treated with androgens during critical periods of perinatal life showed partial or total impairment of female-typical sexual receptivity while displaying male-like copulatory behaviour. The levels of circulating androgens during early ontogeny are, therefore, central in modulating the development of adult sexual behaviour in rats.

The developing foetus and young, however, also possess other organs, apart from the gonads, that can produce and release molecules into the bloodstream, affecting the development of sexual behaviour. The adrenal glands are one such organ, releasing both steroids (e.g. dehydroepiandrosterone, androstenedione) and corticosteroids (e.g. corticosterone, cortisol) not only under pathological, but, more importantly under physiological conditions, especially in response to various stressors (Ward 1972). Adrenal corticosteroids in particular can interfere with testosterone production by the gonads, thus potentially affecting the development of sexual behaviour (Dunlap et al. 1978; see Weinstock 2001 for a recent review of the various effects of prenatal stress on development), whereas adrenal steroids such as androstenedione may affect the process of masculinisation when they are released in circulation.

Ingeborg Ward's (1972) pioneering work on the study of stress interference with sexual development involved prenatally stressing male and female rats by restraining their pregnant mother in semi-circular tubes three times a day for 45 minutes per session during gestation days 14 to 21. Half of the litters that were stressed prenatally were also stressed during their 10 initial postnatal days, along with other individuals that had not been stressed prenatally. When they were about 90 days old, males were tested for heterosexual behaviour by placing them in the company of a female in oestrus. Ward (1972) found that it is prenatal stress that has the greatest effect on adult sexual behaviour. However, prenatally stressed males simply did not show any sexual interest in the oestrous females. Therefore, they appeared to be de-masculinised, but it was unclear whether they were feminised. In order to explore the latter possibility, the males were castrated and given an injection of oestradiol benzoate followed by progesterone, a hormonal treatment that would elicit female-typical behaviour in feminised individuals. Treated and control animals were then exposed to a stud male and what Ward (1972) observed was that 73% of prenatally stressed males showed female-like lordotic behaviour compared with only 36% of control males. Postnatally stressed males showed an intermediate level of response (53%). Therefore, although the feminised sexual response of prenatally stressed males to the presence of a stud male was lower than that of females, it was twice as high as that of non-stressed males and their lordotic behaviour was very female-like (Ward 1972). Ward suggested that this prenatal effect of stress in rats is mediated by adrenal release of the steroid androstenedione, which may compete for binding sites with testosterone. It is interesting to note that Ward did not interpret her results in the context of 'pathology', but attributed the effect to an adaptation, although in her case she leaned towards a group-selectionist interpretation, implicating population self-control at time of food scarcity, that was still popular at the time but that is no longer entertained.

Dahlöf et al. (1977), following an experimental protocol similiar to that of Ward (1972), also tested the effect of more 'natural' stressors such as crowding, in addition to the traditional restraint-stress in rats. Their results showed a significant decrease in heterosexual behaviour among male offspring of mothers that had experienced crowding conditions during pregnancy, compared with offspring of control mothers. Therefore crowding seems to be a stressor that is able to behaviourally de-masculinise male foetuses. In tests of lordotic behaviour, males exposed to both crowding and mother-restraint stress during prenatal life were behaviourally feminised compared with controls.

Rhees & Fleming (1981) also studied the effect of a more 'natural' prenatal stressor such as nutritional stress, in which pregnant rats received only half of the food fed to controls, along with Ward's (1972) original stress protocol. In addition, some of the females were injected daily with 20 units of adrenocorticotropic hormone (ACTH) rather than being subjected to a stressful environment. ACTH is naturally produced in the anterior pituitary and it releases corticosteroids and other steroids from the adrenals when the organism is under stress (see, for example, Bornstein & Chrousos 1999). All three modes of induction of prenatal stress response resulted in behavioural de-masculinisation of male offspring and also resulted in significant feminisation of sexual behaviour (Rhees & Fleming 1981). This was coincident with a significantly reduced size of the sexually dimorphic nucleus of the preoptic area of the hypothalamus (SDN-POA) at 60 days of age in prenatally stressed male rats (Anderson et al. 1985). The SDN-POA is normally larger in males than in females and is an important centre contributing to the control of sexual behaviour (see Chapter 5). That is, prenatally stressed males were not only behaviourally feminised, but they also had a feminised SDN-POA.

Another 'food stressor' such as addition of alcohol to the diet of pregnant females also increased lordosis (feminisation) and decreased ejaculation (de-masculinisation) in their male offspring (Ward et al. 1994). Apart from a reduced SDN-POA, prenatally stressed male rats also developed fewer motoneurons in the spinal nucleus of the bulbocavernosus and the dorsolateral nucleus that control penile erection (Kerchner et al. 1995), but the size of the sexually dimorphic medial amygdala, a brain area also involved in the control of aspects of sexual behaviour, was not affected by prenatal stress (Kerchner et al. 1995). If Sprague–Dawley male and female rats are prenatally heat- and restraint-stressed from gestation day 15 until birth three times a day for 30 min each time, the size of the rAca (i.e. the rostral position of the anterior division of the anterior commissure, a nerve fibre link between the right and left hemispheres) becomes masculinised in stressed females and feminised in stressed males (Jones et al. 1997).

The de-masculinising effect of prenatal stress in male rats could not be replicated by Whitney & Herrenkohl (1977), although they did confirm the feminising effect. Such a feminising effect, known as *Prenatal Stress Syndrome* (Ward 1984), requires an intact anterior hypothalamus, as prenatally stressed male rats subject to electrolytic lesions of their anterior hypothalamus at 180 days of postnatal age displayed a significantly reduced level of lordotic behaviour.

Although prenatal stress can de-masculinise male rats, the effect is far more dramatic if the individual is subsequently reared in social isolation than if it is reared with social stimulation (Dunlap et al. 1978), the latter being the normal condition of rats in the wild. This suggests that there is a degree of resilience (i.e. developmental stability) to withstand the effect of stressors during early ontogeny facilitated by social rearing conditions, although in the work of Dunlap et al. (1978), stress alone accounted for more than 20% of the reduction of ejaculatory capability in male rats.

Ward & Reed (1985) followed up on the experiments of Dunlap et al. (1978) by studying the effect of a range of postnatal social rearing conditions on the development of sexual behaviour in prenatally stressed (pregnant mother restrained and heat-stressed with light) and control male rats. The social conditions in which control and prenatally stressed males were reared since 16 days of age were: (a) total social isolation, (b) caged with a same-aged control (i.e. non-prenatally stressed) male, (c) caged with a prenatally stressed male and (d) caged with a female. At 60 days of age all males were tested for copulatory behaviour with a female. They were subsequently castrated and tested for feminine sexual behaviour in the presence of a stud male. Patterns of heterosexual sexual behaviour indicate that males reared in isolation were the least sexually active (i.e. most de-masculinised) followed by the prenatally stressed males reared with an equally prenatally stressed male, the highest levels of ejaculatory activity with a female were obtained by non-stressed control males reared with a female. Between the latter and the stressed males reared together were the results of the other treatments: controls reared with another control male or with a stressed male, and stressed males reared with a control male or a female. From this it can be concluded that socialisation, even if it involves the company of equally prenatally stressed males, is an important factor in the resilience of ontogenetic programmes against the effects of prenatal stress in rats. However, prenatal stress remains a factor in the partial de-masculinisation of males, even if those males are postnatally reared in the company of a female. What about feminisation? The males that showed the highest level of lordosis were those stressed prenatally and then reared with an equally prenatally stressed male, followed by those reared with a control male. Therefore feminine sexual behaviour in the presence of a stud male is enhanced in male rats by *both* prenatal stress and social experiences with same-sex individuals during rearing. Interestingly, both controls and prenatally stressed males reared with a female show a similar level of lordosis, indicating the very significant level of plasticity displayed by adult male rats regarding the manifestation of masculine and feminine sexual behaviour.

A review of major work carried out on the *Prenatal Stress Syndrome* in rats in the 1970s and early 1980s can be found in Ward (1984). The broad trends suggested by those experimental studies include: (a) prenatal stress affects feminisation/masculinisation, especially in males, but to a degree also in females; (b) the effect of prenatal stress on male sexual behaviour is also variable, ranging from no effect, to behavioural feminisation (i.e. increased lordotic behaviour in the presence of a stud male), to asexuality and bisexuality; (c) external morphology and many behaviours other than lordosis are generally masculinised in prenatally stressed males; (d) there is a sharp decrease in circulating testosterone around days 18 and 19 of development in prenatally stressed males and also lower levels of brain aromatase activity during the same period.

More recent studies have unravelled further details of the mechanisms involved in the *Prenatal Stress Syndrome*. The specific neuroendocrine mechanisms that affect the development of male sexual behaviour in rats include: (a) reduced volume of central nervous system areas controlling sexual behaviour and orientation due to reduced aromatase activity in the foetus' hypothalamic region, (b) alteration in brain monoamines, and (c) a potential reduction of circulating testosterone (see Holson *et al.* 1995 for a review).

The response to stress experienced by a pregnant female may involve an increased release of corticosteroids in circulation that may in turn affect the development of the embryo. Holson *et al.* (1995) specifically tested the potential effect of corticosterone and ACTH in an experimental design involving pregnant females being subject to either of the following treatments: stress, ACTH injections, corticosterone injections, dexametasone (DEX, a corticosterone antagonist) injections or they were left untreated (controls). All injections were administered daily for one week between gestational days 14 and 21. As expected, all treatments increased maternal circulating corticosterone except the DEX group. Interestingly, the only treatments that showed temporarily impaired heterosexual behaviour in male offspring were the prenatal DEX and the prenatal stress, but the effect was not permanent. Similarly, exposure to a stud male after gonadectomy and hormonal priming resulted in elevated lordosis in the DEX group only. Therefore, in this work, DEX seems to be involved in de-masculinising and feminising male rats. Holson *et al.* (1995) interpreted their results in view of the following potential mechanism of development of brain areas involved in control of sexual behaviour: prenatal testosterone masculinises those brain areas, but testosterone secretion by the gonads is suppressed by the joint action of glucocorticoids and the catecholamine noradrenaline, both secreted under stress. Moreover, glucocorticoids themselves also enhance production of adrenaline by the adrenal medulla, meaning that stress should increase circulating corticosteroids and catecholamines by various mechanisms. Thus stress may induce de-masculinisation through the mediation of a variety of stress hormones. DEX may depress testosterone production due to its strong affinity with corticosteroid receptors, whereas stress may activate production of both corticosteroids and catecholamines. ACTH and corticosterone alone seem to be less effective than the joint activation of glucocorticoids and catecholamines achieved

Table 4.2. Effects of prenatal stress on catecholamine contents and functional metabolism of testosterone in the preoptic area and the mediobasal region of the hypothalamus in 10-day-old rats

Measure	Intact		Prenatally stressed	
	Females	Males	Females	Males
POA				
Noradrenaline (nmol/g tissue)	4.39	2.95^a	3.49	4.71^b
Dopamine (nmol/g tissue)	3.98	3.85	3.05^a	5.94
Aromatase activity (pmol oestradiol/h/g tissue)	0.404	0.616^a	0.423	0.385^b
5α-reductase activity (nmol 5α-reduced metabolites/h/g tissue)	14.34	6.16^a	12.06	$18.63^{b,c}$
MBH				
Noradrenaline	1.95	2.56^c	1.73	2.07^c
Dopamine	2.43	4.72^a	3.26^a	3.84^a
Aromatase activity	0.289	0.253	0.313	0.171
5α-reductase activity	8.82	8.25	9.91	$22.45^{a,b,c}$

Significant differences: ^{a}with intact females, ^{b}with intact males, ^{c}with prenatally stressed females.
Modified from Table 1 of Reznikov et al. (2001).

by stress. The levels of stress needed to produce significant effects on development of sexual behaviour, however, seem to be high, which again suggests that ontogeny is to some extent resilient to a degree of environmental perturbations during prenatal life.

In a remarkable work, first published in Ukranian in the journal *Problemy Endokrinolologii* but then translated into English and published in *Neuroscience and Behavioral Physiology*, Reznikov et al. (2001) stressed pregnant female rats through 1 hour daily immobilisation from days 15 to 21 (i.e. the last week) of pregnancy. When offspring were 10 days old their preoptic area and mediobasal region of the hypothalamus were analysed for concentration of noradrenaline and dopamine, and also for aromatase and 5α-reductase activity. Table 4.2 summarises the results of Reznikov et al. (2001).

From Table 4.2 it can be seen that prenatally stressed males are feminised in their POA production of noradrenaline whereas females decrease significantly, and males increase non-significantly, their POA dopamine if prenatally stressed. Aromatase activity in the POA is dramatically reduced in prenatally stressed males to female levels, whereas 5α-reductase activity in prenatally stressed males becomes hyperfeminised. Therefore, prenatal stress tends to feminise not only the POA structure but also the POA function in male rats. With regard to the mediobasal hypothalamus, prenatal stress does not affect the sexual dimorphism present in controls, nor does it affect aromatase activity, but it masculinises dopamine production in females and hypermasculinises 5α-reductase activity in males.

Similar studies of the effects of prenatal stress have also been carried out in another rodent, the guinea pig (*Cavia aperea*), by Sylvia Kaiser, Norbert Sachser and their collaborators in Germany. Sachser & Kaiser (1996) performed an experiment testing the effects of prenatal social stress on daughters' sexual behaviour: pregnant mothers were kept in either a stable or an unstable social environment, achieved by changing or unchanging group composition. They also studied the effects of social stress during lactation. Only prenatal social instability had an effect on female offspring sexual behaviour, with daughters of mothers exposed to social stress displaying a higher level of courtship behaviour typical of males (i.e. they were behaviourally masculinised). Although prenatally stressed females had normal levels of circulating cortisol, they had higher absolute and relative (to body mass) mass of adrenals. In addition, they also had significantly higher levels of circulating testosterone at day 100 of age than daughters of control mothers (Kaiser & Sachser 1998). Prenatally stressed daughters were not only more behaviourally masculinised but they also displayed higher tyrosine hydroxylase (TH) activity; TH availability is a limiting factor in catecholamine production, and catecholamines are released under stress (see Holson et al. 1995). It is very tempting to speculate that female behavioural masculinisation in this case may be an adaptive mechanism to increase fitness in the face of social subordination. Subordinate females that may have reduced reproductive output due to social stress imposed by dominant females may compensate for the decreased current production of daughters by producing more socially competitive offspring: masculinised females, but also feminised males, who may display alternative strategies for survival and reproduction in a competitive social environment. That is, they may be compensating for low production of offspring by increasing production of grand-offspring. Such a hypothesis could be easily tested in studies of guinea pigs and other rodents held in captivity. ACTH treatment seems to be unable to replicate the effects of social stress on the development of sexual behaviour in female guinea pigs (Kaiser et al. 2000), unlike the feminising effect that ACTH treatment may have in male rats (Rhees & Fleming 1981).

Behavioural masculinisation of female guinea pigs born of mothers subject to social stress during pregnancy was again obtained by Kaiser et al. (2003) in a subsequent experiment. In this work, however, the authors also studied upregulation (i.e. expression) of androgen (AR) and oestrogen (ER-α) receptors in the nucleus arcuatus and the MPOA in the hypothalamus, and in the hippocampus (pyramidal layer of the CA1 region), all regions of the limbic system that are involved in control of sexual behaviour and, in the case of the hippocampus, in learning and memory. Females born of mothers subject to an unstable and therefore stressful social environment showed greater upregulation of the androgen receptor in the nucleus arcuatus compared with females born of unstressed mothers; the same was found for ER-α in the nucleus arcuatus and AR and ER-α in the MPOA. Both receptors were also upregulated in the same prenatally stressed females in the hippocampus. Therefore, behavioural masculinisation of prenatally stressed female guinea pigs is associated with production of both androgen and oestrogen receptors in specific areas of the central nervous system traditionally associated with sexual behaviour and learning, thus making the cells in those brain areas susceptible to the action of steroid hormones. Kaiser et al. (2003) also suggest that catecholamines released under stress may have an activational effect on AR and ER-α receptors expression.

The evidence so far points to prenatal stress clearly affecting sexual behaviour and gender role (masculinity, femininity) in rodents. Does it also affect the development of sexual orientation? A recent experiment carried out by Meek et al. (2006) on Swiss Webster mice (Mus musculus), where females were stressed from day 12 of pregnancy by handling them, exposing them to noise and subjecting them to temperature stress, showed that sexually naïve male offspring of stressed females, given the simultaneous choice between a tethered oestrous female and an equally tethered sexually active male, made more visits and spent more time in the male than the female compartment than control male mice. In addition, prenatally stressed male mice were less sexually active with the female than controls. In a test with an untethered sexually active male, prenatally stressed males also displayed a higher incidence of lordosis than did control males.

In sum, experimental work on various rodents suggests that pre- and perinatal stress could be a potentially powerful mechanism for the development of same-sex sexual behaviour and same-sex partner preference in males and females. The effect of stress is at least mediated by the release of corticosteroids, steroids and catecholamines into the bloodstream, which may result in feminisation of male and masculinisation of female areas of the brain that control sexual behaviour. This, in turn may produce changes in sexual behaviour. Although the ontogeny of the various organisms is obviously somewhat buffered against the effect of stressors, as the stress levels become particularly elevated during sensitive periods of development their effects become noticeable, potentially leading to the expression of homosexual phenotypes.

Hormones, such as corticosteroids, that are produced in response to stress are also known to be transported transplacentally from mother to embryo in humans (see Mulder *et al.* 2002 for a review). Maternal stress during pregnancy is associated with increased risk of abortion and, in those offspring that survive, it is associated with developmental effects on brain organisation and functioning (Mulder *et al.* 2002). In particular, prenatal stress has also been suggested as a potential cause for the development of homosexuality in humans.

Stress effects and human homosexuality

Günter Dörner and his collaborators (Dörner *et al.* 1980, 1983) carried out the initial, pioneering works that tested the association between stress experienced by pregnant mothers and subsequent development of a homosexual orientation in their male offspring. Dörner *et al.* (1980) studied 865 homosexual men who were attending clinics for the treatment of sexually transmitted diseases in Germany, and related their sexual orientation to the date of their birth. What the authors found was that a significant majority of the homosexual men were born during the years of the Second World War or in the immediate postwar period, with a peak in frequency detected for the years 1944–1945. Dörner *et al.* (1980) interpreted this trend as being a result of the stress suffered by mothers during pregnancy. They subsequently carried out a study based on interviews of 100 heterosexual and 100 non-heterosexual (i.e. homosexual and bisexual) men and asked them about stressful conditions that their mother might have experienced during pregnancy. Although the result was coincident with that of their previous work, i.e. more non-heterosexual men reported alleged maternal stresses during pregnancy than heterosexual men, this study clearly suffers from methodological problems associated with potential biases in recalling events on the part of mothers.

Dörner and collaborators' studies were savagely criticised by Sigusch *et al.* (1982) in an article published in the *Archives of Sexual Behavior*. With regard to their specific attack on the link between prenatal stress and development of homosexuality, their criticism is worth quoting in full (Sigusch *et al.* 1982: 447–8):

> This reduction of war, an exceptional state of mental, social, and societal emergency, to the effects of hormone activity, suggests that Dörner himself has doubts about the tenability of his 'biological reasoning'. The flimsiness of his arguments is obviously meant to be cushioned by such recourse to a crude sociobiologism. But this does not make them any more scientifically valid.

Presumably, this was supposed to put a nail in the coffin of the prenatal stress hypothesis for the development of human homosexuality. That the critical argument is spurious should be clear from the fact that a hypothesis is not only strengthened by its theoretical underpinning (e.g. 'sociobiology'), but also by its ability to explain past events (e.g. patterns of birth) and the ability to predict future ones (e.g. results of prospective studies). It is this empirical approach – fundamental to the scientific endeavour – that allows seemingly unusual and surprising theories to be given a chance through experimental or observational tests. Dörner did not pull a stress effect on development of sexual orientation out of nowhere, but out of previous studies showing the association of pre- and perinatal androgen levels with sexual behaviour and the interference of corticosterone, a hormone released under stress, with testosterone release in rodent models (see above). The hypothesis predicted the patterns found for men born during wartime. From then on better tests are expected to be carried out controlling for specific stressors acting on mothers at specific times during their pregnancy; with the hypothesis being evaluated after a sufficient amount of empirical evidence of the highest quality becomes available. The preliminary work and theoretical underpinning only provide a framework to evaluate the overall plausibility of the hypothesis; they do not provide a conclusive prove of its validity. Dörner (1983) did reply to the criticism by Sigusch *et al.* (1982) and accepted, among other things, the limits of retrospective studies as tests of prenatal or perinatal stress mechanisms. This controversy also involved important ethical issues and here Dörner's stance of regarding homosexuality as an illness was simply wrong. To be fair, however, Dörner subsequently changed his view on the issue of homosexuality as pathology, regarding it instead as an expression of the heterosexuality–bisexuality–homosexuality continuum found in our species (Dörner *et al.* 2001).

Dörner's initial study was replicated by Schmidt & Clement (1995) who compared the incidence of homosexuality in males born before (1936–1940), during (1941–1945) and after (1951–1955, 1956–1960) World War II in western Germany. The authors did not detect a significant difference in homosexual activities during the period of adolescence in the various cohorts, but when adult ages were considered, homosexual activity was more prevalent in the postwar generations. Schmidt & Clement interpreted this result in a social constructionistic manner, attributing it to the increased sexual liberalisation of the postwar period. Whether the various cohorts studied by Schmidt & Clement differed or not in the relative distribution of *exclusive* homosexuals in adulthood is unknown. More recently, de Rooij *et al.* (2009) have analysed a corpus of 380 men and 472 women born during the Dutch famine of 1943–1947. They found no statistically significant effect of prenatal exposure to the famine on the development of sexual orientation in both males and females. However, they also acknowledge that their results might have been affected by a degree of underreporting of homosexuality among people of the age group that they studied (mean age = 58 years).

Incidentally, the same problem might have affected Schmidt & Clement's study. Clearly, future studies of this kind will require a better control of relevant variables.

Formalising it as the *Maternal Stress* hypothesis (which is the human equivalent model of the *Prenatal Stress Syndrome* proposed in rodents), Ellis *et al.* (1988) took up the challenge represented by the work of Dörner *et al.* They suggested that stress hormones such as corticosteroids and catecholamines, released by the mother into her bloodstream in response to specific stressors, can cross the placenta and decrease production of steroids in the offspring which, in turn, could lead to alternative pathways of brain differentiation if stress is coincident with androgen-sensitive periods of brain development. Such an ontogenetic process could lead to production of adult male phenotypes displaying same-sex sexual preference through, for instance, feminisation of brain areas (e.g. nuclei) that control sexual orientation. Ellis *et al.* (1988) did accompany the formulation of their hypothesis with a very preliminary study of the association of maternal recall of stresses suffered during pregnancy and subsequent development of homosexuality in their offspring, finding an association between the two in the case of stresses suffered during the second trimester of pregnancy.

Bailey *et al.* (1991) also carried out a retrospective study of subjects who were classified in terms of their sexual orientation on the basis of their sexual fantasies alone, a method that is likely to blur differences between categories. Male and female heterosexual subjects were compared with non-heterosexual counterparts in terms of Kinsey sexual fantasy scores, self-rated childhood gender non-conformity and maternally rated childhood gender non-conformity. Their results suggest that for male non-heterosexuals there is no association between the three variables mentioned above and recalled maternal stress during pregnancy. However, for female non-heterosexuals, a small maternal stress effect was detected.

More recently, Entringer *et al.* (2009) have used a retrospective approach to test for effects of prenatal stress on the activity of the hypothalamus–pituitary axis (HPA) of adults. Their work suggests that prenatally stressed individuals respond to a standardised behavioural challenge paradigm (a speech and mental arithmetic task to be performed and filmed in front of an audience for 15 min) by showing a higher level of cortisol increase than controls. This result is relevant because such an effect of prenatal stress on HPA function in postnatal life could potentially affect the probability of development of a homosexual orientation via the mechanisms operating during adrenarche that will be analysed in the next section.

Retrospective studies of this kind, based on recalling of past events, interesting as they are, are well known to suffer from potential biases and it is doubtful whether they can be taken too seriously unless those biases, and also the effect of confounding variables, are satisfactorily controlled. Prospective studies, that involve the follow-up of individuals throughout several stages of their ontogeny, are far better suited for the study of stress effects on the development of sexual behaviour and orientation, and this is obviously the method used in the rodent studies reviewed above.

Hines *et al.* (2002) published a prospective study based on the ALSPAC, the English 'Avon Longitudinal Study of Parents and Children' database. Stress was assessed in the ALSPAC cohort of pregnant women at 18 weeks of gestation through the administration of a questionnaire for stresses experienced during the first half of pregnancy and then again at 8 weeks postnatal for the second half of pregnancy and the immediate postnatal period. Children were then assessed for gender role behaviour at 42 months of age through the Pre-School Activities Inventory (PSAI) that is completed by the child's primary caretaker, usually the mother. The PSAI includes questions aimed at assessing the child's use of sex-typical games and toys and engagement in sex-typical activities. Results indicate that prenatal stress did not affect PSAI scores of boys, but it did affect those of girls. In particular, the more prenatal stress, the more masculinised the girl was according to the PSAI scores. This result is also in accordance with the trend for non-heterosexuality found in prenatally stressed females by Bailey *et al.* (1991) that I mentioned above.

Interestingly, girls in the study by Hines *et al.* also showed a greater tendency towards masculine behaviour if they had older brothers, or older or more educated mothers. Girls with mothers who were more timid or had a smaller social network showed a greater tendency towards feminine gender role behaviour. However, when those variables were controlled, prenatal stress was still able to predict the PSAI score, notwithstanding the fact that the additional variables were important. The work of Hines *et al.* therefore suggests that a suite of ontogenetic factors contribute to the development of 'gender atypical' behaviour, that certainly include both prenatal factors and postnatal social factors involving potential stress effects, but also learning. Unfortunately the study was limited to a follow-up until the age of 3.5 years only; a longer-term study will be needed to see whether any of those children develops a homosexual sexual orientation.

Although Dörner's original model emphasised prenatal stress, early postnatal stresses may also affect the development of some areas of the brain that control sexual behaviour. For instance, a study that may suggest a role of postnatal stress during early development in the causation of a homosexual sexual orientation is that carried out by Frisch & Hviid (2006) using a large dataset of marriages recorded in the Danish Civil Registration System. In Denmark legal marriages can be either heterosexual or homosexual. Increased likelihood of a homosexual marriage was recorded by Frisch & Hviid (2006) among: (a) men with unknown father, (b) women who lost their mother at a young age, and (c) men and women who came from households where parents divorced a relatively short period of time after marrying; all circumstances that might have been the cause of significant stress on the young boy or girl. Whether early postnatal stress may be associated with the tendency to marry, develop a homosexual orientation, or both is obviously something that such data cannot explain on their own.

Some evidence consistent with early postnatal stress effects on the ontogeny of homosexuality can also be found in studies of child negative relationships with parents, including situations where sexual abuse was involved. Early studies carried out by Liddicoat (1956), West (1959) and Westwood (1960) suggested that recalled tensions between son and father are a much more recurrent feature of the early experience of homosexuals, compared with heterosexuals, than attachment to mother (the latter is a classic prediction of psychoanalytic theory). Eva Bene (1965a, b) also provided evidence based on a retrospective study of the early experiences of male and female homosexuals with both mother and father. Bene's work, based on an English sample of 83 homosexual and 84 heterosexual males and 37 homosexual and 80 heterosexual females, strongly supports a postnatal association between homosexuality and family stress. The stress effect seems to derive from two different sources, however: (a) parental violence, mainly from father, but to a much lesser extent from mother as well, or (b) lack of security received from father owing to the latter's weak and insecure personality; such lack of paternal protection and reassurance may make the child more susceptible to the effect of external stressors. Broadly speaking, the same patterns are valid for male and female homosexuals. These data, however, cannot clarify the exact direction of causality: whether it is a homosexual orientation (or gender atypicality) in the offspring that elicits tensions within the family, or vice versa.

More recently, child sexual abuse has been associated with increased likelihood of development of a homosexual sexual orientation in both men and women (see Henderson et al. 2008 and references therein). McConaghy & Silove (1992) have also reported a relationship between male homosexuality and recalling of negative relationships with one or both parents. The pattern is maintained for the case of female homosexuals, but the effect is weaker than in males (McConaghy & Silove 1992). The response of children to an environment of abuse and stress can be variable, however, with individuals growing up developing sexual orientations that may range from heterosexuality to homosexuality and also asexuality (Henderson et al. 2008). Individual differences in the mechanisms to cope with stress (i.e. *developmental stability* or resilience, and degree of *canalisation* of sexual orientation) may explain this variability. Again, care should be taken in the interpretation of reports of sexual abuse suffered by homosexuals when they were young, to determine whether the abuse could have caused or whether it was a consequence of 'gender non-conforming' behaviours in the child (see, for example, Harry 1989). However, even in a context where gender role 'non-conformity' in childhood is the cause, rather than the consequence of stress, what such stress may do in those children is to tip the balance towards the development of homosexuality later in life, at least in some individuals (remember the case of *unidirectional plasticity* that I mentioned at the beginning of this chapter). This issue will be further discussed in the 'Do "sissy" boys and "tomboy" girls become adult homosexuals?' section below.

From an adaptive perspective, the development of a homosexual orientation on the part of the individual who experienced significant stress during specific periods of his/her early life can be seen as a coping strategy to increase survival under stressful environmental circumstances, but also as a strategy to increase inclusive fitness if homosexuality is associated with increased tendency to cooperate. In this context homosexuality would be clearly adaptive. The view of homosexuality as an adaptive strategy to cope under stress is antagonistic to an alternative view that would interpret homosexuality as pathology (see the 'Is homosexuality a pathology?' section of Chapter 10 for my critical views on this issue). Resilient individuals, on the other hand, may develop a heterosexual orientation in adulthood in spite of suffering abuse or other stresses as children, whereas individuals who are non-resilient and who also lack coping mechanisms may develop asexuality. We will see in Chapter 8 how *Reproductive Skew* theory provides a general evolutionary framework for the understanding of homosexual development in some members of a family group, as one potential outcome of interactions with parents and/or siblings that is meaningful in evolutionary perspective.

On the other hand, Kaiser & Sachser (2001) report that sons of socially stressed mothers display delayed behavioural development (e.g. play behaviour typical of juveniles) compared with sons of mothers living in stable social conditions. I will return to the issue of infantilism in the last section of this chapter where I explore the potential role of heterochrony, neoteny in particular, in the evolution of homosexual behaviour.

Given the potential social implications that studies on homosexuality and stress have, it is no wonder that such research tends to elicit strong reactions from one or the other member of the community. It is also perfectly understandable and indeed welcome that close scrutiny of the scientific quality of research should be especially elevated for those issues that may have a significant social impact. Yet, knowing is better than not knowing, and if a reality does not fit our expectations it won't disappear just by turning our sight away. In any event, if and when prenatal and/or postnatal mechanisms involving stress that may result in the development of homosexuality are confirmed in humans, theoretical considerations suggest that such mechanisms could be adaptations evolved in a social environment. Further support for homosexuality and bisexuality as social adaptations will be provided in primates and other taxa throughout this book.

Adrenarche and the development of homosexuality

So far in this section, I have reviewed the experimental evidence for the development of same-sex sexuality based on prenatal and early postnatal effects of stress. In humans, however, sexual maturation is extended over various years and undergoes further changes that include the onset of ovulation in females and production of spermatozoa in males, a process known as gonadarche (spermarche for boys and menarche for girls) that is manifested at puberty between the ages of 9 and 15 years (Herdt & McClintock 2000). Heterosexual males are well known to experience a later onset of puberty than heterosexual females (see, for example, Bogaert *et al*. 2002). Sexual attraction develops slowly before puberty and by the age of about 10 years boys and girls have already experienced their first sexual attraction (McClintock & Herdt 1996). At 10, or before, the gonads are usually not fully mature; however, what does mature at that age are the adrenal glands, hence the naming of adrenarche for that stage of development (McClintock & Herdt 1996; Havelock *et al*. 2004).

Adrenarche is a developmental process that is mainly confined to the Hominoidea (Havelock *et al*. 2004). Starting at about six years old, and therefore well before gonadarche, the adrenals begin to develop and by the age of 10 their steroid secretory activity is in full swing. One of the main steroids secreted by the adrenals is dehydroepiandrosterone (see, for example, Herdt & McClintock 2000), which is a precursor of both oestrogens and testosterone. Thyrotropin and cortisol are also secreted during this period (Ponton & Judice 2004). From this we may predict that stressful events that enhance the secretory activity of the adrenal glands may result in the early release of steroid hormones into the bloodstream that, in turn, could promote an early surge of sexual activity (see Ponton & Judice 2004 and references therein). Moreover, stress at around age 10 that increases the secretory activity of the adrenals may also affect the development of brain centres that control sexual behaviour and orientation and that are sensitive to steroids, at a time when sexual attraction is developing.

Adrenarche, therefore, may be at the centre of a potential mechanism of prepubertal development of homosexuality in humans. Within this mechanism, changes in brain architecture may be modulated by stressful social experiences for instance, thus producing ontogenies leading to homosexuality. However, a pre- or perinatally developed homosexual orientation may also be the cause, not a consequence, of social stress at adrenarche, at least in societies where homosexuals are not tolerated.

The two potential mechanisms of stress action postulated here (pre/perinatal and at adrenarche) are independent and, in theory, they may also act in tandem. For instance, in cases where homosexuality may be mainly caused by prenatal or perinatal processes, stress at adrenarche may affect the probability that a homosexual phenotype could further develop into a transsexual. Alternatively, pre- and perinatal processes affecting gender role (e.g. development of femininity in young boys) may pave the way, in some individuals, to a further development into a homosexual as a result of social stress experienced during adrenarche.

High endocrine activity of adrenals during adrenarche is also associated with precocious gonadarche, an association that could arise through the same molecular signal triggering both processes, whereas a direct interference between the two seems unlikely (see, for example, Reiter & Grumbach 1982).

Alfred Kinsey and collaborators (1948, 1953) had already indicated a tendency for homosexual males to experience an early puberty and also to manifest an earlier interest in sex than heterosexual males. Kinsey also noticed that homosexual women had sexual experiences at earlier ages than heterosexual women. Early onset of puberty was also reported by Bogaert & Blanchard (1996) and Bogaert et al. (2002) among homosexual men compared with heterosexual men, whereas Tsoi (1990) noticed an earlier psychosexual development in male transsexuals. Savin-Williams & Ream (2006) provide evidence that, at least in the women included in their study, homosexuals have an earlier onset of menarche than heterosexuals. Age at sexual maturity was not significantly different between homosexual and heterosexual men in their sample. The above evidence is broadly consistent with an association between the development of homosexuality and an earlier onset of sexual functions.

The earlier onset of menarche found in homosexual women by Savin-Williams & Ream (2006) was not confirmed by Bogaert (1998), Tenhula & Bailey (1998) and also Ostovich & Sabini (2005), but the latter did confirm Kinsey's findings that lesbians have sexual experiences at earlier ages than heterosexuals. The same authors could not find any difference between gay men and heterosexual men, however. Also, Meyer-Bahlburg et al. (1985) compared women who had an 'idiopathic precocious puberty' with a sample of controls and concluded that the two groups did not differ in sexual orientation when they were interviewed at an age of between 13 and 20 years.

Tsoi (1990) described an earlier psychosexual development in transsexual than heterosexual women, but menarche in transsexuals occurred 2 years later than in heterosexuals. Lesbians and bisexual women were also found to have their first sexual experience at younger ages than heterosexual women in an English sample studied by Bogaert & Friesen (2002). The trend was similar, although marginally not significant ($p = 0.09$), for men: homosexuals tended to have their first sexual experience at a younger age than heterosexuals.

In sum, heterosexual males have a later puberty compared with heterosexual females. Also, adrenarche occurs prior to gonadarche. If gonadal steroids are involved in the mechanisms of development of homosexuality around puberty and homosexual males are feminised, then in that case we would expect homosexual males to have an earlier puberty compared with heterosexuals. This is consistent with the results obtained by various authors. The results for homosexual females, however, were mixed, with some suggesting an earlier onset of menarche in homosexual females too. That is, although age of attainment of puberty does correspond to a model of feminised homosexual males (*sexual inversion*), the early puberty of homosexual females follows a *hyperfeminisation* pattern. But in fact early puberty in **both** homosexual males and females and also their precocious interest in sex are more simply consistent with potential effects of stress on adrenarche and then on gonadarche in both sexes via some common, but still hypothetical molecular trigger. By affecting adrenarche, stress may accelerate the expression of sexual behaviour and attraction, and by independently affecting gonadarche it may produce precocious puberty. Precocious puberty in turn may be associated with decreased fertility, at least in some individuals (see, for example, Ibáñez et al. 1999). Which molecule/s released under stress by which organ/s may start such a process? Opioids or some neurotransmitter, for instance, are potential candidates (see, for example, Reiter & Grumbach 1982).

All this suggests that stress and neuroendocrine events occurring just prior to puberty may provide, at least in principle, an additional mechanism to either explain the development of homosexuality in some individuals or perhaps explain the transition from homosexuality to transsexuality in those individuals who have already a homosexual predisposition due to pre- or perinatal processes. The variability in the available data, however, is considerable, suggesting that at best this is only one of the many proximate mechanisms that could explain homosexuality in some individuals.

From an adaptive point of view, what sense does it make to increase sexual activity, start it earlier in life, but direct it to a member of the same sex in response to a situation of elevated stress, social stress in particular, during the period before and around puberty? Of course, the whole process may not be a result of an adaptation at all, but we will see in Chapter 8 that reproductive skew is a common outcome of social interactions within groups. Through aggressive interactions dominant members of the group may sometimes achieve lower reproductive activity in subordinates (e.g. younger members of the group) including, in some extreme cases, reproductive suppression that could be mediated by opioids (see, for example, Bribiescas 2001). In social species, however, adaptations are also expected to evolve in which subordinates may resist attempts by dominants to reproductively suppress them. That is, they are expected to evolve coping strategies to withstand the stresses of sociality. Homosexuality may be one of the possible outcomes of coping under stress: survival may increase in homosexuals owing to decreased competition for reproduction with other members of the group. Same-sex sexual behaviour could then be used to cement social bonds that may also help in survival. The homosexual may still achieve some reproductive success at some point in time if he or she is actually bisexual, and, in groups that also contain relatives, the homosexual may achieve some indirect fitness gains by helping close relatives if homosexuality is associated with release of resources and/or cooperative behaviours.

Fluctuating asymmetry

One currently active line of research, regarding the potential impact of prenatal stress on development of homosexuality, is the study of fluctuating asymmetry (FA). During the early ontogeny of organisms, bilaterally symmetrical structures may undergo processes that alter such symmetry. Small fluctuations away from symmetry could be caused by the effects of environmental stressors on development (see Leamy & Klingenberg 2005 for a recent review) that alter the patterns of cell growth, differentiation and division. Therefore FA measurements could be used as proxy variables to estimate developmental stress. If homosexuality is a result of early ontogenetic processes affected by stress, then it is expected that degree of homosexuality and FA values be positively correlated.

In a recent special issue of the *Archives of Sexual Behavior* dedicated to *Biological research on sex-dimorphic behavior and sexual orientation* (Zucker 2008a), Hall & Schaeff (2008) and Miller et al. (2008) report on two studies of the association between FA and sexual orientation carried out in the USA. The study of Hall & Schaeff (2008) included 97 heterosexual adult females, 59 heterosexual males, 75 homosexual females and 57 homosexual males. What they found was that homosexual men had higher values of FA than heterosexual men in four of the eight traits measured, and also in a measurement that was a composite of various traits. A similar result was obtained in the case of women: homosexuals had higher FA values in five of the eight traits and in the composite. Miller *et al.* (2008) obtained similar results after studying 51 homosexual men, 48 heterosexual men, 27 homosexual women and 41 heterosexual women: ear breadth FA and the values of two composite indices of FA increased with the values of Kinsey scores in men. That is, values of FA are positively correlated with homosexuality. In the case of women, Kinsey scores were positively correlated with FA in the fourth digit. In these studies, although not all markers showed an association between FA and sexual orientation, those that did tended to support the hypothesis that developmental stress may be a causative factor of same-sex sexual orientation. This conclusion, however, has recently been challenged by Martin *et al.* (2008), who described lower values of FA in homosexuals than heterosexuals.

Clearly, more research is needed before we can have a better idea of the potential effects of prenatal and postnatal stress on male and female homosexuality in humans, but we can also safely conclude, from research already carried out, that stress experienced during prenatal development or during the early postnatal development, or perhaps even at ages just prior to puberty, is an important candidate cause for the development of a homosexual orientation, especially through the mediating effects of adrenal steroids, corticosteroids and catecholamines, but perhaps other molecules as well such as opioids and neurotransmitters. For instance, Roper (1996) has proposed an alternative pathway linking perinatal and early postnatal stress with development of homosexuality where stress favours the release of endogenous opioids or serotonin (or both) into circulation. Increased circulating levels of opioids and/or serotonin, in turn, may increase production of prolactin by the anterior pituitary in infants (see, for example, Guyda & Friesen 1973 for human infant production of prolactin) that finally may depress testosterone production by the developing gonads thus leading to feminisation of the male brain. In this regard, more studies of sexual orientation development in a variety of stressful prenatal and postnatal conditions will be clearly welcome.

Developmental effects of endocrine-disrupting environmental chemicals

Many compounds, known as xenobiotics, that we ingest or somehow absorb into our body from the external environment can have a direct effect on our physiology, altering normal biological processes. Some are compounds with the capacity to disrupt endocrine functions and they are therefore collectively known as *endocrine-disrupting chemicals* (EDCs) (Tchernitchin & Tchernitchin 1992; Colborn *et al.* 1993; Ottinger *et al.* 2005; Tabb & Blumberg 2006; Crews & McLachlan 2006). When EDCs ingested by a fertilised female are able to reach the developing embryo (before laying in birds and before parturition in mammals), or when EDCs are ingested by the young through their food after birth, they could potentially interfere with a series of physiological processes, leading to alternative ontogenies. Table 4.3 lists some EDCs and their effects on the development of sexual behaviour and reproductive tissues in birds and mammals. Not all compounds affect male and female sexual behaviour development equally. Moreover, there is a tendency for the EDCs studied to de-masculinise males, but they could also feminise males or masculinise females (Table 4.3).

EDCs can exert their action by modulating (a) steroid hormone metabolism, (b) cell nuclear receptor coactivators, (c) cell nuclear receptor degradation, (d) hormone receptor activity, or (e) DNA methylation in the male germline; in the latter case the EDC could be exerting an intergenerational effect (Tabb & Blumberg 2006; see also Crews & McLachlan 2006 for a work emphasising DNA methylation effects). Among the specific actions exerted by EDCs, oestrogenic and antiandrogenic activities are far more common than antioestrogenic or androgenic functions (McLachlan

2001). McLachlan (2001) lists several EDCs that have oestrogenic activity such as environmental 17β-oestradiol, diethylstilboestrol (DES), ethynyl oestradiol, the fungally originating zearalenone, pollutants such as DDT, PCB, bisphenol A, nonylphenol, kepone and plant products such as the isoflavone genistein, the flavone luteolin, reveratrol and coumestrol. Compounds that have antiandrogenic activity include pesticides such as fenitrothion, linuron and vinclozolin. The source of environmental androgens that could act as EDCs, such as testosterone, can be most varied and not necessarily related to urban or industrial pollution. For instance, the plant steroid stigmasterol can be metabolised into testosterone and released into aquatic environments by bacteria growing on decaying plant material (McLachlan 2001).

The mechanism of action of EDCs mostly involves the same cell receptors that recognise and bind endogenous hormones. Xenobiotics may either compete with the endogenous hormones for binding sites on the receptors or they may even bind some additional receptors on the cell membrane, e.g. the SXR (steroid/xenobiotic) receptor (McLachlan 2001). Steroid receptors could be targeted in this manner by many of those environmental chemicals, including heavy metals such as cadmium, which can bind to the oestrogen receptor α (McLachlan 2001). Below I review studies done on one EDC, diethylstilboestrol, that has been implicated in the development of same-sex sexual behaviour and orientation in humans and birds.

Diethylstilboestrol (DES)

Diethylstilboestrol is a synthetic non-steroidal oestrogen that has been used since its development in 1938 and until as recently as the early 1980s, to treat a series of reproductive conditions in women, including the prevention of miscarriages (Newbold 1993; Kerlin 2005). Its use, however, has since been banned owing to the discovery of carcinogenic and teratogenic effects on embryos. Moreover, DES could have affected both young males and females postnatally through ingestion with food, as it was also used to stimulate growth in cattle (Kerlin 2005). DES has been associated with a series of developmental syndromes in both sexes, including psychosexual effects.

Ehrhardt *et al.* (1985) carried out a study to determine whether there was any association between DES exposure in utero and sexual orientation in women. Their work found higher rates of bisexuality and homosexuality in women who were

Table 4.3. Effects of prenatal action of environmental EDCs on development of sexual behaviour

Compound	Effect	Reference
Diethylstilboestrol	Sexual orientation in human	Ehrhardt *et al.* 1985; Kerlin 2005
	De-masculinisation in male Japanese quail	Viglietti-Panzica *et al.* 2005
Phenobarbital	Feminisation of male hamster behaviour	Clemens *et al.* 1979
Nicotine	De-masculinisation of mounting in male rat	Segarra & Strand 1989
Polychlorinated biphenols (PLCBs)	Possible human female masculinisation	Sandberg *et al.* 2003
	Impaired reproductive behaviour in Japanese quail	Ottinger *et al.* 2005
Vinclozolin	Reduced male sexual behaviour in Japanese quail	McGary *et al.* 2001
p,p'-DDE	De-masculinisation of male Japanese quail	Ottinger *et al.* 2005
DDT	Feminisation of reproductive tissue in male California gull	Fry & Toone 1981
Fadrozole	Masculinisation of reproductive tissues in chicken and turkey	Sanderson 2006

DES-exposed in utero compared with two kinds of control. Although this result is highly suggestive of a developmental effect of DES on sexual orientation, the sample of women studied by Ehrhardt *et al.* (1985) also differed in aspects of their ethnic background (66.7% Jewish in the DES sample vs. 43.3% Jewish and 43.3% Roman Catholic in non-DES sample) that could have biased the likelihood of developing a non-heterosexual sexual orientation regardless of prenatal DES exposure.

Kerlin (2005) studied the contribution of 'DES-sons' (i.e. men suffering various developmental complications due to in utero exposure to DES) to a website of the DES Sons International Network set up for the purpose of his project. What Kerlin (2005) found was that 'DES-sons' reported a high prevalence of male-to-female transsexual orientation and also an elevated frequency of 'gender dysphoria', defined by John Money (1988: 201) as: 'the state, as subjectively experienced, of incongruity between the genital anatomy and the gender-identity/role (G-I/R)'.

In a retrospective study where women with either homosexual, bisexual or heterosexual offspring were asked to recall the use of a series of drugs during their pregnancy, Ellis & Hellberg (2005) reported that 5 of the 19 drugs considered showed a significant association with sexual orientation of offspring:

(a) Mothers of homosexual or the combination of homosexual/bisexual males took significantly less anti-nausea and vomiting medicine than mothers of heterosexual males; however, they took more diet pills and also more gamma globulin.
(b) Mothers of homosexual or the combination of homosexual/bisexual females took more DES than mothers of heterosexual females. They also took more diet pills and more synthetic thyroid medication. In this study, then, DES is associated with development of same-sex sexuality in women. However, in this case the effect of DES can only explain 3.5% of homosexual development in women as only 4 out of 114 mothers of homosexual females took DES during pregnancy.

Ellis & Hellberg (2005) carried out further analyses of their dataset in order to study the effects of the various compounds while controlling for maternal age, maternal education and self-rated ability to recall events that occurred during pregnancy. Once those factors were controlled in a logistic multiple regression, some new patterns emerged, with mothers who took prednisone, an adrenocortical steroid similar to the stress hormone cortisol, having more male homosexual offspring than mothers who did not take prednisone. For female offspring the patterns remained unchanged as far as the effects of diet pill use and synthetic thyroid medication were concerned, but the effect of DES became insignificant. Additional analyses suggested that the effect of those compounds on the development of daughters' homosexuality seemed to occur during the second and third months of pregnancy. Therefore, this work suggests that early development of both male and female homosexuality is sensitive to drug consumption by the mother, especially those that have a link with adrenal function such as prednisone and also thyroidal hormones that can affect the adrenal production of catecholamines. The specific effect of DES on female homosexuality, however, seems to be less strong than initially suggested.

The above effects of DES on homosexuality development could not be replicated by Titus-Ernstoff *et al.* (2003) in a large study of 5600 women and 2600 men that were DES-exposed in utero, although they did find a small trend for DES-exposed women to be more left-handed than controls.

Prenatal effects of DES have also been investigated in other species. Viglietti-Panzica *et al.* (2005) carried out a study of the effect of DES on development of Japanese quail sexual behaviour. After incubating male embryos that received either one of two doses of oestradiol benzoate (25 μg EB/50 μl sesame oil or 10 μg EB/50 μl sesame oil), or 700 μg of DES/50 μl sesame oil, or just 50 μl sesame oil on day three of incubation, they tested the individuals, once they reached sexual maturity, for sexual activity with a sexually receptive female. Both EB and DES prenatal treatments resulted in de-masculinised (i.e. sexually unmotivated by the receptive female) males. Unfortunately, Viglietti-Panzica *et al.* (2005) did not test the males for female-like behaviour in the presence of a sexually active male; neither did they expose them to the simultaneous choice between a sexually motivated male and a female to assess sexual orientation.

In sum, research on the potential effects of DES on development of a homosexual sexual orientation has produced results that remain somewhat ambiguous. A recent study of the effect of progesterone ingested by pregnant mothers through medication, however, has reported some effects on offspring sexual orientation: fetal exposure to progesterone increased the likelihood of homosexuality in offspring assessed at an age of around 23 years (Reinisch *et al.* 2008), and I have already mentioned above the work of Ellis & Hellberg (2005) suggesting a prenatal effect of prednisone and synthetic thyroid medication on the development of homosexuality.

Early social experience and learning

As soon as a bird hatches or a mammal is born, it leaves the maternally controlled environment that contributed to prenatal ontogeny. The postnatal period of development can be, and in most cases is, also affected by maternal influences that range from nutritional (e.g. food brought to the nestling by the mother, or the milk fed to young mammals) to psychological (e.g. social learning), but the postnatal development is also subject to the influences of many other individuals, both conspecifics, especially in the case of social species, and heterospecifics. The developing central nervous system of the newborn will therefore be subject to the action of a series of external influences that will affect its learning and behavioural development in a manner that varies from species to species and also from individual to individual within a species.

Learning is one of the processes whereby the organism develops behavioural patterns. The specific mechanisms through which this is achieved vary from Pavlovian classical conditioning, to Skinnerian instrumental or operant conditioning, but also imprinting, a learning process that occurs during specific sensitive periods in ontogeny (Lorenz 1935; Bateson & Hinde 1987; see a recent review in Hogan & Bolhuis 2005) and the inferential learning stressed by cognitive developmental psychology (Kohlberg 1966; Bandura 1989). In this section I will mainly focus on learning mechanisms that involve interactions with conspecifics (i.e. social learning) and will critically evaluate the potential effects of learning on the early ontogeny of same-sex sexual behaviour.

Juvenile sex play and development of sexual orientation

During early postnatal life, young vertebrates may engage in play activities with conspecifics (i.e. social play) that may include sexual behaviours. Such behaviours are precursors of the adult mating repertoire. Through these play activities, juveniles will supposedly improve their skills in mating, but also use those interactions for socio-sexual functions such as an early establishment of dominance hierarchies or affiliative rapports with other conspecifics in some species. From a cognitive perspective, Allen & Bekoff (1997) have suggested that social play may also afford an opportunity to learn the difference between perception and reality, at least in some taxa, as the specific clues associated with play behaviour may be seen as indicators of the fact that the behaviour is a pretence (e.g. pretence aggression, pretence sexual intercourse).

Diamond & Bond (2003) have recently published a comparative analysis of social play in birds. The first striking finding in Diamond & Bond's (2003) work is that social play is extremely uncommon in birds compared with mammals, being restricted to mainly the Psittacidae, Bucerotidae, Bucorvidae and a few genera of Passeriformes. The association of social play with those avian taxa seems to be related to the possession of traits such as altriciality, sociality and a larger brain size (see Diamond &

Bond 2003 and references therein). To these traits, Diamond and Bond add a delayed age of attainment of sexual maturity among species that display complex social play, a finding that is consistent with similar studies carried out in mammals, where duration of the juvenile period of development is associated with greater play complexity (Joffe 1997; Pellis & Iwaniuk 2000). I will return to this issue when neotenic effects are discussed later in this chapter. This asymmetry between mammals and birds in juvenile social play also coincides with a lower prevalence of homosexual mounting recorded in adult birds compared with mammals (see Chapter 3). Is this association between juvenile social play and adult homosexual behaviour a result of a broad evolutionary relationship between play and development of sexual behaviour, or are they specifically linked causally through learning at the ontogenetic level? That is, would same-sex play mounting at young ages directly lead to the development of homosexual behaviour and eventually to a homosexual orientation in adults?

In birds, learning during early ontogeny may affect sexual orientation, but not necessarily in a permanent fashion. In an experimental study carried out on zebra finches (*Taenopygia guttata*), Adkins-Regan & Krakauer (2000) showed that males that grew in a colony where all adult males had been removed before the study subjects reached their eighth day of post-hatching life showed greater preference for pairing with same-sex conspecifics, as measured by courtship behaviours and consortship, than control birds did, but this preference was not long-lasting. Learning mechanisms might have been involved in this phenomenon if males require the presence of both adult males and females during post-hatching development to initially recognise the 'correct' model for a future sexual partner in reproduction. An early learning component has also been suggested in the development of sexual orientation in Japanese quails (*Coturnix coturnix japonica*) (Nash & Domjan 1991).

Among mammals juvenile sexual play, including same-sex sexual play, is widespread across taxa and is more common in males than females (Tanner *et al.* 2007). However, I will argue here that such early same-sex sexual experiences do not necessarily lead to the development of a homosexual sexual orientation in adults; instead, they seem to facilitate the development of sexual behaviour in general, with most individuals finally developing a heterosexual orientation.

In bottlenose dolphins (*Tursiops* spp.) homosexual activities, especially in the context of play, are more common in juveniles than in adults. Young male calves may engage in homosexual activities that may have socio-sexual functions such as the establishment and reinforcement of coalitions and affiliative bonds (Mann 2006; see also Chapter 7). Among Atlantic walruses (*Odobenus rosmarus rosmarus*), male juveniles have also been reported to mount each other (Sjare & Stirling 1996), while American black bear (*Ursus americanus*) cubs can also engage in sexual play occasionally (Henry & Herrero 1974), with males play-mounting both male and female littermates.

Juvenile same-sex sexual play has also been widely reported in ungulates. In the domestic sheep (*Ovis aries*), sexual play among males occurs from as early as days 2 and 3 of postnatal life and it extends for the whole of the suckling period, during which males mount more actively other lambs of both sexes than females do (Orgeur & Signoret 1984). Orgeur & Signoret (1984) also noted that the frequency of sex-play among lambs reached a maximum before the peak in postnatal circulating testosterone occurs, suggesting that the behaviour, if influenced by testosterone, is probably a result of prenatal organisational effects of the hormone on brain development. The prenatal testosterone effect hypothesis was experimentally tested by Orgeur (1995), who administered an 800 mg testosterone implant to ewes on day 50 of pregnancy. At weeks 4–8 of postnatal life both male and female lambs were observed. Figure 4.2 clearly shows that although the implant did not have a significant effect on male lamb sexual play, it did have a masculinising effect on female lambs, thus suggesting that development of juvenile sexual play is controlled by prenatal testosterone (or one of its metabolites) in this species. Testosterone implants did not have the same effect in males as in females, simply because male lambs are already exposed to high levels of androgens in utero. Sexual behaviour development in O. aries, however, is resilient to changes in social conditions of rearing, as social deprivation during infancy was associated with only minor and transitory effects on mounting activity (Orgeur & Signoret 1984). To complicate the matter further, Zenchak et al. (1981) showed that rams reared in all-male groups developed two kinds of phenotype: those displaying normal levels of heterosexual performance and those who showed little sexual interest in an oestrous female. Indeed, the latter preferred other rams as sexual partners. Same-sex mounting normally occurs in rams in the context of dominance relationships, as it does in other Caprinae such as *Capra hircus* (Orgeur et al. 1990), but some of the rams in the study by Zenchak et al. courted other males only, whether oestrous females were present in the pen or not. That is, those rams had a preponderant homosexual orientation. The potential mechanisms explaining this phenomenon will be addressed in detail in Chapter 5; what I want to stress here is that, given that most males engage in same-sex mounting when they are lambs, the fact that male-oriented rams represented only a small fraction of all the rams reared in all-male groups strongly argues against a significant effect of early same-sex play-mounting experiences in the development of adult homosexuality in this species.

Play-mounting in juveniles has also been described in the pronghorn (*Antilocapra americana*) in both buck and doe fawns, with the former displaying the behaviour at higher frequency than the latter (Kitchen 1974). Frequency of same-sex mounting decreases in males and females as they mature. Muskox (*Ovibos muschatus*) calves also engage in same-sex mounting, which is usually resisted by the mounted individual (Reinhardt & Flood 1983). Mounting is indiscriminate with respect to the sex of the mountee in this species, and it is performed more frequently by male calves. When female calves mount other individuals, however, they do so more frequently with other females than with males (Reinhardt & Flood 1983). Among same-sex calves, mounter and mountee roles are frequently reversed, a common pattern in same-sex sexual behaviour at young ages.

Hemsworth et al. (1978) kept young male pigs (*Sus scrofa*) from 3 weeks of age in either a control monosexual group, or in social restriction (visual but not physical contact with other pigs), or isolation, or permanently in all-male groups from either 3 or 12 weeks of age, and observed sexual behaviour at 32–52 weeks of age. Although piglets reared in permanent all-male groups did perform homosexual mounting with anal intromission, they also developed heterosexual mounting, whereas piglets socially restricted or isolated did not perform as well in heterosexual copulations. This suggests that same-sex sexual experiences at young ages in pigs favour the development of a heterosexual sexual orientation at the adult age rather than an exclusive homosexual orientation.

Studies carried out in rodents have also shed a considerable amount of light on the early ontogeny of same-sex sexual behaviour. Early studies carried out by Beach (1942) indicated that *Rattus* males that underwent normal development until weaning (i.e. at 21 days of age) were not affected in their heterosexual sexual orientation by subsequent rearing in either isolation or all-male groups. However, if the social rearing environment is modified from the early days of postnatal life (e.g. from day 2 until day 94; Hård & Larsson 1968) so that males are either reared in solitude, or in an all-male group or in a heterosexual group, sexual behaviour undergoes significant developmental changes. Hård & Larsson (1968) described a higher frequency of heterosexual mounting in males reared in heterosexual groups than in males reared in monosexual groups, although the latter had higher mounting rates of females than males reared in isolation. However, as males became exposed to females after 95 days of age, those reared with other males tended to approach the heterosexual sexual performance of males reared in mixed-sex groups

Juvenile sex play and development

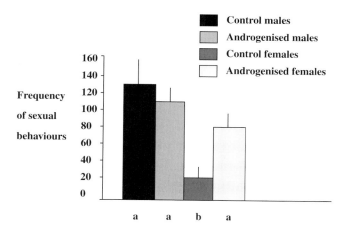

Figure 4.2. Mean frequency per individual per 15 h of male-like sexual play behaviour (± SD) recorded in male and female lambs that were born to either control ewes or ewes that were androgen-treated during pregnancy. Different letters indicate a statistically significant difference. Adapted from Orgeur (1995).

(Hård & Larsson 1968). Lack of experience with females during early development is easily overcome by male rats in terms of their sexual preference when they are adult (Vega Matuszczyk et al. 1994), provided that they come into contact with females at least before 16 months of age (Drori and Folman 1967), a common occurrence in natural conditions. Therefore the ontogeny of heterosexual behaviour seems to be quite resilient to the diversity of partners in early mounting experiences. Moreover, such mounting activities with other males do help in the subsequent adult development of heterosexual sexual behaviour.

Juvenile social play, including sexual play, is widespread among primates and has been recorded in all species studied so far (Lewis & Barton 2006). In a comparative analysis of 12 primate species, Lewis & Barton (2006) found a positive correlation between the size of the amygdala and the hypothalamus, two brain regions involved in sexual behaviour and orientation (see Chapter 5), and social play ($p = 0.005$ and $p = 0.01$, respectively), but not with non-social play ($p > 0.50$). This suggests a close causal connection between social play in the early stages of development and the development of sexual behaviour. Whether it is social play with same-sex peers, peers of the other sex or adults, especially the mother, it is likely to be variable between species. In *Macaca mulatta*, Goy et al. (1974) described a fundamental role of the mother in the development of proper mounting techniques in male infants, but experiences with both mother and peers are even better (Wallen et al. 1981). Studies of the ontogeny of sexual behaviour in macaques (or equally socially complex species), however, must take into account a variety of factors that can modulate the expression of same-sex sexual behaviour: from social conditions at rearing, to dominance relationships, to availability of mates at times of increased sexual activity, and establishment and duration of individual bonds associated with affiliative and cooperative relationships (see, for example, Goldfoot et al. 1984). These issues will be addressed in great detail in subsequent chapters.

Finally, I would like to mention work carried out on our closest living relatives: apes of the genus *Pan*. I will start with a recent comparative work published by De Lathouwers and van Elsacker (2006) on play-mounting in bonobo (*Pan paniscus*) and chimpanzee (*P. troglodytes*) infants. Although bonobos are well known for their frequent same-sex sexual interactions, but also for heterosexual sexual behaviour performed in socio-sexual contexts as we will see in Chapters 5, 8 and 9, the difference between the two species in the behaviour of young individuals seems to be less dramatic than first envisaged. In fact, at least in captivity, play-mounts among bonobo infants are not more frequent than play-mounts among chimpanzee infants (De Lathouwers & van Elsacker 2006). This suggests that the higher frequency of same-sex sexual behaviours observed in adult bonobos is not a direct consequence of a particularly elevated frequency of play-mounting among infants in this species as compared with chimpanzees. Chie Hashimoto (1997) carried out a detailed study of the ontogeny of sexual behaviour in bonobos in the wild. She studied a group at Wamba, Zaïre, composed of seven infants, five juveniles, two adolescents and 16 adults. The overall sex ratio was 1.1 M/F and it was 0.77 M/F among adults. Behavioural observations of the immature individuals indicated that they already start displaying genital contact activities when they are less than 1 year old. Most (78%) of the genital contact occurring between the immatures can be described as sexual play. Sexual play was more frequent in males than females and tended to be performed in the ventro-ventral more than the ventro-dorsal position. Although genital contact occurred in both M–M and F–F dyads, it was more frequent in M–M dyads. The frequency of F–F mounting increases in females as they mature, but the contexts in which same-sex sexual behaviour occurs are different at different ages: play among juveniles and tense situations among adults. This difference suggests that same-sex sexual behaviour is part of the normal heterosexual development of bonobos, as it is in other social mammals, but that the behaviour has been co-opted in this species for specific socio-sexual functions in adults.

From the available evidence of the relationship between juvenile play-mounting and adult sexual orientation in birds and non-human mammals, we can safely conclude that same-sex mounting among juveniles is a stage in the development of the adult heterosexual sexual behaviour, with young males being more frequently engaged in play-mounting than young females. Moreover, it seems unlikely that juvenile same-sex sexual play is a direct and specific cause of adult homosexual preferences in those species. Juvenile sexual play improves the neuromotor and perhaps other aspects of sexual behaviour (whether this will be homosexual or heterosexual in adults) but it does not determine sexual orientation. In the last section of this chapter I will suggest a link between juvenile sexual play and homosexuality through neoteny, but it should be borne in mind that this is an evolutionary, not an ontogenetic or developmental, link.

Juvenile sex play and homosexuality in humans

Juvenile sex play has been also studied in humans. When children of both sexes are reared together in a fully permissive environment they do engage in sex play activities. For instance, Fox (1962) mentioned the case of the Kiryat Yedidim Israeli kibbutz where both boys and girls engaged in sex play activities up until 12 years old, at which age girls became less interested in sexual behaviours performed with their male kibbutzniks, a pattern expected from the action of inbreeding avoidance mechanisms. A similar situation where relatively free sex play occurred in children of up to 11 years, in a social environment where adults were permissive of such behaviour, was also described in the Tallensi of northern Ghana and other native populations around the world (e.g. Trobriand Islanders, Pondo, Tikopia; see Fox 1962 for an early review). Similar trends are also observed nowadays in some European ethnic groups such as the Swedes (Larsson & Svedin 2001). Child sex play should therefore be regarded as a normal occurrence in human populations (Finkelhor 1980; Greenwald & Leitenberg 1989; Friedrich et al. 1991; Sandfort & Cohen-Kettenis 2000; Friedrich 2000; Larsson & Svedin 2002). Children tend to avoid sex play activities with each other only as a response to bans enforced by adults; in the Chiricahua Apaches and many industrialised Western societies, for instance (Fox 1962; Friedrich et al. 1991; Larsson & Svedin 2001).

Proper scientific studies of child sexual play have become available only recently, and they have mainly been motivated by issues of child sexual abuse (Finkelhor 1980; Larsson & Svedin 2001). Finkelhor (1980) interviewed 796 undergraduate students from the New England region of the USA, asking them about their sexual experiences during childhood. A total of 13% of respondents reported having had childhood sexual experiences with siblings (15% of females and 10% of males), although Finkelhor suspects that these values are underestimates. Most of those experiences were of the heterosexual kind, but 16% were M–M (brothers) and 10% were F–F (sisters). Finkelhor interprets the patterns found in a dual fashion: sexual play at very young ages may have a sexual developmental function, but it may also play a socio-sexual role (e.g. in the development of dominance relationships). In fact, respondents in Finkelhor's study were divided as to whether their childhood sex play experiences had been positive (30%) or negative (30%). Interestingly, a substantial 40% of respondents reported that their feelings regarding those experiences were neutral.

Greenwald & Leitenberg (1989) reported that in the USA pre-adolescent sex play behaviours are not only common, but, at least in boys, they are more frequent in a same-sex than in an heterosexual context (boys: 34% heterosexual and 52% homosexual; girls: 37% and 35%, respectively). Greenwald & Leitenberg (1989) also studied hetero- and homosexual child sex play among pre-adolescents in the USA and, following the responses to a questionnaire, they concluded that about 61% of their 19-year-old respondents recalled some kind of sexual experience before they turned 13. Moreover, 28% of them also reported same-sex sexual experiences in the pre-adolescent years, a figure that is almost one order of magnitude higher than the frequency of exclusive homosexuals found among adults in the USA. Although involvement in heterosexual sexual behaviours continues to increase with age, involvement in homosexual sexual behaviours decreases during adolescence (Leitenberg et al. 1989), again suggesting that homosexual sexual experiences at younger ages tend to be more a precursor to the development of heterosexuality than of homosexuality in most individuals. Both kinds of sexual play experiences (heterosexual and homosexual) had the same positive effect on the development of sexual arousal as the child matured (Leitenberg et al. 1989).

A prospective study carried out by Dickson et al. (2003) in New Zealand followed up about 1000 children from the age of 3 to 26 years. They found that 88% of men and 74.1% of women had always been heterosexual, whereas 10% of men and 24.6% of women showed some degree of bisexuality, with a smaller 0.6% of men and 0.0% of women that had always had a homosexual inclination. As expected from a model of ontogenetic development of same-sex sexual behaviour with highest preponderance at younger ages, as is found in other mammals, the percentage of people displaying same-sex sexual attraction decreased with age for bisexually oriented individuals. This suggests that whatever homosexual play experience the individuals had at young ages, it did not affect the development of heterosexual sexual preferences as an adult. However, the percentage of exclusive homosexuals showed a slight increase with age in the work by Dickson et al. (2003): 1.6% of men currently declaring themselves as exclusively homosexual vs. 0.6% declaring themselves as having ever been exclusively homosexual; for women the percentages are 0.8% current and 0.0% ever being exclusively homosexual. It is possible, however, that this trend may be also affected by recent cultural changes occurred in New Zealand society (Dickson et al. 2003).

Although child sex play is a common occurrence in modern Western societies, cross-cultural differences can also be detected. Friedrich (2000), for instance, has reported a greater frequency of sex play activities among Dutch than among American children. The same trend was observed for Swedish children, both boys and girls, compared with their American counterparts (Larsson et al. 2000). It is quite possible that these differences may be explained, at least in part, by adult permissiveness with regard to child sexual play.

Understanding the ontogeny of sexual behaviour in humans necessarily requires a cross-cultural, comparative approach for more reasons than one. Most trivially, if social influences are suspected to play a role in the early development of homosexuality, the specifics of those social interactions may vary from culture to culture and so may the contexts in which a homosexual orientation emerges in boys and girls. Göncü *et al.* (2000), for instance, show how social play in toddlers may differ in some respects between socio-cultural communities as diverse as urban USA or Turkey, a Mayan peasant community in Guatemala and a tribal community in India. Such cross-cultural studies, carried out with a common methodology and appropriate controls, are still in their infancy in the context of the early ontogeny of homosexual behaviour. How much and in what ways do the frequency and modality of sexual play vary among children across cultures? How do those differences and similarities in child sexual and other play correlate with the patterns of homosexual behaviour among adults? If the suggestion that sexual play at young ages is simply a mechanism to refine sexual behaviour and not a mechanism that, per se, can give rise to a homosexual individual is correct, we would expect that more comprehensive and detailed cross-cultural studies of infant same-sex sexual play will show no correlation between the frequency of such play and the frequency of exclusive homosexuals in the adult population. A further analysis of the role of early learning experiences in the development of a homosexual orientation in humans will be carried out in the section 'What is the role of learning in the early ontogeny of human homosexual orientation?' below. In the next section I review the empirical evidence for and against the link between 'gender-atypical' behaviour during childhood and sexual orientation in adulthood.

Do 'sissy' boys and 'tomboy' girls become adult homosexuals?

Weinrich (1985: 322) defines a 'sissy boy' as: 'a boy whose gender non-conformity (dressing in female clothing, desire to be a girl, friendship with girls, feminine role-playing or gesturing, and lack of interest in athletics) is persistent and clear-cut enough to cause adults to take notice'. Likewise, a 'tomboy' is a behaviourally masculinised girl. Early suggestions that children who are behaviourally *gender-atypical* (or *gender-nonconforming*) may, later in life, develop a homosexual sexual orientation were already present in the writings of Krafft-Ebing (1886). The idea was subsequently developed by Richard Green, who, after studying nine boys 'manifesting anomalous gender role development', pointed out that the boys' behaviour was coincident with the reported behaviour, at the same age, of adult male transsexuals (Green 1968). His subsequent studies led him to conclude that at least some 'sissy' boys and 'tomboy' girls may develop into adult homosexuals, although gender-atypical behaviour at young ages is a relatively better predictor of adult homosexuality in boys than in girls (Green 1979).

Early empirical work pointed to the potential role of postnatal stress in the development of sissiness in boys. For instance, Helen Koch's (1956) pioneering work suggested that sissiness in second-born sons increases as the gap between first- and second-born sons decreases, with competitive interactions between brothers also expected to increase when they are more similar in age (Sulloway 1996). Green *et al.* (1985) also suggested an association between sissy boys and a stressful family environment (see also Manosevitz 1970; Bogaert & Blanchard 1996; and Alanko *et al.* 2009b for additional evidence and comments).

Improving on the initial descriptive work carried out on limited sample sizes, Richard Green (1985) used a prospective approach to study sex atypicality in childhood, whereby subjects were evaluated when they were children and then again later during adolescence or in their young adult years. Green's (1985) study found that, of the boys showing cross-gender behaviour in childhood, 43% scored in the homosexual range of the *Kinsey Scale* (5–6) later in life, whereas 25% fell within a bisexual range of 2–4, and a sizeable 32% developed heterosexual sexual orientation (0–1). Two years later, he published his much discussed and oft-cited book *The 'Sissy Boy Syndrome' and the Development of Homosexuality*, in which he articulated his views that human male homosexuality develops from early ontogenetic stages in some individuals and that subsequent social experiences (including psychotherapies) are not always able to change sexual orientation in those same subjects.

Table 4.4 summarises the results of some of the works relating childhood behavioural gender atypicality and subsequent development of sexual orientation. Clearly, with the exception of Money & Russo (1979), the association of childhood gender non-conformity and either homosexuality or bisexuality is not perfect and it is variable between studies. Not all gender-atypical children develop a homosexual/bisexual orientation; in fact, between 6.3% and 62% of gender-non-conforming boys develop into heterosexuals. The same is true for 9.2%–44% of gender-non-conforming girls.

Alanko et al. (2009a) have recently published the results of a retrospective study carried out in Finland, focusing on recalled gender-atypical behaviour among male and female homosexual and heterosexual members of various monozygotic or dizygotic twin pairs. Their sample size included over 3000 participants, who were administered both the Recalled Childhood Gender Identity/Gender Role Questionnaire (Zucker et al. 2006) and the Sell Assessment of Sexual Orientation (Sell 1996). Recalled gender-atypical behaviour was positively correlated with adult homosexual behaviour in both men and women, but the relationship was stronger for men. These authors also found a larger positive correlation in both recalled gender-atypical behaviour and sexual orientation (0.56 and 0.50, respectively) for male monozygotic twins than for female monozygotic twins (0.53 and 0.47, respectively), but also a lower correlation among dizygotic twins (0.27 and 0.25, respectively, for males; and 0.07 and −0.01, respectively, for females). That is, what Alanko and collaborators' work suggests is not only that homosexuals are more likely to recall gender-atypical behaviour as children than heterosexuals, but that both recalled gender-atypical behaviour and sexual orientation display heritable genetic variability that is stronger in males than in females.

Prospective studies such as the one carried out by Richard Green, however, are better suited to

Table 4.4. Percentage of men and women who were gender-non-conforming in their behaviour as children, and their sexual orientation as adults

Heterosexual	Homosexual	Bisexual	Other[a]	Reference
Boys				
32	43	25	—	Green 1985
6.3–27.1	72.9–93.7	—	—	Zuger 1989
≤ 62.5	≥ 37.5	—	—	Lebowitz 1972
—	100	—	—	Money & Russo 1979
40	30	—	30	Davenport 1986
Girls				
44	24[b]		32	Drummond et al. 2008
9.2[c]	45.0	37.5	—	Safir et al. 2003
33.8[d]	42.5	43.8	—	Safir et al. 2003
56.9[e]	12.5	18.8	—	Safir et al. 2003

[a]Various patterns such as asexual or not interested in sex. [b]Homosexual and bisexual combined. [c]High-degree gender non-conformer. [d]Intermediate-degree gender non-conformer. [e]Gender conformer.

investigating developmental effects than retrospective studies. Davenport (1986) followed up 10 gender-non-conforming boys from the age of 2–6 years to 15–27 years. A total of 40% developed a heterosexual orientation; 30% were homosexual/transsexual; the remaining 30% were either masculine but not dating anyone, 'artistic but not effeminate' and heterosexual according to his parents, or heterosexual and dating, also according to the parents. Zuger (1989) studied 55 males for over 30 years who as boys displayed effeminate behaviour. A range of 72.9%–93.7% developed into homosexuals, whereas 6.3%–27.1% became heterosexuals.

Studies of female homosexuality also indicate a degree of childhood gender atypicality (Phillips & Over 1995), although Phillips & Over (1995) also found that some of the lesbians they interviewed did not differ from heterosexual women in recalled childhood gender conformity. In an Israeli sample of women, heterosexuals tended not to have been a 'tomboy' when a child (56.9%), but a sizeable 43% displayed either a high or an intermediate degree of tomboyism (Safir et al. 2003). The degree of tomboyism, however, was higher for bisexuals (81.3%) and for lesbians (87.5%). With regard to sex roles, 'tomboys' tended to develop into either androgynous or feminine women, but the same occurs with women who were not 'tomboys' as girls. Drummond et al. (2008) have recently reported the results of a prospective study carried out in Canada where 25 girls diagnosed with 'gender identity disorder' (GID) were followed from the age of initial diagnosis (3–13 years old) up to a variable age range of 15–36 years old. At follow-up, and in spite of their initial GID diagnosis, 88% of participants did not report any problem with their gender identity, with only 12% being classified as 'gender-dysphoric'. With regard to sexual orientation in behaviour, 44% were heterosexual, 24% were either bisexual or homosexual and 32% were asexual.

Some researchers have simultaneously studied both boys and girls by using the same methodology. Sandberg et al. (1993a) described a total of 22.8% of boys and 38.6% of girls displaying gender-atypical behaviours in their sample; these percentages are much higher than the percentages of adult exclusive homosexual men (about 3.1%; Laumann et al. 1994; Kendler et al. 2000) and exclusive homosexual women (about 2.0% on average; Laumann et al. 1994; Kendler et al. 2000), thus suggesting that child gender-atypical behaviour does not necessarily lead to the development of adult homosexuality.

Madeleine Wallien and Peggy Cohen-Kettenis (2008) have recently published a detailed prospective study of Dutch boys and girls identified as 'gender-dysphoric' at ages of between 5 and 12 years, who were followed up until they were 16–28 years old. Their work is a great improvement on previous prospective studies in that they do not treat the population of 'gender-dysphoric' children as homogeneous. By dividing their sample between those individuals who maintained their 'gender dysphoria' at follow-up ('persisters') and those who were no longer 'dysphoric' ('desisters') they were able to determine which one of the two kinds of 'gender-dysphoric' child was more likely to have developed a homosexual sexual orientation later in life. Overall, only 27% of the initially 'gender-dysphoric' children were still 'dysphoric' at follow up, again suggesting that the so-called 'gender dysphoria' is mainly part of the normal behavioural development of the young in our species. In a detailed comparison of sexual attraction, behaviour, fantasy and identity between 'persisters' and 'desisters' they found some very interesting patterns (see Table 4.5). In general, girls are characterised by all-or-none responses: all 'persisters' are homosexual in attraction, behaviour and fantasy, whereas all desisters are heterosexual for those three traits, but these results are based on extremely small sample sizes and therefore should be regarded as unreliable. Sexual identity, with a slightly larger sample size, shows some variability, with 'persister' girls divided into 88% homosexuals and 12% bisexuals; the three 'desisters' available were all heterosexual in their identity.

The sample sizes for boys are slightly larger. Among boys, the only trait that was associated with 100% of 'persisters' becoming homosexuals was sexual identity. For all the other traits, although a majority of between 83% and 92% was homosexual, a proportion of between 8% and 17% was heterosexual. The variability is even greater in the 'desister' group. This group is not homogeneously heterosexual, which means that homosexuality among males is not associated with gender role inversion. However, a variable percentage of 'desisters' does display heterosexual patterns (19%–44%), whereas among 'persister' males the vast majority shows homosexual patterns. This work clearly suggests that gender non-conformity in childhood is only predictive of homosexuality in a subset of males and females, and that more extreme cases of gender non-conformity are more likely to develop a homosexual orientation later in life. The work, however, remains inconclusive with regard to the exact mechanisms that can explain these results.

Although prospective studies are ideal for the investigation of the development of homosexuality, Rieger et al. (2008) rather cunningly managed to overcome the usual suspicion of bias in the recollection of past events in retrospective studies by means of analysing family videos of interviewees' childhoods and comparing them with their degree of 'gender non-conformity' in adulthood. The authors asked heterosexual and homosexual males and females to rate 21 childhood videos of currently homosexual males, 20 videos of homosexual females, 23 videos of heterosexual males and 26 of heterosexual females, on the basis of a unidimensional scale of more masculine (1) to more feminine (7) behaviours. The results of Rieger et al. (2008) indicate that childhood videos of homosexuals show children behaving in a more gender-non-conforming manner than videos of heterosexuals. Gender non-conformity becomes recognisable in homosexuals between the ages of 3 and 4 years. Interestingly, Rieger et al. (2008: 52) also state that 'past rejections from peers tended to correlate significantly with both childhood and adult gender non-conformity. If this peer rejection experienced by more sex-atypical targets was, in fact, elicited by their gender non-conformity, it would suggest that gender non-conformity persisted into adulthood despite past negative social interactions.' In fact, the result also lends itself to the alternative interpretation that it may well be the stress produced by peer rejection (see the above section on Stress Effects) that, from about the age of 3–4 years, led the gender-non-conforming child through a developmental pathway that eventually resulted in an adult who is both non-conforming in gender role and homosexual in orientation. Such a model would predict that among children who are gender-non-conforming the probability of developing a homosexual orientation in adulthood should be greater in those cases where the child suffered

Table 4.5. Percentage of participants rating themselves on sexual orientation (three dimensions) and sexual identity

Group	Attraction		Behaviour		Fantasy		Sexual identity	
	Boys	Girls	Boys	Girls	Boys	Girls	Boys	Girls
Desisters	$n = 25$	$n = 3$	$n = 13$	$n = 2$	$n = 16$	$n = 1$	$n = 18$	$n = 3$
Heterosexual	44	100	23	100	19	100	27	100
Bisexual	0	0	23	0	25	0	17	0
Homosexual	56	0	54	0	56	0	56	0
Persisters	$n = 12$	$n = 7$	$n = 6$	$n = 3$	$n = 5$	$n = 2$	$n = 9$	$n = 8$
Heterosexual	8	0	17	0	17	0	0	0
Bisexual	0	0	0	0	0	0	0	12
Homosexual	92	100	83	100	83	100	100	88

Adapted from Table 5 of Wallien & Cohen-Kettenis (2008).

specific social stresses, whereas that probability should be lower among those gender-non-conforming children who were socially accepted. In fact, as I have already mentioned, only a variable subsample of the boys and girls who are gender-non-conforming goes on to develop a homosexual sexual orientation in adulthood. On the other hand, those individuals who are subject to a more canalised development of homosexuality, owing to a closer genetic control of the trait, are likely to develop a homosexual sexual orientation under a wide umbrella of possible social environments.

Gender-atypical children may be subject to various social stresses coming from adults (including parents and teachers) and also peers (see, for example, Isay 1990, Landolt *et al.* 2004 and references therein). Such stresses may include sexual harassment, especially when the child is also considered handsome (see, for example, Zucker *et al.* 1993), but broader social stigmatisation coming from peers is even more common (Sandberg *et al.* 1993b), especially in the case of gender-non-conforming boys, whereas gender-non-conforming girls are relatively better tolerated (Zucker *et al.* 1997). Note that the potential for sexual harassment to contribute to the development of a homosexual orientation does not contradict the suggestion, made in the previous section, that sexual play at young ages seems to be a normal component of the behavioural repertoire of young primates. It is when the sexual experience occurs, not as a playful behaviour performed between peers, but as a traumatic or otherwise persistently stressful experience, that I suggest it could affect the development of sexual orientation in children. In my view research on the causative mechanisms that link gender non-conformity in childhood and the development of adult homosexuality should not ignore the potential role of social stress.

That early postnatal social stress may be a contributing factor in the development of homosexuality starting from childhood gender non-conformity is also suggested by the work of Carol Lynn Martin, who showed pervasive adult negative stereotypes about more feminised boys (C. Martin 1990, 1995). Whether, in their attempts to 'cure' the boy of his femininity, a process that may subject the child to significant stresses, adults are enhancing the chances that the child will develop a homosexual sexual orientation, is something that requires specific investigation. In a recent study, Alanko *et al.* (2009b) have also shown that childhood gender-atypical behaviour in both boys and girls is associated with a negative relationship between parent and child. In turn, a homosexual sexual orientation at adult ages is also associated with recalled negative and presumably stressful relationships with parents during childhood, although heterosexual adults also recall having experienced a stressful relationship with parents as a result of their childhood gender atypicality.

The results obtained from studies carried out in humans regarding the long-term effects of same-sex sexual play (see the previous section) and gender-non-conforming behaviours (this section) at young ages are consistent with our review of the effect of same-sex sexual play among young in other vertebrates: same-sex sexual activity among young social mammals is just part of the normal development of the individual. This leads to the conclusion that a child who may be predisposed to develop a same-sex sexual orientation may well engage in same-sex sexual play and also gender-non-conforming behaviours (see Bartlett & Vasey 2006; Bem 2008 provides a recent review), but not all children who engage in same-sex sexual play and gender-non-conforming behaviours develop a homosexual orientation when they become adult. In the words of Kenneth Zucker (2008a: 1362):

> ... it is apparent that there is no one 'natural history' for GID in children: some children show persistence in their gender dysphoria, whereas a large number show clear desistance. Some children differentiate into a homosexual orientation; and others into a heterosexual orientation.

Children who are gender-atypical and do develop into homosexuals when they become adult may do so due to genetic or prenatal developmental mechanisms, such as the ones I have mentioned in this chapter and Chapter 3, but postnatal effects such as those caused by social stress should not be dismissed.

What is the role of learning in the early ontogeny of human homosexual orientation?

Early reviews of the ontogenetic aspects of homosexuality in humans indicated the potential role played by the prenatal organisational actions of hormones and also by postnatal learning (Ehrhardt & Meyer-Bahlburg 1981). Not surprisingly, psychologists have emphasised associative learning processes in the development of homosexuality, whereby eroticisation of specific stimuli (e.g. fantasies involving same-sex individuals) could be reinforced by actual social interactions with individuals of the same sex, in Michael Storms' (1981: 340) words: 'erotic orientation is influenced by an interaction between sex drive development and social development during early adolescence'. In this model, masturbation is understood as a secondary reinforcing mechanism of the same-sex sexual partner preference. Which stimuli are eroticising, and which are not, is supposed to be culturally specific and be subject to social learning.

In his *Erotic Orientation Model*, Storms (1981) suggests that prolonged isosexual interactions (e.g. as they may occur in sexually segregated social groups) should promote same-sex bonding, leading to the development of a homosexual orientation, especially during adolescence, a period of high erotic development. Storms' (1981) model is falsified by the low frequency of homosexual males and females compared with heterosexuals even in societies where sexual segregation among adolescents (e.g. at school) is common (Kinsey *et al.* 1948; Peplau *et al.* 1998). The association of early sexual maturation with a homosexual orientation is real but Storms' (1981) conclusion that such an early onset of sexuality may lead to the development of same-sex sexual preference through associative learning mechanisms, especially in a sexually segregated social context, is not warranted (see Chapter 7). In fact, it is entirely possible that the causation is quite different. Individuals with a predisposition to develop a homosexual orientation because of genetic (see Chapter 3) or pre/perinatal or prepubertal (this chapter) effects might also have an elevated sexual motivation and early onset of puberty. A potential link between elevated sexual motivation, early onset of puberty and same-sex sexual orientation is provided by stress effects, as we have already seen, not by learning. Genes having pleiotropic effects could potentially also contribute to such an association.

This is not to say, however, that a homosexual orientation could not be 'fine-tuned' under the influence of learning or that classical or social conditioning are necessarily irrelevant in all cases. In fact they are not (Udry 2000; Woodson 2002), but conditioning *per se* seems an unlikely mechanism to explain the bulk of cases of the ontogeny of same-sex sexual behaviour in humans and other vertebrates. As we have seen, same-sex sexual behaviour is a normal occurrence among the young of many social vertebrates and we could expect that the development of a heterosexual orientation would be buffered against the effect of those early experiences.

Daryl Bem (1996; see also Bem 2008 for an update) produced a further theoretical elaboration, the *Exotic Becomes Erotic* (EBE) model based on the assumption that 'every child, [gender] conforming or nonconforming, experiences heightened, nonspecific autonomic arousal in the presence of peers from whom he or she feels different' (Bem 1996: 321). Therefore a 'sissy' boy who spends more time interacting with girls will develop a sexual attraction for boys, and vice versa for 'tomboy' girls, whereas boys and girls interacting with members of the same sex will consequently develop a heterosexual sexual orientation according to the EBE theory. The psychological mechanism implied by the EBE theory is obviously very different from the classical or social conditioning mechanisms mentioned above, as the stimulus provided by directly interacting with members of one sex is expected to predispose the individual to sexual orientation towards members of the other sex. Classical conditioning predicts a positive association between sexual segregation and development of homosexual orientation; the EBE model predicts that sexual segregation favours the development of a heterosexual sexual orientation, whereas homosexuality would be more likely to develop in conditions of sexual aggregation.

One of the problems with the EBE theory is that it already assumes a pre-existing bias on the part of some boys (i.e. those who are gender-non-conforming) to behave in a gender-atypical manner, and so it is unable to distinguish the specific contribution to the development of homosexual orientation made by the pre-existing bias (presumably resulting from genetic make-up, early neuroendocrine developmental processes, etc.) and by the subsequent social interactions with members of the same and the other sex. Would 'sissy' boys joining all-male groups in various activities subsequently develop a homosexual orientation regardless? Or would they develop a heterosexual orientation as expected from the EBE theory? In addition, the theory cannot explain the development of masculine male homosexuality, as it focuses only on feminine male homosexuality.

Another aspect of the EBE theory that in my view is unconvincingly developed is the transition from being bullied by members of the same sex (e.g. in the case of 'sissy' boys) to develop a sexual preference for the same kind of individual. Bem (1996) explains the transition through the *Opponent Process* theory (Solomon 1980) that posits that a stimulus causing an initially negative effect in an organism is counteracted by an internally generated positive effect; the specific example given by Bem (1996) is a stimulus causing pain being counteracted by the release of endorphins. When the negative effect diminishes, the individual may experience a euphoric after-effect. Repeatedly evoking the opponent process will cause, according to Bem (1996), a conditioned response in the organism. Therefore, a 'sissy' boy harassed by other boys may transform his initial fear into a conditioned arousal leading to sexual attraction towards boys.

Why the 'sissy' boy carrier of a developmental bias towards same-sex sexual partners would not simply retain his sense of hatred or even fear for heterosexual masculinised boys who mercilessly harassed him in his youth, and seek comfort in the company of like-minded 'sissy' boys, or homosexual boys more accepting of effeminate males, is something that I find hard to comprehend (see also Bos *et al.* 2008).

In fact, Bell & Weinberg (1978: 88) report of white homosexual men (WHM) and black homosexual men (BHM) interviewees who were 'asked what they thought they got out of their first affair, almost two-thirds of the WHMs and nearly half of the BHMs mentioned the warmth and love and understanding they had received ...', whereas in a study of bisexuals Weinberg *et al.* (1994b: 126) mention that

> The last question we asked about disclosure was: 'Based on your own experience, what advice would you give another bisexual?' Most frequently people suggested finding a support group first, especially with other like-minded people ... or searching out close friends and relatives who would be accepting.

Carbone (2008: 315–16) has recently provided additional evidence that 'ridicule and ostracism experienced by gay boys during their development may be experienced by some as a threat to their safety and therefore traumatic'; the effects of such stressful experiences on the gender-non-conforming boys was that they were 'left feeling confused and anxious about their desires ... They were left with the choice to either comply with the requests from their social environment to conform or risk being scorned, abandoned, and perhaps physically harmed'. Such traumatic experiences derived into an aetiology of Post-Traumatic Stress Disorder as the child became an adult, if the individual did not have adequate mechanisms for coping with the stress; but if the individual did enjoy the contact with a 'supportive peer network, family network, and teacher network' then the tendency to develop a more 'positive identity' increased. In particular, benefits were derived from specific contact with other gay youths, or gay teachers and family members (Carbone 2008). Moreover, systematic harassment from peers, especially if carried out from a young age, could lead to the development of asexuality, bisexuality or homosexuality through the postnatal stress route that was reviewed in this chapter.

Bem's (1996) theory has also been criticised by Peplau *et al.* (1998) on the grounds that many homosexual males and females did interact with same-sex peers when they were young. In addition, the tendency of children reared together not to become sexually aroused by each other in adulthood is more easily explained by inbreeding avoidance mechanisms, for instance, than by the EBE theory. Finally, Peplau *et al.* (1998), based on an analysis of cross-cultural data, suggested that sexual segregation at young ages is not associated with development of a homosexual orientation in adults, which is in accordance with the EBE theory but also with a model of developmental buffering of sexual orientation (this chapter).

Incidentally, the tendency for homosexuals to seek comfort with each other, which may be modulated by the level of rejection that homosexuals may experience from the rest of society, may explain the onset of the *egalitarian* system (Cardoso 2005; see Chapter 1) of homosexuality that characterises Western and Westernised societies. Where homosexuals are better integrated into the social frame, *pathic* systems (Cardoso 2005) are more prevalent.

Sexual imprinting that may affect heterosexual mate choice may potentially also bias mate preferences toward same-sex partners (Eibl-Eibesfeldt 1967; Laland 1994). Although sexual imprinting is important in birds and some mammals (Bolhuis 1991; Laland 1994), its relevance in the determination of human sexual orientation remains speculative. However, it is possible that imprinting (and also conditioning) could help explain, at least in part, the vast array of sexual fetishisms described in humans, and therefore at least some cases of same-sex sexual preference could have also been affected to some extent by sexual imprinting or other learning mechanisms (Fitzgerald 2000).

Fetishism and homosexuality

Fetishism has been described in both heterosexuals and non-heterosexuals (homosexuals and bisexuals). It is a term originally introduced in psychology by Alfred Binet in the late nineteenth century 'to refer to a predominant or an exclusive sexual interest in an inanimate object (fetish) or part of the human body' (Weinberg *et al.* 1995: 17). Fetishism is much more prevalent in men than in women (see, for example, Weinberg *et al.* 1995; Långström & Zucker 2005) and its development during ontogeny, around the time of puberty, could be explained by a diversity of learning mechanisms including imprinting as well as classical and operant conditioning (see Weinberg *et al.* 1995 for a concise review). Weinberg *et al.* (1995) point out that fetishism is manifested as a continuum of states from very mild to very intense in the male population. Långström and Zucker (2005) concluded, from their study of transvestic fetishism in a random Swedish sample of 2810 men and women, that the variables most strongly associated with fetishism are: lower thresholds of sexual arousal, early separation from parents in childhood, high frequency of masturbation, use of pornography, and same-sex sexual experiences. That is, fetishism seems to be associated with enhanced sexual activity. Most intriguing of all is the recorded high frequency of fetishisms targeted to feet and toes (podophilia) (see, for example, Scorolli *et al.* 2007). This evidence suggests to me that fetishism may derive from normal processes of sexual behaviour, perhaps associated with sexual selection and mate choice. This would explain the link of fetishism with enhanced sexual activity and also with specific objects, body parts in particular, and it could also explain its manifestation as a continuum of states. Alfonso Troisi (2003) suggested exactly such an evolutionary link between some 'sexual disorders' and sexual selection, which may put the understanding of fetishism in a different light, with mild expressions of fetishism being just a result of the action of perfectly normal mechanisms of mate choice (e.g. foot fetishism may be a very good case in point).

In fact, the sexual selection interpretation of fetishisms seems to be a better explanation of these phenomena than the *Erotic Target Location Error* (ETLE) interpretation put forward by Ray Blanchard (Blanchard 1991; Freund & Blanchard 1993). That parts of the body may become the primary stimulus for sexual arousal is not necessarily a result of 'a misdirection of erotic interest toward body features that are peripheral (e.g., hair or feet) or inessential (e.g., clothing)' (Lawrence 2009:196), as the ETLE model would suggest. Those same parts of the body are also targets of mate choice processes in many individuals, not just fetishist (Binet also seems to have been aware of this issue, according to Frank Sulloway (1979: 286)). Could a sexual selection mechanism also explain the more obsessive aspects of fetishism? I think that it could. In fact, Lawrence (2009: 209) quotes the works of LaTorre (1980), Wilson (1987) and others suggesting that fetishistic behaviours tend to be more common among individuals who have less access to actual sexual partners, suggesting that when specific men become socially isolated, they may tend to redirect their sexual desire towards those stimuli that are usually associated with an actual sexual partner (see also Weinberg *et al.* 1995), the latter association being expected from sexual selection. If this is so, then we might also expect that more socially isolated homosexual men would also tend to develop fetishistic tendencies. This issue was investigated by Martin Weinberg and collaborators (Weinberg *et al.* 1995: 27; see also Weinberg *et al.* 1994a) who studied homosexual (88% of their sample) and bisexual (12%) male fetishists belonging to a US-based organisation, the 'Foot Fraternity', to conclude that 'It seems that, along with distinct personality, social isolation played some role in the development of their fetish interests ...'. Lawrence (2009) also points out, rightly, that specific studies are required to determine the exact direction of causation between fetishism and social isolation: is fetishism a result of sexual inexperience with other individuals, or is a lack of confidence in social interactions a result of fetishism? The original link of fetishism with sexual selection that I postulate, however, is consistent with either of those possibilities. In sum, fetishism is not limited to heterosexuals, but it is also found among homosexual men, suggesting that fetishism is more a consequence of being a man than of expressing a specific sexual orientation. In its mildest forms, fetishism could be a result of sexual selection and partner choice mechanisms that may be retained across sexual orientations.

Contributions of social learning

Notwithstanding the limitations of the above learning mechanisms acting at young ages, it is clear that our brain is quite capable of producing a great degree of behavioural plasticity that has been the target of a large amount of research on the early development of homosexuality. Through language, imitation and various control strategies, individuals that interact socially may be able to influence each other and potentially alter each other's sexual behaviour.

Albert Bandura's (1989) *Social Cognitive* theory emphasises postnatal social determination of gender role and gender typing through learning. Building upon Lawrence Kohlberg's (1966) *Cognitive-Developmental* theory, which focuses on the ability of children to learn gender role stereotypes by observation and imitation, Bandura (1989) also points to the influence of social control (punishments and rewards) on directing the ontogeny of gender role and identity. Gender roles are developed early in ontogeny: by age 2–3 years children already start displaying the gender roles characteristic to their culture, being a result of early social interactions not only with the mother, but also with other individuals with whom the child has social contact (Lewis 1987). Social learning therefore rightly emphasises the diversity of behavioural phenotypes resulting from the plasticity of the brain during development, but some students of social learning seem to treat such plasticity almost as if it were unconstrained, with functions that are only limited by the number and type of external, environmental inputs (Hogben & Byrne 1998). Although a high degree of plasticity is clearly relevant in the case of gender role, the development of sexual orientation does not, in general, display the same degree of elevated plasticity.

Mischel *et al.* (1973) take a more realistic approach, proposing that the potential repertoire of behaviours is also constrained by 'individual competencies' – that a biologist would translate into 'genetic, morphological, physiological constraints' – and not just by external contingencies. The issue of developmental constraints limiting the kind of phenotype produced is especially important for the understanding of how sexual orientation gradually takes shape as the child grows.

That children are not internally unconstrained in the development of their sexual orientation is demonstrated by the relatively low percentage of children raised by lesbian mothers developing a homosexual or bisexual sexual orientation (16%, 4/25) compared with children raised by single heterosexual mothers (0%, 0/20) (Golombok & Tasker 1996). Although Golombok & Tasker's (1996) work does indicate an increased spread of sexual orientations among children living with lesbian mothers than among children living with an heterosexual single mother, which the authors interpreted as increased open-mindedness learned from a more tolerant environment in a lesbian social context, the fact that spontaneous imitation did not seem to produce a better result in terms of matching lesbian mothers and children's sexual orientation, just argues in favour of the role of other processes, in addition to learning, contributing to the ontogeny of sexual orientation. Similar results are obtained for male parents, with Bailey *et al.* (1995) showing that 80% of sons of gay fathers are actually heterosexual. In an earlier work Richard Green (1978) reported 100% heterosexuality in 13 eleven- to twenty-year-old children of homosexual or transsexual parents.

Arguably, some of the clearest evidence for the influence of learning processes contributing to at least some manifestations of homosexuality in humans is that of political homosexuals (e.g. political lesbians) (Peplau *et al.* 1999; Gottschalk 2003). I concur with Peplau *et al.* (1999) in that the mechanisms that trigger the onset of homosexuality may be different at different ages, with political homosexuals providing what is probably the best evidence of the effect of brain plasticity on sexual orientation. Such seemingly 'unbound flexibility', however, is not representative of all cases of homosexual behaviour.

Recent trends show a slow movement towards the convergence of different perspectives on homosexuality into a more integrationist model. The aim throughout this book is to build a case for a biosocial theory of homosexuality. In the context of the ontogeny of homosexual orientation in humans, both Belsky *et al.* (1991) and Udry (2000) concur in that both the phenotypic plasticity displayed by our species and the constraints imposed by our own reality as organisms must be intricate parts of any explicative model of homosexuality. The *Biosocial Model* that will be introduced in Chapter 10 posits to achieve exactly such an integration.

I now turn to another phenomenon that has been described in birds and to a lesser extent in mammals: *other-sex mimicry*. Other-sex mimicry involves aspects of the early ontogeny of individuals in species that display this effect. This section is also a prelude to the last section of this chapter, where heterochrony and neoteny in particular will be reviewed.

Other-sex mimicry: an alternative route to animal homosexuality?

When it is used in the context of a discussion about the evolution of homosexuality, the expression 'other-sex mimicry' may give the wrong impression that I equate homosexual behaviour with gender role inversion. As is repeatedly stressed in this book, this is not the case: homosexuals of any sex can be feminine, masculine or androgynous in terms of their gender role. However, in sexually dimorphic species the issue of cross-gendered phenotypic traits is probably the first to arise and the most obvious and easy to study. This is the case with other-sex mimicry. Other-sex mimicry (more commonly known in the literature as 'opposite sex mimicry') is simply a state whereby a male or a female resembles in behaviour, external morphology, or both, a conspecific of the other sex. This is a situation not uncommon in birds, although it is probably less common in mammals (Lyon & Montgomerie 1986). The phenomenon of other-sex mimicry involves chromosomally and gonadally normal individuals and it is therefore not a case of intersex. Whether it is a result of specific mutations that affect development through the mediation of gonadal steroids or other hormones, or through other mechanisms, is unknown in most species exhibiting the trait.

In birds with colourful males and less colourful females, juveniles of both sexes tend to resemble females (Lyon & Montgomerie 1986). In some species, however, sexually mature males retain their juvenile, female-like plumage, usually, but not always, for an initial period in their life. Retention of juvenile, female-like traits in some sexually mature males is also known from a number of sexually dimorphic primate species. In birds this process is known as *delayed plumage maturation* (Lyon & Montogomerie 1986; Thompson & Leu 1995), a phenomenon that has been reported in more than 200 species of bird, including 36 passerine families, and that involves mainly males, with only 8 families of birds exhibiting female delayed plumage maturation (Thompson & Leu 1995). Evolutionary ecologists have proposed several adaptive hypotheses to explain this phenomenon, including the *Cryptic*, *Winter Adaptation*, *Status Signalling* and *Female Mimicry* hypotheses (Lyon & Montgomerie 1986; Slagsvold & Saetre 1991; Muether *et al.* 1997). It is not my aim to review all those hypotheses, instead I will mainly focus on the two that have received most of the attention: the *Female Mimicry*, which states that female-like, juvenile plumage retention in sexually mature males has the selective advantage of deceiving mature-plumaged males, who are owners of breeding territories, into accepting the female mimics in or close to the territory (Rohwer 1978); and *Status Signalling*, which states that female-like plumage is a signal of subordinate status to be perceived by dominant males, so that subordinates will then avoid aggression from dominant territory owners by signalling their status (Lyon & Montgomerie 1986).

One of the bird species in which the *Female Mimicry* hypothesis has been more thoroughly tested is the pied flycatcher (*Ficedula hypoleuca*), a small passerine studied by various research teams in Europe. The extent of retention of female-like plumage in sexually mature males is variable across local populations of the pied flycatcher throughout its European distributional range (Røskaft *et al.* 1986), and it is also variable with age, with lighter, sexually mature males becoming darker as they grow older (Král *et al.* 1988). It is possible that such a variability among males may be retained in the overall population, at least to some extent, because of diversity among females in their choice of preferred sexual partners (Røskaft *et al.* 1986; Järvi *et al.* 1987), although there may also be additional advantages in populations where both kinds of males are present, as we will see below.

Tore Slagsvold and Glenn-Peter Saetre (Slagsvold & Saetre 1991; Saetre & Slagsvold 1996) carried out a series of experiments in Norway to test the *Female Mimicry* hypothesis in the pied flycatcher. When either a caged territorial male or a female or a female-plumaged male were experimentally exposed to territorial males in the wild the latter responded attacking the caged, dark-plumaged territorial male, but their first reaction was to court both the female and the female-mimic male. That this was the result of a 'mistaken identity' error, was clear from the observation made by Slagsvold & Saetre (1991) that courtship calls directed towards the female-plumaged male decreased with the degree of previous experience of the territorial male with real females.

Saetre & Slagsvold (1996) subsequently carried out an experiment to determine whether female-mimic males reap any advantage from the territory owner's mistake. In pair-wise contests over the ownership of a nest-box, wild-caught males of the two phenotypes were confronted for a short period of time, usually less than one hour in order to avoid injuries. To be precise, three kinds of male were used in the tests: black-plumaged, female-plumaged and intermediate-plumaged mature males in different pair-wise combinations. In a second test, some female-plumaged and intermediate-plumaged males were painted black in order to make them resemble black-plumaged males and a new series of pair-wise contests was carried out on those males. The results of the experiment are summarised in Figure 4.3. It is clear that, in the context of these brief encounters, female-like males have an advantage over both naturally black and painted males in terms of winning access to the nest-box, whereas intermediate and black males had similar levels of wins. This suggests that the colour of the plumage acts as a deceiving trait, perhaps allowing an advantage in terms of 'surprise' in territorial contests. Saetre & Slagsvold (1996) also indicate that the naturally female-plumaged males were the first to attack, which explains why painted males had a slightly higher level of wins than the intermediate males in Figure 4.3. It is possible that the difference between the two female-plumaged males (painted vs. not painted) shown in Figure 4.3 is due to the latter seizing the initiative in response to the artificial black plumage of the opponent. This advantage of female mimics may explain the retention of the trait in the population, even where the trait may not be preferred by females, as those female-mimics may be better able to remain close to black male territories, potentially reaping advantages of survival and future reproductive success. As they grow older, they will develop the black plumage that may attract females. This is an adaptive scenario that can be easily tested in long-term studies of wild populations.

Delayed plumage maturation also occurs in males of the genus *Passerina* (Rohwer 1986; Thompson & Leu 1995; Muether *et al.* 1997; Greene *et al.* 2000). Muether *et al.* (1997) studied Lazuli buntings (*P. amoena*) in Montana (USA) to test several hypotheses for the delayed plumage maturation displayed by this species. Lazuli bunting males retain their juvenile plumage into their

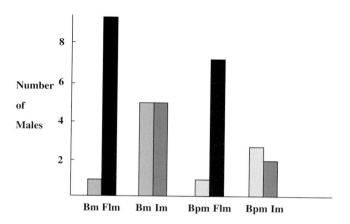

Figure 4.3. Result of contests over the control of a nest box between pairs of male pied flycatchers possessing different plumage coloration. Bars indicate the number of winners. Bm, black males; Flm, female-plumaged males; Im, intermediate-coloured males; and Bpm, black-painted males. Modified from Saetre & Slagsvold (1996).

second year, but their gonads are fully mature and capable of siring offspring. These birds, however, are competitively inferior to fully adult-plumaged males and occupy marginal territories only, but they are also tolerated by the older males. Moreover, this may not be a case of simple female mimicry, as older males seem to be able to distinguish between second-year males and females (Muether *et al.* 1997). What is it driving the evolution of delayed plumage maturation in this species, then? If young sexually mature males are better tolerated by fully mature males as neighbours, and those young males are able to attract a female and breed, even if they may suffer some loss of reproductive success through extra-pair copulations by the neighbouring fully plumaged male (Muether *et al.* 1997; Greene *et al.* 2000) they may still have a better lifetime reproductive success than males who do not reproduce at all during their second year of life. Young, mature-plumaged males may not be able to compete with older and more experienced males, as the latter attack mature-plumaged males more frequently than any other kind of male (Muether *et al.* 1997). Would a marginal increase in reproductive success during the second year of life be enough to select for delayed plumage maturation? We do not know, but in a variable environment with only limited suitable habitat available and a lifespan that is not very long, being allowed to settle in a territory close to a brightly plumaged owner because you are more likely to be tolerated if your plumage is dull can be a life- (and fitness-) saving opportunity if there is nowhere else to go (Greene *et al.* 2000).

Although here I am implicitly interpreting delayed plumage maturation as an ontogenetic programme that has been selected for, delayed development may also be a result of current physiological status resulting from lack of access to food and other resources (Landmann & Kollinsky 1995). In either case the organism is just displaying enough behavioural plasticity to make the best of a bad job under harsh social and environmental circumstances.

If the signal that subordinate males are sending in most of these species is that of being a female, then we would expect delayed plumage maturation to be associated with same-sex mounting in a comparative analysis. Table 4.6 shows a list of bird species taken from Table 2.1, for which I have information about both same-sex mounting and delayed plumage maturation. All species are members of different genera with the exception of *Larus argentatus* and *L. delawarensis*, which, however, differ in their presence of delayed plumage maturation. Given the small sample size and the taxonomic diversity of the sample, the hypothesis will be tested through a Fisher's exact test, taking the species as the unit of statistical analysis. A total of 71% (5/7) of the species displaying delayed plumage maturation also engage in same-sex mounting, whereas 55.5% (5/9) of the species that do not have delayed plumage maturation engage in same-sex mounting. The difference goes in the direction supporting a potential association between delayed plumage maturation and the display of homosexual sexual behaviour, but the trend is statistically not significant ($p = 0.45$).

In most bird species sexual maturation, and therefore fully functional gonads, are attained concomitantly with adult plumage. In species with delayed plumage maturation, however, full development of male-typical plumage is achieved after becoming sexually mature (see, for example, Saetre & Slagsvold 1996). Evolutionarily, such a contrasting pattern could result from simply slowing down the development of some bodily structures (e.g. plumage coloration) while maintaining the rate of gonadal development intact. If this occurred the outcome would be a gonadally mature, but juvenile-looking organism. Such a process is called neoteny and could potentially be an important evolutionary mechanism explaining homosexual behaviour in some taxa. Neoteny will be the focus of the last section of this chapter.

Evolution and ontogeny: is homosexuality a neotenic trait?

In his classic book *Ontogeny and Phylogeny*, Stephen Jay Gould (Gould 1977: 482) defined the process of heterochrony, following De Beer (1930), as 'a phyletic change in the onset or timing of development, so that the appearance or rate of development of a feature in a descendant ontogeny is either accelerated or retarded relative to the appearance or rate of development of the same feature in an ancestor's ontogeny'. That is, new species may evolve through mutations that alter the rate of development of some traits relative to others in relation to the ancestral state. Many such genes that control developmental programmes affecting diverse parts of the body, including the brain, have been already found in vertebrates: e.g. *Otx*, *Hox*, *Wnt*, *Sox* (Parr *et al.* 1993; Burke *et al.* 1995; Prior & Walter 1996), and they are the subject of intense evolutionary study within the new research programme of 'Evo-Devo' (Hall 2003).

The study of heterochrony is a very active field of research and has been the focus of several recent reviews (Smith 2003; Hall 2003; McNamara & McKinney 2005; Webster & Zelditch 2005). The array of potential processes that fit the definition of heterochrony is diverse, as illustrated in Figure 4.4.

Table 4.6. Presence or absence of delayed plumage maturation in bird species with or without same-sex mounting

Species	Delayed plumage maturation	Same-sex mounting	Reference
Philomachus pugnax	N	N	Chu 1994[a]
Uria aalge	N	Y	Chu 1994
Falco tinnunculus	Y	Y	Hakkarainen *et al.* 1993
Tryngites subruficollis	N[b]	Y	Lanctot *et al.* 1998
Pluvialis apricaria	N	N	Chu 1994
Haematopus ostralegus	N	Y	Chu 1994
Himantopus h. himantopus	N	Y	Chu 1994
Tringa totanus	N	Y	Chu 1994
Larus argentatus	N	N	Chu 1994
Larus delawarensis	Y	N	Ryder 1975
Carpodacus mexicanus	Y	Y	Brown & Brown 1988b
Ficedula hypoleuca	Y	N	Røskaft *et al.* 1986
Poephila acuticauda	Y	Y	Langmore & Bennett 1999
Wilsonia citrina	N	N	Lyon & Montgomerie 1986
Chiroxiphia caudata	Y	Y	Foster 1987
Tachycineta bicolor	Y[c]	Y	Lozano & Handford 1995

[a]I used Chu's (1994) low levels 0 and 1 of *Extensiveness of first spring moult* as a conservative criterion of delayed plumage maturation. [b]Males mimicking females have been observed in this species, but the mimicry is behavioural, not morphological (Lanctot *et al.* 1998). [c]Delayed plumage maturation occurs in females in this species (Lozano & Handford 1995).

Paedomorphosis is an evolutionary process in which the development of individuals in a derived species is slowed down compared with its ancestor, whereas *peramorphosis* is an acceleration of development (Gariépy *et al.* 2001). In the context of somatic vs. reproductive structures and functions, peramorphic evolutionary processes that involve acceleration of development of somatic structures, with the timing of gonadal maturation remaining unchanged, are called *acceleration*, whereas those involving an unchanged somatic development throughout speciation but a delayed or slowed down gonadal maturation are called *hypermorphosis*. It can be seen that peramorphosis will be involved in evolutionary changes that delay the onset of reproductive ability relative to the development of other physiological and morphological characteristics in a derived species; in other words, species that evolved through a process of peramorphosis will have adult-looking (compared with the ancestor) developmental stages that nevertheless are slower at developing full reproductive capability. Paedomorphic processes may involve either an accelerated gonadal maturation with rate of somatic development remaining unchanged, a process known as *progenesis*; or delayed somatic development with gonadal development unchanged, a process known as *neoteny* (Raff & Wray 1989; Gariépy *et al.* 2001). In either way, paedomorphosis results in a derived species with individuals attaining reproductive capacity at a developmental stage in which they look like a juvenile of the ancestral species.

Students of avian evolutionary biology have implicated paedomorphosis in the evolution of flightless birds (Chatterjee 1999; Cubo & Arthur 2001) and delayed plumage maturation (Lawton & Lawton 1985). In mammals, paedomorphosis is consistent with the evolutionary patterns of several morphological structures and behaviours found in various groups, including primates (Hafner & Hafner 1984; Doran 1992; German *et al.* 1994; Berge 1998; Chaline *et al.* 1998; Gariépy *et al.* 2001; Cubo *et al.* 2002; Goldberg 2003; Mitteroecker *et al.* 2004). However, a recent comparative study of

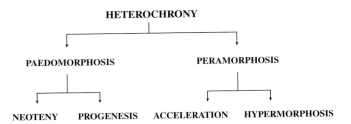

Figure 4.4. Diversity of heterochronic processes.

mammals carried out by Chris Fraley, Claudia Brumbaugh and Michael Marks (Fraley *et al.* 2005) indicates that pair bonding is not associated with their variables measuring neoteny; the same non-significant result was obtained in an analysis of primate data only. In the study by Fraley *et al.* (2005), however, the degree of neoteny was estimated through lifespan, gestation time, age at dispersal and age at puberty measurements, which may have not been sufficiently representative of the true level of neoteny that may affect behaviour in the different taxa. When it comes to understanding the evolution of behaviour in general and homosexual behaviour in particular through neoteny, it is ultimately the brain that we are interested in.

As we will see in Chapter 5, the architecture and functions of the brain are determined to a great extent by ontogenetic processes controlling the production and death of neurons and also by processes controlling the number of synaptic connections among neurons. Whenever the organism undergoes *regressive events* in the development of the brain (Finlay *et al.* 1987), that is, whenever the early developmental pathway of brain organisation involves an overproduction of neurons, dendrites and axons, with subsequent reshaping of brain architecture through apoptosis or programmed cell death as the organism develops, any mutation that slows down apoptotic processes will produce phenotypes that retain greater numbers of neurons and connections compared with the parental generation. In other words, that mutation will produce a phenotype resembling in brain properties the younger developmental stages of the parental generation. Repeated throughout many generations, it may be seen how this process will produce a neotenic descendant possessing a large and complex brain. As we have seen in a previous section of this chapter, same-sex sexual behaviour is often expressed at young ages of social vertebrates and it is part of the normal sexual behavioural development of both males and females, but especially males. If slowed-down nerve cell death allows the retention of a 'juvenile-like' brain into the sexually mature ages, then same-sex sexual behaviour can also be manifested among adults. This is the *Neotenic Theory for the Evolution of Homosexuality* in a nutshell. From a historical perspective, a precursor of this theory was already present in the writings of Sigmund Freud (Freud 1905). I will devote the rest of this section to putting some additional flesh on this fascinating theory.

During development, apoptosis can deplete brain regions of up to 50%–80% of the original number of neurons (Finlay *et al.* 1987). This fact can already give a very graphic idea of the immense amount of room that there is for mutations, or even external environmental factors such as the hormonal milieu in utero or the more general chemical environment to which a developing embryo is exposed, to modulate brain structures through changes in apoptosis alone. Both males and females lose neurons with age, but in some brain regions controlling sexual behaviour males tend to lose fewer neurons than females, apoptosis being slowed down in those regions in males by testicular testosterone produced during prenatal and perinatal life. Altering apoptotic processes during early ontogeny can either feminise male brains or masculinise female brains (Finlay *et al.* 1987); this can be achieved in a continuous manner, producing a gradation of behavioural phenotypes from more masculine to more feminine in either a chromosomally male or chromosomally female individual. An evolutionary neotenic trend that spans across sexes will retain juvenile-like homosexual behaviour in both males and females, but adults of both sexes may still differ in their likelihood of expressing same-sex sexual behaviour if additional processes (e.g. sex-specific steroid modulation of apoptosis) are superimposed on the overall neotenic evolutionary trend. In mammals, the general prediction of the *Neotenic Theory* is that both sexes will be likely to express homosexual behaviour compared with evolutionary ancestors, but males will manifest exclusive homosexual orientation more frequently than females. This is in accordance with what we know about the distribution of homosexuality in human societies (Kirkpatrick 2000). Of course we do not know the prevalence of homosexuality in early hominids, but we can address the evolutionary issues indirectly through comparative analyses of extant species. This is what I will do in Chapters 7 and 8.

The early ontogeny of the mammalian brain is not only characterised by the presence of large populations of neurons, but also by a great extent of diffuse connectivity among those neurons, which then becomes more selective as the organism develops (Finlay *et al.* 1987). Finlay *et al.* (1987) suggested the possibility that neotenic evolutionary processes could also be involved in the retention into the adult stage of large numbers of nerve cell connections that are typical of the juvenile stages of development. If those larger numbers of connections sustain greater diversity and flexibility of sexual behaviours, then they may well be part of another mechanism explaining adult bisexuality and homosexuality through neoteny (see also Bourgeois 1997).

The idea that humans evolved through neotenic processes is not new. Gould (1977) traces the first indication of thoughts about adult humans resembling the young of great apes for evolutionary reasons rather than chance, back to the first half of the nineteenth century. However, it was the Dutch anatomist Louis Bolk (1926) who explicitly made the link between what we now recognise as neoteny and human evolution with his theory of foetalisation of human anatomy. When I started working on this book in early 2006 I had a 'Neoteny' file ready to accumulate references and ideas to develop in the manuscript about the potential links between this evolutionary process and homosexuality, but in early 2007 I realised that 'somebody else had got there first'. Clive Bromhall's (2004) book *The Eternal Child* makes for extremely pleasant reading and a treasure trove of ideas, not all of equal quality, however. His basic neotenic model for the evolution of homosexual behaviour implies the following steps:

(a) aridification of the early hominids' East African habitat (see, for example, de Menocal 1995, 2004) favouring

(b) formation of large groups, which provided the selective context for neotenic evolutionary processes. This can be seen as infantilised adults being better able to withstand the potentially conflicting social life within a group and also being more inclined to cooperate. In addition, increased learning abilities coming with an infantilised brain would have favoured this process.

(c) Sexual selection through female choice would have then accelerated the process of neoteny as more infantilised males would be less aggressive, more cooperative, more intelligent and resourceful and therefore better able to care for offspring, all traits expected to be preferred by females.

Finally,

(d) homosexual behaviour, in Bromhall's (2004) model, would be an unselected evolutionary outcome of the runaway sexual selection for infantilised males.

Moreover, if neotenic processes affect males more than females, as Bromhall (2004) postulates, then the asymmetry in the frequency of exclusive homosexuality observed between men and women could be easily explained.

For the reasons stated above, I obviously agree with the basic insight that neoteny may be an important evolutionary processes that contributed to the production of homosexual phenotypes in humans. The specific mechanism proposed by Bromhall (2004), however, is more debatable. I agree that sociality provides the environment for the display of very complex behaviours and that such a selective environment is likely to favour the further evolution of large and complex brains (Dunbar 1998). But for social groups to be behaviourally complex they do not necessarily need to be large, as the sphere of social interactions is usually limited to a subset of the large group anyhow (see, for example, Shaner & Hutchinson 1990). We will see in Chapter 8 that homosexuality is more likely to be selected at intermediate group sizes. Second, large and complex brains do not only permit complex cooperative interactions, but also complex competitive and manipulative interactions. Both are likely to be useful in the long-term maximisation of reproductive fitness in a social context. I also agree that neoteny may have been accelerated in humans via sexual selection through female choice for 'brainy' males, but I would also add male choice: in a harsh environment smart women are as useful to men's reproductive success as smart men are to women's. Such two-way sexual selection (see, for example, Clutton-Brock 2009) would be favoured by the need for the provision of biparental care. The final step linking homosexuality to neoteny in the way of the former being an unselected outcome of selection for larger and more complex brains is a possibility. Although I agree that same-sex sexual behaviour might have arisen evolutionarily as an exaptation (Gould & Lewontin 1979), I find less convincing the idea that same-sex mounting is a neutral trait (see, for example, Vasey 1996). I will argue in Chapter 8 that there are too many current socio-sexual functions of same-sex sexual behaviour that look like adaptations to social life in mammals, to simply assume a current non-adaptive value for those behaviours across taxa.

Shaner & Hutchinson (1990) also provide theoretical arguments and empirical evidence to support neotenic evolution of humans, and they also correctly emphasise neuroplasticity aspects of neoteny and social bonding, based on previous works by Gerald Edelman (1987) and Helen Fisher (1982).

Apart from humans, the bonobo (*Pan paniscus*) is another species in which same-sex sexual behaviour could be also explained by neotenic evolutionary processes. The genus *Pan* is the closest living relative to our own species and the bonobo shows a remarkable display of same-sex sexual behaviours both in the wild and in captivity, especially among females (see, for example, de Waal 1997). Ben Blount (1990) has suggested that bonobo same-sex sexual behaviour may be a result of a more general paedomorphic (perhaps neotenic) evolutionary process in this species, as could be the case in humans. Blount (1990) includes the following traits in bonobos among those that resemble traits typical of immature or juvenile stages of development in the congeneric chimpanzee (*P. troglodytes*), and that therefore could have evolved through paedomorphosis: (a) female genitalia positioned more anteriorly than in chimpanzees, (b) labia majora retained in the adult individual, (c) more extended intermenstrual interval than in chimpanzees, a feature that characterises adolescent female chimpanzees, (d) shorter period of lactational amenorrhoea, (e) skull resembling that of adolescent chimpanzees (e.g. reduced prognathism, smaller mandibles and more rounded cranium), (f) bonobos are more social than chimpanzees (a high degree of sociability characterises the young of many primates). Enomoto (1990) also argues that the social play displayed by adult bonobos at a study site in Wamba, Zaire, is a neotenic trait as, in primates, social play is typically expressed by juveniles but rarely so by adults. More recently, Lieberman *et al.* (2007) have also suggested that at least some of the characteristics of the bonobo's skull are paedomorphic compared with chimpanzees. In addition, it is also tempting to speculate that the propensity to adopt a ventro-ventral position during sexual intercourse in both heterosexual and homosexual contexts, that is reminiscent of the infant's embrace of its mother, may also be facilitated by neotenic processes in species that exhibit it such as bonobos, humans, Japanese macaques and others. With regard to Japanese macaques, however, although ventro-ventral contact between two individuals may be reminiscent of mother–infant interaction (e.g. unrelated adults may embrace and while they do so the younger member of the pair may hold the nipple of its partner in its mouth; Harold Gouzoules, pers. comm.), sometimes female–female ventro-ventral mounting

> developed among particular pairs of females as a result of both females trying to mount their female partners at the same time. This resulted in a 'dance' routine, but then the two females would tumble to the ground and mount in a v-v position. After this had occurred a number of times, they would go straight to the v-v position (Harold Gouzoules, pers. comm.).

This suggests that ventro-ventral mounting could be also affected by learning processes. Moreover, the *Female Weaning Period* hypothesis of Edwards & Todd (1991; see Table 1.1) suggests that a young male who is being weaned may seek the comfort of an older male as a maternal substitute, whereas the older one may seek sexual comfort from the younger, thus a ventro-ventral embrace between two males of different ages, which may also involve the young male sucking the nipple of the older male, could be explained in this proximate way. In both cases, however, the *Neotenic Theory* suggests that neotenic processes will lower the threshold for the expression and the development (e.g. through learning) of infantile-like behaviours in adults engaged in sexual intercourse with each other.

Both the arguments and current evidence in favour of a neotenic origin of same-sex sexual behaviour in at least some taxa are sufficiently compelling to encourage further research in this field. In fact, neoteny figures prominently in the *Biosocial Model of Homosexuality* that will be introduced in Chapter 10.

Summary of main conclusions

- Developmental processes can be canalised or plastic as a result of natural selection. In social species, canalisation or plasticity of homosexual behaviours may result from species-specific or sex-specific selective processes. Homosexuality tends to be relatively more canalised in men and more plastic in women.
- Hormones, steroids in particular, may have an organisational role on homosexuality during early development (pre-, peri-, postnatal). Both gonadal androgens and oestrogens exert their action during specific sensitive periods, in a species-specific and brain-area-specific fashion. In particular they may exert their organisational role by altering the processes of apoptosis and nerve cell architecture.
- Direct genetic mechanisms and postnatal learning processes interact with hormone-driven mechanisms to potentially modulate the expression of homosexual behaviour as the organism develops.
- Masculinisation/de-masculinisation, feminisation/de-feminisation are distinct processes of brain development and their modification (e.g. by hormone action) may alter gender role and sexual orientation development.
- Proxy variables such as 2D:4D ratio that, at least in part, reflect prenatal exposure to steroids, in general do not covary with homosexuality in a very predictable way, and their usefulness is relatively greater in the case of predicting female than male homosexuality.
- Both the *Birth Order* and the *Family Size* effects are well-established patterns associated with homosexuality, especially in males. Younger brothers are more likely to develop a homosexual sexual orientation in relatively large families.
- Prenatal stress is an important factor that can explain the development of homosexuality in humans and other mammals. Postnatal stress

- effects, including stress suffered at adrenarche, are also of great potential relevance.
- Juvenile same-sex sexual play is part of the heterosexual development of social animals, mammals more than birds, and, as such, is not conducive to homosexuality in adults, including humans.
- Gender non-conformity in childhood (e.g. masculine girls and feminine boys) is not an unequivocal precursor to adult homosexuality. Adult homosexuals tend to have been gender-non-conforming children, but not all gender-non-conforming children develop into adult homosexuals.
- Potentially, in some individuals, the transition from childhood gender non-conformity to adult homosexuality might be even facilitated by postnatal stress effects, perhaps contributed by social (e.g. family) intolerance of gender role non-conformity in children.
- Learning mechanisms can also be involved in the development of gender role and, to some extent, they may also modulate some aspects of the expression of homosexuality, especially in women. However, learning effects are not unconstrained.
- Neoteny is a major evolutionary process that can explain the evolution of homosexual behaviours in some taxa, including anthropoid primates.

In the next chapter I review, but also endeavour to integrate, the endocrinological and neurological mechanisms that can be associated with the expression of homosexual behaviours in adult individuals.

5

The endocrine and nervous systems: a network of causality for homosexual behaviour

Sexual behaviour is expressed in the adult animal thanks to the activities of the motor, sensory, central nervous, peripheral nervous and endocrine systems. The activation of such systems is affected by stimuli coming from the external environment, which also includes the social environment, and by internal signals. The various systems also interact with each other in subtle and sometimes not so subtle ways. In this chapter I specifically explore the role of endocrine and nervous systems in the proximate causation of homosexual behaviour, with emphasis on the adult stage of development. Some aspects of the more complex mental faculties of our species will be also considered. The chapter, however, starts with a transitional link with the previous chapter in the form of a review of human endocrine disorders that have been studied in the context of the early ontogeny of homosexuality.

Human endocrine disorders and homosexuality

Most of the medical conditions that have been studied as means of unravelling the endocrinological mechanisms of human homosexuality are considered disorders in need of a cure or at least of containment of their most damaging effects. From this, however, it should not be concluded, as stated in Chapter 3, that homosexuality is a disorder. In fact I argue in this book that it is not. The usefulness of studying such syndromes lies in the opportunity they afford to study those specific neuroendocrinological mechanisms that are common to both the medical condition and also the manifestation of homosexual behaviour. In this way, the causation of homosexual behaviour can be better understood in its more usual, functional context. This is a simple, but very important point that is worth further elaboration. To illustrate the argument by analogy I will briefly discuss the case of human autoimmune disorders.

The many individuals who are sufferers of an autoimmune disorder will have little problem in recognising it as a disease. Most people will be familiar with terms such as *insulin-dependent diabetes*, *multiple sclerosis*, *myocarditis*, *rheumatoid arthritis* and *systemic lupus erythematosus*, all of which are autoimmune conditions that are objects of medical attention (Jacobson et al. 1997). From this, one may conclude that *autoimmunity* is necessarily a disease state and as such it should be dealt with and studied. Not so: there is nothing intrinsically pathological about autoimmunity. In fact it would be quite easy to envisage perfectly adaptive contexts in which an autoimmune response can be beneficial to the organism. For instance, an immune system that is capable of attacking and destroying self cells is perfectly suited to defend the organism against cancer. In fact, this is one of the hypotheses that have been proposed to understand autoimmunity in an evolutionary and adaptive perspective (Torchilin et al. 2003), so much so that induction of autoimmune responses is envisaged as one of the many strategies that could be used to treat certain types of cancer (Pardoll 1999; Torchilin *et al.* 2003). In a similar manner, although same-sex sexual behaviour can take forms that we may regard as pathological, we cannot conclude from this that the behaviour as such is a result of an abnormal, non-adaptive functioning of the organism. Incidentally, heterosexual sexual behaviour can also take pathological forms (see, for example, Stein *et al.* 2001). As autoimmunity may have evolved in the context of anticancer defence, homosexuality may have evolved in the context of some adaptive functions (e.g. socio-sexual behaviours and others) as argued in this book. At the same time I do recognise the causal complexity of the different cases of homosexual behaviour too, and will argue that there is no one-size-fits-all cause (i.e. a single and simple mechanism) when it comes to understanding same-sex sexual behaviour across mammalian and avian taxa. In what follows, therefore, I review human endocrine disorders affecting sexual behaviour with the objective of understanding the hormonal underpinning of human homosexuality and its potential relevance outside the context of those disorders.

We have seen in Chapter 4 how androgens, oestrogens, corticosteroids, catecholamines and other hormones can play an important role in the early development of sexual orientation in some vertebrates. In fact most of the human disorders studied involve, in one way or another, some alteration of steroid and corticosteroid functions during sensitive periods of brain development that may be subsequently associated with same-sex sexual behaviour in the adult. Several detailed reviews of such disorders are available (Ellis & Ames 1987; Wilson 1999; Dörner *et al.* 2001; Cohen-Bendahan *et al.* 2005) that emphasise altered testosterone biosynthesis (ATB), 5α-reductase deficiency (5α-RD), congenital aromatase deficiency (CAD), complete androgen insensitivity syndrome (CAIS) and congenital adrenal hyperplasia (CAH). The analysis below will mainly focus on those five disorders.

Altered testosterone biosynthesis

The major pathways in steroid biosynthesis are summarised in Figure 5.1. A very good review of ATB and its effects on gender role and identity can be found in Wilson (1999). In brief, testosterone is synthesised in the Leydig cells mainly from its precursor androstenedione, a reaction that is reversible and that is catalysed by a set of isoenzymes of 17β-hydroxysteroid dehydrogenase (17βHSD). The same enzyme also reversibly catalyses the conversion of oestrone into oestradiol, whereas oestrone is produced from androstenedione by aromatisation. Isoenzymes are very similar variants of the same enzyme that, although encoded by different genes, catalyse the same chemical reaction. Therefore mutations in one or more of the 17βHSD isoenzymes may impair production of testosterone (but also oestradiol), although in fact the variable number of isoenzymes available buffers the biosynthesis of testosterone against mutations at one of the loci (Wilson 1999). However, in some cases mutated 17βHSD isoenzymes can sufficiently alter production of testosterone (and/or oestradiol) during early life to affect gender role development during ontogeny (Wilson 1999 and references therein). Genetic males (46,XY) with ATB are usually reared as girls owing to their initial development of female external genitalia. However, with time, ATB males tend to develop normal male external genitalia owing to the ability of non-affected isoenzymes to convert androstenedione into testosterone. In spite of their female gender of rearing at early ages, about half the ATB boys undergo gender role reversal as they grow up, but the other half retain their feminine gender role of rearing, and this may occur even between two brothers: one retaining the feminine gender role of rearing whereas the other changes to masculine (Wilson 1999). From this it is reasonable to conclude that ATB may have a variable effect on the development of gender role that may be associated with the number of mutated isoenzymes or the degree of functional impairment of each one of those isoenzymes: more severe cases would be more likely to undergo gender reassignment or

Human endocrine disorders and homosexuality 161

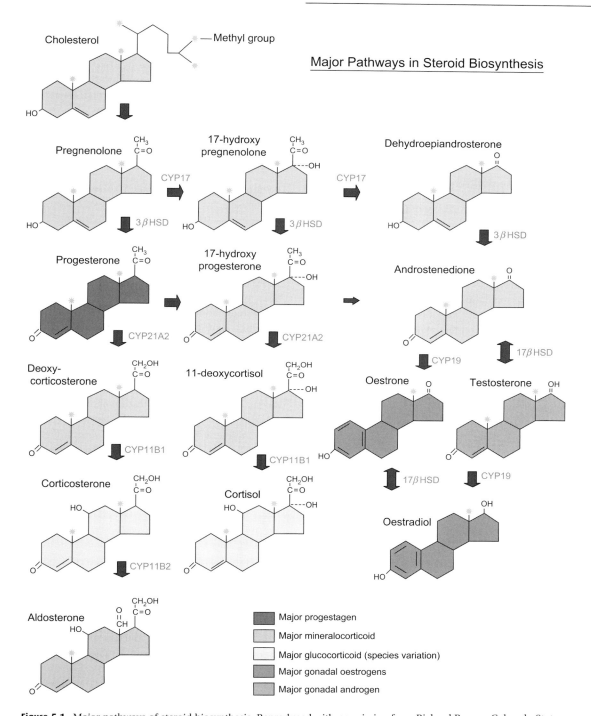

Figure 5.1. Major pathways of steroid biosynthesis. Reproduced with permission from Richard Bowen, Colorado State University.

perhaps develop a same-sex sexual orientation, whereas less severe cases are more likely to develop an heterosexual sexual orientation. Moreover, the effect of ATB may be complementary to other mechanisms, e.g. those involving learning, which may also contribute to differences between siblings.

5α-Reductase deficiency

Testosterone can be eventually reduced into dihydrotestosterone (DHT), a molecule that has a higher affinity for the androgen receptor than testosterone, and that is therefore a more powerful androgen. The reduction of testosterone into DHT is catalysed by two isoenzymes of 5α-reductase (5α-R). Again, as in ATB, the availability of isoenzymes may buffer the organism against mutations at one of the two loci. Whenever DHT deficiency is sufficiently severe during foetal development in males carrying 5α-R mutations, those males may develop a feminine gender role at puberty. However, such DHT deficiency is not always sufficiently elevated and some males carrying mutations at the locus may indeed develop a masculine gender role. That not all males carrying 5α-R mutations develop a feminine gender role (see, for example, Imperato-McGinley et al. 1991; Al-Attia 1996) may be explained by the ability of the mutated isoenzyme/s to retain sufficient functionality and/or by mutations at one of the two isoenzymes only and/or by additional masculinising mechanisms not involving DHT (Wilson 1999). Herdt & Davidson (1988) provide some interesting anthropological data on 5α-R pseudohermaphrodites (i.e. individuals who display external traits of one sex while possessing gonads of the other sex) that include information on sexual orientation and gender role/identity. Out of three adult individuals who had been reared as males, one displayed some degree of homosexual tendencies, subsequently becoming a local shaman. A second individual, after having adopted a feminine gender role and being married to a man, left his village and 'likes to go out with women... (but) does not now engage in heterosexual intercourse' (p. 48).

A third individual was apparently fully heterosexual. Three cases are reported by Herdt and Davidson of adult 5α-R pseudohermaphrodites reared as females; in all three cases, the individuals switched to a masculine gender role at puberty.

Congenital aromatase deficiency

The aromatisation of testosterone into oestrogen is catalysed by an aromatase enzyme complex that includes the aromatase cytochrome P450 (P450arom) coupled to a reductase. P450arom (currently known as CYP19) is highly conserved among vertebrates, especially among mammals, and is encoded by the *CYP19* gene (Conley & Hinshelwood 2001). In a rat model Bakker et al. (1993) showed that aromatisation blockage by using the aromatase inhibitor 1,4,6-androstatriene-3,17-dione (ATD) produces behaviourally de-masculinised and feminised male rats. Aromatase knockout (ArKO) mice have also been produced that carry mutations at the *CYP19* gene and that cannot aromatise testosterone (Bakker et al. 2002a). In a test choice between the odour of an oestrous female or that of a sexually active male, the ArKO mice did not show any preference, whereas control male mice preferred the oestrous female odour (Bakker et al. 2002a), indeed, ArKO mice tended to avoid social contact with conspecifics of both sexes altogether. Their behaviour was more asexual than homosexual (see also DuPree et al. 2004 for a review). The same occurs in male quails treated with the aromatase inhibitor racemic vorozole: they show suppressed sexual response towards a female (Balthazart et al. 1997).

In humans DuPree et al. (2004) reported three cases of males with congenital aromatase deficiency; in all cases their sexual orientation was heterosexual and their gender role was masculine. The same authors also carried out a more comprehensive linkage analysis of a polymorphic TTTA repeat that is closely associated with the *CYP19* locus and found that male homosexuals are not more similar among themselves at that locus than expected from Mendelian inheritance, whereas although the

CYP19 gene seems to be slightly more expressed in homosexual males than heterosexual males the difference is not statistically significant. On the other hand, male-to-female transsexuality has been positively associated with both the mean length of an oestrogen β-receptor gene repeat polymorphism, and also polymorphisms of both the androgen receptor gene and the *CYP19* (Henningsson *et al.* 2005). Hare *et al.* (2009), however, could only confirm the association between male-to-female transsexuality and longer gene polymorphisms at the androgen receptor gene, but not with the oestrogen β-receptor gene or the *CYP19* that was described by Henningsson *et al.*

Carani *et al.* (2005) described the case of a single individual affected by aromatase deficiency. The subject had undetectable circulating oestradiol upon testosterone treatment (Carani *et al.* 1999). An analysis of his sexual behaviour (i.e. frequency of morning penile erections upon waking up), and also gender identity and sexual orientation evaluated through the Bem Sex Role Inventory, indicated that the individual had a male gender identity and a heterosexual sexual orientation; moreover, he experienced an increase in erotic fantasies and libido only upon combined treatment with both oestradiol and testosterone. Although Carani *et al.* (1999) concluded that this case study suggests that aromatase deficiency affects sexual activity but not sexual orientation, as also indicated by the rodent and quail studies reviewed above, in this case some legitimate doubts may remain about the validity of the sexual orientation reported by the subject. The subject was a Catholic priest and we do not know whether this fact may have affected his reporting. Results of the few reported cases of CAD in humans have been reviewed by Rochira *et al.* (2001, 2004) and more recently by Jones *et al.* (2007) and Rochira *et al.* (2008). All reviews conclude that the up to six cases of reported CAD and the one case of oestrogen receptor-α inactivation in men indicate a heterosexual sexual orientation in the subjects displaying the conditions.

In sum, studies of congenital aromatase deficiency, or of the impairment of the oestrogen receptor-α or -β, do not provide strong evidence for an association of the syndrome with a same-sex sexual orientation in humans, although the usual caveat that more studies are needed is especially appropriate in this case.

Complete androgen insensitivity syndrome

Although an organism may have perfectly normal levels of circulating androgens, their effects on target tissues may be negligible if the androgen receptors on those tissues are not functional. Severe modifications of the androgen receptor, resulting, for instance, from mutations at the X-linked locus encoding for the polypeptide, may be conducive to conditions such as complete androgen insensitivity syndrome (CAIS). CAIS chromosomal males may develop female external genitalia and secondary sexual traits, along with female sexual behaviour and gender identity (Macke *et al.* 1993; Melo *et al.* 2003). The effect is less severe in cases where androgen insensitivity is only partial (PAIS) (Wilson 1999). The chromosomally female (XX) counterpart of CAIS is currently known as Mayer–Rokitansky–Küster–Hauser syndrome and produces phenotypic characteristics similar to those of CAIS (Money *et al.* 1984).

CAIS individuals suggest that, at least in humans, the oestrogen receptor, which is functional in these individuals, may not be relevant to the development of male heterosexual sexual orientation and identity (Wilson 1999; Cohen-Bendahan *et al.* 2005; Hines 2006). On the other hand, Collaer & Hines (1995) do quote a case of an androgen insensitivity syndrome male reared as female and who requested sex reassignment to male. This might have been a case of partial androgen insensitivity, however, but it could also be a case that can be explained by mechanisms relying on the oestrogen receptor.

The association between androgen receptors and the development of homosexual sexual behaviour, gender role and gender identity in chromosomal men with male external genitalia and secondary sexual traits remains controversial, however

(Quigley 2002). Family studies of the androgen receptor gene suggest that the androgen receptor trait and homosexual orientation segregate independently and that sequence variation in the androgen receptor gene is not associated with variation in homosexual orientation in most subjects (Macke et al. 1993). A similar study on the oestrogen receptor would be clearly welcome. The trend is even clearer for PAIS, as affected individuals tend to develop a heterosexual sexual orientation irrespective of whether they were reared as boys or girls (Money & Ogunro 1974).

The situation is more clearcut in rats, at both the neuroanatomical and behavioural levels. In rats both androgen and oestrogen receptors are involved in the development of sexual behaviour and gender role. In these rodents, full masculinisation of brain areas controlling various aspects of sexuality, such as the medial amygdala and the sexually dimorphic nucleus of the preoptic area (SDN-POA) of the hypothalamus, require functional androgen and oestrogen receptors (Morris et al. 2005; see Zuloaga et al. 2008 for a recent review).

Congenital adrenal hyperplasia

Congenital adrenal hyperplasia is a condition caused by a deficiency in the 21-hydroxylase enzyme that affects about one in 10 000–15 000 newborns, both males and females (Cohen-Bendahan et al. 2005). A deficient 21-hydroxylase impedes conversion of progesterone and its metabolite 17-hydroxyprogesterone into corticosteroids, whereas the excess progesterone thus formed could be converted into androstenedione that will finally produce both testosterone and oestradiol (Figure 5.1). Therefore CAH individuals tend to be exposed to higher than normal levels of testosterone (and oestradiol) during development (Dörner et al. 2001). CAH girls display a tendency to engage in masculine-typical activities whereas CAH boys tend not to differ from controls (Berenbaum & Snyder 1995; Mathews et al. 2009). Berenbaum & Snyder (1995) have shown that the specific preference of CAH girls to play with 'boy toys' was weak and mainly restricted to a small group of girls. Mathews et al. (2009) compared CAH females and unaffected female relatives (mainly sisters) in some of their behavioural and personality characteristics and found that CAH females were less tender-minded, more aggressive and less interested in infants, all traits that they categorised as being masculine, although the same subjects did not differ with regard to 'dominance'. Women who have been diagnosed with CAH tend to have a female sexual identity and a heterosexual sexual orientation, with a small minority, however, identifying as males (Collaer & Hines 1995; Berenbaum et al. 2000; Berenbaum & Bailey 2003; Cohen-Bendahan et al. 2005). Dittmann et al. (1992) interviewed 34 CAH females and 14 of their sisters and found that CAH individuals tended to have more homosexual relationships and masculine gender role patterns than that of their non-CAH sisters; moreover, their index of sexual orientation indicates that from 11 to 21 years old the degree of homosexuality and especially bisexuality of CAH females was higher than that of their non-CAH sisters and, conversely, the degree of heterosexuality was higher in the sisters. In a study of 43 CAH girls aged between 3 and 18 years old that were reared as girls, Berenbaum & Bailey (2003) found that the CAH girls scored, in their gender identity, intermediate between 'tomboy' and control, with only 11.6% (5/43) of the CAH girls scoring beyond the range of controls. Hines (2006) reports a higher tendency of bisexual sexual orientation among CAH girls. The same result was obtained by Money et al. (1984), with CAH females showing elevated incidence of non-heterosexuality. Bachelot et al. (2007) also report that 22.2% (2/9) of females diagnosed with the classical salt-wasting form of CAH and who also had sexual experience are homosexual.

Meyer-Bahlburg et al. (2008) have recently published the most comprehensive review (see their Table 1 that summarises the major works published between 1968 and 2007) of the association between CAH and homosexual orientation in women, confirming previous reviews (see above) that identified a consistent minority of CAH women displaying a homosexual/bisexual orientation. Meyer-Bahlburg et al. (2008) also report the results of their own work on CAH women attending two New York City (USA) clinics. The women were administered a Sexual Behaviour Assessment Schedule (SEBAS-A) questionnaire and their sexual orientation was rated with the *Kinsey Scale*. Their results indicate that CAH women had greater tendency towards homosexuality and bisexuality than non-CAH controls. CAH may also affect males, but in this case results tend to be mixed. In comparisons between CAH males and their non-affected brothers, they were found to be similar in both gender identity and sexual orientation (Cohen-Bendahan et al. 2005 and references therein). On the other hand, CAH males did show a degree of feminisation in some personality traits in the work by Mathews et al. (2009).

Table 5.1 summarises the main results of studies carried out on human endocrine syndromes. In sum, these studies indicate that mutations at relevant loci affecting endocrine mechanisms may not always have large effects on development of sexual orientation, gender role and gender identity in affected individuals, owing to: (a) availability of multiple loci that can buffer normal development from the early effects of mutations at one of those loci; (b) potential, but still controversial, contribution of oestrogens when androgen receptors are impaired, as occurs in rodents, and (c) effects of learning on sexual behaviour that may be more or less effective in accordance with the severity of the endocrine condition. CAH is to a certain degree an exception, with a small but consistent minority of CAH females showing a comparatively elevated degree of non-heterosexuality compared with non-CAH females. One obvious difference between CAH and the other syndromes mentioned here is that CAH involves an excess of androgens and oestrogens, whereas all the other syndromes involve a potential deficit of production or impairment of action. Potential deficits of specific hormones can be compensated by alternative biosynthetic pathways, or by alternative mechanisms involving other hormones, thus decreasing the impact of the syndrome on development of sexual orientation, whereas excesses, at least in this case, seem relatively more likely to produce an effect on sexual orientation in humans.

Table 5.1. Human endocrine syndromes and their associated effects on gender role

Syndrome	Steroid/s affected during development	Effect on steroid levels/action	Effect on gender role	Reference
Altered testosterone biosynthesis	Testosterone, oestradiol	Lower levels	Variable	Wilson 1999
5α-reductase deficiency	Dihydrotestosterone	Lower levels	Variable	Wilson 1999
Congenital aromatase deficiency	Oestrogen	Lower levels	Negligible	Carani et al. 1999; DuPree et al. 2004
Complete androgen insensitivity syndrome	Androgens (receptor)	Lower action	Variable	Macke et al. 1993; Quigley 2002
Congenital adrenal hyperplasia	Testosterone, oestradiol	Higher levels	Consistent masculinisation in some affected women	Dittmann et al. 1992; Berenbaum & Bailey 2003; Hines 2006; Meyer-Bahlburg et al. 2008; Mathews et al. 2009

Circulating hormones in the adult and homosexual behaviour

That hormones could contribute to the early organisation of homosexual behaviour seems to be clear from the above review and from the additional ontogenetic works discussed in Chapter 4 (see also Adkins-Regan 1988). However, what kind of action do hormones exert in the adult individual that may affect same-sex sexual behaviour? Can hormones *activate* homosexual behaviour in adults? Can they even exert an *organisational* effect in adult organisms associated with plasticity of the central nervous system and learning (see, for example, Moore 1991)?

Hormones and homosexual behaviour in primates

Circulating gonadal steroids that are elevated during breeding clearly influence the motivation to copulate in most vertebrates (see Goy & Roy 1991; Wallen 2001; Wallen & Baum 2002 for reviews). In some species, however, even during periods of elevated mating activity, rates of copulation by females may vary not only according to the phase of their ovarian cycle but also according to the social context (e.g. presence or absence of reproductive competitors) (Wallen 2001), indicating that in fact there is a dual control of mating behaviour: sex-hormonal mediation and presumably direct nervous system control (see also Goy & Roy 1991). Wallen (2001) exemplifies this dual control also in males in a study of rhesus monkeys (*Macaca mulatta*), where adult males may mount other males independently of their circulating levels of androgens, but suppressing adult gonadal androgen production reduces male mounting. In addition, the more experience males have at mounting females, the more they prefer females as sexual partners (Wallen 2001). Goldfoot *et al.* (1978) also suggest a diverse control of copulation behaviour in *Macaca arctoides* as ovariectomised and adrenally suppressed females, that is females who lack two major endocrine organs producing steroid hormones, continue to copulate.

Copulatory behaviour and reproduction can be decoupled in vertebrates, primates in particular. For instance, copulations of any kind may sometimes occur outside the main mating season as well as during mating periods, suggesting a dual activational control of copulations (Crews 1987; Wallen 2001; Ziegler 2007). A behaviour that presumably relies on this dual neuroendocrinological control is *postconception mating*. Postconception mating is expressed by females who engage in copulatory activities with males outside their fertile period, a pattern frequently observed in primates (Hrdy 1974; Saayman 1975; Taub 1980; Ziegler 2007; Engelhardt *et al.* 2007). When this or any other kind of mounting occurs in a homosexual context, socio-sexual explanations such as expression of dominance, but also affiliation and reconciliation, could also apply (Rowell 1972; Ziegler 2007; see also Chapters 8 and 9).

A more specific focus on hormones suggests that in many species of primate the female hormonal cycles influence, sometimes quite dramatically, the social and sexual interactions among members of a group or band (Wallen & Tannenbaum 1999). This is reflected, for instance, in the temporal coincidence between copulations, including homosexual copulations, and circulating levels of sexual hormones (Nieuwenhuijsen *et al.* 1987).

Female primates regularly copulate not only during the oestrous cycle, but also during pregnancy (Rowell 1972; Wallis 1982; Engelhardt *et al.* 2007). In rhesus monkeys, a species that breeds seasonally (Zehr *et al.* 1998), mounting by females is coupled to their ovarian cycle (Pope *et al.* 1987; Wallen & Tannenbaum 1999), with copulations being mainly confined to the follicular phase of the cycle and sharply declining as soon as the corpus luteum starts secreting progesterone (Gordon 1981). The effect of sexual hormones on mounting was clearly demonstrated by Zehr *et al.* (1998) in seasonally breeding rhesus females by implanting ovariectomised females with oestradiol during the non-breeding season. After implants were provided, females initiated sexual interactions with males, with a slight increase in mounting as well. Males, on the other hand, remained relatively uninterested in sexual interactions with females. Moreover, in a previous study of the same individuals, oestradiol treatment of females during the non-breeding season also increased female–female sexual behaviour, an observation that Zehr et al. (1998), however, did not repeat. Therefore, in *M. mulatta* mounting, including same-sex mounting, can be enhanced by circulating levels of sexual hormones (see also Pope et al. 1987). Pope et al. (1987) showed that although circulating oestradiol can control the display of sexual behaviour in females, the specific partner for sexual intercourse also depends on availability: if males are not interested in mounting then females may resort to female–female mounting when they are under the influence of sexual hormone implants (Pope et al. 1987).

In a troop of *Macaca fuscata* studied by O'Neill et al. (2004a) in captivity, the proportion of females engaging in same-sex sexual behaviour varied interannually from 51% to 78%, a higher frequency than that reported in the source population (i.e. Arashiyama B) from Japan: 27%–47%. In part, the very high female-bias in the adult sex ratio occurring in the troop studied by O'Neill et al. could explain this difference in terms of differences in the availability of partners for copulations during periods of increased circulating oestrogens.

O'Neill et al. (2004a) studied hormonal correlates of same-sex sexual behaviour in their *M. fuscata* females. They measured both oestrone (E_1, a metabolite of oestradiol) and pregnanediol (PdG, a metabolite of progesterone) in faeces during the ovulatory cycle. PdG shows baseline levels during the follicular phase of the cycle, with a slight increase during the periovulatory phase, whereas in such a phase E_1 reaches a peak, and finally both hormones return to baseline levels during the luteal phase of the cycle. Of the eight sexually mature females that they observed, five engaged in same-sex sexual behaviour; of those five females, three did so during their cycling whereas two did so during pregnancy. Among those females engaging in homosexual mounting during cycling, the frequency was lowest during the luteal phase (heterosexual mounting was also lowest during that phase), whereas the highest frequency of all kinds of mounting, homosexual and heterosexual, occurred during the follicular and periovulatory phases. This suggests a similar hormonal activational mechanism for copulations with partners of any sex in females of this species.

O'Neill et al. (2004a) report rates of same-sex mounting among females that are very similar to rates of heterosexual mounting during the ovulatory cycle. In pregnant females, however, same-sex mounting coincided with a drop in circulating PdG, suggesting that potential hormonal control of female mounting during this period may be exerted by other hormones such as oxytocin and vasopressin (which will be reviewed towards the end of this section) or by non-hormonal mechanisms. O'Neill et al. suggest that *M. fuscata* females may engage in same-sex sexual interactions because (a) they minimise interruptions from males, (b) there are no available male partners owing to sex-ratio bias, (c) they are a 'novelty', and/or (d) they are purely seeking sexual gratification. In this chapter I will suggest a simpler and, in my view, more effective mechanism that could explain female same-sex sexual behaviour in macaques and perhaps other social primates that I term the *Lock-in Model*.

Gouzoules & Goy (1983) studied 79 female *M. fuscata* held in captivity in a mixed-sex troop, all of them were involved in consortships with the males. Of the 79 females, 39 (49.4%) only formed consortship with male partners which, sometimes, involved the female mounting the male, with a trend for pregnant females to mount the male more frequently than non-pregnant females. A total of 40 (50.6%) females established consortships with female partners and of those 40 females 17 exclusively mounted another female. Therefore 21.5% of females did not engage in mounting with males, but did so with another female. The sex ratio in this troop was 3:1 F:M if only adult males are taken into account and 1.6:1 F:M if all males are taken into account. This means that at any time a minimum of 0.6/1.6 = 37.5% of females did not have access to a male, which could at least partly explain the presence of 21.5% of females that only mounted another female. The fact that specific females who are sexually motivated to engage in copulation may nevertheless be unable to access an equally sexually motivated male partner may be due to dominance interactions between females preventing access to males, but also to unavailability of a male (because all males are engaged with another female or because those available may be unmotivated); or the female being already engaged in copulations with another female when a male becomes available. If a female behaviourally 'locks herself in' a mounting sequence with another female, primed by circulating hormones and motivated by accessibility to a same-sex partner and inaccessibility to an other-sex sexual partner, she may ignore a male during that period even if one becomes available. Such a 'lock-in' behavioural pattern may be adaptive in a heterosexual context in order to achieve ejaculation and fertilisation, and as a male strategy in the context of sperm-competition, but it may seem *prima facie* maladaptive in a homosexual context. However, such an initially maladaptive behaviour could become adaptive in the context of socio-sexual functions of same-sex mounting such as reinforcement of dominance or affiliative relationships.

Bonobos (*Pan paniscus*) are well known for routinely engaging in same-sex sexual behaviour. Females are particularly inclined to rub their genitalia in a behaviour called Genito-Genital rubbing, or GG-rubbing (Figure 5.2). Alan Dixson will further elaborate on bonobos' GG-rubbing in Chapter 9, here I focus on endocrinological correlates of GG-rubbing. In a recent study carried out by Adinda Sannen and collaborators on bonobos held in captivity, GG-rubbing was frequently displayed in contexts of access to food, being also associated with a longer period of proceptivity during the menstrual cycle in this species compared with its closest relative, the chimpanzee (*Pan troglodytes*) (Sannen 2003). The latter result was also reported by Shimizu *et al.* (2003), who found that bonobos had an extended follicular phase – which is associated with increased proceptivity, at least in some primates (O'Neill *et al.* 2004b) – compared with chimpanzees, orang-utans, gorillas and macaques.

Figure 5.2. Female bonobos (*Pan paniscus*) GG-rubbing in the *Lola ya* bonobo sanctuary, Democratic Republic of Congo. Photo courtesy of Vanessa Woods.

Sannen (2003) studied bonobos in various zoos in both Europe and the USA and measured 5α-androstane-17α-ol-3-one, a urinary metabolite of testosterone. Although 5α-androstanolone concentration is higher in males than in females in both bonobos and chimpanzees, the difference between the sexes is much lower in bonobos. On average male bonobos have 37.8 ng/mgCr of 5α-androstanolone, whereas females have 29.2 ng/mgCr, which is similar to the concentration found in female chimpanzees (about 20 ng/mgCr). Male chimpanzees, however, have higher urinary levels of 5α-androstanolone (about 150 ng/mgCr) than male bonobos (Sannen 2003). These results suggest that elevated female same-sex sexual behaviour in bonobos compared with chimpanzees is unlikely to be only a result of different levels of circulating testosterone in females, although lowered levels of testosterone in males are consistent with the less aggressive nature of bonobo males.

Marshall & Hohmann (2005) also measured urinary testosterone in bonobos, but they did so in a wild population and the results reported were restricted to males. Adult males in Marshall & Hohmann's (2005) study had 525 pmol of testosterone/mgCr on average, a value higher that that found in subadult males (309 pmol/mgCr), suggesting that in this species adult male same-sex sexual behaviour is maintained in concomitance with an increase in testosterone production during development. In a study carried out at Lui Kotal, in the Democratic Republic of Congo, Dittami et al. (2008) found that both urinary 17α-androgen and testosterone metabolites were higher in male than female bonobos in the wild, but cortisol metabolites did not differ between the sexes. If GG-rubbing has a socio-sexual function associated with affiliation and coping with social stress in bonobos, it could be predicted that levels of cortisol, a hormone secreted in response to stress, in this species should be lower than in the congeneric chimpanzee in the wild. In fact, Dittami et al. (2008) report levels of urinary cortisol metabolites of about 50 pg/mgCr in male bonobos, whereas Muller & Wrangham (2004) report levels of 281–568 pg/mgCr in the morning and 120–277 pg/mgCr in the afternoon for male chimpanzees at the Kibale National Park in Uganda. In the case of females, bonobos' urinary concentration of cortisol metabolites is around 40 pg/mgCr (Dittami et al. 2008), whereas in adult female chimpanzees sampled at the Kibale National Park the concentration is much higher: 70.6 ng/mgCr for immigrant adult females and 12.3 ng/mgCr for resident adult females (Kahlenberg et al. 2008). Also in partial support for the role of same-sex genital contact in both males and females as a strategy to decrease stress under conditions of potential social tension, Hohmann et al. (2009) have recently shown that bonobos held at the Frankfurt Zoo in Germany do respond to a situation of competitive conflict for clumped food with an increase in both cortisol, as measured from saliva samples, and the general level of aggression. This is followed by an increased rate of genito-genital contact that in turn seems to lead to a decreased relative level of salivary cortisol.

Shimizu et al. (2003) analysed urinary concentrations of oestradiol metabolites (i.e. oestrone conjugates, E_1C), pregnanediol (PdG) and follicle-stimulating hormone (FSH) in captive *Pan paniscus*, *Pan troglodytes*, *Pongo pygmaeus*, *Gorilla gorilla* and *Macaca fascicularis* and compared them with the levels found in humans. All apes shared with humans two peaks in E_1C concentration, one at about mid-cycle and the second in the luteal phase; the pattern was not found in macaques. PdG levels were also similar among apes and humans, although the similarity was closer between bonobo females and women (Shimizu et al. 2003). Although suggestive of a common endocrinological mechanism underlying sexual behaviour in apes and some convergence in the reproductive biology between the two species that display the highest levels of female same-sex sexual behaviour (i.e. bonobos and humans), these trends will need to be investigated with an increased sample size as Shimizu et al. (2003) only studied a very small number of individuals.

The effectiveness of hormones as activators of sexual behaviour depends not only on the concentration of the hormone in circulation but also on other factors such as number and distribution of hormone receptors on target tissues, which may help explain some interspecific differences. On the other hand, Jurke *et al.* (2001) suggest that testosterone in bonobos positively correlates with GG-rubbing. In fact, Sannen (2003) does report a rough association between 17α-androstanolone concentration and GG-rubbing among four female bonobos, although a better association is found between 5α-androstanolone and heterosexual copulations. This suggests that GG-rubbing may be affected by testosterone, but is not under exclusive testosterone control, a conclusion supported by short-and long-term data on 5α-androstanolone concentrations and frequency of GG-rubbing (Sannen 2003). GG-rubbing is better associated with urinary levels of 5α-androstanolone over the longer term, as females that tend to be more engaged in GG-rubbing also have higher concentrations of 5α-androstanolone. The association is lost as soon as we consider fluctuations in GG-rubbing activity within a day (Sannen *et al.* 2005). This suggests that GG-rubbing is likely to be under the dual control of hormones and direct central nervous system mechanisms.

Hormones and homosexual behaviour in other mammals and in birds

Hormonal activation of various aspects of same-sex sexual behaviour in adults has also been studied in rodents. Circulating sex steroids in adult rodents can affect the motivation to engage in sexual behaviour, the expression of feminine/masculine sexual behaviour, and preference for social and also sexual partners of a specific sex. Both testosterone and oestradiol have activational effects on various aspects of male and female sexual behaviour (Goldfoot & Van der werff ten Bosch 1975; Gorski 1987; Slob *et al.* 1987; Woodson *et al.* 2002; Hull & Dominguez 2007), with both mounting and lordosis being elicited by sex steroids in the two sexes, even though lordosis is a more female-typical and mounting a more male-typical behaviour (Gorski 1987).

Although female–female mounting in equids is unusual, when it occurs it may be associated with either non-ovulatory oestrous cycles that may be manifested late in the breeding season, with mares in oestrus mounting non-oestrous females (Asa *et al.* 1979); or with diverse phases of the ovulatory oestrous cycle (Gastal *et al.* 2007). Gastal *et al.* (2007) analysed in detail 15 same-sex sexual interactions that occurred during the mating season among *Equus caballus* mares. Of those interactions, 13 occurred during the follicular phase of the mounting mare, whereas in the other two a mare in the luteal phase mounted another that was in the follicular phase. The frequency of mounting among mares was higher early and late in the season than during the mid-period. Gastal *et al.* (2007) also measured circulating steroids in both the mounting and the standing mares, along with control mares that were not involved in mounting. Although none of the statistical analyses reached significance, there was a consistent qualitative trend for circulating concentrations of testosterone, androstenedione, oestradiol, oestrone and progesterone to be slightly higher in the mounting mares than in controls, with the latter having slightly higher levels than standing mares. Female donkeys (*Equus asinus*), known as jennies, also tend to mount other females when both mounter and mountee are in oestrus, and they tend to do so rather more frequently than is the norm among *E. caballus* mares (Henry *et al.* 1991, 1998).

In female koalas (*Phascolarctos cinereus*) female–female mounting is unusual; however, it has been observed in captivity with the involvement of females in oestrus (Feige *et al.* 2007). In domestic cattle oestrus is also associated with same-sex sexual behaviour in females (see, for example, Cecim & Hausler 1988; Rodtian *et al.* 1996; Medrano *et al.* 1996), eliciting both mountee and mounter behaviour (Williamson *et al.* 1972; Hurnik *et al.* 1975; Rodtian *et al.* 1996). Cows not in oestrus usually avoided being mounted, whereas cows in late pro-oestrus, oestrus or early meta-oestrus would perform mounting activity (Williamson *et al.* 1972; Medrano *et al.* 1996). However, some non-oestrous cows may mount oestrus cows (Williamson *et al.* 1972 and references therein), perhaps as a display of dominance (Hurnik *et al.* 1975; see also Cecim & Hausler 1988). Oestrous cows also have a tendency to associate together and, while they do so, engage in mounting with each other (Williamson *et al.* 1972). One potential adaptive value of female–female mounting, at least for the mounter, is that the behaviour may cause pseudopregnancy in the mountee (see also Chapter 7 for a further discussion of pseudopregnancy effects). Therefore female–female mounting, although it is a proximate expression of the effects of circulating steroids on behaviour, may also have specific adaptive socio-sexual functions such as those associated with female–female competition in this case.

In sum, in both primates and non-primate mammals the activation of same-sex sexual behaviour in adults is proximately controlled by both hormonal and nervous system mechanisms, with the hormonal effects being particularly well documented in females.

Androgens have also been studied in adult birds in the context of same-sex sexual behaviour. Wingfield *et al.* (1982) studied the endocrinological correlates of female–female pairing in Western gulls (*Larus occidentalis*) in California. Birds in this population of *L. occidentalis* showed a tendency to form same-sex female pairs that could be long-lasting and involve various mating behaviours including courtship displays and, occasionally, mounting. This same-sex pair formation was a direct result of a female-biased sex ratio in the population (Wingfield *et al.* 1982; see also Chapter 8). Wingfield *et al.* (1982) measured circulating luteinising hormone, progesterone, dihydrotestosterone, oestrone and oestradiol in female gulls in this population and detected no difference between monosexually (i.e. female–female) and disexually (i.e. female–male) paired females in all hormones measured, with the exception of LH, which continued to increase for a longer period of time in monosexually paired females than in disexually paired females.

This pattern can be explained by the fact that not all those monosexually paired females were in attendance of eggs. Biparental care in this mainly socially monogamous species can explain the formation of monosexual pairs in demographic situations where the adult sex ratio is female-biased, but the endocrinological correlates of the monosexual pairing are indistinguishable from those found in the disexual pairs in females. I will return to the case of monosexual partnerships in female gulls in Chapter 8.

Non-steroid hormones

Although steroid hormones have been the major focus of research on the mechanisms of homosexual behaviour activation in male and female mammals, other hormones can also play an important role. In particular, the nine-amino-acid peptide oxytocin, which is mainly, but not exclusively, produced in the supraoptic and paraventricular nuclei of the hypothalamus, is a molecule of potential great interest (see Carter 1998 for a review).

Oxytocin (OT) can act as an endocrine hormone but also as a neurotransmitter in diverse brain regions such as the cortex, amygdala, ventral hippocampus, ventromedial hypothalamus, bed nucleus of the stria terminalis, anterior olfactory nucleus and others (Arletti *et al.* 1992; Carter 1992). Most of these are centres that may directly or indirectly affect sexual behaviour. OT is also produced in reproductive tissues such as the ovary and testis (Insel *et al.* 1997).

OT has been traditionally associated with activational roles in milk production, uterine contraction, foraging, grooming, parental behaviour, modulation of memory and motor activity (see, for example, Arletti *et al.* 1985). However, in male Wistar rats OT also shortens the ejaculatory latency and increases the number of mounts and, in general, facilitates male sexual behaviour (Arletti *et al.* 1985; Arletti & Bertolini 1985), whereas in female rats OT increases lordosis in a dose-dependent manner (Arletti & Bertolini 1985; Arletti *et al.*

1992). OT acts in synergy with gonadal steroids (oestradiol and progesterone in females, testosterone in males) to elicit sexual behaviours (see, for example, Witt 1999). In fact, the medial preoptic area (MPOA) of the hypothalamus, a region associated with sexual behaviour control and that responds to steroid input, is also linked to oxytocinergic neurons (Caldwell 1992). Both oestrogen and progesterone have the ability to trigger the release of oxytocin by those neurons (Caldwell 1992) and if the MPOA of female Sprague–Dawley rats is infused with an OT antagonist, the treated individuals decrease their receptive sexual behaviour (Caldwell et al. 1994).

The role of OT in the control of sexual behaviour has been reviewed by Carter (1992) and Insel et al. (1997), suggesting that OT can have an important role in the expression of sexual behaviour in humans in conjunction with but also independently of oestrogen and androgen action. Moreover, OT release in humans can be behaviourally conditioned, which may link OT production to specific cognitive stimuli, thus allowing endocrine mechanisms controlling sexual behaviour to operate in synergy with learning (Carter 1992). In rodents, there is substantial evidence that OT is involved in heterosexual partner preference for instance (Williams et al. 1994; Young et al. 1998; Cho et al. 1999; Pitkow et al. 2001; Keverne & Curley 2004).

The above evidence suggests that OT levels coupled with conditioning could be partly responsible for at least some cases of same-sex sexual behaviour in mammals. This is certainly an area that requires further research.

Studies should also focus on arginine vasopressin (AVP), another nonapeptide that is similar to OT. Apart from their function in control of sexual behaviour, both AVP and OT are also involved in the establishment of social bonds and could therefore mediate attachment between same-sex sexual partners (Keverne & Curley 2004). Their association with both the dopamine reward system (see the *Dopamine and serotonin: modulation of homosexual behaviour* section later in this chapter) and the processes of individual recognition mediated by olfactory stimuli, again suggests some potential role for conditioning in the establishment of social bonds between specific individuals, bonds that in various mammals could be mediated by olfactory cues (Young & Wang 2004).

With regard to homosexuality in humans, Purba et al. (1993) did carry out a study of the brain of male patients who had died of AIDS, comparing OT and AVP neurons (i.e. neurons containing receptors for OT or AVP) in the paraventricular nucleus of the hypothalamus of homosexuals and heterosexuals. What they found was that neither the number or cell volume of OT and AVP neurons differed statistically between homosexuals and heterosexuals, although the qualitative trends were for cell volumes for the two kinds of neuron to be larger in homosexuals than heterosexuals, whereas the number of OT neurons tended to be larger and the AVP neuron numbers smaller among heterosexuals than homosexuals.

In sum, same-sex sexual partner preferences could potentially result from: (a) the double action of OT and AVP on both the promotion of social bonding and sexual behaviour, with (b) same-sex sexual partner preference being reinforced by dopamine that (c) could mediate the association of specific same-sex stimuli (e.g. olfactory) with both social bonding and sexual behaviour via the mesolimbic reward system (Young & Wang 2004). A similar model for the proximate mechanisms of same-sex sexuality in humans, women in particular, has been recently suggested by Lisa Diamond (2008b). Direct empirical tests of this model, however, are still lacking.

Sex hormones and the *Lock-in Model* for homosexual behaviour in macaques

As we have already seen, in primates F–F mounting frequently occurs in concomitance with oestrus and pregnancy (Gouzoules & Goy 1983), suggesting an association with neuroendocrine factors. In Japanese macaques (*Macaca fuscata*) held in captivity, over 50% of heterosexual mountings occur during the initial three hours of the day (Hanby & Brown

Figure 5.3. One of the variants of female–female dorso-ventral mounting used by Japanese macaques (*Macaca fuscata*). Photo by S. Kovacovsky, from Vasey & Duckworth (2006).

1974). This may not only coincide with a peak in circulating testosterone during the night (as suggested by data from the congeneric *M. mulatta* (Goodman *et al.* 1974) and *M. radiata* (Kholkute *et al.* 1981) and a peak in circulating androstenedione that, at least in *M. mulatta*, is reached between 6 and 9 am (Goncharov *et al.* 1981), but it may also provide a relatively small window of opportunity for heterosexual copulations, which, in turn, may impinge on the accessibility of other-sex sexual partners depending on the adult sex ratio in the group. In Hanby & Brown's (1974) study colony of Japanese macaques heterosexual mounting, M–F and F–M, but also female homosexual (F–F) mounting, occurred almost exclusively during the mating season in the autumn and winter months (see also Wolfe 1986; Chapais & Mignault 1991; Vasey 2006a), whereas M–M mounting was spread the year round. However, the frequency of M–M mounting was higher just before and just after the peak in frequency of heterosexual mounting (Hanby & Brown 1974). This suggests that in females in particular, but to some extent also in males, same-sex mounting is at least partly modulated by endocrine mechanisms.

The full diversity of sexual partner combination during the breeding season was also observed in wild *M. fuscata* (Enomoto 1974), suggesting that the presence of a degree of variability in sexual partner combinations is not just an artefact of captivity (Figure 5.3). Captivity, however, can dramatically influence the probability and frequency of expression of same-sex mounting (see Chapter 2) especially when the socionomic sex ratio is highly biased (for example, see Figure 8.6). For instance, in a group held in captivity in Italy, Lunardini (1989) observed F–F and M–F mounting also during the non-breeding season.

Hanby & Brown (1974; see also Vasey & Duckworth 2006) also report a pattern of consortship of variable time length between two individuals. In particular, male–female and female–female pairs may protract their exclusive consortship for a variable period from prior to the onset of mounting all through the length of their mounting interaction and extending to a period after mounting as well. Consorting female partners often engage in mutual mounting performed in succession, the so-called *series mounting* (Wolfe 1986; Chapais & Mignault 1991; Vasey 2002a; Vasey & Duckworth 2006; Vasey

2006a). It can be easily seen how an equal degree of sexual motivation between two females that can temporarily access only each other for sexual intercourse may lead to series mounting between the two. Whether series mounting may also be co-opted in socio-sexual processes associated with affiliation/coalition building is perfectly plausible, as sexual and socio-sexual functions of same-sex mounting are not mutually exclusive. Moreover, we would expect that as soon as socio-sexual functions are selected, then there may even be occasions when the need to engage in affiliative behaviours with other females, for instance, may be more important than the need to copulate with a specific male, at a specific point in time, for reproductive purposes (data available for *Pan paniscus* females are supportive of this possibility; Paoli et al. 2006).

O'Neill *et al.* (2004a) suggest that F–F mounting and consortship in *M. fuscata* may be explained by (a) females actively choosing another female sexual partner to prevent interactions with non-preferred males, and/or (b) for 'novelty' or 'sexual gratification', and/or (c) due to a female-biased ratio of preferred consorts. The first hypothesis can explain a consortship and mutual support in defence against males' advances, but does not necessarily explain homosexual mounting. The second hypothesis can explain mounting but, presumably, F–F mounting is not the only way to achieve sensory gratification. The third hypothesis assumes, rather than explaining, preference for specific consorts. In fact, the manifestation of same-sex sexual behaviour in female macaques only requires: (a) neurohormonally primed sexual motivation, (b) sex bias in the availability of willing sexual partners (see also Figure 8.6, which predicts that F–F mounting in *M. fuscata* should reach very low levels when the sex ratio in the troop approaches parity), and (c) a relatively rigid behavioural sequence during which the two partners are unwilling to break a temporary consortship once such short-term consortship is established. I call this the *Lock-in Model*. The *Lock-in Model* can explain not only the observations that females engaging in same-sex sexual behaviours may ignore available males, but also why those same females can copulate with the previously rejected males at a later time, and, if the sex ratio is dramatically female-biased, why a small number of females may be only observed in same-sex sexual interactions. The latter case may be further exacerbated by dominance relationships among females that can translate into socio-sexual behaviours such as mounting, and dominants actively preventing subordinates from accessing males. Additional socio-sexual functions of F–F mounting are also possible, such as establishment of alliances that can explain reciprocal mounting between two females, but such functions will be reviewed in Chapters 8 and 9.

The most controversial part of this model is probably the assumed rigidity of the pattern of consort association during an interval of time that includes mounting. Why should a rigidity of this kind be selected? Why not retain full flexibility and cut short the homosexual interaction as soon as a willing male becomes available? The model requires that there must be a fitness premium in maintaining a consortship for a period of time until the full sequence of behaviours from invitation to mounting, to the actual mounting to ending of the sequence is completed. The benefits of maintaining the behavioural sequence should be clear in a heterosexual context, where mounting is expected to lead to ejaculation and presumably to fertilisation. Cutting the sequence short because a better male happens to pass by is likely to attract a retaliatory reaction from the current male partner. The larger the cost a female may incur from this retaliation and its consequences, the less likely it is that full flexibility will be selected. If selection for flexibility of a specific behaviour is decreased by those costs, then individuals will be more likely to be locked-in during the copulatory sequence, whether they are two females or one male and one female. Additional socio-sexual functions of mounting may simply strengthen the lock-in period. A similar model that interprets the patterns of homosexual behaviour in females as a side effect of their heterosexual behaviour was also proposed by Symons (1979) with regard to explaining partner selectivity in F–F homosexual mounting.

In what follows I provide a detailed analysis and constructive critique of a very impressive body of work produced by Paul Vasey and his collaborators, from the University of Lethbridge in Canada, on same-sex sexual behaviour in female *M. fuscata*. Following the *Lock-in Model* suggested above, my comments and suggestions provide alternatives to some of Vasey and his team's interpretations of their results, but I want to make clear that I do not dispute the quality of their observations as such.

Vasey *et al.* (2008a) have recently reported results of non-mounting sexual behaviours performed by females in either homosexual or heterosexual consortships and found that although most behaviours are not sexually dimorphic some are and in the latter case females behave similarly whether they are in a homosexual or a heterosexual consortship, suggesting that females involved in same-sex sexual interactions are feminised, not masculinised. Vasey *et al.* (2008a) also show a result that it seems to suggest a socio-sexual function of female same-sex mounting in *M. fuscata*. In fact, sexual vocalisations, although they are uttered significantly more frequently by all kinds of females than by males, are also produced slightly more frequently by subordinate (39.7%) than by dominant (26.8%) females in homosexual consortships, the difference being only marginally not significant ($p = 0.054$). That is, subordinate females seem to be adopting a relatively more feminised pattern of use of sexual vocalisations during homosexual consortships than dominant females.

Paul Vasey (see, for example, Vasey & Duckworth 2006, 2008) has proposed a model to explain the proximate causes of same-sex mounting in *Macaca fuscata* based on the rewards, in terms of 'pleasure', accruing to both members of the homosexual consortship. Although this is an entirely plausible proximate mechanism (see the analysis of hedonic aspects of same-sex sexual behaviour among primates in Chapter 9) which is not necessarily alternative to the lock-in and the socio-sexual mechanisms suggested above, it should be further refined in this case, in order to explain why, if the pursuit of immediate pleasure is the major drive behind F–F mounting in *M. fuscata*, those females are not more prone to engage in self-masturbation than they are. Masturbation would provide a more immediate, easily accessible, hassle-free means of experiencing pleasure than the lengthy, perhaps frustrating, and often complex process required in seeking a partner (whether of the same or different sex). A 'pleasure-seeking' model would predict that frequency of same-sex sexual behaviour should be positively correlated with self-masturbation frequency. The positive correlation is expected as pleasure rewards could be obtained similarly through same-sex mounting when a partner is available or self-masturbation when it is not, and pleasure obtained from genital stimulation should not necessarily be biased in favour of any specific means used to achieve it (some data available from a single individual in Rendall & Taylor's 1991 work may be supportive of this). The *Lock-in Model* predicts a negative correlation between self-masturbation and same-sex sexual behaviour, simply because the more females are engaged in homosexual (and heterosexual) consortships the less likely it is that they will respond to their sexual drive through masturbation, as seeking a sexual partner is a greater priority than just seeking immediate pleasure rewards. Testing these predictions, however, will require a control of the socio-sexual functions of same-sex sexual behaviour; for example, the most subordinate female in the group may simply be rejected as sexual partner by most males, and she may also be a non-preferred partner for homosexual mounting by other females, leaving her to engage in a few homosexual interactions and also masturbation (Rendall & Taylor 1991).

Vasey & Duckworth's (2006) work suggests that mounter more than mountee females reap the most rewards in terms of 'pleasure'. This can explain series mounting, but it cannot explain those cases where a specific female is always the mounter and another is always the mountee whenever the two engage in same-sex sexual activity with each other;

such biases would be better explained as a result of dominance asymmetries, for instance (see Chapters 8 and 9). Moreover, affiliative socio-sexual functions also predict a symmetrical pattern of mounter–mountee roles. This possibility can be tested against the 'pleasure' mechanism by studying the association between frequency of same-sex mounting across female dyads and probability of coalition formation between members of the same dyads; if the latter correlation is positive, then series mounting is more than just a 'pleasure-seeking' behaviour. Sexual partner preference in F–F mounting has also been suggested by Vasey (2002a) as further evidence of a pleasure-seeking mechanism. However, active preference motivated by pleasure rewards must be tested against additional 'availability' models, as I explain in the next paragraph, and also against socio-sexual models involving dominance or affiliation, as suggested above.

Vasey & Gauthier (2000) provide a direct test of the same-sex partner preference hypothesis based on pleasure rewards by using a captive colony of *M. fuscata* held in Quebec, Canada. They compared the rate of female homosexual and heterosexual sexual behaviours in the colony during two periods: one in which the operational sex ratio was 0.31 (5/16) M/F and another when the ratio was 0.06 (1/16) M/F. In the latter case only one male was available. They observed an increase in F–F homosexual activities as the sex ratio became more female-biased, whereas heterosexual activity did not show a statistically significant increase. Moreover, they reported a pattern for females engaged in homosexual activities in the highly F-biased group to reject male solicitations. They interpreted their results as evidence of an active homosexual preference on the part of females:

> higher levels of female homosexual activity observed in the context of female-skewed operational sex ratios can be primarily attributed to *female preference for certain same-sex sexual partners relative to certain opposite-sex mates*. The likelihood that these preferences will be expressed simply increases whenever preferred same-sex sexual partners are abundant in the population
>
> (Vasey & Gauthier 2000: 23, italics mine).

Unfortunately, their experimental method did not allow for a test of relative change in the availability of 'preferred' females. In fact the ratio of 'preferred' vs. 'non-preferred' females remained constant because the total number and identity of females remained constant; the only variable that changed was the availability of males. Therefore their results could be easily explained by the *Lock-in Model* through increased relative availability of same-sex partners during periods of elevated sexual activity compared with other-sex partners. The 'preference' effect would be explained by the model as a side effect of females being locked in a behavioural sequence with each other even though a male becomes available during that period of time. In fact, Vasey & Gauthier's (2000) work suggests that the alleged 'preference' disappears as males become more accessible. In order to test for specific 'preference' effects within F–F sexual interactions, a study of preferential same-sex consortship should be carried out first among females, where biases in interindividual interactions are noted, and then an experiment should be performed where the ratio of preferred vs. non-preferred females (as defined by using the above criterion) is changed while keeping the total number of females and the total number of males (i.e. the sex ratio) constant. Vasey & Gauthier's (2000) hypothesis predicts that F–F mounting should increase as the proportion of 'preferred' females increases, whereas the *Lock-in Model* would predict that relative frequency of F–F mounting should not change significantly if the sex ratio remains constant.

In a subsequent work, Paul Vasey showed that whereas all males harassed F–F consortships, only 19% of females did so, and once a consortship between two females was established the bond was defended against the intrusion of a third party (Vasey 2004). The fact that third party females tend to disrupt ongoing F–F consortships far less than males do suggests that females may not be competing among themselves for exclusive access to preferred female individuals. This suggests that the pattern of same-sex consortship observed among females seems to be more a reflection of a common mechanism that stabilises ('locks-in') a current consortship (whether M–F or F–F) for a given period of time, rather than an active preference of females to consort with a specific other female instead of a male as Vasey (2004) seems to suggest.

The *Lock-in Model* can also explain why females who are engaged in a homosexual sexual sequence actively reject sexual advances made by males (Vasey 1998). From an evolutionary perspective, such a rejection would be perfectly adaptive in a heterosexual context as explained above. Moreover, the model can also explain not only the patterns of increased homosexual sexual behaviour between adult females but also that of increased heterosexual sexual behaviour between adult females and sexually *immature* males that occurs when the sex ratio becomes more female-biased (Wolfe 1986): as adult individuals, both males and females, become locked into a sexual interaction, the 'supernumerary' adult females may be more motivated to engage immature males in sexual activities. In addition, the *Lock-in Model* is entirely consistent with the generally feminised behavioural features displayed by females between same-sex mounting periods described by Vasey *et al.* (2008b).

If female *M. fuscata* are actually exerting a preference for same-sex sexual partners in order to achieve pleasure rewards, as Vasey and collaborators suggest, then we may also predict that the frequency of 'orgasm' in females (Troisi & Carosi 1998) should be higher or more easily achieved, at least in some individuals, when the partner is female more than when it is male. This requires specific comparisons of the same individuals engaged in sexual behaviours with male and female partners and an association between the relative levels of orgasmic response in those contexts with relative frequency of sexual partnerships with males and females. The pleasure–reward hypothesis predicts that females should prefer sexual partnerships with individuals with whom they achieve orgasm more easily or at higher frequency, and that there are females who achieve orgasm more easily with other females than with males, after having experienced sexual intercourse with both. I am not aware of any specific test of these predictions.

More recently, Vasey & Pfaus (2005) have reported behavioural results for *M. fuscata* held in captivity in mixed-sex groups of 16 sexually active adult females and 5 sexually active adult males. There were also one sexually inactive adult female and 16 sexually inactive immatures in the group. All females engaged in same-sex mounting during the breeding season, with the proportion of consortships being very similar between homosexual (55%) and heterosexual (45%) combinations. Interestingly, Vasey & Pfaus (2005) mention how 'competition' between males and females is expressed in terms of two females maintaining their already established same-sex mounting sequence against the interference of an interloping male. The point here is whether those two females 'chose' to maintain their homosexual interaction in spite of the presence of an available male, as Vasey & Pfaus (2005) seem to suggest, or whether they were just locked into a relatively stable behavioural interaction not easily disrupted by any conspecific (whether male or female) as the *Lock-in Model* suggests. An active choice mechanism would predict exclusive homosexuality among at least some females, whereas the *Lock-in Model* predicts functional bisexuality. That is, females that engage in homosexual sexual interactions at one time may as easily engage in heterosexual sexual interactions at another. This, apparently, is what occurs in most if not all *Macaca* populations studied so far.

Male–male mounting also occurs in *Macaca fuscata* (Hanby & Brown 1974). Although it seems to be more widespread throughout the year than F–F mounting, it nevertheless peaks in frequency towards the end of the breeding season (Hanby 1974), thus suggesting that endocrinological factors may also affect M–M mounting in this species in addition to the effects of unavailability of sexually receptive females.

Same-sex mounting is also observed in other *Macaca* species. In rhesus monkeys (*Macaca mulatta*) females also engage in same-sex mounting during the breeding season, with the mounter role being more prevalent during the early stages of the menstrual cycle, whereas the mountee role is more frequent in the middle, ovulatory phase of the cycle (Akers & Conaway 1979). Both mounter and mountee females decrease homosexual sexual activity during the luteal phase of the cycle.

Reinhardt et al. (1986) studied same-sex sexual behaviour in male *M. mulatta*, describing an increase in frequency of the behaviour in a troop where the sex ratio was more male-biased (0.83 M/F vs. 0.56 M/F). Same-sex sexual behaviour of male *M. mulatta* was also studied by Gordon & Bernstein (1973) in an experimental setup where they established a mixed-sex group (7 adult M, 46 sexually mature F and 23 juveniles and infants), a small monosexual male group of 8 adult individuals and a larger monosexual male group of 20 adult individuals. Same-sex mounting increased dramatically during the mating season in both mixed-sex and monosexual groups. However in the latter it did so only if the all-male group was caged in visual contact with the mixed-sex group; if there was no such contact, the breeding season did not affect same-sex sexual behaviour frequency in the all-male group. Therefore, in male *M. mulatta* visual stimulation is also important to trigger the onset of sexual activity in the mating season (Gordon & Bernstein 1973), in support of the dual neuro- and endocrinological control of sexual behaviour that I have stressed in this chapter. Support for the *Lock-in Model* in *M. mulatta* comes from Huynen's (1997) study describing increased frequency of same-sex consortships among middle- and low-ranking females, as most of the heterosexual consortships involved high-ranking females. In this case male choice for higher-ranking females may have left subordinate females with no other option but to engage in same-sex sexual interactions under the influence of elevated circulating sexual hormones.

In a group of stumptail macaques (*Macaca arctoides*) studied in captivity by Chevalier-Skolnikoff (1976), males engaged in same-sex sexual behaviours that included mounting and sometimes anal intromission, during periods when all females were sexually unresponsive, being either pregnant or lactating.

In sum, within the genus *Macaca* same-sex mounting, especially among females but also among males, is strongly influenced by: (a) the endocrinological status of the potential interactants, (b) the behavioural patterns of consortship between two individuals (e.g. 'lock-in'), (c) the operational sex ratio and, in general, the relative availability of other-sex compared with same-sex sexual partners, and (d) socio-sexual interactions such as dominance and coalition/affiliation.

In the next three sections I analyse in some detail the cases of *Bos taurus*, *Ovis aries* and *Crocuta crocuta* and the potential role endocrine factors may play in explaining same-sex mounting (or lack thereof) in these species.

Bos taurus: hormones, sexual behaviour and the buller steer syndrome

Unfamiliar male cattle that are joined together in the same group usually engage in aggressive and sexual interactions (Mohan Raj *et al.* 1991). Male bovids or steers that are mounted by other males, the riders, are referred to as bullers (Irwin *et al.* 1979). The frequency of bullers in domestic cattle herds has been variably reported to range from 1.5% to 3.5% (Irwin *et al.* 1979; Edwards 1995; Taylor *et al.* 1997).

Irwin *et al.* (1979) indicated that about 80% (326/409) of bullers in their study had been treated with progesterone and oestradiol, whereas in a study by Jezierski *et al.* (1989) there was a negative correlation between circulating testosterone and progesterone in the buller and the number of mounts he received. Riding behaviour, on the other hand, is associated with elevated circulating testosterone (Jago *et al.* 1997). Edwards (1995) distinguishes between two categories of buller: Type I or 'true buller' and Type II or 'picked-on buller'. The latter, in Edwards' (1995) view, is a result of submission–dominance relationships among males in the herd (see also Klemm *et al.* 1983/84; Blackshaw *et al.* 1997; and Taylor *et al.* 1997), whereas the former is suggested to be a consequence of hormonal treatment of cattle: oestrogen treatment feminises the buller's behaviour and pheromone production, and riders respond to volatile oestrogenic compounds released by the buller with mounting (see also Blackshaw et al. 1997). In fact, phytoestrogens ingested with food, such as coumestrol, also increase the incidence of buller behaviour, with synergistic actions occurring when males also carry an anabolic steroid implant (Edwards 1995). Therefore the buller steer syndrome seems to be a proximate result of an activational (and perhaps organisational when the steer is young) role of oestrogen treatment on male cattle pheromone production and also sexual behaviour: directly in the 'passive' buller and indirectly in the 'active' rider, the latter presumably responding to pheromonal stimuli coming from the buller that are usually associated with oestrous females. However, this cannot be the whole story, as even in herds where many individuals are hormonally treated, only a small percentage of males become bullers. This strongly suggests a hormonal activation of genetically controlled propensities, whether such propensities are behavioural and/or biochemical (e.g. capacity to metabolise implanted oestradiol or progesterone into volatile metabolites that can act as female-like sexual pheromones). It is possible that interactions between genotype and levels and timing of exposure to steroids, in addition to dominance relationships among individuals, may suffice to explain the observed frequencies of bullers in different herds.

Ovis aries: the case of homosexual rams

Same-sex sexual behaviour is not uncommon in wild ovids; although it can be expressed in both males and females, it is more frequent in the former than in the latter (see Chapters 7 and 8). In general, ovids are polygamous and seasonal breeders, but rams are also capable of breeding in direct response to the availability of oestrous ewes (Roselli et al. 2004a). *Ovis* spp. males tend to roam widely and join flocks of females in search of ewes in oestrus. They use olfactory cues and the ungulate-typical flehmen response to detect sexually receptive females (Roselli et al. 2004a). In addition, in some species such as *Ovis canadensis* (Berger 1985) some males may display 'feminised' behaviours such as joining a female flock during periods of sexual segregation, being subordinate to some of the females in the flock and adopting a female-like urination posture in contexts where such a posture is not adopted by males. On the other hand, in *Ovis aries* males may also initiate homosexual mounting following a masculinised mounter pattern, with the male mountee usually resisting such attempts (Pinckard et al. 2000a).

Several domestic sheep (*O. aries*) breeds have been the focus of recent studies of same-sex sexual behaviour in males. The reason why *O. aries* has attracted so much attention is that some rams seem to have an exclusive homosexual sexual orientation; that is they ignore ewes in oestrus preferring to mount other rams, a phenomenon first described by Hulet et al. (1964) and subsequently characterised by Zenchak et al. (1981). This makes *O. aries* only the second mammal known, apart from humans, capable of displaying exclusive homosexuality.

Among the rams studied by various authors, four main phenotypes have been described: heterosexual, bisexual, homosexual and asexual; the last phenotype do not show sexual interest in either ewes or other rams. Table 5.2 summarises the relative frequencies of the different phenotypes found by various authors. Overall, exclusively homosexual rams account for 8.6% (range 7.4%–9.5%) of all rams in the populations studied. Several hypotheses have been proposed to explain such homosexual behaviour in rams, ranging from prenatal hormonal effects, to genetic mutations and postnatal social mechanisms (see Roselli et al. 2004b; Roselli & Stormshak 2009a,b for recent reviews). Here I will review this case and propose an explanatory synthesis.

Price et al. (1988) carried out sexual partner choice experiments to determine how exclusive the ram preference to mount other rams was

Table 5.2. Frequency distribution (%) of sexual preference phenotypes in rams

Heterosexuals	Bisexuals	Homosexuals	Asexuals	Reference
55.6	18.4	7.4	18.5	Price et al. (1988)
74.4		8.5	17.0	Perkins et al. (1992)[a]
55.6	22.0	9.5	12.5	Roselli et al. (2004b)
70.4			29.6	Hulet et al. (1964)[b]

[a]Perkins et al. (1992) did not distinguish between heterosexual and bisexual rams.
[b]Hulet et al. (1964) only distinguished between rams that were 'sexually inhibited' and those that were not, in heterosexual mating tests.

Figure 5.4. Male-oriented ram mounting a female-oriented ram in an experimental setup where the male-oriented ram had a choice of mounting either males or females. Photo courtesy of Charles Roselli.

(Figure 5.4). Test rams were simultaneously exposed to two unfamiliar sexually active rams and two ewes in oestrus. The four stimulus animals were restrained and only exposed their rear and side to the test rams, this was done in order to prevent the onset of agonistic interactions between individuals. Price et al. (1988) reported three kinds of sexually active rams according to their choice of mounting partner: (a) sexually active rams that only mounted ewes, (b) sexually active rams that only mounted other rams, and (c) sexually active rams that mounted both ewes and rams in variable relative frequencies, from those biased towards mounting males, to those preferentially mounting females, to those showing no preference between the two sexes. In addition, Price et al. (1988) also identified 18.5% of the rams that were sexually inactive.

Could this phenotypic diversity be the result of early experience, or lack thereof, with females and/or males during development? Katz et al. (1988) set themselves to test this hypothesis using 25 rams reared with both oestrous ewes and males between the ages of 4.5 months (i.e. after weaning) and 9 months and compared them with 23 rams that had been exclusively exposed to other males after weaning. When all rams were 10 months old they were tested for sexual orientation. The rams reared in disexual groups mounted more and achieved more ejaculations with an oestrous ewe than those reared with other males only, suggesting that post-natal learning experience may play some role in the development of sexual behaviour. However, most, but not all, males later developed a heterosexual mate preference, indicating that sexual orientation was not compromised by those early experiences. The exceptions described by Katz *et al.* (1988) came from two rams out of those reared in a disexual social environment – a result not expected from learning – and also one ram of those reared in a male monosexual social environment that not only developed, but also maintained, a sexual preference for males. Hulet *et al.* (1964) also suggested that learning is not a factor in the development of exclusive homosexuality in rams.

In a subsequent work, Price *et al.* (1994) also confirmed the important effect of early heterosexual experience on the development of sexual performance later in life, but not on sexual orientation. In fact, during their experiment they only found two of their control rams that simply did not show a sexual interest in females but instead preferred to mount other males. Could those males be responding to social dominance interactions with other rams? In their original series of experiments aimed at studying the sexual performance of rams reared in all-male groups, Zenchak & Anderson (1980) discovered that although some of the males could respond with the expected heterosexual behaviour when exposed to a ewe in oestrus, others did not show any interest in the ewe. This motivated them to test the rams for sexual preference between a ewe in oestrus and a heterosexually active ram. In this mate choice experiment heterosexually oriented rams preferred to mount ewes, whereas the rams that showed a low response in the previous experiment preferred to mount other rams. Zenchak *et al.* (1981) controlled for the effect of dominance, which resulted in subordinate rams not having access to the female, and once this was done by removing dominant rams, the rams that exhibited a homosexual orientation continued to do so and ignored the oestrous female. Roselli *et al.* (2004b) also report that dominance is not an explanatory factor for exclusive male homosexual orientation in this species.

Male ungulates rely heavily on their sense of smell to detect females in oestrus. Lindsay (1965) carried out a mate choice experiment comparing intact rams and rams that had undergone olfactory ablation, in their choice between a ewe in oestrus and an anoestrous ewe. These were all adult rams that had developed a heterosexual sexual orientation. Intact rams preferred oestrous females, whereas olfactory impaired rams approached the two kinds of female at random, and simply mounted the female that did not walk away. Although mating success was ultimately similar between the two kinds of ram, they differed in their mating behaviour. In fact, olfactory impaired rams did not show foreplay behaviour in advance of their mounting, nor did they show other typical male courtship behaviours, such as nudging (Lindsay 1965). Although the experiment did not test for choice between a male and a female sexual partner, it is clear that olfactory impaired rams lost their ability to discriminate sexual partners, while retaining their male-typical sexual behaviour of mounting. Although volatile chemicals coming from potential sexual partners seem to play a role in mate choice, they are obviously not sufficient on their own; visual stimuli are probably also needed, as demonstrated by an experiment carried out by Gonzalez *et al.* (1991). In this experiment, sexually experienced heterosexual rams did not respond to the smell of urine, wool or vaginal secretions of oestrous females with a surge in circulating luteinising hormone (LH) and testosterone (T) after the odours were presented without the visual stimulus of an oestrous female.

If olfactory mechanisms are involved, could specific alterations of the neuroendocrinological mechanisms controlling sexual behaviour, especially those associated with brain areas responding to the detection of volatile chemicals, explain the occurrence of homosexual rams? And more generally, are homosexual rams different from heterosexual rams from an endocrinological and a neurological perspective?

We have seen in Chapter 4 how male-typical sexual behaviour in some mammals can be affected by conversion of testosterone into oestrogen in the brain. Pinckard *et al.* (2000b) studied the sexual response of heterosexual rams that they called female-oriented rams (FOR), male-oriented rams (MOR) and sexually inactive rams (SIR) (the last simply do not mount either males or females). Mounting of an oestrous female declined in both FOR and MOR individuals after castration (see also Roselli *et al.* 2004a) and this decline could not be prevented by treatment with oestradiol (E_2). The oestradiol treatment was also ineffective in SIRs. Intact rams of the three types also did not differ in their circulating levels of oestrone (E_1), E_2 and testosterone (T). The authors also measured testicular venous oxytocin (OT) and also did not find any difference among the three types of male. The experiment by Pinckard *et al.* (2000b) suggests that it is not the differences in the basal levels of circulating steroids and OT that may explain differences in sexual orientation of adult rams. However, although MOR and FOR individuals have equal basal levels of T they do not respond equally with a T surge to the presence of a ewe in oestrus (Roselli *et al.* 2002b and references therein). FORs produced higher T levels than MORs and SIRs upon exposure to an oestrous female. This result is expected if MORs and FORs differ, for instance, in the sensory system detecting pheromones.

Stellflug *et al.* (2004) treated FORs, MORs and SIRs with naloxone, an opioid inhibitor that should favour release of testosterone in circulation. Both FORs and MORs responded to naloxone treatment with the expected surge in circulating T, whereas SIRs did not; the same surge was detected in circulating LH. Concentrations of T were indistinguishable between FORs and MORs. However, after exposure to a female in oestrus, MORs did not respond with a surge in LH whereas FORs did. Again, these results point to a potential difference between MORs and FORs at the peripheral sensory level.

Also consistent with peripheral sensory differences between FORs and MORs are the results obtained by Perkins & Fitzgerald (1992), who described a lack of LH surge in MORs exposed to either restrained male or female stimulus individuals, and yet those same rams responded to luteinising hormone releasing hormone (LHRH) treatment with both an LH and eventually a T surge. Perkins *et al.* (1992) also described a similarity between FORs and low-performance males (probably a mixture of MORs and SIRs) in their basal levels of LH, and a similar LH surge upon LHRH injection. In addition, T concentrations did not differ between the two groups of males. Exposure to oestrous ewes elicited investigative behaviour in low-performance rams but, with the exception of one individual, they did not respond sexually to the female presence. In addition, Perkins *et al.* (1992) described how long-term exposure to oestrous females increased LH secretion in FORs, but did not in low-performance males.

Perkins *et al.* (1995) treated four MORs and four FORs with oestradiol benzoate during both the breeding and the non-breeding seasons and detected no change at any time in circulating LH after treatment; that is, both MORs and FORs are equally de-feminised in their mechanisms that control the surge in LH (see Roselli & Stormshak 2009b for a review). However, after analysing the oestrogen receptor concentration in brain regions they found that the number was higher in the amygdala in FORs than in MORs and ewes. There was no statistically significant difference in the number of occupied and unoccupied oestrogen receptors in either the hypothalamus or the anterior pituitary between FORs and MORs. It is known that the amygdala receives sensory inputs from both the primary olfactory system and the vomeronasal organ and innervates central nervous system nuclei associated with control of sexual behaviour. Alexander *et al.* (1999) exposed FORs, MORs and low-performing rams, (the latter probably included mainly SIRs), to either a FOR or a female in oestrus through a fence that did not allow mounting. FORs performed more exploratory sniffs than MORs and low-performing rams towards the test individuals, whereas the latter two performed in a similar manner. In addition, FORs directed more sniffs to the oestrous ewe than to the ram, whereas both MORs and low-performing rams directed more sniffs to the ram than to the ewe, with the proportion being more biased among MORs (about 4:1) than among low-performing rams (about 2:1). This again suggests a difference at the sensory level among the three types of ram. Circulating levels of LH did not differ among the three types of ram before the test; however, they increased in FORs after exposure to the oestrous ewe but not after the exposure to the ram. Although both MORs and low-performance rams did not show a significant change in circulating LH after exposure to ram or ewe, the qualitative trend is the opposite to that found in FORs, with both MORs and low-performance rams showing a slight surge in LH after exposure to a sexually active ram and a decrease after exposure to an oestrous ewe. That is, MORs seem to be perceiving male rams as sexually attractive individuals and their neuroendocrine system responds accordingly.

Endocrinological evidence is also available with regard to stress response to restraint (Stellflug 2006). After subjecting rams of different phenotypes to one hour of restraint stress, Stellflug (2006) found that MOR, FOR and SIR rams responded with a similar surge in circulating cortisol and a similar decrease in circulating testosterone.

Alexander et al. (1993) found that although circulating T and oestradiol were similar between FORs and 'low performance rams', the latter being probably a mixture of MORs and SIRs, FORs had a higher proportion of their oestrogen receptors in the hypothalamic preoptic area (POA) that were occupied than the low-performance rams, although no difference was found for the medial basal hypothalamus and the amygdala. Moreover, low-performance rams had more occupied oestrogen receptors in the anterior pituitary than FORs. Although these results differ somewhat from what Perkins et al. (1995) found (see above) they nevertheless suggest that differences in some areas of the brain may be also important. With regard to circulating hormones, Resko et al. (1996) also stated that FORs had circulating levels of T, E_1 and E_2 statistically similar to those of MORs, a result consistent with the findings of Alexander et al. (1993). In addition, Resko et al. tested for the capacity of testes homogenates to synthesise progesterone (17OH-P) and T in vitro and found that MORs had a decreased biosynthesis capacity compared with FORs (Resko et al. 1996). Importantly, aromatase activity was lower in the POA of MORs than in that of FORs and it was undetectable in the anterior pituitary of both groups (Resko et al. 1996). Therefore, the work of Resko et al. suggests that MORs have testes that produce slightly less T and they also have a diminished aromatase activity in the POA compared with FORs. Similar results to those described by Resko et al. (1996) were also obtained by Roselli et al. (2002a). These results clearly indicate a source of differences between MORs and FORs at the brain level in addition to the peripheral sensory level.

The combined evidence reviewed in the preceding six paragraphs strongly suggests that homosexual rams are likely to be responding sexually to male-typical pheromones rather than to female-typical pheromones, that is, modifications should be probably sought in the biochemistry and cellular biology of the olfactory system. Neural circuitry linking olfactory tissues to the amygdala and other brain regions controlling sexual behaviour seems to be functioning in these rams in a manner similar to that found in female-oriented rams, as does their neuroendocrine system associated with steroid and corticosteroid hormone action. One important exception is the POA, which seems to differ in some of its properties between MORs and FORs.

One intriguing possibility that, to my knowledge, has not been tested empirically in this species is that modifications of peripheral sensory tissues (e.g. due to a mutation) may also contribute to modifications of specific centres in the brain that are connected with those sensory tissues. For instance, the amygdala projects to the hypothalamus, which is well known to be associated with control of sexual behaviour and where, as we will see below, an area known as the ovine sexually dimorphic nucleus (oSDN) is located in the POA; the amygdala also receives innervations from the olfactory bulb. The olfactory bulb, in turn, is neuronally connected with both the vomeronasal organ and the main olfactory epithelium. Using a mouse model, Peter Mombaerts and collaborators (Rodriguez *et al.* 1999; Mombaerts 2006) have shown how alterations of olfactory receptors in the vomeronasal organ may lead to changes in the patterns of coalescence of neuronal axons into specific olfactory bulb structures called glomeruli. Could those changes in olfactory bulb organisation also affect the organisation of regions of the limbic system that control sexual behaviour? Using a Wistar rat model, Wrynn *et al.* (2000) measured Jun (a transcription factor produced in response to neuronal injury) in the amygdala after the rats were subject to unilateral olfactory bulbectomy. The amygdala responded to unilateral bulbectomy by increasing the production of Jun, suggesting that distal neuronal alterations can cause some changes in the neuronal activity of brain centres that may directly control or that may be connected to areas that control sexual behaviour. Although this evidence does not prove the suggestion that changes in peripheral nerve cells of the olfactory tissues can alter the structure of specific brain centres during ontogeny, at least it is consistent with such a proposition.

Could sexual hormones be implicated in the causation of homosexual orientation in rams through an organisational effect during early ontogeny? Sexual differentiation in sheep occurs during prenatal days 30–100 of a 150-day gestation period (Roselli *et al.* 2003). We have already seen how gonadal testosterone can exert its organisational activity on the development of the brain, whether directly or after the aromatisation into oestrogen that occurs in brain tissues (Chapter 4). Prenatal androgen exposure in rams affects age at puberty (earlier in males) and male-typical sexual behaviour (Roselli *et al.* 2006a and references therein) and it could also affect the development of centres such as the medial-POA/anterior hypothalamus (MPOA/AH) that control sexual behaviour. Roselli *et al.* (2004a) found that a sexually dimorphic nucleus that they described in the MPOA/AH region of the brain was larger in FOR than in MOR individuals and also larger in FORs than in ewes. They labelled this region the ovine sexually dimorphic nucleus (oSDN), which could be homologous to similar regions in humans and rodents that I will describe later in this chapter. The oSDN has a high content of aromatase-expressing neurons (Roselli *et al.* 2007). Roselli *et al.* (2007) describe how prenatal testosterone exposure de-feminises and masculinises ewes, which as adults display an increased level of mounting behaviour and aggressiveness. Moreover, the oSDN of females prenatally exposed to testosterone propionate (TP) was larger than that of female controls, whereas the prenatal treatment with TP did not affect oSDN size in males. It is therefore possible that **lack** of early exposure to androgens in utero may have affected the size of oSDN in MORs compared with FORs or that the difference may have developed postnatally under the influence of modifications in the peripheral organs of olfaction, or perhaps even both, as the two mechanisms are not mutually exclusive. On the other hand, a previous study by Alexander *et al.* (2001) did not find a significant difference between FORs and MORs in both cell densities and cell sizes in the medial amygdala, preoptic area, bed nucleus of the stria terminalis and ventromedial hypothalamic nucleus, all brain areas associated with control of sexual behaviour.

If a mutation is the main cause of the development of homosexual rams, then the homosexual trait should be genetically heritable. In their review, Roselli *et al.* (2004b) report a value of genetic variance for the trait of 0.22; sexual performance and motivation also respond to artificial selection. Stellflug & Berardinelli (2002) carried out a series of experiments using Rambouillet sheep with the objective, among others, of determining whether selection for high or low reproductive rates in ewes would affect sexual behaviour and orientation in male offspring, that is, they in fact carried out a test of the *Sexually Antagonistic Selection* hypothesis for the evolution of homosexuality in sheep. They used rams born from lines of ewes held at Montana State University that had been selected for either high or low reproductive rates since 1968. They tested sexual performance of rams with an oestrus-induced ovariectomised ewe; rams that did not mount the ewe were further tested for sexual orientation by concurrently exposing them to two restrained rams and two restrained oestrous females. None of the six rams from the ewe low reproductive rate line behaved in a MOR fashion, whereas one out of four rams from the ewe high reproductive rate line behaved like a MOR. Although the sample sizes are too small to make the results statistically meaningful, it is interesting that the trend is consistent with the *Sexually Antagonistic Selection* hypothesis: homosexual rams are more likely to be born of ewes selected for high reproductive rates.

In sum, it is well known that sexual behaviour in both rams (Stevens *et al.* 1982) and ewes (Gelez & Fabre-Nys 2004) is modulated by olfactory and visual stimuli that can also affect learning processes. Both males and females produce specific pheromones that are sexual attractants to the other sex (Stevens *et al.* 1982; Gelez & Fabre-Nys 2004). From the available evidence reviewed here, it seems plausible that homosexual *O. aries* rams are masculinised males (in fact they are mounters, not mountees) that respond sexually to stimuli coming from other males rather than females. If differences are to be found between heterosexual and homosexual rams, they are more likely to occur in the peripheral sensory tissues involved in odour (e.g. pheromone) detection, with the basic brain organisation and physiology seemingly similar to that of heterosexual rams, with the exception perhaps of centres such as the oSDN of the MPOA/AH. From an evolutionary perspective, male-oriented rams could be a result of *Sexually Antagonistic Selection* processes.

Crocuta crocuta: selection of masculinised females

The spotted hyena (*Crocuta crocuta*) is a large carnivore and member of the family Hyaenidae that lives in mixed-sex, mixed-age social units called clans, which have a fission–fusion social organisation not too dissimilar to that of some primates (Dloniak *et al.* 2006b). Within a clan the social structure is based on dominance of adult females over immigrant males, whereas males establish a dominance hierarchy among themselves on the basis of a queuing system. Dominance among females in the clan is organised according to matrilines. Aggression among males is rare, whereas among females it is common, although such aggression is not reflected in complete control of reproduction by the most dominant female, as all females in the clan breed (Holekamp & Smale 1998; Dloniak *et al.* 2006b). Dominance is most clearly demonstrated in feeding precedence at large carcasses, where the dominant female/s and their offspring have precedence (Frank 1997). Males are the dispersing sex, contributing to a diversity of male individuals within the clan: clan-born males that are related to the philopatric females and immigrants that are not (Dloniak *et al.* 2006b). The sex ratio among adults in the clan is female-biased, usually about 2F:1M (Holekamp & Smale 1998) and although during a given breeding season males mate with multiple females, females may also copulate with more than one male in any oestrous period (Holekamp & Smale 1998), suggesting that the mating system is effectively polygynandrous or promiscuous (Engh *et al.* 2002).

A striking characteristic of spotted hyenas is the highly masculinised external genitalia of females. Females posses an enlarged, fully erectile clitoris, known as a *pseudopenis,* that is similar in size to the male's phallus (Glickman *et al.* 1987). The erect penis and pseudopenis are used by males and females, respectively, in appeasement displays such as the greeting ceremony (Frank *et al.* 1985). The enlarged clitoris of spotted hyenas is an amazing organ as such (Figure 5.5), but if we consider that

Figure 5.5. Female *Crocuta crocuta* showing the pseudopenis. Drawing courtesy of Christine Drea.

females also deliver their offspring through this structure (see, for example, Cunha *et al.* 2003) then Laurence G. Frank's (1997) question: 'Why would evolution create a reproductive organ so hazardous that 9–18% of females die during their first birth, and those that survive lose over 60% of their first-born young?' makes this case a real evolutionary enigma.

Several hypotheses have been suggested to explain the remarkable level of female masculinisation in the spotted hyena (Frank 1997; Muller and Wrangham 2002):

(a) *Mimicry hypothesis* (Kruuk 1972; Muller & Wrangham 2002): females mimic male penile displays as this behaviour may appease dominant females that tend to direct aggressive acts preferentially towards other females.

(b) *Male Infanticide hypothesis* (Kruuk 1972): female clitoridal displays help prevent infanticide by males.

(c) *Siblicide hypothesis* (East *et al.* 1993): androgens that affect female masculinisation also affect cubs, who become aggressive, leading to competition between siblings and potentially to siblicide.

(d) *Chastity Belt hypothesis* (East *et al.* 1993): the female clitoris prevents forced copulations as copulation through the pseudopenis is difficult.

(e) *Competition–aggression hypothesis* (Frank 1996, 1997): strong feeding competition at a carcass selected for high levels of androgen-controlled aggressiveness in both cubs and adults. Masculinisation of female genitalia is an unselected effect of fetal exposure to high levels of androgens.

The past 20 years or so have witnessed the production of a considerable amount of research focusing on the endocrinology of spotted hyenas, and how organisational and activational effects of hormones, androgens in particular, could potentially explain not only the elevated levels of aggression observed among both cubs and adult females in this species, but also the ontogeny and evolution of the pseudopenis.

Female spotted hyenas often produce litters of two cubs. However, the elder sibling usually attacks the second-born within minutes of birth (Frank et al. 1991) with the level of aggression being most intense on the day of birth and falling rapidly thereafter. This behavioural pattern is positively correlated with levels of circulating androgens, which are especially elevated at birth, with plasma androstenedione concentration being as high in male as in female cubs, suggesting a maternal origin of the androgen. In fact androstenedione is the main circulating androgen in adult females, whereas in males testosterone is the main androgen (Frank et al. 1985, 1991). Elevated circulating androgens at birth are not only associated with elevated inter-sibling aggression, but they are also associated with elevated siblicide, with same-sex sibling pairs being more likely to be reduced to one by siblicide than mixed-sex pairs (Frank et al. 1991). Discrimination of sexes could be achieved by olfactory means in cubs, as it is in adults (Drea et al. 2002b).

Surviving cubs gather at a communal den where they engage in social play with other cubs of the clan, with female cubs being more active in social play than males (Pedersen et al. 1990), a bias that may be related to the tendency of males to disperse from their native clan. Play behaviour among cubs also includes play-mounting (Holekamp & Smale 1998). However, play-mounting is mainly performed by male cubs and only for a few weeks, after which mounting is rarely observed until it re-emerges at the adult age, when it is performed in the context of reproduction (Holekamp & Smale 1998; Tanner et al. 2007). Dloniak et al. (2006a) also described higher frequency of play-mounting in male than female cubs, although female cubs that had been exposed to higher levels of androgens during the second half of gestation did show a trend for a more frequent use of mounting during play activities. Rates of aggression, however, were high in both male and female cubs and were also positively correlated with levels of circulating maternal androgens (Dloniak et al. 2006a). That is, ontogenetically masculinised female cubs do not seem to engage much in play-mounting and certainly they do so far less than males. This is the first piece of what I will call the *Spotted Hyena Puzzle*.

What is the role of hormones in the masculinisation of female spotted hyenas? Pregnant females held in captivity show a circulating steroid profile of increased concentrations of progesterone first, followed by an increase in oestradiol and testosterone as the pregnancy progresses (Licht et al. 1992). Androstenedione also increases during pregnancy compared with the non-pregnant state, but its levels in circulation are not correlated with those of the other steroids. Towards the end of pregnancy, circulating androgens in females may reach values similar to those found in males (Licht et al. 1992), whereas androgen levels in non-pregnant females are lower than those detected in males. Male and female foetuses also do not differ in their levels of circulating androgens. Licht et al. (1992) showed that androstenedione produced by the mother's ovaries is metabolised into testosterone and DHT in the placenta. Therefore it is the steroids of placental origin that may contribute to masculinisation of the female fetuses (see also Yalcinkaya et al. 1993; Glickman et al. 1999).

In a series of works published in the same issue of the *Journal of Reproduction and Fertility* (Glickman et al. (1998); Drea et al. (1998); Licht et al. (1998)) Steve Glickman and his collaborators provided strong evidence for female foetuses that still have undifferentiated gonads on days 33 and 48 of gestation to already display a clitoris that is visibly masculinised, being similar to a 50-day-old male foetus' phallus (see also Browne et al. 2006). This is coincident with an elevated production of testosterone and oestradiol from androstenedione that takes place in the placenta (Licht et al. 1998). If androgens of maternal origin are responsible for the observed masculinisation of female embryos then it would be expected that antiandrogen treatment of pregnant females might feminise embryos.

The test was carried out by Drea *et al.* (1998) by treating pregnant females with the antiandrogens flutamide, which blocks the androgen receptor, and finasteride, which inhibits 5α-reductase. The results of the experiment were quite intriguing. Antiandrogen treatment did not produce male offspring with abnormal positioning of the urethral meatus or female offspring possessing a typical mammalian vagina and clitoris, as would be expected from similar experiments carried out in other mammals. Instead, feminisation of females simply produced the typical morphological external sexual structures that characterise females in this species but in a more exaggerated fashion, whereas feminisation of males resulted in the development of a penis having the morphological characteristics of the normal enlarged clitoris that characterises spotted hyena females. Drea *et al.* (1998) correctly concluded that development of the external genitalia in this species must be under a dual control of androgen-dependent and androgen-independent mechanisms (see also Cunha *et al.* 2005). Postnatal penile and pseudopenile growth is also largely independent from androgens, and the clitoris growth is only partly dependent on oestrogens (Glickman *et al.* 1998). These three works provide the second piece of the *Spotted Hyena Puzzle*.

Place *et al.* (2002) followed up those males and females that had been prenatally treated with antiandrogens, into their adult life. In terms of their reproductive capability, females were perfectly able to reproduce, but for males the effect was quite different as the morphological changes apparent in their phallus incapacitated them from normal reproduction, although a couple of treated males were observed mounting females. Control males were all capable of normal breeding. Therefore prenatal antiandrogen treatment decreases sexual activity in males, as is observed in other mammals, whereas females seem to be able to develop relatively normally in spite of their ontogeny being affected by androgens (Place *et al.* 2002). However, their clitoris is shorter and thicker, making delivery more difficult and dangerous to the offspring (Drea *et al.* 2002b).

In spite of external genital masculinisation, and relatively high circulating androgens, female spotted hyenas display heterosexual sexual behaviour, with female–female mounting being uncommon among young and unreported in the wild among adults. However, an analysis of the medial preoptic area and the anterior hypothalamus has led to the discovery of a sexually dimorphic nucleus, named by Fenstemaker *et al.* (1999) the hyena sexually dimorphic nucleus (hSDN), that is two times larger in males than females. If the nucleus is homologous to similar nuclei described in other mammals (e.g. rodents and humans) then the sexual dimorphism is less accentuated in the spotted hyenas. Fenstemaker *et al.* (1999) suggest that exposure to prenatal androgens in females may partly explain the decreased level of sexual dimorphism in the hSDN in this species. The relative masculinisation of the hSDN in female spotted hyenas makes the apparently infrequent female–female mounting in this species very puzzling indeed (see Figure 5.6).

I can now formally state the *Spotted Hyena Puzzle*. Studies so far indicate that in this species:

(a) ontogenetically androgenised females develop a reproductively costly pseudopenis, a clearly masculinised trait;
(b) prenatal inactivation of androgen action does not change the development of female external genitalia to produce typical mammal female genitals, suggesting a direct genetic contribution to the development of the pseudopenis; also,
(c) prenatal androgenisation is not translated into behavioural masculinisation in terms of female–female mounting at all stages of postnatal ontogeny;
(d) high levels of circulating androgens in adult females, especially the parous ones, do not translate into female–female mounting, not even in socio-sexual contexts associated with dominance displays, and yet the establishment of dominance is important among females;
(e) absence of female–female mounting occurs in spite of the relative masculinisation of areas of

Figure 5.6. Female–female mounting in spotted hyenas (left) in captivity compared with male–female mounting (right) (from Fenstemaker *et al.* 1999).

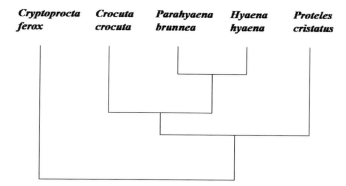

Figure 5.7. Phylogeny of the extant members of the family Hyaenidae according to Koepfli *et al.* (2006).

the female brain that are associated with sexual partner preference in various mammals.

Why is it that a species that should be expected to display a high frequency of female homosexual mounting does not do so, or does so only rarely?

I believe that the solution to the *Spotted Hyena Puzzle* can be found if we take into considerations various levels of analysis. At the very least we must address the issues of origin, maintenance and subsequent modifications of female masculinisation in this species and explain which aspects of female masculinisation seem to be adaptive, which ones seem to be not, and which ones seem to be con- straining the spread of female same-sex mounting behaviour.

Within the family Hyaenidae there are four living species: *Crocuta crocuta*, *Hyaena hyaena*, *Parayaena brunnea* and *Proteles cristatus*, the sister taxon of this clade being the viverrid *Cryptoprocta ferox* according to Koepfli *et al.* (2006) (see Figure 5.7).

According to the fossil record the hyenids had a peak of species diversity in the Late Miocene with about 20–30 species being available at the time. The group, however, seems to have undergone a series of relatively rapid extinctions (Koepfli *et al.* 2006) a process that might have been favoured by the prevalence of hypercarnivory (i.e. exclusive consumption of meat as food) in this group (van Valkenburgh *et al.* 2004). Of the five species included in Figure 5.7, *Crocuta crocuta* is the only one to have external genitals that are masculinised in adult females. However, recent studies carried out in the fossa (*Cryptoprocta ferox*) and the striped hyena (*Hyaena hyaena*) suggest that penile-like clitoridal morphologies do appear, although only transiently, during the development of females (Hawkins *et al.* 2002; Wagner *et al.* 2007). Fossa females develop a penis-like clitoris, which is both enlarged and possesses spines as the male penis does, this structure appears at about 7 months of age and lasts only a few years. Adult and juvenile profiles of both androstenedione and testosterone seem to be similar, suggesting that the appearance and disappearance of the pseudopenis may not be under direct androgenic control (Hawkins *et al.* 2002). Young female striped hyenas also develop transient structures between the postnatal age of one and 18 months that resemble a scrotum and a conical protrusion that looks penis-like (Wagner *et al.* 2007). As for the fossa, and to a great extent the spotted hyena, the expression of these structures in the striped hyena does not seem to be under androgenic control (Wagner *et al.* 2007). In some species it is well known that external female or male genitalia develop before the testis or the ovaries start secreting steroid hormones, suggesting that their development may be under a different kind of control, perhaps a direct genetic control (O *et al.* 1988; Glickman *et al.* 2006). I do not have information regarding development of external genitalia in *Parahyaena brunnea* and *Proteles cristatus*; however, if we put all these pieces of the jigsaw puzzle together it is possible to suggest the following hypothetical evolutionary scenario.

Given the available evidence, it is plausible that the development of external male-like genitalia in females is an ancestral state in the clade that includes hyaenids and the fossa and that genes controlling the trait were maintained throughout speciation, but their degree of expression and perhaps accumulated mutations have been different in different taxa. Co-occurrence of the trait across several species may be the result of an ancient mutation that survived a bottleneck during a period of rapid extinctions of some supercarnivore taxa that occurred probably about 30 mya (Koepfli *et al.* 2006). More recent bottlenecks may also explain the low genetic diversity found in both striped hyenas and brown hyenas (*Parahyaena brunnea*) (Rohland *et al.* 2005). The high reproductive costs of masculinisation in females may have selected for a regression of the trait in most taxa, some of which retain vestigial and transient evidence of external genital masculinisation in females. In spotted hyenas, however, masculinisation was retained in spite of the obvious costs. What are the benefits that can compensate, in terms of fitness, for those costs of hazardous delivery of newborns in *Crocuta crocuta*? I believe that the answer should be found in increased survival of individuals who make it alive out of the maternal womb, and who face a postnatal life in a highly competitive social environment. Enhanced survival starts with elevated aggression of cubs, which is under androgenic influence. This allows the offspring an advantage in a scramble competition for food at a carcass, and the same is valid for adult females (Glickman *et al.* 2006). Supercarnivory in this species could be associated with the ability of groups of hyenas to hunt relatively large prey; this requires behavioural mechanisms of social tolerance as well. In fact, the large female pseudopenis is used in greeting ceremonies and this may suffice in a society where the major tactic to avoid and resolve conflicts is to just move away (Wahaj *et al.* 2001). In addition, all female spotted hyenas in the group seem to be rather promiscuous, engaging in an elevated frequency of heterosexual mounting with various males in the group, which, according to the *Lock-in Model*, would lead to low levels of homosexual female mounting. Therefore high female aggressiveness seems to have selected for conciliatory tactics that, unlike those of primates, do not involve close body contact, whereas elevated sexual behaviour during oestrus easily leads to heterosexual copulation by all individuals (males and females) owing to the effectively promiscuous mating system of this species, thus decreasing the chances for same-sex mounting among oestrous females in both sexual and socio-sexual contexts.

Female spotted hyenas are not behaviourally unable to perform masculinised homosexual mounting, as Figure 5.6 shows, and when they do engage in mounting other females, the behaviour is described by those who have witnessed it as 'dramatic', including 'erections and movements of the clitoris that are similar to (but not exactly the same as) those of the hyena penis during mating' (Steve Glickman, pers. comm.). This suggests that there must be a selection against the widespread use of female–female mounting in this species in socio-sexual contexts and selection for elevated aggressiveness may, as suggested above, be the reason for that. This model could be tested in comparative, experimental studies of the five species mentioned above. If the scenario that I suggest is correct, then the level of hypercarnivory and aggression described in wild populations should be negatively correlated with frequency of same-sex mounting among females in captive groups of equal size and sex ratio of the five species listed in Figure 5.7. A test of animals held in captivity is necessary in order to determine the potential for homosexual female–female mounting once the probability of two females meeting each other while in oestrus is made constant across species, as not all species have the same degree of sociality in the wild.

Circulating hormones and sexual orientation in adult humans

That a physiological increase in circulating gonadal steroids may also increase the motivation to engage in sexual behaviour in adult humans is a well established fact (see Bancroft 2005 for a review). That is, gonadal steroids have a well established activational effect on sexual behaviour in adults, even though in humans and also in most other primates, copulation may also occur independently of gonadal hormonal influence (see Wallen 2001), suggesting a dual neuroendocrine control of copulation as I have already stressed in this chapter. In which ways and to what extent, then, might circulating levels of sexual hormones also affect behaviours that define a homosexual orientation in adult humans?

I start this review with an analysis of the potential effects of the use of anabolic androgenic steroids, asking whether the use of such steroids by adult individuals may affect their sexual orientation. I will then continue with a review of the studies that compared circulating levels of various hormones in homosexuals and heterosexuals, to subsequently review studies of orchiectomy (removal of testes) and oophorectomy (removal of ovaries) and whether these interventions affect sexual orientation.

Does the use of anabolic androgenic steroids by adults affect sexual orientation?

In this era of use of anabolic androgenic steroids (AAS) by persons interested in improving their body build or athletic performance, one would imagine that a plethora of data would be available to test whether alteration of circulating levels of steroids in adults can affect sexual orientation or not. Surprisingly, most works that focus on sexual orientation and AAS address the causal relationship between the two variables from the perspective of whether an already established homosexual sexual orientation favours or not the use of AAS and, if so, what are the psychosocial reasons for this link (e.g. improving self-image; Herzog et al. 1991; Baird 2006), rather than whether an initially heterosexual individual may or may not increase his/her bisexual or homosexual tendencies over time as a result of continued use of AAS.

Dillon et al. (1999) interviewed a sample of 100 individuals (94% males and 6% females) that included 70% heterosexuals, 27% homosexuals and 3% bisexuals. Their results suggest that use of AAS may be an unlikely cause of homosexual/bisexual orientation development in adult males. In fact, non-heterosexual males tended to start their use of AAS at older ages than heterosexuals, they also used fewer types of AAS per cycle and lower dosages. In addition, non-heterosexual males also had fewer cycles of AAS use over the previous 12 months than their heterosexual counterparts (Dillon *et al.* 1999). The results of Dillon *et al.* (1999) are consistent with the hypothesis that in adult males AAS do not affect sexual orientation and that already established non-heterosexual men use AAS for psychosocial reasons. On the other hand, Miller *et al.* (2002) rightly emphasise the paucity of studies directly and explicitly focusing on the potential effects that AAS may have on the development of a homosexual orientation, although they do quote previous research linking use of AAS with increased sexual activity (Miller *et al.* 2002 and references therein). In their own research, they carried out an analysis of data from the Youth Risk Behaviour Survey of the Centers for Disease Control and Prevention of the USA that were collected through the administration of questionnaires to teenage students in their last four years of schooling. In their sample, 2% of females and 4.1% of males reported ever having used AAS. A total of 68.6% of AAS male users had sex during the 3 months previous to the study, whereas only 32% of non-AAS users had sex. Females show a very similar trend: 63.9% of steroid users had sex vs. 35.9% of non-steroid users. Although this work did not specify whether sexual intercourse was with same-sex, other-sex or both kinds of partners, results are consistent with an increase in libido among male and female AAS users, which may predispose them indirectly to a subsequent increase in the frequency of same-sex sexual intercourse under specific social circumstances, e.g. those that favour monosexual group formation (e.g. schools, sport clubs). This should be regarded as a hypothesis in need of specific testing, however.

I suggest that with respect to potential links between AAS use and development of homosexual orientations, it is probably female users of AASs that should be the focus of more research. Yet the literature dealing with AAS is biased towards studies that focus on men rather than women (Gruber & Pope 2000). Although the effect most often reported of AAS use in men is increased level of aggression, effects on sexual behaviour are quite variable, ranging from hypersexuality to hyposexuality (Borges *et al.* 2001). It is only at higher doses of AAS usage that males seem to develop clearly feminising traits such as gynaecomastia (Borges *et al.* 2001; Rashid *et al.* 2007). In women who are users of AASs, however, clear masculinising effects have been reported that include the development of acne, facial hair and coarsening of the voice (Borges *et al.* 2001; Rashid *et al.* 2007). Moreover, AAS female users also tend to develop menstrual irregularities (Borges *et al.* 2001). Gruber & Pope (2000) carried out a study of female AAS users with data coming mainly from the New England region of the USA. Their 75 subjects were subdivided into 33% AAS users and 67% non-AAS users; data were collected mainly through interviews and direct physical examinations. A total of 76% of women reported one or more effects of AAS use (voice change, acne, clitoromegaly, increased facial hair) including 64% of AAS users reporting one or more psychological effects such as labile mood, and general irritability and aggressive behaviour, with only 6.24% of them reporting increased libido. Gruber & Pope (2000), however, also describe that more than 88% of the AAS users reported some level of 'gender identity disorder' both as children and as adults, whereas only 66% of the non-AAS users did. AAS users also expressed a preference to be involved in stereotypical masculine activities and wear stereotypical masculine clothing along with a preference for having male friends. Given these results, the issue of a causal relationship between use of AAS and alteration of sexual orientation remains unclear. Are women with an already established lesbian sexual orientation or a masculinised gender role also more likely to use AAS? Or can AAS modify the sexual orientation of a woman with a well-established prior heterosexual sexual orientation? This is an area where further research would be very welcome.

Does sexual orientation in men and women correlate with levels of circulating hormones?

The issue of AAS notwithstanding, some reviews of the role of hormones in the proximate causation of homosexual behaviour in adult men and women have given a *prima facie* negative result. In an early review of 25 studies that compared circulating testosterone (T) levels in male homosexuals and heterosexuals, Meyer-Bahlburg (1984) reported that 20 of those studies did not find a significant difference between homosexuals and heterosexuals, whereas three studies found that the levels of circulating T in homosexuals were lower than in heterosexuals, and two studies reported evidence for the reverse trend: circulating T was higher in homosexual men than heterosexual men. In a more recent review, Banks & Gartrell (1995) also reported that most of the studies that they reviewed (10/16) found no difference in circulating levels of T between heterosexual and homosexual men and of those that detected a statistically significant difference three indicated that homosexuals had higher circulating levels of T and three found the opposite trend. In a work that was not included in Banks and Gartrell's article, Gladue (1991) did not detect a statistically significant difference in both total and free T between homosexual and heterosexual males. A more recent work by Neave & Menaged (1999) describes a statistically non-significant trend for homosexual men to have slightly higher levels of salivary T than heterosexual men.

Meyer-Bahlburg (1984) also reported the results of seven studies that measured circulating oestradiol. Of those, only two studies found a significant difference between heterosexuals and homosexuals, with homosexual men having higher circulating levels of oestradiol than heterosexual men. Gladue (1991) did not detect any difference in circulating oestradiol between homosexual and heterosexual men. Luteinising hormone (LH) also shows very little difference between homosexual and heterosexual men, with only 3 out of 14 studies reporting a difference, with homosexual men having higher levels of circulating LH than heterosexual men (Meyer-Bahlburg 1984). Circulating levels of follicle stimulating hormone (FSH) also tend not to change with sexual orientation in men in 5 out of 6 studies (Banks & Gartrell 1995); in the single study that showed a statistically significant difference, homosexual men had higher levels of circulating FSH than heterosexual men. With regard to prolactin (PRL), only 25% of studies reported a difference, and those that did, reported an increased level of circulating PRL in homosexual as compared to heterosexual men (Meyer-Bahlburg 1984). Androstenedione concentration is also similar between homosexual and heterosexual men in 5 out of 7 studies, with the two studies that reported a difference indicating that homosexual men had higher levels of circulating androstenedione than heterosexual men. Meyer-Bahlburg (1977), in an earlier review, had also highlighted four studies showing lower values of the androsterone/etiocholanolone ratio (two metabolites of testosterone) in homosexual than in heterosexual men.

Meyer-Bahlburg (1984) also reported two studies that compared circulating cortisol, with one of the studies finding no difference between homosexual and heterosexual men, and the other finding that homosexuals had higher circulating levels of cortisol than heterosexuals. Given the number of non-significant results, both Meyer-Bahlburg (1984) and Banks & Gartrell (1995) cannot be blamed for concluding that circulating levels of the gonadal steroids, corticosteroids and peptide hormones studied seem to show little association with homosexual behaviour in male humans. However, among those studies that did find statistically significant results, there is an intriguing level of consistency, with oestradiol, luteinising hormone, prolactin, androstenedione and cortisol levels being higher in homosexual men than heterosexual men, the former also have higher levels of etiocholanolone and androsterone than the latter. Testosterone is the only hormone that shows some ambiguity, with 40% of works indicating higher levels in homosexual men and 60% of works indicating higher levels in heterosexual men in Meyer-Bahlburg's (1984) review and 50% of works for each case in Banks & Gartrell's (1995) review.

Meyer-Bahlburg (1984) also reviewed two endocrinological studies of lesbians that showed that they had higher levels of circulating T than heterosexual women, but they did not differ in their circulating levels of gonadotropins, cortisol or prolactin. Neave & Menaged (1999) also found a non-statistically significant trend for homosexual women to have higher salivary testosterone levels than heterosexual women. The trends described by both Meyer-Bahlburg (1984) and Neave & Menaged (1999) are consistent with Kozak's (1996) study that detected a positive correlation between circulating free (i.e. unbound to globulins) T and the masculinity dimension of the sex role identity scale as defined in the Bem Sex Role Inventory that was administered to women. That is, more masculinised women also tend to have higher circulating levels of free T. Behaviourally masculinised lesbians (i.e. 'butch' lesbians) were also found to have higher levels of salivary T than 'femme' lesbians by Pearcey et al. (1996) and Singh et al. (1999). Bosinski et al. (1997) studied homosexual female-to-male transsexuals (FMTs) in Germany and described higher levels of circulating testosterone and androstenedione in FMTs than in heterosexual females; moreover, FMTs also seemed to have an adrenal cortex hyperresponsive to adrenocorticotropic hormone. A similar result was obtained by Baba et al. (2007) in a study carried out in Japan. They describe significantly higher levels of androstenedione and luteinising hormone and non-significantly higher levels of testosterone in FMTs with polycystic ovary syndrome, with FMTs being more likely to express the syndrome (58%) than the general female population (5%–10%).

Therefore, according to those studies, masculinised and homosexual women seem to have higher levels of circulating testosterone and other androgens. Other researchers, however, have reached different conclusions. Downey et al. (1987) were unable to detect a statistically significant difference in circulating androstenedione, T and cortisol between lesbian and heterosexual women. A study carried out by Dancey (1990) also did not detect a statistically significant difference between primary, intermediate and secondary lesbians and heterosexual women in their circulating levels of T, androstenedione, progesterone and oestradiol. Gladue (1991) did not detect any difference in circulating T (and oestradiol) between lesbians and heterosexual females either. However, the frequency distribution of sexual activity among lesbians does not seem to follow the same general pattern as for heterosexual women according to Matteo & Rissman (1984): whereas heterosexual women tend to increase their level of sexual activity during the mid-follicular (i.e. postmenstrual) and late luteal (i.e. premenstrual) phases of their cycle (Bancroft et al. 1983), lesbians do not report a decline of sexual activity during menstruation, although they reported a peak in orgasms during mid-cycle and an increase in sexual thoughts during the first 3 post-menstrual days (Matteo & Rissman 1984). The latter result may be due to specific differences in aspects of the endocrinology of lesbians and heterosexuals (e.g. elevated circulating testosterone in the former, as suggested above) or a result of differences in the direct control of sexual behaviour by central nervous system mechanisms that do not depend on endocrine modulation, or both. In any case, these results also suggest that discrepancies in endocrine measurements between homosexual and heterosexual women could be detected, or not, in endocrinological studies depending on the timing of sampling with respect to the stage of the menstrual cycle.

Variability in the results notwithstanding, among women the statistically significant trends regarding differences in testosterone concentration between homosexuals and heterosexuals are more consistent than in the case of men, with homosexuals and masculinised women having higher levels of circulating testosterone than heterosexual and feminised women.

Gooren (1986) administered luteinising hormone releasing hormone (LHRH) followed by conjugated oestrogen to groups of heterosexual and homosexual men and women. Basal levels of LH, T and oestradiol did not differ between the two groups of women, neither did they differ in their LH response. Basal LH, T and oestradiol levels also did not differ between homosexual and heterosexual men, and neither did the T and LH response to conjugated oestrogen administration (Gooren 1986). Interestingly, an increase in LH levels after 4 days of conjugated oestrogen administration that resembles the normal female response was detected in 5 of the 15 (33%) heterosexual men and 11 of the 23 (47%) homosexual men, suggesting a degree of 'feminisation' of the endocrine response to oestrogen even in the heterosexual male population. This intrasexual variability among men may explain early findings of a positive oestrogen feedback in homosexual men (Dörner *et al.* 1975). Hendricks *et al.* (1989) also suggested that the male population may be physiologically variable regarding their ability to produce an oestrogen feedback. For instance, different physiological responses to stress among the male participants in these studies may partly explain the various results obtained (Hendricks *et al.* 1989). Homosexual and heterosexual women did not differ in their LH response to oestrogen administration.

I started this section on the effects of circulating hormones on sexual orientation in adults with an analysis of AAS use. The use of AAS implies an input of steroids into the circulation of the user, however an alternative 'natural experiment' in humans that involves a decrease rather than an increase in circulating gonadal steroids is castration, i.e. the removal of the gonads. In men, the removal of one or both testes is technically known as orchiectomy, whereas in women the removal of one or both ovaries is known as oophorectomy.

Do orchiectomy and/or oophorectomy lead to a homosexual orientation?

Orchiectomy has been practised on men for centuries in diverse cultures for various purposes. I will skip the fascinating historical and cultural background of orchiectomy and refer the reader to some excellent reviews: Wilson & Roehrborn (1999); Aucoin & Wassersug (2006); Johnson *et al.* (2007); Brett *et al.* (2007); Ringrose (2007); Wassersug & Johnson (2007). The basic question I want to address here is whether historical and present cases of orchiectomy throw any light on the potential effects of gonadal removal on the expression of same-sex sexual behaviour in adults.

In an early review of the effects of orchiectomy, Tauber (1940) pointed to studies of injured war veterans as the ones that at the time provided the largest dataset for a study of the effects of orchiectomy on sexual behaviour. Some general trends suggested by Tauber included the onset of impotence (i.e. lack of penile erection) concomitant with an often good level of libido. However in some cases both potency and libido decreased markedly. A diversity of results was also found in cases of patients orchiectomised as part of a treatment for genital tuberculosis and tumours. Tauber did not mention effects of orchiectomy on sexual orientation, but then he did mention that some aspects of patients' reporting could suggest a less than accurate information coming from the patient: 'We know on clinical grounds that although some persons have lost their conscious libido and are also impotent, they may be extremely distressed over the combined deficit' (Tauber 1940: 80). Whether in these circumstances any increase in same-sex sexual attraction or behaviour would have been reported by the orchiectomized individual in the first half of the twentieth century is unclear.

Wilson & Roehrborn (1999) report an increase in gynaecomastia (i.e. enlargement of the breasts) in orchiectomised individuals and also a decrease in size of the prostate gland, but they do not mention effects on sexual orientation. More recently, Johnson *et al.* (2007) have carried out a study of men that requested orchiectomy for reasons not related to a medical condition. The subjects were recruited from the Eunuch Archive and provided responses to a questionnaire that were sometimes supplemented by additional narratives and interviews. Respondents in the Johnson *et al.* study gave various reasons why they decided to undertake orchiectomy, with homosexuality and transition to transsexuality being two of those. Although most of the participants identified themselves as male, a proportion considered themselves to be in an alternate gender (neither male nor female) situation. In this case, however, homosexuality, at least in some individuals, seems to have been antecedent not consequent to orchiectomy. Brett *et al.* (2007), in a companion article, provide some additional light on this issue. They report that the same individuals also listed some side effects of orchiectomy such as loss of libido, the experience of hot flushes and genital shrinkage, and that 22% of respondents mentioned an actual change in sexual orientation.

A recent study co-authored by Alicia García-Falgueras and Dick Swaab (2008) also suggests a potential effect of castration on development of brain areas that are known to covary in their size and structure with sexual orientation. They studied the INAH3 region of the anterior hypothalamus, which, as I will mention in the 'The hypothalamus and homosexuality' section of this chapter, is smaller (i.e. more feminised) in homosexual than in heterosexual men. The authors compared both the volume and number of neurons in this subnucleus between heterosexual males, heterosexual females, intact male-to-female transsexuals and, importantly, castrated male-to-female transsexuals. Individuals in the latter group underwent castration as part of their process of feminisation of external genitalia. Although intact male-to-female transsexuals were feminised in their INAH3 volume and number of neurons (i.e. volume and number smaller than in heterosexual males and similar to heterosexual females), castrated male-to-female transsexuals had INAH3 volume and number of neurons that were intermediate between heterosexual males and females. Although this result is at least compatible with the possibility that castration might have caused neuronal changes in an area of the adult brain that is involved in the control of sexual orientation, it is not conclusive as castrated transsexuals did not have a more feminised INAH3 than intact transsexuals, as one might have expected.

In a recent review, Aucoin & Wassersug (2006) mention a tendency throughout history for some eunuchs to behave bisexually, and they emphasise that historical eunuchs had the ability to retain not only sexual potency but also a heterosexual sexual orientation. If boys were castrated before puberty, however, they developed some feminised morphological traits (Ringrose 2007), but whether they also developed a homosexual sexual orientation as adults remains uncertain, although there are suggestions from the practice of castration in antiquity indicating that some of them might. For instance, in his *Satyricon*, Petronius (who died in 66 c.e.) wrote:

> It is a Persian custom; they abduct young boys, scarcely of years; the mutilating steel condemns them all to lust, and in this bid to stay the hurrying years and delay swift-changing age, Nature seeks her natural way and cannot find it. So for pleasure every man has a minion, with effeminate body and mincing gait, with flowing hair and heaps of novel-sounding clothes, the very things to entice a man
>
> (translated by Paul Dinnage 1998: 97).

Currently some heterosexual men may be subject to androgen-deprivation therapy (ADT) that may involve either surgical or chemical castration as part of a standard treatment for advanced prostate cancer (Wassersug & Johnson 2007). Although men who undergo ADT tend to show decreased libido and evidence of de-masculinisation, it is unclear whether the treatment as such is capable of altering sexual orientation (Wassersug & Johnson 2007).

Some doubts about the validity of reports provided by participants may shadow some of the studies carried out on the effects of chemical castration of sex offenders. Grossman *et al.* (1999) carried out a review of the effects of chemical castration of sex offenders achieved through the administration of the antiandrogens medroxyprogesterone acetate and cyproterone acetate, two synthetic progesterones that reduce the levels of circulating testosterone. Although inmates that received the antiandrogens reported a reduction of libido, erections and ejaculations, in those studies that also used a placebo control the results were mixed and somewhat suggestive of patients' manipulative responses: for example, some contradictions between self report and plethysmography results concerning ability to experience an erection were detected, as were similarities in the trends of hormone-treated and placebo-treated subjects (Grossman *et al.* 1999). Obviously the subjects had a vested interest in providing evidence that they 'were cured' of their rapist tendencies, thus making any conclusion based solely on their testimonials somewhat dubious. Likewise, potential effects of chemical castration on sexual orientation could have been hidden by those subjects.

Much less is known about the effects of oophorectomy on women's sexual orientation. Bilateral oophorectomy is associated with a decrease in libido and a decline in circulating testosterone (Shifren 2002). Testosterone and oestradiol replacement suffices to restore libido, but Shifren (2002) makes no mention of any of the above procedures affecting sexual orientation of women.

In sum, assigning no role at all to current circulating hormones in the causation of homosexual behaviour in sexually mature humans does not seem to be warranted from the available evidence. The above review suggests that there are both endocrine-dependent and endocrine-independent mechanisms that co-contribute to the expression of sexual orientation in different individuals in the population, and I would like to emphasise here the considerable inter-individual variability within sex category, especially males. Notwithstanding any statistical issue relative to those studies that showed no difference in circulating hormones between homosexual and heterosexual individuals (e.g. was the sample size large enough to detect a difference? Are there some other biases or effects of uncontrolled variables?), in cases where a difference was found it tended to be consistent, indicating higher levels of circulating hormones (oestradiol, luteinising hormone, prolactin, androstenedione, cortisol) in homosexual males than in heterosexual males, with testosterone being clearly much more variable between sexual orientation categories. In women, on the other hand, levels of circulating testosterone seem to be more predictably higher in lesbians and masculinised females than in heterosexual and feminised females. This result is in accordance with the studies of AAS use among women that I reviewed above. On the other hand, studies of orchiectomy and oophorectomy, which involve lowered levels of circulating steroids, are inconclusive with regard to potential effects of those treatments on sexual orientation and, if anything, they suggest no effect on sexual orientation if they are carried out on adult individuals. Interestingly, the association between increased circulating hormones and homosexuality in adults, especially women, is consistent with the results obtained for the congenital adrenal hyperplasia syndrome that were reviewed at the beginning of this chapter.

The available evidence suggests that a role of circulating hormones in the expression of homosexual behaviour in adult humans should not be dismissed. In particular, the role of circulating hormones is likely to be variable between individuals within the same category of sexual orientation, therefore studies should be much more detailed than they have been so far, identifying diverse categories of lesbians and gay men that may differ in their physiology. The same is valid in the study of sexual orientation in other species as well; see, for instance, the diverse categories of rams that engage in same-sex mounting described in a previous section. Some do so as part of the normal socio-sexual repertoire of the species; some may do so perhaps as a result of a specific mutation.

With the exception of the case of AAS users and that of chemical castration in humans, I have assumed that the origin of hormones that could potentially alter sexual behaviour and orientation is internal to the adult organism. However, across a vast array of taxa, adult individuals may intake substances from their environment (e.g. with food) that can exert actions similar to those of endocrine hormones or that can disrupt the activity of endogenous hormones, and that therefore could potentially affect sexual behaviour.

Effects of endocrine-disrupting chemicals on the expression of sexual behaviour in adult mammals and birds

In Chapter 4 I reviewed the effects of environmental chemicals that have endocrine-disrupting capabilities on the early sexual development of birds and mammals. Here I focus on the effects of both natural products and industrial pollutants on sexuality at the adult stage.

Natural or synthetic xenobiotics can alter sexual hormone synthesis through their interference with hormone (e.g. androgen and oestrogen) receptors located on target cells (Greim 2004), but non-receptor-mediated mechanisms have also been described (Henley & Korach 2006). Among industrial pollutants that have endocrine-disrupting properties, phthalates, alkylphenolic compounds, organochlorine pesticides, some metals, polychlorinated biphenyls and dibenazodioxins have the capacity to decrease the level of sexual activity with some exceptions, such as methoxychlor, which stimulates mating in species such as *Rattus norvegicus* (Zala & Penn 2004). Decreased sexual activity may be achieved by various routes, including reduction of endogenous production of testosterone and DHT, but also of oestrogen, as is the case with 2,3,7,8-tetrachlorodibenzo-*p*-dioxin (TCDD) in rats (Rosselli *et al.* 2000; Sanderson 2006).

Phytoestrogens are plant compounds that have oestrogenic activity and that are classified into three main classes: lignans, coumestans and isoflavones (Murkies *et al.* 1998; Dixon 2004), whereas oestrogenic compounds produced by fungi are known as mycoestrogens, e.g. lactones. Lignans and isoflavones ingested by humans are metabolised in the gastrointestinal tract into heterocyclic phenols that resemble oestrogens in their chemical structure (Murkies *et al.* 1998). Coumestrol is among the most potent of phytoestrogens, having an effect on human cells in culture that is about 20% the effect of oestradiol, whereas genistein has an 8.4% oestradiol-equivalent effect (Murkies *et al.* 1998). When administered to female rats, coumestrol can suppress the oestrous cycle (Dixon 2004). In fact, by competing with oestrogen for oestrogen receptors and sex hormone binding globulins, phytoestrogens can have either an oestrogenic or an antioestrogenic effect on reproductive and central nervous system tissues (Wang *et al.* 1994; Whitten *et al.* 1995; Dixon 2004) leading to either increased or decreased sexual activity (see, for example, Mani *et al.* 2005; Henley & Korach 2006).

In rats the anteroventral periventricular nucleus (AVPV) of the brain is associated with control of sexual behaviour through its influence on luteinising hormone secretion. The AVPV is sexually dimorphic in rats, with the female's AVPV being larger than the male's (Bu & Lephart 2007). Bu & Lephart (2007) observed that adult male rats kept on an isoflavone-rich diet underwent a significant decrease in their AVPV, i.e. their AVPV became 'hypermasculinised'. The effect was due to isoflavones inducing apoptosis in the AVPV of male rats. In particular, cells undergoing death were the ones expressing oestrogen receptor β, rather than the oestrogen receptor α (Bu & Lephart 2007). This result indicates that, at least in rats, ingestion of specific compounds with food may contribute to the reshaping of areas of the central nervous system that could subsequently affect behaviour, sexual behaviour in particular, including, potentially, sexual orientation. This is possible whenever the central nervous system retains a degree of both structural and functional plasticity into the adult stage of an organism, an issue that will be discussed at length when I analyse the roles of the nervous system in the expression of same-sex sexual behaviour later in this chapter.

Phytosterols are an additional group of compounds that can be ingested with food and that can exert an action on the endocrine system. Natural phytosterols that occur in plant tissues include stigmasterol, β-sitosterol and campesterol, which can be eventually metabolised into androgens, oestrogens or progesterone depending on the specific compound (Sarangthem & Singh 2002). They may also have sex-hormone inhibiting action: β-sitosterol in particular has the capacity to reduce the synthetic activity of steroids by the gonads by decreasing availability of cholesterol (Ling & Jones 1995; Fritsche & Steinhart 1999), the precursor of all steroid hormones (see Figure 5.1). In an experiment carried out in captivity, however, phytosterols added to the diet increased testis size in male field voles (*Microtus agrestis*) with circulating levels of oestradiol and testosterone also increasing in both males and females, but especially males (Nieminen *et al.* 2003). Unfortunately field voles were housed singly and therefore sexual behaviour and orientation were not recorded. Nieminen *et al.* (2004) also gave phytosterols (sitosterol mainly) as food supplements to tundra voles (*Microtus oeconomus*). Although the researchers detected lower circulating concentrations of testosterone in treated than control males, reproductive success was higher among treated pairs; the exact reasons for this effect remain unclear (Nieminen *et al.* 2004).

Female rats may also experience an increase in reproductive activity after ingestion of plant compounds such as 6-methoxybenzoxazoline (6-MBOA), a non-oestrogenic molecule (Butterstein *et al.* 1985). Interestingly, 6-MBOA has a molecular structure that resembles that of melatonin (Butterstein *et al.* 1985), a molecule involved in the synchronisation between reproductive activity and photoperiod (Nelson *et al.* 1998).

Most of the evidence available so far directly points to a potential ability of compounds ingested with food to alter reproductive behaviour. When this alteration implies an increase in sexual activity, then opportunities for the expression of homosexual behaviour may arise in social or demographic circumstances where the sex ratio is biased. A direct example of this effect is provided by Fratta *et al.* (1977), who carried out a study of the consequences of tryptophan deficiency on sexual behaviour of male rats and rabbits, following previous reports of *p*-chlorophenylalanine (PCPA), an inhibitor of tryptophan hydroxylase, inducing homosexual mounting in male rats. Fratta *et al.* (1977) kept both adult rats and rabbits on a tryptophan-free diet that resulted in a marked increase in copulation behaviour in both species. Males were caged in monosexual groups so what Fratta *et al.* (1977) probably observed was an increase in homosexual mounting among the treated individuals that had their sexual activity enhanced and had access only to same-sex sexual partners.

Although some studies also suggest potential direct effects of chemicals ingested with food on the architecture and physiology of brain areas controlling sexual behaviour, specific effects on sexual orientation are still unclear and require further research. Clearly, relatively simple tests of experimental animals kept under different diets and then offered a simultaneous choice of a sexually motivated male and a female in oestrus would provide crucial evidence regarding the potential effect of phytoestrogens, phytosterols, industrial pollutants and other environmental compounds on sexual orientation in adults.

The issue of compounds ingested with food potentially predisposing animals to engage in homosexual behaviour may also suggest the need for a careful re-analysis of same-sex sexual behaviour observed in animals held in captivity (e.g. zoos). Although this issue may not be relevant to all studies, researchers would be nevertheless advised to carefully check what their animals were fed (for example, alfalfa has a high content of phytoestrogens, Seguin *et al.* 2004), as a regular ingestion of compounds that could enhance sexual activity may produce elevated same-sex mounting under conditions of confinement and a biased sex ratio.

The neuroendocrine system and homosexual behaviour

Hormones can mediate heterosexual and homosexual behaviour only if they ultimately exert an effect on the nervous system. As we have seen in previous sections of this chapter, molecules that may have an effect on brain centres involved in control of sexual behaviour in both males and females include testosterone, oestradiol, progesterone, prolactin, oxytocin, vasopressin, cortisol, pheromones, and

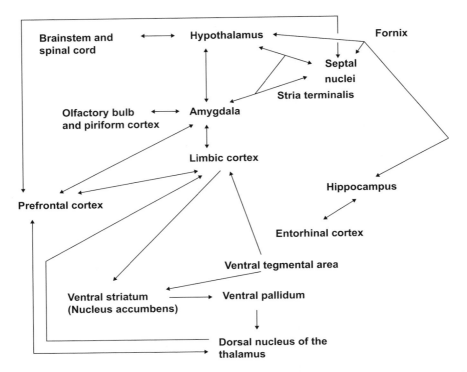

Figure 5.8. Major central nervous system centres involved in control of sexual behaviour in vertebrates. Modified from R. Swenson: www.dartmouth.edu/~rswenson/NeuroSci/chapter_9.html, 2006.

also neurotransmitters such as dopamine, and others (see, for example, Meston & Frohlich 2000). Their action is exerted on cells of specific areas of the brain, the most important of which are summarised in Figure 5.8. I will start this section by briefly reviewing the major brain centres that control sexual behaviour and some of the major molecular and cellular mechanisms that determine brain plasticity, and then consider their specific relevance to the expression of homosexual behaviour.

In mammals, classic areas of the brain that are involved in the control of copulation and, in males, penile erection are the medial preoptic area (MPOA) and the bed nucleus of the stria terminalis (BNST) in the hypothalamus, the medial amygdala and the nucleus accumbens (see, for example, Liu et al. 1997; Pfaus 1999). However, recent studies have shown that central nervous system control of sexual behaviour is likely to be exerted by various areas according to the specific stage in the behavioural sequence from attraction to a mate to copulation to orgasm and, in males, ejaculation. Both male and female rats show activation of a diverse array of brain centres in concomitance with the display of behaviours such as mounting, intromission and ejaculation, but also anogenital investigation. Veening & Coolen (1998) list the medial preoptic nucleus, the BNST, the caudal thalamus and the posterodorsal section of the medial amygdala as some of the areas associated with control of sexual behaviour in rats.

In humans the study of the neurological mechanisms involved in the expression of sexual behaviour is complicated by the many dimensions defining sexual arousal in our species: motivational, physiological, emotional and cognitive (Stoléru *et al.* 1999). Pfaus (1999) reviewed early works that used positron emission tomography (PET) scan technology to detect activated areas of the brain upon perception of sexual stimuli. Such studies indicate an association between sexual arousal (for example, when subjects were exposed to a sexually explicit film) and activation of various cortical areas. Redouté *et al.* (2000) also used PET scanning to determine sexual activation of brain areas of heterosexual men following the perception of stimuli of increasing sexual intensity. They also complemented the PET data with records of penile plethysmography, a method used to measure the degree of erection, and circulating testosterone. Not surprisingly, a film showing heterosexual coitus was perceived by the heterosexual participants as the stimulus eliciting the highest subjective sexual arousal. More interestingly, however, the level of subjective perception was positively correlated with circulating testosterone. The film also activated the following brain areas: claustra, putamens, nucleus accumbens, hypothalamus, thalamus, parietal lobes and central sulcus. A level of cognitive involvement in the response to the stimulus was also inferred through the activation of the right orbitofrontal cortex (Redouté *et al.* 2000).

Holstege *et al.* (2003) also studied right-handed heterosexual men, but they focused on brain area activation during ejaculation. In this case the sexual stimulus was provided not by an erotic film but by manual penile stimulation performed by the subject's sexual partner in life. During ejaculation/orgasm the following brain structures were activated: subparafascicular nucleus, zona incerta, lateral central tegmented field, ventral tegmental area and several thalamic nuclei. Sections of the telencephalon were also activated. Regions of the isocortex such as the Brodmann area also showed activation. Interestingly, the visual cortex was activated bilaterally in spite of the subjects having their eyes closed.

That is, different stages of the sexual act such as arousal and orgasm are associated with the activation of some similar, but also some different central nervous system areas.

Karama *et al.* (2002) also studied the brain's activational response to the exposure to an erotic film as compared with a 'sexually neutral' film, but they did so in both male and female heterosexual, right-handed adults. The authors used functional magnetic resonance imaging (fMRI) to determine the regions of the brain activated by watching the films. Although male participants had a higher level of self-reported sexual arousal than female participants in general, both sexes displayed activation of the medial prefrontal, anterior cingulate, orbitofrontal, occipitotemporal and insular cortices along with the ventral striatum and the amygdala. In addition, males also showed activation of the thalamus and hypothalamus. The activation of the hypothalamus in females was statistically non-significant, but the trend was in the same activational direction as in males.

In sum, sexual stimuli, whether visual or tactile, activate thalamic and hypothalamic centres along with amygdala, nucleus accumbens and striatum, but also various isocortical areas, which is indicative of a strong involvement in sexual behaviour of centres controlling the more complex mental and cognitive faculties. Patterns of brain region activation associated with appetitive sexual behaviour and copulation are similar in mammals and birds (see Goodson *et al.* 2005 for a review).

Sexual behaviour and brain plasticity

The probability of the brain areas mentioned in the previous section being activated by external environmental stimuli, such as watching a couple copulating, is in part modulated by the levels of circulating sexual hormones (e.g. gonadal steroids). Gonadal steroids such as testosterone and oestrogens that reach the cells of the brain, or hormones that are produced by brain cells themselves, exert their activational role on sexual behaviour by interacting with cellular receptors in the relevant brain centres. Both mammals and birds possess regions of the brain that have high levels of steroid receptors. Such regions are collectively known as the 'social behaviour network' and include the medial amygdala (known as the nucleus taeniae in birds), the bed nucleus of the stria terminalis, lateral septum, anterior hypothalamus, ventromedial hypothalamus, preoptic area and various midbrain areas (reviewed in Goodson *et al.* 2005).

Hormonal activation of cellular receptors in those and other centres occurs via three major mechanisms: (a) through a genomic action of classical receptors, (b) through a non-genomic action of classical receptors, and (c) through a non-genomic action of non-classical receptors (see, for example, Lösel & Wehling 2003; Simoncini & Genazzani 2003) (see Figure 5.9). To these, we should also add a ligand-independent action (Mani 2003). The immediate effect of the genomic action of hormones is the transcription of nuclear genes, the secondary transcript being ultimately translated into specific neurotransmitters that will facilitate activation of neuronal circuits in those specific areas of the brain where cells posses receptors for the hormones (Mani 2003). The genomic action of hormones, however, is a slow process, taking hours to unfold (Lösel & Wehling 2003). Neural circuits could be also activated within seconds by non-genomic mechanisms involving membrane receptors and secondary messenger cascades (Kelly *et al.* 1999). For instance, it is well known that oestradiol, progesterone and testosterone can modulate activity of ion channels such as the calcium channel on the cell membrane of neurons (see, for example, Simoncini & Genazzani 2003). The calcium channel, in turn, can modulate neurotransmitter release at neuronal synapses. On the other hand, ligand-independent action involves the activation of transcription factors – which are usually dependent on a specific hormone (e.g. progesterone) – by other molecules, e.g. the neurotransmitter dopamine (Mani 2003). This latter mechanism is extremely relevant for the understanding of neuroendocrine control of homosexuality as it implies mechanisms that may trigger sexual behaviour independently of circulating sex hormones. These mechanisms, in turn, could explain behaviours such as same-sex mounting occurring outside of the breeding season or during periods of the breeding season when circulating levels of sex hormones are low. For instance, Auger *et al.* (1997) described how vagino-cervical stimulation in rodents activates neural progesterone receptors even in the absence of progesterone! Moreover, activation of sexual behaviour outside of the breeding season may be also directly influenced by adrenocortically or even brain-secreted sex hormones that interact with steroid receptors on brain centres (see Goodson *et al.* 2005 for a review). Given the number and complexity of these molecular mechanisms, it is easy to see how individual differences, including gender differences in the distribution of receptors on cells of the nervous system, along with the specific architecture of neuronal circuits and their plasticity during the adult stage of development, could account for at least some aspects of the interindividual variability in the behavioural response to specific sexual stimuli (Kelly *et al.* 1999).

The complexity of nerve cell activation described above points to one traditional source of plasticity in the adult brain function: activation or inhibition of current neuro-pathways. The classic view of brain function is that once the organisational processes that characterise the early ontogeny of the brain are over, the adult brain retains a degree of plasticity, mainly in the extent to which the surviving neuronal pathways are activated or inhibited by internal signals (e.g. hormones and neurotransmitters). This means that individuals may be anatomically similar in their brain circuitry but behave very differently if regions of the brain are diversely activated or inhibited. In this context, homosexual, heterosexual and bisexual behaviours could be associated with the ability of the same brain to produce variable sexual preferences and also to produce both feminine and masculine sexual behaviour, a situation very common in rats and mice (Aron *et al.* 1991). This is an idea first proposed by Krafft-Ebing (1886). Alternatively, male and female brains may be anatomically and functionally different, and therefore diverse sexual orientations may also reflect a large degree of intraspecific diversity in brain morphology. However all this has been

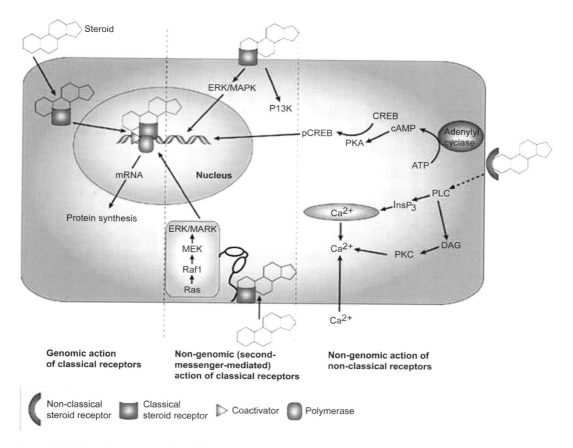

Figure 5.9. Schematic representation of the mechanisms of steroid action that can occur within a cell. The pathways comprise genomic action of classical receptors (left), non-genomic action of classical receptors (middle part), and pathways driven by non-genomic action of non-classical receptors (right). CREB, cyclic AMP response-element-binding protein; DAG, diacylglycerol; InsP$_3$, inositol trisphosphate; MEK, MAPK, PI3K, PKA, PKC and ERK are all kinases; PLC, phospholipase C. From Lösel & Wehling (2003).

postulated on the assumption that the brain does not change after the individual has attained sexual maturity. Results of current research do not warrant such an assumption.

Phenotypic variability in sexual behaviours may be increased if the brain retains a level of plasticity that implies not just activation or inhibition of fixed neuronal circuits, but the actual change in the architecture of cell interconnections in specific areas. Nerve cell interconnections could be increased by increasing the number of synapses between neurons, a process known as synaptogenesis. In female rats, for instance, the number of synapses among neurons in the arcuate nucleus varies daily during the ovarian cycle in relation to the circadian variation in circulating gonadotropins. The arcuate nucleus innervates the medial preoptic area which is well known to control aspects of sexual behaviour (Naftolin et al. 1996). Naftolin et al. (1996) indicate how oestrogen treatment of female rats produces a loss of gamma-aminobutyric acid (GABA) and dopamine synapses in the arcuate nucleus, suggesting that such neurotransmitters may be involved in the variation of synapses during the ovarian cycle in rats. Moreover, recent studies have indicated non-neuronal cell components of the central nervous system such as glial cells (e.g. astrocytes and oligodendrocytes) as possessing a synaptogenesis regulatory role in the adult brain that can contribute to brain plasticity and therefore behavioural plasticity (Nedergaard et al. 2003; Slezak & Pfrieger 2003). Cholesterol is one of the glial factors that contribute to synaptogenesis (Slezak & Pfrieger 2003). Therefore, as suggested by these works, another factor underpinning adult brain plasticity is the ability of nerve cells to undertake synaptogenesis.

Brain plasticity can be achieved through increased synaptogenesis following two major mechanisms: (a) changes in number of synapses without altering interconnectivity among neurons (i.e. no new neurons are innervated by the new synapses), or (b) synaptogenesis involving increased interneuronal connectivity (Stepanyants et al. 2002) (i.e. new neurons are innervated). The latter could be achieved by an increase in the number of dendritic spines or by a modification in the topology of dendrites or of the axonal branches (Stepanyants et al. 2002). Sex steroids such as oestrogens, but also androgens, can modulate the level of synaptogenesis occurring in areas of the brain possessing the appropriate receptors. In sexually mature female rats, for instance, oestrogens can increase synapses in the hypothalamic arcuate nucleus and ventromedial nucleus, the lateral septum and in the midbrain central grey (Matsumoto 1991). In adult Passeriformes androgens can exert a synaptogenic effect in the forebrain (Matsumoto 1991).

The finding that the adult brain retains a degree of synaptogenic capability clearly opens up the possibility for brain and therefore behavioural restructuring during the adult ages of the individual. One immediate and obvious potential consequence of this effect in the context of homosexual behaviour is that post-sexual maturation experiences (including, potentially, learning) may not only activate available neuronal interconnections that were established during early ontogeny, but may actually promote the production of completely new ones!

In a classic experiment carried out on adult mice, Knott et al. (2002) passively stimulated single whiskers for 24 h. Each whisker has a corresponding neuronal representation in layer IV of the somatosensory cortex in the form of a specific assembly of several neurons called a barrel. Continued stimulation of the whisker is associated with an increase in the density of synapses in the associated barrel. This increase is due to the production of new excitatory synapses observed on the dendritic spines (Knott et al. 2002). That is, as peripheral sensory organs are repeatedly stimulated, there is an actual change in the fine architecture of specific areas of the brain (see also the section Ovis aries: *the case of homosexual rams*, above). Therefore, even at the adult stage continuous experiences of a specific kind may be followed by equally specific restructuring of some areas of the brain that could make the repetition of the experience more likely or more efficient in the future. In other words, activity itself can morphologically and functionally reshape neuronal circuitry in the brain (Kleim et al. 1996; Zito & Svoboda 2002; Johansson 2004; Waites et al. 2005). From this, however, it would be naïve to conclude that the plastic capabilities of the brain are structurally unconstrained: they are obviously not. Brain plasticity clearly opens up the possibility of a contribution of learning, but not all individuals will be able to learn and maintain all behaviours (e.g. preference for same-sex mounting) all the time.

In their review of the relationship between experience of adult individuals and brain plasticity, Markham & Greenough (2004) indicated how exposure to stimulus-enriched environments can affect the morphology of brain regions such as the auditory cortex, hippocampus, visual cortex, primary somatosensory cortex, entorhinal cortex, cerebellar cortex, amygdala and basal ganglia, some of which, such as the amygdala, can directly influence sexual behaviour. In particular, Markham & Greenough point to the effects of a stimulus-enriched environment on increasing dendritic arborisation, density of dendritic spines and number of synapses per neuron.

Specific areas of the adult brain can not only modulate their functions through activation/inhibition of established neuronal connections, or the formation of new connections among existing cells as I mentioned above, but also through an actual increase in the number of neurons (neurogenesis) or a decrease in the number of neurons (apoptosis) (Buss *et al.* 2006; Bredesen *et al.* 2006).

Neurons that innervate target cells (e.g. other neurons) can survive only if the target cell provides trophic factors called neurotrophins (e.g. nerve growth factor and brain-derived neurotrophic factor among others; Black 1986; Sastry & Subba Rao 2000; Buss *et al.* 2006). Neurons that are not participant to a network and that therefore have not established active connections (e.g. synapses) with other cells are starved of neurotrophins and therefore die. Such apoptotic processes can be prevented by sexual hormones, for instance, as we have seen in Chapter 4 (see also Meier *et al.* 2000 for a review). Apoptotic processes occurring in the adult brain can therefore provide additional specific plasticity if they are limited to specific areas of the brain. The architecture of the brain is therefore strongly influenced by the tendency of neurons to undergo apoptosis, whereas apoptosis is prevented by survival signals received by the cell from its surrounding environment (Meier *et al.* 2000). Apoptosis affects not only neurons, but also glial cells (Buss *et al.* 2006).

In the adult brain neurons may die, but new ones may also be produced, a process known as neurogenesis. That specific regions of the adult mammal brain are capable of undergoing neurogenesis is a fact that has been recognised only recently (see Markham & Greenough 2004; Ming & Song 2005 for reviews). Knowledge about neurogenesis in the adult avian brain was already well established in the 1980s by the work of Fernando Nottebohm at Rockefeller University (Nottebohm 1985; see Gross 2000 for a review). Neurogenesis in the mammalian adult brain has been described in the lateral ventricle and the subgranular zone of the dentate gyrus in the hippocampus (Ming & Song 2005) and the olfactory bulb (Buss *et al.* 2006), but whether it also occurs in other areas of the adult brain remains to be seen. In the avian brain, neurogenesis has been described in various regions, including the hippocampus and some nuclei of the song system (Rousselot *et al.* 1997; Gross 2000). Current evidence indicates that the new neurons in the adult mammal brain develop from astrocytes, a remarkable finding that suggests that some glial cells act as stem cells for the production of new neurons in the adult brain (Ming & Song 2005). Neurogenesis, like apoptosis, can also be influenced by circulating hormones. For instance, it is inhibited by circulating corticosterone, whereas it is promoted by oestrogen and prolactin (Ming & Song 2005). As for apoptosis, not only neurons but also glial cells can undergo population growth in specific areas of the adult brain (Markham & Greenough 2004).

The objective in the initial paragraphs of this section was to highlight current research on the neuroendocrine mechanisms of sexual behaviour (see also Goy & Roy 1991), and I have emphasised those studies indicating that plasticity of the central nervous system is not confined to specific sensitive periods of pre- and perinatal life, or even the period of postnatal life prior to attainment of sexual maturity, as I stressed in Chapter 4, but also to periods after sexual maturity. Presumably, it is such neurological plasticity that provides the basic mechanisms to produce mental and behavioural plasticity. I have also mentioned, however, that I do not conclude from this evidence that the brain has an *unlimited* capability for learning or that learning mechanisms may overcome the effects that any other process may have on behaviour. Even brain plasticity operates within the boundaries set by the previous ontogeny and the genetics, biochemistry and molecular biology of the cells involved, e.g. neurons. I have also stressed the many interactions between cells of the brain and hormones in producing such plasticity. I will return to the issue of brain plasticity and its specific role in male and female homosexuality when I address how learning, cognition and emotions affect the manifestation of homosexual behaviour in the adult individual, humans in particular. In the next section we will see how the neuroendocrine mechanisms that co-contribute to the causation of sexual behaviour can be organised in specific manners according to the sex of the individual, and that such sexual dimorphism can be both morphologically and functionally variable within and between sexes across species.

Sexual dimorphisms in the adult brain

As we have seen in Chapter 4, the brain undergoes developmental processes during early ontogeny that result in some differences in brain architecture between males and females (see, for example, Swaab *et al.* 2003 for a review of sexual dimorphisms in the hypothalamus). Such processes (e.g. apoptosis) can be controlled by mechanisms that depend on gonadal steroids (see Morris *et al.* 2004b for a review), but also by direct effects of genes located in the sex chromosomes (Agate *et al.* 2003). As we have just seen, in some species both apoptosis and neurogenesis may occur in specific areas of the brain even during the adult stage of development. Table 5.3 summarises a list of sexual dimorphisms in areas of the nervous system that are implicated in sexual behaviour.

Different brain regions in males and females that control sexual behaviour may differ not only in terms of the total number of neurons, neuronal size and/or degree of neuronal interconnectedness, but also in the distribution of steroid receptors in their cells (Scott *et al.* 2004; Dugger *et al.* 2007), which may determine their differential capability to respond to circulating steroids.

The adult male and female brains are therefore not necessarily equivalent, and yet the regions of behavioural overlap between the sexes may be considerable in some species (e.g. in rodents such as *Rattus* spp. and *Mus* spp.). As we have seen in Chapter 4, during development brain regions undergo independent processes of masculinisation or de-masculinisation and de-feminisation or feminisation (Whalen 1974). This means that, under certain circumstances, the brain architecture could retain both masculine and feminine characteristics that could even vary during the adult period of the life of the organism as a result of brain plasticity (Hutchinson 1997). Such overlap in function between male and female nervous system structures and physiology is also seen at the level of response to steroid stimulation, as not only male, but also female neurons have the capability of increasing aromatase production in response to testosterone (Hutchinson 1997). What kind of changes, if any, do these sexually dimorphic brain regions that control sexual behaviour undergo in homosexual individuals? Are homosexual brains structured in a manner typical of individuals of the other sex or are they simply different from both heterosexual males' and females' brains? Is same-sex sexual behaviour a result of a degree of bisexual capability of the brain that is retained in the adult life? How can the experience of the adult individual modulate a plastic brain to produce and maintain brain structures associated with manifestation of homosexual behaviour? I will devote the rest of this neuroendocrinological section to answering these questions.

Homosexual behaviour and the central nervous system

Whenever we adopt a hypothetico-deductive method of scientific enquiry, even in its strong inference version (see Chapter 2), we are always faced with the prospect of producing a set of alternative hypotheses that are strongly biased by our prejudices. For example, *homosexual males are feminised men, therefore their brains must look like those of heterosexual women*. This specific hypothesis was actually first proposed by Dörner *et al.* (1975) but later criticised by Swaab & Hofman (1990) and Gorman (1994). Of course such a hypothesis could be falsified – and we will see that in fact it is – by finding that some sexually dimorphic brain areas are indeed 'hypermasculinised' in

Table 5.3. Sexual dimorphism in some specific nuclei and areas of the brain and spinal cord that influence sexual behaviour

Nucleus/Area	Sexual dimorphism in the adult	Process causing the dimorphism	Species	Reference
Sexually dimorphic nucleus of the preoptic area[a]	M > F	Apoptosis[b]	Mouse, rat, ferret, human, spotted hyena, sheep	Morris et al. 2004b; Roselli et al. 2004a; Fenstemaker et al. 1999
Anteroventral periventricular nucleus[a]	M < F	Apoptosis	Mouse, rat	Simerly 2002
Bed nucleus of the stria terminalis[a]	M > F	Neurogenesis	Human	Chung et al. 2002
Lateral anterior bed nucleus of the stria terminalis[a]	M < F	?	Rat	Cooke et al. 1998
Suprachiasmatic nucleus[a]	M ≠ F[f]	?	Human	Swaab 1995
Ventromedial nucleus of the hypothalamus	M > F	Apoptosis?	Rat	Cooke et al. 1998
Onuf's nucleus[c]	M > F	Apoptosis	Human, dog, rat	Forger & Breedlove 1986
Posterodorsal medial amygdala	M > F	Apoptosis	Rat, mouse	Cooke et al. 2003; Morris et al. 2003
Hyperstriatum[d]	M > F	Apoptosis	Canary, zebra finch	Nottebohm 1980
Nucleus robustus[e]	M > F	Apoptosis	Canary, zebra finch	Gurney 1981
Anterior commissure	M < F	?	Human	Allen & Gorski 1992
Accessory olfactory bulb	M > F	Apoptosis	Rat	Valencia et al. 1992
Vomeronasal organ	M > F	Apoptosis	Rat	Cooke et al. 1998
Locus coeruleus	M < F	Apoptosis and neurogenesis	Rat	García-Falgueras et al. 2005

[a] Located in the hypothalamus
[b] Not in all species
[c] Located in the spinal cord, in the rat it is known as the spinal nucleus of the bulbocavernosus
[d] Also known as the higher vocal centre
[e] Located in the archistriatum
[f] The nucleus is elongated in females and spherical in males, but the volume and number of cells are similar. Additional data are available from Table 1 of Zuloaga et al. (2008).

homosexual males rather than 'feminised'. The problem with the initial formulation, however, remains, as it may retain a biasing effect on our enquiry (Poiani 1995). For instance, if our research into homosexuality is chiefly guided by the gender inversion view of homosexuality, then we may conceivably be prevented from seeing a male who mounts another male or a female who is mounted by another female as homosexuals. That such a bias against the mounter male, or the female mountee, would be unwarranted is clearly exemplified by the case of the homosexual rams studied by Charles Roselli and colleagues that I reviewed in a previous section. Those homosexual rams are mainly mounters, not mountees!

Research programmes on the relationship between central nervous system structure (and function) and same-sex sexual behaviour should start by recognising the full diversity of the patterns of homosexuality along a series of dimensions: male–female, homosexual–bisexual–heterosexual, masculine–feminine, mounter–mountee (see also Whitam 1977). Only then can we propose mechanisms that explain both the possible combinations of phenotypes (e.g. masculine male homosexual, 'butch' vs. 'femme' lesbian) and transitions from one combination to another that may even occur in the same individual during his/her lifetime. Such behavioural diversity is presumably a product, at a proximate level, of the diversity and also plasticity of structures and functions of the central nervous system and the sensory organs associated with it, along with being a result of the interactions of brain centres with genes and their products and, of course, environmental factors during development. With this dynamic scenario in mind, in what follows I will review studies that test the specific association between different areas of the nervous system and sexual orientation.

The hypothalamus and homosexuality

The hypothalamus is a critical part of the diencephalon associated with control of sexual behaviour. In the past 20 years or so it has been the focus of several studies aimed at understanding its role in the causation of same-sex sexual behaviour. Following Dörner's model that views the homosexual brain as mirroring other-sex organisation and function, most studies have initially looked at sexually dimorphic regions and nuclei within the hypothalamus, such as the paraventricular nucleus, the sexually dimorphic nucleus of the preoptic area and others (see Figure 5.10 and Table 5.3) and compared them between homosexual and heterosexual males and females.

Arguably, the most intensely studied areas of the hypothalamus in the context of same-sex sexual behaviour across various taxa are what in rodents are known as the sexually dimorphic nucleus and the medial preoptic area (SDN–MPOA). In humans the SDN is known as the interstitial nucleus of the anterior hypothalamus 3 (INAH3) (Figure 5.10). The initial anatomical works of Gorski et al. (1978, 1980) in rats and the experimental work of van de Poll & van Dis (1979) in rats and Rodriguez-Sierra & Terasawa (1979) in guinea pigs (but see also previous experiments carried out by Brookhart & Dey 1941), paved the way for the subsequent surge in research on the association between SDN–MPOA and sexuality in a variety of taxa.

Gorski et al. (1978, 1980) determined that the sexual dimorphism of the MPOA in rats – the area being larger in adult males than females – is established, at least in part, perinatally and it is largely independent of circulating levels of sexual hormones in adult individuals. In addition, the sexual dimorphism in the MPOA is not explained by body size sexual dimorphism in rats (Gorski et al. 1978). As soon as a sexual dimorphism in a brain area known to control sexual behaviour was found, the opportunity was open to carry out experimental manipulations of such an area to study correlated alterations in sexual behaviour, including sexual partner preference.

Van de Poll & van Dis (1979) carried out an experimental destruction of the caudal section of the MPOA and the rostral section of the anterior hypothalamus in adult male rats and observed a decrease in masculine sexual behaviour, but also a slight increase in feminine behaviour (i.e. lordosis). Hennessy et al. (1986) obtained even more dramatic results after more extensive damage to the POA in male rats, which subsequently showed lordosis levels indistinguishable from those of females. In the work of Paredes et al. (1998), bilateral destruction of the MPOA/anterior hypothalamus (MPOA/AH) of adult male rats also increased the level of consortship with other males and produced a lack of sexual interest in females, but did not produce the level of feminised sexual behaviour reported in previous studies. These experiments taken together strongly suggest that the degree of damage suffered by the sexually dimorphic nucleus, the medial preoptic area and the anterior hypothalamus is associated with varying degrees of de-masculinisation and feminisation of male rats' sexual behaviour,

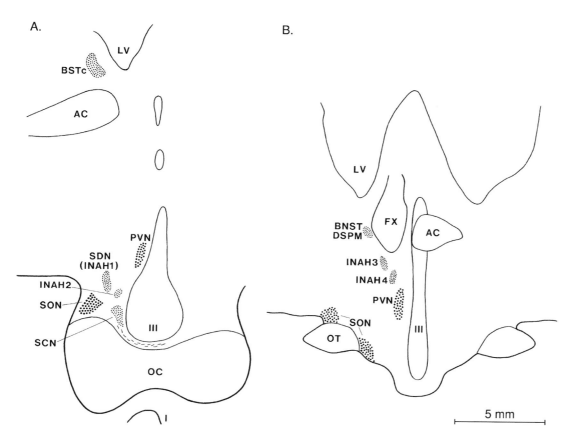

Figure 5.10. Scheme of the sexually dimorphic structures in the hypothalamus of humans. (A) is more rostral than (B). III, third ventricle; AC, anterior commissure; BNST-DSPM, darkly staining posteromedial component of the bed nucleus of the stria terminalis; FX, fornix; I, infundibulum; INAH1–4, interstitial nuclei of the anterior hypothalamus 1–4; LV, lateral ventricle; OC, optic chiasm; OT, optic tract; PVN, paraventricular nucleus; SCN, suprachiasmatic nucleus; SDN, sexually dimorphic nucleus of the preoptic area (= INAH-1); SON, supraoptic nucleus. The AC, BSTc, BNST-DSPM, INAH2, 3, 4, SCN and SDN are different in men and women. The SCN and INAH-3 differ according to sexual orientation. From Swaab (2003).

indicating a seemingly continuous transition from one state to another, a result expected from a capability of at least some regions of the adult rat brain to control both feminine and masculine sexual behaviour. A similar experiment involving MPOA lesions was carried out by Rodriguez-Sierra & Terasawa (1979) on male and female guinea pigs to conclude that, as in rats, the MPOA can control copulation and lordosis in *both* males and females.

Experiments have also been carried out in rodents (e.g. rats) to study early developmental effects on SDN–POA size and its control of sexual behaviour. Rhees *et al.* (1990) showed that the SDN–POA of males reaches its sexually dimorphic typical size by day 5 of postnatal life in both males and females, a process that is dependent on aromatisation of pre- and neonatally circulating testosterone into oestradiol (Rhees *et al.* 1990; Houtsmuller *et al.* 1994) that affects the sex-specific degree of apoptosis and synaptic patterning in the area (Amateau & McCarthy 2004). Treatment with either testosterone propionate in males and females (Rhees *et al.* 1990) or the aromatase inhibitor ATD in males (Houtsmuller *et al.* 1994) can alter not only the size of SDN–POA during development but also the expression of feminine and masculine sexual behaviour accordingly.

Woodson *et al.* (2002) studied Long-Evans rats that were either prenatally, postnatally or both prenatally and postnatally treated with testosterone propionate (TP). Consortship preference tests for an oestrous female were carried out on adult female rats of the three treatment groups. Only female rats treated both pre- and postnatally with TP showed not only a preference to consort with other females, but also an increase in the degree of preference with experience (Woodson *et al.* 2002), a result that is presumably influenced by learning processes. When the same TP-treated females were tested in the presence of an oestrous female after they were also treated with oestradiol benzoate and progesterone, which in control females activate feminine sexual behaviour, the pre- and postnatally testosterone-treated group displayed the least decrease in their preference for the female. In the presence of a stud male, the pre- and postnatally TP-treated females displayed the least lordotic behaviour. However, following a period of increased sexual experience, the postnatally-only TP-treated group also increased lordotic behaviour. These results suggest that in the development of sex role and perhaps sexual orientation in female rats, apart from the well known pre- and perinatal effects of steroids that I described in Chapter 4, there is also a clear effect of learning processes that take place during the adult stages of the individual. The size of the SDN–POA of females mirrors their degree of masculinisation, with pre- and postnatally TP-treated females having a larger SDN–POA than the other categories and with controls having the smallest SDN–POA (Woodson *et al.* 2002). The effect of learning processes on the specific relationship between SDN–POA volume and preference for a male or a female sexual consort in female rats was dramatically demonstrated by Woodson *et al.* (2002) by showing a tendency for naïve female rats not only to increase their preference for female partners as a function of the size of their SDN–POA, but also to increase such preference following previous experience with oestrous females.

The bisexual potential expressed by alterations of MPOA in male and female rodents was also shown in experiments carried out on cats. Hart & Leedy (1983) performed lesions of the MPOA/AH on adult male cats, which were subsequently given oestrogen treatment to test them for female-like sexual behaviour in the presence of stud males. Treated males showed a decreased level of copulatory behaviour and an increased level of proceptivity and receptivity towards stud males. They also displayed female-like sexual behaviour, but this did not reach levels equivalent to those of females. Indeed, operated males were described by Hart & Leedy (1983) as being able to display both male-typical and female-typical behaviour, a mixture that fits better a description of bisexual sexual behaviour. Feminisation of sexual behaviour in males was also observed in ferrets after lesions of the POA/AH by Cherry & Baum (1990), but only in males that sustained bilateral and relatively large lesions of the area. Similar results were also obtained by Paredes & Baum (1995) after carrying out a lesion of MPOA/AH of adult male and female ferrets: only males that sustained extensive bilateral lesions of the MPOA/AH showed a preference to approach a stud male, whereas females treated in the same manner did not show any change in their preference to approach a stud male (see also Kindon *et al.* 1996).

In ferrets, Michael Baum and collaborators have identified a sexually dimorphic nucleus in the preoptic area of the anterior hypothalamus that they named the male nucleus of the POA/AH (MN-POA/AH) and that is larger in males (Tobet *et al.* 1986). The modified sexual preference that male ferrets show after bilateral damage to this area is mediated by changes in the response to olfactory cues, as operated male ferrets not only approached male rather than female odours, but they also showed female-typical patterns of neural activation in the medial POA in response to male odours (Alekseyenko *et al.* 2007).

Genitally masculinised female spotted hyenas also show an attenuated sexual dimorphism in their sexually dimorphic nucleus (hSDN) of the MPOA/AH (Fenstemaker *et al.* 1999), whereas the male-oriented rams studied by Charles Roselli and his collaborators have a sexually dimorphic nucleus (oSDN) in their MPOA/AH that is intermediate between that of the female-oriented rams and ewes. In primates other than humans an area has been described in the anterior hypothalamus of both *Macaca fuscata* (Vasey & Pfaus 2005) and *Macaca mulatta* (Byne 1998) that has been called AHdc and is presumably homologous to the SDN-MPOA, hSDN, MN-POA/AH and oSDN. AHdc is sexually dimorphic in these primate species, with males having larger areas than females, but females that frequently engage in homosexual mounting do not have a larger AHdc (Vasey & Pfaus 2005).

The centres that control sexual behaviour in birds are chiefly the hypothalamic preoptic area, the periaqueductal grey and the nucleus intercollicularis (Ball & Balthazart 2004). In the preoptic area, the medial preoptic nucleus is sexually dimorphic (Balthazart *et al.* 2003; Ball & Balthazart 2004), but apparently no equivalent studies to those carried out in mammals have been performed in birds in order to determine the role of the medial preoptic nucleus in same-sex sexual behaviour.

Swaab & Fliers (1985) were the first to describe a sexual dimorphism in the human brain in an area of the hypothalamus that is homologous to the SDN-POA of rodents. Subsequently, Allen *et al.* (1989) studied the preoptic area of the human hypothalamus and described four groups of cells that they named the interstitial nuclei of the anterior hypothalamus (INAH 1–4). Simon LeVay (1991) was the first to compare the INAH of heterosexual and homosexual males. Although LeVay (1991) studied the brains of homosexual patients who died of HIV–AIDS, thus casting some legitimate concerns about the true representation of his data for the morphology of that same region in healthy brains, the results suggest that the INAH 3 of homosexual men is smaller and more similar in size to the female's INAH 3 than to that of heterosexual men. The size of the other INAHs (i.e. 1, 2 and 4) did not differ between homosexual and heterosexual males. Swaab & Hofman (1990) and Swaab *et al.* (1995) however, did not report differences in this sexually dimorphic nucleus' cell number between heterosexual and homosexual men. On the other hand, the sexual dimorphism in the INAH 3 was confirmed by Byne *et al.* (2000), with the INAH 3 of men being, on average, 50% larger than the INAH 3 of women, after correction for brain size, a difference due to a larger number of neurons found in the male INAH 3.

In view of the criticism that LeVay's (1991) work was subject to because of his use of brains from homosexual HIV-infected persons that were compared with non-HIV-infected heterosexuals, Byne *et al.* (2001) replicated this work using brains from homosexuals who were HIV positive and compared them with those of heterosexuals who were either HIV negative or HIV positive, having acquired the infection through drug use. Byne *et al.* (2001) work confirmed the sexual dimorphism in the INAH 3 and also revealed no statistically significant effect of HIV infection on the size of that specific nucleus, although there was an effect of HIV infection on the INAH 1, with HIV-positive males having an INAH 1 that was 8% larger than that of HIV-negative males. Byne *et al.* (2001) found that the volume of the INAH 3 of homosexual males was intermediate between that of heterosexual males and females, although the difference was not statistically significant. The result did not change after correcting the size of INAH 3 for the effect of brain mass. A recent work carried out by García-Falgueras & Swaab (2008) indicates that the volume of the INAH 3 of male-to-female transsexuals is completely feminised, being significantly smaller than that of heterosexual men and indistinguishable from that of heterosexual women. The same feminisation pattern was detected by García-Falgueras & Swaab in the number of neurons in the INAH 3.

The hypothalamic suprachiasmatic nucleus (SCN) has also been compared between homosexual and heterosexual men. Swaab & Hofman (1990) studied the SCN of three groups of men: homosexuals who died of AIDS and two control groups, one composed of heterosexuals who also died of AIDS and the other of heterosexuals who died of diverse causes that were not related to AIDS. The SCN of homosexual men was 1.73 times larger and contained 2.09 times as many cells as that of heterosexual men who died of multiple causes. This difference was not due to the effect of AIDS, as the SCN of those heterosexual men who had died of AIDS was similar in both total cell number and volume to that of heterosexual men who did not die of AIDS. We will remember from Table 5.3 that the SCN was found to be elongated in females and spherical in males, although the volume and total number of cells did not differ between the sexes (Swaab 1995). Swaab & Hofman (1990) also found a female-like shape in the SCN of homosexual males, which had a longer rostrocaudal axis than the SCN of heterosexual males. Therefore, although the shape of the SCN of homosexual men seems to be feminised, its size is more consistent with a 'third sex' pattern, being larger than that of both heterosexual men and women (see also Swaab et al. 1995).

The ventromedial nucleus (VMN, also abbreviated VMH) is widely regarded as a centre of control of sexual behaviour (Pfaff & Sakuma 1979; McClellan et al. 2006). In rodent studies, electrostimulation of the VMN releases the female-typical lordosis behaviour, whereas lesions to the VMN inhibit the display of lordosis (Pfaff & Sakuma 1979 and references therein). VMN lesions also disrupt female sexual behaviour in ferrets, sheep and cats (Robarts & Baum 2007 and references therein). This control of female sexual behaviour occurs in spite of the VMN being larger in males than females (Matsumoto & Arai 1983), and also in spite of regions of the VMN that are especially rich in sex steroid receptors (oestrogen, but also androgen receptors, Segovia & Guillamón 1993) developing larger number of synapses in males than females (Matsumoto & Arai 1986). Such sexual dimorphism, however, varies cyclically in rodents in response to different stages of the female reproductive cycle. In Wistar rats the VMN is usually 25% larger in adult males than in adult females, but the difference increases to 35% when females are in dioestrus and decreases to 16% when they are in proestrus (Madeira et al. 2001).

The VMN receives projections from the preoptic area of the hypothalamus and also from thalamic and epithalamic areas, along with the amygdala and the medial central grey (McClellan et al. 2006). In turn, the VMN sends projections back to the amygdala, the preoptic area (the medial POA more specifically), the anterior hypothalamus, the bed nucleus of the stria terminalis, the central grey, the zona incerta and also the perpendicular nucleus (McClellan et al. 2006). That is, the VMN is clearly connected with the major centres that control sexual behaviour, including those such as the amygdala that have major links with olfactory areas of the brain (see, for example, Robarts & Baum 2007).

Very controversial studies carried out in male humans, involving surgical lesions of the VMN, mainly coincide in that such interventions decrease sexual drive (Müller et al. 1973; Dieckmann et al. 1988), with Müller et al. claiming that lesions produced in the right VMN of 12 homosexual patients also decreased the specific homosexual fantasies and desire in eight of them.

In sum, studies of various areas of the hypothalamus clearly falsify Dörner's model of a feminised brain in homosexual males (see also Swaab 1995; Swaab & Hofman 1995). Although some areas of the hypothalamus such as the SDN–MPOA tend to be smaller, and therefore more female-like in male homosexuals, in various species the size-feminisation of the SDN–MPOA of individuals displaying same-sex sexual behaviour is not necessarily complete, as the nucleus tends to be intermediate in size between that of heterosexual males and heterosexual females. It is likely that variability in hormone-dependent apoptotic processes may contribute to such diversity of SDN–MPOA development. In other areas of the hypothalamus, homosexual individuals do not have structures that are intermediate in size between those of heterosexual males and females. In terms of the size of the area, for instance, the SCN is larger in homosexual males than in heterosexual males and females, who, in turn, do not differ. The situation for the SCN of homosexuals could be better described as that of a 'third sex'. It is possible that this kind of third-sex pattern involving larger numbers of neurons may be a result of extended neurogenesis in homosexuals (or attenuated apoptosis), a situation that would fit a more general neotenic model for the evolution of the human brain and of homosexuality. As far as the VMN is concerned, more studies are required to understand its potential role in the manifestation of same-sex sexual preference in humans.

Beyond the hypothalamus

Although the hypothalamus has been a major focus for neuroendocrine studies of homosexuality, we have seen that other areas of the nervous system are also involved in controlling sexual behaviour. Their potential association with homosexuality will be reviewed in this section.

The bed nucleus of the stria terminalis

A forebrain limbic area that has been studied in the context of sexual orientation is the bed nucleus of the stria terminalis (BNST) (Figure 5.10). The BNST is involved in control of mating behaviour, receiving neuronal inputs from both the accessory olfactory bulb and the medial amygdala and, in turn, projecting to both the medial preoptic area and the ventromedial hypothalamus (see, for example, Aste *et al.* 1998). Homologous regions have been described in both bird and mammal brains (Aste *et al.* 1998). Zhou *et al.* (1995) compared the central part of the BNST (BNSTc) – that is 44% larger in heterosexual men than heterosexual women – between heterosexual and homosexual men and found that the BNSTc was slightly larger in homosexual men (2.81 mm^3) than heterosexual men (2.49 mm^3) although the difference was not statistically significant. In addition, Zhou *et al.* (1995) also measured the BNSTc of male-to-female transsexuals, finding that they had a statistically non-significantly smaller volume of their BNSTc (1.30 mm^3) than women (1.73 mm^3). In a later study, Kruijver *et al.* (2000) found that the number of somatostatin-expressing neurons in the BNSTc of male-to-female transsexuals was also smaller than in both heterosexual and homosexual men, but indistinguishable from heterosexual women. A different pattern was found in a single brain of a female-to-male transsexual, where the number of somatostatin-expressing neurons in the BNSTc was in the heterosexual male range (Kruijver *et al.* 2000). Interestingly, Chung *et al.* (2002) described a continuous neuronal population growth in the BNSTc of men from the pre/perinatal stages of development to well into the adult ages, whereas in women the BNSTc stops its growth at around puberty. This suggests that either neurons in the BNSTc of homosexual men's and female-to-male transsexuals' brains are produced at a higher rate than in heterosexual men and women, respectively, or that they are produced at a similar rate but for a longer period of time, the latter being a pattern that would be consistent with a neotenic evolutionary mechanism (see Chapter 4). Be that as it may, homosexual men seem to have a 'hyper-masculinised' BNSTc, whereas that of male-to-female transsexuals is feminised. The BNSTc of a single female-to-male transsexual seems to be masculinised.

The anterior commissure and the corpus callosum

The anterior commissure (AC) is a region of the brain that is probably involved in brain lateralisation and, presumably, also in the control of some aspects of cognitive skills, as it seems to affect transfer of olfactory, visual and auditory information between the two hemispheres (Allen & Gorski 1991). The AC is only subtly sexually dimorphic, with females having a larger AC than males on average, but a substantial region of overlap exists between the two sexes (Allen & Gorski 1991). Allen & Gorski (1992) studied the AC of heterosexual men, homosexual men and heterosexual women and found that the AC area was 18% larger in homosexual men than heterosexual women and it was also 34% larger in homosexual than in heterosexual men. Although these results should be taken with caution as the homosexual men died as a result of AIDS, it seems that, in this case, the AC of those homosexual men was 'hyperfeminised'. Lasco et al. (2002) have compared the size of the AC among heterosexual males, heterosexual females and homosexual males while controlling for the effect of HIV infection and found that AC size was not affected by sex and sexual orientation. Therefore the issue of whether there is an association between AC and sexual orientation still requires further investigation.

The AC is an ancient neuronal route of interhemispheric communication that is present in all vertebrates (see, for example, Zeier & Karten 1973; Bruce & Butler 1984; Echteler 1984; Rilling & Insel 1999; Moreno & González 2004). Another major neuronal link between the two hemispheres, which, however, is unique to placental mammals, is the corpus callosum (CC) (Aboitiz & Montiel 2003). As for the AC, the CC is supposed to be involved in brain lateralisation and cognitive functions by means of enhancing interconnectivity between left and right hemispheres (Aboitiz & Montiel 2003). The CC was first compared between men and women by De Lacoste-Utamsing & Holloway (1982), who described a larger splenium – the posterior section of the CC – and also a larger total callosal area relative to brain mass in women than men. A similar difference in sexual dimorphism was obtained by Holloway et al. (1993) for splenium size relative to either brain size or specific sections of the CC, after analysing brains preserved in fixative. The sexual dimorphism, however, was not observed by Going & Dixson (1990) after analysing preserved brains from relatively aged individuals. Sex differences in size and shape of the CC were found by Clarke et al. (1989), but only for brains preserved in various fixatives, whereas results of magnetic resonance imaging (MRI) scans performed on volunteers did not produce the same sexual dimorphism. Similar negative MRI results were obtained by Weis et al. (1989). Against the MRI results of Clarke et al. (1989) and Weis et al. (1989) stands the work of Allen et al. (1991), who did find some sexual dimorphism in CC after carrying out MRI analyses. They calculated various 'bulbosity coefficients' for the splenium: (a) average width of the splenium relative to the posterior fourth fifth of the CC, or (b) relative to the posterior half of the CC, or (c) relative to total CC. All those bulbosity coefficients were significantly larger in females than in males. Jäncke et al. (1997) compared the CC area relative to the forebrain volume between males and females and found that the ratio is larger in the latter. This result, however, simply flows from a negative relation between CC area relative to forebrain volume with brain size, and the tendency for women to have relatively smaller absolute values of brain size than men. In fact, for men with brain size similar to women's, the ratio is also within the female range. Results of studies of the CC therefore seem to point to a great intrasexual variability in CC size and shape that somewhat blurs a possible sexual dimorphism. In fact, across primates, not all species show sexual dimorphism in the CC (De Lacoste & Woodward 1988). This poses some interesting questions as to what we might expect from comparisons of CC between homosexuals and heterosexuals. Emory et al. (1991) studied a group of 10 male-to-female transsexuals and a group of 10 female-to-male transsexuals and compared the total size of the CC, the brain/CC ratio, area and shape of the splenium between them and samples of 20 male and 20 female heterosexuals. MRI analyses indicate that none of the variables measured differed among groups, suggesting that sexual orientation may not be directly associated with changes in CC. This conclusion, however, has been recently challenged by Witelson et al. (2008), who compared the CC of 12 homosexual and 10 heterosexual right-handed men using MRI and found that homosexual men had a larger isthmus, a section of the posterior half of the CC adjacent to the splenium, than heterosexual men.

Olfactory epithelium and vomeronasal organ

Sexual behaviour in most mammals is variably influenced by olfactory stimuli. The two major areas of the nervous system that transduce olfactory information are the main olfactory epithelium and the vomeronasal organ. The former sends neuronal projections into the main olfactory bulb, whereas the latter, which is the main organ for perception of pheromones (Novotny 1987), sends projections into the accessory olfactory bulb. Other organs, however, are also involved in the transduction of olfactory information: the septal organ, the Grueneberg ganglion and free nerve endings that are found in the trigeminal system (see Baum & Kelliher 2009 and Baum 2009 for recent reviews).

Both the main olfactory epithelium and the vomeronasal organ are involved in the neuronal pathways associating olfactory stimuli and sexual behaviour in the brain via links with the amygdala, the stria terminalis and finally the anterior hypothalamus (Savic 2002). In rats, the accessory olfactory bulb is larger in males than females, a difference that is established postnatally, being influenced by circulating testosterone and oestradiol (Valencia et al. 1992). Can alterations of the olfactory bulb produce same-sex sexually oriented phenotypes? Potentially they can. For instance, Hart & Leedy (1983) demonstrated how olfactory bulbectomised male cats respond by increasing sexual receptivity to a stud male without losing their ability to also respond sexually to a female: that is, they displayed bisexual sexual behaviour.

The vomeronasal organ of the mouse has been implicated in the mediation of many pheromone-dependent effects on the species' reproductive biology, such as the *Whitten effect* (synchronisation of oestrus), the *Bruce effect* (blockage of pregnancy), the *Vandenbergh effect* (acceleration of pregnancy) and the *Le–Boot effect* (oestrus suppression) (Novotny 1987). In theory, it could be easily seen how a pheromonal communication system that evolved in the context of heterosexual sexual mate recognition and sexual selection could be implicated in homosexual sexual behaviour. A mutation at the locus/loci that determine the kind of pheromone produced could explain, for instance, why a male can be perceived as a 'female' by other males and therefore be mounted. Conversely, a mutation in the cells of the vomeronasal organ may lead to the transduction of a male-typical pheromone signal as if it were a female-typical pheromone signal, again producing effects such as male–male mounting. And so, we can move step by step upstream along the neuronal pathway from the vomeronasal organ (and/or the olfactory epithelium) to the brain centres that are involved in the control of sexual behaviour and orientation, to see how specific mutations can produce a behavioural response that favours same-sex over other-sex mounting based on the interpretation of olfactory cues originating from an individual to be those of a sexual partner. Such a mechanism could explain the obligatory homosexual phenotypes found among the masculinised but male-sexually oriented rams studied by Charles Roselli and his collaborators (see, for example, Roselli et al. 2004b).

Traditionally, humans have been regarded as a microsmatic species, that is, a species possessing a poor sense of smell, as opposed to macrosmatic taxa such as canids and ungulates (Kohl et al. 2001). In particular, there is a still ongoing debate as to whether the vomeronasal organ that is present prenatally in humans is still maintained and, if so, to what extent it is functional in adults (see Meredith 2001 for a review). These issues notwithstanding, the role of olfaction has recently been reappraised in studies of human sexual behaviour. Among the substances identified as sexual pheromones in humans are secretions from the apocrine glands of the skin that may interact with the skin microflora to produce active pheromones, and the short-chained fatty acids secreted from the vaginal barrel also known as 'copulins' (see Zhou & Chen 2008 for recent experimental work and Kohl et al. 2001 for a review).

To the best of my knowledge it was Daniele Oliva who, in a two-page article published in

Neuroendocrinology Letters (Oliva 2002) first suggested the hypothesis that homosexual men could be sexually attracted to other men as a result of male pheromone action. This would in turn stimulate the hypothalamus and higher isocortical centres controlling sexual behaviour and orientation. Oliva's paper was followed by a comment by Fink & Neave (2002) suggesting that the hypothesis was well worth empirical testing. I certainly agree! In fact, two independent research teams, one from the Karolinska Institute in Sweden and the other from the Monell Chemical Senses Center in the USA, have recently reported the results of experiments testing the hypothesis that homosexual males and females may differ from heterosexuals in their preference for odours from diverse kinds of individuals: males and females, homosexuals and heterosexuals (Savic *et al.* 2005; Martins *et al.* 2005; Berglund *et al.* 2006).

Savic *et al.* (2005) studied the effect of two putative human sex pheromones on sexual activation of men differing in their sexual orientation: the steroids 4,16-androstadien-3-one (AND) and oestra-1,3,5(10),16-tetraen-3-ol (EST), the former being a metabolite of testosterone that is present in the sweat of men, whereas the latter is a metabolite of oestrogen that can be detected in the urine of pregnant women (Savic *et al.* 2005; Sergeant *et al.* 2008 and references therein). Subjects were 12 heterosexual men, 12 heterosexual women and 12 homosexual men, all healthy, right-handed and HIV-negative. They were exposed to either odourless air (control), AND, EST or OO (ordinary odours: lavender oil, cedar oil, eugenol and butanol) and their brains scanned to produce either magnetic resonance imaging (MRI) or positron emission tomography (PET) images. The authors defined two 'regions of interest' to be scanned and compared across individuals: (a) the region activated by AND in heterosexual women that covers the preoptic and ventromedial nuclei of the hypothalamus, and (b) the region activated by EST in heterosexual men, comprising the dorsomedial and the paraventricular nuclei of the hypothalamus. That is, the response of each individual was contrasted against the typical heterosexual response of men and women. In both the case of AND and EST the regions activated in homosexual males corresponded to those activated in heterosexual females rather than heterosexual males (see Figure 5.11): mainly the anterior hypothalamus in the case of AND, with a peak in the preoptic area, and the left amygdala and piriform cortex in the case of EST, with some inclusion of the anterior hypothalamus as well in the latter case (Savic *et al.* 2005). Therefore, in male homosexuals, male pheromones seem to activate areas of the brain associated with sexual behaviour in a manner that resembles the response typical of the female brain, supporting Dörner's model of a feminised pattern in homosexual males as far as brain stimulation by pheromonal inputs is concerned. Interestingly, Savic *et al.* (2005) also indicate that differential perception of such pheromones must operate at a subconscious level simply because homosexual and heterosexual males, but also heterosexual females, did not differ in their verbal rating of AND and EST odours in terms of familiarity, intensity, pleasantness and irritability.

Similar results were obtained by Berglund *et al.* (2008) in a study of non-homosexual male-to-female transsexuals (MTFT). MTFT activated the anterior hypothalamus only when smelling AND, not EST, thus mirroring the pattern found in heterosexual women.

Berglund *et al.* (2006) also carried out a study of pheromones in heterosexual and homosexual women following the same protocol as Savic *et al.* (2005). In contrast to the pattern found in homosexual men, homosexual women did not respond to AND and EST by activating different brain centres. In fact, in both cases activity was detected in the amygdala, the pyriform cortex and the insular cortex. In contrast to this, heterosexual women activated the anterior hypothalamus, after smelling AND, and the classic olfactory regions that were also activated in homosexual women when smelling EST. Interestingly, homosexual women activated a cluster of neurons in the anterior hypothalamus after smelling EST that heterosexual men also activated in response to the same

The central nervous system   217

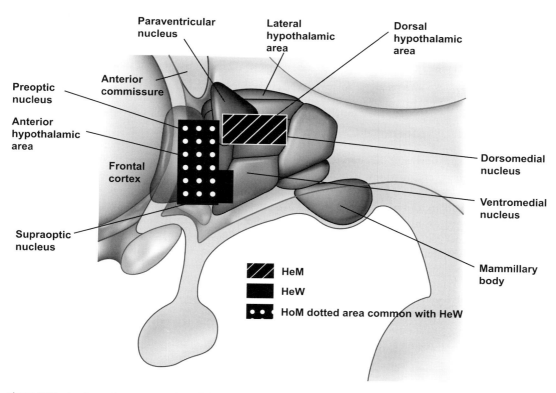

Figure 5.11. A schematic representation of the areas of the hypothalamus, with emphasis on the areas activated by pheromones in heterosexual men (HeM), heterosexual women (HeW) and homosexual men (HoM). From Savic *et al.* 2005.

olfactory stimulus. In sum, lesbians did not respond to male-typical odours such as AND by activating the preoptic area as heterosexual women did and, in fact, they seem to process both AND and EST stimuli in a manner that is more consistent with the brain processes observed in heterosexual men than heterosexual women (Berglund *et al.* 2006), although they did so in a manner that was not as clearcut as for the case of homosexual men activating areas similar to those of heterosexual women (Savic *et al.* 2005). That is, for the relationship between pheromones and sexual orientation, Dörner's model fits the patterns found among both gay men and lesbians, but the fit is better in the former than in the latter.

Martins *et al.* (2005) tested six heterosexual men and six homosexual men along with six heterosexual women and six homosexual women for preference for armpit odours from individuals of those same four categories. Odours were presented in pairs for all possible combinations. Interestingly the preferences of lesbians and heterosexual females did not differ: both preferred armpit odours of heterosexual males and heterosexual females over those of gay males and lesbians. However, the two male categories differed significantly, with heterosexual men preferring odours from other heterosexual men and also from homosexual women, whereas homosexual men preferred odours from heterosexual women and other homosexual men. Apart from the obvious conclusion that males and females of both sexual orientations are capable of distinguishing armpit odours from different individuals and do make a choice when they are asked to, the results of Martins *et al.* (2005) do not lend themselves to a straightforward interpretation. Perhaps the most illuminating trend is that sexual orientation in men seems to be reflected in a clear difference in terms of olfactory preferences, whereas among women, sexual orientation is not associated with any shift in olfactory preference. These results may be reflecting the action of two parallel mechanisms: (a) olfactory cues used to establish *social* consortships, and (b) olfactory cues used to establish *sexual* consortships. Establishing which case is which, however, is made more complicated by the possibility of bisexuality. Does the preference of heterosexual males for armpit odours of other heterosexual males reflect a social consortship preference, or an underlying bisexual tendency? Clearly, more detailed information will be needed from participants in order to interpret the significance of these results. In contrast to Martins *et al.* (2005); Sergeant *et al.* (2007) reported a tendency for heterosexual women to prefer the odour of homosexual men impregnated in a T-shirt, and they even preferred the odour of a clean, unused T-shirt, describing those odours as more pleasant, sexy and preferable than the odour of T-shirts that had been used by heterosexual men. Qualitatively, the trend remained the same during the period of higher fertility of the menstrual cycle. In interpreting these results we should also bear in mind that odours perceived consciously may well produce a response that differs from that of volatile chemicals perceived subconsciously (see the work of Savic *et al.* 2005 above), the former presumably being more constrained by learning experiences in specific cultural contexts.

In a recent review of the association between olfactory functions and sexual orientation, Sergeant *et al.* (2008) also discuss the issue of whether lesbians show an increased level of menstrual synchronisation than heterosexual women or not, a difference expected from the closer physical female–female association among lesbians as compared with heterosexuals. Pheromonally mediated reproductive synchronisation in women could be more effective between lesbian partners, who can engage in very intimate physical contact. Studies that tested this hypothesis, however, are still inconclusive, with two works indicating no enhanced levels of menstrual synchronisation between lesbian partners (Matteo & Rissman 1984; Trevathan *et al.* 1993), whereas one indicates that such synchronisation does occur (Weller & Weller 1992).

The Onuf's nucleus

The spinal cord is also part of the central nervous system and within the spinal cord the Onuf's nucleus, a cluster of neurons located in the sacral region of the spinal cord of cats, dogs and primates that control the activity of muscles used in copulation such as the bulbocavernosus and the ischiocavernosus (Forger & Breedlove 1986), is sexually dimorphic, being larger in males than females (Forger & Breedlove 1986; Pullen *et al.* 1997, but see Seney *et al.* 2006 for an exception). The nucleus receives direct neuronal inputs from the hypothalamus; and some studies of the B-50 (GAP-43) protein that can affect synaptogenesis and axonal growth indicate that the Onuf's nucleus may retain the capacity for neuroarchitectural reorganisation even in adult individuals (Nacimiento *et al.* 1993).

Although no studies are available that compare the Onuf's nucleus of homosexuals and heterosexuals, given its high degree of plasticity we may predict that the size of the Onuf's nucleus may be directly correlated with the degree of sexual activity. In those instances when homosexual males are, on average, more sexually active than heterosexual males it is possible that their Onuf's nucleus may also differ, with the Onuf's nucleus of homosexual men being hypermasculinised. A relative masculinisation of the Onuf's nucleus may be also expected in 'butch' lesbians compared with 'femme' lesbians and heterosexual women. Future studies will be able to test these predictions.

In rats, the spinal nucleus of the bulbocavernosus (SNB) is also sexually dimorphic (larger in males) having functions that overlap to a degree with those of the Onuf's nucleus (Rampin *et al.* 1997). Nordeen *et al.* (1985) have shown that the sexual dimorphism of the SNB is dependent on perinatal action of androgens, which control the degree of apoptosis experienced by SNB motoneurons. Female rats treated perinatally with androgens develop a masculinised SNB. Whether SNB masculinisation is also *specifically* associated with behavioural masculinisation in females is unknown. In adult rodents the SNB, like the Onuf's nucleus in humans, also retains a degree of morphological plasticity that in rodents is associated with seasonal changes in reproductive activity (Hegstrom *et al.* 2002).

Additional functional and morphological studies of central nervous system and homosexuality

Most of the above studies have mainly focused on anatomical aspects of the differences and similarities between central nervous system centres of individuals of varying sex and sexual orientation, whereas other studies, such as those involving responses to pheromones, have focused on functions of the brain. Following the functional approach, Reite *et al.* (1995) recorded the 100 ms latency field component (M100) evoked by a computer-generated auditory stimulus in homosexual and heterosexual men, the field being measured through magnetoencephalography. In comparison between men and women, the M100 is sexually dimorphic, being hemispherically asymmetrical in men, but more symmetrical in women. Reite *et al.* (1995) provide evidence for a greater interhemispheric symmetry in homosexual men than heterosexual men, suggesting a feminisation of brain laterality in male homosexuals.

Ivanka Savic and Per Lindström from the Karolinska Institute in Stockholm (Savic & Lindström 2008) have also carried out a study of cerebral (and cerebellar) asymmetry, although they used MRI and PET scanning. The degree of asymmetry was measured through an Asymmetry Index = (Right − Left)/ (Right + Left), and a PET study of functional connectivity also involved the right and left amygdalae in homosexual and heterosexual men and women. In the PET study they measured regional cerebral blood flow in a situation of rest and passive smelling of odourless air.

Although the values of the asymmetry index did not differ across sexes and sexual orientations for the cerebellar hemispheres, cerebral asymmetry did differ, being significantly more accentuated in heterosexual men than heterosexual women. Among homosexuals, cerebral asymmetry was feminised in men and masculinised in women.

Similar results of feminisation of homosexual men and masculinisation of homosexual women were obtained in the analysis of functional connectivity by Savic and Lindström. Heterosexual women and men differ in their connectivity from the amygdala: right and left, the anterior cingulate, subcallosum, hypothalamus, putamen, caudate and parts of the agranular insular cortex, although they displayed similarities in parts of the temporal neocortex, as homosexuals also did.

The patterns described are convincing, although the authors' further assertions that the variables that they measured are 'unlikely to be directly affected by learned patterns and behavior' (Savic & Lindström 2008: 9403) and that 'although repetitive sex- (or sexual orientation-) specific preferred strategies may, theoretically, have influenced the results, such systematic effects have, to the best of our knowledge, not been reported, and seem unlikely' (p. 9407) seem somewhat premature, as the specific effect of learning was not studied in this work. Studies of this kind that consider potential learning effects, or, more broadly, effects that rely on plasticity of the adult brain, could be carried out relatively easily, especially in women, by recruiting volunteers who, after having identified themselves as heterosexuals are in the process of considering a change of sexual orientation to homosexuality. Volunteers of this kind could be followed up and PET scans obtained, following strict safety protocols to protect the volunteers' health, in order to determine variations in functional connectivity with time as compared with heterosexual controls.

If functional interhemispheric symmetry is affected by the structure of both the anterior commissure and the corpus callosum, which are the major pathways of interhemispheric communication, then the results obtained by Reite *et al.* (1995) and Savic & Lindström (2008) are consistent with those obtained by Allen & Gorski (1992) in the anterior commissure and perhaps with those of Witelson *et al.* (2008) in the corpus callosum (see above): increased interhemispheric anatomical connectivity in homosexual males, perhaps leading to increased functional interhemispheric symmetry, seems to be a feminised trait.

Following a different functional approach, Rahman *et al.* (2003) studied the prepulse inhibition (PPI) of the startle response in healthy homosexuals and heterosexuals of both sexes. PPI consists of a reduction in the startle response (e.g. eye blink) to a strong stimulus whenever it is preceded by a weak stimulus. In rats, the PPI depends on neural activity at the limbic and cortico-pallido-striato-thalamic levels, with particular involvement of the hippocampus, nucleus accumbens, ventral pallidum, amygdala, striatum, thalamus and other centres (Rahman *et al.* 2003 and references therein) that is, areas of the central nervous system that are also involved in the control of sexual behaviour. Rahman *et al.* (2003) described a lower inhibition of the startle response in heterosexual women than heterosexual men. As for the homosexual subjects, Rahman *et al.* found that, whereas homosexual women displayed patterns of PPI that were masculinised, homosexual men did not differ statistically from heterosexual men, although the qualitative trend was for the PPI of homosexual men to be feminised.

I will mention neurobiological studies of sexual orientation that focus on isocortical regions, including the prefrontal cortex, in subsequent sections of this chapter. Here I would like to highlight a study that compared the distribution of white and grey matter in cortical areas of male and female homosexuals and heterosexuals by using MRI scanning (Ponseti *et al.* 2007). Although men have, in general, a larger amount of grey and white matter than females, owing to their larger body mass and therefore larger brain size, women have a larger grey matter/white matter ratio and their layer of grey matter in various cortical areas is also thicker than in men (Ponseti *et al.* 2007 and refs. therein). When men and women of diverse sexual orientation were compared, Ponseti *et al.* (2007) found that specific regions of grey matter that differed between the sexes were not different between homosexual and heterosexual men. Differences, however, were found between homosexual and heterosexual women. Heterosexual women had clusters of grey matter of increased concentration compared with homosexual women in the left ventral premotor cortex, the temporo-basal cortex, and particularly in the left perirhinal cortex. These results suggest that: (a) in general, the size of specific cortical areas does not vary with sexual orientation in both sexes, a finding that runs against Dörner's model, and (b) where a difference between sexual orientations was found, as in women, such difference went in the direction expected from Dörner's model, with homosexual females being masculinised.

In sum, studies of non-hypothalamic centres also falsify Dörner's model as a *general* explanatory mechanism for the neurobiology of homosexuality. Although some of the areas studied or brain activation patterns observed seem to be feminised in homosexual males or masculinised in homosexual females, a result that supports Dörner's model, others are hyperfeminised or hypermasculinised and still others do not seem to be affected by sexual orientation.

Central nervous system, hypersexuality and homosexual behaviour

In this chapter I have already mentioned the amygdala as an important component of the central nervous system circuitry that controls sexual behaviour (see also Baird *et al.* 2003a). The amygdala is located in the temporal lobe and is formed by three areas: the basolateral nuclear group, the centromedial group and the cortical nucleus (Salamon *et al.* 2005). The centromedial group is connected to the bed nucleus of the stria terminalis, whereas the cortical nucleus receives projections from the olfactory bulb and the olfactory cortex. The amygdala is sexually dimorphic, being larger in males than females, and its stimulation may produce effects that range from penile erection to orgasm (Salamon *et al.* 2005). The medial amygdala (MeA) is connected to the MPOA region of the hypothalamus and if lesions are caused to the MeA of males the result is a modification of their copulatory ability (Dominguez *et al.* 2001). The size of the MeA increases in response to elevated circulating androgens in both adult male and female rats, an effect that could be due to both direct androgen action or oestrogen action of aromatised testosterone, as the MeA is rich in both androgen and oestrogen receptors (Cooke *et al.* 1999).

Klüver & Bucy (1939) were the first to detect a radical increase in sexual activity in bilaterally temporal lobectomised male monkeys (*Macaca mulatta*, *M. fascicularis* and *Cebus capucinus*). Hypersexuality in their treated subjects was manifested not only in terms of increased heterosexual and homosexual sexual activity when in the company of other individuals, but also in increased masturbation when alone. Schreiner & Kling (1953) also observed an increase in sexual activity in cats subject to lesions to their rhinencephalon, a region of the central nervous system that connects to the limbic system and the amygdala. Hypersexuality in this case led to homosexual mounting, often reciprocal, when males were caged in monosexual groups.

Schreiner & Kling (1956) also carried out lesions of the amygdala in several other species apart from cats: agoutis (*Dasyprocta agouti*), lynxes (*Lynx rufus*) and rhesus monkeys (*M. mulatta*). They observed a postoperational reduction of aggression and increased mounting behaviour, including attempted heterospecific copulations. Lesions of the amygdala region do not seem to affect sexual orientation as such, just sexual activity, as two operated male cats given the choice to mount a female would prefer to do so rather than mounting each other, the one left out of the heterosexual mounting however, would then mount the other male in a trio copulatory chain (Green *et al.* 1957). In cats, however, hypersexuality may also occur in intact males, being susceptible to modifications through learning and conditioning (Michael 1961). This issue was explored in more detail by Aronson & Cooper (1979) through an experiment in which they produced lesions to the amygdala of male cats and compared their behaviour with that of control males. After sequentially exposing both control and operated cats to a still toy, a moving toy, a rabbit, a male cat and a female cat, controls showed a clear preference for the female cat, whereas operated males were significantly less selective in their choice of object to mount. Interestingly, all males did show some variability in the object to mount in tests carried out preoperatively, lesions to the amygdala simply increased the variance in their pattern of choice. However, the total amount of time spent mounting remained unchanged pre- and postoperatively, suggesting that the kind of lesions effected on the amygdala by Aronson & Cooper (1979) were more likely to affect sexual partner discrimination (they used the expression 'decreased selectivity') than the total level of sexual activity.

Male stump-tailed macaques (*Macaca arctoides*) subject to amygdalectomy also engage in autofellatio and homosexual masturbation, behaviours that had not been observed in those individuals preoperatively. Heterosexual masturbation also increased postoperatively (Kling & Dunne 1976). Interestingly, Kling & Dunne (1976) observed an increase in sexual behaviour postoperatively among the control individuals too, suggesting that the general pattern of sexual behaviour of all individuals within a group can be modulated by the specific behaviour of a subset of the members of the group through imitation.

In general, lesions of the temporal lobe and amygdala tend to lead to hypersexuality more effectively in males than in females (see, for example, Green *et al.* 1957). Moreover, hypersexuality can be also induced by lesions to the hippocampus (Lathe 2001b) and by inhibiting the ability of the pituitary to secrete prolactin, a hormone that exerts a depressing effect on sexual behaviour (Krüger et al. 2003b).

In humans, bilateral damage to the amygdala may be followed by the expression of hypersexuality, but this is not always the case (Hayman et al. 1998). Baird et al. (2003a,b) studied a group of both male and female patients who had undergone temporal lobe resection (TLR) as part of their treatment against epilepsy. Although TLR was associated with a slight increase in sexual activity of men, in women the effect was to slightly decrease sexual activity. However, Baird et al. (2003a) also described an intriguing pattern whereby the larger the size of the amygdala contralateral to the operated side, the larger the increase in sexual activity that was measured, as if the tissue left untouched responded to the operation by overcompensating.

Increased sexual activity per se, however, does not necessarily translate into increased probability of engaging in homosexual behaviour. Mikach & Bailey (1999) compared women who had a very high lifetime number of sexual partners (25–200) with a group of women who had an average number of lifetime sexual partners (0–10). Although the authors found that some of the traits they measured (e.g. waist-to-hip ratio) were 'masculinised' in women who had a higher number of partners, both kinds of woman were heterosexual.

Therefore hypersexuality, of the kind that could be mediated by the amygdala and the temporal lobe, may explain homosexuality mainly if the access to sexual partners of the other sex is somewhat restricted (e.g. in cases of biased sex ratio) and same-sex sexual partners are accessible. In humans, in general, homosexual males and females are not necessarily sexually hyperactive; some are, but there is a large degree of variability (Bell & Weinberg 1978).

Hypersexuality is also an extremely important factor to control for when comparing the neurochemistry of individuals differing in sexual orientation, because if sexual orientation covaries with the degree of sexual activity in those individuals, then their neurochemical differences may be due to the latter rather than the former. A good example of this issue is provided by the work carried out in India by Kurup & Kurup (2002). They measured serum concentration of digoxin, a molecule that affects neuron production of nitric oxide (NO) in the hypothalamus. NO can eventually affect penile erection. They also measured serum NO directly and serotonin among other molecules. They compared the levels of those molecules in circulation among promiscuous heterosexual men, non-promiscuous heterosexual men, homosexual men, bisexual men and heterosexual controls. Each one of those three molecules had higher concentrations in circulation in homosexuals and bisexuals than controls and non-promiscuous heterosexuals, suggesting a potential correlation with sexual orientation. This conclusion, however, is not warranted, as promiscuous heterosexual men also had higher concentrations of the three molecules than controls and non-promiscuous heterosexuals. Moreover, their levels of circulating digoxin, NO and serotonin did not differ from those of the homosexuals and bisexuals. It seems parsimonious to conclude, therefore, that it is the similarity in the level of sexual activity (e.g. frequency of penile erections) that may explain elevated concentrations of the three molecules in homosexuals and bisexuals, rather than their sexual orientation as such.

Future studies focusing on the potential roles that the amygdala and the temporal lobe – but also other central nervous system centres that may control hypersexuality – may have in sexual orientation, may be specifically designed to also test for sexual partner preference, as was done in cats by Aronson & Cooper (1979), in order to distinguish sexual partner discrimination from the effects of a generalised increase in sexual activity.

Dopamine and serotonin: modulation of homosexual behaviour

We have seen at the beginning of this neurological section how sexual hormones contribute to control sexual behaviour during periods of heterosexual mating, the onset of circulating levels of hormones being usually relatively slow. However, once sexual hormones have primed the central nervous system for sexual behaviour, very rapid sexual responses may then be triggered by the release of specific neurotransmitters. Dopamine is a major excitatory neurotransmitter released by neurons in centres controlling copulatory and other sexual behaviours, whereas serotonin generally has inhibitory effects on neurons. Regulation of dopaminergic innervation in central nervous system centres such as the hypothalamus is controlled by oestradiol and androgens (see Giuliano & Allard 2001; Hull et al. 2004; Paredes & Ågmo 2004 for reviews). In addition, dopamine could also activate sexual motivation and reward through its release in the nucleus accumbens (Paredes & Ågmo 2004) and presumably in the amygdala and other areas of the limbic system (Morgane et al. 2005). Pomerantz (1990) showed how male rhesus monkeys treated with apomorphine, a dopamine agonist, reacted to the view of a stimulus female by masturbating. Masturbation increased in a dose-dependent manner with apomorphine concentration up to a maximum of 100 μg/kg. Males were tested in isolation; it was therefore not possible to determine whether homosexual behaviour could be also released by apomorphine.

An important emerging paradigm of dopamine action on central nervous system centres is that its traditional role as mediator of rewarding effects of diverse environmental, including sexual, stimuli through its action on the limbic system is probably too narrow. Dopamine and dopaminergic neuronal networks in the medial frontal cortex may be also involved in a wide range of learning processes (Wise 1996). We have already seen, for instance, how dopamine could mediate the social bonding and also sexual partner preference between same-sex individuals (see the 'Non-steroid hormones' section above).

Serotonin, although traditionally described as an inhibitor of sexual behaviour in both males and females, has much more diverse roles than the traditional paradigm may suggest. For instance, serotonin antagonists or agonists have inhibitory or activational effects depending on the brain area concerned, but also depending on concentration (see Gorzalka et al. 1990 for a review). Kinnunen et al. (2004) specifically tested the hypothesis that the brain is differentially activated in homosexual and heterosexual men through serotonin action by administering fluoxetine to the subjects. Fluoxetine is a selective serotonin reuptake inhibitor. That is, it prolongs the return to the cell of the serotonin released into the synapse, thus extending the effect of serotonin on neurons. What Kinnunen et al. (2004) measured after fluoxetine administration was glucose metabolic changes in the brain by using fluorodeoxyglucose positron emission tomography (FDG-PET). Large areas of the limbic system were activated in both homosexual and heterosexual subjects following treatment. However, their results also suggest that the brain response to fluoxetine differs between homosexual and heterosexual men, with the former exhibiting a smaller reduction of glucose metabolism in the hypothalamus than the latter. Moreover, differential activation also occurred in other areas. For instance, homosexuals displayed increased activity in the prefrontal association cortex upon fluoxetine treatment, whereas heterosexuals did not show any change in the same region. On the other hand, heterosexuals showed specific activity in the lateral anterior cingulate, bilateral hippocampus/parahippocampal gyrus and the cuneate gyrus.

That is, homosexuals and heterosexuals may not only differ in the total number of neurons in various central nervous system areas, as I already mentioned in previous sections, but they may also differ in the distribution of specific kinds of neuron, e.g. dopaminergic, serotoninergic.

In an interesting recent work carried out by a Brazilian research team from São Paulo, Habr-Alencar et al. (2006) studied the effect of chronic fluoxetine administration on both homosexual and heterosexual sexual behaviour in male Wistar rats. Adult male rats were treated with fluoxetine for 21 days, after which they were orchiectomised and two weeks later subjected to oestradiol, followed by progesterone treatment, and exposed to a sexually experienced stud male. This test for homosexual sexual behaviour was also repeated at days 50 and 65 from the initial fluoxetine treatment in order to determine learning effects. A parallel experiment was also carried out exposing fluoxetine-treated males to an oestrous female to test for heterosexual sexual behaviour. These same males were also tested for homosexual behaviour at each stage. In the test for homosexual behaviour of males who lacked heterosexual sexual experience, the number of males exhibiting female-typical lordosis in the presence of a stud male increased with time of exposure to the stud male. However, the number of fluoxetine-treated individuals that displayed lordosis remained constant over time and at numbers lower than those of control males. Moreover, those fluoxetine-treated males that did display lordosis did so at a relatively higher frequency (higher values of the lordosis quotient = (number of lordoses/mount) × 100) than control males, especially after they had accumulated some sexual experience. This suggests that only a subset of the males display lordosis in response to fluoxetine, and that the homosexual behaviour seems to be additionally modulated by experience effects (see Figure 5.12). Figure 5.13 shows the effects that heterosexual experience has on homosexual behaviour in males. In this case, the higher the heterosexual experience, the higher the number of individuals exhibiting lordosis in a homosexual context. This is true for both fluoxetine-treated and control males, but the effect is significantly higher in the treated group. The lordosis quotient also increased dramatically in the fluoxetine-treated group but only after extensive heterosexual experience and with great inter individual variability (Figure 5.13).

In sum, both dopamine and serotonin mediate important learning effects in the context of sexual orientation. Dopamine activates sexual behaviour via the amygdala and it also affects social bonding, and at the same time it mediates learning through the activation of the reward system controlled by the nucleus accumbens. Serotonin also seems to promote processes of learning in the display of homosexual behaviour in adult male rats and such learning effect is boosted by previous heterosexual experience. However, these processes are not common to all individuals. Clearly, in rats, there is a subset of individuals that are especially inclined to develop a serotonin-induced homosexual behaviour, a result that is consistent with studies carried out in humans where homosexual and heterosexual men show some differences in their serotonin-associated neurophysiology. These results may suggest a scenario where genetic predispositions manifested throughout development may affect learning processes that in turn can further modulate the expression of homosexual behaviour in the adult individual.

Orgasm, sexual arousal and homosexuality

Although one would imagine that, at least in humans, it could be possible and relatively straightforward to identify the defining characteristics of what we call orgasm, a recent review by Mah & Binik (2001) lists 25 different definitions that follow 'biological', 'psychological' or 'biopsychological' perspectives. The so-called 'biological' definitions of orgasm tend to emphasise the neuromuscular correlates of the phenomenon, whereas the 'psychological' definitions tend to emphasise climactic subjective experiences. 'Biopsychological' perspectives in turn emphasise both neuromuscular and psychological subjective experiences (Mah & Binik 2001). I agree with Mah & Binik (2001) that, as far as the human experience is concerned, a definition is needed that can encapsulate the diversity of phenomena involved in orgasm. For instance, their *Multidimensional Model* of human orgasm includes three defining dimensions: the *sensory* that emphasises physiological events, the *evaluative* that emphasises sensations such as pleasure, satisfaction or even pain, and the *affective* that focuses on more complex mental experiences such as joy, intimacy and love. Of course, problems will arise as soon as we wish to apply such a model to species other than humans. Almost inevitably, at least given our current knowledge of the minds of other animals, only a minimalistic definition of orgasm may

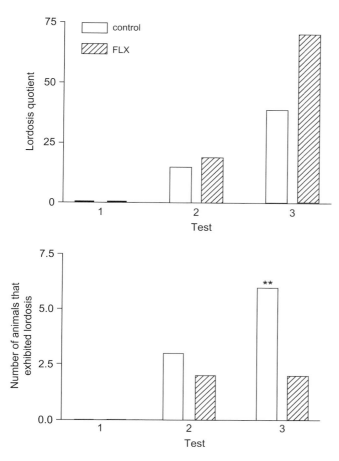

Figure 5.12. Effects of same-sex sexual experience on homosexual behaviour of male rats undergoing long-term treatment with fluoxetine (FLX), a serotonin reuptake inhibitor, as compared with controls. $*p<0.05$. From Habr-Alencar et al. 2006.

be applied for interspecific comparative purposes and such a definition will be more likely to emphasise measurable sensory aspects of orgasm or neurophysiological manifestations (e.g. as measured through fMRI and PET scanning techniques) of whatever mental process is being experienced by the animal (Fox & Fox 1971).

Several hypotheses have been proposed to explain the evolution of orgasm. The *Upsuck* hypothesis (Fox et al. 1970) posits that muscular movements associated with female orgasm facilitate transport of sperm through the female's reproductive tract. Baker & Bellis (1993b) provide some empirical support for the *Upsuck* hypothesis in humans. The *Upsuck* hypothesis emphasises the potential role of orgasm in ensuring fertilisation (see also Allen & Lemmon 1981); other hypotheses, however, focus on sperm competition (see, for example, Bellis & Baker 1990) and female mate choice (see, for example, Troisi & Carosi 1998). In birds, for instance, several species have evolved male organs and structures used in copulation that

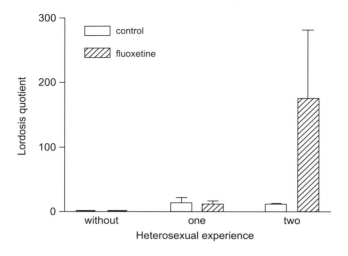

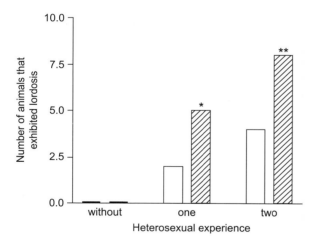

Figure 5.13. Effects of heterosexual experience on homosexual behaviour of rats undergoing long-term treatment with FLX as compared with controls. From Habr-Alencar *et al.* 2006.

seem to have a function in sperm competition and that could stimulate females (and males) during copulation (Briskie 1998). A rather spectacular example of this is provided by the phalloid organ of male red-billed buffalo weavers (*Bubalornis niger*) studied by Winterbottom *et al.* (2001).

The African genus *Bubalornis* comprises two species: *B. niger* and *B. albirostris*, both of them possessing a phalloid organ. The phalloid organ is an appendix located anteriorly to the cloaca. It consists of connective tissue, it is not erectile and it does not encapsulate the sperm duct (Winterbottom *et al.* 2001). Both sexes posses a phalloid organ, but that of males is much larger than that of females. *B. niger* is a colonial species that apparently has a polygamous mating system. Copulation bouts last an unusually long time in this species; presumably, the phalloid organ may be associated with this protracted copulation. Winterbottom *et al.* (2001) carried out a series of detailed observations of copulations among individuals held in captivity, describing how the males appear to experience a form of 'orgasm' during copulation, a state that was achieved after multiple and lengthy copulations. The same orgasm-like response was achieved in males upon direct experimental stimulation of the phalloid organ, which also caused ejaculation. Both orgasm and ejaculation could be achieved in males that were already sexually motivated; in males that were not sexually motivated only orgasm, not ejaculation, was achieved experimentally.

Winterbottom *et al.* (2001) speculate that the phalloid organ and the associated 'orgasm' may have evolved in this species as a result of sperm competition. Interestingly, in captivity where opportunities to escape an interaction with a conspecific are limited, they also observed 118 forced copulations, of those, 49 were between two males. The sex ratio of the captive colony was 2.16 M:F (13 males and 6 females). If the phalloid organ is a stimulatory organ that evolved in response to sperm competition, then the instances of M–M homosexual mounting observed in captivity can be easily explained in this species by the *Lock-in Model* through: (a) the lengthy copulation of a heterosexual pair that prevents other individuals from accessing an other-sex partner, (b) the male-biased sex ratio, (c) the confinement, which prevented mountees escaping. Seeking the sensory rewards presumably obtained by stimulation of the phalloid organ at the most should provide the motivation to find a sexual partner (although masturbation is also an alternative), but of itself does not necessarily establish a bias towards homosexual copulations.

Some authors have interpreted female orgasm simply as a side effect of selection for male orgasm (Symons 1979). Most empirical studies, however, tend not to support such a non-selected hypothesis, as I will show below (see also Puts & Dawood 2006 for a recent review).

Physiologically, orgasm is a state that is achieved as a result of increasing sexual arousal (Chivers 2005), which, in turn, is a process dependent on the activation of brain centres of 'pleasure' such as the cingulate cortex, the nucleus accumbens, the prefrontal cortex and others (Berridge 2003). In humans, some of the brain areas activated by watching an erotic film differ and some others overlap between heterosexual men and women. Karama *et al.* (2002) provide fMRI evidence for men showing greater hypothalamic and thalamic activation than women while watching a sexually explicit film. Such elevated activation in men was congruent with their reported subjective level of sexual arousal during the experiment. Various brain regions, however, were activated in both men and women: e.g. amygdala, ventral striatum, anterior cingulate cortex, insular orbitofrontal cortex, medial prefrontal cortex and occitotemporal cortex.

Homosexual and heterosexual men also activate the same brain regions after each one is exposed to a sexual stimulus consistent with the subject's sexual orientation (i.e. male–male explicit sexual interactions for homosexual men and female–female explicit sexual interactions for heterosexual men) (Safron *et al.* 2007). Using fMRI, Safron *et al.* (2007) found that those regions activated in men by the sexual stimulus specific to their sexual orientation are: rostral anterior cingulate (Brodmann areas 24 and 32), basolateral/medial amygdala, hippocampal complex, medial dorsal thalamic nucleus, hypothalamus, medial orbifrontal cortex, nucleus accumbens/subcallosal cortex, sublenticular extended amygdala and visual cortex (Brodmann areas 17 and 18). Additional brain regions that are associated with sexual arousal and that are equally activated in homosexual and heterosexual men following exposure to appropriate erotic films were also described by Hu *et al.* (2008) using fMRI: middle prefrontal gyrus, postcentral gyrus and bilateral temporal lobe, thalamus, insula vermis, occipital cortex, left precuneus, parietal cortex and cerebellum.

A similar coincidence in activation of common brain areas in homosexual and heterosexual men in response to a visual sexual stimulus concordant with their sexual orientation was reported by Paul *et al.* (2008) using MRI imaging. In this study the common areas were the hypothalamus and the orbifrontal, occipital, temporal and parietal cortices. The magnitude of hypothalamic activation, however, was lower in homosexual men than in heterosexuals. Paul *et al.* (2008) suggest that the pattern of hypothalamic activation in homosexual men is a mixture of female and male typical heterosexual responses.

Ponseti *et al.* (2006) have recently carried out an experiment of sexual arousal of homosexual and heterosexual women and men where the sexual stimulus was limited to showing photographs of aroused male and female genitalia. In this way the authors could control for the complicating effects of exposure to a variety of stimuli such as faces, body movements, voices, and sexually arousing body parts other than genitals. Results of fMRI imaging again suggest an activation of a common set of brain areas that include the ventral striatum, centromedian thalamus, and bilateral ventral premotor cortex after exposure to the appropriate sexual stimulus across sex and sexual orientation categories.

In sum, sexual arousal as such may not necessarily require engaging broad brain areas that are specific to a sexual orientation. In general, similar areas are activated in individuals of different sexual orientations, provided that they are exposed to the sexual stimulus concordant with their orientation. Specificity may be probably sought in the internal anatomy and physiology of those centres that identify a particular kind of individuals as object of sexual desire. Those centres are presumably more likely to be directly associated with decoding of environmental inputs: visual, olfactory, and auditory chiefly (see also the case of homosexual rams in a previous section). Moreover, such specificity may be even found at the level of the peripheral sensory organs. This possibility notwithstanding, Hu *et al.* (2008) did also identify some brain regions in homosexual and heterosexual men that were exclusively activated: the left angular gyrus, right pallidum and left caudate nucleus in homosexual men and the bilateral lingual gyrus, the right parahippocampal gyrus and the right hippocampus in heterosexual men.

A remarkable set of very detailed experiments investigating sexual arousal in males and females of different sexual orientation has been recently carried out by Meredith Chivers, currently at the Department of Psychology of Queen's University, Canada, and her collaborators. Their results strongly indicate that the stimulus-specificity of sexual arousal in men detected across sexual orientations in the above brain activation studies is also backed up by evidence from genital arousal and conscious perception. Below I provide a review of these studies.

Chivers *et al.* (2004) showed sexually explicit films to participants of the four sex (M, F)/sexual orientation (homosexual, heterosexual) categories and also measured genital arousal through plethysmography in males and vaginal pulse amplitude in females. Their results suggest that homosexual men were more aroused by watching two males having sex than by watching two females, the reverse was true for heterosexual men. This marked difference in sexual arousability as a function of sexual orientation described in men was not observed in women. In fact, Chivers *et al.* (2004) report that heterosexual and homosexual women showed an equal level of sexual arousal while watching the F–F or M–M sexually explicit films, and their level of arousal was intermediate between that shown by heterosexual males and homosexual males. It seems therefore that female humans are less specific in terms of the sex that can elicit their sexual arousal, whereas males are more specific as a function of their sexual orientation, being sexually aroused only if at least one member of their preferred sex is present (Mavissakalian *et al.* 1975; Sakheim *et al.* 1985). These results are consistent with the larger reported frequency of male than female exclusive homosexuals in humans (Kirkpatrick 2000), and with the higher level of sexual plasticity and bisexual potential found in women than in men (see, for example, Diamond 2008a,b).

Chivers & Bailey (2005) went on to test whether the pattern described in their previous study (Chivers et al. 2004) was an outcome of a greater generalised degree of sexual arousability in women than in men. On this occasion, apart from exposing the participants to sexually explicit scenes involving M–M, F–F or F–M human pairs, they also included a film displaying a sex scene between a male and a female bonobo (Pan paniscus). The control 'neutral' stimulus was a film showing a landscape or primates engaging in non-sexual behaviour. Interestingly, Chivers & Bailey (2005) found that the level of sexual arousal, as measured by vaginal pulse amplitude, elicited in women by the non-human sexual stimulus was greater than that elicited by the control stimulus, whereas their responses to the different human sexual stimuli did not differ significantly. Their subjective assessment of sexual arousal, however, showed a different pattern, with the heterosexual sexual scene eliciting a greater degree of subjectively evaluated sexual arousal. The responses obtained from men were more specific. Male participants were not aroused by the non-human sexual stimulus, and their maximal sexual arousal was elicited by F–M and F–F stimuli. In addition, their subjective evaluation of sexual arousal was consistent with the results obtained from penile plethysmography. This result again supports the apparent lower degree of specificity in the kind of subjects that can cause sexual arousal in at least some human females; a pattern that could be associated more with a plastic ability to potentially shift between sexual orientations, than with plasticity within a given sexual orientation.

More recently, Chivers et al. (2007) have carried out a very detailed sexual arousal experiment of both homosexual and heterosexual men and women who were exposed to 90 s film clips showing either: non-human sexual activity (a male and a female bonobo mating), female non-sexual activity (nude exercise), female masturbation, female–female intercourse, male non-sexual activity, male masturbation, male–male intercourse, female–male copulation and a control scene (landscape accompanied by a relaxing music soundtrack). As expected from the hypothesis that female sexual arousability is dependent on the characteristics of the act observed rather than on the specificity of the actors, both the homosexual and heterosexual women increased their genital response continuously from the exercise scene to the masturbation scene to the intercourse scene, whatever combination of sexes were involved in sexual intercourse. Moreover, even the bonobo sexual scene elicited a slightly higher response than the landscape scene. Males, on the contrary, showed patterns of sexual arousability that were more dependent on the characteristics of the specific actors involved in the scene. Heterosexual males were more aroused by scenes involving females, whereas homosexual males were more aroused by scenes involving males; in both cases arousability increased with the erotic intensity of the scene. Other stimuli elicited similar responses from heterosexual and homosexual males, including the human heterosexual erotic scene. In the latter case, presumably each participant was focusing on the member of the pair that was most sexually stimulating to them according to their sexual orientation: the male for homosexuals and the female for heterosexuals. Neither kind of male participants was sexually stimulated by either the bonobo sexual scene or the landscape scene (Chivers et al. 2007). I will return to the issue of women's greater ability than men to be plastic in terms of their sexual orientation in my discussion of Baumeister's work in the section 'Behavioural plasticity and sexual orientation' below.

Eventually, an increasing level of sexual arousal may culminate in orgasm. This, in men, may or may not be concomitant with ejaculation: orgasm may occur in the absence of ejaculation and vice versa (see, for example, Mah & Binik 2001). Orgasm is associated with a surge in circulating prolactin in both men and women (Haake et al. 2002; Krüger et al. 2003b, 2005). In an experimental study carried out in Germany with the participation of 10 healthy, heterosexual adult men, Krüger et al. (2003a) measured circulating adrenaline, noradrenaline, oxytocin, prolactin, cortisol, vasopressin, luteinising hormone, follicle-stimulating hormone and

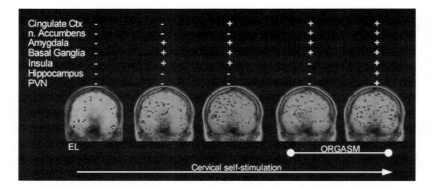

Figure 5.14. The figure shows an activation sequence of forebrain components as orgasm developed in a woman during continuous cervical self-stimulation over an 8-min period. Initially, none of the seven brain regions was activated. This was followed by a gradual increase in brain regions activation that continued during orgasm. Over this period, first the medial amygdala, basal ganglia, and insula showed activation. This was followed by activation of the cingulate cortex; at orgasm, the nucleus accumbens, paraventricular nucleus of the hypothalamus and hippocampus became activated. In addition, the activation of insula and basal ganglia became more extensive at orgasm. From Komisaruk *et al.* (2004).

testosterone from blood samples taken at two-minute intervals for 40 min while the subjects were watching a sexually 'neutral' documentary (initial 10 min), followed by 20 min of a sexually explicit film and then 10 minutes of the documentary again. While the subjects were watching the sexually explicit film they were asked to masturbate to orgasm during the last 10 min of the period. At orgasm, both adrenaline and noradrenaline sharply peaked in concentration, oxytocin also started a sharp rise that peaked a few minutes later. Prolactin experienced a more gentle raise, with levels remaining high compared to control individuals. Cortisol, vasopressin, luteinising hormone, follicle-stimulating hormone and testosterone levels did not change in association with orgasm (Krüger *et al.* 2003a). Orgasm achieved during actual sexual intercourse rather than masturbation is also associated with an increase of circulating prolactin in both men and women (Exton *et al.* 2001).

I mentioned in a previous section of this chapter that homosexual men have higher circulating concentrations of prolactin than heterosexual men in those studies that were able to detect a statistically significant difference between the two (Meyer-Bahlburg 1984). Whether elevated levels of prolactin in homosexuals are a potential cause or a consequence of their sexual orientation (e.g. elevated frequency of orgasmic experiences) remains to be determined (see, for example, Haake *et al.* 2002).

In women, brain areas activated during orgasm include the paraventricular and supraoptic nuclei of the hypothalamus (Komisaruk & Whipple 2005), but also the medial amygdala, the cingulate and insular cortices, the hippocampus, paraventricular nucleus and insula along with areas of the parietal and frontal isocortex (Komisaruk *et al.* 2004) (see Figure 5.14). In men, orgasm is associated with activation of the mesodiencephalic transition area that includes the midline, ventroposterior and intralaminar thalamic nuclei, the suprafascicular nucleus, the zona incerta, the lateral central tegmental field and the ventral tegmental area (Holstege *et al.* 2003) (see Figure 5.15). Areas of the isocortex are also activated, especially regions of the Brodmann area in the right hemisphere (Holstege *et al.* 2003, Figure 5.16). Therefore although some of the brain areas that are activated in men and women during orgasm seem to be different, others are similar, such as isocortical areas, thus paving the way for potential associations of orgasm with more complex cognitive processes.

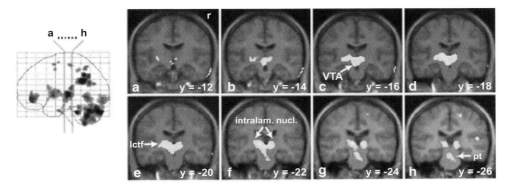

Figure 5.15. Strong activation in the mesodiencephalic transition zone and some isocortical areas in men during an ejaculation/orgasm event. Increased activation is represented in coronal sections (a–h) through the brain. The vertical lines on the glass brain on the left indicate the orientation and location of the sections. Activations are superimposed on the averaged MRI of the volunteers. The activated cluster contains the ventral tegmental area (VTA, sections a–d). The midline thalamic nuclei are located slightly more caudally (sections d–f). The lateral central tegmental field (lctf; sections c–f) and the zona incerta are located lateral to this area. The activated region extends dorsally into the intralaminar nuclei (intralam. nucl.; sections d–h) and the ventroposterior thalamus. Note also the activation in the medial pontine tegmentum (pt; sections g and h). Positioning $y = -14$ means 14 mm posterior to the anterior commissure, r, right side. From Holstege *et al.* (2003).

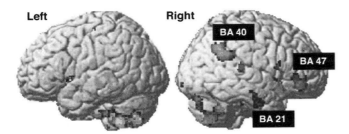

Figure 5.16. Activated areas in the cerebral cortex of men during an ejaculation/orgasm event. Note that the cortical activations are almost exclusively on the right side. BA, Brodmann area: BA 40 is in the parietal cortex, BA 47 is in the inferior frontal gyrus and BA 21 is in the inferior temporal cortex. From Holstege *et al.* (2003).

In sum, orgasm per se is not necessarily associated with or conducive to homosexuality. Sensory sexual gratification may be achieved through heterosexual intercourse, homosexual intercourse or even solitary masturbation. Factors other than the achievement of orgasm will determine which one of the available options is followed by any individual at any specific point in time. Although both males and females can achieve orgasm in some species, including humans, in our species male sexual arousal is more specific than in females, being more dependent on sexual orientation in the former than the latter. Women are more sexually aroused by the observation of sexual acts, independently of their sexual orientation, whereas men are sexually aroused by the sex of the actor involved: men for homosexuals, women for heterosexuals. As far as men are concerned – but, depending on the stimulus, women as well – homosexuals and heterosexuals seem to activate broadly similar areas of the brain during sexual arousal in response to the sexual stimulus concordant with their sexual orientation. However, differences in brain anatomy and physiology could arise at a more fine-grained level within those broad areas. During orgasm various brain areas are activated in men and women, some seem to be sex-specific, others are common to the two sexes.

I now turn to the issue of masturbation and whether the still common belief that masturbatory acts lead to the development of a same-sex sexual orientation receives any support from the available interspecific evidence.

Masturbation and homosexuality

Masturbation could be defined as sexual stimulation achieved by autosexual means (self-masturbation or automasturbation) or with the aid of a partner (allomasturbation). In the latter case the sexual stimulation is achieved by means other than genito-genital or genito-anal contact (e.g. through manual stimulation). Masturbation is not an uncommon phenomenon in mammals (Baker & Bellis 1993a), but it does not seem to be very frequent in birds. Ewen & Armstrong (2002) have described copulatory displays of male stitchbirds (*Notiomystis cincta*), a passerine species endemic to New Zealand, with inanimate objects in the wild. Although the authors did not venture as far as describing such behaviours as *masturbation*, it is certainly a possibility. *Redirected mountings*, such as mounting of inanimate objects, have been also observed in the dovekie, *Alle alle*, where 1.5% of mountings are redirected (Jakubas & Wojczulanis-Jakubas 2008). Jakubas & Wojczulanis-Jakubas (2008) provide additional examples of redirected mounting in birds and also reported that 0.3% of mountings in the dovekie are male–male.

In general, we may expect that bird species that show evidence of orgasm (e.g. *Bubalornis niger*, Winterbottom *et al.* 2001) may be more likely to also display masturbation. If so, it is possible that the phenomenon is indeed restricted to a handful of bird species. On the other hand, as we will see below, if masturbation is an adaptation selected in the context of renewal of sperm (Zimmerman *et al.* 1965), then it could be much more frequent in birds than has been reported so far and, in fact, it may not be directly linked to orgasm but to ejaculation.

Although it is not my aim here to carry out a thorough review of the taxonomic distribution of masturbation among mammals, I will mention some published examples that clearly indicate that the behaviour is widespread across taxa. Masturbation has been described in male baiji, or Chinese river dolphins (*Lipotes vexillifer*) in captivity – sadly, this is a species that has been recently declared extinct (see www.baiji.org). A male baiji was observed rubbing his protruded penis against the wall or bottom of the pond where he was held, a pattern that was coincident with seasonally circulating peaks of testosterone and also lack of a partner (Chen *et al.* 2001, 2002). Male equids also masturbate by rubbing their erected penis against the abdomen, a behaviour that is frequently observed in all-male bachelor bands (McDonnell & Haviland 1995). In some taxa (e.g. *Rattus* spp.) males may even experience more or less rhythmic pulses of spontaneous ejaculation that can be further enhanced by masturbation (Kihlström 1966).

Although masturbation is more frequently reported in male mammals, females in some species are also known to engage in masturbation. For instance, Ishikawa *et al.* (2003) have described masturbation in groups of unmated female Hokkaido brown bears (*Ursus arctos yesoensis*) held in captivity. Masturbation was observed during the oestrus period and until 9–23 days after, therefore masturbation was associated with both endocrine activity and lack of a male sexual partner. Obviously, self-masturbation in females cannot be directly linked to ejaculation, but it could still be associated with orgasm (i.e. pleasure).

Primates have been the special focus of many observations and studies of masturbation (e.g. Mootnick & Baker 1994; Dixson 1998 and references therein) and, because of their close phylogenetic relationship with humans, primates are an important model for the evolutionary study of the potential links between masturbation and human homosexuality. Among the baboons (*Papio* spp.) male masturbation is common (Bielert & van der Walt 1982 and references therein). Male Chacma baboons (*Papio ursinus*), for instance, not only masturbate, but also tend to eat their own ejaculate (Bielert & van der Walt 1982). In an experimental setup where male Chacma baboons were individually caged and visually and olfactorily exposed to a female, Bielert & van der Walt (1982) observed that the highest level of male masturbation was associated with the maximum perineal swelling of the female, which also coincided with increased levels of circulating testosterone in males. Frequency of male masturbation was also positively correlated with the levels of circulating oestradiol in hormone-treated females (Bielert & van der Walt 1982). This experimental work on Chacma baboons therefore shows that male sexual arousal caused by the presence of a female in oestrus may lead to male masturbation if access to a partner is denied.

Nieuwenhuijsen *et al.* (1987) carried out a study of a captive group of stumptail macaques (*Macaca arctoides*), a species that does not show a marked seasonality in sexual activity. This lack of seasonality in mating was also paralleled by a lack of seasonality in masturbation in males. Moreover, male masturbation was negatively correlated with dominance rank, whereas mounting frequency was positively correlated with dominance. This suggests that masturbation in this species is affected by the neuroendocrine mechanisms that control mating, but it is modulated by socio-sexual factors related to dominance that prevent access to an other-sex or even a same-sex sexual partner. In this non-seasonally breeding species, however, masturbation was not closely coupled with circulating levels of testosterone (Nieuwenhuijsen *et al.* 1987). Linnankovski *et al.* (1993) also carried out a study of stumptail macaques in captivity describing greater frequency of masturbation in males when they could establish eye contact with a female than when they could only view or even perform investigation of the perineal area of the female, but without the ability to establish eye contact. This suggests that an additional factor eliciting masturbation in males is the access to specific stimuli coming from the female, such as 'eye contact', indicating willingness to be mounted. Again, this result is consistent with the hypothesis that masturbation is just the best-of-a-bad-job alternative under a situation of sexual arousal and lack of physical access to a partner for copulation.

Mootnick & Baker (1994) studied masturbation behaviour in eight gibbon species (genus *Hylobates*) held in captivity. Although they reported masturbation in both males and females, females engaged in masturbation at a much lower frequency than males. Masturbation in this genus is performed both by individuals that are caged alone and also by individuals that are caged in the company of others (Mootnick & Baker 1994). If masturbation is the result of a best-of-a-bad-job as suggested above, then the specific mechanisms may differ from one situation to another. In particular, being caged alone obviously leads to masturbation as one of the very few options available when the animal is sexually aroused. However, masturbation is also an option for an individual even when it is caged in the company of other individuals if either: (a) all potential partners are already engaged with somebody else in sexual behaviour (see the *Lock-in Model*) or (b) the individual is excluded from sexual interactions with others due to dominance effects or due to those others exercising sexual partner choice (see Thomsen *et al.* 2003 and references therein).

Although masturbation is commonly observed in captivity and its frequency may well be modulated by the specific conditions in the holding facilities (for example, sex ratio, patterns of circulating sexual hormones over time), it is a behaviour that has been also observed in the wild (see, for example, Thomsen & Soltis 2004 for a list of primate species). Thomsen & Soltis (2004) studied a population of free-ranging Japanese macaques (*Macaca fuscata*) on Yakushima Island (Japan) in a demographic situation of virtual sex-ratio parity among adults. *M. fuscata* are seasonal breeders and both males and females mate with multiple partners.

Frequency of masturbating activity among males reached a peak when at least one of the troop females was in oestrus, and decreased with increase in dominance status. Therefore masturbation in this species is also a result of a general effect of sexual arousal and lack of opportunities to engage a partner in sexual activity, either because of potential partners being already engaged with somebody else or due to rejection. The situation, however, seems a bit more complex than that in this population of Japanese macaques. Although males masturbated when in full view of an oestrous female (37.0% of cases) or soon after an oestrous female passed by (13.0%) or in the presence of a mating pair (17.6%), they also masturbated in the presence of other masturbating males (9.3%). Although the first three categories are consistent with a straightforward model of masturbation as the best-of-a-bad-job option, masturbating in the presence of other masturbating males is an indication of a more generalised mechanism of sexual arousability which, paradoxically, does not lead to homosexual sexual behaviour, rather it seems to lead to 'social masturbation'. Something is obviously preventing those males from freely engaging in male–male mounting under conditions of high sexual arousal and lack of willing sexual partners of the other sex. It is possible that socio-sexual implications of same-sex mounting in males, e.g. dominance assertion, may prevent those males from freely engaging in homosexual mounting in this case. That is, if male–male mounting conveys a message of challenge or assertion of dominance, then it may be resisted by some individuals, leading to alternative expressions of sexual behaviour in situations of arousal such as masturbation. Females were not observed engaging in masturbation (Thomsen & Soltis 2004), but then they always had access to a male whenever they were sexually aroused during oestrus.

Masturbation behaviour has also been studied in wild Temminck's red colobus (*Procolobus badius temminckii*) (Starin 2004). The Temminck's red colobus lives in groups with a female-biased sex ratio, with females being polyandrous and showing mate choice preference for the α-male. Starin (2004) reported observations of masturbation in both male and possibly female Temminck's red colobus at low frequencies. In the case of males, only five instances were recorded, all involving the α-male who masturbated during the breeding season on occasions when no female was present or during aggressive encounters between troops.

Alan Dixson will further extend the analysis of primate masturbation in Chapter 9.

Although masturbation could be interpreted as the best-of-a-bad-job in circumstances of elevated sexual arousal but inaccessibility of a sexual partner, the question remains as to why some males masturbate to ejaculation in solitude or even in the presence of other males. Also, why is it that males seem to be more prone to masturbate than females in circumstances when both are in reproductive readiness and have equal inaccessibility to sexual partners? A potential answer to this question was provided by Baker & Bellis (1993a) who, following Zimmerman *et al.* (1965), suggested that male masturbation is an adaptation to sperm competition as it allows a male to renew its spermatozoa and therefore improve the chances of fertilisation when opportunities arise. This is the so called *Sperm Age* hypothesis (Baker & Bellis 1993a).

Baker & Bellis (1993a) tested the *Sperm Age* hypothesis in humans. Although they found an increase in the rate of male masturbation as the interval between copulations with a partner increased, a pattern predicted by the hypothesis, the rate of copulation within human couples was about once every three days, but the sperm can remain competitive for about 7–9 days, which suggests that the rate of copulation per unit of time is higher than that strictly expected from the *Sperm Age* hypothesis in humans. Thus, although masturbation may help in flushing out uncompetitive sperm, other factors may also play a concomitant role. Thomsen *et al.* (2003) carried out a comparative analysis of masturbation patterns in primates and found that, in the wild, 81.8% of the primate species living in multimale–multifemale groups show masturbation activity, whereas only 38.8%

of species living in other social conditions do. Although the data of Thomsen *et al.* (2003) do suggest that effects such as captive and pathological conditions may affect the rate of masturbation, they do so on a background where masturbation seems to be a normal component of the sexual repertoire of those species. The results of Thomsen *et al.* (2003) are consistent with the *Sperm Age* hypothesis, as masturbation was more frequent in mating systems where sperm competition is expected to be higher. However, in multimale–multifemale groups where copulations are biased in favour of the most dominant males owing to both male–male competition and female mate choice, chances are that subordinate males incapacitated to access females or even male sexual partners during the breeding season may resort to masturbation instead as the only outlet left for their sexual drive. Comparative tests of the *Sperm Age* hypothesis require a more fine-grained understanding of the copulatory activities (heterosexual and homosexual) of the individuals in the group and who exactly resorts to masturbation, how often and under what circumstances.

The above body of work suggests that there is no direct causal link between masturbation and both same-sex sexual behaviour and a homosexual sexual orientation in both birds and non-human mammals. Masturbation simply seems to be a result of the activity of mechanisms involving both hormones and brain centres that control sexual arousal under circumstances of lack of access to a willing sexual partner of any kind.

Among humans masturbation is common in both males and females, but especially males. In the latter it is associated with elevated circulating levels of various androgens, pregnenolone and dehydroepiandrosterone in particular (Puris *et al.* 1976). In Western societies, masturbation does not seem to have been an issue from a cultural point of view until the beginning of the eighteenth century (Laqueur 2003; see also Gagnon 2005). The link between masturbation and homosexuality was not even explicit in texts condemning masturbation published in the late nineteenth century (Hunt 1998). Even in Victorian England the main concern with masturbation, e.g. in the context of monosexual groups in schools, was more related to 'the anxiety... that self-abuse would lead to a habit of dangerous indulgence in sensual pleasure with the consequent erosion of self-control and self-discipline' (Hunt 1998: 606) than to the development of a homosexual sexual orientation. In his *Libido Sexualis*, Albert Moll (1897) was explicit in pointing out that same-sex sexual behaviour and mutual masturbation not only occur frequently before puberty but do not lead to the development of adult homosexuality. The idea that masturbation may lead to homosexuality first appeared at the beginning of the twentieth century (1903–1925) according to Martin (1993), in concomitance with the advent of psychoanalysis and also the early studies of conditioning, although Krafft-Ebing did mention the possibility of a causal link between masturbation and the development of homosexuality at least in some cases (Sulloway 1979:284)

Interestingly, Martin (1993) mentions that the first explicit link between homosexuality and masturbation concerned lesbianism rather than male homosexuality, an association that was rather suspicious (to put it mildly) as it was made in the context of a rejection of the emerging movement of women's rights (Martin 1993). More recently, Richard Alexander (1971, 1974, 1975), from an evolutionary biological perspective, and Michael Storms (1981) from a psychological perspective have suggested that conditioning could be a mechanism linking masturbation with the development of a homosexual sexual orientation, especially in men (see also Pfaus *et al.* 2003; James 2004). I argued in Chapter 4 that such a link seems unlikely, in fact the relative frequency of male homosexuals (about 3%; Herdt 1981; Laumann *et al.* 1994; Kendler *et al.* 2000) is dramatically lower than the frequency of males who masturbate (estimated between 60% and 95%; Kinsey *et al.* 1948; Janus & Janus 1993; Laumann *et al.* 1994; Pinkerton *et al.* 2002). The same applies to females: about 50%–60% masturbate (Leung & Robson 1993) vs. 1.5%–2.5% homosexuality (Laumann *et al.* 1994;

Kendler *et al.* 2000). My conclusion is that there is no causal link that necessarily derives homosexuality from masturbation in any of the mammals studied so far, including humans. In a sexual, rather than socio-sexual, context, same-sex sexual behaviour and masturbation are two different consequences of sexual arousal.

Learning processes in adults and their role in homosexual behaviour

We have seen in previous sections of this chapter how the central nervous system retains a variable degree of morphological and functional plasticity even in adult individuals, with the extent of such plasticity being quite variable from species to species and across specific regions of the brain. I have mentioned how behavioural plasticity in adults with a complex brain could reflect the ability of the organism to switch on and off already established alternative neuronal networks. This, however, would provide only a partial picture of the complex reality of brain plasticity, as I have stressed in previous sections. Additional brain plasticity is contributed by the actual ability of neurons and other cells of the brain to modify their structure, number and interactions in adult birds and mammals. Here, plasticity results from a process that may include some basic genetically encoded instructions (e.g. 'grow dendritic spines in response to steroid hormones') but no genetically predetermined end result, such as which specific behaviour the new network of connections will produce. The specific behaviour may be subsequently selected on the basis of the activity or inactivity of the various synapses.

Complex behavioural processes resulting from the action of brain mechanisms that can be more or less plastic include learning, cognition and emotions, all of them being capable of affecting and being affected by sexual behaviour and sexual orientation in particular. In this section we turn our attention to learning mechanisms operating in adult individuals and the role they play in the manifestation of homosexuality in humans and non-human animals.

Both early and more recent reviews of human homosexuality have emphasised the role of learning in the development of sexual orientation (Meyer-Bahlburg 1977; Woodson 2002) and, more generally, in the modulation of processes that rely on the plasticity of brain functions as they are expressed throughout the lifetime of the individual (Byne & Parsons 1993). I should stress from the outset that what I emphasise here is that learning mechanisms, although relying on brain plasticity, are also constrained by genetics and various degrees of canalisation of developmental processes, and therefore the presence of a role model for the learning of homosexual behaviour, such as having a gay father, is no guarantee of development of a homosexual sexual orientation in a son (see, for example, Bailey *et al.* 1995).

Potentially, the activity of brain centres involved in appetitive and consummatory sexual behaviour that are sensitive to specific reinforcers via classical (i.e. Pavlovian) or instrumental conditioning may explain the preference for a homosexual sexual partner over a partner of the other sex if either voluntary or automatic operant sexual behaviour strengthens the association between a sexual partner of the same sex and sexual reward and satisfaction (Robbins & Everitt 1996). In principle, any learning process could be involved in the onset, development and/or modulation of a homosexual sexual orientation, including imprinting, classical conditioning and instrumental (operant) learning (Pfaus *et al.* 2001; Woodson 2002). Major brain centres that are involved in these mechanisms of incentive, motivation and reinforcement of sexual partner preference mainly include areas of the limbic, striatal and pallidal regions: the amygdala, septum and prefrontal cortex in particular, but also the ventral striatum and nucleus accumbens, which receive inputs from the amygdala, the hippocampus and the prefrontal cortex (Robbins & Everitt 1996).

That sexual partner preference can be conditioned by early experience in both birds and mammals has been amply demonstrated by empirical work (see review in Pfaus *et al.* 2001). Does this evidence therefore demonstrate that organisms are born as a blank slate over which experience carves the specific characteristics of sexual orientation? The evidence I have already reviewed in this and also Chapters 3 and 4 demonstrates that this is not the case. What the evidence supporting the effects of learning on sexual orientation demonstrates is simply that there is a dimension of homosexuality that is dependent on the plasticity of the nervous system and that this dimension is subject to specific modulation by experiences with the environment external to the organism. In this context, the external environment 'determines' homosexuality only insofar as it is providing the trigger for the unfolding of specific processes internal to the organism: switching on and off neuronal networks, triggering processes that affect neuroarchitecture (apoptosis, neurogenesis, dendritic arborisation). Which kind of specific process will be activated by which kind of specific environmental stimulus in which kind of individual will be a function of the biology of such an individual, which, in turn, will be constrained by its genes, its previous development, its previous experiences and also the evolutionary history of the taxon. I regard this *integrationist biosocial* approach (see Chapter 1) as a central tool in our understanding of the role of learning in homosexuality. An *integrationist biosocial* approach could explain, for instance, why homosexual experiences during the early periods of development may not necessarily lead to the development of homosexual behaviour in all the individuals that experienced them. We have seen in Chapter 4 how those early homosexual experiences may be adaptive in the context of the development of heterosexual, not homosexual, sexual behaviour in most individuals; and we have also seen in this chapter and also in Chapters 3 and 4 how a subset of those individuals may indeed develop a homosexual sexual orientation as a result of specific genetic and ontogenetic predispositions, enhanced and fine-tuned by early experiences and learning.

The interactive nature of biosocial processes is quite evident as we compare the frequency distribution and the behavioural and cognitive specificity of homosexuality across human cultures. Although different ethnic groups may differ in the relative frequency distribution of specific alleles, it is highly unlikely that all cultural differences are simply a reflection of underlying genetic differences. In the same way in which even genetic clones can develop somewhat different phenotypes according to the environment of rearing, individuals in different ethnic groups differing in their cultural traditions may be expected to experience diverse learning processes that may affect the development of homosexual behaviour. If so, diverse cultures are expected to display homosexuality in a diversity of behavioural modalities and patterns. Intercultural convergence may perhaps reflect common and highly canalised developmental processes, intercultural divergence may in part reflect the plasticity of our brain in action.

An early review by Ford & Beach (1951) reported adult homosexual behaviour being present in about 64% of societies in one form or another. Table 5.4 is modified from Kirkpatrick (2000) and it lists a small sample of widely geographically distributed past and present human societies and the characteristics of adult homosexual behaviour described in them.

The proposal that culturally determined homosexual behaviour resulting from brain plasticity is superimposed upon less ontogenetically plastic (i.e. more canalised) processes is clearly demonstrated by the case of ritualised homosexuality among the Sambia of Papua New Guinea. We have seen in Table 5.4 how homosexual practices were quite widespread in the traditional cultures of Melanesia, especially among men. Sambia young boys were traditionally expected to spend a period of no less than 10 years in a homosexual relationship with an older boy. During this relationship, they were under precise instruction to avoid and even fear women (Baldwin & Baldwin 1989). The same-sex partnership also involved homosexual behaviour such as ritualised 'insemination' through oral semen transfer from the older to the younger member of the

Table 5.4. A sample of past and present cases of adult homosexual behaviour across cultures

Continent/Society	Sex	Concurrent with heterosexual behaviour?	Frequency in the population
Africa			
Siwah (Libya)	M	Yes	c. 95% (?)
Azande (Sudan)	F, M	Yes	Common
Dahomey (Nigeria)	F, M	?	Common
Mpondo (South Africa)	M	Yes	Common
Europe			
Classical Athens	F, M	Yes	Common
Early Roman Empire	F, M	Yes	Common (?)
Dinaric (Serbia)	M, F?	?	Unknown
Florence, 15th century	M	Yes	> 50% (?)
Americas			
Lakota (USA)	M	Yes	Unknown
Mohave (USA)	F, M	Yes	Limited
Oceania			
Precolonial Tahiti	M	Yes	Common
Big Nambas (Melanesia)	F, M	Yes	c. 100%
Marind-anim (Melanesia)	M	Yes	c. 100%
Sambia (Melanesia)	M	Yes	c. 100%
Asia			
China, 700–400 b.c.e.	M	Yes	Unknown
China, 1865–1965 c.e.	F	No	Limited
Japan, 16th, 17th century	F, M	Yes	> 50% (?)
Pukhtun (Pakistan)	M	Yes	Unknown

Adapted from Kirkpatrick (2000), who also reports an extensive list of references.

pair (Herdt 1991). Herdt (1991) suggests that the practice was a way to strengthen male–male bonds in a society where warfare was frequent, whereas Creed (1984) stresses the function of ritualised homosexuality as mediator of young subordination to the elders. Men, however, finally married heterosexually upon reaching an adult age and carried on with their heterosexual relationship from then on (Baldwin & Baldwin 1989; Herdt 1991). Baldwin & Baldwin (1989) suggest that the transition from the early homosexual experience to the later heterosexual life was made possible by the use of 'aversive control' in the rites used by the homosexual pair and also by the fact that the young boy was forcibly taken from his mother to enter the homosexual relationship in an experience that was likely to be traumatic (but see Knauft 2003 for the case of the Gebusi, where ritualised homosexuality was apparently not traumatic for the younger boy). I suggest that an additional dimension should be added to this explanation. That is, as it is the case in many other social mammals (see Chapter 4), same-sex sexual behaviour (e.g. in the form of sexual play) is very common, especially among males, and not only perfectly consistent with adult heterosexuality but also a promoter of a better heterosexual performance in adult life. Thus, the development of an adult heterosexual orientation in spite of homosexual relationships at younger ages is not only possible but is in fact the norm among social mammals. This might be the result of an ancestral genetic programme that evolved in coadaptation with social life. In spite of this general trend for the development of an adult heterosexual sexual orientation, Herdt (1981, 1991) estimated that about 2%–3% of men in the Sambia society did develop a homosexual orientation in adult life. This result is remarkably consistent with the estimated 3.1% prevalence of homosexuality found among men in the USA (Laumann *et al.* 1994; Kendler *et al.* 2000).

What is it about the Sambia ritualised homosexual practices among boys and living in an urbanised society in modern USA that makes the probability of developing a homosexual orientation in adult life among men similar? Or is this coincidence just a demonstration of the independence of homosexuality from environmental influences? At this point the reader may like to remember the warning sign raised in Chapter 4 against the dangers of a manicheistic world-view! I suggest that the above results seem to be consistent with a model for the development of a homosexual orientation in human males that is probably dependent on some pre-existing genetic biases that can explain the low frequencies of exclusive homosexuals. Such biases, however, may be ontogenetically modulated by learning, through the early sexual experiences with members of the same sex. This can explain why those who develop a homosexual orientation among the Sambia seem to have a predilection for younger males, a preference that could be established, at least in part, through conditioning (Baldwin & Baldwin 1989). In addition, Baldwin & Baldwin (1989) also suggest the potential relevance of a mechanism of sexual orientation development that involves the effects of stress during early ontogeny (see Chapter 4 for the review of this pathway).

Diverse cultural experiences can also impinge on the specific behaviours, feelings, imagery and ideologies associated with same-sex sexuality via learning. Thus whether it is culturally acceptable to maintain very small personal distances in physical interactions with members of the same sex (Evans & Howard 1973), or whether it is acceptable for members of the same sex to touch each other (Roese *et al.* 1992; Crawford 1994) and especially, how the concepts of 'masculinity' and 'femininity' are categorised intraculturally, if at all, will probably make a difference in the likelihood that diverse forms of homosexuality will be developed or expressed in the first place, maintained over time, and socially accepted. As I stressed in Chapter 2, same-sex 'touching' is not an unequivocal indication of homosexuality, but homosexuals may or may not touch each other in public depending on cultural influences. I now turn to the important concepts of *masculinity* and *femininity* and how they may relate to the cultural diversity and similarities found among homosexuals and heterosexuals.

Femininity, masculinity and homosexuality

The analysis of homosexuality in males and females is bound to be confronted with and also constrained by our perception of masculine and feminine gender roles. This is valid for studies carried out both in humans and in other species where 'male-typical' and 'female-typical' behaviours can be identified. Terman & Miles (1936) were probably the first to devise a tool for the measurement of masculinity and femininity in humans: the Attitude Interest Analysis Survey (AIAS). The AIAS uses a bipolar approach, with masculinity and femininity being two opposites of the same continuum. It was Sandra Bem (1974), however, who later introduced the measurement of femininity and masculinity as two independent variables co-occurring in the same individual, an approach that is not 'bipolar' but 'bi-dimensional' (see also Klein *et al.* 1985; Kauth 2000, 2002, 2005). Her *Bem Sex-Role Inventory* (BSRI), by combining both masculine and feminine traits, allows for the identification of a large gradation of androgynous states in individuals (see also Constantinople 2005).

A turning point in the cross-cultural study of femininity and masculinity in human societies was arguably represented by the pioneering work carried out by Maccoby & Jacklin (1974). They concluded that men tend to be more assertive or dominant than women, men are also more aggressive and less anxious, whereas women and men are indistinguishable in terms of their levels of self-esteem. Maccoby & Jacklin's (1974) dataset was re-analysed by Feingold (1994) using more sophisticated meta-analytical tools, simply to reach the same conclusions. In addition, Feingold (1994) replicated the meta-analysis carried out by Hall (1984) and also analysed meta-analytically some previously published standardised tests of personality. Again, the results showed the same general pattern initially described by Maccoby & Jacklin.

More recently, Costa et al. (2001) have carried out an analysis of published works that compared masculinity and femininity across cultures using the Revised NEO Personality Inventory (NEO-PI-R) of Costa & McCrae (1992). The NEO-PI-R considers five broad factors: neuroticism (e.g. anxiety, depression, anger, etc.), openness to experience, extraversion (e.g. dominance and love), agreeableness (e.g. submission and love) and conscientiousness. The published data that were included in their analysis originated from 24 countries across Asia, Europe, Africa, and both North and South America. Surprisingly, no work from Oceania was apparently included. Costa et al. (2001) found consistently higher levels of neuroticism and agreeableness among women, but also of warmth, positive emotions, gregariousness, openness to aesthetics, feelings and actions, whereas women scored lower than men in assertiveness, seeking excitement and openness to ideas. Costa et al. (2001) also indicate that gender differences are especially marked in American and European cultures, whereas they are less prominent among Asian and African cultures. They interpret this trend as being a consequence of the relatively greater level of individualism prevalent among Western, industrialised cultures compared with the relatively higher level of collectivism found in more rural cultures.

Although these works seem to be somewhat consistent in their general findings (e.g. in that assertiveness is described as a masculine trait and gregariousness as feminine), the level of cross-cultural variability is significant. Moreover, even within a culture the level of androgyny in males and females may be quite variable, and it may also change as the individual matures (Block 1973), so much so that authors such as O'Neil & Egan (1992: 108) refer to 'the illusions of gender-role stereotypes' and state that individuals undertake a 'gender-role journey' during their lifetime. I also believe that the comment of Costa et al. (2001) on potential discrepancies in the level of gender role polarisation between rural and the more urbanised and economically developed societies (what Ross & Wells 2000 call the *modernist fallacy* in studies of homosexuality) is symptomatic of a far more important and general issue in the understanding of the adaptiveness of masculinity and femininity on the one hand and of the ways in which the culture-specific aspects of gender role may impinge on the perception and expression of homosexuality. Incidentally, cross-cultural studies must be very careful to control for *modernist fallacy* effects and also for ethnic biases in their samples, before claiming to have described human species-specific patterns of behaviour. For instance, Lippa (2008b) has recently published a cross-cultural study based on a large sample coming from 53 countries. However, data were obtained from participants using the internet (a method that is sensitive to *modernist fallacy* effects; see also Ross et al. 2005), and the bulk of data (84%) actually came from four countries only: United Kingdom (45%), USA (29%), Canada (5%) and Australia (5%), all strongly characterised by a common Anglo-Saxon–Celtic cultural heritage. Returning to the point raised by Costa et al., relatively little is known, even in Western societies, about the concepts of femininity and masculinity in rural settings (Little 2002; Little & Panelli 2003) where relatively hard labour may require the involvement of whoever is available in the household, male or female. It is in such societies, where work commitment from both males and females is essential for the survival or well-being of the social group (e.g. the family unit), that gender roles may be subordinated to or at least constrained by the achievement of the broad economic objectives that will ensure such well-being. Translated into evolutionary biological terms, this concept may be rephrased as: the requirement of biparental care to achieve the

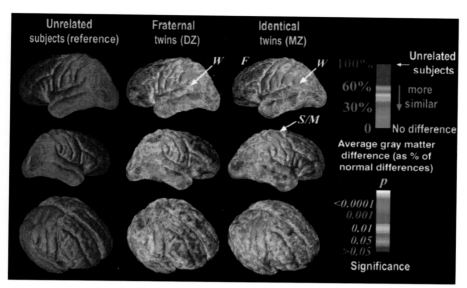

Figure 5.17. Differences in the quantity of grey matter at each region of the cortex were computed for identical and fraternal twins, averaged and compared with the average differences that would be found between pairs of randomly selected, unrelated individuals (blue, left). Colour-coded maps show the percentage reduction in intrapair variance for each cortical region. Fraternal twins are less similar to each other than genetically identical twins in a large anatomical band spanning frontal (F), sensorimotor (S/M) and Wernicke's (W) language cortices, suggesting strong genetic control of brain structure in these regions (the significance of these effects is shown on the same colour scale). From Thompson *et al.* (2001).

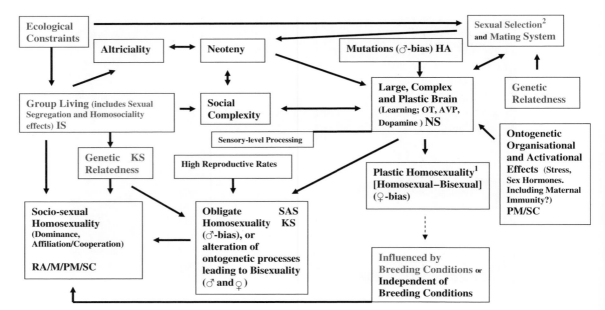

Figure 10.1. A *Biosocial Model* for the evolution and maintenance of homosexual behaviour in birds and mammals. [1]Not highly canalised ontogenetic processes, brain plasticity in adults. [2]Includes all sexual variables that in the *Synthetic Reproductive Skew Model of Homosexuality* were represented by Pair Bond, Sex Ratio, Reproductive Physiology, Mate Choice and EPFs. Variables in colour are components of the *Synthetic Reproductive Skew Model of Homosexuality*: red, core *Reproductive Skew* factors; purple, mainly sexual factors; blue, mainly sociosexual factors. IS, interdemic selection; KS, kin selection; RA/M, reciprocal altruism/mutualism; PM, parental manipulation; SC, sibling–sibling conflicts; NS, natural selection; SAS, sexually antagonistic selection; HA, heterozygote advantage; OT, oxytocin; AVP, arginine vasopressin. Mutations may affect brain centres, peripheral sensory tissues or both, but note also that mutations may also affect other components of the model such as sociality, neoteny, altriciality and reproductive rate. Continuous-line arrows may indicate either (a) proximate causation effects (for example, stress during ontogeny may affect the structure and function of a complex and plastic brain) or (b) ultimate causation effects (for example, sexual selection, evolutionary neotenic processes and a complex social environment may all contribute to the evolution of large and plastic brains). The dotted-line arrow indicates categorisation; for example, plastic homosexuality may be of two kinds, one that is manifested under breeding conditions (e.g. breeding season) and another that is independent of breeding conditions (e.g. non-breeding season).

maximisation of individual fitness given certain environmental constraints may make sex-specific roles subordinated to such a goal. If Darwinian fitness (or economic wealth) is increased by means of a division of labour (e.g. men in the field, women at home), then sex roles will be more clearly defined within such a society (see, for example, Brandth 2002). If fitness (or wealth) is increased by means of an overlap in roles between the sexes (e.g. both males and females have to work in the field to produce enough food) then sex roles may overlap (see, for example, Ross & Wells 2000). In the end, the brain plasticity that is likely to underscore the behavioural plasticity usually found in humans and in other social vertebrates may impinge on the flexibility of gender roles. This model suggests that masculinity and femininity are dynamic traits that may be combined in diverse ways in different individuals, or at different times during the ontogeny of the same individual, in response to social and other environmental circumstances.

That masculine and feminine traits can be combined in the same person, thus giving rise to a more complex gender role/identity, is clearly demonstrated by well-known examples from the anthropological literature (see Lang & Kuhnle 2008 for a recent review). For instance, the *nádleehí*, also known as *berdache*, of Navaho societies (*me-xo-ga*, in the Omaha language, Williams 1993) are one such example. *Nádleehí* are males who take up some well-defined feminine roles, thus identifying themselves as a composite gender that combines characteristics typical of males and females. This means that masculinity and femininity are not seen as mutually exclusive in Navaho culture (Epple 1998) and can be combined in a single person. Similar examples are also found in other cultures, e.g. the *hijras* of Southern India (Weston 1993), the *kathoey* of Thailand (Jackson 1997), the *bantut* of the southern Philippines (Johnson 1995), the *jōgappa* of southern India (Bradford 1983), the *ibne* of Turkey (Cardoso 2009), the *acault* of Burma (Coleman et al. 1992), the *xanīth* of Oman (Wikan 1977), the *muxes* of Mexico (Chiñas 1995), the *aikāne* of eighteenth century Hawaii (Morris 1990), the *ghilmān* of eighteenth to early twentieth century Iran (Najmabadi (2001), the *wakashu* of eighteenth century Japan (Saeki 1997), the *kinaidoi* of ancient Greece (Winkler 1990) and the *cinaedi* of ancient Rome (Gleason 1990). The *virgjeresha* or 'sworn virgins' of Albania are a very good case of apparently heterosexual women displaying an ability to adopt masculine stereotypical behaviours in response to specific social circumstances (Young 2000; Dickrermann 2002). Other examples were mentioned in Chapter 1 (*paneleiros*, *fa'afafine*). In ancient Greece the depiction of male figures combining both masculine and feminine traits was representative of poetry and creativity (Frontisi-Ducroux & Lissarrague 1990).

The previous examples include cases that vary in sexual orientation (e.g. exclusive homosexuals, bisexuals or even the heterosexual *virgjeresha*), what they all have in common is the mixture of gender roles.

As a further illustration of the combinatorial potential of gender roles I would also like to briefly indulge in the telling of a little story that is a favourite of mine. I was born in central Italy, in the Etrurian town of Cerveteri, known to the ancient Romans by the name of Caere. In his beautifully written book *Daily Life of the Etruscans*, the French historian Jacques Heurgon (1961) comments on a passage from the Roman historian Livy where Tullia, the daughter of the Etruscan king of Rome Servius Tullius, in an imaginary dialogue tells her brother-in-law and future king Tarquinius Superbus that he did not deserve the gentle and timid wife he had as she 'lacked feminine audacity' (*muliebri cessaret audacia*). According to Heurgon, such a small, apparently innocuous passage has perplexed all commentators of Livy's work. The perplexity comes from the criticism, from a woman to another woman (her sister), for lacking 'feminine audacity': after all, audacity was not supposed to be a typical feminine attribute (see, for example, the findings mentioned above about *assertiveness* being regarded as a 'masculine' attribute in some modern societies). Most of those commentators have just assumed that somebody corrupted the text at some stage in the process of copying it. Heurgon disagrees and his comment is rather illuminating, arguing that the expression was exactly what Livy intended it to be. There was no error in transcription, it is just that the woman speaking was an Etruscan and Livy was illustrating the role of women in Etruscan society, which was very different from what it was in the Latin culture of Rome:

> Tullia was not in revolt against her own sex; she does not disown her sisters; she does not consider herself to be an exceptional being free from all feminine weakness. Only she has particular ideas regarding the feminine temperament which are not incompatible with a sort of specific audacity, energy and ambition, the *audacia muliebris* which animates Etruscan women
>
> (Heurgon 1961: 88).

Livy's story of Tullia is more than a colourful anecdote. In fact, in ancient Greece, some philosophical schools did recognise the coalescence of both masculine and feminine gender roles in males and females. In the words of the Athenian philosopher of the second century c.e. Polemo:

> You may obtain physiognomic indications of masculinity and femininity from your subject's glance, movement, and voice, and then, from among these signs, compare one with another until you determine to your satisfaction which of the two sexes prevail. For in the masculine there is something feminine to be found, and in the feminine something masculine, but the name 'masculine' or 'feminine' is assigned according to which of the two prevails
>
> (Gleason 1990: 390).

Although strong male femininity or strong female masculinity may be associated with a homosexual sexual orientation in some cases, more often than not various degrees of femininity and masculinity in males or females characterise homosexuals, heterosexuals and bisexuals. So, it is legitimate to ask: where exactly do homosexuals fit in this masculinity–femininity bi-dimensional space (Constantinople 2005)?

Homosexuals are recognised as a category beyond the traditional masculinity–femininity dichotomy in many cultures, being also cross-culturally diverse in their specific characteristics (Weston 1993). To a certain extent, the cross-cultural diversity of gender roles among homosexuals and between homosexuals and heterosexuals is affected by the wider perception of femininity and masculinity in each of those cultures. Lippa & Tan (2001) for instance, studied heterosexual and homosexual men and women from three different ethnic groups: Asian Americans, Hispanic Americans and White Americans. They found that participants who came from relatively more gender-polarised cultures, such as their specific sample of Asians and Hispanics, tended to show a larger degree of differences in gendered traits between homosexuals and heterosexuals than those coming from relatively less gender-polarised cultures such as the White Americans. Hegelson (1994) also studied a sample of people from the USA who were asked to describe in writing what they regarded as the salient characteristics of a masculine male, a masculine female, a masculine person, a feminine female, a feminine male and a feminine person. According to the responses received, the four most frequent traits characterising feminine females were: caring (37.5% of respondents), good manners (35.9%), wears a dress (34.4%) and long hair (29.7%). Interestingly, among the 23 characteristics listed by the respondents, 'homosexual' did not appear. That is, in the perception of the people interviewed a feminine female is unlikely to be homosexual! Feminine males were associated with 21 traits, with the top four being: homosexual (48.3%), thin (46.6%), insecure (31.0%) and emotional (31.0%). Clearly, the situation drastically changed in the case of stereotypical non-concordance between the biological sex and gender role in males. In this case a feminine male is clearly expected to be homosexual. However, what is remarkable about the comparison between the stereotypical characteristics of feminine females and feminine males is that they actually differ. That is, a feminine homosexual male is not expected to be equivalent to a feminine heterosexual female: the former is mainly seen as 'insecure' and 'emotional', whereas the latter is mainly seen as 'caring' and having 'good manners'. That is, even from the perspective of the respondents' stereotypes, homosexual men are a distinct category of gender, rather than a state in a bipolar continuum.

Masculine males are perceived as muscular (71.4%), like sports (52.9%), tall (51.4%) and self-confident (50.0%). None of the respondents included 'homosexual' among the 22 characteristics mentioned for masculine males. Again, gender role concordance is stereotypically associated with heterosexuality, a bias that permeates the perception of homosexuality in many cultures where the focus is on the 'mountee' male rather than the 'mounter' as the homosexual individual. Masculine females, on the other hand, were characterised as liking sports (64.7%), muscular (50.6%), short hair (37.1%) and dress casually (38.8%). In the case of masculine females, being homosexual is ranked 11th among the 21 traits mentioned. The stereotype in this case leans towards a clear identification of lesbians with traits typically associated with the other sex, although masculine females were not as immediately perceived as homosexuals as feminine males were. That is there is a greater cultural tolerance for gender roles variability in females than in males in the USA sample of college students that participated in this study (a similar result was obtained by C.L. Martin 1990).

K. L. Johnson et al. (2007) also studied the response of college students in the USA, but with respect to various stimuli representing males and females varying in waist-to-hip ratio, body motion (more swaggering or more swaying) and sexual orientation. A walking motion involving swaying hips is stereotypically perceived as a feminine motion, whereas walking while swaggering the shoulders is stereotypically perceived as masculine, with walking motion being a more important clue to assessment of gender than waist-to-hip ratio. Subjects judged that homosexuals are those individuals who are female and who shoulder swagger or those who are male and hip sway as they walk, although motion was a clue that was more consistently and also more reliably used in the assessment of male than of female sexual orientation. All this could be easily attributed to the effect of current social stereotypes, as K. L. Johnson et al. (2007: 332) obviously realise, so much so that homosexuals could even mould their walking patterns to the stereotype in order to enhance the public perception of their sexual orientation.

In spite of the stereotypical identification of homosexual males with a feminised gender role and the stereotypical identification of homosexual females with a masculinised gender role, a pattern known as *Gender Inversion* (Lippa 2008a), recent meta-analytical studies indicate levels of masculinity and femininity in lesbians and gay men, respectively, that are intermediate between those of heterosexual men and heterosexual women, a pattern known as *Gender Shift* (Lippa 2008a). In fact, a recent study of gender role in homosexual men carried out in Italy by Zoccali et al. (2008) indicates that homosexual men tend to view themselves as androgynous.

In sum, gender roles are considerably more variable within and between sexes and sexual orientations (see, for example, Connell & Messerschmidt 2005; Leszczynski & Strough 2008) than some previous authors have been willing to grant. Moreover, in studies of gender role care should be taken to also consider the effect of culturally entrenched stereotypes: although these may remain stable over a short (few years) or medium (few decades) time span, they may well change over a longer (few centuries) time frame. In addition, such stereotypes may be variable, at any point in time, across ethnic groups or across subcultures within the same ethnic group.

Behavioural plasticity and sexual orientation

Does the cultural perception of a wide range of gender roles within a specific sexual orientation in women – which I mentioned in my review of Hegelson's work – reflect a more profound degree of sexual plasticity in women than in men? In order to answer this question a clear distinction should be made from the outset. There are at least three major aspects of sexual plasticity that should be distinguished from each other, and their interrelationships should be properly studied and understood. First, there is the plasticity in sexual behaviour within a heterosexual sexual orientation: number of partners in a lifetime, frequency of sexual intercourse per partner, temporal frequency distribution of the number of partners and sexual activity at different timescales, degree of sexual partner choice and constraints on choice, etc. Second, there is plasticity in sexual behaviour within a homosexual sexual orientation, with variables to be studied mirroring the ones that I have just listed for heterosexual relationships. Finally, there is plasticity in terms of the likelihood of transitions from one sexual orientation category to another: homosexuality, heterosexuality, and diverse gradations of bisexuality; these could be experienced by different individuals or even the same individual more than once during his/her lifetime.

A decade ago Roy Baumeister (2000) published an influential work, highly prized by Peplau & Garnets (2000), on the level of sexual plasticity in males and females. In this work, Baumeister builds his argument starting from the premise that women's sexuality is more under 'cultural' influence than men's sexuality, which, in turn is supposed to be more under 'biological' influence. More specifically, Baumeister's thesis is that 'female sex drive is more malleable than the male, indicating higher average erotic plasticity' (Baumeister 2000: 348). In Baumeister's model, the expectation of females having higher erotic plasticity than males could be theoretically explained in the following manner: (a) females' sexual plasticity could be an adaptation to male power and sexual control: by being more flexible and accepting of unwanted sexual relationships females may make such relationships more bearable; (b) flexibility may be inherent to a species-specific female role. Here Baumeister may be emphasising what in the evolutionary literature is known as 'female mate choice'. So far, I do not see why, following (a) and (b), males should be regarded as less plastic than females. Males may be both as unfussy or opportunistic, in their sexual behaviour – thus demonstrating high plasticity – as females, and they can also be selective of whom to choose for reproductive purposes (e.g. to establish a family). Baumeister then continues with (c) females, having a 'weaker sexual drive' than males, may also be more malleable in terms of the specific object of their desire. Here it is unclear what 'weaker sexual drive' means. If it means that females are only aroused during specific periods of time whereas male sexual arousability is, on average, relatively more widespread through time (e.g. during the day or during the year), then this only leads to a difference in sex-specific patterns of sexual activity over time: more continuous in males, more discontinuous in females. Would this also lead to more sexual malleability in females? If males can be continuously aroused and are not engaged with another female, they will then be available whenever their sexual partner is aroused, therefore there will be no need for the female to expand the set of potential objects of her desire on these grounds alone. However, women may expand the set of partners if the preferred partner is not available during the periods when they are sexually aroused. Exactly the same, however, is expected of men! Whichever way we see it, this aspect of plasticity within a heterosexual context is a property of both men's and women's sexuality.

Baumeister's (2000) argument that the sexual revolution that characterised the USA in the 1960s and 1970s, and that saw women modify their sexual behaviour to a greater extent than men, supports his hypothesis of greater sexual plasticity in women than in men is less than convincing. Indeed, the argument is more supportive of the hypothesis of similar heterosexual sexual plasticity in men and women. In fact, what the 1960s and 1970s produced was the cultural freedom that allowed women to openly express their full heterosexual behavioural repertoire. Men did not change their sexual behaviour simply because they were already expressing their heterosexuality without great impediments, e.g. in terms of number of sexual partners. In fact, male and female sexual behaviours are more convergent in societies that are more sexually permissive and egalitarian (Reiss 1980; Weinberg *et al.*

2000). On p. 352 Baumeister (2000) also quotes an example that, again, he misinterprets as evidence of women's heterosexuality being more plastic than men's:

> When asked about their current frequency of sexual activity and their current preferences for frequency of sexual activity, the wife's answers indicated that their marital practices corresponded almost precisely to the amount of sex they wanted, whereas the men reported a significant gap between what they wanted and what they were able to have ... Possibly, women succeeded better than men at adjusting their expectations into line with what they were getting, which could be another manifestation of plasticity ...

If men can be sexually aroused on command, whereas women are restricted in their sexual arousability by specific endocrine states and/or psychological states (e.g. absence of stress) for instance, then necessarily women will be far more likely to achieve fulfilment of their sexual needs than men. To put it bluntly: he will be always ready when she is, but she may not be always ready when he is. This has nothing to do with *greater* plasticity of women's sexuality as it is expressed in a heterosexual context. Moreover, in this specific case it is men who will probably have to resort to the plasticity of their brain in order to find an alternative to the rejected proposal for sexual intercourse: decrease sexual arousal and do something else, keep sexual arousal high and masturbate, keep sexual arousal high and find an alternative partner, etc. On the other hand, as suggested by Baumeister (2000) a woman may have reasons to accede, at least from time to time, to the sexual advances of her partner even though she may not be highly motivated. In sum, **both** heterosexual men and women are behaviourally plastic, broadly speaking, as far as engaging in sexual activities is concerned.

From an evolutionary point of view, heterosexual sexual plasticity in women may have been selected under different adaptive contexts (e.g. as a means to secure parental care from more than one man, or increase fertility or select specific genes, Geary 2000; Schuiling 2003; Gangestad *et al.* 2005; Geary 2006; Kardum *et al.* 2006; Pillsworth & Haselton 2006) from men's heterosexual sexual plasticity, which may have been selected as a means to increase the number of women fertilised, but also to establish social bonds with various females (Pillsworth & Haselton 2006 and references therein). But in the end, both sexes may well be operating in an evolutionary environment where heterosexual sexual plasticity can be adaptive to both males and females (Greiling & Buss 2000; Simmons *et al.* 2004; Schuiling 2005; Pillsworth & Haselton 2006; Koehler & Chisholm 2007). Behavioural plasticity, including plasticity of sexual behaviour, is a result of brain plasticity; broadly speaking, overall brain plasticity is a trait that characterises our (and some other) species, not any particular sex. In the words of Havelock Ellis (1946: 34): 'Within certain limits... the feminine type must have a tendency to adapt itself to the ideals of men, and the masculine type to the ideals of women'. This conclusion, however, is a broad statement and it does not imply, as we will see in what follows, that every single specific aspect of sexuality will be equally plastic in a comparison between the sexes.

In discussing Baumeister's work I have so far referred to plasticity *within* a specific sexual orientation. Current empirical evidence indicates that heterosexuality is plastic in both male and female humans. Do the arguments change as we focus on the transitions, back and forward, through different categories of sexual orientation (homosexual–heterosexual–bisexual) in the adult ages? Indeed they do, and in this case I agree with Baumeister that adult females seem to be more plastic than males (see also the section *Orgasm, sexual arousal and homosexuality* above). We will see in Chapter 9 that across primates females are significantly involved in same-sex sexual behaviour. In humans, however, males are more frequently exclusively homosexual than females (Kirkpatrick 2000). This, however, does not falsify the hypothesis of higher plasticity among females in specific aspects of sexuality as females are more likely to change sexual orientation over their lifetime (see, for example, Peplau *et al.* 1999; Peplau & Garnets 2000; Kinnish *et al.* 2005; Diamond 2007), whereas males tend to establish their sexual orientation relatively early in life and maintain it thereafter. One specific example that clearly demonstrates human female plasticity in terms of transitions from one sexual orientation to another, and which was discussed by Baumeister (2000) and developed by many others, is that of *political lesbians*. Political or cultural lesbians (see, for example, Taylor & Rupp 1993; Jackson & Scott 2004), who embrace a same-sex sexual pattern of behaviour after a conscious process of decision making and who may abandon or return to a lesbian sexuality more than once in their lifetime (Peplau *et al.* 1999), are clearly demonstrating that the causation of their sexual behaviour involves plastic brain mechanisms. In the end, political or cultural lesbianism is possible only because a woman's brain is plastic with regard to transitions across sexual orientations.

In an extremely interesting work, Lisa Diamond (2008a; see also Diamond 2008b) followed 79 non-heterosexual women over a 10-year period – from late adolescence to early adulthood – and tracked changes in both sexual behaviour and sexual attraction. Her results clearly indicate that women are ontogenetically plastic with regard to sexual orientation and that changes to and from homosexuality, bisexuality and heterosexuality can and do occur.

The plasticity of our brain undoubtedly lies at the core of all these issues: understanding the dynamics of lifetime changes in sexual orientation (Peplau *et al.* 1999; Peplau & Garnets 2000; Diamond 2007) is next to impossible if we treat our brain simply as a rigid 'blackbox'. The brain responds to environmental inputs according to its own, internally determined rules, the brain is neither a 'blackbox', nor just a mirror of the external world. Presumably, brain plasticity is a result of evolutionary processes that involved natural selection, sexual selection, genetic drift and neoteny, and once plasticity became a fixed trait in the population, as it is in many social mammals such as rodents and primates, then, as we have seen, the gates became wide open for learning to be involved, as one of many processes, in the modulation of all behaviours in adult individuals, including same-sex sexual behaviours.

Brain, emotions and homosexuality

Helen Fisher (1998) identified three interconnected emotion–motivation systems that characterise the mammalian brain – and presumably the brain of some other taxa as well – in the execution of biological functions such as reproduction, parenting and mating: attraction, lust and attachment. The relationships between sexual attraction and homosexuality will be more specifically reviewed later in this chapter in the section 'Influences of sexual selection on homosexuality', whereas various relationships between lust and homosexuality have been already analysed in the section 'Orgasm, sexual arousal and homosexuality'. Additional analyses will be provided by Alan Dixson in Chapter 9, where the relevance of pleasure in same-sex sexual interactions in humans and other primates will be discussed. Here I focus mainly on attachment. In addition I also review works that have studied the relationship between aggressiveness and empathy and sexual orientation to finish with an analysis of the adaptive value of androgynous gender roles as emotional coping strategies adopted in stressful situations.

By definition, social animals live in close physical contact with each other. When the contact is actively maintained, rather than being a side effect of environmental conditions such as living in a patchy habitat, then a degree of attachment may occur. More complex interactions will arise among the members of the group when the attachment is not simply diffused across the whole membership of the group, but is specific between particular individuals. When the attachment involves conspecifics of the same sex, those individuals should be regarded as being homosocial. Homosocially bonded individuals may or may not engage in sexual activities with each other, but if they do then they can be regarded as homosexuals.

Homosociality and homosexuality can be distinguished not only by definition, but also because the neurological structures and processes associated with them are somewhat different. That is, brain centres activated by what in human terms we call romantic love and we refer to in other species as mate choice or mate attraction, differ to some extent from those controlling sexual drive, which, in turn also differ but are interactive with centres controlling attachment and therefore sociality (see, for example, Bartels & Zeki 2000; see also Fisher *et al.* 2006 for a review). Such interactions may produce outcomes such as sexual attachment.

Sexual attachment between individuals of the same species may be affected by emotive states modulated by neurotransmitters (e.g. dopamine), neuropeptides (e.g. oxytocin, arginine vasopressin) and endogenous opiates (Broad *et al.* 2006). Fisher *et al.* (2002) provided experimental evidence to suggest that dopamine can mediate some aspects of romantic attraction in humans, at least in a heterosexual context. In the analysis by Fisher *et al.*, sex drive, romantic attraction and attachment are all obviously linked to reproductive success via fertilisation and mate guarding, but also through parenting. Such basic mechanisms need not change in the context of homosexual sexual partnerships. That is, the emotional basis of romantic attachment may be the same or similar across sexual orientations, it is just the object of such attachment that obviously changes. This, in principle, may require a less dramatic differentiation in brain function and structure between homosexual, bisexual and heterosexual individuals than we may otherwise expect. Alternatively, brain processes controlling attachment between homosexuals of one sex may mirror those of heterosexuals of the other sex (*gender inversion*) or they may be completely different from those of male and female heterosexuals (*third gender*), in which case the homosexual relationship may reflect a more specific specialisation of the brain.

The expression of attachment and therefore pair bond formation among adults (i.e. positive affective bonds; Burgdorf & Panksepp 2006) is controlled by brain centres such as the amygdala, the prefrontal cortex, the mesolimbic nucleus accumbens and the ventral pallidum (Carter 1998; Young & Wang 2004; Broad *et al.* 2006) and it may have derived from other prosocial behaviours such as mother–offspring bonding (Bowlby 1969; Hazan & Shaver 1987; Panksepp *et al.* 1997; Carter 1998; Esch & Stefano 2005).

Attachment is also responsive to stress, in the sense that it is under stress that social animals seek each other's company (Esch & Stefano 2005), whether they are sexual partners, social partners or mother and offspring, although I should also point out that bonding may break down under conditions of elevated chronic stress (Esch & Stefano 2005).

From an adaptive perspective, if the benefit of positive affective bonds such as those leading to long-term social attachment between two or more individuals is that of increasing inclusive fitness through parental care, alloparental care and/or cooperation in mutual defence or acquisition of resources (see, for example, Fraley *et al.* 2005), then affective bonds usually expressed in a heterosexual context may well retain most of their adaptive value when expressed in a homosexual context. In the case of exclusive homosexuality, social attachment would be adaptive not only in the context of increased survival, but also in that of increased opportunities to help relatives (see Chapters 3 and 8; similar arguments have been independently put forward by Lisa Diamond, 2008b). Interestingly, it is when the broad social interactions become dysfunctional, as in the case of societies where homosexuals are discriminated against, where the adaptive value of exclusive homosexuality becomes less apparent, attachment between two homosexual individuals becomes more common and distancing from family appears to be paradoxical. We have already seen in Chapter 4 how homosexual, but also bisexual, men and women may seek the company of similarly sexually oriented persons as a strategy to cope under social stress caused by intolerance (Bell & Weinberg 1978; Weinberg *et al.* 1994b; Carbone 2008).

Emotional attachment between two persons, two friends for instance, whether of the same or of a different sex, is a phenomenon separate from romantic love, as illustrated by the results of Johnston & Bell (1995) who reported a similar degree of attachment between heterosexual (98%) and homosexual (89%) males with females during adolescence. Attachment is also dependent on other aspects of the social interaction. For instance, studies of attachment among lesbians have emphasised a greater level of satisfaction within the relationship when the interaction is more egalitarian (Peplau et al. 1978). When the symmetry in 'power' between the members of a relationship decreases, attachment between the two partners also decreases (Caldwell & Peplau 1984). Consequently, a decreased level of attachment (e.g. anxious attachment, avoidant attachment, see Grau & Doll 2003) and increased asymmetry in the relationship leads to lower levels of satisfaction in the relationship (Horne & Biss 2009); conversely, the more the homosexual relationship is egalitarian, the more stable the levels of attachment are and also the more the relationship is associated with greater levels of satisfaction among homosexual women. This dynamic process is expected to be conducive to a choice of partners that are more secure and with whom a more stable relationship could be established. This is in fact the case in both lesbians and gay men, where couples that are more likely to be accommodating of each other's individualities and therefore less likely to engage in interactions that are 'destructive' of the social bond also display a greater degree of attachment (Gaines & Henderson 2002). Although all the above features of same-sex attachment need not necessarily change in heterosexual relationships, other characteristics, such as jealousy, may be different and show patterns of gender discordance across sexual orientations.

Indeed, the stability of attachment in a homosexual relationship could be jeopardised by jealousy, as is also the case for heterosexuals. In this regard, the patterns of jealousy seem to be more specific to the different sexual orientations. The expression of jealousy appears to be somewhat masculinised among lesbians and feminised among gay men (*gender inversion*). According to Dijkstra et al. (2001): homosexual females, like heterosexual males, tend to emphasise sexual infidelity as a reason to be jealous, although heterosexual males also regard emotional infidelity as a cause of concern; in the case of homosexual males, they emphasise emotional infidelity as heterosexual females do.

In the evolutionary perspective of this book, it is also interesting to speculate that attachment – that presumably evolved originally in the context of mother–infant bonding (Carter 1998) – may have been subsequently co-opted in the evolution of sociality through the mediation of neoteny in some taxa, with neoteny being reinforced by the benefits of cooperation accruing to affectively bonded individuals. This is a scenario that is consistent with the view of homosexuality as a neotenic trait (Bromhall 2004, see Chapter 4). The results of comparative analyses of mammals, including primates, that I mentioned in Chapter 4 (Fraley et al. 2005) do not support this possibility, as those authors did not find an evolutionary correlation between neoteny and pair bonding. However, as I already remarked in that chapter, their measurement of neoteny may not be sufficiently specific, as they only used a limited number of life-history traits whereas perhaps brain size corrected for body mass might have been more appropriate.

In sum, attachment presumably evolved from mother–infant relationships in mammals; although it is a property of homosexual relationships, as it is of heterosexual relationships, some aspects of attachment such as jealousy suggest a degree of gender inversion in both male and female homosexuals.

Aggressiveness, empathy and homosexuality

We have seen in Chapter 3 and in the previous section that one potential selective value of homosexuality could be found in the aid directed by homosexuals towards relatives, in societies where homosexuals are fully integrated into the social framework and their relatives are not discriminated against by the rest of the community. If homosexuals develop a cooperative personality as predicted by kin selection, then we may also expect them to be in general less aggressive and more empathic than heterosexuals. The available evidence supports this prediction. Blanchard *et al.* (1983), for instance, found that heterosexual males score significantly higher on a Physical Aggressiveness Scale than both homosexuals and transsexual homosexuals, the latter two, in turn score higher than heterosexuals in a Feminine Gender Identity Scale. The same results for aggressiveness were obtained by Gladue & Bailey (1995), with heterosexual males self-reporting higher levels of physical aggressiveness than both homosexual males and heterosexual females; the latter two did not differ. Verbal aggression is also more elevated in heterosexual males than in either heterosexual females, homosexuals (males and females) or bisexual females, with heterosexual females being also more verbally aggressive than homosexual females and more physically aggressive than homosexual males (Wrench 2002). Dickins & Sergeant (2008) have recently confirmed the trends for aggressiveness, which is associated with social dominance as well, being higher in heterosexual males than females, with the latter scoring higher than homosexual males. Finally, Sergeant *et al.* (2006) not only confirmed the lower levels of physical aggression in homosexual than heterosexual males, but also reported higher levels of empathy in the former than the latter.

The above patterns are clearly consistent with the expectations from kin selection: more empathic and less aggressive homosexuals will be more likely to perform cooperative behaviours towards other members of the resident group, which in traditional human societies usually includes various kinds of kin.

Androgyny and emotional coping

Emotional aspects of sexuality may also explain some degree of psychological androgyny detected in some societies, especially as males and females age (see, for example, McCabe 1989). Psychological androgyny in both men and women reiterates not only the brain plasticity that I emphasise here, but also a level of combinatorial capacity of gender roles that I have already mentioned in a previous section. Psychologists express the benefits of androgyny in terms of 'coping'. Evolutionary biologists would agree with such a view, although we would perhaps express ourselves in terms of androgyny being a strategy to maximise inclusive fitness in situations of environmental, including social, stress and uncertainty. Androgyny as a coping strategy is obviously a hypothesis, but a testable one.

Data in support of androgyny being a coping strategy to withstand situations of stress come, for instance, from Baucom & Danker-Brown (1979), who showed experimentally that androgynous men and women are less susceptible to situations of learned helplessness (an analogue for depression). In a study of college students, Shaw (1982) found that androgynous subjects tended to perceive stressful life events as 'less undesirable' (especially in the case of women), and androgynous individuals of both sexes also regarded themselves as happier than masculine or feminine participants. Shaw, however, also suggests that such a difference may also be affected by differential access to social support systems across gender roles. In a short-term longitudinal study spanning 8 weeks, Roos & Cohen (1987) also showed that androgynous individuals were more resilient to stress than masculine or feminine participants. As in the study of Shaw (1982), Roos & Cohen (1987) also indicate the mediation of social support as a relevant moderator of stress in androgynous persons. More recently, Isaac & Shah (2004) studied Indian couples; their work indicates that high levels of androgyny are associated with less distressed individuals and better mental adjustment between heterosexual couples, whereas Crawford (1994) found that androgynous individuals feel in general more comfortable with social physical closeness, including actual touch, than non-androgynous individuals. As for McCabe's (1989) study, Isaac & Shah (2004) also suggest a trend for couples to develop a higher level of androgynous behaviours with time. Moreover, Branney & White (2008) suggest that the inability of individuals (e.g. men) to escape socially imposed bipolarism of gender roles, e.g. if they are prevented from adopting more androgynous behaviours, may be one of the causes of male depression. Evidence against the hypothesis that androgyny is an adaptation to coping comes from Hamilton (2001) who did not find a difference between androgynous and non-androgynous males and females in their scores of 'reason for living'.

Homosexuality and cognitive processes

In the previous section I discussed the potential links between emotional states and same-sex sexual orientation and some aspects of gender role. Although emotions such as attachment and pleasure are known to interact with cognitive processes (e.g. by means of improving cognitive functions (Esch & Stefano 2004)) in this section I focus on the specific association between cognition and homosexuality.

Some psychologists see cognitive processes as contributing to the development of sexual orientation mainly in terms of their providing the reinforcement of the sex labels accepted in the community in which an individual grows up. The community establishes the gender schemata that the developing individual will have to learn. In Bem's (1981: 355) words 'a schema is a cognitive structure, a network of associations that organizes and guides an individual's perception' and through following those schemata the developing individual organises the learning experience that is relevant to sexual behaviour. In this context, then, sexual orientation is presumably determined by both the degree of learning plasticity of the individual and the specific constraints represented by the prevalent sex types accepted in the community. Interestingly, it is when the sexual orientation developed by any specific individual (e.g. homosexuality) is in contradiction with the sex schemata prevalent in his/her community (e.g. adults are expected to be heterosexual) that the limits of brain plasticity and learning may become most apparent. That is why other students of sexual orientation such as Maccoby & Jacklin (1974) tend to emphasise a genetic predisposition towards homosexual behaviour that, nevertheless, is susceptible to further modulation by learning processes. Thus, although cognitive processes are undoubtedly linked to learning and therefore brain plasticity, they are also constrained by those developmental processes of the brain that are more canalised. When the development of a specific sexual orientation is refractory to the sex schemata currently in place in a specific community, various outcomes are possible including ostracism or abandonment of the native community by the atypical individual, or retention of the atypical individual within the community due to change in the prevalent schemata, or retention of the individual without change in the schemata, leading to stress and coping.

Cognitive mechanisms that sustain the development of sexual schemata may have evolved through sexual selection and may also have played a role in the evolution of homosexuality, especially through the mediation of neotenic processes. For instance, Perrett et al. (1998) carried out a study of attractiveness for feminised and masculinised male and female Caucasian and Japanese faces and found that the general trend was for participants of both ethnic groups to prefer feminised faces of both males and females. Such a bias could be an adaptive outcome of a preference for younger individuals (see also the next section: 'Influences of sexual selection on homosexuality'). Younger-looking, yet adult individuals could be regarded as more healthy and therefore more appropriate partners either for reproduction, in the case of individuals of the other sex, or for cooperation, in the case of individuals of the same or other sex. If so, then social and sexual preference for youthful-looking individuals may have at least reinforced any neotenic evolutionary process that might have been already in progress in humans (Bromhall 2004; see Chapter 4).

Once a diversity of sexual orientations has evolved however, would individuals also differ in terms of their cognitive characteristics and could those characteristics adaptively explain the maintenance of the homosexual trait in the population? Moreover, do cognitive characteristics in homosexuals follow a gender-concordant or a gender-discordant (e.g. *gender inversion, third gender*)

Homosexuality and cognitive processes

Table 5.5. Comparisons of cognitive abilities between men and women of diverse sexual orientation

Study	Task	Heterosexuals	Heterosexuals vs. homosexuals[a] Males	Females
Wegesin (1998a,b)	Spatial	M > F	Gender-shifted	Same as heterosexuals
	Verbal	M < F	Gender-shifted (in choice of lexicon); same as heterosexuals (in semantic processing)	Same as heterosexuals
McCormick & Witelson (1991)	Spatial	M > F	Gender-shifted	
	Verbal	M < F	Gender-shifted	
Willmot & Brierley (1984)	Verbal	M < F	Hyperfeminised	
Sanders & Ross-Field (1986)	Spatial: Vincent Mechanical Diagrams	M > F	Gender-inverted	
	Water Level test	M < F	Gender-inverted	
Tuttle & Pillard (1991)	Spatial	M > F	Gender-shifted	Same as heterosexuals
	Verbal	M < F	Gender-shifted	Same as heterosexuals
Gladue (1990)	Spatial: Water jar test	M > F	Gender-shifted	Hyperfeminised
	Mental rotation test	M > F	Gender-shifted	Same as heterosexuals
Hall & Kimura (1995)	Spatiomotor: Throw-to-target Task	M > F	Gender-shifted	Gender–shifted
	Purdue Pegboard	M < F	Same as heterosexuals	Same as heterosexuals
Rahman & Wilson (2003b)	Spatial: Mental rotation test	M > F	Gender-shifted	Gender-shifted
	Benton Judgment of Line Orientation test	M > F	Gender-shifted	Same as heterosexuals
Alexander & Sufka (1993)	Spatial	M < F	Third gender	
	Verbal	M > F	Gender-shifted	
Rahman et al. (2005)	Verbal	M > F	Same as heterosexuals	Gender-shifted
	Spatial: Mental rotation, Cardinal direction	M > F	Gender-shifted	Hyperfeminised
	Landmark navigation, Left–right direction	M < F	Gender-shifted	Same as heterosexuals
Rendall et al. (2008)	Verbal (phonetics): Voice pitch	M < F	Same as heterosexuals	Slightly gender-shifted
	Formant frequencies	M < F	Same as heterosexuals	Gender-shifted

[a] Gender-shifted: the trait value for homosexual males or females is somewhere between the value for heterosexual males and females. Gender-inverted: the value for the trait in homosexuals is indistinguishable from the value of the trait for heterosexuals of the opposite sex. Third gender: the value of the trait for homosexuals does not lie in a linear continuum that includes the values of heterosexual males and females. Hyperfeminised traits go beyond the average value of the heterosexual females (smaller, when heterosexual females have smaller values than heterosexual males and larger, when heterosexual females have larger values of the trait than heterosexual males).

pattern? Table 5.5 summarises the results of various cognitive tests carried out by diverse authors comparing the performance of male and female heterosexuals and homosexuals.

Wegesin (1998a,b) studied both spatial and verbal abilities in heterosexual and homosexual males and females. Heterosexual males perform better than heterosexual females in the Vandenberg and Kuse 3-D mental rotation test, whereas heterosexual females perform better in verbal tasks. In homosexuals, males perform in a manner that resembles more the heterosexual female pattern than the heterosexual male pattern, whereas lesbians responded in a manner similar to that of heterosexual females. Interestingly, with regard to verbal tasks gay men tended to resemble heterosexual females in terms of their choice of lexicon, but their semantic processing was similar to that of heterosexual men. This may suggest that diverse cognitive processes may be under diverse influences in male homosexuals. Tasks such as mental rotation may be controlled by central nervous system mechanisms that are probably less labile than 'choice of lexicon'. Could the latter be more affected by learning than the former? If so, choice of lexicon would be more likely to conform to a cultural expectation of a feminised male homosexual.

McCormick & Witelson (1991) also carried out both spatial and verbal ability tests for right-handed homosexual men, heterosexual women and heterosexual men. For the spatial tests, the scores of homosexual men were intermediate between those of heterosexual men and heterosexual women. The same trend, although not significant, was found for the verbal tests: the scores of homosexual men were also intermediate between those of heterosexual women and heterosexual men. However, Willmott & Brierley (1984) found that in a verbal IQ test male homosexuals scored higher than female heterosexuals, who, in turn scored higher than male heterosexuals, that is, male homosexuals were hyperfeminised for this particular cognitive task. Sanders & Ross-Field (1986) also tested homosexual men, heterosexual men and heterosexual women for visuo-spatial abilities and found that for most tasks (Vincent Mechanical Diagrams test and one of the two water level tests they used) heterosexual males performed better than both homosexual males and heterosexual females, the latter two performed similarly. However, in a different water level test, heterosexual males performed worse than both homosexual males and heterosexual females, with the latter two continuing to perform similarly. Tuttle & Pillard (1991) tested homosexual and heterosexual males and females for spatial, verbal and mathematical abilities using the Primary Mental Abilities test (PMA) and the Wechsler Adult Intelligence Scale (WAIS). Their statistically most significant result was for PMA scores for spatial relations to be higher in heterosexual males than heterosexual females. However, they also found marginal trends for homosexual males to score higher than heterosexual males in WAIS similarities, vocabulary and block design tasks, that is, homosexual males tended to score higher than heterosexual males in both verbal but also some spatial cognitive skills. Interestingly, homosexual and heterosexual females did not differ (Tuttle & Pillard 1991).

Gladue et al. (1990) compared the spatial performance of heterosexual and homosexual males and females, in a water jar test and a mental rotation test. In the water jar test homosexuals, both male and female, performed more poorly than heterosexuals, with men tending to perform slightly better than women but not significantly so. In the mental rotation test men did better than women overall, but the effect of sexual orientation was significant for males only, with homosexual men scoring lower than heterosexuals. No difference in mental rotation capabilities was detected between homosexual and heterosexual women.

Hall & Kimura (1995) tested homosexual and heterosexual males and females for two spatiomotor tasks: a Throw-to-Target Task, where males usually outperform females, and the Purdue Pegboard, where females usually outperform males. In the Throw-to-Target Task homosexual men were outperformed by heterosexual men whereas homosexual women outperformed heterosexual women. That is, homosexual men showed an indication of feminisation, whereas homosexual women were masculinised in their throwing skills. Sexual orientation, on the other hand, did not affect performance in the Purdue Pegboard. More recently Rahman & Wilson (2003b) have studied the effect of sexual orientation in men and women on the performance of tasks that are known to be dependent on the activity of the parietal cortex: the Mental Rotation (MR) test and the Benton Judgment of Line Orientation (BJLO) test. In the BJLO test men, in general, scored better than women; and heterosexual men also achieved scores that were higher than those of homosexual men. However, no effect of sexual orientation was detected among women for this test. Homosexual men and heterosexual women did not differ either, suggesting a feminisation of the cognitive abilities involved in this test among homosexual men. As for the BJLO test, men also scored higher than women in the MR test. Again, in this cognitive ability homosexual men achieved lower scores than heterosexual men, whereas the reverse was true for women: homosexual women achieved higher scores than heterosexual women. Therefore in the MR test homosexual men tend to display a feminised pattern whereas homosexual women show a masculinised set of abilities.

Alexander & Sufka (1993) measured brain activity during cognitive tasks involving spatial or verbal skills in men (homosexuals and heterosexuals) and heterosexual women. They carried out a spectral power analysis of electroencephalograms (EEG) within the range of alpha frequencies (i.e. 8–12 Hz) as subjects were involved in the tasks. They specifically recorded EEG activity from temporal, parietal, frontal and occipital areas in both hemispheres. At the hemisphere level, individuals differed in their level of activation (reduced alpha power) or inhibition (increased alpha power) of the different areas targeted. In concomitance with the performance of verbal tasks, the left hemisphere tended to be only slightly activated in all individuals, whereas activation differed dramatically between men of diverse sexual orientation in the right hemisphere. Whereas heterosexual males were highly activated, homosexual males were inhibited. In contrast, the level of activation in heterosexual women was low in both hemispheres. With regard to the spatial task, the two hemispheres also responded differently depending on sex, and also sexual orientation for men. Heterosexual males showed inhibition of alpha activity in both hemispheres during spatial tasks, but the level of inhibition was higher in the left hemisphere. Homosexual males, on the other hand, showed increased alpha activity in both hemispheres, with the right hemisphere being more activated than the left. Heterosexual females had a slight activation of the left and inhibition of the right hemisphere. These results clearly indicate that, in terms of activation or inhibition of neuronal activity in specific areas of both hemispheres during either verbal or spatial tasks, homosexual males are not feminised, on the contrary, they show a pattern, that is neither masculine nor feminine (Alexander & Sufka 1993), that could be described as 'third gender'.

When navigating, men and women seem to use different strategies: women tend to rely more on the use of environmental descriptors or landmarks, whereas men tend to rely more on Euclidean coordinates for their orientation (Rahman *et al.* 2005 and references therein). Rahman *et al.* (2005) carried out a study of various cognitive tasks in heterosexual and homosexual males and females, testing for vocabulary, mental rotation, road map direction and spatial location skills. Their results were mixed in terms of both sex and sexual orientation effects. For instance, men scored higher than women in vocabulary, mental rotation capabilities, and usage of cardinal direction in navigation, whereas women scored higher than men in usage of landmarks for navigation and also in the use of left–right directions. Homosexuals, on the other hand, displayed a mixed bag of results: homosexual women scored higher than heterosexual women in vocabulary, whereas homosexual men used more landmarks than heterosexual men; homosexuals of both sexes made more mistakes in the road map test than heterosexuals, but there was no effect of sexual orientation on spatial location tests. This suggests that homosexuals are not unequivocally characterised by gender-non-concordant cognitive characteristics; some cognitive skills are indeed gender non-concordant whereas some others are not.

Finally, a recent study carried out by Rendall *et al.* (2008) in Canada, has focused on phonetic aspects of language. They compared voice pitch and formant (the latter describes the peak in the spectrum of frequency of a sound), among 34 adult heterosexual males, 33 heterosexual females, 29 homosexual males and 29 homosexual females, all native speakers of Canadian English not displaying any obvious regional accent. In general, both voice pitch and formant frequencies were lower in men than in women, as expected from previous studies. With regard to sexual orientation effects, pitch did not differ between homosexual and heterosexual men, whereas it was slightly lower for homosexual than for heterosexual women, although the difference was not statistically significant. That is, lesbians are mildly gender-shifted in mean voice pitch. Formant frequencies were also similar between homosexual and heterosexual men, whereas heterosexual women showed significantly masculinised values of formant frequencies. Rendall *et al.* (2008) also analysed vowel sounds across sexes and sexual orientations; although both sexes exhibited some differences in some vowel sounds between sexual orientations, differences among women were more numerous than among men (3/10 sounds for men vs. 7/10 sounds for women).

The results of the above studies (see Table 5.5) clearly indicate that Dörner's model – that was originally proposed for brain organisation – does not have a general applicability when it comes to cognitive skills. Dörner's model would predict feminised cognitive abilities for homosexual men and masculinised cognitive abilities for homosexual women (i.e. a *Gender Inversion* pattern), but this was not what all the studies that I reviewed above found. Some cognitive skills are indeed gender discordant in homosexuals, but others are gender concordant. Sexual dimorphism has been detected in many cognitive tasks; for instance, males tend to outperform females in spatial abilities whereas the reverse tends to be true for verbal abilities. However, the pattern cannot be generalised to all kinds of verbal or spatial skills, as women do outperform men in some spatiomotor tasks such as the Purdue Pegboard (Hall & Kimura 1995). Even after considering the variability in sexual dimorphisms in cognitive tasks, the trends are variable as far as sexual orientation effect is concerned. With regard to spatial cognitive abilities homosexual men can be either feminised or gender-shifted (i.e. their scores lay between those of heterosexual males and females), gender-inverted (i.e. their scores are similar to those of heterosexual females), third sex or, as in the case of the Purdue Pegboard (Hall & Kimura 1995), there may be no effect of sexual orientation on performance. With regard to verbal skills homosexual men can be masculinised, gender-shifted or even hyperfeminised but, interestingly, nobody has ever reported homosexual men to be verbally hypermasculinised.

Most studies that included comparisons between women of different sexual orientation seem to suggest that verbal and spatial skills tend not to differ between heterosexual women and lesbians. More studies were carried out with regard to spatial skills, however, and they show some degree of variability, with some studies indicating a masculinised cognitive ability in lesbians, others suggesting hyperfeminisation, whereas still others did not detect an effect of sexual orientation. With regard to tests of verbal skills, lesbians either did not differ or were gender-shifted when compared with heterosexual women. Although data are still limited it seems that specific cognitive characteristics have a more diverse association with sexual orientation in men than women and that such diversity encompasses, but also goes beyond, the restrictive model of homosexual men being cognitively feminised males. This is consistent with a multicausal model of homosexuality in men.

Influences of sexual selection on homosexuality

Ever since the publication of Darwin's (1871) *The Descent of Man and Selection in Relation to Sex*, sexual selection has been an important focus of evolutionary research. Sexual selection, or selection for increased matings, is a major process that can explain the evolution of mating systems and secondary sexual traits across a variety of taxa, humans included. In particular, sexual selection may have affected the evolution not just of anatomical and physiological human traits, but also of aspects of our psychology (Symons 1979). In the context of this study of homosexuality, I have mentioned sexual selection through female choice for caring males (i.e. males that are more cooperative with their partner and that also provide more paternal care may be preferred sexual mates) that may lead to selection of 'feminised' males who, in turn, may also express various degrees of same-sex sexual preference (see Chapter 3). Two-way sexual selection (males selecting for specific traits in females and vice versa) was also mentioned in Chapter 4 in the context of the evolution of feminised traits in males and masculinised traits in females, and also for the potential role of neoteny in the evolution of same-sex sexual behaviours in some taxa, humans in particular. In Chapter 4 I also mentioned how sexual selection mechanisms could explain fetishisms among homosexual men. In this chapter it could be easily seen how sexual selection may be involved in the processes leading to the evolution of same sex-sexual behaviour via the mechanisms of sexual arousal, learning, emotions, cognition and preferences for masculinised or feminised traits that I reviewed in previous sections. Sexual selection will also have a prominent place in the models for the evolution of homosexuality that will be developed in Chapters 8 and 10.

In this section, I will briefly review recent developments concerning the role of sexual selection in partner choice in humans in general, what kind of effects sexual selection may have on the maintenance of same-sex sexual preferences, and whether adult homosexuals and heterosexuals of both sexes differ, and in what way, in their criteria for choice of sexual partner, apart from the obvious preference for a partner of the same or the other sex.

Humans are an altricial species and devote a great deal of energy and time to parental care over various years. Such costly parental care (parental investment) involves both males and females. On the other hand, women are more limited in their ability to reproduce (e.g. due to pregnancy and lactation) than men. Asymmetry between males and females in reproductive capability may potentially lead to the adoption of very different mating strategies in the two sexes. In particular, it is commonly believed that females are relatively choosier than males. This suggests that in a heterosexual context the choice of partner for purely sexual relations should show the highest levels of gender differences: men should tend to accept a variety of partners for sexual relationships, whereas women are expected to be relatively more selective. However, when it comes to becoming involved in a more long-term partnership, which may also include caring for children, then both men and women are expected to be highly selective, even though the criteria for selectivity may differ between the sexes: in some cases it may be wealth for women and beauty/youth for men (see, for example, Kenrick *et al.* 1990; Jankowiak *et al.* 1992; Regan *et al.* 2001; see also Silverthorne & Quinsey 2000 and references therein). An equal degree of selectivity between men and women for long-term sexual partners, but less selectivity in men than women for short-term sexual partners, is known as the *Qualified Investment Model* (Kenrick *et al.* 1990). Although these patterns of heterosexual partner choice may vary somewhat cross-culturally – and as such they do not imply that in a heterosexual context women are not as behaviourally plastic as men: being *choosy* does not mean being *inflexible*, it just means *discriminating among the available alternatives* (see the comments made in the context of Baumeister's (2000) work in a previous section of this chapter) – some of the preferences (e.g. having *good financial prospects*) are largely conserved across cultures (Buss *et al.* 1990).

Some physical traits have greater importance than others in mate choice. In most cultures, for instance, the face is the most continuously exposed body part that is both complex and variable enough to convey subtle and reliable (*sensu* Zahavi 1975) information regarding some characteristics of the individual. Such information could then be used in the process of partner choice. Faces convey information about men and women regarding developmental processes, health, nutritional status, endocrine activity, age and many other characteristics. Although some trends are relatively well established, e.g. men prefer younger-looking women, whereas women prefer similar-age-looking or older-looking men (Buss 1989), the response that some of those traits elicit from a potential partner may be also culturally variable. For instance, some ethnic groups may prefer 'feminised' faces in males, whereas other may prefer 'masculinised' faces (Johnston 2006). In other circumstances, the preference may be based on the level of bilateral symmetry in the faces (for example, Langlois & Roggman's 1990 result, showing preference for 'averaged' faces, was probably an outcome of increased symmetry).

Gillian Rhodes (2006) has recently reviewed the many studies carried out on human sexual preference based on face characteristics. She identifies three major sexually selected preferences as far as human faces are concerned: (a) preference for bilateral symmetry, (b) preference for *averageness* (usually produced in these studies by computer-blending characteristics of faces from various individuals), and (c) preference for sexual dimorphism ('feminine'-looking faces in females and 'masculine'-looking faces in males). As mentioned above, the motivation for being attracted to a specific kind of face may also vary: from sexual, to partnership in coalitions, but also quality of the potential partner as caregiver. Rhodes carried out meta-analyses to test for those three (a–c) sexually selected preferences in humans and found that sexual partner preference increases with degree of facial averageness and symmetry. With regard to sexual dimorphism, she found that males prefer females with more 'feminine' faces, and females prefer males with more 'masculine' faces, although this result followed only after natural faces, rather than blended faces, were used.

In sum, heterosexual humans clearly display traits that both males and females use in their choice of a sexual and child-rearing partner. Moreover, those same traits could also be used for social partner choice.

If sexual selection has occurred in the heterosexual context, what might its role be in same-sex sexual partner choice? Do homosexuals follow the heterosexual patterns of mate choice that are characteristic of their sex but applied to same-sex partners? Do they show 'inverted' criteria to evaluate traits for their choice of partner? Do they use completely different ('third-sex') criteria? How does this vary between males and females? The rest of this section will be devoted to addressing these issues.

Donald Symons (1979) suggested that sexual behaviour is typical of males for homosexual men and typical of females for homosexual women, it is only the sex of the targeted partner that is obviously different between homosexuals and heterosexuals of each sex. I will call this the *Symons model*. That is, as far as their broad sexual behaviour is concerned, including the criteria used to choose among alternative partners of the appropriate sex, the *Symons model* predicts that homosexuals and heterosexuals are consistent within each sex. Homosexual men are expected to have many partners, especially in short-term relationships, to focus on sexual activity and orgasm, and physical attractiveness of partners and to prefer younger partners, although they are also expected to express a degree of choosiness as their heterosexual counterparts do. Homosexual women are expected to prefer more stable relationships and emphasise social bonds, but also to enjoy sexual activities and orgasm as heterosexual women do, although women of any sexual orientation are expected to emphasise hedonistic aspects of the relationship less than men (Symons 1979).

Empirical tests of the *Symons model* have revealed a greater level of complexity than first suspected. For instance, Jankowiak *et al.* (1992) showed that, in accordance with the *Symons model*, in a choice among various photographs, heterosexual men preferred younger women and homosexual men preferred younger men. However, although heterosexual women preferred older men, homosexual women preferred younger women. Bailey *et al.* (1994) carried out a study of mating psychology of men and women across sexual orientations. As predicted by sexual selection theory, men have more interest in uncommitted sex and visual sexual stimuli, give less importance to partner status, are more sexually than emotionally jealous, prefer younger partners, give greater importance to partner physical attractiveness and emphasise the frequency of sexual behaviour more than women (see also Buunk *et al.* 2002). These between-sex differences are larger than differences between sexual orientations within each sex, which *prima facie* seems to support the *Symons model*. However, for women, two statistically significant results were obtained: homosexual women (a) show greater interest in visual stimuli and (b) give less importance to partner's status than heterosexual women. That is, the statistically significant results indicate a pattern of female-homosexual masculinisation in the criteria for partner choice, not feminisation as Symons would have predicted. Two statistically significant results were obtained in the comparisons between homosexual and heterosexual men: (a) homosexuals give more emphasis to frequency of sexual behaviour and are more *emotionally jealous* than heterosexual men, whereas the latter (b) tend to have a greater preference for a younger partner and have a greater degree of *sexual jealousy* than homosexual men. That is, homosexual men show both patterns of 'hypermasculinisation' (frequency of sexual behaviour) and feminisation (emotional jealousy), again indicating that they are not following a consistently masculinised pattern of choice. A similar mixture of patterns, in some regards consistent and in some others inconsistent with the expectations from heterosexual individuals of the same sex, were obtained by Regan *et al.* (2001) in a questionnaire study of homosexual men and women.

In a recent study where homosexual and heterosexual male and female participants were asked to judge still photographs and a video showing individuals of both sexes displaying typical feminine or masculine patterns, lesbians preferred more feminine women, thus displaying the typical pattern of heterosexual men (*gender inversion*) (Rieger *et al.* 2009a). Gay men showed no particular preference for either feminised or masculinised males, a pattern that is also sex-atypical and consistent with heterosexual women's choices (*gender inversion*) in this study. Therefore the results of Rieger *et al.* run counter to the *Symons model*.

On the other hand, some data in support of the *Symons model* are provided by Sergios & Cody (1986), who showed that male homosexuals put great emphasis on the physical attractiveness of their sexual partner. In a study carried out in Canada, Silverthorne & Quinsey (2000) showed that males always preferred younger partners irrespective of their sexual orientation, and females preferred older partners, also irrespective of their sexual orientation. The tendency for heterosexual women to prefer older men is also matched by a similar preference of homosexual women to prefer older women in the study by Kenrick *et al.* (1990), but only at younger ages, at older ages homosexual women invert their preference (younger partners preferred), showing a masculinised pattern of choice. The same change in preferred age of sexual partner as homosexual women mature was obtained by Hayes (1995). However, for males Hayes' work does support the *Symons model*: older gay men preferred younger sexual partners. Additional support for the *Symons model* comes from other studies of heterosexuals and homosexuals carried out by Freund *et al.* (1973) on aversion to nude photographs and Harris (2002) on jealousy, although a study by Dijkstra *et al.* (2001), also on jealousy, did not support the model.

More recently, Gobrogge *et al.* (2007) have carried out a study of sexual preferences between homosexual and heterosexual males using advertisements from a website that attracted individuals from the USA and also Canada. Although homosexual men tended to have a greater tendency to place advertisements for purely sexual encounters (39%) than heterosexual men (27%) and the latter tended to place more advertisements seeking longer-term relationships (73%) than homosexual men (61%), both homosexual and heterosexual men become highly selective regarding the age range of their potential partners when they are seeking long-term relationships, whereas both increase the age range of potential partners when seeking sexual encounters; these results are expected from the *Symons model*.

In sum, although some studies reject the *Symons model* of sex-typical choice of mate characteristics within each sexual orientation, other studies support the model. One of the variables, among many others, that could possibly affect the outcome of results in these studies is the age of the respondent. In fact, when age is considered patterns of choice within a sexual orientation may vary, especially in women. That is, some homosexual women, especially older ones, do show a 'gender-inverted' pattern of sexual partner choice. Among men, patterns are also variable and indicative of at least two partner choice strategies among homosexuals: (a) one that follows a masculinised (or even hypermasculinised) pattern consistent with the *Symons model* and (b) another that is more consistent with 'gender inversion'.

Most of the above studies were carried out on static signals (e.g. facial traits as observed in photographs or fixed verbal messages as they appear in advertisements), but humans convey non-verbal information about themselves in more dynamic ways as well. For instance, information about a person may be obtained from brief instances of expressive behaviour, known as *thin slices* of behaviour (Ambady & Rosenthal 1992; Ambady *et al.* 1999, 2000). Ambady *et al.* (1999) found that observations of behaviour conveyed information, additional to the static sources of information, about the sexual orientation of a person. Eye-gaze is one such behaviour. Cheryl Nicholas (2004) introduced the concept of 'Gaydar', a term originally coined by the gay and lesbian community, to describe those *thin slices* of behaviour such as eye-gaze that operate as 'gay identity recognition devices'. Using *gaydar* one individual is supposed to be better able to distinguish between a homosexual and a heterosexual person in an encounter. However, as already noted by Rieger *et al.* (2010), the use of eye-gaze as a communication device directed to a potential sexual partner is common to both homosexuals and heterosexuals, and therefore *gaydar* may only require the use of broad communicative abilities, common to all sexual orientations, but targeted to the appropriate sex, a mechanism consistent with the *Symons model*. However, *thin slices* of behaviour, such as particular movements, gesticulations, positions of parts of the body, and others, specific to a given sexual orientation could be even more effective *gaydar* signals (see, for example, Rieger *et al.* 2010). In addition, once *thin slices* of communicative behaviour subject to potential detection by any recipient are in place, the possibility remains open for a homosexual person to conceal such signals in situations where the chances of attracting a partner are low but those of attracting discrimination are high (Sylva *et al.* 2010).

On the basis of the topics reviewed in the above sections dealing with nervous system control of same-sex sexual behaviour, we could in principle conclude that homosexual preferences could be achieved by at least three broad and somewhat independent proximate neuroendocrine mechanisms:

(a) Similarity between homosexual and heterosexual individuals in the brain centres controlling sexual arousal, pleasure (see the previous section 'Orgasm, sexual arousal and homosexuality' in this chapter), romantic love, attachment, cognition and identification of secondary sexual traits, but differences in the centres (e.g. peripheral sensory tissues: visual, olfactory, auditory) identifying a target individual of a specific sex as a potential 'sexual partner'. This mechanism requires an important degree of critical information processing at the sensory level (Lettvin *et al.* 1959) that determines sexual partner preference (see also Martin-Alguacil *et al.* 2006 for an emphasis on sensory-level processing).

(b) Environmental inputs are processed similarly in homosexuals and heterosexuals at the sensory level, but their integration in brain centres that produce sexual responses are different. This may result from sexual differences in central neuroanatomy (e.g. size of specific nuclei) and/or neurophysiology (e.g. differential neurotransmitter production or expression of neurotransmitter receptors).

These two mechanisms – (a) and (b) – are what we may call 'more static' mechanisms, that simply require neuroendocrinological processes and structures that are established during early ontogeny and are fully operational during the reproductive ages of the individual. However,

(c) inter-individual modifications of both peripheral sensory information processing and brain neuroactivity may also occur at the adult stage of development as a result of brain plasticity (a 'more dynamic' mechanism).

Variability in those three mechanisms may at least partly explain, at a proximate level, the variability in sexual orientation patterns observed within and between sexes.

Brain, complex social behaviour and homosexuality

I conclude this chapter on the neuroendocrine system and homosexual behaviour in adult individuals with a broader inquiry into the evolution of the brain. In particular, I ask in which specific ways increased brain size and complexity in birds and mammals may have favoured the evolutionary emergence of homosexual behaviour. Before I address this specific question, however, I will briefly review our current knowledge of brain evolution and its association with complex and plastic behaviour more broadly.

In general, the evolution of the brain among vertebrates has been the subject of the same kind of general dynamics that have affected the evolution of other bodily structures and functions, with natural selection likely to have been one of the major mechanisms implicated in brain evolution, although not the only one. This, of course, requires that brain structure (including size) and function be heritable. There is mounting evidence suggesting heritability of some aspects of brain organisation in humans coming from very detailed twin studies that have used functional magnetic resonance imaging (fMRI), positron emission tomography (PET), electroencephalography (EEG) and magnetic resonance imaging (MRI) techniques (see Thompson *et al.* 2002 for a review). Such studies have reported realized heritability (h^2) estimates of 0.90–0.95 for the middle frontal regions of the brain near Brodmann areas 9 and 46 (Thompson *et al.* 2001), 0.94 (Bartley *et al.* 1997) to 0.66 (Wright *et al.* 2002) for total brain volume, 0.80 for the corpus callosum (Sullivan *et al.* 2001, cited in Thompson *et al.* 2002), and 0.88 for cerebellar volume (Posthuma *et al.* 2000). Paul Thompson and collaborators at the Department of Neurology, UCLA, have provided some startling evidence of similarity in grey matter distribution in the human isocortex consistent with genetic control of at least the frontal cortex and those cortical areas involved in language (Thompson *et al.* 2001; see Figure 5.17). Candidate genes involved in brain development have been identified (see Toga & Thompson 2005 for a review) such as the brain-derived neurotrophic factor (*BDNF*) gene and the catechol-*o*-methyltransferase (*COMT*) gene. The *BDNF* gene plays an important role in synaptic plasticity (see, for example, Soulé *et al.* 2006), whereas the COMT gene plays a pivotal role in dopamine metabolism (see, for example, Malhira *et al.* 2002).

Convergence is a recurrent pattern in brain evolution (Nishikawa 1997), suggesting that compartments of the brain may evolve in concordance with each other. This idea has been encapsulated in

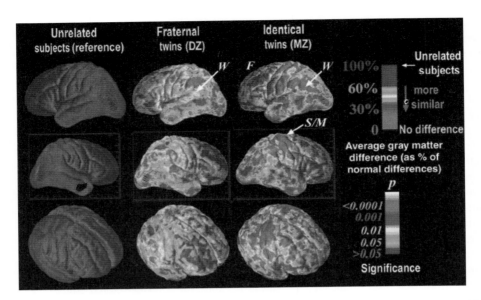

Figure 5.17. Differences in the quantity of grey matter at each region of the cortex were computed for identical and fraternal twins, averaged and compared with the average differences that would be found between pairs of randomly selected, unrelated individuals (left). Colour-coded maps show the percentage reduction in intrapair variance for each cortical region. Fraternal twins are less similar to each other than genetically identical twins in a large anatomical band spanning frontal (F), sensorimotor (S/M) and Wernicke's (W) language cortices, suggesting strong genetic control of brain structure in these regions (the significance of these effects is shown on the same grey scale). From Thompson *et al.* (2001). See also colour plate.

what is known as the *Developmental Constraints* theory (Finlay & Darlington 1995). On the other hand, in many instances different compartments (e.g. areas, nuclei) may have evolved at different rates in different lineages, thus giving rise to specific adaptations in each of those lineages, even in cases where there may be a common evolutionary trend towards an overall increase in brain size over time. This is known as the *Mosaic* or *Modular* theory of brain evolution (Barton & Harvey 2000; Iwaniuk *et al.* 2004; Lefebvre *et al.* 2004). There is no reason to believe that only one of those two theories can apply to all cases of brain evolution. They are both plausible and chances are that some lineages may show clear evidence of mosaic brain evolution (see, for example, Iwaniuk *et al.* 2004), whereas others may have undergone brain evolution that is better explained by developmental constraints (Finlay & Darlington 1995). On the other hand, the nature of brain tissues is such that changes in neuronal interconnectivity, cellular receptors and ion channel distribution on the cell membrane, and distribution and abundance of neurotransmitters that may be undetected in comparisons of gross brain morphology (e.g. total volume), may none the less be responsible for major changes in behaviour and therefore major modifications in the adaptive evolution of the lineage (Nishikawa 1997).

The isocortex, also known as the neocortex (Northcutt & Kaas 1995), is one of those brain areas that shows not only increase in size following increase in body size within a lineage (e.g. primates), but also lineage-specific rates of increase that are not expected from body size variation (e.g. within the genus *Homo*, Nishikawa 1997). This again suggests that both developmental constraints and mosaic evolution may have played a role in brain evolution. The isocortex has evolved independently in various groups of mammals (Northcutt & Kaas 1995), suggesting that behaviours that may be, at least in part, under isocortical influence, such as same-sex sexual behaviour, may or may not share common neurological mechanisms or even common adaptive functions across taxa, a concept that has been already stressed in this chapter and will continue to be stressed throughout the rest of the book. Convergences of function (i.e. analogies) may and indeed have occurred, homologies have also been found in both structures and function, but divergences have occurred too (see Lefebvre et al. 2004 for a recent review).

That the brain seems to have evolved under the effect of natural selection is strongly suggested by comparative studies of the functional evolution of the vertebrate brain indicating:

(a) loss or reduction of olfactory neurocentres in mammals living in aquatic environments (Johnson et al. 1994);
(b) adaptive evolution of main olfactory bulb, hippocampus and auditory nuclei associated with specific foraging habits (Hutcheon et al. 2002);
(c) increase in brain size (corrected for body mass) with level of altriciality in birds (Iwaniuk & Nelson 2003); and
(d) decreased forebrains in long-distance migratory birds (Winkler et al. 2004).
  After adjusting for the effect of body mass,
(e) gestation length is associated with larger brains in ungulates (Pérez-Barbería & Gordon 2005), whereas
(f) complexity of ecological niche is positively correlated with relative size of the isocortex in mammals (Jolicoeur et al. 1984).

Therefore brain size and complexity are correlated with the complexity of the selective environmental challenges faced by the organism, a trend that is clearly expected from the adaptive role of a large and complex brain. In fact, Lefebvre et al. (2004) point out how areas of the brain that are involved in complex integration of information, such as the isocortex and striatum in primates and the hyperstriatum ventrale and neostriatum in birds, tend to have greater size in species that display high frequency of production of novel behaviours. Similar results had been obtained by Keverne et al. (1996) and Reader & Laland (2002) in primates; see Reader (2003) for a review. The ability to produce novel behaviours is likely to be very adaptive in a complex and unpredictable environment. If the ability to produce novel behaviours is genetically heritable (Reader 2003), then it will be a trait that could be transmitted intergenerationally and it could also be subject to natural selection.

In species that live in more or less permanent groups, chances increase for complex interindividual relationships to be established: cooperation, alliances, competition, dominance, deception, manipulation, reciprocation, communication, and so forth. Such a complex social environment may have provided the selective framework for the evolution of large and complex brains, this is known as the *Social Intelligence* hypothesis (Byrne & Whiten 1988). The *Social Intelligence* hypothesis has been supported by comparative analyses carried out in primates (Barton 1996; Joffe & Dunbar 1997; Barton 1998), carnivores (Gittleman 1986) and ungulates (Pérez-Barbería & Gordon 2005), although the specific evolutionary dynamics linking sociality and evolution of brain size are likely to differ among those taxa (Pérez-Barbería et al. 2007b). The hypothesis was also supported by comparative analyses carried out in birds. Burish et al. (2004) have recently shown that the degree of social complexity across avian taxa is positively correlated with relative (to total brain volume) size of the telencephalon (forebrain).

As soon as a trait such as a large and complex brain has evolved under the selective pressure of a complex social environment or a variable and unpredictable habitat, or both, the same trait, through its manifested phenotypic expressions – behaviour in this case – could also be subject to sexual selection. For instance, increased social and foraging skills afforded by a large and complex brain could be traits selected by a potential sexual partner in her (but presumably also his) mate choice. In fact, Pawłowski et al. (1998) found that, using a comparative analysis, in primate species with larger relative volume of the isocortex mating success of males decreases as a result of shared paternity. Pawłowski et al. interpret this result as indicating increased conflict among males in group-living species possessing a large and complex brain or, alternatively, an increased tendency to form alliances and pay for it through shared paternity. Both of these are perfectly reasonable possibilities; however, the authors seem to have overlooked the additional effect of female mate choice: 'smart' females may well be selecting males with diverse behavioural capabilities (skills, personality, etc.) as the fathers of their various offspring even in circumstances where a dominant male is capable of controlling copulations (e.g. through aggression and mate guarding) most, but not all, of the time.

Returning to our main topic of interest: is homosexual behaviour a trait that might have emerged from the evolution of a complex brain, and if so, how? Obviously, the performance of same-sex mounting does not necessarily require a particularly complex brain, as exemplified by the various case studies of invertebrates that were mentioned in Chapter 3. However, homosexual behaviour due to 'mistaken identity', as often occurs in invertebrates, is certainly not what characterises same-sex mounting in most birds and mammals, primates in particular. We have seen in this section how evolutionary trends towards a larger and more complex brain are associated with group living and the complexities of social life. Such complexities involve not only behaviours that fall within the competition and cooperation rubrics (e.g. alliances, deception, etc.), but also those that may be relevant to sexual selection, particularly when it involves mate choice. In addition, from an ontogenetic point of view, a large and complex brain capable of both *producing* and adaptively *responding to* a complex social environment is more likely to be developed in altricial rather than precocial species (Iwaniuk & Nelson 2003). The slow development typical of altricial species may in turn be subject to further modification by neotenic processes (see Chapter 4). The mutual influence of social complexity and brain complexity through manifest behaviours may have driven the neotenic processes that have affected our and other species, with sexual selection playing a role of accelerator of such processes. In this evolutionary model, same-sex sexual behaviour can emerge at more than one stage in association with a complex central nervous sytem. For instance,

(a) same-sex mounting may be a behaviour originally selected in the context of socio-sexual functions (dominance–cooperation) in a complex society, a process that could have been facilitated by a complex brain;
(b) same-sex mounting may be a behaviour that emerges in social animals early in their ontogeny and that may be maintained in adult life as a result of neotenic processes that are driving the evolution of a complex brain;
(c) once a complex and plastic brain has evolved, animals may retain a degree of flexibility in their behaviour, including sexual behaviour, that allows them to engage in same-sex sexual interactions under specific circumstances that can vary from individual to individual and from time to time (e.g. *political lesbians* in humans).

All these possibilities are consistent with basic models of genetic heritability of the homosexual trait (because of heritability of some aspects of brain structure, see above) but also with the emergence of less canalised ontogenetic processes leading to the expression of same-sex sexual behaviours that involve plastic properties of the brain. Those evolutionary mechanisms will be synthesised, along with the proximate mechanisms listed at the end of the previous section, into an *integrationist biosocial model* that will be introduced in Chapter 10. The model will summarise the various causal pathways involving both proximate (e.g. developmental) and more ultimate (e.g. evolutionary) mechanisms that alone or in various combinations may explain the emergence of same-sex sexual behaviour across taxa.

Summary of main conclusions

- Studies carried out on human endocrine syndromes indicate that excesses of circulating steroids may have a greater developmental effect on sexual orientation and gender role than deficits.
- Activation of same-sex sexual behaviour in adult individuals is under the dual control of sexual hormones (e.g. gonadal steroids) and direct neurological mechanisms that are especially evident during the mating periods of the year.
- Depending on the species, in adult social mammals same-sex sexual behaviour may have various socio-sexual functions associated with dominance and cooperation, but also coping with stress under conditions of social conflict.
- Both oxytocin and arginine vasopressin are involved in the control of social bonding and sexual behaviour and their link with the dopamine reward system may explain learning effects in the establishment of same-sex sexual relationships.
- A *Lock-in Model* is proposed to explain some aspects of same-sex mounting in female macaques.
- Traditionally, circulating levels of sex hormones have not been associated with sexual orientation in adult humans. However, a more critical analysis of the available evidence suggests that in cases where there is a difference, a consistent trend was described for higher levels of circulating oestradiol, luteinising hormone, prolactin, androstenedione and cortisol in homosexual than heterosexual men, with concentrations of circulating testosterone being less predictable between sexual orientation categories in men. Women, on the other hand, show more predictably higher levels of circulating testosterone in lesbians and masculinised females than in heterosexual and feminised females.
- The case of homosexual *Ovis aries* rams first described by Hulet *et al.* (1964) and then characterised by Zenchak *et al.* (1981) represents one of the few examples of exclusive homosexuality known in non-human animals. The available evidence suggests that homosexual rams are masculinised males that respond sexually to stimuli (e.g. pheromonal) coming from other males rather than from females.
- Variable sexual partner preference may reflect variability in brain structure and physiology in terms of: (a) establishment of a diversity of alternative neuronal circuits during pre-adult life that are then differentially activated or inhibited; (b) actual change of the brain circuitry due to neurogenesis and apoptosis during adult life; (c) actual change of the brain circuitry due to synaptogenesis in adult life. Mechanisms (a), (b) and (c) are not mutually exclusive and together may account for the observed plasticity of sexual orientation in adult humans, women in particular.
- Some regions of the brain differ (e.g. in size) between adult males and females, but not all regions of the brain that are involved in the control of sexual behaviour are feminised in male homosexuals and masculinised in female homosexuals.
- Sexual orientation also affects brain activation by putative sex pheromones in humans. In general, male and female homosexuals tend to show a response to putative human pheromones that is typical of heterosexuals of the opposite sex, but the pattern is stronger for male than female homosexuals.
- Hypersexuality controlled by the amygdala and the temporal lobe may also mediate the expression of same-sex sexual behaviour in social situations of a biased sex ratio.
- Dopamine activates sexual behaviour through its action in the amygdala and it is also involved in the promotion of social bonding, while at the same time, it mediates learning processes through the reward system controlled by the nucleus accumbens. In this way changes in dopamine levels may affect learning, leading to the expression of homosexual (not just homosocial) bonds. Serotonin is also involved in the processes of learning that affect the expression of homosexual sexual behaviour.
- In general, sexual arousal may not necessarily require brain centres that are specific to a sexual

orientation. A similar diversity of centres tends to be activated in individuals of different sexual orientations provided that they are exposed to a relevant sexual stimulus. In women, however, sexual arousability tends to be relatively less specific than in men.

- Sexual arousal may eventually culminate in orgasm. At orgasm, adrenaline, noradrenaline, oxytocin and, more slowly, also prolactin, increase in circulation. Orgasm per se is not necessarily associated with or conducive to homosexuality.
- Masturbation is mainly an outcome of sexual arousal expressed in circumstances of lack of access to a willing sexual partner of any kind. There is no proven causal link that masturbation leads to the development of a homosexual sexual orientation in any bird or mammal species, including humans.
- Learning, through the mechanisms of motivation, incentive and reinforcement of sexual partner preference, contributes to a partial explanation of the within- and between–culture variability in homosexuality in humans. Moreover, masculine and feminine gender roles – which are independent from homosexual, heterosexual and bisexual sexual orientations – can also vary cross-culturally as a result of the combined effects of experience, development and perhaps genetic predispositions. Different gender roles can be also combined to different degrees in the same person, with the extent of behavioural androgyny being also variable cross-culturally. Through learning, gender roles may be moulded by individuals of various sexual orientations in order to adapt themselves to a current cultural or subcultural stereotype.
- Women display greater sexual orientation plasticity than men, both between individuals and within an individual over his/her lifetime.
- The emotion–motivation systems of attraction, lust and attachment sustain the basic functions of reproduction, parenting and mating in mammals and presumably other vertebrates. Centres controlling attraction, lust and attachment need not differ, necessarily, between homosexuals and heterosexuals except in those specific aspects that determine the sex of the preferred sexual partner. Attachment may also have additional functions in coping with stress. Here, androgynous personalities may be better endowed with the emotional tools to withstand the effects of stressful experiences.
- Cognitively, men show a more diverse association between sexual orientation and both verbal and spatial skills than women. Overall, however, both homosexual men and homosexual women **do not** show a clear-cut gender inversion pattern in their cognitive capabilities.
- Patterns of sexual partner choice among homosexuals and heterosexuals do reflect the action of sexual selection processes. Such patterns are consistent both with sex-typical choices and with specificities associated with sexual orientation, especially with regard to the age of preferred partner among homosexual women.
- The evolution of a large and complex brain may result from processes of natural and sexual selection favoured by neoteny and the complexities of social living. This evolutionary framework is the basis to understanding the use of same-sex sexual behaviours in sexual and socio-sexual contexts in social birds and mammals.

So far I have emphasised the interactions occurring among genetic (Chapter 3), endocrinological and neurological mechanisms during early (Chapter 4) or adult (this chapter) ontogeny in the causation of homosexual behaviour. One crucial piece of information that is still missing, however, is the potential role that the immune system may play. It is well known that components and functions of the immune system influence and are influenced by both endocrine and neurological factors (Solomon 1987), so much so that it would be most appropriate to speak of a neuroimmunoendocrinological system rather than of three separate systems. The next chapter critically reviews the idea that the immune system may affect the development of a homosexual sexual orientation in humans.

6

Immunology and homosexuality

In placental mammals, offspring develop in the maternal womb for the initial stages of their life. This implies a risk that the foetus will be recognised as a foreign body by the maternal immune system. For instance, foetal expression of molecules encoded in the genes of paternal origin, which may be different from maternal molecules, may trigger an immune response from the mother. Some specific targets of maternal immunity include products of paternal major histocompatibility (MHC) genes that can be expressed by the outer layer of cells (the trophoblast) surrounding the blastocyst. In fact, foetal MHC molecules are major antigens (Bainbridge 2000). Other important antigens of foetal origin that are paternally inherited, such as human leukocyte antigens (HLA), are responsible for the production of specific immunoglobulin G (IgG) antibodies in about 30% of mothers in humans (Luppi 2003; see also Goulmy *et al.* 1977; Gangestad *et al.* 1996).

A maternal immune reaction against her own offspring may lead, in the most extreme cases, to abortion (Thellin & Heinen 2003). Some major maternal immune mechanisms leading to rejection of the foetus include T helper cells 1 and 2 imbalance (Th1/Th2) favouring Th1 cells, which produce specific cytokines, e.g. tumour necrosis factor alpha (TNFα), interleukin 2 (IL2) and gamma interferon (IFNγ), among others. Such Th1 cytokines can then activate natural killer (NK) cells and inflammatory macrophages, which, in turn, can kill other cells, those of the embryo in this case. Cells of the trophoblast can also increase their production of the pro-thrombinase fgl2 as a result of Th1 cytokine action, with fgl2 having the ability to produce a thrombosis in the placenta, thus shutting down blood flow, which may ultimately lead to loss of the foetus (see Thellin & Heinen 2003 for additional mechanisms also leading to foetal rejection). If, on the other hand, the Th1/Th2 imbalance favours Th2, then the pregnancy is carried on normally (Szekeres-Bartho 2002). Whenever the foetus is capable of surviving a maternal immune attack, chances are that the newborn may display a phenotype that reflects the results of that attack during its early development. In particular, when the immune reaction targets foetal brain tissues, specific changes in the offspring's behaviour may be expected (Gualtieri & Hicks 1985; Ader *et al.* 1990).

The central nervous system enjoys a substantial degree of *immune privilege*, meaning that cells of the brain are sheltered from contact or from the consequences of contact with components of the immune system (see, for example, Niederkorn 2006). For instance, both neurons and glial cells produce a cytokine Fas ligand (FasL) that interferes with inflammation processes caused by immune reactions. The brain is also capable of inhibiting the activity of Th cells. In addition, maternal pregnancy-associated glycoproteins (PAG) can interfere with MHC functions (Roberts *et al.* 1996), whereas local maternal secretion of corticosteroids may help dampen immune responses (Priddy 1997), all of which may contribute to defending the foetal brain tissues.

Maternal immune response to paternal antigens may also be prevented, as would be expected from natural selection, by paternal action. Such paternal inhibitory effects on maternal immune activity include suppressing the production of antigen-presenting cell subsets, but also introducing immunomodulatory factors such as cytokines and progesterone with the seminal fluid (see Seavey & Mosmann 2006; Poiani 2006 for reviews). Moreover, the foetus also has its own immune system that can be activated, for example, in preparation for delivery or in response to the presence of pathogens in the amniotic fluid (Berry et al. 1995) and that could protect the developing organism from external immune attack. Such defence mechanisms, however, are not necessarily perfect and when they fail foetal tissues such as those of the brain may indeed be exposed to maternal immune action and therefore undergo changes, with potential consequences for behaviour.

If a maternal immunity effect exists, we would predict that mortality will be higher in male than female foetuses, that surviving male offspring are more likely to develop an exclusive homosexual sexual orientation than females, and that, if the prenatal immunity effect is a heritable condition in the mother, homosexuals should be especially frequent in extended families among brothers and first cousins through the maternal line.

Just over twenty years ago, Gualtieri & Hicks (1985) suggested an *Immunoreactive* theory for the higher levels of developmental disorders observed in male than in female humans, whereas female humans are relatively more susceptible to autoimmune disorders (see also Jacobson *et al.* 1997). They proposed that maternal immunoreaction against male offspring would be able to explain, not only male-biased developmental disorders due to the higher immunogenicity of male (XY) than female (XX) foetuses, but also the *parity effect*; i.e. increased likelihood of developmental disorders with birth order (e.g. Lalumière *et al.* 1999), this being a result of the sensitisation process undergone by the maternal immune system with subsequent pregnancies. The *Fraternal Birth Order* (FBO) effect that will be analysed below is a special case of the *parity effect*.

Gualtieri & Hicks hypothesised that the *parity effect* may be mediated by the H-Y antigen: a minor histocompatibility complex protein encoded in the Y chromosome that is implicated in the differentiation of the embryonic gonadal anlage into testes (see Müller 1996 for a review). H-Y antigen mediation of the *parity effect* was hypothesised by Gualtieri and Hicks to proceed in the following manner:

(a) H-Y antigenicity would be associated with an increase in the size of the placenta,
(b) which would increase the development of the firstborn,
(c) this would lead to an increase in maternal immunity against H-Y,
(d) which would affect the development of subsequent male children in particular (see also Singh & Verma 1987).

Gualtieri and Hicks supported their model with data showing patterns consistent with the FBO effect (which they actually called the *antecedent brother* effect). The FBO effect was tested by Flannery & Liederman (1994) using a sample of 11 578 mother–child pairs; the children were followed up until the age of 7 years and the distribution of neurodevelopmental disorders noted. What the authors found was that only one of the neurodevelopmental disorders that they studied showed a clear FBO effect: mental retardation. This suggests that even though the mechanism may not be a general explanation for various developmental conditions of the central nervous system, some aspects of the development of behaviour may well be explained by the *Immunoreactive* theory.

Immunity and Fraternal Birth Order effect in the development of a homosexual orientation in males

Eliot Slater (1962) was the first to explicitly point out an association between male homosexuality and late birth order within a sibship. In fact, he had proposed the idea for the first time at a symposium four years before in 1958 (Slater 1962). Slater introduced the so-called *Slater's Index* to measure the FBO effect (number of proband's older brothers/number of all his brothers). On the other hand, Blanchard *et al.* (1996) and LeVay (1996) trace back to Theo Lang (1940) the initial finding that homosexual men tend to live in families with a male-biased sex ratio among siblings. Almost twenty years following Slater's initial insight, Malcolm MacCulloch & John Waddington (1981) proposed an explicit immune theory (*Maternal Antibody* theory) for the development of a homosexual orientation, which attempted to explain both male and female human homosexuality and also the FBO effect in the incidence of homosexuality. In their view, maternal immune attack would affect second- and subsequent-born children in healthy mothers as a consequence of increased immune sensitisation following the first pregnancy. According to MacCulloch & Waddington (1981), some mothers may produce abnormally elevated titres of antitestosterone antibodies, thus leading to higher production of homosexual male offspring, or of antiprogesterone antibodies, thus leading to production of homosexual female offspring. In addition, MacCulloch and Waddington postulated an immunological mechanism, acting at the level of the adult individual, in which homosexuals would produce elevated titres of antitestosterone (males) or antiprogesterone (females) autoantibodies.

The general idea of an immunological mechanism determining human homosexuality was subsequently taken up by Ray Blanchard and colleagues, and by others as well, and developed into a currently very active research programme testing what it is now known as the *Maternal Immune* hypothesis (MIH) (Blanchard & Klassen 1997). The current focus is mainly on characterising the FBO effect observed among male homosexuals, specific empirical studies focusing on the postulated immunological mechanisms are still in their infancy, as will be mentioned in this chapter.

A specific mechanism of prenatal immune interactions leading to the development of a homosexual sexual orientation was proposed by Blanchard & Klassen (1997), who suggested that the FBO effect observed among homosexual men could be explained by the H-Y immune mechanisms that may mediate the *parity effect* (Gualtieri & Hicks 1985). This specific H-Y mechanism, however, can only explain male, not female, homosexuality.

What is the evidence for the FBO effect in male homosexuality? After controlling for parental age, Blanchard & Bogaert (1996) demonstrated an FBO effect in a Canadian sample of homosexual males and suggested a potentially explanatory role for immunity but also for genetic inheritance. Their estimates indicate that the effect of older brothers on the development of a homosexual orientation in younger brothers is frequency dependent, with the odds of being homosexual increasing by 33% for each additional older brother. A similar result of about 38% was obtained by Bogaert (2003b), although Ellis & Blanchard (2001) reported a slightly smaller value of 28%. Blanchard & Bogaert (2004) calculated the probability of development of a homosexual orientation due specifically to FBO at 28.6%, whereas Cantor *et al.* (2002) estimated it to be in the range between 14.8% and 15.2%. Blanchard *et al.* (1996) also studied this effect in a sample of Dutch homosexual males. Among homosexual probands, the proportion of males in the sibship was 0.57 whereas among non-homosexual (i.e. heterosexual and bisexual) probands the proportion of males in the sibship was 0.53. The expected proportion of males in the Dutch population was 0.51, which did not differ from the proportion found among non-homosexuals but was significantly different from that found in homosexuals (Blanchard *et al.* 1996). This effect could not be explained by differences in the total number of siblings between homosexual and non-homosexual probands as the average sibship size was similar, the difference was more in the age distribution of sibs around the age of the proband.

Blanchard *et al.* (1998) analysed British and US data and found that homosexual men tended to be younger in their sibship than heterosexual men; moreover, homosexual men also tended to come from male-biased sibships. The effect size, calculated as the mean value of the number of older brothers for the homosexual men included in the analyses minus the mean value of older brothers for heterosexual men divided by the pooled standard deviation of those two means, found in their data by Blanchard et al. (1998) was relatively small: 0.26.

After analysing data from 97 gay men sampled in the south-eastern USA, Purcell et al. (2000) found that the sex ratio among siblings in the case of gay men (0.56) was more male-biased than the ratio expected from the population at large (0.51), with gay men being more likely to be younger siblings than was expected from a random distribution; this effect was specific for male rather than female sibships. Bogaert (2000) analysed data available in the *National Health and Social Life Survey* of the USA (Laumann et al. 1994) and found that, in accordance with the FBO effect, the older brother tended to be more heterosexual, whereas the likelihood of developing a same-sex sexual orientation increased in later-born brothers. More recently, Côté et al. (2002) carried out a study of male inmates in a Canadian prison who had committed at least one offence of a sexual nature and found that the values of their Most Deviant Index score were best predicted by the number of older brothers, the values of the score being also positively correlated with the length of the birth interval between probands and their next younger brother. Bogaert (2003b) also analysed data from a British national probability sample to conclude that the number of older brothers was a good predictor of the development of a same-sex sexual attraction.

Number of older siblings was also positively associated with probability of homosexuality in a recent work that analysed a Canadian sample (Bogaert & Liu 2006). King et al. (2005), however, after carrying out a study of the effect of family size on male sexual orientation in England, concluded that although their sample of homosexual men had more brothers they also had more sisters than heterosexual men, with the effect of each kind of sibling being independent of the other. Moreover, King et al. (2005) also point out that gay men tend to come from families that have larger numbers of members across various categories of kinship: paternal aunts and uncles, paternal cousins, older brothers, older sisters, maternal complete generation (aunts, uncles, cousins, siblings), although the family size effect was stronger for the paternal side than for the maternal side of the kinship. Interestingly, adopted males were especially likely to report a homosexual sexual orientation (King et al. 2005b). Rahman (2005b) has also described an FBO effect in an English sample of 80 heterosexual and 80 homosexual males, where homosexuals tended to have more older brothers than heterosexuals, but the FBO effect did not correlate with psychological gender.

Blanchard et al. (2006; see also Blanchard & Lippa 2008) have recently pointed out an interaction between two early ontogenetic correlates of homosexuality: FBO effect and handedness. In their work they concluded that the FBO effect was significant for right-handed men only. Blanchard et al. interpreted their results in terms of a potential mutual cancelling out of the FBO effect and left-handedness on the development of homosexuality (see Chapter 4). However, it is also possible that if the FBO effect is mediated by social interactions among siblings rather than an immune mechanism (see below in this chapter), then its effect will be more obvious in cases where the plasticity of relevant brain structures is fully retained after birth, and this may be more the case in right-handed than in left-handed individuals owing to constraining differences in their early ontogeny (see Chapter 4). Blanchard (2008b) also seems to be leaning towards such a differential plasticity mechanism. Future studies should focus on which specific cellular and molecular mechanisms link, on the one hand, non-right-handedness and early ontogenetic canalisation of sexual orientation, and on the other more plasticity in the development of sexual orientation in right-handed individuals. The discovery of such a mechanism will presumably be able to explain the development of a same-sex sexual orientation due to both pre- and postnatal effects. We will return to the issue of postnatal effects in the last part of this chapter.

McConaghy *et al.* (2006) have carried out a retrospective study of awareness of homosexual feelings among men who currently describe themselves as either homosexual or heterosexual. With the well-known proviso that such personal recall studies may be subject to biases, McConaghy *et al.* (2006) provide evidence for an FBO effect on such awareness of homosexual feelings in men. Bogaert (2006a) has recently published an analysis of four samples of men that also included one sample of individuals from blended families (i.e. families where some of the siblings are genetically unrelated). There was variability in this latter sample in terms of the amount of time of contact between non-related siblings during their early ontogeny, although the exact age distribution of the non-related children when they joined their foster-family was not made explicit by Bogaert (2006a). In testing a sociological theory for the FBO effect in opposition to the specific immunological theory entertained so far, Bogaert (2006a) predicted that purely social interactions between males would produce an FBO effect on the development of a homosexual sexual orientation independently of the genetic relatedness between the siblings, moreover, such a sociological mechanism also predicts an irrelevant effect of the number of older biological brothers on the development of a homosexual sexual orientation of a proband, if the proband and his brothers were reared apart. What Bogaert (2006a) found was that the development of a homosexual sexual orientation was only affected by the number of older biological brothers reared with or not reared with the proband, a result that supports the MIH. Bogaert's (2006a) work was enthusiastically reviewed by Puts *et al.* (2006) in the same issue of the *Proceedings of the National Academy of Sciences of the USA* where the original work was published.

We should obviously welcome better and more precise tests of any reasonable hypothesis that can explain the development of a homosexual sexual orientation, and Bogaert's (2006a) work certainly represents an improvement on previous research and a step in the right direction. In my view, however, there is at least one major methodological issue that affects Bogaert's (2006a) results and therefore their interpretation. The issue, which was already mentioned in the 'Birth order and family effects' section of Chapter 4, is the uncertainty, or the variability among samples, regarding: (a) the exact time of separation between biological siblings and (b) the exact timing of union (from the point of view of the age of both individuals) between genetically unrelated stepbrothers. As explained in Chapter 4, from an early ontogenetic perspective timing of events is of crucial importance. For instance, if there is a discrete sensitive period during the development of the brain where social interactions (e.g. social stress or others) are more likely to affect the development in one direction or another regarding sexual orientation, then what matters is the precise timing of these events and not just whether separation or co-habitation between siblings and stepsiblings has occurred. If genetic brother and proband were separated *after* a postnatal sensitive period where development of brain centres controlling same-sex sexual behaviour are most susceptible to social interactions with siblings, then the FBO effect may have well been the result of social–developmental mechanisms, not of potential prenatal immune-mediated effects. On the other hand, if a non-genetically related stepbrother joins the family after that *sensitive period*, then he will have no social effect on the proband's development of a homosexual sexual orientation and the foster-sibling himself may not be affected by this union if he was old enough. These patterns, however, will occur not only because of the lack of prenatal immune effects in non-biological siblings, but also because of the mistiming of their social interactions with respect to the postnatal sensitive period/s. All these issues are left unresolved in Bogaert's (2006a) paper; a more detailed analysis that explicitly takes the timing of ontogenetic events into account should be welcomed.

In the past ten years or so, Ray Blanchard (Jones & Blanchard 1998; Blanchard 2001, 2004, 2008a) has published successive reviews of the FBO effect in studies of human homosexuality. Across nine studies reviewed by Jones & Blanchard (1998), values of the Slater's Index were higher for homosexual males: 0.524 (all sibs), 0.572 (brothers), 0.520 (sisters) compared with heterosexual males: 0.452 (all sibs), 0.480 (brothers) and 0.464 (sisters); i.e. homosexual males tended to be born later in their sibship compared with heterosexual males and, although the values of the index were highest for the number of brothers, they were also higher for the number of sisters. Three years later, in 2001, Blanchard published a second review where five additional studies were included, for a total of fourteen. The trend remained unchanged, with homosexual men being born later in their sibship. Subsequently, Blanchard (2004) proposed a more detailed immunological mechanism that could explain the birth order effect in the ontogeny of homosexuality. This issue will be reviewed in a different section below.

Green (2000) studied the FBO effect among male transsexuals of various sexual orientations: asexual, heterosexual, homosexual and bisexual. Homosexual male-to-female transsexuals tend to be born later within their sibship; they also have more siblings, and Green estimates that the likelihood of being homosexual among male transsexuals increases by 40.5% for each older brother. A similar pattern of later born male-to-female homosexual transsexuals was obtained by Poasa *et al.* (2004) from a study of Samoan *fa'afafine*, although they also detected both a fraternal and a sororal birth order effect in their sample.

The above evidence suggests that the FBO effect in the development of a homosexual sexual orientation in males is not an artificial phenomenon or a result of analytical biases. Its exact importance and, above all, the causal mechanisms that can explain it, however, are still wide open for discussion and investigation as it will be argued in the rest of this chapter. For instance, Camperio-Ciani *et al.* (2004) have recently reported that only 6.7% of the variance in sexual orientation is explained by the effect of the number of older male siblings in their sample, whereas Francis (2008), after analysing data from the National Longitudinal Study of Adolescent Health of the USA, concluded that the FBO effect on male homosexuality was only mild.

Fraternal and Sororal Birth Order effect in the development of a homosexual orientation in females

Blanchard & Klassen's (1997) H-Y immune mechanism can only explain an FBO effect for the development of a same-sex sexual orientation in males. In their original formulation of the hypothesis, however, MacCulloch & Waddington's (1981) immunological mechanism was supposed to apply to both male and female homosexuality. Therefore it is appropriate to ask whether there is a sororal birth order (SBO) effect or a FBO effect in the distribution of female homosexuality within a family.

Bogaert (1997) provided initial evidence that neither the number of older male nor that of female siblings accounts for the development of a lesbian sexual orientation (see also Bogaert 2003b). Moreover, the tendency in British and USA data, if anything, suggests that lesbians tend to be older within their sibships whereas for male homosexuals the reverse pattern was found (Blanchard *et al.* 1998), although some effects of the sibship sex ratio may occur as lesbians tended to come from male-biased sibships. The same pattern was described by Zucker *et al.* (1998), who found that girls described as having a 'gender identity disorder' were older within their sibship, especially with regard to their sisters (SBO effect), although not their brothers. The same trend could not be confirmed by Bogaert (2000) in a separate study, however. Instead, what Bogaert (2000) found was that female same-sex sexual orientation was more likely in women whose brothers had an influence on them in terms of teaching them about sex. Whether there is a causal relationship here, or whether already established lesbians seek advice from their brothers in order to better mould a masculine gender role and identity, is unknown.

McConaghy *et al.* (2006) measured awareness of homosexual feelings by the time probands were 15 years old, and found that homosexual feelings were more preponderant among women who were older in their sibship; however, for those who had older siblings, those tended to be brothers more than sisters. More recently, Francis (2008) has studied family and other correlates of homosexuality and bisexuality in young women in the USA and found that the probability of developing a homosexual orientation decreased with having one older brother or adopted brothers, whereas the probability of developing a bisexual sexual orientation decreased if there is an older brother in the family, or an older sister, or a younger sister or a sister of the same age.

In sum, although a greater diversity of research teams studying a greater diversity of ethnic groups across a larger spectrum of countries would be very welcome, the available evidence suggests that there is a consistent FBO effect on the development of a homosexual orientation in men (17 of the studies reviewed here), whereas an SBO effect on male homosexuality has been reported in only three studies so far (i.e. Jones & Blanchard 1998; King et al. 2005; Poasa et al. 2004). In the case of lesbians, evidence for a *parity effect* is much less clear-cut. The FBO effect has been found in one study of lesbians (McConaghy et al. 2006), whereas previous studies had not found such an effect (Bogaert 1997, 2003b). Three studies also mention the absence of any SBO effect in lesbians (Bogaert 1997, 2000, 2003b) although others have described an SBO effect, but one that goes in the opposite direction from either the FBO in lesbians or both the FBO and SBO in gay men; that is, lesbians tend to be older within their sisterships, not younger (Zucker et al. 1998; Blanchard et al. 1998; McConaghy et al. 2006). Moreover, Francis (2008) found a consistent decrease of homosexuality or bisexuality with the presence of siblings. We will return to these trends in the discussion of the mechanisms of the FBO/SBO effects at the end of this section.

Parental age effects

To what extent are parity effects a result of parental age? Clearly, the younger individual in a large sibship not only has a number of elder siblings but she or he was also born of an older mother. Moreover, the individual may also have an older biological father, if we assume monogamy and no extra-pair fertilisations, compared with the firstborn. Although parental age may have an effect on studies of the FBO/SBO phenomenon that is more methodological than anything else (for example, a sample of individuals that is biased towards younger parents will have an over-representation of firstborn, and the same may happen when the population under study is increasing rather than decreasing (Hare & Price 1969; Price & Hare 1969)), I will not focus on these methodological issues, which, eventually, can be addressed by limiting the study to full sibships, e.g. by only studying families with parents who are postreproductive. The main focus will be on potential biosocial effects of parental age and special emphasis will be put on studies where the above methodological problems have been reasonably controlled.

It was Eliot Slater (1962) who first pointed out the potential role of parental (maternal in his case) age effects on the development of homosexuality in humans. In fact, his article was titled *Birth order and maternal age of homosexuals*. Abe & Moran (1969) subsequently pointed out that paternal age was also positively associated with the development of a homosexual orientation in offspring. Ten years after Abe & Moran's (1969) study, Hare & Moran (1979) confirmed that the parental age effect detected in the distribution of homosexuality among male siblings was valid for both mother's and father's age in their sample, but no parental age effect of any kind was detected for lesbian offspring.

Anthony Bogaert and his collaborators have recently focused their attention on the study of *parity* and *parental age* effects in the distribution of homosexual orientation and on the interactions between the two effects (Bogaert & Cairney 2004; Bogaert 2006a; Bogaert & Liu 2006). Bogaert & Cairney (2004) analysed two datasets: one from Canada and the other from the National Comorbidity Survey (NCS) of the USA. Interestingly, in the NCS sample they found that, in males, probability of developing a homosexual orientation increases with the number of older siblings in young mothers and fathers, but it actually decreases in older mothers and fathers. For lesbians, the probability increases when father is older, especially in elder siblings. The Canadian sample only included gay men and the analysis of these data suggests that although the probability of being gay increases with the number of older siblings, the effect is stronger when parents (both father *and* mother) are young and weakens when parents are older. Bogaert (2006a) also confirmed a trend for gay men to come from families where the mother is younger. After analysing a Canadian sample of male homosexuals from Ontario, Bogaert & Liu (2006) concluded that the likelihood of being gay increases with the number of older siblings (brothers and sisters together) but that the effect varies with both maternal and paternal age, being stronger when parents are younger and weaker when they are older. Countering this trend, Frisch & Hviid (2006) reported an increase in the likelihood of homosexual marriage with an increase in maternal age, but admittedly their sample may not be representative as only a subset of all homosexuals formalise their relationship with a marriage, even in a country such as Denmark where homosexual marriage is legal.

In sum, recent studies suggest that, in general, homosexuality in men tends to be associated with a younger mother but also a younger father, whereas the *parity effect* detected for homosexuals decreases as both parents get older. In the more limited evidence available for lesbians, the effect whereby the older sister is more likely to develop a lesbian orientation seems to be increased by having an older father. Interestingly, these conclusions spring from studies carried out in the 2000s. Results of studies carried out by Slater, Abe, Hare and Moran in the 1960s and 1970s reached a different conclusion: the probability of developing a homosexual orientation in sons increased with the age of parents. Methodological issues notwithstanding, this difference may suggest the possibility that factors other than immunity could explain the parental age effect and perhaps the *parity effect* as well.

Can immunology explain the development of homosexuality in males and/or females?

The mechanism originally proposed by MacCulloch & Waddington (1981) involved the production of either individual autoantibodies or of maternal antibodies against (a) testosterone, in the case of male homosexuality and (b) progesterone, in the case of female homosexuality. Such mechanisms could potentially explain the FBO and SBO effects. In fact it is possible to experimentally induce an elevated production of antisteroid (e.g. antitestosterone or antiprogesterone) antibodies in animals if the steroid is, for instance, conjugated to a protein. When unconjugated, however, steroids are believed to be poorly immunogenic owing to their small molecular size; peptides are in general far more immunogenic than steroids (Kuwahara *et al.* 1998). Kuwahara *et al.* (1998) studied a woman who had elevated levels of circulating antitestosterone autoantibodies, an immune response that was apparently mediated by immunoglobulin G. The subject was affected by hypergonadotropic hypogonadism; that is, she had poorly developed gonads in spite of displaying high production of luteinising hormone and follicle-stimulating hormone. Antioestrogen antibodies have also been detected in women with autoimmune diseases such as lupus erythematosus (Counihan *et al.* 1991) or who had been taking oral contraceptives (Bucala *et al.* 1987); the latter may also suggest potential roles of oestrogens (e.g. phytoestrogens) ingested with food (see Chapter 5) as immunogens. Ródenas *et al.* (1998) have also reported a woman with a dermatitis produced by an autoimmune reaction against progesterone. Therefore, although immune reactions against self or foetal steroids may occur in women that could affect their offspring's development, they seem to be too rare to account even for the roughly 3% levels of male homosexuality described in humans. Whether such a mechanism could account for a small proportion of that 3%, however, remains to be determined.

The model proposed by Gualtieri & Hicks (1985), on the other hand, could potentially explain the

FBO effect and the bias in male homosexuality in humans; the evidence, however, is less than impressive. If the H-Y antigen was responsible for eliciting a maternal immune attack against male fetuses we would expect Klinefelter syndrome (KS) males (47/XXY), who have increased levels of circulating antibodies, a characteristic that is typical of females (XX) (Oktenli *et al.* 2002), but also have a Y chromosome that encodes for the H-Y antigen, to be especially prone to developing a homosexual orientation owing to a concomitant maternal and self-immune attack on developing tissues. Such a propensity towards homosexuality should be especially evident in second- and later-born KS males. Available evidence, however, suggest that among homosexual males KS individuals are not more represented than in the general population (Kessler & Moos 1973). Moreover, although some studies have characterised KS males as being less masculine (e.g. using the Bem Sex Role Inventory; Bancroft *et al.* 1982) the same subjects do not qualify as more feminine, using the appropriate BSRI criteria. Bancroft *et al.* (1982) suggest that XXY boys score low in 'masculinity' traits simply as a result of a generalised low self-esteem, which, incidentally, may also affect their romantic relationships with members of the other sex. In their review, Diamond & Watson (2004) do stress the general 'feminine' behavioural characteristics of KS subjects but, again, here 'feminine' may be just the reflection of their insecurity, shyness and gentleness of character, hardly a set of traits that characterise all XX females or that are absent from all XY males in Western societies. With respect to the specific issue of homosexuality, Diamond & Watson (2004) point out that the prevalence of homosexuality among KS individuals is not different from the prevalence found among the general population.

If KS males do not seem to show higher levels of homosexuality than XY males it may just be because of the excess X chromosome. In that case, a better test of Gualtieri & Hicks' (1985) hypothesis will be afforded by the study of XYY trisomic males, who have a supernumerary Y chromosome. Studies of XYY individuals, however, either do not mention any effect of the karyotype on same-sex sexual orientation (Kessler & Moos 1973), or mention very low frequencies of XYY males among those men attending a gender identity clinic: 1/136 (0.7%) (Blanchard *et al.* 1987).

Ray Blanchard (2004; see also Blanchard 2008a) has recently proposed a series of more detailed immunological mechanisms that could explain the development of a homosexual orientation and the birth order effect on male homosexuality. For instance, antibodies may bind to and inactivate male-specific molecules, perhaps molecules of the membrane of cells found in specific areas of the CNS, whose inactivity may alter sexual behaviour; e.g. the H-Y antigens mentioned above, but also protocadherin 11 Y-linked and neurologin 4 Y-linked (Blanchard 2004). He also indicated the possibility of a Th1/Th2 imbalance that I have already mentioned at the beginning of this chapter. Countering these immunological mechanisms, however, recent evidence put forward by James *et al.* (2003) suggests that in humans and rats it is multiparous females that are more tolerant of male skin grafts than reproductively naïve females, a result that falsifies a major prediction of the MIH (see also James 2006 for additional criticisms of the MIH). An additional falsification of the MIH comes from a study carried out by Tavares *et al.* (2004) in Porto Alegre, Brazil, where they studied sexual orientation among only children and, for those coming from families with more than one child, first-born and not-first-born in the sibship. Against the prediction of the MIH, it was the only children who reported the highest prevalence of non-heterosexual sexual orientation, followed by first-children and not-first-children. A similar result was obtained by Frisch & Hviid (2006) who studied homosexual and heterosexual marriages in Denmark. Denmark was the first country that, in 1989, made same-sex marriage legal. After analysing a very large dataset from the Danish Civil Registration System, Frisch & Hviid (2006) reported that men and women who grew up as only children had same-sex marriage rates 11% and 15% higher, respectively, than men and women with siblings. For individuals who grew up in families with more than one child, however, Frisch & Hviid (2006) did report a birth order effect where younger children (males and females) are more likely to marry homosexually. Of course we do not know whether the tendency to marry may also be affected by birth order effects, thus introducing an additional complication in the interpretation of Frisch and Hviid's results. Additional arguments against the MIH have been provided by Neil Whitehead (2007) in what is probably the fiercest critique of the MIH so far.

In a recent work, Piper *et al.* (2007) have studied *parity effects* on maternal production of H-Y-specific $CD8^+$ T cells. Although they found a trend for more mothers who had two or more boys to also have a more elevated H-Y-specific CD8 immune response (50%, 4/8) compared with mothers who only had one boy (32%, 6/19), the difference was not statistically significant. However, only women who gave birth to at least one male also had H-Y-specific $CD8^+$ T cells, but such an H-Y-specific population only represented between <0.01% and 3% of the overall $CD8^+$ population of T cells (Piper *et al.* 2007). Piper *et al.* (2007) also point out that there are cellular mechanisms, perhaps involving $CD4^+ CD25^+$ regulatory T cells, that could suppress maternal immune reactions against fetal H-Y antigens.

So far in this chapter immunity has been mainly analysed in the context of prenatal immune mechanisms potentially affecting the development of homosexuality. Is there an association between immunity and homosexuality in adults? Geschwind & Galaburda's (1985, 1987) model (see Chapter 4) suggests an association not only between developmental instability and both left-handedness and homosexuality, but also between such instablity and a higher susceptibility to infection. If so, left-handed male homosexuals should be especially at risk of contracting an infectious disease as their immune system may be less efficient. Becker *et al.* (1992) tested this hypothesis by comparing left-handed, right-handed and mixed-handed homosexual and bisexual men that were either HIV-seronegative or HIV-seropositive. Their results are generally unsupportive of an association between immunity and handedness. HIV-seronegative male homosexuals tended to be as left-handed as HIV-seropositives (6% vs. 7%). Moreover, the rates of seroconversion rates were similar between right- (16.0%) and left-handed (20.0%) individuals, whereas the probability of developing AIDS did not differ statistically between the two groups (23.1 % for right-handers vs. 20.2% for left-handers). Finally and importantly, the incidence of reported autoimmune diseases or allergies did not differ according to handedness (Becker *et al.* 1992). Left-handed males tended to be more likely to develop a homosexual sexual orientation compared with males in the general population, but this was not associated with an increased probability that homosexual males would also display medical conditions associated with immune system activity (see also Marchant-Haycox *et al.* 1991). What the immune activity of the mothers was during pregnancy (Blanchard 2008a), however, is not known.

Immunity and the parental age effect

We have seen above that recent studies indicate that younger parents (mother and father) are more likely to produce a homosexual male offspring, whereas lesbians are more likely to be born to couples where the father is older. The MIH predicts that such a pattern should be the result of a more elevated immune activity in adults (at least females) at younger ages. The effect of a younger father on the offspring's homosexuality does not seem to fit comfortably with an immunological hypothesis, although arguments could always be produced for a possible immune effect of the father on the foetus via paternal molecules introduced with the semen into the female's reproductive tract throughout at least part of the pregnancy (e.g. Poiani 2007). A similar argument, however, would not be able to explain the statistical association of a lesbian daughter with an older father.

Are male and female adult humans more immuno-active at younger ages? Production of antibodies, or humoral immunity, tends to decline with age, although this does not seem to result from an increased immune deficiency at older ages; rather, it seems more likely to be the product of an age-related increase in *immune dysregulation* (Le Maoult *et al.* 1997). In fact, in both humans and mice levels of autoantibodies increase with age due to decreased antibody affinities and increased levels of polyreactivity against a variety of antigens, including self antigens (LeMaoult *et al.* 1997; Castle 2000). The thymus does involute by the age of 40, although this does not mean that at that age individuals become completely immunodeficient (Castle 2000). One consequence of aging, however, is to limit the clonal expansion capability of T cells (Castle 2000; Goronzy & Weyand 2005; Sadighi Akha & Miller 2005), especially that of naïve T cells, but not so much memory T cells (Haynes 2005). Therefore, although aging may affect some aspects of both humoral and cellular immunity (Allman & Miller 2005) in the direction predicted by the hypothesis, other aspects of immunity such as *dysregulation* suggest increased probability of immune attack at older ages. It remains to be seen whether high levels of immune activity at younger reproductive ages of parents are sufficient to explain the development of homosexuality in offspring.

Maternal autoimmunity, which results from a hyperactive immune system, has certainly been associated with the development of at least some mental disorders in offspring; for example, systemic lupus erythematosus in the mother is associated with learning disorders in offspring, sons in particular (Ross *et al.* 2003; Tincani *et al.* 2006; see Flannery & Liederman 1994 for a review). Current hypotheses suggest that antiphospholipid and/or anti-Ro/La antibodies produced by a mother suffering from an autoimmune condition may explain those specific effects on the development of the child's brain (see, for example, Neri *et al.* 2004a,b). One advantage of the molecular mechanism involving antiphospholipid and/or anti-Ro/La antibodies, when applied to the MIH, is that it is not specific to offspring of one sex only (e.g. males). Whether the same mechanisms may be implicated in the development of homosexuality or not, however, can only be determined by future empirical studies. Chances are, however, that if maternal autoimmunity is a factor in the development of homosexuality in their offspring, such mechanisms will be able to explain only a fraction of the cases of homosexuality, as I have already suggested.

Non-immunological explanations for birth order and parental age effects

In my view, at this point in time, the MIH for homosexuality is still on rather shaky ground. Could the well-established trend for male homosexuals to be younger in their sibship and be born to younger parents be explained by an alternative mechanism not involving the direct effect of maternal immunity on brain development?

Slater (1958) proposed a model based on early learning experiences whereby sex play among brothers may condition the younger one to develop a homosexual orientation. We have already seen how this specific mechanism is unlikely, as such, to explain the development of a homosexual orientation across the board, unless the behavioural experience just reinforces an existing predisposition (e.g. a genetic one) to develop a homosexual orientation or, as I will suggest below and have already discussed at length in Chapter 4, the experience is also concomitant with or preceded by early stress effects. In this context then, conditioning could be retained as a potential factor in the causation of homosexuality but only as one of the variables operating within a more complex mechanism.

On the other hand, if learning is a major mechanism for the spread of homosexual behaviour in a population, contagion effects should be evident within families. Jones & Jones (1992: 149) define a *contagious behaviour* as one where 'one person is more likely to exhibit it when a relevant other person has already done so'. Such an effect relies on various potential mechanisms (including genetic inheritance) but here I stress one of the mechanisms listed by Jones & Jones (1992), namely behavioural contagion based on learning. The *Behavioural Contagion* hypothesis predicts family clustering of the homosexual trait and also a sex-biased distribution (Jones & Jones 1994, 1995), but as such it does not predict a bias in the distribution of the trait towards the younger members of the sibship. For a younger brother to develop a homosexual orientation through learning, an older brother with a similar orientation is required to teach him; this means that under the conditions of a behavioural contagion mechanism homosexuality will be widespread within sibships but clustered among different sibships. The observed FBO effect runs against the first of those two predictions.

The problem may be apparently solved by Bem's (1996) cognitive model *Exotic Becomes Erotic* (EBE) that was discussed in Chapter 4, if younger brothers are more likely to develop a perception of themselves as being different from their elder brothers. This, according to Bem's EBE model, should lead to the development of a same-sex sexual attraction in the younger brother within a brothership. I argued against the EBE model as a general mechanism for the development of a homosexual orientation in Chapter 4 and its application to the FBO effect does not seem to be any more successful. Although a younger brother could well develop a tendency to differentiate himself from his elder brothers, including a differentiation in his sexual orientation, this is certainly not the only outcome of interactions among brothers as it is often the case that the younger member of a sibship in fact takes his elder siblings as raw models, thus becoming similar to, rather than different from, them, a process emphasised by social learning theory (Bandura 1977; Slomkowski *et al.* 2001; Kowal & Blinn-Pike 2004; Pomery *et al.* 2005; Snyder *et al.* 2005; Whiteman *et al.* 2007). We will see below that interactions among brothers can be of both the competitive kind – leading to some degree of differentiation – and the cooperative kind, leading to some degree of similarity between siblings. On the other hand, this latter social learning mechanism will be a special case of behavioural contagion and therefore, although it can explain clustering of homosexuality across families, it cannot explain the FBO effect. It is also unclear why younger parents should be associated with increased likelihood of their children developing a homosexual orientation through learning. At the most a younger, and presumably more open-minded, parent may be less likely to interfere with the development of a homosexual orientation in the child, but s/he may not necessarily be the initial raw model for such development through learning.

We seem to have reached a point where both the FBO and the younger parent effects in the distribution of homosexuality cannot be easily explained by a so-called 'biological' mechanism, neither can they by various so-called 'social' mechanisms only. At this stage, we should try to bridge the divide between the 'social' Scylla and the 'biological' Charybdis and see whether a biosocial approach can help us understand those patterns.

I will start by analysing a recent series of models that could be used to achieve such a biosocial integration, to then provide a very brief sketch of the synthesis that will be further developed in the subsequent chapters and fully explained in Chapter 10. The models were originally developed by various authors especially to explain family effects on personality development, but I will extend them to the specific case of homosexuality.

Family dynamics

In 1995 Frank Sulloway published an article in the journal *Psychological Inquiry*, soon to be followed by the publication in 1996 of his well-known book *Born to Rebel: Birth Order, Family Dynamics and Creative Lives* (Sulloway 1995, 1996), where he laid down the foundations of what has become a controversial, but in my view promising, biosocial theory of the FBO, family size and parental age effects. Sulloway (1995) strictly operates within a Darwinian framework by identifying what he calls the 'four great Darwinian conflicts' in human behaviour: parent–offspring conflict (Trivers 1974), differential parental investment by the two sexes (Williams 1966; Trivers 1972), sibling–sibling competition (Trivers 1974) and sexual selection (Darwin 1871).

Siblings growing together are usually envisaged as sharing the same environment, potentially leading to the development of some phenotypic similarities among them. However, Sulloway (1995, 2007a) rightly points out that the fact of growing together may in itself be also a cause of phenotypic differentiation between siblings during development, with some of the mechanisms leading to such differentiation being driven by conflicts with parents over parental investment and conflicts among the siblings over appropriation of such parental investment (see also Koch 1960). In an evolutionary ecological analogy: competition may lead to habitat partitioning in the short-term and character displacement in the long-term. Birth order establishes an asymmetry between siblings that can be measured in terms of age and, at least potentially, in terms of size but also in terms of social power and privilege (Sulloway 1995). Such asymmetry may well affect the family dynamics and outcomes of within-family conflicts.

Sulloway (1995, 1996) bases his analysis of birth order effects on McCrae & John's (1992) five-factor model of personality structure that includes: neuroticism (N), extraversion (E), openness to experience (O), agreeableness (A) and conscientiousness (C), popularly known among psychologists as the *Big Five*. As a result of the dynamics of both confluence and conflict of interests operating within a family social milieu, first-born offspring are expected to be characterised by personality traits of extraversion (dominant, assertive) and conscientiousness (respectful of parental authority), whereas later-born siblings are expected to be characterised by higher levels of openness (untraditional, rebellious) and agreeableness (approachable, popular). As for neuroticism, the distinction between first- and later-born seems to be less clear-cut, but first-born siblings are expected to score higher on this trait as they may be more jealous in their effort to protect the privileges they enjoy (Sulloway 1995). In a meta-analysis of the published literature, Sulloway (1995) supported this model indicating a significant effect of birth order for most of those personality traits, although the effect was particularly strong for openness and conscientiousness (see Table 6.1).

Table 6.1. Results of birth-order studies according to the five personality dimensions

Behavioural Domain	Outcome
Extraversion	
First-borns are more extraverted, assertive, and likely to exhibit leadership qualities	5 confirming (17%) 6 negating 18 with no difference
Agreeableness/Antagonism	
Later-borns are more agreeable, popular and easygoing	12 confirming (39%) 1 negating 18 with no difference
Neuroticism (or emotional instability)	
First-borns are less well-adjusted and more anxious, neurotic, fearful, and likely to affiliate under stress	14 confirming (29%) 5 negating 29 with no difference
Openness	
First-borns are more conforming, traditional, and closely identified with parents	21 confirming (49%) 2 negating 20 with no difference
Conscientiousness	
First-borns are more responsible, achievement-oriented and organised	20 confirming (44%) 0 negating 25 with no difference

(modified from Table 1 of Sulloway 1995)

As a result of sibling competition (leading to *de-identification*, or maximisation of behavioural differences among adjacent siblings), parental manipulation and investment (the latter may vary with parental age: smaller for younger parents, larger for older ones), siblings carve for themselves *family niches* that are dependent on birth order often leading to sibling personality stereotypes (Sulloway 2001, 2007a,c). Importantly, such family niches are a result of both *prenatal* and *postnatal* effects of the family environment on the individual's development.

Sulloway's more specific *Family Dynamics Model of Radical Behaviour* (Sulloway 1996: 197) is partly consistent with what we have already seen regarding homosexuality in terms of birth order and postnatal stress effects. For instance, if we replace the word 'rebellious' in Sulloway's model with the word 'homosexual' we obtain: (1) 'Compared with firstborn, laterborns are more *homosexual*', (2) 'High [parent–offspring] conflict increases *homosexuality*'; although his model also predicts that (3) Larger sibships are less *homosexual*, whereas in reality we know the contrary to be the case. The discrepancy with prediction (3) may be if both competitive and cooperative effects within a sibship are taken into account. For instance, stresses suffered by the younger siblings as a result of interactions with their elder siblings may potentially affect the development of same-sex sexual preferences of the former, as explained in Chapter 4. If a homosexual phenotype develops, cooperative behaviours may be also expressed by the homosexual, as expected from kin selection (see Chapter 3). This would increase the adaptiveness of homosexuality in large sibships. Sulloway's prediction (3), however, will be rescued in the *Synthetic Reproductive Skew Model of Homosexuality* that I will introduce in Chapter 8. This model predicts that frequency of homosexuality would reach a maximum at some intermediate group size and then decrease for very large groups.

This, in a nutshell is Sulloway's (1995, 1996, 2001) *Family Dynamics Model*. The specific phenotypic outcome of such family effects is mediated and constrained by learning but also by the individual's genetic make-up and influences of stress on behavioural development. Thus, for poorly canalised ontogenetic programmes the family environment will be a major influence on the development of specific behavioural phenotypes, whereas for more canalised ontogenetic programmes the initial genetic make-up will provide a greater limit to the outcomes of that behavioural development (for example, this may perhaps explain the case of *resilient children* mentioned in Chapter 3; Cove *et al.* 2005, and the variable outcomes of sibling–sibling interactions leading to de-identification in some cases and identification in others reported by Whiteman *et al.* 2007). The specificities of age differences among siblings and also parental age, differences in body size and general health, but also social power and privilege as established by cultural traditions, are some of the factors that will determine niche differentiation among siblings in Sulloway's model.

Sulloway (2001) provides empirical support for his *Family Dynamics Model* through a meta-analysis of published works, concluding that birth order accounts for 4.1% of variance in sibling personality, an effect that is probably more important than it seems if we consider that sex accounted for 2.1% of variance in the same analysis. Jefferson *et al.* (1998) have independently supported Sulloway's *Family Dynamics Model* with analyses that focus on personality differences among adult sibs. A weaker support for the model is obtained when families that include adopted children are analysed (Beer & Horn 2000), which emphasises the need to also include pre- or perinatal effects (e.g. those caused by stress, as I have just mentioned) on brain and therefore personality development, but also postnatal (e.g. prepubertal) effects. As already indicated, timing of events is everything: what matters is not only whether a factor does occur but also when it occurs in the context of the ontogeny of the individual.

Somewhat variable results were also obtained by Rohde *et al.* (2003) in a large study of university students from Germany, Israel, Norway, Austria, Russia and Spain. Against the *Family Dynamics Model*, the last-born child in a family of three children was considered to have been most favoured by parents, as expected from parental investment allocation with increasing parental age. However, it was also found that the last-born was the most rebellious, in apparent contradiction to perceived parental favouritism. This last result fits within the *Family Dynamics Model*. The finding of Rohde *et al.* (2003) suggests that parent–offspring conflict may be resolved in favour of the youngest sibling as parents age, but sibling–sibling conflict was resolved in favour of the elder sibling. Recent research that also supports the predictions of the *Family Dynamics Model* includes that by Healey & Ellis (2007), Dixon *et al.* (2008) and Wang *et al.*

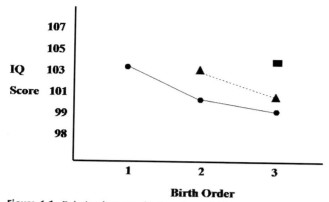

Figure 6.1. Relation between birth order and IQ score. Mean IQ scores for male conscripts according to birth order and number of elder siblings who died in infancy (age < 1 year). Circles, no siblings died; triangles, one sibling died; squares, two siblings died. Scores are adjusted for parental education level, maternal age at birth, sibship size, birth weight, and year of conscription. Modified from Kristensen & Bjerkedal (2007).

(2009), although other studies do not support the model (see, for example, Wichman et al. 2006).

Sulloway (2007a) does acknowledge that personality traits that develop during childhood and adolescence, when the individual tends to be more closely influenced by the family environment, may undergo further modifications as a result of novel experiences in adulthood, when individuals may be more under the influence of the social environment at large. Such further developments are expected from the continued plastic ability of brain structure and function as described in Chapter 5, and Sulloway (2007a) also suggests that the 'behavioural toolkit' that we learned as children in the family environment can be continually modified throughout the individual's lifespan. I would add, however, that such plasticity is *not* unlimited or unconstrained.

The developmental mechanisms emphasised in the *Family Dynamics Model* are indeed expected to retain a degree of plasticity through natural selection, as a family environment of various siblings has clearly not been constant throughout our evolution as a social species. Here I am referring to the likelihood that the high levels of child mortality experienced by our species until relatively recently (Sulloway 2007c) would have produced sudden and unpredictable changes in positions within a sibship – a *functional birth order* as opposed to a *biological birth order*, to use Sulloway's (1996) distinction; for example, the middle-born in a trio of brothers could have suddenly become the eldest (for the first time in his life) or the youngest (for the second time) upon the death of the elder or the younger brother, respectively. An ability of the middle brother to respond to such changes in the family demographics would certainly be adaptive (Sulloway 2007c). This hypothesised brain plasticity mechanism has recently been investigated by Kristensen & Bjerkedal (2007), who carried out a test of birth-order effects on IQ scores in a sample of Norwegian conscripts. Families of one, two and three children were studied; for the families with more than one child, cases were compared where the first-born (in two-child and three-child families) or the first-born and the second-born (three-child families) had died. Brain plasticity would allow second- and third-born children to quickly and adaptively adjust their phenotype to the new social environment in the case of death of elder brothers. Figure 6.1 shows their results, which indicate a clear trend for a monotonic decrease in IQ scores with increasing birth order. Younger siblings, however, are capable of re-adjusting their phenotype and reaching IQ scores equivalent to those of first- borns if elder siblings suddenly die. This clearly indicates: (a) adaptive plasticity of the brain and (b) that the youngest brother is just doing the best-of-a-bad-job in terms of IQ development in the competitive environment where older brothers are present. Sulloway (2007b) interprets Kristensen & Bjerkedal's (2007) results following the niche partitioning expected in the interactions among siblings that flows from his *Family Dynamics Model* and also in terms of Zajonc, Markus & Markus' (1979) *Confluence Model*. Following the suggestion throughout this book that human female homosexuality is more plastic than male homosexuality, it would be interesting to determine to what extent the development of women's sexual orientation may be particularly responsive to current changes in sibship order and sex composition. We have seen in this chapter that homosexuality in women tends to be associated with older sisters: would younger sisters increase their probability of developing a homosexual sexual orientation if the older sister were no longer part of the family? Does the exact age of separation matter?

In the previous paragraph I mentioned the *Confluence Model*. This model was put forward by Zajonc *et al.* (1979) to explain the apparent patterns of decline in IQ scores with family size and birth order. The structure of the model comprises a depressing effect of newborns on the average intellectual maturity of the family, the effect being especially important in the case of small families and families where the inter-birth interval is larger, up to a point. However, whenever the elder sibling is old enough to start teaching the younger, the elder benefits in terms of mental (e.g. IQ) development through the training experienced with teaching the younger siblings, whereas the younger brother benefits by learning from his elder. This means that, depending on age differences and relative stage of development, all members of the family may become resources, in terms of intellectual development, for each other. Therefore, the *Confluence Model* predicts a dynamic change in the intellectual environment of sibships as they grow older and/or the family expands or contracts, a prediction that perfectly fits the results of Kristensen & Bjerkedal (2007). The *Confluence Model* has recently come under attack (Wichman *et al.* 2006), provoking a strong and convincing rebuff from Zajonc & Sulloway (2007). In my view the *Family Dynamics* and the *Confluence* models are complementary, with the former mainly emphasising competitive interactions within the family and the latter mainly emphasising cooperative interactions. The complementarity of *Family Dynamics* and *Confluence* models suggests that they could both be subsumed into a more general model.

Given that both Sulloway and Zajonc root their models into broad evolutionary processes such as parent–offspring conflict, differential parental investment by the two sexes, sibling–sibling competition, sexual selection and cooperation, I suggest that the interindividual postnatal variability in behaviour and various aspects of personality highlighted by the *Family Dynamics* and *Confluence* models may also be relevant to the understanding of the development of homosexuality in men and women.

More specifically, patterns of sexual orientation development of offspring in a family environment, which are affected by interactions with both siblings and parents (but also other relatives and non-relatives in both humans and many other social vertebrates), could be better understood by a more synthetic model that can account for both cooperative and competitive dynamics, and that involves both prenatal mechanisms and also those postnatal developmental mechanisms associated with stress and to some extent learning. Such a synthetic (*integrationist* or *biosocial*) model will be fully developed in Chapter 10 following the application of *Reproductive Skew Theory* to homosexuality that will be carried out in Chapter 8.

Could a model of cooperation/competition that synthesises both the *Family Dynamics* and the *Confluence* models for within-family interactions explain the FBO effect and the young-parents effect of human homosexuality? I suggest that it can, especially through the mediation of ***stress effects on development on the one hand and coping and, more generally, fitness-enhancing mechanisms that evolved to withstand the stresses and complexities of social life on the other***. If same-sex sexual behaviour is adaptive under a variety of social circumstances, such as avoiding competition with more aggressive and dominant members of the extended family or social group (e.g. the elder siblings) or to enhance cooperation with the rest of the extended family, or both, specific alleles may have well been selected that make the expression of such behaviour more likely and reliable, especially when it is expressed in bisexual individuals or in self-identified homosexual individuals who, however, do reproduce. In traditional societies where mutual help is paramount for survival, kin selection would have further increased the chances of homosexual phenotypes becoming adaptive (see Chapter 3). That homosexuality among younger sons is especially common for younger parents, as shown in studies carried out in the 2000s, may also be a result of a particularly stressful family environment in situations where parents are less experienced. In the 1960s, however, homosexuality in sons was associated with an older father. If postnatal stress is an important factor, we would expect father–son interaction to have been more stressful in the 1960s, at least in some societies, if the father was older, and more stressful in the 2000s if the father is younger, this is a prediction that should be easily testable. Further suggestions indicating an important role of stress in the development of human homosexuality comes from Francis' (2008) work on male and female adolescents and young adults. Among males in Francis' study, the percentage of homosexuals was higher in families with no father or mother, among African–Americans, and among those with less than twelfth grade of education. The percentage of male bisexuals increased in mother-only families, father-only families, in families without father and mother and among African–Americans. In females, homosexuality increased in mother-only families and families without mother and father, whereas bisexuality increased in girls growing up in father-only families, and in those with less than twelfth grade education.

In sum, the amalgamation of both the *Family Dynamics* and the *Confluence* models stresses the need to consider the combined effect of cooperation and competition in the development of personality traits associated with sexual orientation and gender role.

I will further develop these ideas into an initial model in Chapter 8 that will be finally expanded into a *Biosocial Model of Homosexuality* in Chapter 10. On the other hand, the specific immune mechanisms reviewed in this chapter may be, in general, less important, although more detailed studies should be welcome (Blanchard 2008a). In fact, it seems that it is probably time for research effort on the MIH to be shifted from demonstrating birth order effects to empirically testing the specific immunological mechanisms that have been proposed.

Summary of main conclusions

- Maternal immunity may affect the development of embryos, with males and later-born offspring expected to be more sensitive to this effect.
- There is a clear trend for homosexual men to be younger members of a sibship, especially with regard to their brothers: this is the fraternal birth order effect (FBO). Sororal birth order effects (SBO) on male homosexuality are less frequent.
- FBO and SBO effects are less consistent in lesbians across studies than they are in gay men, although some trends indicate that lesbians tend to be older, not younger, within their sistership.
- Early studies indicated a positive association between parental age and development of homosexuality in men. More recent studies, however, suggest the opposite; i.e. that homosexuality is negatively correlated with age of parents. These contradictory trends may be explained by postnatal social effects, such as stressful social interactions associated with parental age. Such social interactions may have undergone modifications in the course of a few decades in some Western societies.

- The role of immunity in the development of a homosexual sexual orientation remains controversial, but more specific studies investigating the various potential immunological mechanisms should be welcomed.
- Sulloway's *Family Dynamics Model* and Zajonc, Markus and Markus' *Confluence Model*, taken together, represent the first step towards the assembling of a *Biosocial Model of Homosexuality* that integrates genetic, developmental and learning effects.

In the next chapter the focus shifts to a supra-individual level of analysis. There I review the potential effects that segregation between the sexes may have on the evolution of homosexual behaviour.

Sexual segregation effects

In some species of animals, including birds and mammals, the sexes may segregate for part or even most of the year (Ruckstuhl & Neuhaus 2005). Although a widely accepted definition of sexual segregation remains somewhat elusive (Bowyer 2004), owing in part to the diversity of behavioural patterns and mechanisms involved in this phenomenon, its essential features could be encapsulated in the following definition: *sexual segregation is a greater distancing in space or time between members of different sexes than between members of the same sex*. Such spatiotemporal distancing between the sexes may occur at all possible scales (Catry *et al.* 2005): from the micro (few hours or few metres) to the macro (many days or kilometres), and it will encompass both social and non-social species. Moreover, this definition also allows the use of appropriate statistical tools (e.g. multidimensional scaling) in order to determine whether a population does indeed display sexual segregation at any specific spatiotemporal scale or not.

Along the time axis, we should also consider the possibility that sexual segregation may vary ontogenetically, with taxon-specific tendencies for sexual segregation to be more prevalent at some ages than at others. For instance, larger and older adult beluga whale (*Delphinapterus leucas*) males segregate from females in summer in the Canadian high Arctic, but younger males do not (Loseto *et al.* 2006). Obviously, when individuals in a population are not sexually segregated, it means that they are *sexually aggregated* (Bowyer 2004). Following the above definition of sexual segregation, in species such as the Douglas's squirrel (*Tamiasciurus douglasii*) (Koford 1982) where members of a local population hold individual territories that are not clumped into sex-specific clusters but instead male and female territories are spatially interspersed, we cannot talk of sexual segregation. Rather, species such as *T. douglasii* show *individual* segregation irrespective of sex (i.e. they are individually segregated but sexually aggregated).

When a population is sexually segregated, same-sex interactions may become more likely, including interactions of the sexual kind; hence any study of same-sex sexual behaviour should naturally consider the potential effects of sexual segregation.

As is often the case in evolutionary biology, the study of sexual segregation dates back to Darwin's first edition of *On the Origin of Species* (Darwin 1859, see also Bowyer 2004). Hypotheses for sexual segregation must explain not only why individuals maintain relatively smaller distances from members of the same than from members of the other sex, but also why members of each sex sometimes form monosexual groups (i.e. distinct social units composed of individuals of the same sex) rather than wandering as single individuals (Calhim *et al.* 2006). Although sexual segregation does not necessarily imply segregation into monosexual groups (which is known as *Social Sexual Segregation*, Bon & Campan 1989; Conradt 1998), because females and males may move around alone, each in a sex-specific habitat (*Habitat Sexual Segregation*) or microhabitat (*Spatial Sexual Segregation*, Geist & Petocz 1977; Conradt 1998), it is *Social Sexual Segregation* that is most relevant for the potential expression of same-sex sexual interactions.

The two patterns, sexual segregation and constitution of monosexual groups, may or may not be a result of a common cause or causes. Group formation is often the result of high costs of solitary living and high benefits accruing to individuals that live in close proximity to each other, regardless of the sex-specific composition of the group (see Winnie & Creel 2007 for a list of costs and benefits of group living), whereas sexual segregation has many potential causes that are independent of sociality. However, sociality and sexual segregation do co-occur in some species (Table 7.1).

Moreover, there is also the possibility that sexual segregation may just be, in some cases at least, an unselected or random effect of the confluence of two conditions: a tendency among individuals of one species to form groups and a highly biased sex ratio in the population (Conradt 1998, see also Thirgood 1996). On the other hand, in highly polygamous species where single males control a group of females – as is the case in harem polygyny – and actively exclude other sexually mature males during periods when females are sexually receptive, females are segregated into groups, whereas bachelor males that do not control a harem are left with no choice but to form male monosexual groups, if benefits of group living exceed costs, or move around alone, if costs of group living exceed the benefits. In such a case sexual segregation will be a product of specific selective processes rather than mere chance.

Although sexual segregation may occur along a spatial axis (*Environmental Sexual Segregation*: this includes *Habitat* and *Spatial Sexual Segregation*), it may also manifest itself along a temporal axis, with conspecifics of different sexes being sympatric but active at different times of the day for at least part of the year (*Temporal Sexual Segregation*; Mooring *et al.* 2003, see Table 7.2).

Everything else being equal, it seems intuitively appealing to think that in sexually segregated species homosexual encounters should be more frequent, relative to heterosexual encounters, than in sexually aggregated species, whereas sexually segregated species that form monosexual groups should be more likely to manifest same-sex sexual behaviours than sexually segregated species that are not social. This would be expected, in the first instance, from the increased opportunities for same-sex sexual interactions as the sex ratio of available potential partners is biased towards same-sex individuals. Obviously everything else is not always equal, and other ecological, social and demographic variables that will be analysed in this chapter and, in greater detail, in Chapter 8, not to mention potential individual-specific genetic, neuroendocrinological and ontogenetic effects analysed in previous chapters, apart from sexual segregation, can and indeed do play an important role in the expression and selection of same-sex sexual behaviour in birds and mammals.

When an individual has mainly access to same-sex sexual partners as a consequence of sexual segregation, she or he has three choices as far as sexual behaviour is concerned: to skip sexual interactions altogether (asexuality), to engage in homosexual sexual interactions (homosexuality) or to engage in autoerotic sexual activities such as self-masturbation (autosexuality). Which one of those behaviours will be displayed will be a function of individual propensities and partner response at a proximate level, and the history of selection for the expression of sexual behaviours in social contexts at the evolutionary level.

Sexual segregation may be also favoured by specific and active preferences for consorting with same-sex conspecifics, preferences that may have evolved for reasons quite unrelated to homosexuality. In addition, increased opportunities for homosexual behaviours in species displaying social sexual segregation may also lead to the evolution of specific sociosexual functions of male–male (M–M) and/or female–female (F–F) sexual interactions related to, for instance, dominance or cooperation (e.g. alliances).

In this chapter, I first review the different hypotheses and mechanisms of sexual segregation

Table 7.1. Same-sex mounting, sexual segregation and other potentially correlated variables in birds and mammals

Species	Sex Involved	Same-Sex Mounting	Sexual Segregation	Mating System	Sociality	Body Mass (kg)	Reference
Birds							
Columba livia	M?	Y	N	M	C	0.556	Brackbill 1941; Hetmanski & Wolk 2005; Soman et al. 2005
Phoebastria immutabilis	M	Y	Y	M	C	2.5	Fisher 1971; AnAge Database 2007
Philomachus pugnax		N	Y	L	T	0.137	Dit Durrell 2000; AnAge Database 2007
Phoenicopterus chilensis	M > F	Y	N	M	C	3	del Hoyo et al. 1996; King 2006; Animal Diversity Web 2007
Phoenicopterus ruber ruber	M = F	Y	N	M	C	2.5	del Hoyo et al. 1996; King 2006; AnAge Database 2007
Phoenicopterus ruber roseus	F	Y	N	M	C	2.5	del Hoyo et al. 1996; King 2006; AnAge Database 2007
Anas platyrhynchos		N	Y	M	C	1.048	Lebret 1961; AnAge Database 2007
Cygnus atratus	M	Y	N	M	C	6.225	Braithwaite 1981; Brugger & Taborsky 1994; del Hoyo et al. 1996; Animal Diversity Web 2007
Branta canadensis		N	N	M	C	4.55	Koplam 1962; del Hoyo et al. 1996; Animal Diversity Web 2007
Anser c. caerulescens		N	N	M	C	2.807	Prevett & MacInness 1980; Quinn et al. 1989; Gregoire & Ankney 1990; Dunn et al. 1999; AnAge Database 2007
Anser anser	M	Y	N	M-FPy	C	3.309	Huber & Martys 1993; del Hoyo et al. 1996; Kotrschal et al. 2006; AnAge Database 2007
Bubulcus ibis	M	Y	N	M	C	0.39	Fujioka & Yamagishi 1981; del Hoyo et al. 1996; AnAge Database 2007
Ardea cinerea	M	Y	N	M	C	1.443	Ramo 1993; del Hoyo et al. 1996; Lekuona & Campos 1998; AnAge Database 2007
Egretta caerulea	M	Y	N	M	C	0.396	Werschkul 1982; del Hoyo et al. 1996; Animal Diversity Web 2007
Gallinula tenebrosa	M	Y	N	Pya	G	0.525	Garnett 1978; del Hoyo et al. 1996
Porphyrio porphyrio melanotis	F > M	Y	N	Pya	G	0.895	Craig 1980a,b; Jamieson & Craig 1987; del Hoyo et al. 1996

Table 7.1. (cont.)

Species	Sex Involved	Same-Sex Mounting	Sexual Segregation	Mating System	Sociality	Body Mass (kg)	Reference
Birds							
Alca torda	M	Y	N	M	LC	0.617	Wagner 1996; del Hoyo et al. 1996; Animal Diversity Web 2007
Uria aalge	M	Y	Y	M	C	0.94	Hatchwell 1988; del Hoyo et al. 1996; Nagy 2001
Melanerpes formicivorus	(M = F)?	Y	N	Pya	G	0.082	MacRoberts & MacRoberts 1976; del Hoyo et al. 1996; Nagy 2001
Dinopium benghalense	M?	Y	N	M	T	0.109	Neelakantan 1968; del Hoyo et al. 1996; Hutchins et al. 2002
Calypte anna	M	Y	Y	Py	T	0.0045	Stiles 1982; del Hoyo et al. 1996; Nagy 2001
Gypaetus barbatus	M	Y	N	M	T	5.07	del Hoyo et al. 1996; Bertran & Margalida 2003; AnAge Database 2007
Falco tinnunculus	M	Y	N	M	T	0.211	Olsen 1985; del Hoyo et al. 1996; Korpimaki et al. 1996; Nagy 2001
Chionis minor		N	N	M	C	0.505	del Hoyo et al. 1996; Bried et al. 1999; Garamszegi et al. 2005
Tryngites subruficollis	M	Y	N	L	T	0.05	Prevett & Barr 1976; Myers 1979; del Hoyo et al. 1996; Lanctot et al. 1998; Loesch et al. 1999
Pluvialis apricaria		N	N	M	T	0.214	Parr 1992; del Hoyo et al. 1996; AnAge Database 2007
Haematopus ostralegus	F	Y	Y	M	T	0.48	Heg & van Treuren 1998; Ens 1998; Catry et al. 2005; AnAge Database 2007
Himantopus h. himantopus	F	Y	Y	M	LC	0.185	Kitagawa 1988; del Hoyo et al. 1996
Tringa totanus	M	Y	N	M-FPy	LC	0.149	Hale & Ashcroft 1983; AnAge Database 2007
Struthio camelus australis		N	N	Py	SS	88.3	Sauer 1972; Bertram 1980; Nagy 2001
Dromaius novaehollandiae		N	N	M-Pa-Pr	SS	52.5	Pople et al. 1991; Coddington & Cockburn 1995; Hough et al. 1998; Animal Diversity Web 2007
Rhea americana		N	N	HPy	SC	23	Codenotti & Alvarez 1997; Fernandez et al. 2003; Animal Diversity Web 2007

Table 7.1. (cont.)

Species	Sex Involved	Same-Sex Mounting	Sexual Segregation	Mating System	Sociality	Body Mass (kg)	Reference
Birds							
Pygoscelis adeliae	M	Y	Y	M-Pa(14%)	C	3.99	Hunter *et al.* 1995; Davis *et al.* 1998; Catry *et al.* 2005; Nagy 2001
Phalacrocorax carbo sinensis		N	N	M	C	3.629	Kortlandt 1995; del Hoyo *et al.* 1996; AnAge Database 2007
Agapornis roseicollis	F	N	N	M	C	0.0548	Dilger 1960; del Hoyo *et al.* 1996; Animal Diversity Web 2007
Aratinga canicularis	F	N	N	M	SC	0.07	Hardy 1963; del Hoyo *et al.* 1996
	M	Y		M			Buchanan 1966
Brotogeris versicolurus		N	N	M	SC	0.071	Arrowood 1988; del Hoyo *et al.* 1996; AnAge Database 2007
Eolophus roseicapillus		N	N	M	SC	0.3	Rowley 1990; del Hoyo *et al.* 1996; AnAge Database 2007
	F	Y					Rogers & McCullock 1981
Scopus umbretta		N	N	M	T	0.355	Cheke 1968; Prange *et al.* 1979; del Hoyo *et al.* 1996
Larus ridibundus	M	Y	N	M	C	0.26	van Rhijn & Groothuis 1987; del Hoyo *et al.* 1996; Animal Diversity Web 2007
	M	Y		M	C		van Rhijn & Groothuis 1985
	M	N		M	C		van Rhijn 1985
Larus atricilla		N	N	M	C	0.275	Hand 1985; del Hoyo *et al.* 1996; AnAge Database 2007
	M	Y		M	C		Noble & Wurm 1943
Larus argentatus		N	N	M	C	1	Shugart *et al.* 1987; del Hoyo *et al.* 1996; AnAge Database 2007
		N		M	C		Shugart *et al.* 1988
		N		M-OPy	C		Shugart 1980
Larus occidentalis	F	Y	N	M	C	0.761	Hunt *et al.* 1984; del Hoyo *et al.* 1996; AnAge Database 2007
	F	Y		M	C		Hunt & Hunt 1977
		N		M	C		Pierotti 1981
		N		M	C		Hunt *et al.* 1980
Larus novaehollandiae	M	Y	N	M	C	0.188	Mills 1994; del Hoyo *et al.* 1996; Iwaniuk & Nelson 2002
Larus delawarensis		N	N	M	C	0.5	Conover 1984; del Hoyo *et al.* 1996; Animal Diversity Web 2007
		N		M	C		Conover & Hunt 1984
		N		M	C		Fox & Boersma 1983

Table 7.1. (cont.)

Species	Sex Involved	Same-Sex Mounting	Sexual Segregation	Mating System	Sociality	Body Mass (kg)	Reference
Birds							
		N		M	C		Kovacs & Ryder 1985
		N		M	C		Kovacs & Ryder 1981
		N		M	C		Conover 1989
		N		M	C		Lagrenade & Mousseau 1983
Larus californicus		N	N	M	C	0.6	Conover 1984; del Hoyo et al. 1996; Animal Diversity Web 2007
Rissa tridactyla		N	N	M	C	0.386	Coulson & Thomas 1985; del Hoyo et al. 1996; Nagy 2001
Menura novaehollandiae	M	Y	N	Py	T	0.9	Smith 1988; del Hoyo et al. 1996; Rahn et al. 1985
Centrocercus urophasianus	F	Y	N	L	T	2.5	Gibson & Bradbury 1986; del Hoyo et al. 1996; Nagy 2001
Gallus gallus	F	Y	N	Py	SC	2.58	Guhl 1948; Banks 1956; del Hoyo et al. 1996; AnAge Database 2007
Fringilla coelebs		N	N	M	T	0.02	Marjakangas 1981; Linstroem 1989; Sheldon 1994; Sheldon & Burke 1994; Maciejok et al. 1995; Browne 2004; AnAge Database 2007
Carpodacus mexicanus	M	Y	Y	M	SS	0.02	Belthoff & Gauthreaux 1991; McGraw & Hill 1999; Lindstedt et al. 2007; Dhondt et al. 2007; Animal Diversity Web 2007
Ficedula hypoleuca		N	N	M	T	0.013	Slagsvold & Saetre 1991; Nagy 2001
Lanius collurio		N	Y	M	T	0.030	Fornasari et al. 1994; Jakober & Stauber 1994; Herremans 1997; Tryjanowski & Reuven 2002; AnAge Database 2007
Cyanistes caeruleus		N	N	M-FPy	T	0.011	Kempenaers 1994; Råberg & Stjernman 2003
		N		M-FPa	T		Kempenaers 1993
Corvus monedula		N	N	M-FPy	T	0.249	Roell 1979; AnAge Database 2007
Callaeas cinerea wilsoni		N	N	M	T	0.225	Flux et al. 2006; Higgins et al. 2006
Euplectes franciscanus	M	Y	N	Py	C	0.019	Craig 1980, 1982; Nudds & Bryant 2000
Poephila acuticauda	M	Y	N	M	C	0.015	Langmore & Bennett 1999; Higgins et al. 2006

Table 7.1. (cont.)

Species	Sex Involved	Same-Sex Mounting	Sexual Segregation	Mating System	Sociality	Body Mass (kg)	Reference
Birds							
Wilsonia citrina		N	Y	M	T	0.010	Niven 1993; Stutchbury 1994; Catry *et al.* 2005; Animal Diversity Web 2007
Chiroxiphia caudata	M	Y	N	L	T	0.025	Sick 1967; del Hoyo *et al.* 1996
Mionectes oleagineus		N	N	L	T	0.013	Westcott 1992; Westcott & Smith 1994; del Hoyo *et al.* 1996; Animal Diversity Web 2007
Perissocephalus tricolor		N	N	L	T	0.334	Trail 1990; del Hoyo *et al.* 1996
	M	Y	N	L	SC		Snow 1972; del Hoyo *et al.* 1996
Rupicola rupicola	M	Y	N	L	T	0.21	Trail & Koutnik 1986; Trail & Adams 1989; del Hoyo *et al.* 1996
Pseudonigrita arnaudi	M	Y	N	Pm	C	0.019	Collias & Collias 1980; Schluter & Repasky 1991; Perrins 2003
Philetairus socius	M	Y	N	M	C	0.025	Collias & Collias 1978; Nagy 2001; Covas *et al.* 2006
Tachycineta bicolor	M	Y	N	M	C	0.020	Lombardo *et al.* 1994; del Hoyo *et al.* 1996; Nagy 2001
Riparia riparia	M	Y	Y	M	C	0.014	Carr 1968; del Hoyo *et al.* 1996; Nagy 2001
Hirundo pyrrhonota	M	Y	Y	M	C	0.023	Brown & Brown 1996; Meek & Barclay 1996; AnAge Database 2007
Paradisaea raggiana		N	N	L	SC	0.215	Frith 1982; Beehler & Pruett-Jones 1983; Frith & Cooper 1996; McNab 2005
Notiomystis cincta	M	Y	N	M-Pa-Py-Pya	T	0.03	Castro *et al.* 1996; Ewen & Armstrong 2002; Ewen *et al.* 2004; Higgins *et al.* 2006
Lichenostomus melanops cassidix	M	Y	Y	M	T	0.023	Franklin *et al.* 1995; Moysey 1997; Higgins *et al.* 2006
Mammals							
Suncus murinus	M	Y	Y	Py	SO	0.085	Khokhar 1991; Dellovade *et al.* 1995; Matsuzaki 2004; Animal Diversity Web 2007
Dugong dugon	M	Y	Y	L-FPr	FSC	569	Anderson 1982, 1997, 2002; Animal Diversity Web 2007
Loxodonta africana		N	Y	Py	T	4540	Buss & Smith 1966; Laws 1969; Stokke & du Toit 2002; Animal Diversity Web 2007

Table 7.1. (cont.)

Species	Sex Involved	Same-Sex Mounting	Sexual Segregation	Mating System	Sociality	Body Mass (kg)	Reference
Mammals							
Felis catus	M>F	Y	N	Py	LG	3	Yamane 1999, 2006; Molsher 2006; Brown et al. 2007
Panthera leo persica	F	Y	N	Py	SC	199	Chavan 1981; Animal Diversity Web 2007
	M	Y		Py	SC		Pati 2001
Panthera leo leo	M	Y	Y	Py	SC	199	Cooper 1942; Animal Diversity Web 2007
	F	Y	Y	Py	SC		Cooper 1942
	M	Y	Y	Py	SC		Schaller 1972
Acinonyx jubatus		N	Y	Py	SC	45	Caro & Collins 1987; Earle 1987; Clutton-Brock 1989; Broomhall et al. 2003; Bissett & Bernard 2007
Helogale undulata rufula		N	N	Pa	SC	0.27	Rasa 1979; Rood 1980; Silva & Downing 1995
	M≫F	Y	Y	Pya	SC		Rasa 1977; Rood 1980
Canis familiaris	M≫F	Y	N	Pya	SC	35.5	Daniels & Bekoff 1989; Pal et al. 1999; Animal Diversity Web 2007
Canis lupus		N	N	M	SC	43	Derix et al. 1993; Morand & Poulin 1998; Jedrzejewski et al. 2001; Scandura 2004
Chrysocyon brachyurus	M	Y	N	M	SO	21.5	Kleiman 1972; Dietz 1984; Bandeira de Melo et al. 2007; Animal Diversity Web 2007
Speothos venaticus	M	Y	N	M	T	6	Kleiman 1972; Animal Diversity Web 2007
		N		M	T		Drüwa 1983
Ursus arctos horribilis		N	Y	Pya	T	233	Craighead et al. 1969; Wielgus & Bunnell 1995; Craighead et al. 1995; Morand & Poulin 1998; McLoughlin et al. 2002
Desmodus rotundus		N	Y	HPy	C	0.032	Wilkinson 1984, 1985; Park 1991; Gomes & Uieda 2004; Animal Diversity Web 2007
Myotis lucifugus	M	Y	N	Pr	SC	0.009	Thomas et al. 1979; Agosta et al. 2005; Psyllakis & Brigham 2006; Animal Diversity Web 2007

Table 7.1. (cont.)

Species	Sex Involved	Same-Sex Mounting	Sexual Segregation	Mating System	Sociality	Body Mass (kg)	Reference
Mammals							
Balaena mysticetus		N	Y	Py	SC	42500	Würsig et al. 1993; Richardson et al. 1995; Cosens & Blouw 2003; Animal Diversity Web 2007
Stenella longirostris	M	Y	N	Py-Pya	SC	65	Norris & Dohl 1980; Perrin & Mesnick 2003; Karczmarski et al. 2005; Silva et al. 2005; Animal Diversity Web 2007
Tursiops truncatus	M > F	Y	Y	Py	SC	175	Smolker et al. 1992; Wells & Norris 1994; Mann 2006; Animal Diversity Web 2007
		N			SC		Connor & Smolker 1995; Félix 1997; Samuels & Gifford 1997; Lusseau et al. 2003
	M	Y			SC		Shane et al. 1986
	M	Y			SC		Caldwell & Caldwell 1977; Östman 1991; Irvine et al. 1981
Odobenus rosmarus	M	Y	Y	Py	SC	1069.5	Sjare & Stirling 1996; Weckerly 1998
	M	Y	Y	Py	SC		Miller & Boness 1983
		N	Y	Py	SC		Miller 1976
	M	Y		Py	SC		Miller 1975
Halichoerus grypus	M	Y	Y	Py	SC	194	Backhouse 1960; Weckerly 1998; Breed et al. 2006
Mirounga angustirostris		N	Y	Py	SC	1600	Le Boeuf 1974; Weckerly 1998
Neophoca cinerea	M ≫ F	Y	Y	Py	SC	178.5	Marlow 1975; Weckerly 1998
Phocarctos hookeri	M	Y	Y	Py	SC	227	Marlow 1975; Weckerly 1998
Rattus norvegicus	(M = F)?	Y	Y	Py	SC	0.32	Barnett 1958; Animal Diversity Web 2007
Oryctolagus cuniculus	F	Y	N	Py-Pr	SC	2.25	Cowan 1987; Albonetti & Dessì-Fulgheri 1990; Morand & Poulin 1998
Notomys alexis	M	Y	Y	M	SC	0.035	Happold 1976; Dewsbury & Hodges 1987; Animal Diversity Web 2007
Pseudomys albocinereus	M	Y	Y	Pr?	SC	0.032	Happold 1976; Bubela & Happold 1993; Nagy 2001
Microcavia australis		N	N	Pya	SC	0.275	Rood 1972; Ebensperger et al. 2006; Animal Diversity Web 2007

Table 7.1. (cont.)

Species	Sex Involved	Same-Sex Mounting	Sexual Segregation	Mating System	Sociality	Body Mass (kg)	Reference
Mammals							
Marmota flaviventris	F	Y	N	Py	SC	3.5	Armitage 1998; Oli & Armitage 2003; Animal Diversity Web 2007
Marmota caligata	F	Y	N	M-Py-Pya	SC	6.35	Barash 1974; Kyle *et al.* 2007; Animal Diversity Web 2007
Marmota olympus	F > M	Y	N	Py-Pya	SC	6	Barash 1973; Animal Diversity Web 2007
Tamiasciurus douglasii	M	Y	N	Pr	T	0.226	Smith 1968; Koford 1982; Animal Diversity Web 2007
Tamiasciurus hudsonicus	M	Y	N	Pr	T	0.196	Smith 1968; Morand & Poulin 1998; Haughland & Larsen 2004
	F	Y		Pr	T		Layne 1954
Sciurus carolinensis		N	N	Pr-Py	T	0.55	Thompson 1977, 1978; Koprowski 1992, 1993; Morand & Poulin 1998
Macropus giganteus		N	Y	Pr	SC	49	Poole 1973; Grant 1973; Poole & Catling 1974; Kaufmann 1975; Jarman 1991; Weckerly 1998
			Y		SC		Jarman & Southwell 1986
Macropus rufogriseus banksianus				Pr		16.85	Jarman 1991; Weckerly 1998
		N	Y	Py	SC		Johnson 1989a,b
Macropus agilis	M	Y	N	Pr	FSC	15	Jarman 1991; Dressen 1993; Stirrat & Fuller 1997; Weckerly 1998; Blumstein *et al.* 2003
Macropus parryi	M	Y	N	Pr	SC	13.5	Kaufmann 1974; Jarman 1991; Weckerly 1998
Macropus rufogriseus rufogriseus				Pr		16.85	Jarman 1991; Weckerly 1998
	F	Y	N	Py	SO		La Follette 1971; Strahan 1983
Aepyprymnus rufescens	F	Y	N	Py	SO	2.48	Johnson 1980; Frederick & Johnson 1996; Animal Diversity Web 2007
		N					Frederick & Johnson 1996
Sminthopsis crassicaudata		N	Y	M	SC	0.016	Morton 1978a,b; Nagy 2001
Dasyurus hallucatus		N	N	Pr	T	0.25	Schmitt *et al.* 1989; White 1990; Oakwood 2002

Table 7.1. (cont.)

Species	Sex Involved	Same-Sex Mounting	Sexual Segregation	Mating System	Sociality	Body Mass (kg)	Reference
Mammals							
Kobus leche kafuensis		N	Y	L-FHPy	SC	85.5	Nefdt 1995; Weckerly 1998
Kobus vardoni		N	Y	L-HPy	SC	71.5	Rosser 1992; Weckerly 1998; Yearsley & Perez-Barbería 2005
	F	Y		L-HPy	SC		de Vos & Dowsett 1966
Kobus ellipsiprymnus		N	Y	Py	SC	210	Wirtz 1983; Weckerly 1998
	F≫M	Y	Y	Py	SC		Spinage 1982a,b
Gazella thomsoni	M	Y	Y	Py	SC	23	Walther 1978a,b; Weckerly 1998
Adenota kob thomasi		N	Y	Py	SC	174	Ledger & Smith 1964; Leuthold 1966; Animal Diversity Web 2007
	F	Y	Y	Py	SC		Buechner & Schloeth 1965
Antilope cervicapra	M≫F	Y	Y	HPy	SC	37.5	Dubost & Feer 1981; Animal Diversity Web 2007
Antilocapra americana	M	Y	Y	Pm	SC	49.7	Gilbert 1973; Maher 1997; Weckerly 1998; Carling *et al.* 2003
	M≫F	Y					Kitchen 1974
Odocoileus hemionus columbianus	F	Y	Y	Pm	SC	80.75	Wong & Parker 1988; Bowyer *et al.* 1996; Weckerly 1998; Ruckstuhl & Neuhaus 2002
Odocoileus virginianus	M	Y	Y	Pm	SC	65.75	Hirth 1977; Weckerly 1998; Ruckstuhl & Neuhaus 2002; Sorin 2004
Alces alces		N	Y	Py	SC	566.5	Miquelle *et al.* 1992; Ballenberghe & Miquelle 1993; Weckerly 1998
Cervus elaphus	M	Y	Y	Py	SC	297.5	Lincoln *et al.* 1970; Weckerly 1998
	F	Y	Y	Py	SC		Guinness *et al.* 1971
Cervus elaphus roosevelti	M	Y	Y	Py	SC	297.5	Harper *et al.* 1967, Rue 1989; Weckerly 1998; Ruckstuhl & Neuhaus 2002; Williams *et al.* 2002; Lung & Childress 2007
Cervus nippon	M	Y	Y	L	SC	36.01	Chapman *et al.* 1984; Weckerly 1998; Borkowski 2000; Ruckstuhl & Neuhaus 2002; Bartoš *et al.* 2003

Table 7.1. (cont.)

Species	Sex Involved	Same-Sex Mounting	Sexual Segregation	Mating System	Sociality	Body Mass (kg)	Reference
Mammals							
Dama dama	M	Y	Y	L	SC	57	Weckerly 1998; Holečková et al. 2000; Apollonio et al. 2003; Yearsley & Pérez-Barbería 2005; Focardi & Pecchioli 2005
Elaphurus davidianus	(M = F)?	Y	Y	HPy	SC	186.5	Schaller & Hamer 1978; Wemmer et al. 1983; Weckerly 1998; Jiang et al. 2004; Yearsley & Pérez-Barbería 2005
Muntiacus reevesi	M	Y	N	Py	SC	13.5	Barrette 1977; Chapman et al. 1997; Weckerly 1998; Yearsley & Pérez-Barbería 2005
Rangifer tarandus	(M = F)?	Y	Y	Py	SC	87	Bergerud 1974; Weckerly 1998; Ruckstuhl & Neuhaus 2002; Røed et al. 2005
Axis axis	M	Y	Y	Py	SC	69.5	Schaller 1967; Albes 1977; Moe & Wegge 1994; Weckerly 1998; de Silva & de Silva 2001
Pudu puda	M	Y	N	M?	SO	9	Mooring et al. 2002; Fisher et al. 2002; Yearsley & Pérez-Barbería 2005; Bartoš & Holečková 2006
Giraffa camelopardalis	M	Y	Y	Py	SC	900	Coe 1967; Weckerly 1998
	M > F	Y					Pratt & Anderson 1985
	M	Y	Y		SC		Innis 1958
	M	Y	Y		SC		Leuthold 1979
Vicugna vicugna	M?	Y	Y	Py	SC	50	Koford 1957; Animal Diversity Web 2007
Tayassu tajacu	M	Y	N	Py	SC	30	Dubost 1997; Morand & Poulin 1998
Tayassu pecari	(M = F)?	Y	N	Py	SC	32.5	Dubost 1997; Animal Diversity Web 2007
Phacochoerus aethiopicus	F	Y	Y	Pr	SC	55	Somers et al. 1995; Silva & Downing 1995
Bison bonasus	F	Y	Y	Py	SC	610	Krasiński & Raczyński 1967; Animal Diversity Web 2007
	M	Y	Y	Py	SC		Cabón-Raczyńska et al. 1987
	F	Y	Y	Py	SC		Jaczewski 1958
Bison bison	M	Y	Y	Py	SC	565.5	Rothstein & Griswold 1991; Weckerly 1998

Table 7.1. (cont.)

Species	Sex Involved	Same-Sex Mounting	Sexual Segregation	Mating System	Sociality	Body Mass (kg)	Reference
Mammals							
	M	Y	Y	Py	SC		Reinhardt 1985
	M > F	Y	Y	Py	SC		Vervaecke & Roden 2006
	M	Y		Py	SC		Reinhardt 1985
Bison bison athabascae	M	Y	Y	Py	SC	565.5	Komers *et al.* 1994; Weckerly 1998
Bos indicus	M ≫ F	Y	N	Py	SC	755	Reinhardt 1983; Animal Diversity Web 2007
Bubalus bubalis	F	Y	Y	Py	SC	725	Tulloch 1979; Nowak 1999; Animal Diversity Web 2007
Ovibos moschatus	M > F	Y	Y	Hpy	SC	232.5	Reinhardt & Flood 1983; Reynolds 1993; Cote *et al.* 1997; Weckerly 1998; Ruckstuhl & Neuhaus 2002
Equus caballus	M > F	Y	Y	Py	SC	1150	Feist & McCullough 1976; Animal Diversity Web 2007
Equus przewalskii	F	Y	Y	Py	SC	250	Boyd 1991; Animal Diversity Web 2007
		N					Dierendonck *et al.* 1996
Equus quagga		N	Y	Py	SC	275	Schilder & Boer 1987; Animal Diversity Web 2007
	M	Y	Y	Py	SC		Schilder 1988
Equus zebra zebra	M	Y	Y	Py	SC	350	Penzhorn 1984; Morand & Poulin 1998
Rupicapra pyrenaica ornata		N	Y	Py	SC	37	Lloyd & Rasa 1989; Rasa & Lloyd 1994; Pepin *et al.* 1996; von Hardenberg *et al.* 2000; Ruckstuhl & Neuhaus 2002; Animal Diversity Web 2007
		N					Lovari & Locati 1993
Pseudois nayaur	M	Y	N	Py	SC	47	Wilson 1984; Weckerly 1998; Lovari & Ale 2001; Cao *et al.* 2005
Ammotragus lervia	M ≫ F	Y	Y	Pr	SC	99.5	Habibi 1987a,b; Weckerly 1998
	M	Y					Katz 1949
Ovis orientalis musimon	F > M	Y	Y	Py	SC	45	McClelland 1991; Silva & Downing 1995
Ovis aries	M	Y	Y	Pr	SC	23	Orgeur *et al.* 1990; Weckerly 1998; Pemberton *et al.* 1999; Ruckstuhl *et al.* 2006
Ovis canadensis	M	Y	Y	Pr	SC	185	Geist 1968; Hogg 1987; Hass & Jenni 1991; Shackleton 1991; Weckerly 1998; Pelletier & Festa-Bianchet 2006

Table 7.1. (cont.)

Species	Sex Involved	Same-Sex Mounting	Sexual Segregation	Mating System	Sociality	Body Mass (kg)	Reference
Mammals							
Capra hircus	M	Y	Y	Py	SC	45	Orgeur et al. 1990; Saunders et al. 2005; Ruckstuhl 2007; Animal Diversity Web 2007
Cebus capucinus	M	Y	N	Pm	SC	3.267	Robinson & Janson 1987; Perry 1998; Weckerly 1998
Propithecus verreauxi	M	Y	N	Pya	SC	3.6	Richard 1974; Weckerly 1998
Saimiri sciureus		N	Y	Py	SC	0.763	Mitchell 1994; Weckerly 1998
	F	Y					Talmage-Riggs & Anschel 1973
	M	Y					Ploog et al. 1963
Nasalis larvatus	F = M	Y	Y	Py	SC	15.19	Yeager 1990; Weckerly 1998; Boonratana 2002
Miopithecus talapoin	M	Y	Y	Py	SC	1.25	Rowell 1973; Blaffer Hrdy & Whitten 1987; Weckerly 1998
Cercopithecus aethiops	M > F	Y	Y	Pm	SC	4.05	Gartlan 1969; Weckerly 1998; Whitten & Turner 2004
Hylobates lar	M	Y	N	M	SC	5.5	Edwards & Todd 1991; Weckerly 1998
Presbytis entellus	F	Y			SC	12.8	Sommer 1988; Weckerly 1998
	F	Y	Y	Py	SC		Srivastava et al. 1991
	M > F	Y	Y	Py	SC		Sommer et al. 2006
Papio ursinus	M	Y	N	HPy	SC	19.41	Hall 1962; Weckerly 1998; Altmann & Alberts 2003
Papio cynocephalus hamadryas	(F = M)?	Y	N	HPy	SC	13.85	Kummer 1995; Weckerly 1998
Papio cynocephalus anubis	M	Y	N	Pr	SC	19.5	Smuts & Watanabe 1990; Weckerly 1998
	M	Y		Pr	SC		Smuts 1987
	M ≫ F	Y		Pr	SC		Owens 1976
Theropithecus gelada	M ≫ F	Y	Y	HPy	SC	17.05	Bernstein 1975; Kawai et al. 1983; Weckerly 1998
Macaca fuscata	M ≫ F	Y	N	Pr	SC	10.1	Hanby & Brown 1974; Weckerly 1998
	F	Y	N	Pr	SC		Vasey & Duckworth 2006
	F	Y	N	Pr	SC		Vasey 2006a
	F	Y	N	Pr	SC		Vasey 2002a
	F	Y	N	Pr	SC		Vasey & Gauthier 2000
	F	Y	N	Pr	SC		Vasey et al. 1998
	F	Y	N	Pr	SC		Vasey 1998
	F	Y	N	Pr	SC		Vasey 1996
	M	Y	N	Pr	SC		Hanby 1974
	F	Y	N	Pr	SC		Wolfe 1986

Table 7.1. (cont.)

Species	Sex Involved	Same-Sex Mounting	Sexual Segregation	Mating System	Sociality	Body Mass (kg)	Reference
Mammals							
	F	Y	N	Pr	SC		Takahata 1982
	(F = M)?	Y	N	Pr	SC		Enomoto 1974
	M	Y	N	Pr	SC		Takenoshita 1998
	F	Y	N	Pr	SC		Lunardini 1989
	F	Y	N	Pr	SC		Chapais & Mignault 1991
	F	Y	N	Pr	SC		Rendall & Taylor 1991
	F	Y	Y	Pr	SC		Corradino 1990
Macaca arctoides	F	Y	N	Pm	SC	7	Slob *et al.* 1986; Blaffer Hrdy & Whitten 1987; Silva & Downing 1995
	(M > F)?	Y			SC		Chevalier-Skolnikoff 1976
	F > M	Y			SC		Chevalier-Skolnikoff 1974
		N			SC		Gouzoules 1974
Macaca mulatta	F	Y	N	Pm	SC	4.6	Akers & Conaway 1979; Weckerly 1998
	M	Y			SC		Reinhardt *et al.* 1986
	F	Y			SC		Fairbanks *et al.* 1977
	M	Y			SC		Gordon & Bernstein 1973
	M > F	Y	Y		SC		Bernstein *et al.* 1993
	F	Y			SC		Loy & Loy 1974
	F	Y			SC		Kapsalis & Johnson 2006
	F	Y			SC		Huynen 1997
Macaca nemestrina	F	Y	N	Pm	SC	8.5	Giacoma & Messeri 1992; Weckerly 1998
	(M = F)?	Y			SC		Oi 1990
Macaca radiata	M	Y	N	Pm	SC	5.15	Silk 1994; Weckerly 1998
Macaca tonkeana	(M = F)?	Y	N	Pm	SC	12	Thierry 1986; Lindefors 2002
Macaca nigra	M	Y	N	Pm	SC	8	Silva & Downing 1995; Reed *et al.* 1997
	(F = M)?	Y			SC		Dixson 1977
Pan paniscus	F ≫ M	Y	N	Pr	SC	35	Weckerly 1998; Hohmann & Fruth 2000
	F ≫ M	Y	N	Pr	SC		Fruth & Hohmann 2006
Pan troglodytes	F	Y	N	Py?	SC	45	Weckerly 1998; Firos Anestis 2004
	N	N		Py?	SC		Nishida 1997
	F	Y	N	Py?	SC		Yerkes 1939
Gorilla gorilla	M	Y	Y	Py	SC	126.5	Weckerly 1998; Yamagiwa 2006
	M	Y	Y	Py	SC		Yamagiwa 1987
	F	Y	Y	Py	SC		Harcourt 1979
	M ≫ F	Y	Y	Py	SC		Nadler 1986
	M	Y	Y	Py	SC		Robbins 1996

Table 7.1. (cont.)

Species	Sex Involved	Same-Sex Mounting	Sexual Segregation	Mating System	Sociality	Body Mass (kg)	Reference
Mammals							
	F	Y	Y	Py	SC		Fischer & Nadler 1978
Pongo pygmaeus	M	Y	N	HPy	T	53	Weckerly 1998; Fox 2001, 2002
Homo sapiens	M>F	Y	Y	M-FPm	SC	57	Kinsey *et al.* 1948; Brown *et al.* 2007
	M>F	Y	Y	M-FPm	SC		Kinsey *et al.* 1953
	M	Y	Y	M-FPm	SC		Hensley 2000 (*Prison)
	F	Y	Y	M-FPm	SC		Hensley 2000 (*Prison)
	M	Y	Y	M-FPm	SC		Ashworth & Walker 1972 (*Boarding school)
	F	Y	Y	M-FPm	SC		Ashworth & Walker 1972 (*Boarding school)

Mating systems: facultative promiscuity (FPr), monogamy (M), lek (L), facultative polygyny (FPy), polygynandry (Pya), polygyny (Py), polyandry (Pa), harem polygyny (HPy), occasional polygyny (OPy), polygamy (Pm), promiscuity (Pr), facultative polygamy (FPm), facultative polyandry (FPa), facultative harem polygyny (FHPy). Sociality: colonial (C), territorial (T), group-living (G), loosely colonial (LC), semi-social (SS), social (SC), solitary (SO), facultatively social (FSC), loose group (LG), flock (F). References for Sex Involved and Same-sex Mounting can be found in Table 2.1.

proposed for birds and mammals. This will be followed by an extensive review of sexual segregation across both mammals and birds, with some comments on various evolutionary links that may exist between sexual segregation and same-sex mounting. I then suggest evolutionary models specifically linking sexual segregation and homosexual behaviours, to finally test the models through comparative analyses in birds and mammals separately.

List of hypotheses proposed to explain sexual segregation

The hypotheses that have been proposed to explain sexual segregation may be broadly grouped into five categories: *Stochasticity, Sociality, Predation, Foraging* and *Reproduction*. Below I list the hypotheses with a very brief explanation and include examples of species whose sexual segregation has been attributed to the mechanism implied by the hypothesis. The examples are taken from references included in the literature review that follows and are not intended to be exhaustive.

Stochasticity hypotheses

Stochastic Effect: Sexual segregation is the result of a random association of individuals in

Table 7.2. Patterns of spatiotemporal distribution of the sexes

DISTRIBUTION OF SEXES	
Sexual segregation	**Sexual aggregation**
(I) *Environmental Sexual Segregation*:	
(a) Habitat sexual segregation	
(b) Spatial sexual segregation	
(II) *Temporal sexual segregation*	
(III) *Social sexual segregation*	

groups, under conditions of biased sex ratios (Conradt 1998).
Example: possibly *Inia geoffrensis*.

Sociality hypotheses

Social Factors (Main *et al.* 1996, Le Pendu *et al.* 2000b, Wolf *et al.* 2005): Sexual segregation is determined by social interactions among individuals such as females herded by a male in a harem, whereas males excluded from breeding form bachelor groups during the mating season, where they may engage in the establishing of dominance relationships or improving specific skills (e.g. fighting; see also the *Social Preference* hypothesis mentioned by Bon & Campan (1996) and Ruckstuhl & Neuhaus (2000)). Females with dependent offspring may subsequently segregate themselves from all males during lactation in order to protect the developing young.
Example: *Loxodonta africana*.
A more subtle version of this hypothesis is known as
   Coercion Defence (Brereton 1995): This hypothesis posits that females not only spatially segregate, but also form coalitions within a larger social unit in order to defend themselves against males' harassment. This hypothesis is also related to the *Social Dominance* and *Behavioural Incompatibility* hypotheses.
   Example: *Tursiops sp.*
   Social Dominance (Weckerly *et al.* 2001, Catry *et al.* 2005): One sex is dominant over the other and the dominant sex excludes the other sex from specific areas.
   Example: *Falco sparverius*.
   Behavioural Incompatibility: In sexually dimorphic species, juvenile females avoid agonistic and sexual acts of males (Bon & Campan 1996).
   Example: *Ovis gmelini*.

Predation hypotheses

Predation Risk: Gestation and offspring rearing increase sensitivity to predation in females and dependent offspring, thus promoting sexual segregation (Bon & Campan 1996, Ruckstuhl & Neuhaus 2000):
Example: *Ovis canadensis*.

Foraging hypotheses

Activity Budget (Ruckstuhl & Neuhaus 2000, Ruckstuhl & Kokko 2002): Different energetic constraints affecting activity budgets may result in sexual segregation (see also Bon & Campan 1996, Calhim *et al.* 2006). The hypothesis has been criticised by Yearsley & Pérez-Barbería (2005).
Example: *Capra hircus*.
This hypothesis is related to:
   Forage Selection: Gestation and offspring rearing determine sex-specific physiological constraints promoting sexual segregation (Main & Coblentz 1990, Ruckstuhl & Neuhaus 2000).
   Example: *Alces alces*.
   Sexual Dimorphism – Body Size (Main *et al.* 1996, Stewart 1997, Le Boeuf *et al.* 2000, Pérez-Barbería & Gordon 2000 and references therein): Sexual body size dimorphism determines sex-specific energetic constraints, leading to habitat sexual segregation that follows a patchy distribution of appropriate food items.
   Example: *Bos bison*.
Specialisation (Catry *et al.* 2005): The two sexes are specialised in the use of specific resources that can affect not only spatial, but also temporal segregation (e.g. differences in arrival time at breeding grounds).
Example: *Thalassarche melanophrys*.
Gastrocentric (Barboza & Bowyer 2000, 2001; Bowyer 2004): Originally proposed to explain

sexual segregation in ruminants. Sexes segregate as a result of their different body size, minimum quality of food required, digestive retention (proportional to rumen size in ruminants), and sex-specific reproductive requirements.

Example: *Odocoileus virginianus*.

Reproduction hypotheses

Reproductive Strategy (Bergerud & Gratson 1988, Gruys 1993, Main *et al.* 1996): Females prefer habitats that increase survival of themselves and their immature offspring, whereas males remain on the breeding grounds to defend breeding sites or move to areas where they can find appropriate food.

Example: *Ovis canadensis*.

Sex (Main & Coblentz 1990, Wielgus & Bunnell 1995): Females and their dependent young segregate into a specific area away from males to avoid infanticide or sexually motivated aggression from males when females are not reproductively receptive.

Example: *Ursus arctos*.

Moult Migration (Catry *et al.* 2005): The sex that moults earlier in migratory bird species at the end of the breeding season is also ready earlier to undertake seasonal migration. Sex-specific departure time will produce sexual segregation.

Example: Many anatids.

Male Intervention (Lyons *et al.* 1992): Inter-male aggression prevents males from physically interacting with females, thus maintaining spatial sexual segregation. Spatial sexual segregation through a male intervention mechanism may manifest itself at very small scales.

Example: *Saimiri sciureus*.

Mating System: In polygamous mating systems (e.g. those based on harems), the sexes may be largely segregated during mating independently of any other factor. In fact, sexual segregation is particularly pronounced in cervids and bovids, which are predominantly polygynous (Main *et al.* 1996).

Example: *Syncerus caffer*.

Indirect Paternal Care: Males may leave preferred habitats to females and dependent young of their own accord as an indirect form of parental care, as decreased competition for food with females and their offspring will increase the probability of survival of those young. This hypothesis predicts that only males who participated in copulations will tend to leave the food-richest sites to females and dependent offspring.

Example: possibly *Myotis daubentonii*.

Altruistic (or, more appropriately, Kin Selected): Main & Coblentz (1990) proposed a similar hypothesis to the *Indirect Parental Care* hypothesis that they called the *Altruistic* hypothesis. The difference between the *Indirect Parental Care* and the *Altruistic* hypotheses is that in the latter all males will segregate from females and their dependent offspring provided that they are related to them, but independently of whether the male participated or not in copulations.

Example: None available.

Competition Avoidance (e.g. Mooring *et al.* 2003): A subtle version of the *Altruistic* hypothesis, which posits that sexually mature males will actively segregate from females and their offspring not only to enhance fitness of the young, which may be their own (*Indirect Parental Care*) or genetically related in a broader sense (*Altruistic*), but also to enhance survival of females, i.e. their sexual partners in future mating interactions.

Example: None available.

Although the predictions (all or some) of each one of the above hypotheses are necessarily unique, the hypotheses themselves are not necessarily alternative. In fact, some of them may well be complementary; for instance, sexual segregation may be a result of the combined effects of sex-specific foraging constraints and predator avoidance. Moreover, it will be noticed that some of the hypotheses are just subtle variations around a common theme; e.g. sex-specific physiological constraints.

A review of sexual segregation studies

In the simplest models, sexual segregation results from different constraints faced by the two sexes and individual decisions to move to areas where – or be active at times when – the specific needs of individuals of each sex are satisfied. Whenever preferred environmental conditions are separated in space or time, sexes will become segregated (Ruckstuhl & Kokko 2002). One such constraint is body size. Although body size dimorphism in mammals can be an evolutionary outcome of niche specialisation between the sexes, sexual selection or intersexual competition for food (or a combination of those) (Hedrick & Temeles 1989), once body size dimorphism has evolved it may reinforce sexual segregation (Pérez-Barbería et al. 2002). The more niche-specialised the sexes become, the more segregated they will be during specific periods of the year.

In a comparative analysis of the Artiodactyla (even-toed ungulates), Pérez-Barbería & Gordon (2000) supported the *Sexual Dimorphism – Body Size* hypothesis to explain sexual segregation in this group. However, they also suggested a more sophisticated evolutionary model (Pérez-Barbería & Gordon 2000: 675) linking body size, sexual dimorphism, alimentary constraints and mating system:

> *occupancy of open habitats increases the aggregation of individuals* and favours polygyny, and sexual dimorphism evolves because a larger body size confers advantages during fights between males for breeding access to females in polygynous mating systems. Subsequently, sexual dimorphism favours sexual segregation because of the differences between the sexes in the efficiency of selecting and digesting food. However, sexual segregation could also favour increased levels of polygyny [italics mine].

Thus the benefits conferred in open habitats by group formation (e.g. antipredator defence) may trigger a cascade of evolutionary events leading to polygyny and sexual size dimorphism to finally result in sexual segregation in some species.

After carrying out a comparative analysis of ungulates, Ruckstuhl & Neuhaus (2002) concluded that sexual segregation in this group is mainly a result of sexual differences in activity budgets, whereas risk of predation and selection for particular food items are additional factors. We will see below that social sexual segregation cannot be just described as a side effect of independent individual decisions to move to a preferred, patchy habitat. More often than not, group formation will have its own potential selective relevance independent of sexual segregation (e.g. as an antipredator strategy). In this context, sexual segregation will result from factors that determine the monosexual composition of the group but not necessarily the tendency to form groups. Both processes, however, can be linked, whenever it can be demonstrated that individuals actively prefer members of the same sex in the process of group formation (see, for example, Pérez-Barbería et al. 2004). Such an active preference must be experimentally demonstrated, however, as Pérez-Barbería et al. (2004) did, as it cannot be deduced from the patterns of monosexual group formation itself. Monosexual groups will always be necessarily but passively formed as a result of, for instance, sexual differences in physiological needs and food preference plus group formation in an environment where the distribution of the preferred food is patchy (Ruckstuhl & Neuhaus 2002). Why should a species evolve individual active preference to join conspecifics in monosexual groups? We will see in the review below that in many species time spent in a monosexual group confers additional advantages in learning, establishing dominance hierarchies and, in some species, establishing affiliative bonds that reinforce useful coalitions in intra- and interspecific competition, antipredator defence and food acquisition (these and other factors will be incorporated into an initial evolutionary model of homosexuality in Chapter 8).

Caprinae

The Caprinae (sheep, goats and others) is a subfamily of the Bovidae. Among the Caprinae, *Capra hircus*, studied on the Isle of Rum in Scotland by Calhim *et al.* (2006), manifests sexual segregation more commonly in spring and summer, except during the rut period. The study by Calhim *et al.* supports the *Activity Budget* hypothesis for *C. hircus*, although they suggest that an element of social avoidance may be also playing a role in sexual segregation in this species at their study site. Neuhaus & Ruckstuhl (2002a) studied Alpine ibex (*Capra ibex*) in France in summer (Figure 7.1). Although differences in activity budget or habitat preference between the sexes could explain sexual spatial segregation in the Alpine ibex, other factors are also likely to contribute to sexual segregation in this and other species of Caprinae (Neuhaus & Ruckstuhl 2002a). For instance, Pérez-Barbería *et al.* (2007a) carried out a controlled experiment testing for activity budget effects in Soay sheep (*Ovis aries*) and found no supporting evidence.

During periods when the sexes are segregated, caprids may display same-sex mounting. It has been suggested that female–female mounting in *C. hircus* may have evolved to signal female readiness to mate to males from a distance (Shearer & Katz 2006). Such behaviour would undoubtedly make adaptive sense in sexually segregated species such as *C. hircus*. However, what Shearer & Katz (2006) also found was that the mounting female is not always in oestrus, whereas the mountee is. If so, what is the mounter female communicating to the distant males? In this case it seems more plausible to think that the mounter is not necessarily communicating with distant males, instead she may be primarily involved in reproductive interference with the mountee. It is well known that in many social mammals dominants attempt to interfere with the reproductive capacity of subordinates, sometimes leading to reproductive suppression of subordinates (see, for example, Albonetti & Dessì-Fulgheri 1990; Creel *et al.* 1995; Woodroffe & MacDonald 1995; Asa & Valdespino 1998). In particular, F–F mounting may cause 'pseudopregnancy' in the mountee, i.e. a prolonged dioestrous or luteal phase (see, for example, Asa & Valdespino 1998). In fact, in *C. hircus*, a species where pseudopregnancy has been widely reported, oestrus can be modified, when the mounter is a male for instance, owing to the mechanical effects of mounting and intromission rather than the effects of fluids secreted by the male accessory glands (Romano & Benech 1996). Female mounting of another female may well produce similar results. This could also explain why Shearer & Katz (2006) recorded some degree of F–F mounting in *C. hircus* even in the close presence of a male.

Pérez-Barbería *et al.* (2004) studied sexual segregation in Soay sheep in Scotland using an experimental maze setup that allowed an individual to have free choice of alternative associates. The authors showed that both males and females prefer to associate with individuals of the same sex. This social preference for a same-sex partner is expected, for instance, from the *Mating System* hypothesis. In fact, such same-sex preference in this species did not change during the rut period, when many males may not have access to females, in spite of the capability for promiscuous mating behaviour, as described for the *O. aries* population in St Kilda, Scotland (Coltman *et al.* 1999). In spite of the common tendency described in both males and females to form monosexual groups, ewe groups in *O. aries* are socially more stable than ram groups, perhaps reflecting the relatively greater relatedness among females in a flock than among males in bachelor groups (Pérez-Barbería & Gordon 1999).

Predation risk can also explain sexual segregation and group formation in the Caprinae. For instance, in the Dall's sheep (*Ovis dalli dalli*) studied by Corti & Shackleton (2002) in the Canadian province of Yukon, the authors described a pattern of spatial sexual segregation in late spring and early summer whereby females with lambs occupy more secure sites, with higher cover, whereas all-male groups were larger and stayed away from cover. Different susceptibility to predation between females with

Figure 7.1. Monosexual groups of bachelor male (A) and female (B) Alpine ibex (*Capra ibex*). Female groups may also contain young males that have body mass similar to females. As males grow to a larger size they either form bachelor groups of similarly aged males or join groups of older males. Photos taken by Peter Neuhaus.

lambs and males could explain this pattern. In *Ovis canadiensis*, however, Ruckstuhl (1998) suggested that sexual social segregation can be explained as a side effect of independent individual decisions of males and females at their study site in Alberta, Canada. According to Ruckstuhl, the sexes are expected to have different activity time budgets as a result of their sexual body size dimorphism and this alone would be sufficient to determine differences in activity patterns. Although such factors could indeed explain sexual segregation in *O. canadensis*, with the additional potential influence of antipredator defence to explain group formation, social interactions between and within sexes also remain a valid hypothesis. Mooring *et al.* (2003) studied desert bighorn sheep (*Ovis canadensis mexicana*) in New Mexico. *O. c. mexicana* is segregated in the study area into all-male groups, female groups with both dependent offspring and yearlings, and mixed mating groups of sexually mature males, females and immatures. They suggest that this pattern of sexual segregation can be explained by a combination of *Reproductive Strategy* and *Predation Risk* hypotheses.

Mouflons (*Ovis gmelini*) in France distribute themselves following a pattern very similar to that of *O. c. mexicana*: female groups with young, all-male groups and mixed-sex groups (Cransac & Hewison 1997). Cransac & Hewison (1997) found that all-male group formation in mouflons may be favoured by an ontogenetic pattern of close interactions between males throughout their postnatal development. Thus, although sociality may be a selective response to environmental factors such as risk of predation, formation of all-male groups may be enhanced by early social interactions among males. In polygynous species where chances of becoming a breeder early in life are slim for males, mechanisms that favour group living such as the development of early male–male affiliative bonds could have a significant selective advantage in survival. Le Pendu *et al.* (2000) studied *O. gmelini* in Germany in winter and also described social sexual segregation, which is presumably driven by social factors as it can occur even within mixed-sex social groups. Le Pendu *et al.* (2000) suggest that outside the rut females tend to avoid sexual or agonistic behaviours of males, hence the sexual segregation. Rams tend to interact among themselves rather frequently in a fashion that is not just affiliative but also agonistic (Le Pendu *et al.* 2000).

In sum, sexual segregation in the Caprinae seems to be driven by differential nutritional requirements of the two sexes associated with differences in body size and also with gestation and lactational constraints faced by females. Those same asymmetries between the sexes may make them differentially vulnerable to predators. Antipredator defence could also explain group formation in the segregated sexes, facilitated by the tendency of individuals to seek the company of conspecifics, with some preference bias for same-sex conspecifics. This tendency to seek the company of same-sex conspecifics, however, seems to be as strong in females as it is in males among the Caprinae, with some variability across species. Once sexual segregation has been established, same-sex mounting may occur in both all-male and all-female groups. Evidence suggests that in the Caprinae same-sex mounting has, at the very least, a socio-sexual function associated with dominance and interference with reproduction.

Other bovids

Sexual segregation is also common among bovids other than the Caprinae. In the kudu (*Tragelaphus strepsiceros*), a browsing antelope studied in South Africa by Du Toit (1995), there is social sexual segregation, especially outside the rut, with males forming unstable all-male bachelor groups and females associating with other females, calves and subadults in more stable group compositions. In this species the *Antipredator* hypothesis is more likely to explain sexual segregation. On the other hand, the *Sexual Dimorphism – Body Size* hypothesis seems to explain sexual segregation in wild bison (*Bos bison*) in Kansas (Post *et al.* 2001), whereas feral cattle (*Bos taurus*) in Doñana, Spain, spatially segregate by body size irrespective of sex: larger males vs. females and smaller males, presumably as a result of dietary constraints (Lazo & Soriguer 1993).

Turner *et al.* (2005) studied the African buffalo (*Syncerus caffer*) in KwaZulu–Natal, South Africa. *S. caffer* segregates into all-male groups, groups of females with young, and mixed herds. Mature males in mixed herds are involved in costly reproductive activities that prevent them from feeding adequately. However, they regularly transfer to all-male herds, where they spend more time feeding and recuperating body condition (Turner *et al.* 2005). In fact bulls in mixed herds spent more time moving, courting females and scanning than feeding. In a species such as this, which displays breeding activity throughout the year, the sustained costs of mating (e.g. in terms of loss of body mass) could jeopardise lifetime reproductive success, unless breeding is intercalated with periods of more sustained foraging. Joining all-male groups affords such an opportunity. Why should bulls improve body condition by joining all-male groups rather than foraging alone or perhaps staying in the mixed herd but ignoring females in oestrus? Turner et al. suggest that one potential advantage for bulls that join all-male groups is the establishment of dominance relationships with other males. An established dominance hierarchy with potential reproductive competitors could save the bull valuable time and energy while he is attending the females in the mixed herd. This is a reasonable adaptive scenario, which, however, should be treated as a hypothesis requiring specific testing.

In sum, sexual segregation in bovids, including the Caprinae, follows a general pattern that is also shown by other artiodactyls such as the cervids (see below), of a process driven by energetic and alimentary constraints and sociality favoured by antipredator defence, with the addition of more active preferences for joining same-sex groups driven by social benefits such as establishment of dominance relationships.

Cervids

Although sexual segregation is especially prominent among polygynous ungulates in general, it is particularly pronounced in the cervids (Stewart et al. 2003). Stewart et al. (2003) studied white-tailed deer (*Odocoileus virginianus*) in Texas and concluded that sexual segregation in this species is in part explained by the *Gastrocentric* hypothesis, as larger-bodied males prefer habitats of abundant, although low-quality food (e.g. graminoids), whereas females prefer habitats of higher-quality food (e.g. forbs and shrubs) (Stewart et al. 2003). The study by Jenks et al. (1994) of *O. virginianus* in Oklahoma and Arkansas also supports the *Gastrocentric* hypothesis for this species. However, it is clear that the *Gastrocentric* hypothesis is unlikely, on its own, to fully account for sexual segregation patterns in *Odocoileus*, as indicated by Weckerly's (1993) work on the black-tailed deer (*O. hemionus columbianus*) in California. Sexual segregation in this species is not associated with differences in food requirements determined by body size sexual dimorphism (Weckerly 1993).

Kie & Bowyer (1999) carried out an experimental study of the effect of predation on sexual segregation in a Texas population of *O. virginianus*. They observed that, after removal of predators population density increased and sexual segregation decreased throughout the year, a differentiation of diet among the sexes followed decreased sexual segregation. Main & Coblentz (1996) carried out a study of Rocky Mountain mule deer (*O. hemionus hemionus*) in Oregon that supported the *Reproductive Strategy* hypothesis, as sexual segregation seemed to be best explained by females' avoidance of predators (e.g. coyotes) and food and water access limitations imposed by dependent offspring, whereas males were free to maximise their food intake in appropriate habitats. In a population of fallow deer (*Dama dama*) in Italy, females more than males segregated in response to the presence of potential predators (in this case human visitors), especially when they had fawns (Ciuti et al. 2004). Thirgood (1996) studied *D. dama* in England where the species displays social sexual segregation that is more pronounced in females than males, mirroring the pattern found in Italy. Female group size changed according to season (larger in winter and spring) and habitat (larger in open than closed habitats) suggesting a potential antipredator response. Interestingly, males tended to occur more often in all-male groups as the proportion of males in the population increased (Thirgood 1996). Decreased sexual segregation when proportion of males is low can be easily explained by benefits of group living (e.g. antipredator defence) overcoming benefits of sexual segregation. Similarly, when the sex ratio is heavily biased towards one sex, group formation may lead to (partial) sexual segregation as a result of purely stochastic processes.

Bonenfant et al. (2004) studied European red deer (*Cervus elaphus*) in France and Norway. The authors described pronounced habitat sexual segregation during the calving season, but otherwise the tendency was for the sexes to be socially sexually segregated. Harem size was smaller in their two forested habitats compared with the open habitat of the species on the Isle of Rum (Scotland). This suggests that the size of bachelor groups is also likely to be larger on the Isle of Rum during the rut and therefore it may be expected that same-sex sexual behaviours in males are also more likely to be manifested there. In a study of Roosevelt elk (*Cervus elaphus roosevelti*) carried out in California, Weckerly et al. (2001) showed that spatial sexual segregation was mainly explained by food competition. In fact, mixed-sex groups were more likely to form when food was abundant. However, females also displayed a higher degree of social cohesion than males regardless. Winnie & Creel (2007) studied the potential effect of predation by wolves (*Canis lupus*) on *C. elaphus*'s tendency to form groups in Montana (USA). Risk of wolf predation affected group size in a sex-specific manner: mixed herds of females with young decreased in size in response to increased predation risk from wolves, whereas bull-only herds tended to increase in size, leading to an overall convergence in group size. However, this effect per se does not explain sexual segregation in this species. Winnie & Creel (2007) suggest that sexual segregation in winter may be explained by the need of bulls to recover body mass lost during the rut, whereas group formation could be explained by the threat from wolves.

Thamins (*Cervus eldi thamin*) studied by McShea et al. (2001) in Myanmar also sexually segregate according to habitat, apparently as a result of predator avoidance and specific nutritional requirements imposed by lactation in females. Hirotani (1990) studied reindeer (*Rangifer tarandus*) in a semi-domestic setting in Finland. Although the author describes greater social associations between females than males that are maintained throughout most of the year, at the individual level those relationships are unstable and group membership is fluid. In the case of the huemul (*Hippocamelus bisulcus*) of southern Chile, Frid (1999) described that although huemul groups are mainly mixed-sex, females seem to have a greater propensity to maintain same-sex social bonds than males. Group formation in this species is interpreted as an antipredator response (Frid 1999).

Sexual segregation has also been studied in the Alaskan moose (*Alces alces gigas*). Miquelle et al. (1992) studied a wild population of this species, showing that social sexual segregation is more pronounced in winter than at other times of the year. However, all-male groups are always present, although at varying frequencies: from 6.5% of all group types (all-male, all-female and mixed) during the rut to 41.2%–42.7% in winter. All-male groups during the rut are expected whenever some males are unable to compete for access to females (owing to age or body condition constraints) but associate with other males owing to the benefits of group living (e.g. antipredator defence). This situation can set the stage for homosexual interactions between males in all-male groups driven by neuroendocrine conditions during the rut. This, however, would not be in contradiction with additional sociosexual functions of same-sex sexual interactions in all-male groups during the rut associated, for instance, with establishment of dominance hierarchies. Interestingly, Miquelle et al. (1992: 23) mention that 'Groups composed of small males were extremely rare (1% of observations). When observed in groups, small males were more likely to be associated with females than were large males in winter..., and summer..., but not during the rut...or postrut'. Association of small males with female groups outside the rut is consistent with the *Forage Selection* and *Predation Risk* hypotheses: similarity of body sizes between small males and females allow them to feed in the same area, whereas group formation may be favoured by antipredator defence, which is of benefit to both male and female members of the group. During the rut, however, small males are more likely to be excluded from the female group by larger males as the former will be treated as reproductive competitors by the latter. That small males are reproductive competitors to large males is suggested by their tendency to court non-receptive females (Miquelle *et al.* 1992: 42). Therefore, in spite of preferentially joining female groups, small male moose are neither feminised nor do they display a homosexual sexual orientation.

Group living can be an antipredator strategy through dilution of predation risks, also known as the *selfish herd effect* (Hamilton 1971) or through active cooperation in defence, or both. When selfish herd effects prevail, however, individuals will have to balance the positive dilution effects with the negative effects that living in larger groups imply in terms of increased detectability to predators, but also increased food competition, increased potential for harassment from conspecifics, and transmission of pathogens. This balance can explain why females with calves tend to be less social (Miquelle *et al.* 1992).

In sum, the major mechanisms for sexual segregation in cervids mirror, to a great extent, those proposed above for the Caprinae and other bovids: sexual body-size dimorphism, and also gestation and lactation in females, impose alimentary limitations that are different in both sexes thus driving the spatial or habitat sexual segregation. Predation may also help explain the differential preference of males and females with calves for different habitats and, in addition, it may help explain group living in both sexes. In cervids, females show a greater tendency to seek the company of same-sex individuals than males thus suggesting the presence of an active rather than passive mechanism of monosexual group formation. In both bovids and cervids, it is clear that same-sex mounting can have, at the very least, a function in the establishment and maintenance of dominance hierarchies.

Giraffids and equids

In the giraffids Caister *et al.* (2003) described social sexual segregation within herds of Niger giraffes (*Giraffa camelopardalis peralta*). *G. c. peralta* does not have a defined breeding season, and male–male competition can occur the year round. According to Caister *et al.*, spatial sexual segregation could be explained in this species by the *Reproductive Strategy* and *Predation Risk* hypotheses. Masai giraffes (*G. c. tippelskirchi*) were studied by Ginnett & Demment (1999) in Tanzania. The authors suggest that spatial sexual segregation in this subspecies is better explained by the *Reproductive Strategy* hypothesis and, given that it is females with their dependent offspring who seem to prefer specific habitats not used by males and females without offspring, the *Predation Risk* hypothesis could also have explanatory value.

Other ungulates such as the perissodactyls (horses, zebras, donkeys and the like) have also been studied. The equid plains zebra (*Equus quagga*) is a non-ruminant, sexually monomorphic for body size, yet polygynous species, with a stallion controlling a harem of up to six females at a study site in Namibia (Neuhaus & Ruckstuhl 2002b). This species can breed throughout the year; Neuhaus & Ruckstuhl (2002b) observed only one bachelor group of seven males.

On the basis of this review of sexual segregation in ungulates, it is reasonable to suggest that a general model for sexual segregation in this taxon should consider sex-specific alimentary requirements related to physiological differences between the sexes (e.g. body size and also lactation in females) favouring habitat and spatial segregation. In addition, predation is likely to be a major selective pressure favouring group living, thus leading to social sexual segregation. Group living may increase selection towards polygyny – which is quite common among ungulates – as females clumped into groups will be more easily defendable from other males. This in turn may drive body size differences between the sexes even further via sexual selection (e.g. male–male dominance interactions). As soon as polygyny is established, sexual segregation will occur during the mating period as well, as males unable to control groups of females will have the alternative to either wander alone or join other bachelors in an all-male group as an antipredator strategy and as a strategy to establish dominance relationships with other males (i.e. potential competitors). Such dependence on group living may further select for specific propensities to seek the companionship of members of the same sex. Group formation seems to be relatively more characteristic of females than males in ungulates. A female-biased propensity towards sociality and affiliative behaviour is also widespread among primates (Silk *et al.* 2006a,b) although it is not totally absent in males either (Hill 1994; Silk 1994; van Hoof & van Schaik 1994). This hypothetical evolutionary scenario for sexual segregation, which can be reconstructed on the basis of the available evidence, predicts that in ungulates same-sex sexual behaviour could be a consequence of *both* sexual and socio–sexual mechanisms operating during breeding periods in particular, and facilitated by the formation of monosexual groups.

Bears, polecats, marsupials and rodents

Although sexual segregation is very common among ungulates, the trait is also displayed by other mammals. Sexual segregation in grizzly bears (*Ursus arctos*) in Alberta, Canada, seems to be a result of a tendency for females to avoid attack from potentially infanticidal immigrant males in some localities. In habitats where older males have higher survival rates, male immigrant influx is lower and sexual segregation is also low; sexual segregation increases in areas where influx of migrant males increases as a result of higher mortality of resident older males (Wielgus & Bunnell 1995).

In the Abruzzo region of central Italy, European polecats (*Mustela putoricus*) show temporal sexual segregation without monosexual group formation, with males being nocturnal whereas females are diurnal and crepuscular (Marcelli *et al.* 2003). The authors suggest several hypotheses to explain this pattern of temporal sexual segregation. Given that males are larger than females temporal sexual segregation may be the result of within species niche partitioning resulting from interference competition between individuals of different body sizes. Alternatively, the smaller-bodied females may have shifted to a more diurnal habit in order to avoid nocturnal predators such as foxes. Although in this species there is no monosexual group formation, temporal sexual segregation in the same habitat increases the chances that if a conspecific encounter occurs, it will be of a monosexual kind.

Some marsupials also show sexual segregation. MacFarlane & Coulson (2005) studied sympatric populations of western grey (*Macropus fuliginosus*) and red (*M. rufus*) kangaroos in Victoria (southeastern Australia). Whereas *M. fuliginosus* is a seasonal breeder, *M. rufus* is a continuous breeder in the area. Both *Macropus* species differ in their degree of social sexual segregation. Whereas *M. fuliginosus* sexually segregates outside the mating season, the continuously breeding *M. rufus* sexually segregates throughout the year. Large males just move throughout the habitat in search of females in oestrus.

In an experimental study carried out on *Mus musculus domesticus*, a social species that does not undergo sexual segregation in the wild on a regular basis (Terranova & Laviola 1995), Terranova *et al.* (2000) showed that pups housed in mixed-sex (i.e. disexual) groups manifested a social preference to interact with pups also reared in a disexual social environment over pups reared in monosexual groups. It is interesting to note, however, that upon a choice experiment for social partnership, males always preferred females, irrespective of their former prepubertal sexual segregation, whereas females who were prepubertally reared in sexual segregation with other females preferred females in tests of social interaction. From this it could be concluded that experimental sexual segregation in this species tends to affect females' social partner choice more than males', a result that mirrors the female social partner preference for other females found in species where sexual segregation is the norm, such as ungulates.

Bats, elephants and pinnipeds

Among the Chiroptera, Indian false vampires (*Megaderma lyra*) show partial spatial sexual segregation at roosting sites, at least when females are pregnant or lactating (Goymann et al. 2000). Bat species in Europe and North America show a trend for sexual segregation by altitude, with males being found more frequently at higher and females at lower altitudes (Russo 2002 and references therein). This pattern of sexual segregation is already apparent in young of both sexes in the vespertilionid *Myotis daubentonii*, a bat species studied by Russo (2002) in the Abruzzo region of central Italy. Body condition of males is better at lower altitudes, yet they move towards higher altitudes from very young ages. Although lactating females have been suggested to aggressively exclude males from better foraging areas, males seem to offer very little resistance, thus suggesting that other factors may be at play (e.g. parental facilitation or kin selection).

Broders & Forbes (2004) studied the northern long-eared bat (*Myotis septentrionalis*) and the little brown bat (*M. lucifugus*) in Canada. Sexes seem to be segregated by habitat in *M. lucifugus*, whereas there is social segregation in *M. septentrionalis*, with females having a greater tendency to roost in all-female groups, whereas males tended to roost alone. Entwistle et al. (2000) studied the brown long-eared bat *Plecotus auritus* in Scotland and found less sexual segregation at higher latitudes, suggesting that the thermoregulation benefits of communal roosting may decrease the benefits of sexual segregation. It is possible that in bats monosexual group formation is particularly developed in females.

In the body size sexually dimorphic and polygynous African elephant (*Loxodonta africana*; order Proboscidea), bulls control family units of females (cows) and their young, whereas bulls not in musth wander singly or in small all-male groups (Stokke & du Toit 2002). Stokke & du Toit (2002) proposed a *Social Factors – Reproductive Strategy* hypothesis to explain spatial segregation between family units and all-male bachelor groups in African elephants, with dominant bulls chasing subordinates away from females, while females are restricted by the need to stay close to water, needed by the developing young.

Among the pinnipeds, foraging habitat sexual segregation was described in northern elephant seals (*Mirounga angustirostris*) in California by Le Boeuf et al. (2000). *M. angustirostris* is one of the most sexually size-dimorphic mammal species known, with males weighing 10–15 times more than females. Le Boeuf et al. (2000) suggest that spatial sexual segregation during foraging is likely to be a result of different food requirements determined by the sexual dimorphism in body size. The hypothesis was also supported by Stewart (1997), who described sexual segregation by body size: larger adult males vs. females and young (less than 2 years old) males, during migrations in *M. angustirostris*. Sexual segregation is also found in harbour seals (*Phoca vitulina concolor*). Kovacs et al. (1990) studied this species in New Brunswick, Canada, where *P. v. concolor* shows social sexual segregation during the breeding season, with females and their pups staying separated from adult male and juvenile aggregations. Wolf et al. (2005) studied Galápagos sea lions (*Zalophus californianus wollebaeki*), a non-migratory pinniped that lives the year round in colonies. Sexual segregation during the mating period can be easily explained in this species by its polygynous mating system, where dominant bulls control a harem of females and males with no harem form all-male bachelor groups. After giving birth, females and their pups segregate themselves from all males during lactation (habitat sexual segregation). Habitat sexual segregation decreased during the non-reproductive period in the study by Wolf et al. (2005), but spatial sexual segregation within the common habitat was significant. Thus in this species, social factors associated with mating system and subsequent protection of pups during lactation against conspecific harassment are sufficient to explain sexual segregation. However, both males and females may accrue some specific benefits from being members of monosexual groups. Wolf et al. reject the unselected hypothesis that group formation is simply the side effect of limited habitat availability. According to those authors, potential adaptive advantages of group living in this species include establishment of dominance hierarchies and improvement of fighting skills in males. Why females should gather in all-female groups, especially during non-breeding periods and after lactation, remains to be determined. However, emphasis on dominance may perhaps be overrated in this and perhaps other species of pinnipeds, as living in monosexual groups can also afford opportunities for cooperation to both males and females (for example, sea lions cooperatively hunt anchovy schools (Packer & Ruttan 1988); see also cooperative foraging in Steller sea lions *Eumetopias jubatus* (Gende *et al.* 2001).

Cetaceans

Sexual segregation is also a common pattern in cetaceans. Among the mysticetes, North Atlantic whales (*Megaptera novaeangliae*), like many other baleen whales, sexually segregate during annual migrations (Stevick *et al.* 2003). In the odontocetes, river dolphins or botos (*Inia geoffrensis*) were studied in Brazil by Martin & da Silva (2004). Botos are not very social, probably forming only short-term associations with conspecifics; adults also have no known predators according to Martin & da Silva (2004). The population sex ratio is very close to even but sex ratios across a sample of habitats were variable: from female-biased to male-biased to even. Although lactating females may be under specific energetic and alimentary constraints to dwell more in some habitats than others, it is unclear whether apparent cases of sexual segregation in this species could simply be explained by chance associations.

The highly social bottlenose dolphin (*Tursiops* spp.) has been extensively studied around the world, especially in the USA and Australia. Smolker *et al.* (1992), working on the well-studied population of *Tursiops* sp. at Shark Bay, Western Australia, describe how in this sexually monomorphic species individuals associate into parties (*Social Sexual Segregation*) of variable sex composition: 49% of the parties were of mixed-sex composition, 27% were of females only and 24% of males only. Male monosexual associations were more stable than female monosexual associations (Smolker *et al.* 1992). Interestingly, of the three kinds of parties, it was the male–female association that was most transient. This suggests that male–female associations may occur mainly in the context of fertilisation, which, of course, does not require necessarily a long-term association, whereas same-sex parties may be cooperative alliances beneficial to males in sexual competition to control females in oestrus and to females in defence against harassing males (Smolker *et al.* 1992). However, other functions related to cooperative feeding or defence against predators may also play a role in the establishment and maintenance of same-sex alliances. Establishing a reliable cooperative alliance obviously requires time; in fact temporally stable same-sex associations in *Tursiops* already occur among subadults (Smolker *et al.* 1992; Connor *et al.* 1999). Connor *et al.* (1999) also report more labile alliances of brief duration, which they call 'superalliances', among male *Tursiops* at Shark Bay, alongside the more stable ones. The function of these 'super-alliances' is unknown. Again, *Tursiops* species could be a good example for the study of the association between social bonds and same-sex sexual interactions, the latter being likely to occur in both sexual and socio-sexual contexts.

In the sexually dimorphic beluga whale (*Delphinapterus leucas*) sexually mature males are typically larger than sexually mature females (Loseto *et al.* 2006). Loseto *et al.* (2006) studied *D. leucas* in the Canadian high Arctic in summer (Northwest Territories) describing a trend for habitat segregation by sex associated with body size: relatively small females, with or without calves, and small (probably subadult) males used mainly offshore and shallow open habitats, whereas slightly larger females, with or without calves, and medium-sized males (presumably sexually mature) used ice edge habitats, while large males are restricted to closed- and mixed-ice habitats. The authors suggest that large males are mainly constrained by access to food whereas smaller males and females are constrained in their spatial distribution by both food availability and predation avoidance.

Birds

Sexual segregation also occurs in birds (reviewed by Catry *et al.* 2005). Catry *et al.* (2005) indicate that there is a common pattern of sexual segregation in many migratory birds during non-nesting periods of the year. They suggest that the major factors explaining this sexual segregation are likely to be niche divergence between the sexes, which often differ in body size, and social dominance of one sex over the other leading to habitat sexual segregation. In addition, some migratory bird species may show a sex-specific phased pattern of return to the breeding grounds. For instance, in many species males arrive first (protandry) whereas in other, especially the 'sex-role reversed' species, such as some charadriids, females arrive first (protogyny) (Mills 2005). It would be expected that in protandrous species same-sex sexual behaviour during the initial period of the breeding season would be biased towards males, whereas in protogynous species it may be biased towards females as a result of the temporary sexual segregation occurring during a period of increased sex hormone circulation and activation of sexual behaviour.

Sexual segregation in foraging areas has been described in albatrosses, e.g. the black-browed albatross *Thalassarche melanophrys* and the grey-headed albatross *T. chrysostoma* during the breeding season in subAntarctic South Georgia. Sexual segregation occurs in both species during incubation, but it is not as clear during brood rearing. The larger body size of males allows, or perhaps constrains, them to feed in either Antarctic or subAntarctic waters, where strong winds provide sufficient lift for sustained flight over long distances. Females, with lower wing loading, can better exploit resources in more northern regions where winds are less strong (Phillips *et al.* 2004). In the sexually size-dimorphic (males larger than females) northern giant petrel (*Macronectes halli*) also inhabiting the South Georgia, males forage along the coast, whereas females venture further out to sea during incubation (González-Solís *et al.* 2000).

Lopez Ornat & Greenberg (1990) studied habitat sexual segregation in seven passerine species on the Yucatan peninsula of Mexico and found significant sexual segregation in five: American redstart (*Setophaga ruticilla*), common yellowthroat (*Geothlypis trichas*), magnolia warbler (*Dendroica magnolia*), northern parula (*Parula americana*) and hooded warbler (*Wilsonia citrina*), with a sixth species, the green warbler (*Dendroica virens*) also possibly showing habitat sexual segregation. The only species that they studied that did not show habitat sexual segregation was the yellow warbler (*Dendroica petechia*), a migratory species. The pattern for the five cases showing significant habitat sexual segregation was for males to occupy more mature habitats. Sexual habitat segregation has been also described in American kestrels (*Falco sparverius*) (Mills 1976, cited in Lopez Ornat & Greenberg 1990) and Eastern great reed warblers (*Acrocephalus orientalis*) (Nisbet & Medway 1972, cited in Lopez Ornat & Greenberg 1990). In *F. sparverius* females are larger than males, a common pattern in raptors. Although some studies indicate the occurrence of habitat sexual segregation in this species (see, for example, Mills 1976, cited in Lopez Ornat & Greenberg 1990; Smallwood 1987, cited in Arnold & Martin 1992), in a locality in south-western Ontario Arnold & Martin (1992) did not find evidence for habitat sexual segregation. They suggest that the lower population densities at higher latitudes decrease intersexual competition for food, leading to sexual aggregation.

Gruys (1993) studied willow ptarmigans (*Lagopus lagopus alexandrae*) in British Columbia during the non-breeding season, detecting habitat sexual segregation whereby most males remain in subalpine habitats and females with the majority of juveniles move to boreal forest habitats. Gruys (1993) suggests a *Reproductive Strategy* hypothesis to explain sexual segregation in this species during the non-breeding season, as females and their immature offspring will transfer to a habitat where survival is maximised (e.g. in terms of protection against predators), whereas males remain on the breeding grounds where they defend specific sites in order to maximise reproductive success in the following breeding season.

In sum, sexual segregation in birds does not seem to be very common, although it does occur in several taxa. Foraging specialisation between the sexes is a frequent cause of sexual segregation among sedentary species, whereas different timing of departure to, arrival at and location of overwintering areas may also explain sexual segregation among seasonally migratory species.

Sexual segregation in primates

Although, in general, sexual segregation has not been a strong factor in the social evolution of primates (Di Fiore & Rendall 1994), some species do exhibit various degrees of sexual segregation. Squirrel monkeys (*Saimiri sciureus, S. boliviensis boliviensis*), for instance, are characterised by sexually segregated social groups that occur both in the wild and in captivity (Williams & Abee 1988; Lyons *et al.* 1992). Lyons *et al.* (1992) showed that in *S. sciureus* held in captivity sexual segregation is more likely to be maintained by intermale aggression preventing males from interacting with females (*Male Intervention* hypothesis). Other examples of social and spatial sexual segregation can be found in talapoin monkeys (*Miopithecus talapoin*, Rowell 1973) and bonnet macaques (*Macaca radiata*, Handen & Rodman 1980). Among Old World primates and the Hominoidea in particular, sexual segregation, when it occurs, is mainly manifested in terms of small-scale, spatial segregation rather than larger-scale habitat segregation.

Sexual segregation also occurs in many contexts in human societies, especially, but not exclusively, among children and adolescents (Lipman-Blumen 1976; Maccoby & Jacklin 1987; Maccoby 1994, 2002; Pellegrini 2004; Pellegrini *et al.* 2005). Children younger than about 30 months are mainly sexually aggregated in their social activities with peers, but from the age of about 30–36 months they start leaning towards same-sex peers (Maccoby 2002; Fabes *et al.* 2003). It is possible that juvenile humans sexually segregate as a female avoidance strategy of males' rough play (Pellegrini 2004, but see Maccoby (1998) and Maccoby & Jacklin (1987) for a criticism of this interpretation), or as an opportunity for both sexes to learn and develop sex-distinctive skills that will be useful later in life, a process known as *sex-typing* (see, for example, Maccoby 2002).

Eleanor Maccoby (2002) suggests that between the ages of 3 and 8–9 years old, boys tend to play aggressive/competitive games and girls preferentially play collaborative discourse games, a sex difference that, although it may be, at least in part, driving sexual segregation in children, is not enough as a causative explanation, as the region of overlap between the sexes is considerable (Maccoby 1990). Maccoby's studies suggest that the sex of the child itself may be already a major causative factor for the sexual segregation tendencies in children and adolescents (Maccoby & Jacklin 1987; Maccoby 1990, 2000). However, sexual segregation of young humans is also reinforced or encouraged by adults (Pellegrini 2004) and, in fact, it is unclear whether boys and girls would be more willing to interact with each other (i.e. decreased sexual segregation) if adult intervention in this regard were eliminated, through fully permissive parenthood, throughout the children's development and not just during brief periods in an experimental setting (e.g. see the case of egalitarian societies with high levels of permissiveness) (Reiss 1980, Weinberg *et al.* 2000). Moreover, initial sexual segregation at relatively young ages may be subsequently driven to greater extremes by a runaway process resulting from peer pressure, as the child grows older and his/her socialisation becomes more sophisticated (see, for example, Maccoby 1994; Martin *et al.* 1999).

Pellegrini *et al.* (2005: 201) interpret human sexual segregation during juvenile periods of development as an adaptation 'to learn and practice skills consistent with their respective adult reproductive roles... males learn competitive and aggressive skills useful in securing and maintaining status in their juvenile and adult groups. Females, on the other hand, learn nurturing skills'. But then the authors also realise that successful rearing of offspring in humans requires a degree of coordination between male and female (unlike the case of precocial taxa such as ungulates). Such successful coordination is unlikely to develop from long-term sexual segregation (Kalmijn 2002 makes a similar point). Moreover, to what extent is parental perception of gender roles promoting reinforcement of behaviours in children that conform to such roles and discouraging those that do not conform? Although some experimental studies provide evidence against the role of adult reinforcement in children's sexual segregation (Martin & Fabes 2001), more detailed and continuous studies of parental behaviour at home should be conducted, starting at a very young age of the child. In my opinion, studies of sexual segregation in humans should focus much more on which role adults play in enforcing such segregation, either through subtle reinforcements (e.g. reprimanding girls who play rough and tumble games with boys, or discouraging nurturing games that boys may play with girls) or through the active establishment of monosexual playgroups (e.g. sport teams, gatherings of mothers with their boy or girl children in separate groups, single-sex schools).

One distinct possibility that should be the matter of more serious research is that, to a degree, sexual segregation among children may be a relatively recent phenomenon that may be driven by: (a) parental concerns about the potential for early sexual experiences as boys and girls mingle together in an unsupervised setting, (b) parental encouragement of behaviours deemed useful in each sex for a successful adult life, (c) parental concern that if boys are brought up to behave 'aggressively', then girls would be better to stay on their own to learn the appropriate 'nurturing' skills of their sex while avoiding disruptive 'aggressive' interactions with the boys. The tendency of parents to especially protect girls (e.g. from potential sexual molestation) may also explain some differences in parental attitudes regarding freedom of movement of the child. In particular, Maccoby (1990) mentions that boys tend to spend more time playing in public places such as streets, whereas girls tend to play more at home; it is not difficult to see how this pattern may result simply from parental intervention aimed at protecting girls. By contrast, Margaret Mead (1935) describes in the Arapesh of Papua New Guinea that 'Small children are not required to behave differently to children of their own sex and those of opposite sex. Four-year-olds can roll and tumble on the floor together without anyone's worrying as to how much bodily contact results.' (p. 47). 'The Arapesh have no fear that children left to themselves will copulate, or that young people going about in adolescent groups will experiment with sex...' (p. 89), and it is only because the Arapesh married when they were still very young that 'As the little girl approaches puberty, her parents-in-law increase supervision of her, both for her sake and for the sake of her boy husband' (p. 89). Among the Trobrianders of Melanesia, Malinowski (1932: 44–51) also reported sexual aggregation in young children concomitant with freely expressed sexual behaviours that were unimpeded by adults.

Studies of apparently spontaneous sexual segregation in children in school playgrounds clearly show that such a segregation is a matter of degree, with Pellegrini *et al.* (2005) mentioning that both boys and girls were part of 'mostly' (the authors originally used the word in quotes) segregated groups 80% of the time, with 'mostly' segregated groups being defined as those where 60% or more members of the group were of the same sex. In a study by Fabes *et al.* (2003), children played with other children of both sexes in about a quarter of their interactions, whereas Martin & Fabes (2001) report that, for children about 4 years old, 55% of girls played with children of either sex, whereas the frequency for boys was 65%.

La Freniere *et al.* (1984) suggested an interesting mechanism to explain a degree of sexual segregation observed among preschool children. They studied groups of 1–6 year olds in day care centres in Canada. After correcting for the effects of biased sex ratios, they found that children up to 17 months of age displayed no sexual segregation. A degree of preference to play with children of the same sex began at about 27 months, with such preference increasing up to preschool years, by which time children were 20% more likely to affiliate with a same-sex partner than expected by chance. The preference for same-sex peers, however, was more marked for boys than for girls at the older ages. La Freniere *et al.* (1984: 1962) also noted a degree of variability in these trends 'with some children showing no sex-related preferences and a few children showing a preference for opposite-sex peers'. Therefore, more than a preset tendency to choose same-sex social partners, children seemed to be inclined to choose whichever partner was more compatible with them, be that partner a male or a female (see also Goodenough 1934).

My suggestion is that humans are likely to follow the general primate pattern of fluid association among juveniles, with a broad tendency for some degree of sexual segregation, but with the choice of social partners being also influenced by individual interactions. In addition, the intervention of adults and the effects of peer pressure may also discourage sexual aggregation and encourage sexual segregation in some human societies. Although parents in some societies may be concerned about boys and girls learning the appropriate skills expected from their sex, thus encouraging sexual segregation, I also suggest that parents may be concerned about early heterosexual sexual experiences of their children, girls in particular. This predicts that in more sexually permissive societies, or in societies differing in the levels of sexual permissiveness at different times of their history, and in those less sex-typed societies where skills for life expected to be developed by boys and girls overlap (professional development, including military careers, sharing of household duties, etc.) juvenile sexual segregation should be relatively less pronounced than in more gendered and sexually less permissive societies.

What about sexual segregation in adult humans? Adult humans sexually segregate in various social contexts, some involving more or less individually driven sexual segregation (e.g. friendship networks) others involving externally constrained sexual segregation (e.g. at work). Early reviews of adult sexual segregation emphasised sociological aspects of resource control. In particular, in societies where one sex tends to control resources, members of that sex also tend to interact preferentially among themselves (see, for example, Lipman-Blumen 1976). This is valid for both men and women. In Lipman-Blumen's (1976) view, adult homosociality, although it is reinforced by current common interests, also directly derives from ontogenetic processes that start at much younger ages (see above and also Kalmijn 2002). However, adult homosociality and therefore sexual segregation is a fluid trait in humans. This is shown, for instance, by the prospective study carried out by Matthijs Kalmijn in the Netherlands (Kalmijn 2002). Kalmijn followed up young men and women from the age of 18–26 to 26–34 years, from 1987 to 1995, and interviewed the participants at three different stages. One very enlightening result of Kalmijn's study is that adult sexual segregation in the context of friends' networks is strongly influenced by issues of opportunity. For instance, as sexual segregation at work decreases, networks of friends become more sexually aggregated. In contrast, a sexually segregated workplace is associated with increased sexual segregation of friendships. This workplace effect may explain why sexual segregation in young adults increases with age, although Kalmijn also correctly points out that as individuals become married, their ability to befriend an individual of the other sex becomes more limited. As already mentioned above for the case of child sexual segregation, Kalmijn's results also suggest a degree of interindividual variability in the propensity to sexually segregate dependent on personality and compatibility. In particular, sexual aggregation in terms of friendships increases in men as a result of, at least in part, social isolation, whereas in women it increases as a result of increased self-esteem.

In sum, in broad agreement with Eleanor Maccoby I suggest that an integrationist, biosocial approach to sexual segregation is probably the approach that will most likely provide a full explanation of homosociality in humans, both young and adults. Although there is a broad tendency for boys and girls to segregate sexually as they age, such a tendency seems to result from a mixture of sex-specific propensities, individual characteristics and social constraints.

The above review suggests that the patterns and mechanisms of sexual segregation seem to show some similarities, but also some important differences, between birds and mammals. First of all, sexual segregation has not been as frequently reported in birds as it has in mammals. Second, although trends for habitat sexual segregation driven by feeding differences between the sexes is a common theme in species of the two vertebrate classes, social sexual segregation is definitely more a mammalian than an avian characteristic. Across mammals, females show an especially developed trend towards joining monosexual groups, a pattern that is referred to as homosociality. In humans, sexual segregation among juveniles is a matter of degree, with parental intervention and peer pressure probably playing a role in enforcing such sexual segregation, alongside more ontogenetically canalised mechanisms and individual decisions based on compatibility. Although sexual segregation during childhood tends to be associated with early same-sex sexual experiences, those experiences do not lead, per se, to the development of a homosexual sexual orientation in adults as argued below.

Does homosociality at young ages lead to adult homosexuality? Although sexual segregation has been suggested as a potential factor in the development of human homosexuality (Goy & Goldfoot 1975: 414), Peplau et al. (1998) argue that sexual segregation across human cultures is not associated with the development of adult obligate homosexuality or even attitudes towards homosexuals. For instance, ritualised homosexuality among some Melanesian cultures was associated with sexual segregation but this did not lead to exclusive homosexuality (Herdt 1984b). None the less, when boys or girls are raised in sexually segregated groups they may rehearse-play sexual behaviours in a homosexual context (Goldfoot et al. 1984; Berndt & Berndt 1992: 195, see also Chapter 4). Moreover, among young who are sexually segregated (e.g. in boarding schools), same-sex sexual behaviour is frequent. The point however is that such early homosexual experiences usually do not have, by themselves, long-lasting consequences on the sexual orientation of all the individuals involved, especially if the experience is consensual and therefore not traumatic (Ashworth & Walker 1972).

Ellis & Ames (1987: 243) indicated that experimental interventions in which young monkeys, and to some extent rats as well, are reared in same-sex groups have the effect of increasing same-sex sexual preferences in adulthood. Although biological differences between species could explain differences in behaviour between those species and humans, it is important to notice that forced rearing in monosexual groups from a young age not only exposes individuals to same-sex interactions during periods of learning, but it also may expose them to specific stressors. I have already mentioned in Chapter 4 how stresses suffered at young ages are potential factors in the development of adult homosexuality. Thus experiments set to test early learning mechanisms on development of homosexuality in monosexual groups must be able to control for the effects of social stressors (see Chapter 4).

What about sexual segregation in adult humans? Is adult sexual segregation associated with homosexual experiences and if so do those experiences lead to a change in sexual orientation from heterosexuality to homosexuality? Same-sex sexual experiences are common in prison and, perhaps contrary to some prejudices, they are more prevalent in a consensual context (2%–65% of inmates engaged in consensual homosexual behaviour, see Hensley 2000 and references therein) than in the context of homosexual rape (1.3%–20%; Hensley 2000 and references therein), although instances of rape could be more underreported than instances of consensual homosexual experiences.

One of the first studies to address the issue of whether homosexual behaviour, expressed in sexually segregated adult humans, may have long-lasting effects on sexual orientation was carried out by Edward Sagarin on prison inmates in the USA (Sagarin 1976). Although his sample size is too small and, above all, too biased, as he himself admitted, his results suggest that inserter individuals who usually adopted a dominant stance ($n = 5$) all returned to their former heterosexual orientation after they served their term in prison, whereas those insertee (passive) individuals that he followed up ($n = 4$) did adopt a homosexual sexual orientation once they were released. Unfortunately, given the small sample size and the lack of a long-term follow-up of those individuals it is difficult to make much of these results. Hensley & Tewksbury (2002) have published the most updated review of same-sex sexuality in both women and men in a prison setting that I am aware of. Their results suggest that about 21%–86% of female inmates engage in same-sex sexual behaviours. A defining characteristic of female inmate homosexuality, however, is its affiliative socio-sexual aspect expressed in terms of alliances and cooperation: 'Companionship, security, a sense of belonging, and interdependence were reasons given for participation in these partnerships' (p. 229). Coercion and dominance was associated with same-sex sexual behaviour in only a small proportion of cases, probably much less than 20%. With regard to homosexual sexual interactions among male inmates, Hensley & Tewksbury indicated that both affiliative and coercive homosexuality are expressed, the latter being more frequent than is the case in women (14%–60%). More importantly, most of those inmates regarded themselves as heterosexuals (78%–80%). It seems clear from these studies that although homosexual behaviour is a common occurrence in the context of sexual segregation in prison in humans, it is not always coercive and, above all, it does not seem to be able to change the participants' sexual orientation.

Therefore, homosociality sets the stage for individuals living in groups to potentially also express same-sex sexual behaviours. However, such homosexual experiences may not lead to the development of an exclusive homosexual sexual orientation, whether they occurred in young or in adults – if they did, most ungulates would be exclusively homosexual, for instance. Species that evolved social sexual segregation, however, may also have evolved same-sex sexual behaviours relevant to their sociality (e.g. same-sex mounting as expression of dominance or affiliation); such behaviours tend to be expressed within a functionally bisexual sexual orientation.

I will devote the last part of this chapter to testing models for the potential evolutionary association between sexual segregation and same-sex mounting in birds and mammals. This phylogenetic approach differs from the ontogenetic approach that I just used in the analysis of homosexuality and homosociality, especially in humans.

In what follows I first introduce two models for the causal evolutionary links between sexual segregation and the incidence of same-sex sexual behaviour, to subsequently test those models using comparative analyses. More detailed information about the phylogenies and the comparative method used is provided in Chapter 2.

Evolutionary models for the potential link between sexual segregation and same-sex mounting

The above review of the empirical work carried out on sexual segregation suggests at least two major evolutionary models for the potential causative association between same-sex sexual behaviour and sexual segregation. In the first model, which could be called the *Socially Driven Model*, the mating system (e.g. polygyny) determines a degree of sexual segregation that may then set the stage for the grouping of the segregated sexes into social units, with same-sex mounting occurring within the monosexual groups as a result of hormonally driven sexual behaviour or a response to socio-sexual functions, or both. The second model, inspired by what we know from ungulates, could be labelled the *Ecologically Driven Model*. In this model sex-specific alimentary constraints may determine a degree of habitat or spatial sexual segregation that may then give rise to sociality in response to predation pressure. Once monosexual social groups are formed the evolution of a polygamous mating system (e.g. polygyny) becomes more likely, thus opening up the opportunity for members of the same sex that are excluded from heterosexual contact (e.g. males in bachelor groups) or those that live in groups with a highly biased sex ratio (e.g. females in harems) to engage in homosexual sexual behaviours in response to neuroendocrinological states and/or specific socio-sexual functions, especially during the breeding periods of the year (see Figure 7.2).

Testing of the above two models was carried out through comparative analyses of independent contrasts (Felsenstein 1985) (see Chapter 2). The phylogenies used are those appearing in Figure 2.4 (birds) and Figure 2.5 (mammals). Calculation of phylogenetically independent contrasts was carried out using the PDAP program (Midford *et al.* 2005), which runs in the *Mesquite* program of Maddison & Maddison (2006) as explained in Chapter 2. All analyses were run twice: once after setting all branch lengths of the phylogenetic tree equal to one (this approximates a punctuational mode of character evolution) followed by a second run of the program with branch lengths set following Grafen's (1989) method. Table 7.1 shows the variables used in these analyses: body mass (in kg), mating system (facultative promiscuity, monogamy, lek, facultative polygyny, polygynandry, polygyny, polyandry, harem polygyny, occasional polygyny, polygamy, promiscuity, facultative polygamy, facultative polyandry, facultative harem polygyny), sociality (colonial, territorial, group-living, loosely colonial, semi-social, social, solitary, facultatively social, loose group, flock), same-sex mounting and sexual segregation. Categorical variables were numerically coded, and intraspecific variability was factored in by assigning a mean value of the trait for a species based on the values for that species that were reported by various authors (see Table 7.1), whenever more than one value was available. Coding of mating system categories ranged from 1 to 5 spanning from monogamy (1) up to promiscuity and polygynandry (5). In this case, the value of the code for mating system increases with the qualitative increase in the number of heterosexual sexual partners involved. Sociality was coded from 1 to 3, with solitary or pair territoriality being coded 1 and sociality/coloniality being coded 3; facultatively social, semi-social or loosely group-living species were coded 2. Same-sex mounting was either observed (code 2) or not (code 1), with intermediate values also being possible owing to intra-specific variability. The coding was the same for sexual segregation: 2 if observed and 1 if not.

The procedure was to first correlate mating system, sexual segregation, sociality and same-sex mounting contrasts with log (body mass + 1) contrasts and subsequently correlate mating system, sexual segregation, sociality and same-sex mounting contrasts with each other.

Birds

The bird dataset produced 71 phylogenetically independent contrasts. Table 7.3 shows the results of the Pearson's product-moment correlations for the standardised independent contrasts, all probabilities are one-tailed.

It is clear from Table 7.3 that the results of the comparative analyses for the birds dataset are consistent no matter what method of branch length assignment is used. In both cases only two correlations are either significant: Mating System vs. Sociality and Sociality vs. Same-sex mounting, or approach significance: Mating System and Sociality for branch lengths set to 1 ($p = 0.05$). This suggests that evolutionary changes towards an increased

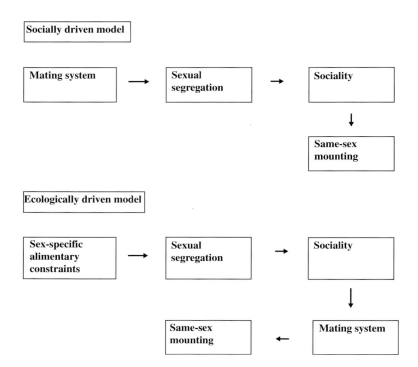

Figure 7.2. The *Socially Driven Model* emphasises the role of mating system in the evolution of sexual segregation, sociality and ultimately same-sex mounting. The *Ecologically Driven Model* considers sexual segregation as a precondition for the evolution of sociality and consequently mating system and same-sex mounting.

level of polygamy are associated with evolutionary changes towards an increased level of sociality on the one hand, and, on the other, an evolutionary trend towards increased levels of sociality is associated with an increased level of same-sex mounting. Although the analyses as such cannot determine causality, just co-occurrence of evolutionary trends between variables, these results for birds are more consistent with the *Socially Driven Model* than with the *Ecologically Driven Model*. The *Socially Driven Model* predicts a stronger positive correlation between mating system and sociality than between mating system and same-sex sexual behaviour, as the hypothesised path of causality is shorter between the former two than the latter two (see Figure 7.2; Wright 1921), and it also predicts a strong positive correlation between sociality and same-sex sexual behaviour. Both of these predictions are confirmed by the results. Two crucial predictions of the model, however, are not supported: the positive association between mating system and sexual segregation and also that between sexual segregation and sociality. In both cases correlations between independent contrasts are not significant. This suggests that, after all, same-sex sexual behaviour in avian taxa does not seem to be associated with sexual segregation; rather, it seems to be more closely associated with sociality and mating system in the manner suggested by the *Socially Driven Model*. This apparent link will be further explored in Chapter 8. The model therefore requires a crucial modification, with the paths to and from sexual segregation being eliminated (see Figure 7.3).

Table 7.3. Pearson's product-moment correlations (*r*) between independent contrasts of same-sex mounting, log-body mass, mating system, sexual segregation and sociality in birds

	r	*p*[a]
Grafen's (1989) branch length method		
Body Mass[b] vs. Mating System	0.041	0.363
Body Mass vs. Sexual Segregation	0.039	0.372
Body Mass vs. Sociality	−0.016	0.446
Body Mass vs. Same-sex Mounting	−0.060	0.306
Mating System vs. Sexual Segregation	−0.083	0.243
Mating System vs. Sociality	0.336	0.001**
Mating System vs. Same-sex Mounting	0.030	0.398
Sexual Segregation vs. Sociality	0.010	0.465
Sexual Segregation vs. Same-sex Mounting	0.054	0.325
Sociality vs. Same-sex Mounting	0.200	0.045*
All branch lengths set to 1		
Body Mass vs. Mating System	0.056	0.318
Body Mass vs. Sexual Segregation	−0.125	0.147
Body Mass vs. Sociality	0.068	0.283
Body Mass vs. Same-sex Mounting	−0.004	0.486
Mating System vs. Sexual Segregation	−0.072	0.273
Mating System vs. Sociality	0.195	0.050
Mating System vs. Same-sex Mounting	0.079	0.252
Sexual Segregation vs. Sociality	−0.045	0.351
Sexual Segregation vs. Same-sex Mounting	0.067	0.285
Sociality vs. Same-sex Mounting	0.196	0.048*

[a] One-tailed.
[b] In fact, log(body mass + 1).

Mammals

The phylogeny of mammals produced a total of 106 standardised phylogenetically independent contrasts. Correlations are summarised in Table 7.4.

Analyses of independent contrasts in mammals show a rather more complex situation than that found among birds. First of all, in mammals sexual segregation does play an important role, in contrast with the situation we encountered in birds. For instance, evolutionary changes towards increased body mass are consistently associated with evolutionary changes towards increased sexual segregation, a result that is in accordance with the *Foraging* and *Reproduction* hypotheses for sexual segregation that were reviewed in this chapter. Evolutionary changes towards increased body mass are also associated with evolutionary changes towards increased sociality – at least when branch lengths are assigned following Grafen's (1989) method – a trend that is consistent with, for instance, the selective effect of ecological factors such as predators. This immediately suggests that, for mammals, the

FIGURE 7.3. The *Modified Socially Driven Model* retains all the components of the original model except sexual segregation.

Table 7.4. Pearson's product-moment correlations (r) between independent contrasts of same-sex mounting, body mass, mating system, sexual segregation and sociality in mammals

	r	p^a
Grafen's (1989) branch length method		
Body Mass^b vs. Mating System	0.075	0.218
Body Mass vs. Sexual Segregation	0.221	0.011**
Body Mass vs. Sociality	0.176	0.034*
Body Mass vs. Same-sex Mounting	-0.076	0.216
Mating System vs. Sexual Segregation	-0.152	0.058
Mating System vs. Sociality	-0.045	0.321
Mating System vs. Same-sex Mounting	0.166	0.042*
Sexual Segregation vs. Sociality	0.434	0.028×10^{-4}***
Sexual Segregation vs. Same-sex Mounting	-0.152	0.057
Sociality vs. Same-sex Mounting	-0.058	0.274
All branch lengths set to 1		
Body Mass vs. Mating System	0.042	0.331
Body Mass vs. Sexual Segregation	0.253	0.008**
Body Mass vs. Sociality	0.108	0.133
Body Mass vs. Same-sex Mounting	-0.150	0.061
Mating System vs. Sexual Segregation	-0.125	0.098
Mating System vs. Sociality	-0.003	0.484
Mating System vs. Same-sex Mounting	0.073	0.227
Sexual Segregation vs. Sociality	0.310	0.0005***
Sexual Segregation vs. Same-sex Mounting	-0.158	0.051
Sociality vs. Same-sex Mounting	0.048	0.311

^{a}One-tailed.
^{b}In fact, log(Body Mass + 1).

Ecologically Driven Model may be more likely to apply (see Figure 7.2). The results involving body mass contrasts are supportive of the first link in the *Ecologically Driven Model*, i.e. that sex-specific alimentary constraints set the stage for sexual segregation in mammals. The direct association between sexual segregation and sociality that follows from the model is also consistently supported by the results of the comparative analyses, independently of the method used to assign branch lengths. Finally, the expected link between mating system and same-sex mounting was also found, although only for the case of branch lengths set using Grafen's (1989) method. The only part of the model that did not receive support from the comparative analyses was the hypothesised positive association between sociality and mating system. Therefore, although the *Ecologically Driven Model* for the evolution of same-sex mounting seems to be quite promising in the case of mammals, more studies will be required in order to determine the specific evolutionary association between sociality and mating system, as highly social species could evolve a diverse array of mating systems in response to internal (e.g. degree of relatedness) and external (e.g. ecological constraints) factors, as we will see in Chapter 8 in the context of the review of *Reproductive Skew* theory. On the other hand, neither the original nor the modified *Socially Driven Model*, which look more promising in the case of birds, is well supported by the comparative analyses of the mammalian data.

In sum, sexual segregation seems to be a far less common phenomenon in birds than it is in mammals. Among the avian genera included in Table 7.1, 22.2% (14/63) display sexual segregation, whereas 77.7% (49/63) do not; the opposite is true for mammals, with 48 out of the 72 monotypic genera (66.6%) showing sexual segregation, whereas only 33.3% (24/72) of the monotypic genera among mammals do not display sexual segregation ($\chi^2_1 = 24.96$, $p < 0.0001$). In addition, among the mammalian genera included in Table 7.1, there are four that are polytypic, that is that include both reported cases of sexual segregation and also cases of sexual aggregation: *Macaca*, *Macropus*, *Panthera* and *Helogale*. The evolution of sexual segregation was expected to facilitate the evolution of same-sex sexual behaviour (mounting in this case) by simply increasing the availability of same-sex partners during periods of elevated sexual activity, or by producing the social context for the expression of socio-sexual behaviours (e.g. those associated with dominance and/or affiliation). The results show that this is the case in mammals. At this stage, it is important to reiterate that the association between sexual segregation and same-sex mounting in mammals that was obtained in the analyses should be interpreted in an evolutionary perspective. That is, sexual segregation provides the selective background for the evolution of same-sex mounting in various contexts: dominance, affiliation, or the effect of neuroendocrine mechanisms that are active during the mating season. The results should not be interpreted in an ontogenetic perspective: i.e. that sexual segregation at young ages necessarily results in the development of obligate homosexuality in adults. This is not what these results imply.

Contrary to the results obtained for mammals, sexual segregation is not an important part of the evolutionary causal network leading to same-sex mounting in birds. This does not mean that if specific conditions of sexual segregation are induced in an avian population, same-sex mounting will not be observed, what it means is that sexual segregation does not seem to have been an important selective factor in the evolution of same-sex mounting in most of the bird species included in the analyses. Perhaps this is because sexual segregation is relatively uncommon in birds in the first instance, and when it does occur it is confined to the non-breeding periods of the year when birds usually do not copulate. In mammals, on the other hand, sexual segregation leading to sociality may not only affect the evolution of specific mating systems – although which end mating system may be achieved within a social species is clearly open to many and very diverse possibilities, as we will see in Chapter 8 – but the more polygamous or promiscuous that mating system is, the analyses suggest that the more likely it is that such species will also display higher levels of same-sex mounting.

Summary of main conclusions

- Sexual segregation is more common in mammals than birds and mainly results from physiological (e.g. alimentary), ecological (e.g. predation) and social selective factors.
- Social sexual segregation is particularly conducive to the expression of same-sex sexual behaviours.
- Same-sex mounting during periods of social sexual segregation results from neuroendocrinological states associated with mating periods and evolutionary trends associated with the establishment of dominance hierarchies and/or coalitions.
- Although in humans there is a trend for boys and girls to initially aggregate sexually and then start to segregate sexually at about 2–3 years old, throughout development the choice of social partner seems to be an outcome of sex-specific propensities, individual characteristics unrelated to sex and social constraints.
- Importantly, homosociality, as expressed during periods of sexual segregation in humans, is not conducive to the development of an exclusive homosexual sexual orientation through learning processes.

- The association of adult homosexual behaviour with homosociality in mammals is a result of evolutionary processes that result in adaptation in a bisexual context.

The above themes will be further developed in the next chapter, where I will address in greater detail the social, life history and ecological theatres of animal homosexual behaviour.

The social, life history and ecological theatres of animal homosexual behaviour

We have seen in Chapter 7 how variables such as mating system, sociality and, for mammals at least, sexual segregation can affect, directly or indirectly, the evolution of homosexual behaviour across species. For the bird species that were used in those analyses, same-sex mounting is positively associated with sociality and polygamy (see also MacFarlane *et al.* 2007), whereas in mammals same-sex mounting can increase with polygamy and be indirectly associated with sexual segregation and sociality.

This chapter explores the potential effects that a series of behavioural, ecological and life history variables exert on homosexual mounting. Apart from the variables that were already analysed in Chapter 7 in the context of sexual segregation, here I also examine the relationship between same-sex mounting and the specific sex involved (males, females or both), the adult sex ratio, dominance, affiliative behaviours, genetic relatedness, group size, ecological constraints on breeding, pair bonding, extra-pair or extra-group heterosexual copulations and, in birds, plumage sex-dichromatism. In mammals I will consider the effect of body mass sex-dimorphism. The aim is not to determine the effect of each one of those variables in isolation but to do so in the context of a broader theoretical framework. I will therefore start this chapter by introducing *Reproductive Skew* theory, as this provides an important starting point for a general evolutionary theory of homosexuality. Some necessary modifications that the theory should incorporate, to gain in generality, will be highlighted, and I will then briefly describe some empirical tests of the theory that have been carried out in birds and mammals. A qualitative model will subsequently be produced that directly addresses the link between the modified *Reproductive Skew* theory and homosexual behaviour; finally, I will test this model through comparative analyses. The chapter will also analyse some case studies in more detail and end with a section that deals with the potential effect of homosexual behaviour on the transmission of sexually transmitted pathogens.

Reproductive Skew Theory

Whichever sex is involved, homosexual behaviour is a *prima facie* evolutionary paradox in so far as its performance does not necessarily lead to immediate achievement of fertilisation in taxa such as birds and mammals. Whether the individuals involved in same-sex sexual behaviour are indeed bisexuals (sequential or not) or whether they are and always have been exclusive homosexuals will also make a difference in terms of the potential evolutionary mechanisms that can explain same-sex sexuality. Bisexuality presumably leads to a potentially higher reproductive success than strict lifetime homosexuality, although in both cases additional Darwinian fitness could be also achieved through helping the reproductive effort of close relatives, directly through the provision of alloparental care or indirectly through the release of resources. Transmission rate of own genes through the combined modes of direct and indirect fitness is encapsulated in the concept of *inclusive fitness* and the evolutionary process that maximises *inclusive fitness* is called *kin selection*. I will therefore start this introduction to *Reproductive Skew* theory by discussing the seminal work of Bill Hamilton (1963, 1964a,b) on inclusive fitness and kin selection.

As already mentioned, an individual's genes could be transmitted to the next generation in at least two major ways: (a) directly by producing offspring, and also (b) indirectly by helping close relatives increase the production of their offspring beyond what they would have been able to achieve on their own. Inclusive fitness is measured by taking into account both the direct and the indirect mode of transmission of genes to the next generation. It was John Maynard Smith (1964) who introduced the concept of *kin selection* as the evolutionary process that maximises inclusive fitness. This followed the publication of an article by Hamilton in the *American Naturalist* (Hamilton 1963) where he first introduced what has become known as *Hamilton's Rule*. *Hamilton's Rule* encapsulates the essential features of kin selection:

$$B/C > 1/r$$

where B represents the additional offspring produced by a recipient of help beyond what it would have produced unaided, C represents the offspring that the donor of help did not produce, and r is the coefficient of genetic relatedness between the recipient and the donor of help. An equivalent expression of *Hamilton's Rule* that will be used later in this section is:

$$Br - C > 0$$

That is, when inclusive fitness is taken into account, forfeiting reproduction could be selected if it is associated with a sufficient increase in close relatives' reproductive success. This kin selection process could potentially explain the apparent decrease in reproductive output of homosexual individuals. For instance, if by not reproducing homosexuals make more resources (e.g. food) available to their close relatives, who could then use those resources to increase their reproductive success, then homosexuality could be explained on these grounds alone. Moreover, non-reproducing individuals may even provide active help towards the reproductive effort of close relatives, but this is not strictly necessary; releasing resources by forfeiting reproduction may suffice. Kin selection, however, fails to explain decreased reproductive activity associated with helping non-kin, and in the case of homosexuality it also fails to explain why non- (or less) reproductively active individuals should engage in same-sex sexual behaviours such as mounting. In order to overcome the first objection – 'why should some individuals help non-kin?' I will introduce *Reproductive Skew* theory, which incorporates elements of kin selection but also considers interactions between non-kin (see also Trivers 1971; Alexander 1974, and game-theoretical models inspired by John Maynard Smith's seminal work of 1982 *Evolution and the Theory of Games*). The second objection, 'why should some individuals engage in same-sex sexual behaviours?' will be partly addressed by further developing *Reproductive Skew* models to incorporate same-sex mounting in a socio-sexual context (e.g. as an expression of dominance or affiliation/cooperation), and it will also require a modification of *Reproductive Skew* theory in order to incorporate purely sexual aspects of same-sex mounting; for example, by including elements of sexual selection and mate choice into the theory. I will subsequently introduce the *Synthetic Reproductive Skew* theory for the evolution of homosexuality. This theory will be a very important stepping stone and a prelude to an even more general model that will be developed in Chapter 10.

A historical overview of *Reproductive Skew* Theory

Modern *Reproductive Skew* theory has its immediate precursors in the pioneering works of Mary Jane West-Eberhard (1975), Steve Emlen (1982a,b) and Sandra Vehrencamp (1983a,b). West-Eberhard (1975) mainly operated within the framework of kin selection but she also emphasised the role of ecological constraints in the modulation of dominance relationships occurring among relatives in a social unit. Dominance could eventually affect the partitioning of reproduction among members of a group; the expression *Reproductive Skew* refers to exactly such partitioning of reproduction. Ecological constraints on dispersal and independent breeding were also highlighted by Emlen (1982a) as important environmental conditions for retention of offspring within a family unit. As soon as a group is formed owing to a lack of opportunities for dispersal, the individuals in the group almost inevitably engage in social interactions that could be a manifestation of more profound conflicts over the allocation of resources: reproducing or not, providing help or not. Such interactions may have various consequences when occurring among adults: if a subordinate has nowhere else to go, then it could be allowed to stay within the natal territory by more dominant individuals provided, for instance, that the subordinate helps in return. In the case of young offspring, however, adults could interfere with their ontogeny in order to manipulate their development and therefore the offspring adult phenotype, thus making them less (or more) likely to disperse and/or provide help (Emlen 1982b; see also Chapter 4). Selection may also be expected for young, developing individuals to resist potential manipulations (Zahavi 1974). On the other hand, whenever interests are convergent, both dominant and subordinate may easily engage in cooperative interactions.

Vehrencamp (1983a,b) was among the first to introduce a *Reproductive Skew* model using an optimisation approach. Her model highlighted the conflicts between dominant and subordinate over dispersal, reproduction and allocation of resources that ultimately result in differential reproductive output, such difference being the chief measurement of reproductive skew. In Vehrencamp's (1983a,b) model, reproductive skew increases when group breeding is more productive than solitary breeding and when the degree of genetic relatedness between dominant and subordinate increases.

As for Emlen's (1982b) model, Vehrencamp's (1983a,b) also considers the potential effect of dominants' manipulation of subordinates' ontogeny. It was not until the 1990s, however, that *Reproductive Skew* theory really took off as a major model of social interactions in vertebrates.

Emlen (1995) applied *Reproductive Skew* theory to the context of evolutionary dynamics within family groups of vertebrates. As already mentioned, in family groups offspring may delay dispersal (for example, because of ecological constraints) and stay at home in close association with their parents, waiting for an opportunity to become breeders. The higher the probability of successful offspring dispersal, the higher 'incentive to stay' parents should offer their offspring in order to retain them in the family and benefit from their help (Reeve & Ratnieks 1993). In this context, four major parameters are relevant in order to understand the degree of reproductive skew that may be expressed among members of the group: (a) the benefits (in terms of inclusive fitness) that dominants enjoy if the subordinate stays compared with the alternative of the subordinate leaving, (b) the benefits accruing to the subordinate if it decides to leave compared with the alternative of it staying, (c) the asymmetry in terms of dominance between the members of the family group and (d) the degree of genetic relatedness among them. Reproductive skew between members of a group is expected to increase as the differences in their dominance status increase, chances of successful dispersal are low for the subordinate and they are more closely related.

Recent reviews of *Reproductive Skew* theory identify two broad views for the evolution of interactions within groups. On the one hand, the *Compromise* or *Tug-of-War* approach emphasises the struggle occurring among members of a social unit for the control of limited resources, and on the other the *Transactional* approach emphasises the sharing of reproduction among members of a group that results from cooperation among them (Reeve *et al.* 1998). Here we can see some parallels between the *Reproductive Skew* models developed by evolutionary biologists and the *Family Dynamics* and

Confluence models developed by psychologists that were introduced in Chapter 6. The parallelism is expressed in terms of the emphasis on either competitive or cooperative interactions. We will see later in this chapter how competitive and cooperative views are indeed complementary. *Transactional* models, in turn, can take the form of either *Concession* if subordinates require an incentive to stay in the group or *Restraint* if subordinates limit their share of reproduction in order to avoid retaliation from the dominant (Johnstone 2000; Magrath & Heinsohn 2000). A proposal for a synthesis between *Concession* and *Restraint* models has recently been put forward by Buston *et al.* (2007). When dominants from different family groups compete for the services of subordinates, the *Concession* model may be modified to take the form of a *Bidding Game* (Reeve & Keller 2001).

Reeve & Keller (2001) proposed a general formulation for *Reproductive Skew* models that is an extension of *Hamilton's Rule*. Let us imagine the realistic possibility that within a social unit individuals could adopt more than one behaviour (evolutionary biologists use the term *strategy*) when confronted with specific circumstances. Let us also assume that there are two alternative strategies to choose from: *i* and *j*. *Hamilton's Rule* is used to decide which one of those two alternative strategies will be favoured by natural selection. For instance, strategy *i* will be selected over *j* if:

$$(Pi - Pj) + r(Ki - Kj) > 0$$

where P is the individual's reproductive output associated with the strategy, whereas K is the other party's reproductive output if either of the two strategies is performed, and *r* is the coefficient of relatedness between the interactants. Within this framework, if strategy *i* is 'defer reproduction and behave homosexually' (i.e. P*i* = 0), then the strategy will be selected over an alternative strategy *j* (e.g. 'behave heterosexually and reproduce') if

$$r(Ki - Kj) > Pj$$

That is, if the net gain in reproductive output enjoyed by close relatives when the individual behaves homosexually, corrected by the coefficient of relatedness between the two, is larger than the reproductive output of the homosexual had she or he decided to engage in heterosexual sexual behaviour and reproduce instead, then homosexuality will be selected. Note that this formulation does not necessarily predict any helping action from the part of the homosexual (e.g. caring for a relative's offspring); in fact, across social species not all members of a group engage in alloparenting, even when they do have the opportunity to do so (see, for example, Poiani 1994). It will suffice if the resources that could have been used for reproduction, such as food and others, are saved for the reproductive benefit of relatives. Moreover, the formulation does not even preclude the male or female homosexual from actually engaging in very frequent same-sex sexual behaviour, as the energy required by both sexes for mounting and, in the case of males, semen production if homosexual mounting culminates in ejaculation, is usually only a small fraction of that required for the overall reproductive effort. That is, the usual objections that *Kin Selection* has little to do with evolution of homosexuality because: (a) homosexuals may not necessarily perform active helping behaviours directed to kin preferentially and (b) same-sex sexual behaviour is a waste of time and energy that would be expected to be used to help kin are not necessarily warranted from a theoretical perspective (see also Chapter 3). On the other hand, if non-reproductive homosexuals also help in the breeding effort of their close relatives (e.g. by bringing food to the young, or providing protection), then the likelihood of the homosexual trait persisting over time will increase even further (see, for example, Johnstone 2008).

The above formulation has been extended by Reeve & Emlen (2000) for the case of an *N*-individual group size to give:

$$Pi - Pj + \sum_{m=1}^{N-1} r_m (K_{m,i} - K_{m,j}) > 0$$

where $K_{m,i}$ is, in this application of the model to homosexuality, the reproductive output of the mth group member if a group mate behaves homosexually, $K_{m,j}$ is the reproductive output of the mth group member if the mate behaves heterosexually instead, and r_m is the coefficient of relatedness between the two.

Under the conditions we are considering in this model, selection of homosexual behaviour should therefore occur when

$$\sum_{m=1}^{N-1} r_m \left(K_{m,i} - K_{m,j}\right) > Pj$$

This predicts a higher threshold of independent reproductive output where homosexuality could be selected if the group is of an intermediate size, especially for a group formed by close relatives. In other words, an individual is more likely to develop a homosexual sexual orientation in intermediate groups of close relatives where its lack of reproduction has a more significant effect on the reproductive success of the rest of the group. In very large groups the values of K are probably too small; in very small groups they are presumably larger but they add up across a smaller number of group members.

In most of the above models, the level of reproductive skew and hence, in my application of *Reproductive Skew* theory to homosexuality, the probability that one or more members of the group will be homosexual should increase as the:

(1) dominant and subordinate are more closely related,
(2) subordinate has fewer chances of successful dispersal,
(3) subordinate's activities within the natal group, including skipping reproduction but also helping, can significantly increase the production of offspring by those group members who reproduce, and
(4) ability of the subordinate to resist the behavioural imposition or manipulation by the dominant decreases.

Reeve & Keller (1995) extended the modelling to those cases where the dominant and the subordinate are asymmetrically related to each other's offspring; for example, in the case that the subordinate is the offspring of the dominant then she or he will be more closely related to subsequent dominant's offspring (if they are full sibs, $r = 0.5$ on average) than the dominant is to the subordinate's offspring (i.e. $r = 0.25$). In this case, associations of parents with offspring should lead to greater reproductive skew, and hence we may predict a higher probability of development of homosexuality in the offspring if parents are still reproductive (e.g. if they are still young, see Chapter 6). In interactions between close relatives, however, the expression of same-sex sexual behaviour may be also constrained by inbreeding avoidance mechanisms, as I will explain in the 'Mate choice and incest avoidance' section below.

I have mentioned above that *Compromise* and *Transactional* models are indeed complementary: the former emphasises competitive and the latter cooperative interactions. This means that they could well be synthesised into a more comprehensive model. The *General Model of Skew* of Johnstone (2000) and the *Bordered Tug-of-War* model proposed by Reeve & Shen (2006) are two such attempts. In both cases the competitive interactions within the group over allocation of reproduction are evolutionarily constrained by the achievement of some reproduction within the group. This can be illustrated by the extreme case where excessive competition leads to no reproduction by any member of the group; in such a case any sample of unique alleles represented in that group will become extinct. This implies what is intuitively quite obvious, that there is a cap to the spread of reproductive suppression and, by extension, to the spread of exclusive homosexuality. In the extreme case of groups formed by exclusive homosexuals only, such groups can only last in time if they are replenished by the immigration of individuals produced outside the group by heterosexuals and/or bisexuals. If, on the other hand, groups contain both reproducing (heterosexuals, bisexuals) and some non-reproducing (e.g. exclusive homosexuals), but highly cooperative individuals, then they may be less likely to suffer local extinction and more likely to produce dispersers. Broad evolutionary processes of *interdemic selection* (see, for example, e.g. Wade & McCauley 1984), for instance, could account for the evolution of both competitive (e.g. aggressive behaviour, partitioning the access to reproduction within the group) and cooperative (e.g. affiliative behaviour, increasing group competitiveness against other groups) tendencies within societies (Nonacs 2007), including those cooperative and competitive interactions mediated by same-sex mounting. Therefore a synthetic approach to reproductive skew applied to the evolution of homosexuality, with the addition of *interdemic selection*, predicts that same-sex sexual behaviour is more likely to be maintained at high frequencies in the context of bisexuality, where the expression of same-sex sexual behaviour does not negate reproduction, whereas exclusive homosexuality should occur at low frequency in the population, with such a low frequency becoming stable if homosexuality is associated with increased cooperation.

The above *Reproductive Skew* models assume *complete control* from the part of the dominant in that the level of skew may be controlled, for instance, through the provision of *staying incentives* to subordinates in order to delay dispersal and *peace incentives* given to subordinates in order to avoid aggressive conflict with the dominant. Tim Clutton-Brock (1998) called this the *Concession* model of *Reproductive Skew* theory. However, he also suggested that in some circumstances control of subordinates' reproduction may not be complete. In situations where control is *incomplete*, subordinates may reproduce, not as a concession from the part of the dominant, but as a result of the dominant's inability to control them. The *Incomplete Control* model is a viable alternative to explain cases of lower skew than expected from concession mechanisms (Cant 1998; Emlen *et al.* 1998; Reeve *et al.* 1998). In my extension of *Reproductive Skew* theory to homosexuality, incomplete control should be associated with a lower degree of exclusive homosexuality among the members of the group than that expected from complete control mechanisms. Incomplete control becomes more likely as groups reach very large sizes. Table 8.1 summarises the various reproductive skew models reviewed here.

Queuing to breed

In some animal societies, breeder status may be achieved through an age-based queuing system, in which the eldest subordinate is the one becoming breeder upon the death of the dominant. If subordinates in such groups have higher survival rates than the breeder they may stay in the group, queuing for a breeding opportunity, without any need for high incentives (Kokko & Johnstone 1999; Ragsdale 1999). Higher subordinate than dominant survival rates will ensure that subordinates down the queue have a chance of breeding before they die. Stable queuing systems with high turnover, therefore, may decrease the probability of development of homosexual phenotypes. Such queuing systems, however, are not always stable (Johnstone *et al.* 1999). In fact, in some circumstances it may pay a subordinate to challenge an immediate dominant (Cant & Johnstone 2000). For instance, as the queue becomes too long a subordinate may have little choice but to disperse or challenge if it is to achieve direct fitness gains through reproduction (Ridley & Sutherland 2002). Alternatively, such subordinates may exact some indirect fitness gains via kin selection in the way I mentioned above. If group membership changes unpredictably (e.g. owing to very high mortality), however, it will pay subordinates to retain reproductive readiness no matter where they are in the queue (Zink & Reeve 2005), thus lowering the probability of development of an exclusive homosexual phenotype even further. Linear, often age-based, dominance hierarchies are very common among social vertebrates (e.g. *Macaca arctoides*, Nieuwenhuijsen *et al.* 1987; *Junco hyemalis oreganus*, Jackson 1988; *Equus caballus*, *Rangifer tarandus*, Rothstein 1992; *Gallus gallus domesticus*, de Vries 1998; *Metriaclima zebra*, Chase *et al.* 2002; *Ovis canadensis*, Adams 2005), although in some species coalitions between group members may disrupt the linearity of dominance (see, for example, Silk 1993). In sum, high chances of

Table 8.1. A taxonomy of *Reproductive Skew* models

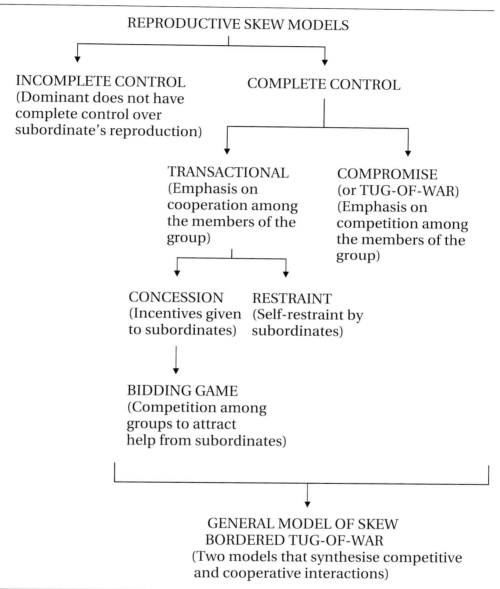

disruption of dominance or fixed, short-term patterns of orderly attainment of breeding status should be associated with lower levels of *obligate* or *exclusive* homosexuality. However, same-sex mounting, expressed within a bisexual sexual orientation in the context of affiliation and maintenance of competitive coalitions, may be expressed as a strategy to 'jump the queue'.

Early development effects

The dominant, which is usually older than the subordinate, has the advantage of interacting with subordinates from the time when the latter are young. This confers an advantage on the dominant in terms of perhaps manipulating the behavioural development of subordinates, especially during their early ontogeny (see Chapters 4 and 6), so that a subordinate may be less likely to disperse and more likely to stay and help once she or he has become an adult (Crespi & Ragsdale 2000). This idea was first proposed by Richard D. Alexander (1974), who labelled it *Parental Manipulation*. Crespi & Ragsdale (2000) suggest that *Parental Manipulation* is a more efficient strategy to achieve retention of offspring than *Concessions* (see also Emlen 1982b; Vehrencamp 1983a,b). The reason for a greater efficiency of *Parental Manipulation* compared with *Concessions* is simply that it is energetically and, in terms of fitness costs, cheaper for the dominant to adjust the early behavioural development of a young subordinate than to concede sufficient reproductive incentives once that subordinate is a fully grown adult (Crespi & Ragsdale 2000). The ability of dominants to manipulate the subordinates' ontogeny, however, will be limited by the degree of plasticity of such ontogeny (see Chapter 4 and also Beekman *et al.* 2006). We have seen in Chapter 4 that stress suffered during early development could contribute to biasing the ontogeny of an individual towards a homosexual phenotype. Therefore family members (e.g. parents and/or siblings) could potentially affect homosexuality in the young through this route, with the outcome, however, being somewhat variable depending on the genotype of the young members of the family and also on the effect of prenatal developmental processes that are independent of maternal stress.

Mate choice and incest avoidance

Reproductive skew is not only affected by dominance, ecological constraints to dispersal and relatedness. Subordinate males, for instance, may also have little chance of breeding if females do not prefer them as mates. Such female choice may lower the threshold conditions for the evolution of homosexuality in males and this may occur in both monogamous and polygamous mating systems (see, for example, Kokko 2003) depending on the level of bias in the sex ratio. That is, even if a female is polygamous or engages in extra-pair copulations, if she does not copulate with all males indiscriminately but exerts a choice of partners; and if the sex ratio is sufficiently male-biased, then there will be some males who do not have access to heterosexual copulations and who therefore may be inclined to establish homosexual partnerships. The same is obviously true in the case that the sex ratio is female-biased and males also exert some mate preference; in this case homosexual behaviours may be expressed in females.

In addition to mate choice, when the group is formed by closely related individuals, incest avoidance may also decrease the probability of heterosexual matings, thus, again, lowering the threshold for selection of homosexual partnerships (Kokko 2003; Magrath *et al.* 2004). Emlen (1996) had already recognised the limitations of the above four-parameter model of *Reproductive Skew* based on costs and benefits accruing to dominant and subordinate and genetic relatedness between them, and suggested that *incest avoidance* should also be included as an additional variable affecting the probability of reproduction of subordinates within a family group. The effect of incest avoidance would be to increase the level of reproductive skew in a group, as subordinates will tend not to mate with their parents or siblings. On the other hand, if extra-group fertilisations occur, then the level of relatedness within the group may decrease, thus leading to a lower level of reproductive skew (Emlen 1996; Reeve & Keller 1996). Paradoxically, although incest avoidance may increase the chances of the expression of homosexual behaviour owing to increased reproductive skew, it may also decrease the likelihood of homosexual intercourse between closely related members of a social unit (e.g. a family), if the incest avoidance mechanisms that are adaptive in the heterosexual context remain active in the homosexual context as well. Therefore the combined effects of reproductive skew and incest avoidance lead to the prediction that homosexuals, when they occur, will tend to preferentially engage in sexual intercourse, especially when they are adult, with non-kin or with kin that are not first degree (e.g. cousins; see Adam (1985) for some examples from the anthropological literature).

Tests of *Reproductive Skew* Theory

Many empirical tests of *Reproductive Skew* theory have been carried out in social birds and mammals. Although it is not my aim to provide an exhaustive review of such works, I would like to highlight the following trends in the results:

(a) Members of the group establish dominance hierarchies that are also expressed in terms of differential reproductive success: dominants reproduce more than subordinates (e.g. *Sericornis frontalis*, Whittingham *et al.* 1997; *Porphyrio porphyrio*, Jamieson 1997; *Suricata suricatta*, Clutton-Brock *et al.* 1998, 2001; *Turdoides squamiceps*, Lundy *et al.* 1998; *Papio cynocephalus*, Alberts *et al.* 2003; *Macaca mulatta*, Widdig *et al.* 2004).

(b) Reproductive skew among group members of the same sex increases as the degree of genetic relatedness increases among them (e.g. *Sericornis frontalis*, Whittingham *et al.* 1997; *Porphyrio porphyrio*, Jamieson 1997; *Turdoides squamiceps*, Lundy *et al.* 1998; *Cryptomys damarensis*, Cooney & Bennett 2000; *Corcorax melanoramphos*, Heinsohn *et al.* 2000; *Dacelo novaguineae*, Legge & Cockburn 2000; *Crocuta crocuta*, Engh *et al.* 2002; *Guira guira*, Macedo *et al.* 2004; but see Haydock & Koenig 2003 for more complicated patterns in the acorn woodpecker, *Melanerpes formicivorus*).

(c) The fewer males there are in the group, the higher the level of extra-group copulations carried out by females (e.g. *Sericornis frontalis*, Whittingham *et al.* 1997) thus suggesting that reproductive skew can be modulated by female mate choice (*Crocuta crocuta*, Engh *et al.* 2002; *Papio cynocephalus*, Alberts *et al.* 2003; *Cyanocorax morio*, Williams 2004; *Macaca mulatta*, Widdig *et al.* 2004).

(d) The higher the level of ecological constraints on dispersal, the higher the level of reproductive skew (e.g. *Porphyrio porphyrio*, Jamieson 1997).

(e) Reproductive skew is usually higher among members of the philopatric sex (e.g. *Suricata suricatta*, Clutton-Brock *et al.* 1998; *Corcorax melanoramphos*, Heinsohn *et al.* 2000; *Melanerpes formicivorus*, Haydock & Koenig 2002; *Cyanocorax morio*, Williams 2004; see the list of various bird species in Vehrencamp 2000).

(f) Reproductive skew decreases with increase in group size (e.g. *Guira guira*, Macedo *et al.* 2004).

(g) Reproductive skew increases when genetic relatedness with members of the other sex in the group increases (e.g. *Suricata suricatta*, Clutton-Brock *et al.* 2001).

(h) Subordinates can reduce reproductive skew by collaborating in coalitions (e.g. *Papio cynocephalus*, Alberts *et al.* 2003).

Therefore, broadly speaking the empirical evidence supports the various improvements that *Reproductive Skew* theory has been incorporating since the initial formulations put forward by Mary Jane West-Eberhard, Steve Emlen and Sandra Vehrencamp more than 20 years ago.

The *Synthetic Reproductive Skew Model* of *Homosexuality*

The above review of theory and empirical evidence suggests a series of properties for a *Synthetic Reproductive Skew Model of Homosexuality*. In what follows, I provide a narrative of how the *Synthetic Reproductive Skew Model of Homosexuality* is set to link the various variables of interest in a causal network, the specific effects that the variables are expected to have on each other and the manner in which such interactions may select for the expression of homosexual behaviour. The *Synthetic Reproductive Skew Model of Homosexuality* consists of the amalgamation of three major theoretical aspects of animal homosexuality: (a) *Reproductive Skew Theory* modified to take into account aspects of mate choice, inbreeding avoidance and queuing to breed (Figure 8.1); (b) sexual aspects of homosexual behaviour (Figure 8.2; see also Vasey 2006a) and (c) socio-sexual aspects of homosexual behaviour (Figure 8.3; see also Wickler 1967). Those three 'modules' of the synthetic theory are amalgamated together in the manner shown in Figure 8.4. Note how the inclusion of both conflict and cooperation in this model also provides a synthesis of the *Family Dynamics* and *Confluence* models that were introduced in Chapter 6 in the context of human homosexuality.

The links between diverse variables in Figure 8.4 are either directional causal links (arrows) or non-causal (i.e. correlational) associations between variables (straight lines). For instance, a polygamous mating system may directly lead to a biased sex ratio among breeders, thus favouring same-sex sexual behaviour in the breeding season, as the sex that has less access to heterosexual partners may form monosexual groups and engage in same-sex mounting primed by their reproductive neuroendocrinological status. Therefore the mating system is linked to a biased sex ratio through an arrow. The '+' sign inside the 'Biased Sex Ratio' box indicates the direction of the association between that variable and same-sex sexual behaviour. That is, as the sex ratio becomes more biased, same-sex sexual behaviour is expected to become more likely or frequent.

Providing a realistic mathematical formulation for all the relationships summarised in Figure 8.4 in a single model will not be easy and it is not my aim to embark on such an endeavour. Interesting advances are being made in the right direction (see, for example, Gavrilets & Rice 2006; Camperio-Ciani *et al.* 2008a) but are still far from producing a comprehensive formal model of the kind that is required. I do realise that this decision may frustrate some readers, but I hope that this will motivate them to try their skills at producing a proper and comprehensive formal model. Be that as it may, we can also argue that advances in knowledge can certainly be made even before a theory is mathematically formalised, and there are plenty of classic examples to support this view: evolution by natural selection (Darwin 1859), the handicap principle (Zahavi 1975), or the transmission theory for the evolution of parasite virulence (Ewald 1994), to mention just a few. My suggestion is that we need a mathematical formulation that is at the same time realistic and that takes into account the complexity of the phenomenon. Oversimplifying the problem in order to make it more easily treatable mathematically will be tantamount, to paraphrase the old Islamic tale of Mullah Nasruddin, to looking for a lost ring where you did not drop it simply because there you have plenty of light.

The model summarised in Figure 8.4 is formulated in a sufficiently general manner that it can be relevant for the understanding of both mammalian and avian homosexual behaviour. In spite of its qualitative formulation, the *Synthetic Reproductive Skew Model of Homosexuality* clearly makes the following broad predictions, which will be tested by using comparative analyses:

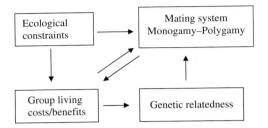

Figure 8.1. Four core variables that play an important role in *Reproductive Skew Theory* are included: ecological constraints, mating system, group living and genetic relatedness.

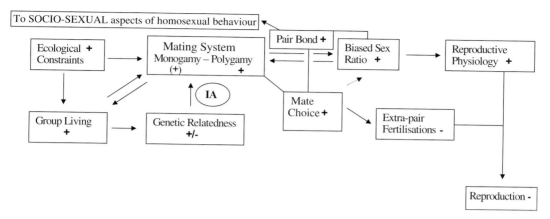

Figure 8.2. Core *Reproductive Skew Theory* variables and sexual variables. Note that the hatched arrow that links mate choice with sex ratio indicates that although mate choice may not necessarily affect the adult sex ratio as such it does affect the operational sex ratio, that is, the sex ratio of individuals involved in mating behaviours.

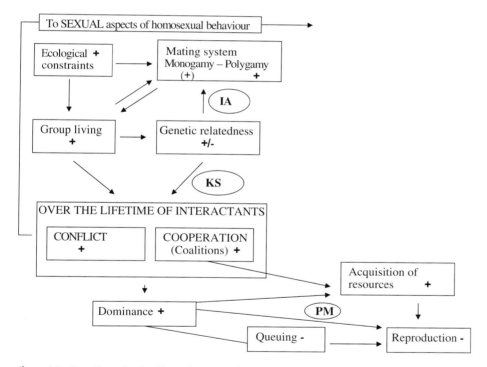

Figure 8.3. Core *Reproductive Skew Theory* variables and sociosexual variables.

(1) Ecological constraints to dispersal favour philopatry (Emlen 1982a) and set the stage for the expression of homosexual behaviour in either a sexual or a socio-sexual context as individuals of the same sex are more likely to encounter each other.

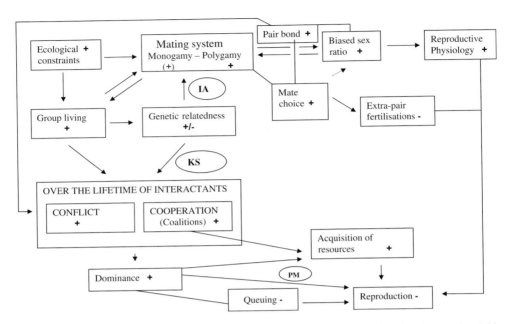

Figure 8.4. *Synthetic Model* that includes core *Reproductive Skew Theory*, sexual and socio-sexual variables. Signs (+, −) indicate the expected evolutionary correlation between the variable and same-sex mounting. IA, inbreeding avoidance; KS, kin selection; PM, parental manipulation.

(2) Philopatric conspecifics may organise themselves into social groups when benefits of group living exceed costs (Koenig & Pitelka 1981); again, close interactions among group mates may give raise to either sexual or socio-sexual expressions of homosexuality. The frequency of homosexual behaviours should be higher at intermediate group sizes, as opportunities for heterosexual copulations increase at very large group sizes (see, for example, LeBoeuf 1974) thus decreasing selection pressures for homosexual mounting.

(3) In social species with reduced natal dispersal, group members may include closely related individuals, produced by the reproductive activity within the group (Brown 1987; Emlen 1982a). Although group members may tend to avoid same-sex sexual behaviours with close relatives (e.g. adult siblings), as a result of selection for inbreeding avoidance in a heterosexual context (hence the predicted negative association of relatedness with homosexual interactions), less closely related members of a group (e.g. cousins) may engage in cooperative interactions that could be mediated by affiliative behaviours such as grooming but also same-sex reciprocal mounting (hence the predicted positive association between genetic relatedness and homosexual behaviour in those circumstances). Whether a mainly positive or negative correlation is selected will probably depend on the level of inbreeding depression caused by consanguineous heterosexual matings (O'Grady *et al.* 2006): higher levels of inbreeding depression will shift the balance towards a negative correlation between genetic relatedness and both heterosexual and, by extension, same-sex mounting.

(4) Ecological constraints, group living and inbreeding avoidance may all contribute to shape the mating system of the species (Bennett & Owens 2002), which of course may

range from monogamy, at one end of the spectrum, to polygamy/promiscuity at the other end. Potentially, both monogamy and polygamy may give rise to homosexual behaviours in social mammals and birds, but we should expect a trend towards higher levels of same-sex mounting in polygamous species, where the operational sex ratio is more biased.

(5) Five major elements of reproductive biology associated with the mating system can co-contribute to the manifestation of same-sex sexual behaviours from a purely sexual perspective. (a) If the mating system is such that there is a biased sex ratio among breeders (e.g. polygyny will produce an excess of non-breeding males if the adult sex ratio in the population is around 1:1), (b) those individuals not engaging in heterosexual sexual behaviour during the mating period (a period of increased circulation of sexual hormones in the blood) may be prone to copulate with alternative individuals such as same-sex partners. (c) Pair bonding behaviour that may have been originally selected in a heterosexual context (Cézilly *et al.* 2000) (or that might have been initially selected in the purely social context of homosociality; see prediction (6) below), may continue to operate in the new homosexual context. (d) Mate choice that may have also been selected in a heterosexual sexual context (Hasselquist & Sherman 2001) may also operate in the new homosexual context during the periods of breeding. All the above factors (a)–(d) contribute to the expression of same-sex sexual behaviours because they either limit the availability of heterosexual partners during periods of elevated sexual activity or promote bonding of same-sex partners for short or even long periods. Once a bond is established, sexual behaviour *may* follow. However, (e) opportunities for extra-pair fertilisations may decrease the likelihood of homosexuality in this purely sexual context. That is, this sexual module of the model clearly predicts that homosexuality during the breeding season is less likely to occur if all sexes have free access to members of the other sex for copulations and more likely to occur if such access is restricted.

(6) The reproductive skew module of the model links up with the final, socio-sexual module through group living and genetic relatedness, whereas the sexual module links directly with the socio-sexual module through pair-bonding. Outside the sexual context, same-sex pair bonding is an expression of homosociality; within the sexual context, it is an indication of homosexuality. Within a group, it is predicted that same-sex mounting may be associated with specific conflictive (e.g. competition) or cooperative (e.g. coalitions) interactions. The link with the sexual module through pair bonding also indicates that same-sex sexual behaviour may be preferred in the socio-sexual context of conflict over an alternative aggressive behaviour or in the context of cooperation over grooming behaviour under the influence of the reproductive condition of the individuals. In other words, depending on the context, same-sex mounting may be a socio-sexual manifestation of conflict or cooperation, especially during breeding periods of the population, i.e. periods of elevated neuroendocrinological readiness to mount.

(7) Same-sex sexual behaviour, by favouring coalition formation in its affiliative socio-sexual function or by reinforcing dominance in its competitive socio-sexual function, may lead to acquisition of resources that could then be used for reproduction (here bisexuality is predicted if both partners finally reproduce) or to the reproduction of one of the partners in the relationship (here exclusive homosexuality of one of the partners is a clear possibility, whereas the reproducing partner will obviously be defined as bisexual).

(8) Parental manipulation (Trivers 1974) may also contribute towards biasing the development of offspring towards a homosexual phenotype under the general effect of the inclusive fitness costs and benefits of having homosexual offspring.
Finally,

(9) A dominance hierarchy based on short queues for breeding status (Ridley & Sutherland 2002),

along with easy access to reproduction, are expected to decrease the extent of homosexuality in the population.

A comparative test of the *Synthetic Reproductive Skew Model of Homosexuality*

Comparative analyses will be carried out in the manner described in Chapters 2 and 7, using the same set of species and phylogenetic trees. I start with analyses of the avian dataset and continue with the analyses of the mammalian dataset.

Birds and the *Synthetic Reproductive Skew Model of Homosexuality*

Table 8.2 displays the dataset used in the analyses. The variables Body Mass, Mating System, Sociality and Same-Sex Mounting have already been defined and their codification (if applicable) used in the comparative analyses as explained in Chapter 7. Sex Involved in same-sex mounting might be males, females, or both. When both sexes display same-sex mounting, their relative frequencies might be similar or different; in the latter case males or females may be the sex with the higher reported level of same-sex mounting. Therefore this variable was entered in the analysis of independent contrasts, coded as follows: no same-sex mounting reported (code 1), M only (2), M > F (3), M = F (4), F > M (5) and F only (6); that is, the higher the value of the variable, the higher the relative level of female involvement in same-sex mounting. Adult Sex ratio was subdivided into five states ranging from heavily female-biased adult sex ratios (M << F, code 1) to M < F (2), M = F (3), M > F (4) and M >> F (5); that is, the larger the value of the variable, the more male-biased the adult sex ratio.

Information on homosexual mounting being explicitly performed in a socio-sexual context of dominance (Dominance Mount) was recorded as N (code 1) (the authors explicitly describe a lack of same-sex mounting behaviour when discussing the behavioural repertoire associated with dominance), Y (code 2) (same-sex mounting is described as a socio-sexual manifestation of dominance), Ya (code 3) (as for Y but in this case the authors explicitly state that the dominant individual is always the mounter). Affiliative Mount describes cases of same-sex mounting performed in contexts of affiliation, which usually manifests itself as bouts of reciprocal mounting. Codification in this case is either absence (N, code 1) or presence (Y, code 2). Degree of genetic Relatedness within the group, flock or colony was categorised as either Non-related (N, code 1), Low (2), Medium (3) or High (4).

The Social Unit involved could vary from a family group to a flock or a colony according to the species; here direct values of number of individuals observed were used and they were entered in the analysis after logarithmic transformation. Ecological Constraints are very difficult to measure for such a diverse number of species; I therefore used a proxy variable that, at least in part, reflects the degree of ecological constraints faced by the population. Species typically characterised by a Cooperative Breeding system (CB, code 2) are also typically under elevated ecological constraints on independent breeding (such constraint is often *limited dispersal*, sometimes *high predation*, etc.), whereas Non-cooperatively Breeding species (NCB, code 1) are comparatively less ecologically constrained (e.g. fewer impediments to young individuals to dispersing and breeding independently). Plumage Sexual Dichromatism in birds will be codified as 1 for sexually monochromatic species, 2 for species with only a mild sexual dichromatism and 3 for sexually dichromatic species. Heterosexual Pair Bonding may be present (Y, code 2), absent (N, code 1) or for a limited number of species, it may be absent but social bonds may nevertheless occur between dominant and subordinate males at a lek (~Y, code 1.5). Finally, I collected published information about the levels of Extra-pair Copulations (EPC). In this regard, for each species I was able to obtain information about either EPC or Extra-pair Paternity (EPP), but usually not both. Obviously, raw values of EPC and EPP cannot be mixed in the

Table 8.2. Life history and behavioural variables used in the comparative test of the *Synthetic Reproductive Skew Model of Homosexuality* in birds

All observations were made when birds were in breeding condition. Same-sex Mounting: C = copulations. Dominance Mount: Ya = mounter is always dominant. Affiliative Mount: RM = Reciprocal Mount. Relatedness: within group, flock or colony. Social Unit Size: number of individuals in group, flock or colony. Ecological Constraints: CB (cooperative Breeder, high ecological constraints), NCB (Non-cooperative Breeder, low ecological constraints). Plumage Sexual Dichromatism: monomorphic (M), dimorphic (D). EPCs = Extra-pair Copulations, see text for detailed explanation for the qualitative categorisation of different variables (EPC; EPP = Extra-pair Paternity). n.a. = not applicable due to unreported same-sex mounting for that species. (a) Large values of social unit size usually refer to breeding colonies. References for Sex Involved and Same-sex Mounting can be found in Table 2.1, whereas references for Mating System, Sociality and Body Mass can be found in Table 6.1.

Bird Species	Sex Involved	Same-sex Mounting	Adult Sex Ratio	Mating System	Sociality	Body mass (kg)	Dominance Mount	Affiliative Mount	Relatedness	Social Unit Size	Ecological Constraints	Plumage Sexual Dichromatism	Pair Bonding	EPCs	References
Columba livia	M?	Y (2 instances)	?	M	C	0.55	?	Y? (RM)	Low (probably)	169	NCB	M	Y	L (1% EPP)	Brackbill 1941; Murton et al. 1972; Robichaud et al. 1995; Moller 1997
Phoebastria immutabilis	M	Y	M≫F	M	C	2.5	?	?	?	~90	NCB	M	Y	?	Fisher 1971; Pitman et al. 2004
Philomachus pugnax	M	N	?	L	T	0.13	n.a.	n.a.	L	?	NCB	D	N	?	Thuman & Griffith 2004; Lank et al. 2002
Phoenicopterus chilensis	M > F	Y	M<F	M	C	3	N	N	L?	55	NCB	M	Y	M (9.57% EPC)	Farrell et al. 2000; King 2006
Phoenicopterus ruber ruber	M = F	Y (2%–3% of all C)	M>F	M	C	2.5	?	?	L?	?	NCB	M	Y	M?	Shannon 2000; King 2006
Phoenicopterus ruber roseus	F	Y (15.6% of all F)	M<F	M	C	2.5	?	?	L?	8600–19900(a)	NCB	M	Y	?	Cézilly & Johnson 1995; King 2006
Anas platyrhynchos	M	N	M>F	M	C	1.04	n.a.	n.a.	L?	274	NCB	D	Y	L (3% EPP)	Titman & Lowther 1975; Bossema & Roemers 1985; Spottiswoode & Moller 2004
Cygnus atratus	M	Y	M>F	M	C	6.22	?	Y	L	22	NCB	M	Y	H (15% EPP)	Braithwaite 1981; Kraaijeveld & Mulder 2002; Kraaijeveld et al. 2004
Branta canadensis		N	M=F	M	C	4.5	n.a.	n.a.	L?	150	NCB	M	Y	H (10%–12% EPP)	Klopman 1962; Abraham et al. 1999; Hundgen et al. 2000; Kalmbach 2006
Anser c. caerulescens		N	M<F	M	C	2.80	n.a.	n.a.	L?	10000	NCB	M	Y	L (2% EPP)	Rockwell et al. 1983; Lank et al. 1989; Schwagmeyer et al. 1999

Table 8.2. (Cont.)

Bird Species	Sex Involved	Same-sex Mounting	Adult Sex Ratio	Mating System	Sociality	Body mass (kg)	Dominance Mount	Affiliative Mount	Relatedness	Social Unit Size	Ecological Constraints	Plumage Sexual Dichromatism	Pair Bonding	EPCs	References
Anser anser	M	Y	M>F	M-Fpy	C	3.30	?	Y	High	120	NCB	M	Y	L?	Huber & Martys 1993; Hirschenhauser et al. 2000; Frigerio et al. 2001; Kortschal et al 2006
Bubulcus ibis	M	Y (3.3% of all C)	?	M	C	0.39	Ya	?	L?	?	NCB	M	Y	H (29.2% EPC)	Fujioka & Yamagishi 1981; Spottiswoode & Møller 2004
Ardea cinerea	M	Y (1% of all C)	?	M	C	1.44	Y	?	L?	808–978	NCB	M	Y	L	Wiese 1976; Lekuona 2002; Ramo 1993
Egretta caerulea	M	Y (3.1% of all EPC)	?	M	C	0.39	Y	?	L?	400–4000	NCB	M	Y	H	Meanley 1955; Werschkul 1982; Olmos & Silva e Silva 2002
Gallinula tenebrosa	M	Y (0.7% of all C)	M>F	Pya	G	0.52	Y	?	L	3–7	CB	M	?	?	Garnett 1978; Putland & Goldizen 2001; Crowley & Magrath 2004
Porphyrio porphyrio melanotis	F>M	Y (6.8–14% of all C)	M>F	Pya	G	0.89	Ya	?	H	14	CB	M	Y	L	Craig 1979, 1980; Jamieson & Craig 1987a,b; Lambert et al. 1994; Spottiswoode & Møller 2004
Alca torda	M	Y (9% banded M mountee; 66% mounter)	M>F	M	LC	0.61	Ya	?	?	~600	NCB	M	Y	VH (80% of F)	Wagner 1991, 1992, 1996; Sandvik et al. 2005
Uria aalge	M	Y (0.8% of all C)	M>F	M	C	0.94	?	?	Medium	900	CB	M	Y	M (1.6–9.9% EPC)	Birkhead 1978; Birkhead & Nettleship 1984; Hatchwell 1988; Birkhead et al. 2001; Friesen et al. 1996b

Species														References
Melanerpes formicivorus	M = F?	Y				0.08	N			5	CB		Y	MacRoberts & MacRoberts 1976; Koenig & Mumme 1987; Dickinson et al. 1995; Haydock & Koenig 2002; Spottiswoode & Moller 2004
Dinopium benghalense	M?	Y (single obs.)	?	M	T	0.10	?	Y (RM)	?	Single or pairs	NCB	Mild-D	Y	?	Ali 1961; Neelakantan 1968
Calypte anna	M	Y (single obs.)	M<F	Py	T	0.004	?	?	L	Single or pairs	NCB	D	N	H?	Wolf & Stiles 1970; Ewald & Rohwer 1980; Stiles 1982; Wethington et al. 2005
Gypaetus barbatus	M	Y (7.35% of all C)	?	M	T	5.07	Y	Y (RM)	M?	3	CB	M	Y	L	Bertran & Margalida 1999, 2003; Kimball et al. 2003; Godoy et al. 2004
Falco tinnunculus	M	Y	?	M	T	0.21	?	?	L	3	CB	D	Y	L (2% EPC)	Olsen 1985; Korpimäki 1988; Schwagmeyer et al. 1999; Kimball et al. 2003
Chionis minor	M	N	M<F	M	C	0.50	n.a.	n.a.	L	61 (island pop.)	NCB	M	Y	N	Burger 1981; Bried et al. 1999
Tryngites subruficollis	M	Y	M>F	L	T	0.05	Ya	?	L	?	NCB	M	N	L	Myers 1979; Lanctot & Weatherhead 1997; Lanctot et al. 1997, 1998
Pluvialis apricaria		N	M>F	M	T	0.21	n.a.	n.a.	L?	206	NCB	M	Y	?	Parr 1980
Haematopus ostralegus	F	Y	M<F	M	T	0.48	N	Y (RM)	Not related	3	NCB	M	Y	L (2% EPC)	Ens 1998; Heg & van Treuren 1998; Owens & Hartley 1998; Schwagmeyer et al. 1999
Himantopus h. himantopus	F	Y (very low freq.)	M<F	M	LC	0.18	?	n.a.	L?	26	NCB	D	Y	L (2.2% EPC)	Kitagawa 1988; Cuervo 2003
Tringa totanus	M	Y	M>F	M-FPy	LC	0.14	?	n.a.	L?	300	NCB	M	Y	?	Hale & Ashcroft 1982; 1983

Table 8.2. (*Cont.*)

Bird Species	Sex Involved	Same-sex Mounting	Adult Sex Ratio	Mating System	Sociality	Body mass (kg)	Dominance Mount	Affiliative Mount	Relatedness	Social Unit Size	Ecological Constraints	Plumage Sexual Dichromatism	Pair Bonding	EPCs	References
Struthio camelus australis		N	M<F	Py	SS	88.3	n.a.	n.a.	L	3–9	CB	D	N	VH (68.9% EPP)	Sauer 1972; Kimwele & Graves 2003
Dromaius novaehollandiae		N	M>F	M-Pa-Pr	SS	52.5	n.a.	n.a.	L	3–22	NCB	M	N	VH (73% EPC)	Coddington & Cockburn 1995; Hammond et al. 2002
Rhea americana		N	M>F	Hpy	SC	23	n.a.	n.a.	H	up to 12	CB	M	N	?	Raikow 1969; Codenotti & Alvarez 1997; Bouzat 2001; Fernandez et al. 2003; Fernandez & Reboreda 2003
Pygoscelis adeliae	M	Y	M≫F	M-Pa (14%)	C	3.99	?	Y? (RM)	L	9	NCB	M	Y	M (6.05 EPC)	Davis et al. 1998; Owens & Hartley 1998; Pilastro et al. 2001; Spottiswoode & Moller 2004; Shepherd et al. 2005
Phalacrocorax carbo sinensis		N	M>F	M	C	3.62	n.a.	n.a.	L	1900–10000	NCB	M	Y	M (10% EPC)	Moller & Birkhead 1993; Kortlandt 1995; Goostrey et al. 1998; Frederiksen & Bregnballe 2001
Agapornis roseicollis	F	N	?	M	C	0.05	n.a.	n.a.	?	?	NCB	M	Y	?	Dilger 1960
Aratinga canicularis	F	N	?	M	SC	0.07	?	?	?	9	?	M	Y	?	Hardy 1963
Brotogeris versicolurus	M	Y (single obs.)													Buchanan 1966
Brotogeris versicolurus		N	M>F	M	SC	0.07	n.a.	?	?	10	NCB	M	Y	?	Arrowood 1988
Eolophus roseicapillus		N	M<F	M	SC	0.3	?	Y (RM among M Imm)	?	18	?	M	Y	H	Rogers & McCullock 1981
	F	Y													Rowley 1990

Species														References
Scopus umbretta		N	?	M	T	0.35	?	?	?	?		Y	?	Cheke 1968; Neuchterlein & Storer 1989; Du Plessis *et al.* 1995
Larus ridibundus	M	Y	M>F	M	C	0.26	Y (but some reciprocal)	?	M	500	NCB	M	Y	?	van Rhijn & Groothuis 1985
		Y (10.5% of all pairs)													van Rhijn & Groothuis 1987
		N													Prévot-Julliard *et al.* 1998
Larus atricilla		N	M>F	M	C	0.27	?	?	?	10000	NCB	M	Y	H	Noble & Wurm 1943
	M	Y													Burger & Burger 1975; Burger & Shisler 1978; Hand 1985
Larus argentatus		N	M<F	M	C	1	n.a.	n.a.	?	~210- 5000	NCB	M	Y	L	Shugart 1980
				M-OPy M										(0.5% EPC)	Shugart *et al.* 1987, 1988
Larus occidentalis	F	Y (60% of all pairs)	M<F	M	C	0.76	?	?	?	~1200	NCB	M	Y	N	Schwagmeyer *et al.* 1999; Schreiber 1970; Hunt & Hunt 1977
		Y													Hunt *et al.* 1980
		N													Pierotti 1981; Hunt *et al.* 1984
		N													Spottiswoode & Møller 2004
Larus novaehollandiae	M	Y (10% of all? EPC)		M	C	0.18	?	?	?	?	NCB	M	Y	?	Mills 1994
Larus delawarensis		N	M<F	M	C	0.5	n.a.	n.a.	L	~5000	NCB	M	Y	?	Kovacs & Ryder 1981; Lagrenade & Mousseau 1983; Conover & Hunt 1984; Brown 1998
Larus californicus		N	M<F	M	C	0.6	n.a.	n.a.	?	~5000	NCB	M	Y	?	Pugesek 1983; Conover 1984; Conover & Hunt 1984b
Rissa tridactyla		N	M<F	M	C	0.38	n.a.	n.a.	L	500	NCB	M	Y	L (0.6% EPC)	Coulson & Thomas 1985; Helfenstein *et al.* 2004

Table 8.2. (*Cont.*)

Bird Species	Sex Involved	Same-sex Mounting	Adult Sex Ratio	Mating System	Sociality	Body mass (kg)	Dominance Mount	Affiliative Mount	Relatedness	Social Unit Size	Ecological Constraints	Plumage Sexual Dichromatism	Pair Bonding	EPCs	References
Menura novaeho llandiae	M	Y	M=F	Py	T	0.9	?	?	L	1–3	NCB	D	N	H	Kenyon 1972; Lill 1979; Smith 1988
Centrocercus urophasianus	F	Y	M>F	L	T	2.5	Y	?	L	5–100 (at lek)	NCB	D	N	H	Gibson & Bradbury 1986; Weatherhead & Montgomery 1995; Gibson *et al.* 2005
Gallus gallus	F	Y (11.9% of all F)	M≪F	Py	S	2.58	Ya	?	L	45	NCB	D	N	H (40% EPC)	Guhl 1948; Collias & Collias 1996
Fringilla coelebs		N	M=F	M	T	0.02	N	?	L?	~30	NCB	D	Y	H (7.75–17% EPC)	Marler 1955; Marjakangas 1981; Sheldon 1994; Schwagmeyer *et al.* 1999; Jabłoński & Matyjasiak 2002
Carpodacus mexicanus	M	Y (2 copul. only)	M≫F	M	SS	0.02	Y (but Sub mounts Dom)	?	L?	~4 (roost)	NCB	D	Y	M (8% EPC)	McGraw & Hill 1999; Schwagmeyer *et al.* 1999; Altizer *et al.* 2004b; Dhont *et al.* 2007
Ficedula hypoleuca		N	M=F	M	T	0.01	n.a.	n.a.	?	?	NCB	D	Y	H (12.8% EPC)	Alatalo *et al.* 1981; Slagsvold & Saetre 1991; Schwagmeyer *et al.* 1999
Lanius collurio		N	M>F	M	T	0.03	n.a.	n.a.	?	?	CB	D	Y	M (5.26% EPP)	Carlson 1989; Herremans 1997; Owens & Hartley 1998; Griffith *et al.* 2002; Spottiswoode & Moller 2004
Cyanistes caeruleus		N	M<F	M-FPy M-FPa	T	0.01	n.a.	n.a.	L	~3	NCB	D	Y	H (15% EPP; mean value)	Kempenaers 1994 Blakey 1996; Telleria *et al.* 2001; Griffith *et al.* 2002; Akçay & Roughgarden 2007

Species															References
Corvus monedula			N	M>F	M-FPy	T	0.24	n.a.	n.a.	L	3	NCB	M	Y	N	Röell 1979; Owens & Hartley 1998; Schwagmeyer et al. 1999; De Kort et al. 2006
Callaeas cinerea wilsoni			N	M>F	M	T	0.22	n.a.	n.a.	M?	2 or more	NCB	M	Y	L?	Meenken et al. 1994; Hudson et al. 2000; Flux et al. 2006; Higgins et al. 2006
Euplectes franciscanus	M		Y	M<F	Py	C	0.01	?	?	?	Large?	NCB	D	?	?	Craig 1974, 1980, 1982
Poephila acuticauda	M		Y	?	M	C	0.01	Ya	?	L?	20-30	NCB	M	Y	?	Burley & Symanski 1998; Langmore & Bennett 1999
Wilsonia citrina			N	M>F	M	T	0.01	n.a.	n.a.	M	80 (local pop. size)	NCB	D	Y	H (27% EPC)	Niven 1993; Stutchbury 1994; Stutchbury & Evans Ogden 1996; Schwagmeyer et al. 1999; Griffith et al. 2002
Chiroxiphia caudata	M		Y	?	L	T	0.02	Ya	N	H	2-6 M	NCB	D	~Y	L	Sick 1967; Foster 1984; Mercival et al. 2007
Mionectes oleagineus			N	M>F	L	T	0.01	n.a.	n.a.	?	2-6 M at lek	NCB	M	N	H	Westcott & Smith 1994; Westcott 1997a
Perissocephalus tricolor			N	?	L	T	0.33	Ya	?	?	8 M at lek	NCB	M	~Y	L	Westcott 1997b; Snow 1972; Trail 1990
Rupicola rupicola	M		Y	M>F	L	T SC	0.21	Y (but Sub mounts Dom)	?	?	55 M at lek	NCB	D	N	L	Trail 1985, 1990
Pseudonigrita arnaudi	M		Y (2 copul. only)	?	Pm	C	0.019	Y	?	?	2-9	CB	M	Y	?	Collias & Collias 1990
Tachycineta bicolor	M		Y	M>F	M	C	0.02	Ya	?	L	?	NCB	M	Y	VH (51% EPC)	Lombardo et al. 1994; Schwagmeyer et al. 1999; Conrad et al. 2001; Griffith et al. 2002

Table 8.2. (*Cont.*)

Bird Species	Sex Involved	Same-sex Mounting	Adult Sex Ratio	Mating System	Sociality	Body mass (kg)	Dominance Mount	Affiliative Mount	Relatedness	Social Unit Size	Ecological Constraints	Plumage Sexual Dichromatism	Pair Bonding	EPCs	References
Riparia riparia	M	Y	M=F	M	C	0.01	?	?	?	60–80	NCB	M	Y	H (14.37% EPP)	Peterson 1932; Carr 1968; Szep 1995; Moller 1997; Griffith *et al.* 2002; Spottiswoode & Moller 2004
Hirundo pyrrhonota	M	Y	M>F	M	C	0.02	?	?	L	350–2200	NCB	M	Y	H (23.7% EPP + CBP)	1988a; Brown & Brown 1996
Paradisaea raggiana		N	?	L	SC	0.21	n.a.	n.a.	?	?	NCB	D	N	?	Frith & Cooper 1996
Notiomystis cincta	M	Y	M>F	M-Pa-Py-Pya	T	0.03	N	N	L?	15–18	NCB	D	Y	H (35% EPP)	Ewen *et al.* 1999; Ewen & Armstrong 2002; Spottiswoode & Moller 2004
Lichenostomus melanops cassidix	M	Y (single obs.)	M=F	M	T	0.02	?	?	L?	Small	NCB	M	Y	?	Akçakaya *et al.* 1995; Franklin *et al.* 1995

same analysis as they represent two different variables. In particular, levels of EPP tend to be lower than the recorded levels of EPC when, for instance, there is a high degree of female choice. In this case, not all males copulating with a given female have the same chance of fertilizing her eggs. Sperm competition could also explain such patterns of differential paternity. If, on the other hand, EPCs are very difficult to observe, then the values of EPP may be higher than the recorded values of EPCs. In order to minimise these problems and yet use the full dataset in the comparative analyses I divided the percentages of EPC and EPP into 5 broad levels: No EPC (or EPP) (0%, code 1), Low (>0%–3%, code 2), Medium (>3%–10%, code 3), High (> 10%–40%, code 4) and Very High (>40%, code 5).

All observations were reported during periods of breeding activity for the species.

The predictions 1–9 of the model listed in the previous section suggest a series of results expected in the specific comparative tests. These expected results are summarised in Appendix 2.

In Table 8.3 I report the correlations between phylogenetically independent contrasts of the relevant variables, following the methods of phylogenetic tree branch length assignment of Grafen (1989) and the punctuational model of evolution that assigns equal branch lengths (of value 1) across the tree. None of the variables is significantly correlated with log (Body Mass + 1) contrasts (results not shown) with the exception of Adult Sex Ratio contrasts, which significantly decrease with log (BM + 1) contrasts (Pearson's product-moment correlation $r = -0.238$, $p = 0.039$, $n = 54$) at least when branch lengths were assigned by using Grafen's (1989) method. That is, in birds, evolutionary trends towards increasing body mass seem to be associated with an evolutionary trend towards a female-biased adult sex ratio. However this is unlikely to explain per se the association between same-sex mounting and adult sex ratio that I describe below, simply because same-sex mounting is not associated with body mass. In Table 8.3 all probabilities are one-tailed.

In general, the sign of the coefficient of correlation tends to be concordant between the two methods of branch length assignment, although Grafen's method produced the highest number of significant results, which are also consistent with the significant results obtained with the punctuational evolution method. Overall, five statistically significant correlations were obtained, all of them supporting the specific predictions of the *Synthetic Reproductive Skew Model of Homosexuality* (see also Appendix 2).

The five statistically significant results suggest that:

(a) When the size of the social unit becomes very large over evolutionary time, same-sex mounting seems to decrease.
(b) Evolutionary changes towards increased plumage sexual dichromatism are also associated with evolutionary changes towards increased polygamy.
(c) As the adult sex ratio becomes more male-biased, same-sex mounting also increases.
(d) Evolutionary shifts towards increased sociality are also associated with increases of same-sex mounting.
(e) As sociality increases, the specific level of dominance same-sex mounting also increases.

These results are consistent with a concomitant sexual and socio-sexual evolutionary context of homosexual behaviour in birds, as is stressed in the synthetic model. Highly social species tend to display same-sex mounting, especially in the context of establishing and maintaining dominance relationships. However, same-sex mounting also increases, in males for instance, as the sex ratio becomes more male-biased. Same-sex mounting in the supernumerary sex may be a result both of dominance interactions and of purely sexual interactions in circumstances where individuals of the other sex are scarce. Among the statistically non-significant trends, 7 tend to support and 6 tend to falsify the specific predictions of the model (see Appendix 2).

Table 8.3. Pearson's product-moment correlations (r) between standardised independent contrasts of various life-history and behavioural variables in birds

	n^a	r	p^b
Grafen's (1989) branch length method			
Ecological Constraints vs. Same-sex Mounting	69	− 0.019	0.43
Log-Social Unit Size vs. Same-sex Mounting	59	− 0.287	0.01**
Relatedness vs. Affiliative Mount	9	0.31	0.46
Relatedness vs. Same-sex Mounting	49	0.184	0.09
Mating System vs. Same-sex Mounting	71	0.030	0.39
Mating System vs. Plumage Sexual Dichromatism	71	0.389	0.0003***
Adult Sex Ratio vs. Same-sex Mounting	54	0.301	0.01**
Adult Sex Ratio vs. Sex Involved	54	− 0.079	0.28
Adult Sex Ratio vs. Mating System	54	− 0.063	0.32
Pair Bonding vs. Same-sex Mounting	69	− 0.001	0.49
EPCs vs. Same-sex Mounting	50	− 0.065	0.32
Sociality vs. Same-sex Mounting	71	0.200	0.045*
Sociality vs. Dominance Mount	23	0.570	0.001***
Mating System vs. Dominance Mount	23	− 0.222	0.14
Mating System vs. Sex Involved	71	0.149	0.10
Sociality vs. Affiliative Mount	11	0.052	0.43
All branch lengths set to 1			
Ecological Constraints vs. Same-sex Mounting	69	0.036	0.38
Log-Social Unit Size vs. Same-sex Mounting	59	− 0.261	0.02*
Relatedness vs. Affiliative Mount	9	0.071	0.42
Relatedness vs. Same-sex Mounting	49	0.139	0.16
Mating System vs. Same-sex Mounting	71	0.079	0.25
Mating System vs. Plumage Sexual Dichromatism	71	0.343	0.001***
Adult Sex Ratio vs. Same-sex Mounting	54	− 0.024	0.42
Adult Sex Ratio vs. Sex Involved	54	0.099	0.23
Adult Sex Ratio vs. Mating System	54	− 0.005	0.48
Pair Bonding vs. Same-sex Mounting	69	− 0.008	0.47
EPCs vs. Same-sex Mounting	50	− 0.020	0.44
Sociality vs. Same-sex Mounting	71	0.196	0.048*
Sociality vs. Dominance Mount	23	0.403	0.025*
Mating System vs. Dominance Mount	23	− 0.215	0.15
Mating System vs. Sex Involved	71	0.171	0.07
Sociality vs. Affiliative Mount	11	0.175	0.29

[a]Number of contrasts.
[b]One-tailed.

Before we turn to the comparative test of the *Synthetic Reproductive Skew Model of Homosexuality* in mammals, I address a case that is commonly reported as an example of 'homosexual behaviour' in birds: the case of female–female nesting in gulls (and other species). This case is interesting for at least two main reasons: first, it illustrates the effect of life history and demographic variables on same-sex nesting, and second it illustrates how an exceedingly broad definition of homosexuality may lump together all sorts of behaviours that have little to do with sexual orientation.

Is joint nesting an indication of homosexuality?

Among the species listed in Table 8.2, same-sex nesting has been reported, for instance, between females of *Phoebastria immutabilis*, *Chionis minor*, *Haematopus ostralegus*, *Himantopus himantopus*, *Larus argentatus*, *L. novaehollandiae*, *L. delawarensis*, *L. californicus* and *Rissa tridactyla* and between males of *Wilsonia citrina*, *Rhea americana*, and *Callaeas cinerea wilsoni* (see Table 7.1 for references and the recent work of Young *et al.* (2008) for *Phoebastria*). In this study, however, only those species that also displayed same-sex mounting were classified in the analyses as engaging in same-sex sexual behaviour (see Table 8.2). Most cases of joint nesting occur in species where the behaviour has been specifically selected (e.g. in the context of a polygamous mating system) or where the species is monogamous and displays both biparental care and strong pair bonds. In the latter situation same-sex nesting may occur among supernumerary individuals whenever the adult sex ratio is biased (Conover 1989).

Studies carried out in gulls have been especially informative in this regard. Females can form pair bonds with other females and become involved in joint nesting during the breeding season when the adult sex ratio is female-biased in western gulls, *Larus occidentalis* (Hunt *et al.* 1980; Pierotti 1981; Fry *et al.* 1987), California gulls, *L. californicus* (Conover & Hunt 1984a,b; Fry *et al.* 1987), ring-billed gulls, *L. delawarensis* (Conover & Hunt 1984a,b), herring gulls, *L. argentatus* (Shugart *et al.* 1987) and kittiwakes, *Rissa tridactyla* (Coulson & Thomas 1985). Alternatively, female–female pairs may be the result of a former polygynous trio (female–female–male) that has lost the male (e.g. through death) (Conover *et al.* 1979); e.g. ring-billed gulls *L. delawarensis* (Lagrenade & Mousseau 1983).

Long-term pair bonding in gulls, which may have been originally selected in the context of monogamy, may be subsequently retained in the context of polygyny and therefore it can be observed among two females, whether they paired as a result of a biased sex ratio or as a result of the loss of the male in an originally polygynous trio (Kovacs & Ryder 1981). The important role of pair bond formation was experimentally demonstrated by Conover & Hunt (1984a), who produced experimental short-term 'widows' during the incubation period in *L. delawarensis* and observed that none of them re-mated, thus suggesting that female–female pairs that are formed anew are likely to be formed during the period of pair bond formation, before egg laying. If such pair bond behaviour was evolutionarily selected within the context of heterosexual pairing, then individuals engaging in same-sex consortships may display a reproductive physiology that is not substantially different from that of heterosexual pairs. In fact, Kovacs & Ryder (1985) report how females in same-sex pairs have reproductive endocrinological traits similar to those of females in male–female pairs, except for higher levels of circulating progesterone and lower levels of circulating cholesterol in F–F pairs. The pattern of elevated progesterone can be easily explained by the prolonged incubation experienced by F–F pairs if eggs are infertile, whereas the lower levels of cholesterol may also be due to the elevated production of progesterone, which is a metabolite of cholesterol (Kovacs & Ryder 1985).

Long-term stability of F–F pairs could be adaptively explained if the reproductive success of the members of the pair is higher than if they remained unpaired on the face of the biased sex ratio in the population. This has been recently demonstrated in the Laysan albatross (*Phoebastria immutabilis*) nesting on the Hawaiian island of Oahu (Young *et al.* 2008). There, the proportion of females in the adult population is 59% and 31% of nests are attended by F–F pairs. Nests of monosexually paired females do contain the one-egg clutch typical of this species, the egg being laid by one of the two members of the pair, and reproductive success, although lower than in disexual pairs, is none the less higher than in the case of non-nesting individuals. Given that the members of the pair are not related, and that only one egg per season is incubated, the long-term stability of this association could be explained if, over various breeding seasons, the members of the pair somehow share their reproductive success. Young *et al.* report some F–F pairs being together for 8 years: in one case that was recorded on the island of Kauai, the two females had been nesting together for 19 years.

Whenever females do indeed engage in copulations with each other, then I regard the use of the expression 'homosexual behaviour' as appropriate (e.g. *Larus occidentalis* in the studies carried out by Hunt & Hunt 1977; Hunt *et al.* 1984).

My argument against considering same-sex pair bonds as an unequivocal evidence of homosexual behaviour, unless actual sexual behaviour (e.g. mounting) does occur, also impinges on the understanding of homosexuality in humans. For instance, in a situation where not only child but also adult mortality is high, as it has been the case for most of the evolutionary history of our species, relatives' cooperative caring of children who lost both of their parents is a behaviour that I expect to be adaptive as a result of kin selection. Relatives have cared for parentless children regularly (see, for example, Fauve-Chamoux 1996; van Solinge *et al.* 2000) although in some societies legal provisions impeded a two-female guardianship of an orphaned child (e.g. the nineteenth-century Netherlands; van Solinge *et al.* 2000). However, if not legal, at least *de facto* same-sex guardianship might have occurred as two sisters or two brothers came to live together upon the death of a spouse and cared for their children or even the orphaned child of a relative. Labelling two persons of the same sex 'homosexuals' safely on the grounds that they are living together and collaborating in the rearing of a child seems preposterous. In a cross-species comparative perspective, it is not the social bond and shared parental care that uniquely defines homosexuality, it is the engaging in same-sex sexual behaviours (see also Vasey 2006a). Therefore those species that show same-sex sharing of parental duties and that even establish strong same-sex pair bonds should not be described as homosexuals unless the members of the same-sex consortship do in fact copulate or engage in mounting behaviour. In brief, a different term should be introduced to describe same-sex sharing of parental duties and performance of any other non-sexual activities between individuals of the same sex. In Chapter 7 I introduced the term *homosociality*. Homosexuality and homosociality are independent; each one may occur without the other and they may also occur together. This distinction is important in the **evolutionary** study of homosexuality.

We now turn to the comparative analysis of the mammalian dataset.

Mammals and the *Synthetic Reproductive Skew Model of Homosexuality*

The complete dataset used in these analyses is shown in Table 8.4. Body Mass, Mating System, Sociality, Sex Ratio, Same-Sex Mounting, Dominance Mount, Affiliative Mount, Degree of Genetic Relatedness and Ecological Constraints were all treated and codified where applicable as in the case of birds. The variable Sex Involved in same-sex mounting was also codified as in the case of birds. However, given that several species of mammals were the subject of various studies showing a degree of intraspecific variability in the sex displaying same-sex mounting, I took such variability into account by assigning a value to the species that is the arithmetic mean of the codes for this variable across studies. For instance, for *Macaca mulatta* 8 published works were included: 5 indicating female homosexual behaviour, 2 reporting male homosexual behaviour and one where homosexual behaviour was more frequent in males than females. For *M. mulatta* then the variable Sex Involved took the value of $[(5 \times 6) + (2 \times 2) + 3]/8 = 4.6$. Whereas for birds I used the variable Plumage Sexual Dichromatism, for mammals Body Mass Sexual Dimorphism (BMSD) was used:

$$BMSD = BMM / BMF$$

Table 8.4. Life history and behavioural variables used in the comparative test of the *Synthetic Reproductive Skew Model of Homosexuality* in mammals

Observations were made during breeding periods of the year. Species: the number in brackets refers to the list of references below. Same-sex Mounting: C = copulations; a number in brackets after Y (yes) or N (no) indicates the number of different works that support such evidence. Dominance Mount: Ya = mounter is always dominant. Social Unit Size: where indicated, the value is the arithmetic mean of more than one kind of social unit; e.g. lek, bachelor herd and territorial harem. Ecological Constraints: some species are polytypic for this trait. Extrapair Copulations: unlabelled results are qualitative estimates of EPC, more specific quantitative values are available for Extra-pair Copulations (EPC), Extra-pair Offspring (EPO), Extra-group Paternity (EGP), Extra-group Copulations (EGC), Extra-alliance Paternity (EAP), Extra-unit Paternity (EUP), Non-αMale Paternity or Copulations (NαP/C), Peripheral Lek Male Copulations (PLMC), Copulations by Males from Outside the Resident Population (CMORP), Females Engaged in Extra-harem Copulations (FEHC). n.a., Not applicable owing to unreported same-sex mounting for that species.

Mammal species	Sex Involved	Same-sex Mounting	Adult Sex Ratio	Mating System	Sociality	Body Mass (kg)	Dominance Mount	Affiliative Mount	Relatedness	Social Unit Size	Ecological Constraints	Sexual Dimorphism	Social Bonding	EPCs
Suncus murinus (1)	M	Y	0.81	Py	SO	0.08	?	?	?	?	NCB	1.52	?	H
Dugong dugon (2)	M	Y	1	L-FPr	FSC	569	?	?	L	1.3	NCB	0.88	N	?
Loxodonta africana (3)		N	0.88	Py	T	4540	n.a.	n.a.	M	10	CB	1.76	Y	M
Felis catus (4)	M>F	Y	1.84	Py	LG	3	Y	N	H	~5	NCB	1.2	Y	H (23.8% EGC)
Panthera leo persica (5)	M, F	Y (0.9% of all C)	0.45	Py	SC	199	Ya(M)	?	M	~9	CB	1.32	Y	?
Panthera leo leo (6)	M, F	Y	0.44	Py	SC	199	?	Y	M	14	CB	1.5	Y	?
Acinonyx jubatus (7)		N	2.6	Py	SC	45	n.a.	n.a.	H	3	NCB	1.22	-Y	H
Helogale undulata rufula (8)	M>>F	Y(1), N(1)	2	Pa, Pya	SC	0.27	N	N	H	12	CB	1.1	Y	N
Canis familiaris (9)	M>>F	Y (M1.9%, F0.4% of all C)	1.35	Pya	SC	35.5	Ya	N	L	1.23	NCB	1.03	Y	H
Canis lupus (10)		N	1.3	M	SC	43	n.a.	n.a.	H	~5	CB	1.15	Y	L
Chrysocyon brachyurus (11)	M	Y	?	M	SO	21.5	?	?	L?	1.5	NCB	1.03	N	?
Speothos venaticus (12)	M	Y(1), N(1)	?	M	T	6	?	Y	H	~3	CB	?	Y	?
Ursus arctos horribilis (13)		N	0.82	Pya	T	233	n.a.	n.a.	L	2	NCB	1.42	N	H

Table 8.4. (*cont.*)

Mammal species	Sex Involved	Same-sex Mounting	Adult Sex Ratio	Mating System	Sociality	Body Mass (kg)	Dominance Mount	Affiliative Mount	Relatedness	Social Unit Size	Ecological Constraints	Sexual Dimorphism	Social Bonding	EPCs
Desmodus rotundus (14)		N	0.64	Hpy	C	0.03	n.a.	n.a.	M	10	CB	0.83	Y	?
Myotis lucifugus (15)	M	Y	2	Pr	SC	0.009	?	?	L	6	NCB	1.04	Y	H
Balaena mysticetus (16)		N	0.66	Py	SC	42500	n.a.	n.a.	H	5	NCB	0.91	N	?
Stenella longirostris (17)	M	Y	0.83	Py-Pya	SC	65	?	?	M	56	CB?	1.14	–Y	L? (EGC)
Tursiops truncatus (18)	M>F, M	Y (3), N(1)	1.02	Py	SC	175	Ya	Y(M)	M	12.8	CB	1.03	Y	H (17.6% EAP)
Odobenus rosmarus (19)	M	Y(3), N(1)	0.21, but all M in haul out sites	Py	SC	1069.5	Ya	Y?	L?	487.7	NCB	1.48	N	L? (NαP)
Halichoerus grypus (20)	M	Y	0.57	Py	SC	194	?	?	M?	~1000	CB, NCB	1.5	N?	VH (62.5% EHO)
Mirounga angustirostris (21)		N	0.45	Py	SC	1600	n.a.	n.a.	M?	8600	CB, NCB	2.54	N	VH (44% NαP)
Neophoca cinerea (22)	M>>F	Y	0.33	Py	SC	178.5	N	Y	H?	200	CB, NCB	3.62	N	?
Phocarctos hookeri (23)	M	Y	0.5	Py	SC	227	N	Y	H	1000	NCB	2.32	N	?
Rattus norvegicus (24)	(M = F)?	Y	1.8	Py	SC	0.32	Ya(M)	?	?	?	CB	1.47	?	?
Oryctolagus cuniculus (25)	F	Y	0.66	Py-Pr	SC	2.25	Y	N	M	~4	NCB	1.05	?	M
Notomys alexis (26)	M	Y	~1.11	M	SC	0.03	Y	Y	?	≥ 4	?	0.83	Y	?
Pseudomys albocinereus (27)	M	Y	?	Pr?	SC	0.03	Y	Y	?	3	?	?	Y	?
Microcavia australis (28)		N	1.48	Pya	SC	0.27	n.a.	n.a.	?	21	CB	~1	N	?
Marmota flaviventris (29)	F	Y	0.43	Py	SC	3.5	Y	?	M	5	NCB?	1.39	Y	L?

350

Species													
Marmota caligata (30)	F	Y (33.3% of all C)	0.55	M-Py-Pya	SC	6.35	?	H	12	CB	1	Y	M (6% EPR)
Marmota olympus (31)	F>M	Y	0.66	Py-Pya	S	6	?	?	9	CB?	1.35	Y	?
Tamiasciurus douglasii (32)	M	Y	?	Pr	T	0.22	?	L	2	NCB	1.03	N	VH (54.5% NaP)
Tamiasciurus hudsonicus (33)	M, F	Y	1.02	Pr	T	0.19	?	L	~4	NCB	1.88	N	H
Sciurus carolinensis (34)		N	1	Pr-Py	T	0.55	n.a.	H(F)	2.6	NCB	-1	Y	?
Macropus giganteus (35)		N	0.49	Pr	SC	49	n.a.	L	3.7	NCB	1.88	N	?
Macropus rufogriseus banksianus (36)		N	?	Pr, Py	SC	16.85	n.a.	?	2.2	NCB	1.68	N?	H (~35% NaP)
Macropus agilis (37)	M	Y (Single obs.)	?	Pr	FSC	15	?	?	10	NCB	1.72	?	?
Macropus parryi (38)	M	Y (1.3% of all C)	0.57	Pr	SC	13.5	N	?	6.2	NCB	1.72	N	?
Macropus rufogriseus rufogriseus (39)	F	Y (Frequent)	0.65	Pr, Py	SO	16.85	Y	?	1.5	NCB	1.68	?	?
Aepyprymnus rufescens (40)	F	Y(1), N(1)	0.4	Py	SO	2.48	?	M	1.2	NCB	0.9	N	H
Sminthopsis crassicaudata (41)		N	0.66	M	SC	0.01	n.a.	L?	~3	NCB	-1.03	N	H?
Dasyurus hallucattus (42)		N	0.9	Pr	T	0.25	n.a.	?	1	NCB	1.34	N	H
Kobus leche kafuensis (43)		N	0.74	L-FHPy	SC	85.5	n.a.	?	20 (lek, bach. herd, territory)	NCB	1.3	?	M
Kobus vardoni (44)	F	Y(1), N(1)	0.46	L-HPy	SC	71.5	?	?	4.3	NCB	1.16	N	?
Kobus ellipsiprymnus (45)	F>>M	Y(1), N(1)	0.62	Py	SC	210	?	?	5.9	NCB	1.32	?	M (9% FEHC)
Gazella thomsoni (46)	M	Y (Infrequent)	0.51	Py	SC	23	Ya	?	4.6	NCB	1.2	N?	?
Adenota kob thomasi (47)	F	Y(1) (6.9% F involved), N(1)	0.33	Py	SC	174	?	?	285 (mixed sex, bach. herd)	NCB	?	N	M

Table 8.4. (*Cont.*)

Mammal species	Sex Involved	Same-sex Mounting	Adult Sex Ratio	Mating System	Sociality	Body Mass (kg)	Dominance Mount	Affiliative Mount	Relatedness	Social Unit Size	Ecological Constraints	Sexual Dimorphism	Social Bonding	EPCs
Antilope cervicapra (48)	M>>F	Y	0.7	HPy	SC	37.5	Y	?	?	13	NCB	1.08	?	M (10% PLMC)
Antilocapra americana (49)	M, M>>F	Y (9.4% of all C)	0.64	Pm	SC	49.7	Ya	?	L?	9 (territory, bach. herd)	NCB	1.18	N	H (≥ 22% EPO)
Odocoileus hemionus columbianus (50)	F	Y	0.78	Pm	SC	80.75	?	?	?	2	NCB	1.34	?	?
Odocoileus virginianus (51)	M	Y	0.51	Pm	SC	65.75	Y	?	L	56	NCB	1.64	-Y	M (≥5.3% EPO)
Alces alces (52)		N	0.63	Py	SC	566.5	n.a.	n.a.	L	≤ 36	NCB	1.26	N?	M
Cervus elaphus (53)	M, F	Y	0.65	Py	SC	297.5	N	N	L	7	NCB	1.32	N	L
Cervus elaphus roosevelti (54)	M	Y	0.74	Py	SC	297.5	Y	?	L	34	NCB	1.32	Y	?
Cervus nippon (55)	M	Y	0.38	L	SC	36.01	?	?	H	~13	NCB	1.3	N	H (27.2% NαC)
Dama dama (56)	M	Y	0.65	L	SC	57	N	?	L	5.8	NCB	1.42	N	M
Elaphurus davidianus (57)	(M = F)	Y	0.72	HPy	SC	186.5	?	?	?	58	NCB	1.34	N	L (2% EPC)
Muntiacus reevesi (58)	M	Y	0.66	Py	SC	13.5	?	?	?	1.5	NCB	0.8	N?	L?
Rangifer tarandus (59)	(M = F)?	Y	0.61	Py	SC	87	?	?	L?	~20	NCB	1.44	N	M (3.6% NαP)
Axis axis (60)	M	Y	0.85	Py	SC	69.5	?	?	?	3.3	NCB	1.78	N?	M (~4% NαP)
Pudu puda (61)	M	Y	?	M?	SO	9	?	?	?	1.5	NCB	1	N?	?
Giraffa camelopardalis (62)	M, M>F	Y	2.5	Py	SC	900	Y(M)	?	L?	4	NCB	1.56	N(M), Y(F)	L?
Vicugna vicugna (63)	M?	Y	0.24	Py	SC	50	?	?	?	8.5	NCB	1.08	Y	N (~0% EHC)
Tayassu tajacu (64)	M	Y	1	Py	SC	30	?	?	H?	10	CB	1.12	Y	L?
Tayassu pecari (65)	(M = F)?	Y	0.63	Py	SC	32.5	?	?	?	20	?	1.02	Y	?
Phacochoerus aethiopicus (66)	F	Y	0.33	Pr	SC	55	?	?	?	3.1	CB	1.37	?	?
Bison bonasus (67)	F, M	Y	0.76	Py	SC	610	?	?	H	18	NCB?	1.69	?	?
Bison bison (68)	M, M>F	Y (M: 45% of all C)	0.11	Py	SC	565.5	Y(F), N(M)	Y(M)	L	21	NCB, CB	1.64	N	H

Species													
Bison bison athabascae (69)	M	Y	0.76	Py	SC	565.5	?	L	15	NCB	1.64	N	H
Bos indicus (70)	M >>> F	Y	0.03	Py	SC	755	N	?	30	?	?	Y	?
Bubalus bubalis (71)	F	Y	1.01	Py	SC	725	?	?	33	CB?	1.19	N?	?
Ovibos moschatus (72)	M > F	Y	0.73	HPy	SC	232.5	N	?	18.1	?	1.46	Y(F)	?
Equus caballus (73)	M > F	Y	0.98	Py	SC	1150	Ya	M	3.4 (harem, bach. group)	NCB	1.07	Y?	L
Equus przewalskii (74)	F	Y(1), N(1)	0.88	Py	SC	250	?	H(F)	5.2 (harem, bach. group)	?	1.09	N	M?
Equus quagga (75)	M	Y(1), N(1)	0.52	Py	SC	275	Ya	L	26	NCB	1.13	Y	L
Equus zebra (76)	M	Y	0.41	Py	SC	350	Y, N	Y?	L	2.2 (harem, bach. group)	NCB	1.08	?	?
Rupicapra pyrenaica ornata (77)		N	0.3	Py	SC	37	n.a.	M(F)	22	?	1.48	?	?
Pseudois nayaur (78)	M	Y	1.05	Py	SC	47	Y	?	26.2	NCB?	1.34	N	L
Ammotragus lervia (79)	M>>F, M	Y	0.3	Pr	SC	99.5	Y?	?	7	?	1.44	N	?
Ovis orientalis musimon (80)	F > M	Y (M: 2.6%, F 7.5% of all C)	0.25	Py	SC	45	Y(M)	?	25	?	1.48	N	?
Ovis aries (81)	M	Y	1.2	Pr	SC	23	Ya	H?	3.2	?	1.3	N	M
Ovis canadensis (82)	M	Y	0.67	Pr	SC	185	Ya	L	40	CB, NCB	1.42	N	H
Capra hircus (83)	M	Y	0.42	Py	SC	45	N?	?	5.2	?	1.44	Y(F), N(M)	H (17.2% CMORP)
Cebus capucinus (84)	M	Y	0.66	Pm	SC	3.26	Y?	M	21	CB	1.44	N?	L? EGC
Propithecus verreauxi (85)	M	Y	0.87	Pya	SC	3.6	?	L	4.7	CB	1.04	Y	H (~25% EGP)
Saimiri sciureus (86)	F, M	Y(2), N(1)	1	Py	SC	0.76	Ya	M	60	CB	1.26	?	?
Nasalis larvatus (87)	F = M	Y (M: 12.5%, F: 12.5% of all C)	0.35	Py	SC	15.19	N?	?	12	-CB	2.04	?	?
Miopithecus talapoin (88)	M	Y	1.2	Py	SC	1.25	?	?	45	?	1.26	Y	?
Cercopithecus aethiops (89)	M > F	Y	0.53	Pm	SC	4.05	?	H	24	CB	1.24	Y	M (9% NoC)

Table 8.4. (Cont.)

Mammal species	Sex Involved	Same-sex Mounting	Adult Sex Ratio	Mating System	Sociality	Body Mass (kg)	Dominance Mount	Affiliative Mount	Relatedness	Social Unit Size	Ecological Constraints	Sexual Dimorphism	Social Bonding	EPCs
Hylobates lar (90)	M	Y	1	M	SC	5.5	N	Y?	H	3.4	CB, NCB	1.06	Y	H (12% EPC)
Presbytis entellus (91)	F, M>F	Y	0.14	Py	SC	12.8	Ya	Y	L	19.5	CB	1.46	Y	VH (72% NαC)
Papio ursinus (92)	M	Y	0.74	HPy	SC	19.41	Y	Y	M	50	CB	1.66	Y	H (37% NαC)
Papio cynocephalus hamadryas (93)	(F = M)?	Y	0.48	HPy	SC	13.85	Y(F)	Y(M)	?	41	CB, NCB	1.94	Y	VH (81.2% EUP)
Papio cynocephalus anubis (94)	M, M>>F	Y	0.7	Pr	SC	19.5	Ya	Y(M)	M	97.2	CB, NCB	2	Y	VH (53.6% NαC)
Theropithecus gelada (95)	M>>F	Y (M: 25%, F: 2% of all C)	0.37	HPy	SC	17.05	N	Y	M?	30	NCB	1.5	Y	L?
Macaca fuscata (96)	M>>F, F, M, (F=M)?	Y (19% of all C, 59.7% F involved)	0.35	Pr	SC	10.1	N(5), Y(2)	Y	H(F), L(M)	89.8	CB	1.18	Y	VH (78% NαC)
Macaca arctoides (97)	F, (M>F)?, F>M	Y(3), N(1) (M: 9%, F: 16% of all C; 80% F involved)	0.29	Pm	SC	7	Ya(M,F)	Y(M,F)	H?	22	CB, NCB	1.45	Y	L
Macaca mulatta (98)	F, M, M>F	Y (67.5% F involved)	0.27	Pm	SC	4.6	Y	Y	L	53.7	CB	2.06	Y	VH (62.5% NαC)
Macaca nemestrina (99)	F, (M=F)?	Y	0.16	Pm	SC	8.5	Ya(F)	Y(M,F)	?	36	?	1.42	Y	H NαP
Macaca radiata (100)	M	Y	0.69	Pm	SC	5.15	Y	Y	M	24	CB	1.78	Y	?
Macaca tonkeana (101)	(M = F)?	Y (M: 0.2%, F: 0.2% of all C)	1	Pm	SC	12	N?	Y?	?	~17	?	1.65	Y?	?
Macaca nigra (102)	M, (F=M)?	Y	0.2	Pm	SC	8	N	Y(M,F)	?	97	?	1.8	?	?

Species											
Pan paniscus (103)	F>>M	Y (M: 11.7%, F: 96.4% of all C)	0.41	Pr	SC	35	Y(F)	Y(M,F)	H(M), L(F)	37	Y	L EGC
Pan troglodytes (104)	F	Y (High frequency)	0.33	Py?	SC	45	?	Y	H	82	Y	−M (7%EGP, L EGC, 54% N≠C)
Gorilla gorilla (105)	M>>F, M, F Y		0.77	Py	SC	126.5	Y	Y	L	10.8	Y	L
Pongo pygmaeus (106)	M	Y (0.9% of all C)	0.54	HPy	T	53	Y?	Y?	M	2	N	M (10% N≠P)
Homo sapiens (107)	M>F, M, F Y		0.94	M-FPm	SC	57	Y?	Y	L	~150	Y	M (~10% EPP)

References: (1) Louch *et al.* 1966, Symonds 1999, Stockley 2003; (2) Marsh 1995, Aketa *et al.* 2001, Gales *et al.* 2004, McDonald 2005; (3) Buss & Smith 1966, Armbruster & Lande 1993, Weckerly 1998, Archie *et al.* 2006a,b, Hollister-Smith *et al.* 2007; (4) Yamane 1998, 1999, Natoli *et al.* 2001, Denny *et al.* 2002, Kaeuffer *et al.* 2004, Yamane 2006; (5) Ravi & Johnsingh 1993, Pati 2001; (6) Schaller 1972, Bertram 1975, Packer & Pusey 1982; (7) Caro & Collins 1987, Marker & Dickman 2003, Marker *et al.* 2003, Hayward *et al.* 2006, Gottelli *et al.* 2007, Marker *et al.* 2008; (8) Rasa 1977; (9) Daniels & Bekoff 1989, Heusner 1991, Pal *et al.* 1998, 1999, Mertens 2006; (10) Lehman *et al.* 1992, Głowaciński & Profus 1997, Jędrzejewski *et al.* 2005; (11) Kleiman 1972, Fletchall *et al.* 1995; (12) Kleiman 1972, IUCN 2005; (13) Eberhardt & Knoight 1996, Reed-Eckert *et al.* 2004; (14) Wilkinson 1985, 1986, Delpietro & Russo 2002; (15) Thomas *et al.* 1979, Saunders & Barclay 1992; (16) Schell *et al.* 1989, Würsig *et al.* 1993, Skaug & Givens 2007; (17) Norris & Dohl 1980, Dayaratne & Joseph 1993, Perrin *et al.* 2005, Silva *et al.* 2005; (18) Shane *et al.* 1986, Östman 1991, Fish 1993, Félix 1997, Krützen *et al.* 2003, Lusseau *et al.* 2003, Krützen *et al.* 2004, Mann 2006; (19) Miller 1975, Salter 1979, Miller & Boness 1983, Sjare & Stirling 1996, Weckerly 1998; (20) Coulson & Hickling 1964, Riedman 1982, Tinker *et al.* 1995, Weckerly 1998, Ambs *et al.* 1999, Wilmer *et al.* 1999; (21) Le Boeuf 1974, Riedman 1982, Campagna & Lewis 1992, Weckerly 1998; (22) Marlow 1975, Riedman 1982, Weckerly 1998, Campbell *et al.* 2008; (23) Marlow 1975, Weckerly 1998, Childerhouse & Gales 2001; (24) Stroud 1982, Stewart & German 1999, Yamada 1999, Hayes 2000; (25) Daly 1981, Swihart 1984, Cowan 1987, Ebensperger & Cofré 2001; (29) Armitage *et al.* 1976, Johns & Armitage 1979, Frase & Hoffmann 1980, Schwartz *et al.* 1998, Oli & Armitage 2003; (30) Barash 1974, Carranza 1996, Kyle *et al.* 2007; (31) Barash 1973, Armitage 1981, Blumstein & Armitage 1999; (32) Smith 1981, Koford 1982; (33) Layne 1954, Steele 1998, Lane *et al.* 2007; (34) Thompson 1978, Carranza 1996, Koprowski 1996; (35) Kaufmann 1974, Jarman & Southwell 1986, Jarman 1991, Zenger *et al.* 2003; (36) Johnson 1989a, b, Jarman 1991, Watson *et al.* 1992; (37) Stirrat & Fuller 1997, Weckerly 1998; (38) Kaufmann 1974, Jarman 1991; (39) La Follette 1971, Kaufmann 1974, Fleming *et al.* 1983, Jarman 1991; (40) Johnson 1980, Jarman 1991, Frederick & Johnson 1996, Pope *et al.* 2005; (41) Morton 1978a,b; (42) Schmitt *et al.* 1989, Oakwood 2002, Griffiths & Brook 2005; (43) De Vos & Dowsett 1966, Nefdt 1995, Weckerly 1998, Isvaran 2005; (44) De Vos & Dowsett 1966, Weckerly 1998, Corti *et al.* 2002; (45) Spinage 1982, Wirtz 1983, Weckerly 1998; (46) Walther 1978a,b, Borner *et al.* 1987, Weckerly 1998, Ezenwa 2004; (47) Buechner & Schloeth 1965, Leuthold 1966, Modha & Eltringham 1976, Isvaran 2005; (48) Dubost & Feer 1981, Loison *et al.* 1999a, Isvaran & Jhala 1999; (49) Gilbert 1973, Kitchen 1974, Riedman 1982, Weckerly 1998, Berger & Gompper 1999, Carling *et al.* 2003, Stephen *et al.* 2005; (50) Weckerly 1998, Berger & Gompper 1999, Bender 2000; (51) Mathews & Porter 1993, Weckerly 1998, DeYoung *et al.* 2003, Sorin 2004, Bartoš & Holečková 2006; (52) Ballenberghe & Miquelle 1993, Weckerly 1998, Berger & Gompper 1999, Wilson *et al.* 2003; (53) Gibson & Guinness 1980, Grant *et al.* 1992, Weckerly 1998, Berger & Gompper 1999, Slate *et al.* 2000, Bartoš & Holečková 2006; (54) Weckerly 1998, Berger & Gompper 1999, Polziehn *et al.* 2000; (55) Feldhamer 1980, Weckerly 1998, Berger & Gompper 1999, Endo & Doi 2002, Okada *et al.* 2005; (56) Moore *et al.* 1995, Weckerly 1998, Apollonio *et al.* 1998, Berger & Gompper 1999a, Say *et al.* 2003, Bartoš & Holečková 2006; (57) Weckerly 1998, Berger & Gompper 1999, Chunwang *et al.* 2004; (58) Yahner 1979, Weckerly 1998, Loison *et al.* 1999, Berger & Gompper 1999; (59) Fancy *et al.* 1990, Weckerly 1998, Berger & Gompper 1999, Roed *et al.* 2002, Holand *et al.* 2007; (60) Barrette 1991, Raman 1997, 1998, Weckerly 1998, Berger & Gompper 1999;

Notes to Table 8.4 (*cont*). (61) Carranza 1996, Loison *et al.* 1999; (62) Innis 1958, Coe 1967, Leuthold 1979, Weckerly 1998, Le Pendu *et al.* 2000a, Bercovitch *et al.* 2006; (63) Koford 1957, Bosch & Svendsen 1987, Wheeler 2006; (64) Byers & Bekoff 1981, Dubost 1997; (65) Mayer & Wetzel 1987, Dubost 1997, Altrichter *et al.* 2001; (66) Somers *et al.* 1995, Caro 1999, Ruckstuhl & Neuhaus 2002; (67) Krasiński & Raczyński 1967, Caboń-Raczyńska *et al.* 1987, Mysterud *et al.* 2001, Luenser *et al.* 2005; (68) Riedman 1982, Reinhardt 1985, Meagher 1986, Weckerly 1998, Wilson & Strobeck 1999, Vervaecke & Roden 2006; (69) Fuller 1960, Gates & Larter 1990, Wilson & Strobeck 1999, Ruckstuhl & Neuhaus 2002; (70) Reinhardt & Reinhardt 1981, Reinhardt 1985, Orihuela & Galina 1997; (71) Murphey *et al.* 1991, Berger & Gompper 1999, Heinen & Singh 2001; (72) Reinhardt & Flood 1983, Heard 1992, Weckerly 1998, Berger & Gompper 1999, Aastrup 2003; (73) Feist & McCullough 1976, McDonnell & Haviland 1995, Asa 1999, Berger & Gompper 1999, Heitor *et al.* 2006; (74) Boyd 1991, Groves 1994, FAO 1995, van Dierendonck *et al.* 1996; (75) Schilder & Boer 1987, Schilder 1988, Ginsberg & Rubenstein 1990, Bowland *et al.* 2001, Neuhaus & Ruckstuhl 2002b; (76) Penzhorn 1984, 1988, Lloyd & Rasa 1989, Rasa & Lloyd 1994; (77) Lovari & Locati 1993, Nowak & Paradiso 1983, Loison *et al.* 1999a,b; (78) Wilson 1984, Weckerly 1998, Berger & Gompper 1999, Lovari & Ale 2001, Li *et al.* 2007; (79) Gray & Simpson 1983, Habibi 1987, Weckerly 1998, Cassinello *et al.* 2004; (80) McClelland 1991, Le Pendu *et al.* 1995, Berger & Gompper 1999; (81) Shackleton & Shank 1984, Weckerly 1998, Berger & Gomooer 1999, Preston *et al.* 2001, Coltman *et al.* 2003; (82) Geist 1968, Riedman 1982, Hogg 1984, Berger 1985, Hogg 1987, Festa-Bianchet 1991, Weckerly 1998, Pelletier & Festa-Bianchet 2006; (83) Coblentz 1980, Shackleton & Shank 1984, Mysterud 2000, Kessler 2002; (84) Fedigan 1993, Manson 1997, Weckerly 1998, Perry 1998, Manson 1999. Manson *et al.* 1999, Crofoot 2007; (85) Richard 1974, Riedman 1982, Jolly 1998, Weckerly 1998, Lawler *et al.* 2003, Kappeler & Schäffler 1998, Lawler *et al.* 2003, Kappeler & Schäffler 2008; (87) Kohda 1985, Bennett & Sebastian 1988, Yeager 1990, Weckerly 1998, Murai 2004; (88) Rowell 1973, Weckerly 1998; (89) Struhsaker 1967, Johnson *et al.* 1980, Dewsbury 1982, Cheney & Seyfarth 1982, Mitani *et al.* 1996, Mitani & Watts 1997, Weckerly 1998; (90) Harvey & Clutton-Brock 1981, Riedman 1982, Edwards & Todd 1991, Grant *et al.* 1992, Mitani & Watts 1997, Brockelman *et al.* 1998, Weckerly 1998; (91) McKenna 1979, Dewsbury 1982, Grant *et al.* 1992, Mitani *et al.* 1996, Weckerly 1998; (93) Riedman 1982, Grant *et al.* 1998, McKenna 2005, Sommer *et al.* 2006; (92) Hall 1962, McKenna 1979, Dewsbury 1982, Mitani & Watts 1997, Weckerly 1998, Zinner *et al.* 2001, Swedell 2002, Yamane *et al.* 2003; (94) Owens 1976, Dewsbury 1982, Riedman 1982, Melnick 1987, Smuts 1987, Smuts & Watanabe 1990, Mitani *et al.* 1996, Mitani & Watts 1997, Weckerly 1998; (95) Bernstein 1975, Alvarez & Cónsul 1978, Riedman 1982, Kawai *et al.* 1983, Mitani & Watts 1997, Weckerly 1998, Dixson 1999; (96) Hanby & Brown 1974, McKenna 1979, Dewsbury 1982, Takahata 1982, Gouzoules & Goy 1983, Wolfe 1986, Melnick 1987, Lunardini 1989, Corradino 1990, Hill 1994, Vasey 1996, Weckerly 1998, Vasey *et al.* 1998, Takenoshita 1998, Vasey & Gauthier 2000, Vasey 2002, O'Neill *et al.* 2004a,b, Vasey & Duckworth 2006, Vasey *et al.* 2006; (97) Chevalier-Skolnikoff 1974, 1976, Estrada & Sandoval 1977, Riedman 1982, Bauers & Hearn 1994, Hill 1994, Smith & Jungers 1997, Mitani & Watts 1997; (98) Conaway & Koford 1964, Loy & Loy 1974, Fairbanks *et al.* 1977, Akers & Conaway 1979, McKenna 1979, Dewsbury 1982, Reinhardt *et al.* 1986, Melnick 1987, Hill 1994, Huynen 1997, Weckerly 1998, Kapsalis & Johnson 2006; (99) Bolwig 1980, Oi 1990, Giacoma & Messeri 1992, Hill 1994, Gust *et al.* 1996, Weckerly 1998; (100) Simonds 1974, Hill 1994, Silk 1994, Weckerly 1998, Cooper *et al.* 2004; (101) Thierry 1986, Thierry *et al.* 1990, 1996, Smith & Jungers 1997, Aujard *et al.* 1998; (102) Dixson 1977, Smith & Jungers 1997, Reed *et al.* 1997; (103) Furuichi 1989, Parish 1994, 1996, de Waal 1997, Mitani & Watts 1997, Weckerly 1998, Gerloff *et al.* 1999, Hohmann *et al.* 1999, Hohmann & Fruth 2000, Hohmann 2001, Hohmann & Fruth 2003, Fruth & Hohmann 2006; (104) Dewsbury 1982, Riedman 1982, Ely & Ferrell 1990, Nishida 1997, Mitani & Watts 1997, Weckerly 1998, Vigilant *et al.* 2001, Firos Anestis 2004; (105) Riedman 1982, Yamagiwa 1987, Robbins 1996, Mitani & Watts 1997, Weckerly 1998, Chapais & Berman 2004, Robbins *et al.* 2005, Yamagiwa 2006; (106) MacKinnon 1974, Mitani & Watts 1997, Weckerly 1998, Fox 2001, Goossens *et al.* 2006; (107) Dunbar 1993, Smith & Jungers 1997, Lummaa *et al.* 1998, Muscarella 2000, Ross & Wells 2000, Kirkpatrick 2000, Bamshad *et al.* 2004, Platek & Shackelford 2006. References for Sex Involved and Same-sex Mounting can be found in Table 2.1, whereas references for Mating System, Sociality and Body Mass can be found in Table 6.1.

where BMM = adult male body mass and BMF = adult female body mass. The only exception was *Balaena mysticetus* where body lengths, rather than body masses, were used to calculate sexual dimorphism. Note that highly sexually dimorphic species could be associated with BMSD values either larger or smaller than 1, depending on whether the larger sex is the male or the female. In the vast majority of the mammals included in this analysis, however, it is the males that are larger than females (94.3% of species). Given that the majority of mammals are also polygamous, bonding in this case will refer to social bonding between specific individuals. Note that this is different from mere sociality, as individuals may have the tendency to form groups without regard to the specific individual composition of the group. 'Social Bond' as used here refers to preferential associations between specific individuals and is coded as either present (Y, code 2), absent (N, code 1) and ~Y (code 1.5) when of variable occurrence (i.e. some individuals may and some others may not engage in specific consortships). In species such as *Giraffa camelopardalis*, where males have been reported not to form social bonds among themselves, whereas females do, and where both male and female same-sex sexual behaviour occurs, the value for the social bond variable will be the arithmetic mean of codes 1 and 2. Finally, for the avian analyses I used the variables Extra-pair Copulations and Extra-pair Paternity. Given that only a minority of mammals are socially monogamous, I used a more diverse set of variables to measure the degree of accessibility to heterosexual sexual partners. The variables were taken directly from the published works that were the source of the data: Extra-pair Copulations (EPC), Extra-pair Offspring (EPO), Extra-group Paternity (EGP), Extra-group Copulations (EGC), Extra-alliance Paternity (EAP), Extra-unit Paternity (EUP), Non-αMale Paternity or Copulations (NαP/C), Peripheral Lek Male Copulations (PLMC), Copulations by Males from Outside the Resident Population (CMORP), and Females Engaged in Extra-harem Copulations (FEHC). Variables were categorised into 5 broad levels: No EPC (or EPO, EGP, EGC, EAP, EUP, NαMP/C, PLMC, CMORP, FEHC) (0%, code 1), Low (> 0%–3%, code 2), Medium (> 3%–10%, code 3), High (> 10%–40%, code 4) and Very High (> 40%, code 5).

All observations were reported during periods of breeding activity for the species. I first proceeded to carry out correlations between phylogenetically independent contrasts of all the variables used and log-body mass (log-BM) contrasts. Table 8.5 summarises these results. All comparative analyses of the mammalian dataset were carried out with the assignment of branch lengths that follows Grafen's (1989) method. This method gives parallel results to the method of punctuational evolution but has greater ability to detect significant differences. Five variables (standardised independent contrasts) display a significant correlation with log-BM contrasts: Same-sex Mounting, Adult Sex Ratio (both negatively correlated with log-BM contrasts), and Sociality, log-transformed Social Unit Size (log-SUS) and Sexual Dimorphism, which are positively correlated with log-BM contrasts. Therefore I first proceeded to control for the effects of evolutionary changes in body mass on the evolutionary changes of those five variables. Following Harvey & Pagel (1991) I calculated residuals: observed value minus the expected value from a least-squares regression between each variable and log-BM contrasts. I then correlated those residuals against log-BM contrasts, to make sure that, as expected, they were no longer correlated. In fact, in all cases they were not, so the residuals could be used for further analyses. Residuals will be identified by the suffix '-R' (e.g. Same-sex Mounting-R).

Five statistically significant correlations were obtained (see Table 8.6). Four of them directly supported the predictions of the *Synthetic Reproductive Skew Model of Homosexuality*; the fifth did not support the model, but further analyses carried out to test Predictions 8 and 9 suggest a reassessment of that result.

The four out of five statistically significant results that are in agreement with the *Synthetic Reproductive Skew Model of Homosexuality* indicate that:

Table 8.5. Pearson's product-moment correlations between log-body mass standardised independent contrasts (independent variable) and contrasts for various life history variables in mammals

Dependent variable	r	p	n
Sex Involved	0.0017	0.49	106
Same-sex Mounting	−0.1644	0.045*	106
Mating System	0.0757	0.21	106
Sociality	0.1760	0.034*	106
Adult Sex Ratio	−0.2290	0.01*	99
Dominance Mount	0.1260	0.17	54
Affiliative Mount	0.2204	0.08	39
Relatedness	−0.1401	0.12	68
log-Social Unit Size	0.1729	0.038*	104
Ecological Constraints	0.0279	0.39	91
Sexual Dimorphism	0.1822	0.032*	102
Social Bonding	0.0586	0.29	90
EPCs	−0.0033	0.48	66

Branch lengths set using Grafen's (1989) method.

(a) At larger sizes of the social unit the level of same-sex mounting tends to decrease as expected if both opportunities for heterosexual mounting increase and the ability of dominants to control sexual behaviour of subordinates decreases at very large group sizes.

(b) Same-sex mounting is more likely to occur in association with more polygamous mating systems, in which some males may not have access to females and socio-sexual interactions may involve members of the same sex (e.g. females in harems and males in bachelor groups) (see Figure 8.5).

(c) Same-sex mounting is more likely to occur among less closely related individuals; this is expected from inbreeding avoidance mechanisms selected in a heterosexual context that remain active in a homosexual context.

(d) As the adult sex ratio becomes more male-biased, males are more often involved in same-sex mounting, as expected from both the availability of same-sex partners for sexual intercourse and the increased competition among males for females that may promote socio-sexual interactions associated with dominance. In fact, although not significant, there is a trend for dominance mounting to be associated with polygamous mating systems. The association between adult sex ratio and same-sex mounting is also expected as a within-species phenomenon and it is nicely supported by a meta-analysis of the information available for one species, *Macaca fuscata*. Figure 8.6 indicates that as the sex ratio becomes less female-biased within *M. fuscata* troops the percentage of females that become involved in same-sex mounting decreases ($r^2 = 0.715$, $n = 6$, $p = 0.033$). That is, in *M. fuscata* the involvement of females in same-sex mounting is clearly associated with the scarcity of heterosexual sexual partners in a continuous manner. Interestingly, Figure 8.6 predicts that *M. fuscata* females will dramatically reduce homosexual mounting when the sex ratio reaches values close to parity. In fact, Joseph Soltis (pers. comm.), who studied a population of *M. fuscata* on Yakushima Island where the sex ratio varied from 1 to 1.08 F/M (Thomsen & Soltis 2004), reports that at the time 'female-female mounting among Japanese macaques was rare ... but that it did occur. I believe that I only observed it once'. Of course, the trend shown in Figure 8.6 could be also attributed to a continuous variation in some genetic attribute of the different populations that may explain the diverse frequencies of females involved in same-sex mounting, with the association of mounting with sex ratio being spurious. This may well be the case, but to support this hypothesis it will not be sufficient to prove that the different populations are genetically different, but that they differ genetically in such a way so that the continuous pattern shown in Figure 8.6 can be explained by those genetic differences more efficiently than the alternative hypothesis of sex ratio bias can. For instance, Vasey & Jiskoot (2009) have recently reported an association between female same-sex sexual

Table 8.6. Pearson's product-moment correlations between standardised independent contrasts of various life history and behavioural variables in mammals

	n^a	r	p^b
Grafen's (1989) branch length method			
Ecological Constraints vs. Same-sex Mounting-R	91	−0.303	0.0017**
Log-Social Unit Size-R vs. Same-sex Mounting-R	104	−0.231	0.008*
Relatedness vs. Affiliative Mount	30	−0.062	0.36
Relatedness vs. Same-sex Mounting-R	68	−0.545	0.000001***
Mating System vs. Same-sex Mounting-R	106	0.210	0.015*
Mating System vs. Sexual Dimorphism-R	102	0.023	0.40
Adult Sex Ratio-R vs. Same-sex Mounting-R	99	0.045	0.32
Adult Sex Ratio-R vs. Sex Involved	99	−0.209	0.018*
Adult Sex Ratio-R vs. Mating System	99	0.127	0.10
Social Bonding vs. Same-sex Mounting-R	90	−0.018	0.43
EPCs vs. Same-sex Mounting-R	66	0.021	0.43
Sociality-R vs. Same-sex Mounting-R	106	−0.104	0.14
Sociality-R vs. Dominance Mount	54	−0.072	0.30
Mating System vs. Dominance Mount	54	0.115	0.10
Mating System vs. Sex Involved	106	0.070	0.11
Sociality-R vs. Affiliative Mount	39	−0.162	0.16

R indicates that the variable used is contrast residuals from a regression with log-body mass contrasts.
[a] Number of contrasts.
[b] One-tailed.

behaviour and a specific A1 haplogroup in wild populations of *M. fuscata* on the Japanese island of Honshu. However, they did not seem to have controlled for the effect of differences in the adult sex ratio across populations. Their hypothesis of specific genetic differences explaining the distribution of same-sex sexual behaviour would be strengthened if the adult sex ratio is proven not to co-vary with homosexual behaviour and presence of A1 haplotypes. This issue can be illustrated by another recent report published by Vasey & Reinhart (2009) where they described homosexual interactions occurring between two adult *M. fuscata* females during a brief observation period of 50 minutes at the Primate Research Institute of Kyoto University (Inuyama, Japan). Those females pertain to the Wakasa-B group, which is also from Honshu Island. However, the adult sex ratio in this captive population was 0.2 (3M/15F), which, according to Figure 8.6, should be associated with over 80% of females engaging in same-sex sexual interactions. These issues notwithstanding, I should also mention that the two hypotheses are not necessarily alternative: different populations of Japanese macaques may well be genetically predisposed to express different levels of same-sex sexual behaviour among females, with the opportunity to express such behaviour being dependent on the sex ratio. In this case the populations would probably show curves of same-sex mounting vs. sex ratio differing in slope, intercept or both. My point is that both mechanisms can explain some of the behavioural patterns observed and therefore should be specifically tested.

Some ethnographic data also support the association of same-sex sexual behaviour with both

Figure 8.5. Male–male mounting in fallow deer (*Dama dama*). Photo courtesy of Luděk Bartoš.

polygyny and a male-biased adult sex ratio in traditional human societies. For instance, Gilbert Herdt (1984b: 13) associates, at least in part, the occurrence of ritualised homosexuality in New Guinea with both a male-biased sex ratio and polygyny.

The fifth statistically significant result runs against the prediction of the model. It was expected that Same-sex Mounting would be positively associated with cooperative breeding, the proxy variable for Ecological Constraints. However Ecological Constraint contrasts and Same-sex Mounting contrast residuals were significantly negatively associated. In order to understand this result better, I will need to refer to the results of tests 8 and 9 (see Table 8.6). In test 8 cooperative breeders did indeed show a very weak tendency to include species that display same-sex mounting, after limiting the analysis to monotypic taxa, whereas in test 9 a combined sample of cooperative breeders and species that live in stable social groups indicate that the relative frequency of same-sex mounting increases with the size of the social unit. Although both results were not statistically significant, it is possible that among cooperative breeders same-sex mounting is more likely to be displayed if individuals (e.g. subordinates) do not have access to heterosexual copulations and they are not reproductively suppressed; this is more likely to occur when the social unit is of 'intermediate' size, as predicted by the model and, more generally, by *Reproductive Skew Theory*. When the social unit is very large, however, reproductive suppression continues to decrease and control of dominant over subordinates decreases too, hence lower levels of homosexuality are expected (see result of "Test A" of Prediction 2 in Appendix 2).

Among the statistically non-significant trends, 7 tended to support and 6 tended to falsify the specific predictions of the model. See Appendix 2 for additional comments on all the results, both those that were significant and those that were not.

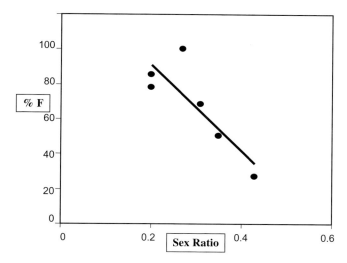

Figure 8.6. The percentage of female *Macaca fuscata* involved in same-sex mounting (%F) decreases linearly as the sex ratio (M/F) in the troop becomes less female-biased ($r^2 = 0.715$, $p = 0.033$). Data are from Gouzoules and Goy (1983), Wolfe (1986; Arashiyama West & Arashiyama B), Lunardini (1989) & Vasey (2002a).

So far I have used a comparative analytical approach that is bivariate; that is, I have tested specific predictions of the model through correlations between two variables (expressed in terms of phylogenetically independent contrasts) at a time. Although the model does make specific predictions about the kind of evolutionary changes expected in bivariate analyses, it also suggests that the links between the various factors involved are likely to be multivariate. In the next section I will carry out a multivariate test of the *Synthetic Reproductive Skew Model of Homosexuality* by using path analysis.

Path analyses

The method of path analysis was already introduced and explained in Chapter 2; here I emphasise that, given that standardised independent contrasts are used in the analyses, the total effect of the model should be interpreted as ***the evolutionary change expected in same-sex mounting following evolutionary changes in the variables of the model as they relate to each other in the manner specifically described by the various causal paths***. In other words, a high and positive value of the total effect of the model means that the model is capable of explaining a high percentage of the variance in the evolutionary increase of same-sex mounting.

For each taxon (i.e. birds or mammals in this case) I produce three alternative models, calculate their total causal effect and then compare the total causal effect across the three models. The first model is a simplified version of the *Synthetic Reproductive Skew Model of Homosexuality* that I introduced in Figure 8.4; the second model is restricted to variables that are associated with sexual behaviour (i.e. I eliminated from the *Synthetic Model* the mainly socio-sexual variables such as dominance mount, affiliative mount and pair/social bond). The third model focuses on socio-sexual variables, and it will be produced by eliminating sex ratio and EPC from the *Synthetic Model*. The effect of body-mass-standardised independent contrasts was controlled in the analyses by using residuals whenever body mass was correlated with other variables. We should also note that the effect of reproductive physiology was broadly controlled in the datasets because all data were gathered during mating periods of the species. That reproductive condition has an important effect on same-sex mounting can be deduced from the fact that such behaviour is relatively unusual outside mating periods across taxa.

The simplified *Synthetic Reproductive Skew Model of Homosexuality* that was used in these analyses is shown in Figure 8.7. In the case of the avian dataset, however, I did not have enough data to also include affiliative mounting in the multivariate analyses; this perhaps already suggests that affiliative same-sex mounting may be relatively uncommon in birds. Figure 8.8 shows the values of path coefficients and error terms for the *Synthetic Model* as it is applied to the bird dataset. The first result worth mentioning is the very small error term for the same-sex mounting dependent endogenous variable. An error term of 0.048 means that the coefficient of determination for the multiple regression between same-sex mounting contrasts and sex ratio contrasts, extra-pair copulation contrasts, relatedness contrasts, dominance mount contrasts and pair bond contrasts is 0.952, indicating that 95.2% of the variance in same-sex mounting contrasts is explained by the combined effect of the above five variables in birds. Accordingly, the total effect of the model is quite large, being 0.913. The strongest paths in the model are represented by the positive association between sociality contrasts and dominance mount contrasts; dominance mount contrasts and same-sex mount contrasts; and sex-ratio contrasts and same-sex mount contrasts, all patterns consistent with the model. The version of the model that is restricted to variables that have a strong sexual component is shown in Figure 8.9 along with the values of the path coefficients and error terms. The total effect of the alternative model that emphasises interactions that have a strong sexual component is 0.597. The mainly socio-sexual version of the model is shown in Figure 8.10 and its total effect is 0.648.

The above path analyses of the bird dataset indicate a much stronger total effect of the *Synthetic Model* compared with either of the two other alternatives: the mainly sexual and the mainly socio-sexual models. That is, in birds, evolutionary trends towards the display of same-sex mounting are a result of a combination of factors that are both sexual and socio-sexual. It can be seen in Figure 8.8 that the contribution of dominance mount contrasts to the multiple regression with same-sex mount contrasts is quite large.

The paths for the *Synthetic Model* in mammals are shown in Figure 8.11. In this case there is enough information to also include affiliative mounting in the model. The total effect of the model is 0.395, a result that suggests a lower explanatory success of this model in mammals compared with birds. The strongest path in the model is the negative association between relatedness contrasts and same-sex mounting contrasts, a result predicted by the model. In mammals there is a similar number of negative and positive path coefficients, whereas in birds there was a majority of positive path coefficients. One consequence of this difference is that in mammals the effects of alternative paths to same-sex mounting tend to cancel each other out. The multiple regression between same-sex mounting contrasts-R and all the variables involved in the last step of the model explains 43.6% of the variance in the evolutionary changes towards same-sex mounting in mammals. When a mainly sexual model is considered (see Figure 8.12) the total effect is −0.231. That is, the overall model seems to predict evolutionary trends toward decreased same-sex sexual behaviour. However, as we will see below, in mammals there is a greater involvement of females in same-sex mounting than there is in birds; this leads to a re-interpretation of the negative partial regression coefficient between sex ratio contrasts-R and same-sex mount contrasts-R shown in Figure 8.12. Such a relationship should be seen as indicating a positive association between a female-biased sex ratio and same-sex mounting in mammals. With this reinterpretation, the total effect becomes 0.235, still lower than the total effect of the *Synthetic Model*. The mainly socio-sexual version of the model is shown in Figure 8.13. Total effect here is also negative and of small value: −0.074. This is mainly due to the

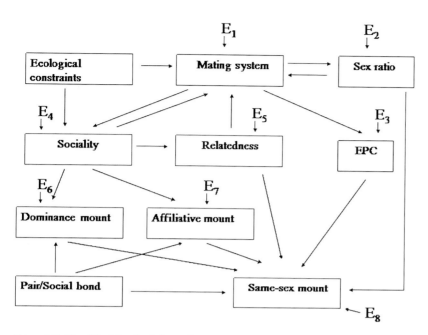

Figure 8.7. Simplified *Synthetic Reproductive Skew Model of Homosexuality*. E, residual error term or disturbance term.

negative association between social bond contrasts and same-sex mount contrasts-R ($\beta = -0.180$). If social bond contrasts are eliminated from the model, the total effect becomes positive but very small (0.053). If we just focus on the last step of the model for both the mainly sexual and mainly socio-sexual alternatives, the coefficient of determination for the former is 26.5% and for the latter is 39.5% (40.7% if social bond contrasts are eliminated). That is, whether we consider the total effect of the path analysis or just the multiple regression of the last step, the *Synthetic Model* has a greater explanatory value than either of the other two alternative models. It should be noted, however, that the *Synthetic Model* is comparatively less successful at explaining evolutionary changes towards same-sex mounting in mammals than it is in birds. Clearly, same-sex mounting in mammals is a far more complex behaviour than it is in birds, involving a diversity of factors whose importance varies from one taxon to another. In fact, in the section 'Direct comparisons between birds and mammals' (see below) I will show that same-sex sexual behaviour is associated with different behavioural and life-history syndromes in birds and mammals, presumably reflecting different causative evolutionary scenarios.

Full correlation matrix

In the previous sections I built a model that takes into account a specific subset of all possible interactions among the variables considered in this work (Figure 8.4) and also tested a subset of those interactions by using path analyses. Here I produce the full correlation matrix between phylogenetically independent contrasts for the 13 behavioural, ecological and life-history variables listed in Tables 8.2 and 8.4. In this case, however, the critical value α will be set by the Dunn–Šidák method, which controls for the number of tests carried out, as the full set of all possible correlations was used. Therefore, given that the number of tests was 78, for an α value of 0.05 it follows that the corrected value $\alpha' = 0.00065$.

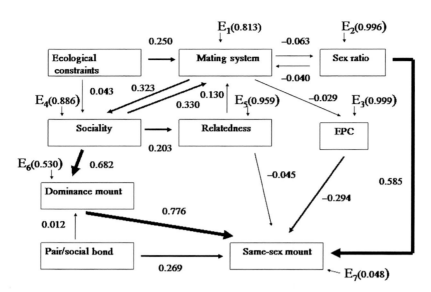

Figure 8.8. Simplified *Synthetic Reproductive Skew Model* tested with bird data.

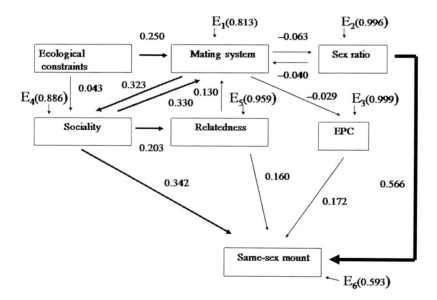

Figure 8.9. Simplified *Synthetic Reproductive Skew Model* restricted to core and sexual variables and tested with bird data.

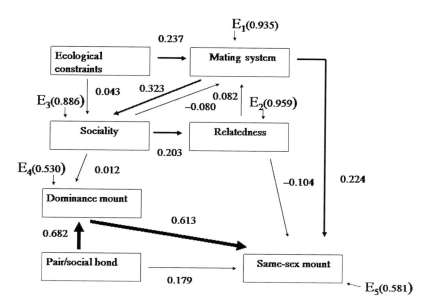

Figure 8.10. Simplified *Synthetic Reproductive Skew Model* restricted to core and socio-sexual variables and tested with the bird data.

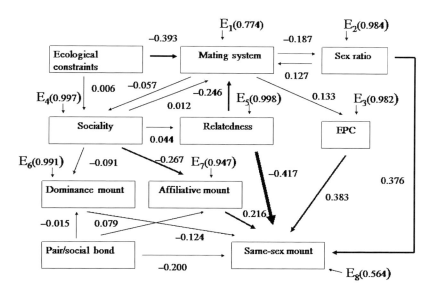

Figure 8.11. Simplified *Synthetic Reproductive Skew Model* tested with mammal data.

366　**Social, life history and ecological factors**

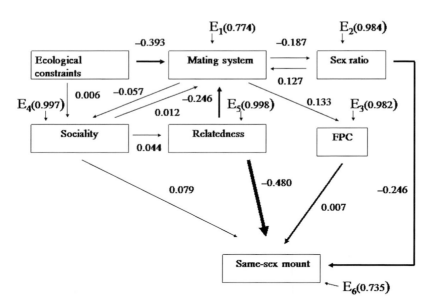

Figure 8.12. Simplified *Synthetic Reproductive Skew Model* restricted to core and sexual variables and tested with mammal data.

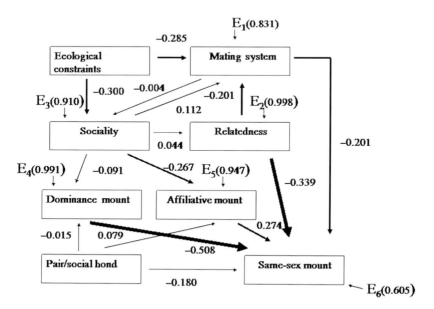

Figure 8.13. Simplified *Synthetic Reproductive Skew Model* restricted to core and socio-sexual variables and tested with mammal data.

Table 8.7 summarises the full set of Pearson's product-moment correlations between phylogenetically independent contrasts for the avian dataset. Under the restrictive conditions of the Dunn–Šidák correction only two correlations are statistically significant: one that I have already analysed between Plumage Sexual Dichromatism contrasts and Mating System contrasts, and a new one between Same-sex Mounting contrasts and Sex Involved contrasts. It appears then that in birds evolutionary trends towards the expression of same-sex mounting are positively associated with evolutionary trends towards a greater involvement of females in same-sex mounting. This result could be linked to the trend (uncorrected $\alpha = 0.04$, see Table 8.7) for same-sex mounting to be positively associated with Sociality and (marginally not significantly using uncorrected α) Dominance Mount. That is, in social birds where females establish dominance hierarchies, such dominance tends to be expressed through same-sex mounting, e.g. *Porphyrio porphyrio*.

Results of the Pearson's product-moment correlations between phylogenetically independent contrasts for the mammalian dataset are summarised in Table 8.8. Three correlations retain their significance in spite of the application of the Dunn–Šidák correction: evolutionary trends towards increased same-sex mounting are associated with evolutionary trends towards decreased relatedness among interactants, a result that was already mentioned in the context of the initial test of the *Synthetic Reproductive Skew Model of Homosexuality*. Two other results that are significant, however, were not part of the previous testing of the model: (a) evolutionary trends towards increased same-sex mounting in mammals are associated with evolutionary trends towards increased involvement of females in same-sex mounting, and (b) evolutionary trends towards increased ecological constraints on independent breeding, measured here as increased levels of cooperative breeding (helping), are associated with evolutionary trends towards increased levels of social bonds. These three results suggest that both the sexual and the socio-sexual aspects of same-sex mounting may interact, in that closely related mammals may avoid same-sex mounting, presumably enacting mechanisms of inbreeding avoidance that are selective in heterosexual contexts, and yet both competitive (e.g. breeding competition) and affiliative (e.g. establishment of social bonds) interactions, especially among members of the philopatric sex (i.e. females in most mammals), can also be expressed through same-sex mounting under certain circumstances. Interestingly, among some primates (e.g. Hominoidea) females tend to be the dispersing sex (Lawson Handley & Perrin 2007); this suggests that same-sex mounting in female mammals may be adaptive to resident and/or dispersing individuals, depending on the species. In fact, this may explain the results of Hensley & Tewksbury (2002) on female prison inmates that I mentioned in Chapter 7. Female inmates, who have usually not met before, encounter each other in prison and there they tend to establish social affiliative partnerships that may also involve same-sex sexual behaviour. The same occurs among wild and captive female bonobos, as we have already seen in Chapter 5.

Additional results for both birds and mammals that were either significant or marginally not significant at the uncorrected α are available in Appendix 3.

Direct comparisons between birds and mammals

In the previous sections of this chapter the analyses of the avian and mammalian datasets have run in parallel. Here I present the results of a direct comparison between the two classes of vertebrates for each one of the variables. Phylogenetic effects are controlled by taking the genera as independent data (Pagel & Harvey 1988). For the case of categorical variables, monotypic genera produced only one entry in the dataset, whereas polytypic genera produced as many entries as there are different types (for example, a genus that includes both monogamous and polygamous species would produce two entries for the Mating System variable, one in the

Table 8.7. Pearson's product-moment correlations between standardised independent contrasts of all the behavioural and life history variables for the birds dataset included in this chapter

Contrasts were calculated using Grafen's (1989) method to set branch lengths. Values shown are r, P (N).

	Same-sex Mounting	Adult Sex Ratio	Sociality	Log-Social Unit Size	Plumage Sexual Dichromatism	Sex Involved	Mating System	Dominance Mount	Affiliative Mount	Relatedness	Ecological Constraints	Pair Bond	EPCs
Adult Sex Ratio	0.301, 0.01* (54)	—											
Sociality	0.200, 0.04* (71)	−0.067, 0.31 (54)	—										
Log-Social Unit Size	−0.287, 0.01* (59)	−0.018, 0.45 (48)	0.119, 0.18 (59)	—									
Plumage Sexual Dichromatism	−0.071, 0.27 (71)	−0.080, 0.28 (54)	−0.16, 0.09[a] (71)	0.051, 0.35 (59)	—								
Sex Involved	**0.465, 0.00004* (71)**	−0.079, 0.28 (54)	0.19, 0.056[a] (71)	−0.12, 0.17 (59)	0.026, 0.41 (71)	—							
Mating System	0.03, 0.39 (71)	−0.063, 0.32 (54)	0.334, 0.002* (71)	−0.04, 0.37 (59)	**0.380, 0.0003* (71)**	0.149, 0.10 (71)	—						
Dominance Mount	0.316, 0.07[a] (23)	−0.057, 0.41 (15)	0.570, 0.001* (23)	−0.08, 0.36 (22)	−0.543, 0.003* (23)	0.392, 0.032* (23)	−0.222, 0.14 (23)	—					
Affiliative Mount	0.342, 0.15 (11)	0.425, 0.14 (8)	0.052, 0.43 (11)	−0.53, 0.04* (11)	0.033, 0.46 (11)	0.174, 0.30 (11)	0.081, 0.40 (11)	−0.373, 0.11 (6)	—				
Relatedness	0.184, 0.09[a] (49)	0.358, 0.01* (41)	0.203, 0.08[a] (49)	0.004, 0.48 (45)	−0.195, 0.09[a] (49)	0.229, 0.05[a] (49)	0.106, 0.23 (49)	0.446, 0.02* (20)	0.31, 0.46 (9)	—			
Ecological Constraints	−0.019, 0.43 (69)	0.127, 0.18 (53)	0.121, 0.16 (69)	−0.03, 0.39 (57)	0.062, 0.30 (69)	0.051, 0.33 (69)	0.241, 0.022* (69)	−0.241, 0.13 (23)	0.075, 0.41 (11)	−0.092, 0.28 (40)	—		

Table 8.7. (*Cont.*)

	Same-sex Mounting	Adult Sex Ratio	Sociality	Log-Social Unit Size	Plumage Dichromatism	Sexual Sex Involved	Mating System	Dominance Mount	Affiliative Mount	Relatedness	Ecological Constraints	Pair Bond	EPCs
Pair Bond	−0.001, 0.49 (69)	0.05, 0.36 (52)	0.016, 0.44 (69)	−0.02, 0.41 (58)	−0.239, 0.02* (69)	−0.022, 0.42 (69)	−0.266, 0.013* (69)	0.202, 0.18 (22)	0.262, 0.21 (11)	0.063, 0.33 (48)	0.09 0.23 (67)	—	
EPCs	−0.065, 0.32 (50)	0.019, 0.45 (42)	−0.109, 0.22 (50)	−0.008, 0.47 (46)	0.034, 0.40 (50)	−0.102, 0.24 (50)	−0.029, 0.42 (50)	0.206, 0.19 (20)	−0.521, 0.06[a] (10)	−0.092, 0.28 (40)	−0.01, 0.47 (50)	−0.125, 0.19 (50)	—

Bold, * = significant at the Dunn–Šidák corrected α, where the 0.05 level is equivalent to 0.00065; *, significant at the uncorrected α of 0.05. [a] Marginally non-significant result at the uncorrected α (0.1 > *p* ≥ 0.05).

Table 8.8. Pearson's product-moment correlations between standardised independent contrasts of all the behavioural and life history variables for the mammals dataset included in this chapter

Contrasts were calculated using Grafen's (1989) method to set branch lengths. Values shown are *r*, *P* (*N*). R = residuals from a regression with log Body Mass contrasts.

	Same-sex Mounting-R	Adult Sex Ratio-R	Sociality-R	Log-Social Unit Size-R	Sexual Dimorphism-R	Sex Involved	Mating System	Dominance Mount	Affiliative Mount	Relatedness	Ecological Constraints	Social Bonds	EPCs
Adult Sex Ratio-R	0.045 0.32 (99)	—											
Sociality-R	−0.104 0.14 (106)	−0.04 0.32 (99)	—										
Log-Social Unit Size-R	−0.233 0.008* (104)	−0.16 0.05[a] (95)	0.262 0.003* (104)	—									

369

Table 8.8. (*cont.*)

	Same-sex Mounting-R	Adult Sex Ratio-R	Sociality-R	Log-Social Unit Size-R	Sexual Dimorphism-R	Sex Involved	Mating System	Dominance Mount	Affiliative Mount	Relatedness	Ecological Constraints	Social Bonds	EPCs
Sexual Dimorphism-R	−0.065 0.25 (102)	−0.02 0.41 (97)	−0.066 0.25 (102)	0.148 0.07[a] (100)	—								
Sex Involved	**0.391** **1 × 10⁻⁵*** **(106)**	−0.20 0.018* (99)	−0.131 0.08[a] (106)	−0.117 0.11 (104)	−0.035 0.36 (102)	—							
Mating System	0.210 0.015* (106)	0.127 0.10 (99)	−0.053 0.29 (106)	−0.108 0.13 (104)	0.023 0.40 (102)	0.070 0.11 (106)	—						
Dominance Mount	−0.418 0.0008* (54)	0.054 0.35 (53)	−0.072 0.30 (54)	0.178 0.10 (53)	−0.236 0.045* (52)	0.053 0.34 (54)	0.115 0.20 (54)	—					
Affiliative Mount	0.197 0.11 (39)	−0.41 0.005* (37)	−0.162 0.16 (39)	0.150 0.17 (39)	−0.186 0.13 (37)	−0.005 0.48 (39)	−0.092 0.28 (39)	0.429 0.004* (37)	—				
Relatedness	**−0.545** **1 × 10⁻⁶*** **(68)**	−0.10 0.20 (65)	0.044 0.35 (68)	0.154 0.10 (68)	0.053 0.33 (67)	0.149 0.11 (68)	0.307 0.005* (68)	−0.267 0.05[a] (38)	−0.062 0.36 (32)	—			
Ecological Constraints	−0.303 0.0017* (91)	−0.29 0.002* (85)	0.017 0.43 (91)	0.325 0.0009* (89)	0.197 0.031* (89)	−0.096 0.18 (91)	−0.202 0.026* (91)	−0.016 0.45 (44)	0.221 0.10 (34)	0.314 0.005* (65)	—		
Social Bonds	−0.018 0.43 (90)	0.029 0.39 (85)	0.068 0.25 (90)	0.110 0.15 (90)	0.038 0.36 (86)	0.106 0.15 (90)	−0.130 0.11 (90)	−0.042 0.38 (47)	−0.076 0.33 (35)	0.161 0.10 (64)	0.382 **0.00029*** (77)	—	
EPCs	0.021 0.43 (66)	−0.08 0.25 (65)	−0.014 0.45 (66)	−0.195 0.05[a] (65)	−0.105 0.20 (65)	0.063 0.30 (66)	0.133 0.14 (66)	0.0091 0.47 (35)	0.140 0.23 (28)	−0.091 0.26 (52)	−0.135 0.14 (62)	−0.135 0.14 (63)	—

Bold *, significant at the Dunn-Šidák corrected α, where the 0.05 level is equivalent to 0.00065; *, significant at the uncorrected α of 0.05. [a] = Marginally non-significant result at the uncorrected α (0.1 > p ≥ 0.05).

Monogamy and the other in the Polygamy categories). In the case of continuous variables, the mean value for the genus was entered. In order to categorise the sex ratio and also mammal sexual dimorphism, a procedure was needed that maximises the use of the datasets and that also makes the direct comparisons between birds and mammals easier (see Tables 8.2 and 8.4). This was achieved by setting the interval $0.8 \leq x \leq 1.2$ as the one demarcating the category M = F. That is, sex ratio and mammal sexual dimorphism values are categorised as: '1' and 'males indistinguishable from females', respectively, if the value is larger than or equal to 0.8 and smaller than or equal to 1.2. The values for Social Unit Size were log-transformed as the variance increased with the mean.

Table 8.9 summarises the results of the analyses. The first striking result is that most of the tests are significant (9/13) with an additional one being marginal ($p = 0.063$). Such a result is very unlikely to be due to chance alone.

Same-sex mounting tends to be more prevalent among mammals than birds. This is in association with a higher prevalence of polygamous mating systems in mammals, a higher degree of sociality, and a higher level of ecological constraints to independent breeding; all these results are predicted by the *Synthetic Reproductive Skew Model of Homosexuality*. Mammals are also more sexually dimorphic than birds, consistent with their higher level of polygamy, as predicted by the model. Females are more often engaged in same-sex mounting in mammals than birds; the bias in favour of males is stronger in birds. This asymmetry fits very well with the significantly female-biased sex ratio found in mammals, whereas the adult sex ratio is biased towards males in birds. This clear association between sex involved in same-sex sexual behaviour and sex ratio is expected from the model. With regard to genetic relatedness, mammals tend to be more closely related within social groups than birds and yet they also tend to display a higher level of same-sex mounting; this suggests that, after all, same-sex mounting may not be completely hampered by the levels of actual genetic relatedness most prevalent within mammal social species through the action of inbreeding avoidance mechanisms, perhaps because social groups are formed by relatives of various degrees, not just first-order relatives.

The model also predicted a convex upwards relationship between social unit size and same-sex mounting. Mammals, the taxon with the higher prevalence of same-sex mounting, tend to also live in significantly smaller groups than birds. This result supports the model if the sizes of social units for both classes of vertebrates are at the large end of the size spectrum, as in this case we would be in the descending (negative slope) section of the curve. There, relatively larger group sizes are associated with a smaller degree of same-sex mounting than relatively smaller group sizes, exactly the result I obtained. The mean value of social unit (e.g. a colony) size in the bird sample is 1141.2 individuals, whereas for mammals it is 172.3 individuals, suggesting that we may indeed be dealing with regions of the social unit size range that are towards the larger end of the spectrum. However suggestive it may be, this aspect of the model still requires more specific testing by using precise same-sex mounting vs. social unit size curves.

Four results seem to falsify the predictions of the model: The occurrence of (a) dominance mount and (b) affiliative mount are not significantly different between birds and mammals in the sample. The sample sizes here are relatively small, however, as I could not gather information for those variables for the vast majority of species. (c) Although birds display relatively less same-sex mounting than mammals their level of social bonding is much higher. Most of the social bonding considered in the bird species however is represented by monogamous relationships between males and females, whereas for mammals the kind of interindividual bonding refers more to intragroup social relationships, and we have already seen that same-sex mounting is more associated with polygamy than monogamy, as predicted by the model. (d) Levels of EPCs tend to be slightly higher in mammals than birds, against the expectations of the model, but the difference is not statistically significant.

Table 8.9. Direct comparisons of social and life-history traits between birds and mammals

In order to control for phylogenetic effects, categorical variable data for monotypic genera were entered only once for the genus, whereas if the genus was polytypic it produced as many entries into the dataset as different types were present. For continuous variables, the mean value for the genus was entered. For both sex ratio and mammal sexual dimorphism, M = F corresponds to values between 0.8 and 1.2 ($0.8 \leq x \leq 1.2$). Values for social unit size were log-transformed before they were entered in the analyses. The values reported here are only representative of this sample of avian and mammalian taxa and may not be representative of broader trends for both orders; for example, the percentage of cooperative breeders (Ecological Constraints) among birds is most likely overestimated.

Variable	Birds	Mammals	Test
Same-sex Mounting	60.2%	75.5%	$\chi^2_1 = 3.45, p = 0.063$
Sex Involved	74.4%(M), 16.2%(F), 9.4%(M + F)	39.7%(M), 8.8%(F), 51.5%(M + F)	$\chi^2_2 = 20.60, p < 0.0001$
Adult Sex Ratio	M > F 51.6% M = F 10.0% M < F 38.4%	10.6% 28.0% 61.4%	$\chi^2_2 = 28.20, p < 0.0001$
Mating System	68.2% Monogamous	8.9% Monogamous	$\chi^2_1 = 50.95, p < 0.0001$
Sociality	62.5% Social	91.0% Social	$\chi^2_1 = 15.13, p < 0.0001$
Dominance Mount	82.6%	66.6%	$\chi^2_1 = 0.68, p = 0.409$
Affiliative Mount	75.0%	70.9%	$\chi^2_1 = 0.00, p = 1.00$
Relatedness	23.4% Above 'Low' category	57.6% Above 'Low' category	$\chi^2_1 = 11.18, p = 0.001$
Social Unit Size	1141.2 ± 2810.0 (SD)	172.3 ± 1008.0 (SD)	$t_{125} = 3.05, p = 0.003$
Ecological Constraints	19.6% Cooperative Breeders	38.7% Cooperative Breeders	$\chi^2_1 = 5.07, p = 0.024$
Dimorphism/Dichromatism	65.5% Monochromatic	42.8% Monomorphic	$\chi^2_1 = 6.00, p = 0.014$
Pair/Social Bonding	80.3%	49.3%	$\chi^2_1 = 12.47, p < 0.0001$
EPCs	58.8% Above 'Low' category	68.9% Above 'Low' category	$\chi^2_1 = 0.814, p = 0.367$

In sum, direct comparisons between the two classes of vertebrates across the behavioural and life history variables considered in the analyses tend to be consistent with the predictions of the *Synthetic Reproductive Skew Model of Homosexuality*, and they also indicate that birds and mammals are characterised by different syndromes as far as same-sex sexual behaviour is concerned: same-sex mounting tends to be more prevalent among mammals in association with their higher levels of polygamy, sociality, ecological constraints and also because in mammals both sexes are more likely to be involved in same-sex mounting compared with birds, in which the involvement is more biased towards males.

What we may conclude from all the previous comparative analyses of the *Synthetic Reproductive Skew Model of Homosexuality* is that the available data are in broad agreement with the predictions of the model and the model in its complete version – which includes reproductive skew, sexual and sociosexual components – provides a more accurate description of the data than any of the more specialised sexual or socio-sexual alternatives. In addition, we could also safely conclude that although mammals and birds share some similarities they are also different in terms of which one of the specific variables included in the model is most relevant to explain the evolution of same-sex sexual behaviour in those taxa.

In the final sections of this chapter I address two specific behavioural, life history and ecological issues: (a) whether homosexuality in humans is associated with an interest in having children or not, and (b) the extent to which transmission of sexually transmitted diseases (STDs), as a cost of same-sex sexual behaviour, may limit the spread of such behaviour.

Homosexuality in humans and interest in having children

We have seen in previous chapters that strong homosexual pair bonds are associated with lowered reproductive success (see, for example, Chapter 3; see Bobrow & Bailey 2001; Rahman & Hull 2005), a pattern that gives substance to regarding homosexuality as an evolutionary paradox. In the case of humans, however, we could push our enquiry even further and ask whether persons who are in a homosexual relationship may also have a lower desire to have children than heterosexuals.

Of course, homosexuals may bring children into the relationship after having had them in a previous heterosexual partnership (see, for example, Crosbie-Burnett & Helmbrecht 1993). What we also want to know, however, is whether childless homosexuals who are engaged in a relationship may or may not subsequently develop an interest in having children within such a relationship (e.g. through adoption, artificial insemination if legally feasible, or the help of a consenting other-sex partner in reproduction). Are the desires to have children as strong among homosexual partners as they are among heterosexual partners? What are the differences and similarities between the sexes across sexual orientations?

In a recent, very interesting study carried out in The Netherlands, Henny Bos, Frank van Balen and Dymphna van den Boom from the Department of Education of the University of Amsterdam (Bos et al. 2003) compared Parenthood Motives (happiness, parenthood, well-being, continuity of interaction with children over time, identity and social control), Reflection (measured as the frequency of thinking about having children), and Strength of Desire to have children between 100 lesbian mother families, and 100 heterosexual families, mostly from urban areas. In general, the variables measured tend to differ consistently between women (whether heterosexual mothers, lesbian biological mothers or lesbian social mothers) and heterosexual fathers, with women considering happiness, parenthood, reflection and intensity of desire more important than men, whereas they consider well-being, identity and social control less important than men do. The most obvious discrepancy between lesbian and heterosexual mothers is that, for the former, continuity of interaction with children is less important than it is for the latter. In terms of their gender patterns of response, lesbians tend to be hyperfeminised for happiness, parenthood (lesbian biological mothers only), reflection, intensity of desire, identity and well-being, whereas they tend to be masculinised for continuity and even hypermasculinised for the case of continuity in the case of lesbian social mothers: lesbian social mothers show an interest in the long-term aspects of the relationship with the child that is even lower than that expressed by the biological father. This may simply be a consequence of the lack of genetic relatedness between the lesbian social mother and the child, a pattern expected from kin selection.

Overall, these results clearly indicate that lesbianism is not associated with a desire to avoid reproduction. Interestingly, Bos et al. (2003) also carried out a set of correlations between Parenthood Motives and both Reflections and Strength of Desire in female homosexual and heterosexual families. Their results suggest that heterosexual families show a positive correlation between reflection and social control, whereas for lesbian families the correlation is not significant. That is, not surprisingly for an urban Western European setting, heterosexual partners are under greater social pressure to have children than lesbian partners.

The correlations between Strength of Desire and Parenthood Motives indicate that both heterosexual and homosexual families show strong personal motivations such as happiness, parenthood and well-being. Again, this seems to be indicative of similarly strong personal motivations for lesbians and heterosexual females to reproduce, in spite of heterosexual females being under a more intense social pressure to do so. A similar conclusion was also reached by Siegenthaler & Bigner (2000).

Do homosexual males differ from heterosexual males in their interest in having children? Unfortunately, on this issue the information regarding homosexual males is rather scanty. Black *et al.* (2000) carried out a study of demographic data for gay men and lesbians available from the general Social Survey and the National Health and Social Life Survey of the USA. What they found was that the frequency distributions of number of children for gay and heterosexual men differed more than the frequency distributions for lesbians and heterosexual women (Table 8.10). That homosexual male partnerships have fewer children than homosexual female partnerships, however, may just be a consequence of women establishing a lesbian relationship after they were involved in a heterosexual relationship during which they had children. The tendencies of courts to give preferential custody of children to the mother over the father may further enhance the difference. Shireman (1996) analysed data from the National Committee for Single Parents of the USA and also found that more female single parents tend to adopt a child than men. Although data on sexual orientation of those single parents are not available, it is possible that a non-insignificant proportion of them may be homosexuals, as children are less likely to be given in adoption to homosexual couples than to single parents, who will obviously not disclose their sexual orientation.

In an early study, Bigner & Jacobsen (1989) compared the responses on a Value of Children scale that 33 gay fathers gave with those of 33 heterosexual fathers and found that, when gay men seek to achieve fatherhood, their motivation is usually elevated. However, not all gay men are equally motivated to become parents. In fact, many express the desire not to become a father (see, for example, Beers 1996). In a sample of 50 gay men studied by Stacey (2006), 16% were childless by choice, whereas the rest were already fathers (76%) or expressed the desire to become one (8%).

Table 8.10. Number of children in homosexual and heterosexual male and female partnerships

	Homosexual (%)	Heterosexual (%)
Men		
No children	94.8	63.8
1 child	3.0	18.1
2 children	1.2	11.0
≥ 3 children	1.1	7.1
Women		
No children	78.3	63.8
1 child	12.6	18.1
2 children	5.0	11.0
≥ 3 children	4.1	7.1

Adapted from Table 6 of Black *et al.* (2000).

In sum, although the desire to have children within a partnership seems to be similar for homosexual and heterosexual women, it seems to be lower in men in general and it is also lower in homosexual than in heterosexual men. Although this suggests that in humans male, but not so much female, homosexuality is associated with a decrease in reproductive output resulting from psychological states associated with a homosexual orientation, more detailed studies of the specific motivations of lesbians and gay men to have or not to have children are needed that control for the legal barriers faced by homosexual men and women in the formation of a family. In fact, Shernoff (1996) indicates that social difficulties to parenthood that male homosexuals face, rather than any lack of desire to have children, is a major obstacle to becoming fathers, at least among some gay men. In addition, given the biological ability of males to reproduce until much older ages than females, it is also expected that males may be under lower pressure to have children at any given point in time than females. These possibilities notwithstanding, I have also argued in Chapter 3 that the lower interest of homosexual males in their own reproduction could be a kin-selected trait if it helps the reproductive effort of close relatives. Kin selection would predict that gay men would express affection towards children (see Murray 1996, for instance) but not as strong an interest in their own reproduction as that expressed by heterosexual men.

Same-sex sexual behaviour and STD transmission

Parasites can be powerful selective factors, not only for the evolution of obvious traits that protect organisms against infection such as the immune system, but also for behaviours such as grooming, sociality, use of specific habitats, eating particular food items, and others (see, for example, Clayton & Moore 1997). In particular, whenever a behavioural pattern enhances the probability of transmission of parasites and pathogens, there may be a selective pressure against that behaviour. Sexual intercourse may be associated with the transmission of venereal or sexually transmitted diseases (STDs). Given that in most species the frequency of same-sex sexual intercourse is lower than that of heterosexual sexual intercourse, questions arise whether STDs, or some STDs, could be more easily transmitted in homosexual rather than heterosexual intercourse and, if so, whether they could be exerting a selective pressure against the spread of same-sex sexual behaviour in birds and mammals.

In this section I review the available evidence regarding sexual transmission of parasites and pathogens during same-sex sexual intercourse in some invertebrates and vertebrates, with a special focus on primates, humans in particular. Unfortunately, I have been unable to find published evidence of same-sex sexual transmission of pathogens in birds, but I will mention a couple of examples from invertebrates. The aim is to tackle some of the issues listed in the previous paragraph in order to obtain a better understanding of the role STDs may have in the maintenance or not of same-sex sexual behaviour.

In a recent review of the factors affecting the risk of parasite transmission in mammals, Altizer et al. (2003) identify a series of social, sexual, demographic, environmental and evolutionary variables that are set to modulate rates of STD transmission via the heterosexual route. In general, they conclude that ST pathogens tend to be rather specialised and relatively less likely to cause host mortality compared with non-ST pathogens, whereas the rate of transmission of ST pathogens is obviously sensitive to the mating system, with promiscuous heterosexual matings being more favourable to sexual transmission of parasites. In general, ST pathogens are well adapted to persist in small host populations as their spread is not dependent on the availability of large numbers of hosts, and their effect on host fitness is usually relatively low. However, as rate of parasite transmission increases, such as is the case in situations of elevated promiscuity, selection for increased ST pathogen virulence is expected. In general, the same patterns of ST parasite transmission and evolution of virulence are predicted if sexual intercourse is homosexual rather than heterosexual; differences may arise, however, if the two patterns of sexual intercourse are associated with differences of mating frequency or with modalities of intercourse (e.g. insertive genito-anal copulation among male mammals) that may favour transmission of some specific pathogens.

Among invertebrates at least two cases of parasite transmission through M–M copulations have been documented or suggested. Anders Møller (Møller 1993) studied the entomopathogenic fungus *Entomophthora muscae*, which is lethal in the domestic fly *Musca domestica*. Whenever an infected fly dies its abdomen becomes swollen; if the dead fly is a male, its enlarged abdomen makes it resemble a female. A passive conspecific with a swollen abdomen becomes a sexually attractive individual to a healthy male, which may then copulate with the dead male. Such attraction to infected individuals may be also enhanced by other cues in addition to the swollen abdomen (Møller 1993). Therefore, through M–M copulation *E. muscae* may be transmitted from a dead host to a live one. Strictly speaking *E. muscae* can be transmitted both heterosexually and homosexually as the healthy male may copulate either with an infected male or a female; the point I want to make, however, is that this is a case where same-sex sexual transmission of a parasite does occur, although the circumstances are obviously unusual owing to the involvement of a dead individual. Among monarch butterflies (*Danaus plexippus*) spores of the protozoan *Ophryocystis elektroscirrha* can be transmitted from males to females via heterosexual copulations (Altizer *et al.* 2004b). Although most matings are heterosexual, about one third of the copulations are M–M and such copulations can last as long as heterosexual ones (Oberhauser & Frey 1999). Therefore, there is at least the potential in this species for the pathogenic protozoan to be sexually transmitted through same-sex sexual intercourse.

Same-sex sexual transmission of pathogens has been relatively better documented in vertebrates, although patterns are not always clear, especially in studies carried out on wild animals. Among domestic pigs (*Sus scrofa*) some males that are held in monosexual groups after the onset of puberty may develop anal ulcerations (McOrist & Williamson 2007). Such anal ulcerations are associated with the presence of bacteria such as *Streptococcus dysgalactiae* and *Clostridium perfringens*. McOrist & Williamson (2007) suggest that anal trauma followed by bacterial infection was caused in the pigs by homosexual mounting. Levtin *et al.* (1983) also suspect that, at least in part, homosexual mounting may have been the cause for the spread of an acquired immunodeficiency syndrome in a captive colony of *Macaca cyclopis* in Massachusetts, USA. Even more indirect evidence for the association between same-sex mounting and sexual transmission of parasites come from some studies carried out in wild apes. It will be remembered that all apes show same-sex mounting (see, for example, Table 8.4); at the same time it is well known that apes engage in behaviours such as swallowing plant materials that have low nutritional value. Such behaviour has been interpreted as self-medication (see, for example, Huffman 1997, 2003). Some such plants contain compounds that have fungicidal activities and that can inhibit the well-known sexually transmitted yeast *Candida albicans* (Huffman 2003). Huffman (1997) also reports *Pan troglodytes verus* from Western Africa consuming plants such as *Palisota hirsuta* and *Eremospatha macrocarpa*, which are used by local human populations in the treatment of venereal diseases.

In contrast to the above positive, although still mainly indirect, evidence for sexual transmission of pathogens through homosexual intercourse, in a comparative test across primate species Nunn *et al.* (2003) did not detect an association between heterosexual mating promiscuity and parasite diversity. It will be remembered that in the bivariate comparative analyses of the mammalian dataset, evolutionary trends towards polygamy were associated with evolutionary trends towards increased same-sex mounting, therefore from the work of Nunn *et al.* (2003) we could probably deduce that at least parasite diversity may not be associated with same-sex mounting across a variety of taxa. This, however, should be regarded as a hypothesis that still requires more direct testing. Rothschild & Rühli (2005) also provide some indirect evidence for the apparent lack of an STD cost of same-sex mounting in *Pan* species. The well-known pattern for *Pan paniscus* to display greater frequency of same-sex sexual behaviour than *Pan troglodytes* (see Chapter 9) is not associated with an equivalent difference in infectious arthritis, which in humans is linked to genital infections by *Chlamydia* and *Mycoplasma*, two well-known sexually transmissible micropathogens. If same-sex sexual intercourse is not associated with costs of STD transmission in most primates, then we would expect the behaviour to at least have one fewer impediment to its expression (see, for example, the widespread occurrence of G–G-rubbing in female *P. paniscus*).

The most firm conclusion that we can draw from the above review, however, is that pathogen transmission during same-sex sexual intercourse in non-human animals is a research area that is still in its infancy.

The situation is rather different in humans, a species that is the subject of considerable research on sexual transmission of pathogens via both heterosexual and homosexual routes. Far from being a problem of the past, or one circumscribed to developing countries, STDs are an important cause of health problems around the world, with same-sex sexual intercourse being a well-established route of STD transmission, especially among men (Fenton & Lowndes 2004). In this section I will sometimes refer to same-sex sexual intercourse as a 'risk factor' for STD transmission. I wish to make quite explicit that the 'risk' involved is not in having a homosexual sexual orientation, but in engaging in specific sexual practices that enhance transmission of STDs. STD transmission among gay men and lesbians will be reduced by making sexual practices safer, not by requesting homosexuals to become heterosexual! Exactly the same argument for increased safety in sexual intercourse is valid for STDs transmitted heterosexually.

STD transmission among people engaging in homosexual sexual intercourse (usually referred to in the medical literature as 'men who have sex with men' or 'women who have sex with women', expressions that are descriptive of sexual behaviour and avoid the issue of gender identity associated with the terms 'gay' and 'lesbian') is enhanced by an elevated number of sexual partners and also by engaging in unprotected sex (see, for example, Wolitski *et al.* 2001; Fenton & Lowndes 2004). Increased promiscuity, accompanied by a lack of barriers that prevent contact between tissues of infected and uninfected partners, are two major causes of recurrent outbreaks of HIV-AIDS among homosexual populations around the world, with multiple infections by various ST pathogens tending to produce a particularly damaging effect on the health of infected people (Røttingen *et al.* 2001).

Among males, same-sex sexual behaviour is a risk factor for STD transmission among both adolescent (see, for example, Zenilman 1988) and adults (see, for example, Stolte & Coutinho 2002). One important factor that affects STD infection in male homosexuals is the perception that diseases such as HIV-AIDS 'have a cure'; this perception tends to increase the risks taken in terms of number of sexual partners and engaging in unsafe sexual practices. This pattern is exactly what an evolutionary biologist would expect whether same-sex sexual behaviour in men has mainly socio-sexual functions or is a result of purely sexual motivational states, or both: variation in costs or perceived costs of the behaviour contributes to modulating its frequency. Responding to increased STD risk by becoming more selective and/or more monogamous is a pattern expected in the context of heterosexual behaviour, as it is in homosexual behaviour. As soon as the perception of risk is relaxed, the impediments for polygamy/promiscuity and broadening of partner selectivity fall in the context of same-sex sexual behaviour, as they do for heterosexual behaviour (see, for example, Ciesielski 2003).

In recent years great focus has also been given to STDs transmitted among homosexual women (Bauer & Welles 2001). Although a great deal of the evidence about STD transmission among lesbians is circumstantial or correlational in nature, some of the trends are worth mentioning. Bacterial vaginosis has been suggested as being one pathology that may originate from sexual contact between two women (Berger *et al.* 1995; Fethers *et al.* 2000). In their recent review of this condition, Marrazzo *et al.* (2002) highlighted the association of bacterial vaginosis in homosexual women with larger numbers of lifetime female sexual partners, a trend expected from sexual transmission of the cause/s of the condition. Studies that use better controls in the comparison of homosexual and heterosexual women also indicate that homosexual women tend to have higher prevalence values of bacterial vaginosis than heterosexual women, although the difference is quite variable across studies (Morris *et al.* 2001). Campos-Outcalt & Hurwitz (2002) have also reported a case of syphilis, an STD caused by *Treponema* bacteria, in women that could potentially be explained by F–F sexual transmission. If bacteria could be transmitted sexually between women, viruses could also. Marrazzo *et al.* (1998) have suggested that at least some of the cases of human papillomavirus infection reported among lesbians may be due to direct sexual infection during same-sex sexual intercourse. Other viruses such as HIV, hepatitis B and herpes simplex can also be transmitted sexually between lesbians (Marrazzo 2000; Marrazzo et al. 2003; Bailey et al. 2004).

Marrazzo (2000) suggested that bisexuals could be a bridge for the transmission of STDs between homosexuals and heterosexuals (see also Matteson 1997), a possibility that is supported by the high ST pathogen infection rates in women and men who are bisexual (Matteson 1997; Bauer & Welles 2001; Millet et al. 2005; Pinto et al. 2005). Weinberg et al. (1994b) also indicate an elevated prevalence of bisexuals who have contracted an STD at least once in their life: 67% in the case of bisexual men compared with 46.3% for heterosexuals and 72.9% for homosexuals, and 61.1% for bisexual women compared with 43.2% for heterosexuals and 37% for homosexuals. The bias towards bisexuals is especially accentuated among women, whereas bisexual men have a slightly lower prevalence value than homosexual men in the Weinberg et al. study.

That homosexual sexual behaviour can be a conduit for the transmission of STDs is quite clear from the above review. The question, however, is whether same-sex sexual intercourse is a more efficient method of transmission of at least some ST pathogens than heterosexual intercourse. This seems to be the case for at least some micropathogens in men. Both HIV and hepatitis B virus (HBV) infections have been reported to be relatively more prevalent among homosexual than heterosexual men. HIV: 1%–35% in heterosexual men vs. 16%–37% in homosexual men (Smith 1991; Fennema et al. 1998; Nicoll & Hamers 2002); HBV: 5.9%–41.8% in heterosexual men vs. 38.7%–68.8% in homosexual men (Dietzman et al. 1977; Mele et al. 1988; Gilson et al. 1998) after controlling for alternative routes of infection such as endovenous injections. Corona et al. (1991) have detected only a slightly higher level of anti-hepatitis C virus antibodies in homosexual men (2.9%) than heterosexual men (2.8%) who were not drug users, although in this case the difference is almost negligible. Judson et al. (1980) also detected trends for homosexual men to have higher prevalence of gonorrhoea (30.31% vs. 19.83%), early syphilis (1.08% vs. 0.34%) and anal warts (2.90% vs. 0.26%) than heterosexual men. Therefore, at least in humans it is possible that homosexual intercourse may carry some increased risks compared with heterosexual intercourse in terms of transmission of specific micropathogens.

So far, I have regarded the relationship between homosexuality and STDs as one where the sexual behaviour of the host affects transmisson rate of the pathogen. But pathogens may also affect host behaviour directly, and through their action on the host's central nervous system increase their rate of transmission; in which case we may think that homosexuality itself could be a result of parasitism. This is the so called *Pathogenic Theory of Homosexuality* first proposed by Cochran et al. (2000). Although the hypothesis should not be dismissed without proper testing, it seems to me that a potential vertical (i.e. parent–offspring) transmission of such postulated infectious agents causing homosexuality is contradicted by birth order effects in the development of homosexuality and also by the heterosexuality of most parents of homosexuals. On the other hand, if the postulated infectious agent is envisaged as being horizontally transmitted (e.g. through sexual or other body contacts) then it is not clear why the high prevalence of pathic systems around the world (see Chapter 1) does not produce a higher rate of exclusive homosexuals through the transmission of the alleged infectious agent from the homosexual to the bisexual/heterosexual partner. Instead, the percentage of exclusive homosexuals is low and roughly constant across societies.

I conclude this section with the proposal of a model for the spread and maintenance of STDs in a population as a result of transmission across sexual orientations. Figure 8.14 provides a diagrammatic description of the model that builds on Marrazzo's (2000) suggestion (see above). The arrows in the model that indicate the directionality of transmission are likely to have different thickness, that is, probabilities of transmission are likely to vary across the model and across infectious agents as well, but I will keep the arrows at constant size in order to focus on more general patterns. The model not only predicts (a) STD transmission from male homosexuals to female and male

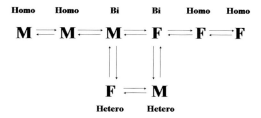

Figure 8.14. Model of STD transmission across sexes and sexual orientations in human populations.

heterosexuals via male bisexuals, and (b) STD transmission from female homosexuals to male and then female heterosexuals via female bisexuals but also, interestingly, indirect transmission of STDs between exclusive gay men and exclusive lesbians via a bisexual/heterosexual network. I am not aware of any such case of indirect STD transmission from an exclusive gay man to an exclusive lesbian from the literature; but if it is not available already I predict that it will soon be described. Some results that are consistent with this model have been recently published by Tao (2008), showing that some viral STDs are more prevalent in bisexual women than lesbians. Mukandavire et al. (2009) have recently published an epidemiological model that also suggests a synergistic role of heterosexuality, homosexuality and bisexuality in the spread of HIV-AIDS.

I do realise that the model could be taken as a justification for discrimination against bisexuals, as occurred during the early years of the HIV-AIDS pandemic in the 1980s (Weinberg *et al.* 1994b). My position in this regard is that scientific models, once they become empirically corroborated, should be made public for the benefit of any interested party, and their application be the subject of the public scrutiny expected in a democratic society. As already mentioned in this section, if a risk factor is detected in specific cases, then that knowledge should be used to educate the concerned parties regarding safety measures recommended during sexual intercourse, not to 'educate' them to change their sexual orientation! I will explain in the last section of Chapter 10 that any discrimination based on sexual orientation should be rejected. Knowledge about risk of STD transmission will simply inform all sexually active individuals, whatever their sexual orientation, about the need to adopt an effective strategy of their choice in order to minimise the risk of infection to themselves and others.

Summary of main conclusions

- My initial evolutionary model of same-sex sexual behaviour is the *Synthetic Reproductive Skew Model of Homosexuality*, which is a synthesis of: (a) *Reproductive Skew Theory*, modified to take into account aspects of mate choice, inbreeding avoidance and queuing to breed, (b) sexual, and (c) socio-sexual aspects of homosexual behaviour.
- Tests of the model, by comparative analyses of independent contrasts in birds and mammals, indicate that the statistically significant results tend to support the model in both taxa.
- More specifically, among birds same-sex mounting is negatively associated with size of the social unit and positively associated with a male-biased sex ratio, sociality and the expression of dominance. Among mammals same-sex mounting is negatively associated with social unit size, relatedness and ecological constraints to breeding, whereas it is positively associated with polygamy. In addition, as the sex ratio becomes more male-biased, males become more involved in same-sex mounting.
- In addition, evolution towards the expression of same-sex mounting is also associated with a greater involvement of females in same-sex mounting in both birds and mammals.
- Path analyses using independent contrasts suggest that the *Synthetic Reproductive Skew Model of Homosexuality* fits the data better than more limited versions that are restricted to either the more sexual or the more socio-sexual aspects of same-sex sexual behaviour. Also, the model fits better the avian than the mammalian dataset,

suggesting that same-sex mounting is a more complex behaviour in mammals than in birds.
- In a direct comparison between birds and mammals, same-sex mounting is more prevalent in mammals than in birds, with mammals being characterised by higher levels of polygamy, sociality, ecological constraints, sexual dimorphism and intragroup relatedness, higher involvement of females in same-sex mounting and higher levels of female bias in the adult sex ratio than birds.
- In humans, the desire to have children seems to be similar for homosexual and heterosexual women, but it seems to be lower in men in general and it also tends to be lower in homosexual than in heterosexual men.
- Finally, although in humans same-sex sexual transmission of pathogens does occur among homosexual and bisexual men and women, it is unclear how frequent this modality of sexual transmission is in other species, and also how many of those human pathogens display an exclusive same-sex sexual transmission.

In the next chapter Alan Dixson will carry out a review of same-sex sexual behaviour in primates that will include not only mounting but also other behaviours that I have not emphasised so far. Chapter 9 will be the final stepping stone before I introduce the *Biosocial Model of Homosexual Behaviour* in the last chapter.

Homosexual behaviour in primates

Alan Dixson

Any attempt at understanding human homosexuality must also include a thorough analysis of same-sex sexual behaviour in the other primates. It is this comparative approach that can uncover those aspects that are common to all or some primates and those that are unique to any particular species. The theory of evolution predicts similarities of traits among closely related taxa due to common descent; it also predicts the possibility of evolutionary convergence among not so closely related taxa if their evolutionary history has unfolded under similar environmental conditions. A further prediction concerns divergences in traits, including behaviour, due to phylogenetic effects, drift and adaptive responses to diverse environmental, including social, circumstances. In this chapter I firstly consider homosexual behaviour as it applies to the non-human primates. This comparative approach may help us to better understand those aspects of human homosexuality that are unique and those that are shared with our closest living relatives. With these thoughts in mind, I start with an introduction to primate diversity and evolution.

Primate diversity and evolution

More than 300 extant species of primate have been described; the precise numbers depend upon the propensity to either 'split' or 'lump' taxa in various classification schemes. All these species, however, may be assigned to one of six major groupings or superfamilies of the Order Primates: the lemurs of Madagascar (Lemuroidea), the galagos and lorises of Africa and Asia (Lorisoidea), the tarsiers of SE Asia (Tarsioidea), New World monkeys (Ceboidea), Old World monkeys (Cercopithecoidea) and the apes and humans (Hominoidea). Phylogenetic relationships between these six superfamilies are shown diagrammatically in Figure 9.1. George Gaylord Simpson (1945) divided the Order Primates into two suborders: the Prosimii (lemurs, lorises, galagos and tarsiers) and the Anthropoidea (monkeys, apes and humans). Because the tarsiers occupy an intermediate position, and display anatomical features found in both these suborders, some authorities prefer to include them with the monkeys, apes and humans, in the Haplorhini, while assigning the lemurs and lorisines to the suborder Strepsirhini. In the account which follows, I adhere to Simpson's scheme, and use the terms prosimian and anthropoid when discussing the evolution and possible functional significance of homosexual behaviour in the various superfamilies of the Order Primates.

For those readers who are less familiar with primatology, it may be helpful to provide a little background information about the behavioural and reproductive biology of the prosimians and anthropoids. The majority of prosimians (Figure 9.2) are nocturnal, relatively small-bodied, arboreal primates. This generalisation applies to all the lorisines and tarsiers, and to the majority of the Malagasy lemurs. Although some lemurs, such as the indris and ringtailed lemur, are diurnal and live in social groups, these species are derived from

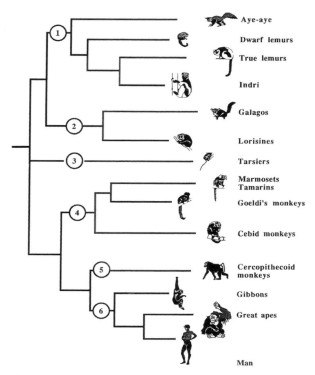

Figure 9.1. Phylogenetic relationships between the six superfamilies of extant primates: 1, Lemuroidea; 2, Lorisoidea; 3, Tarsioidea; 4, Ceboidea; 5, Cercopithecoidea; 6, Hominoidea. Traditionally, superfamilies 1, 2 and 3 (the lemurs, lorises, bushbabies and tarsiers) have been placed in the suborder Prosimii, whereas superfamilies 4, 5 and 6 (the monkeys, apes and humans) have been assigned to the Anthropoidea (Simpson 1945). From Dixson (1998), after Martin (1990).

nocturnal ancestors, and retain traits (such as a reflecting layer or tapetum lucidum in the retina) that arose in nocturnal environments (Martin 1972, 1990). The nocturnal prosimians tend to be non-gregarious, so that the sexes occupy individual, overlapping, home ranges with a dispersed mating system (Dixson 1995). Examples include the greater and lesser galagos of Southern Africa, the African potto and angwantibo, and the slow and slender lorises of Asia (Nekaris & Bearder 2007). These animals make extensive use of olfactory communication in social and sexual contexts. Females exhibit restricted periovulatory periods of sexual receptivity, or oestrus, in contrast to the anthropoids, in which oestrus is lacking (Dixson 2009).

Among the anthropoids, the New World and Old World monkeys (Figure 9.3) show many similarities in their morphology and behaviour, owing to their similar modes of life and the effects of parallel evolution. Thus, they all live in groups, with complex social organisations; many inhabit rainforest, and feed upon fruits, leaves or insects. However, all the New World monkeys are arboreal and some (such as the spider monkeys, howlers and woolly monkeys) have prehensile tails, whereas the Old World forms constitute a more diverse array, including many terrestrial forms (e.g. baboons, geladas and patas monkeys), as well as arboreal monkeys (e.g. guenons, mangabeys, colobus monkeys, proboscis monkeys and many others).

Figure 9.2. Examples of prosimian primates. Left: (lower) The ruffed lemur (*Varecia variegata*); (middle) Lesser bushbaby (*Galago senegalensis*); (upper) The potto (*Perodicticus potto*). Right: A tarsier (*Tarsius spectrum*). After Schwartz (1987) and Clark (1962).

The New World monkeys, like the prosimians, often make use of urine and cutaneous glandular secretions during communication in social and sexual contexts. Such scent-marking behaviour is rare among the Old World monkeys, occurring for example in the mandrill, which possesses a sternal cutaneous gland. However, it is among the Old World forms, such as baboons, talapoins, red colobus, mangabeys and macaques, that one encounters the, oestrogen-dependent, sexual skin swellings, which enlarge and become pink, or red, during the follicular phase of the ovarian cycle. Such swellings also occur among the apes, in female chimpanzees and bonobos. It is also relevant to mention here that all the Old World monkeys and apes exhibit a menstrual cycle, homologous with the human menstrual cycle, whereas very few New World monkeys (e.g. spider monkeys and capuchins) exhibit menstruation.

Turning to the Hominoidea, it has long been thought that the gorilla (Figure 9.3) and chimpanzee might be most closely related to *Homo sapiens* and that humans originated in Africa (Darwin 1871). Modern research has confirmed Darwin's insights, and molecular studies point to a closer common ancestry between humans and chimpanzees, occurring approximately 7–8 million years ago. The chimpanzee and gorilla share numerous phenotypic traits, however, including specialisations of the hand and wrist for terrestrial, quadrupedal (knuckle-walking) locomotion. The social and mating systems of the apes are diverse. The chimpanzee and the closely related bonobo live in flexible fusion–fission communities with

Figure 9.3. Examples of anthropoid primates. Left: (lower) An Old World monkey, the baboon (*Papio sp.*); (upper) A South American capuchin (*Cebus sp.*); Right: A silverback male gorilla (*Gorilla gorilla*) beating its chest. After Clark (1962).

multimale–multifemale mating systems. The larger gorilla, by contrast, is highly polygynous, and massive silverback males compete to associate with small groups of females. In SE Asia, the orang-utan is most unusual among anthropoids in being non-gregarious. Orang-utans occupy very large individual home ranges, and the ranges of adult males overlap extensively with areas occupied by females. Male orang-utans employ loud vocalisations (long calls) and other displays (tree-swaying and snag-crashing) during intrasexual and intersexual communication. The orang-utan is the largest arboreal mammal (adult males may weigh 100 kg), and it is thought that ecological constraints have determined its dispersed social organisation and mating system. By contrast, the Asiatic gibbons, which are much smaller and have historically been referred to as the 'lesser apes', all live in cohesive family units. Although 'extra-pair copulations' have been described (e.g. in the white-handed gibbon, *Hylobates lar*; Palombit 1994) it is thought that most sexual interactions occur within monogamous family groups.

Primate homosexual patterns: phylogenetic considerations

There are qualitative and quantitative differences between the various primate superfamilies, and between the prosimians and anthropoids, in the expression of homosexual patterns of behaviour. Mounting and hindquarter presentation postures are prominent features of homosexual behaviour among the non-human primates: some examples are provided in Figure 9.4. Thus, the propensity for males to mount other males, females to mount females, and members of the same sex to present to one another, has been documented among macaques, baboons, mandrills, talapoins and many other Old World monkeys, as well as among the apes (Vasey 1995; Wallen & Parsons 1997; Dixson 1998). Mounts are commonly accompanied by pelvic thrusting movements (by female mounters, as well as by males). Males may exhibit penile erections during homosexual mounts, and intromissions and ejaculations (although unusual) have been recorded during homosexual mounts in

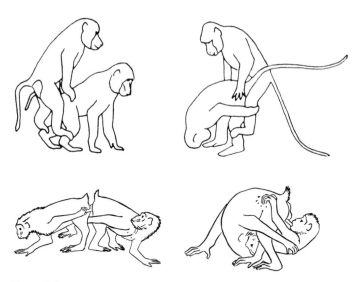

Figure 9.4. Male–male socio-sexual mounts and presentations in anthropoid primates. Upper left: guinea baboons (*Papio papio*). Upper right: bonnet macaques (*Macaca radiata*). Lower left and right: stump-tail macaques (*Macaca arctoides*). After Dixson (1998).

some species (e.g. stump-tailed macaques and orang-utans).

The prosimians are noteworthy for a relative absence of these kinds of homosexual mounting patterns. The sifaka (*Propithecus verreauxi*) may occasionally exhibit male–male mounting, as cited by Bagemihl (1999) in his comparative survey of homosexual behavior in animals. However, among the 13 genera of Malagasy lemurs currently recognised, *Propithecus* provides the only example of same-sex mounting among adults. For the nine genera of the Lorisoidea, and for *Tarsius*, no examples have been documented to my knowledge. Although negative evidence cannot be final, and examples of homosexual mounting in prosimians may come to light in the future, it seems justifiable to conclude that such behaviour must be relatively rare among the prosimian primates.

Among the New World monkeys, same-sex mounts have been recorded in captive and in free-ranging groups of several species. Thus, male–male and female–female mounts have been observed in groups of squirrel monkeys (*Saimiri sciureus*), as well as in capuchins (genus *Cebus*). In the marmosets and tamarins such behaviour has rarely been observed, but it has been recorded among males and females of four tamarin species (*Saguinus fuscicollis*, *S. oedipus*, *S. geoffroyi* and *S. mystax*; Bagemihl 1999). Thus, homosexual mounting has been recorded in only 3 of 19 (or 16%) of the New World primate genera. This is not to say that other forms of homosexual behaviour, such as genital display, are lacking in New World monkeys; I shall return to this topic in a moment.

Homosexual mounting is much more commonly observed among the Old World monkeys (Cercopithecoidea) and apes and humans (Hominoidea) than in other primates. I have been able to locate adequate information, from field studies and observations of captive groups, for 17 of the 24 currently recognised genera of Old World monkeys and apes (see Table 9.1). For 13 of these (76%) male–male mounting occurs in at least one species in the genus. Female–female mounting has also been recorded in 12 (70%) of these genera. In some cases, such as among the macaques and baboons, most

Table 9.1. Presence or absence of male–male and female–female mounting in the various genera of Old World monkeys and apes

Genus	Male–male (M–M) and female–female (F–F) mounts
Macaca	M–M, F–F reported for most species
Papio	M–M, F–F reported for most species
Cercocebus	M–M, F–F C. atys; C. torquatus
Lophocebus	M–M, F–F L. albigena
Mandrillus	M–M, F–F but not between mature males
Theropithecus	M–M, F–F reported
Allenopithecus	Insufficient data
Miopithecus	M–M, F–F reported
Erythrocebus	Same sex mounting absent?
Chlorocebus[a]	M–M, F–F reported
Cercopithecus	?
Colobus	?
Procolobus/Piliocolobus	Reportedly absent in P. badius
Semnopithecus[b]	M–M, F–F reported
Presbytis	?
Trachypithecus	?
Pygathrix	?
Rhinopithecus	M–M?, F–F reported in R. roxellana
Nasalis	M–M, F–F
Simias	Insufficient data
Hylobates	M–M reported in one species
Pongo	Younger M–M, not observed between females
Pan	M–M, F–F most frequent in P. paniscus
Gorilla	M–M, F–F reported

[a]*Cercopithecus aethiops* is currently known as *Chlorocebus aethiops*.
[b]*Presbytis entellus* is currently known as *Semnopithecus entellus*.

(if not all) species engage in same-sex mounting. The typical 'double foot clasp' mounting posture, with the mounter using its feet to grasp the calves of the mountee, occurs in these species, as well as in talapoins, mandrills, langurs and some others. There are sex differences in the exact form of mounting and thrusting patterns, however, and I return to this subject below when the functions of homosexual mounts are considered.

This brief survey of homosexual mounting in the Order Primates leads us to conclude that such behaviour is likely to represent a conserved trait among the Cercopithecoidea and Hominoidea, and was probably present in the common ancestors of both these superfamilies of the Old World anthropoids. The same conclusion applies to homosexual presentations, since hindquarter presentations also occur in those species where males and females engage in homosexual mounting. The situation in the New World anthropoids (Ceboidea) is different, as relatively few species have been shown to exhibit homosexual mounting activity. Some New World monkeys have also evolved complex patterns of genital display, which occur in communication between members of the same, as well as the opposite, sex. Whether it is appropriate to designate these as 'homosexual behaviour' is a moot point. Thus in the squirrel monkeys (genus

Saimiri), males and females use distinctive, stereotyped genital display postures (Figure 9.5). Males may 'jab' the partner with the penis during these displays. Among the callitrichids (marmosets and tamarins), a different type of genital display posture occurs, which involves hindquarter presentation and tail-arching (Figure 9.5). Marmosets and tamarins do not use hindquarter presentations as sexual invitations, however, so that it seems unlikely that their genital displays derive from sexual invitational postures. In both sexes of these New World monkeys, the external genitalia (the pudendal pad of the female and the male's scrotum) are richly supplied with cutaneous scent-marking glands. Both sexes also employ urine in marking displays, and olfactory cues play an integral role in social dominance as well as in sexual contexts (Epple 1980; Epple *et al.* 1993). The genital display may thus derive from behavioural patterns important in olfactory communication. Some other New World monkeys also engage in same-sex and opposite-sex genital displays (e.g. in howlers (*Alouatta*) and in capuchins (*Cebus*)), although these are still poorly documented (see Dixson 1998 for a review).

Stereotyped genital displays do not occur in the Old World anthropoids, except for the use of penile erection as part of precopulatory invitational behaviour by male chimpanzees and bonobos (Goodall 1986; Kano 1992) and orang-utans (Nadler 1988). However, genital touching, manipulation and orogenital contacts do occur during homosexual and heterosexual interactions in a variety of monkey and ape species. Touching, or handling, of the external genitalia has been recorded as part of dominance, greeting and reassurance behaviour between male macaques, baboons, mandrills and chimpanzees, and probably occurs in other species as well. Stimulation of the penis may also be more prolonged and purposeful, as for example during interactions between male bonobos. Genital contacts may also occur between female monkeys, as for example during mutual lateral embracing in Celebes (Sulawesian) macaques (*Macaca nigra*), stump-tails and grey-cheeked mangabeys (Dixson 1998). Whether such contacts constitute mutual

Figure 9.5. Genital displays in New World monkeys. Upper: Two male squirrel monkeys (*Saimiri sciureus*) display in the 'closed counter-position'. The partners display their erect penes but do not make eye contact. Lower: A male marmoset (*Callithrix jacchus*) adopts a genital presentation posture. After Dixson (1998).

genital inspection or active genital stimulation is unclear. In *M. nigra*, for example, females may sniff and briefly touch the partner's vulval area during mutual embraces. Bagemihl (1999) presents a drawing of these macaques engaging in what he describes as mutual (manual) masturbation during the embrace. This was not reported in the original paper (Dixson 1977) cited by Bagemihl, however, and I include the original (unmodified) drawing here (Figure 9.6). The requirement for caution in interpreting same-sex embracing as sexual behaviour also applies to the New World monkeys. Members of some New World genera (e.g. *Ateles*, *Brachyteles* and *Lagothrix*) engage in face-to-face embraces. In *Ateles*, males or females often sniff the neck region of the partner during such embraces, and there are specialised cutaneous glands situated in this area. In such cases, embracing appears to be some form of greeting, or affiliative behaviour, rather than having a primarily sexual function, and it may derive from behavioural patterns associated with mutual olfactory inspections by conspecifics.

In the next section I examine the possible functions and derivation of homosexual patterns of behaviour in monkeys and apes in more detail. Does homosexual behaviour function for social communication, for example, and are there consistent relationships between social structure, and the expression of homosexual interactions, between males and between females?

Functions and derivation of homosexual behaviour

In 1914, the great evolutionary biologist Julian Huxley published an account of the courtship behaviour of the great crested grebe (*Podiceps cristatus*). His paper is widely recognised as the forerunner of later developments in ethology and, as such, it introduced the concept of ritualisation as a contributory mechanism underlying the evolution of courtship displays. Huxley noted that grebes exhibited behavioural patterns derived from non-sexual activities, but which had become ritualised and incorporated into their complex courtship sequences (Figure 9.7). Almost two decades later, when Zuckerman (1932) studied the sexual and social behaviour of hamadryas baboons (*Papio hamadryas*) on 'Monkey Hill' at London Zoo, he noted that both sexes displayed patterns of sexual behaviour, such as hindquarter presentations and mounts, in non-reproductive contexts. These presentations and mounts often occurred in situations where submissiveness and dominance were communicated, rather than any intent to engage in copulation. He noted that amongst male baboons 'one assumes the feminine role and is mounted by the other' and that, likewise, female baboons sometimes present to, or mount, other members of the same sex. Further studies of primate behaviour, involving a much broader range of species, revealed that same-sex (i.e. homosexual) interactions occurred in a variety of contexts, as a social 'greeting' and prior to affiliative interactions (such as grooming), during 'reconciliation' after aggression and so forth.

Figure 9.6. Mutual lateral embracing and genital investigation between two adult female Celebes (Sulawesian) macaques (*Macaca nigra*). After Dixson (1977).

Figure 9.7. Examples of the postures adopted by male and female great crested grebes (*Podiceps cristatus*) during their courtship displays. After Huxley (1968).

Wolfgang Wickler (1967) subsequently developed the concept of socio-sexual mimicry among non-human primates. He proposed that socio-sexual behaviour had facilitated the evolution of morphological traits in the male sex, to resemble those present in females. He cited the example of the hamadryas baboon, in which the male's red rump resembles the female's sexual skin swelling. Wickler proposed that the male's colourful rump might serve to reduce the likelihood of inter-male aggression, especially when displayed as a mimic of female sexual skin, during male–male presentations. This ingenious idea has received little support, however, as more evidence has become available over the years. Thus, in many species of monkeys and apes which engage in homosexual presentation and mounting, males lack homologues of the female's sexual skin. In talapoins, for example, the adult male has a blue perineum and scrotum, in contrast to the female's pink swelling. In baboons other than the hamadryas, males do not have perineal structures resembling those of females. Nor do male chimpanzees or bonobos. The brilliant coloration of the male mandrill's face, rump and genitalia does not closely correspond with the swelling of the female. Moreover, because the physiological mechanisms that produce cutaneous coloration are limited in the primates (i.e. vascular changes causing reddening and dermal melanin causing blue and greenish hues) it is inevitable that sexual skin, when present, may sometimes be similar in the two sexes (Grubb 1973).

Setting aside Wickler's hypothesis of socio-sexual mimicry, evidence that patterns of sexual behaviour are incorporated into same-sex activities has grown steadily over the years as well as the realisation that such patterns may be expressed in young animals, well before puberty. Thus, infant or juvenile male macaques frequently mount each other as well as engaging in 'rough and tumble play'. The expression of such mounting behaviour is important for the development of social relationships and for the successful expression of sexual patterns during later life (Harlow 1971). Mounting frequencies are sexually dimorphic in species such as the rhesus macaque, and prenatal androgenisation of the brain plays a most important role in the development of this dimorphism (Goy & McEwen 1980). If pregnant female rhesus monkeys are treated with long-acting androgens (such as testosterone propionate) then their female offspring are to some degree masculinised. These pseudohermaphroditic females (which often have a well-developed scrotum and penis) show frequencies of mounting and rough and tumble play which are, on average, intermediate between those of normal females and males (Goy 1978).

The conclusion that homosexual behaviour is at least partly concerned with the ritualised expression of sexual patterns in a variety of social contexts is supported by studies of various Old World anthropoids. It is not the case, however, that simplistic explanations such as 'mounting = dominance; presenting = submission' can adequately account for the functions of homosexual behaviour. Thus, it was shown many years ago that dominance hierarchies constructed by measuring agonistic relationships in groups of Old World monkeys do not necessarily correlate with hierarchies constructed from measurements of mounting behaviour in the same groups. Irwin Bernstein (1970) was unable to confirm such correlations based on measurements of aggressive–submissive relationships and mounting interactions in captive groups of vervets, mangabeys, geladas and macaques (three macaque species were examined). The same was subsequently shown to be true in captive groups of talapoins (*Miopithecus talapoin*). However, a closer examination of social communication in talapoins revealed that when presentations and mounts occurred immediately after an aggressive episode, the vast majority of presentations (80%) were made by submissive monkeys and that mounting (87%) was by dominant aggressors (Dixson *et al.* 1975). This function of socio-sexual behaviour to calm aggressive interactions would now be considered part of reconciliation behaviour, which has been extensively studied in monkeys and apes

(DeWaal 1989). Fine-grained analyses of behaviour in anthropoids also indicate that presentation–mounting behaviour serves a number of roles in relation to social communication. In free-ranging Hanuman langurs, for example, Srivastava et al. (1991) showed that female–female mounting is positively correlated with dominance rank, especially between animals that are close to one another in the rank order (Figure 9.8). More recent studies of wild Hanuman langurs by Sommer et al. (2006)

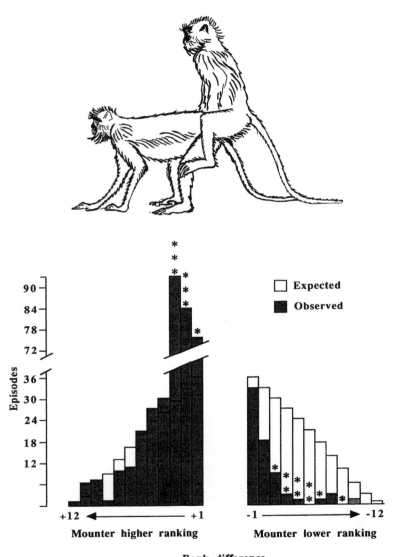

Figure 9.8. Relationships between female rank and socio-sexual mounting in free-ranging Hanuman langurs (*Semnopithecus entellus*). Mounters are higher ranking than mountees in the majority of cases, especially when partners are close to one another in rank. $*p < 0.05$; $**p < 0.01$; $***p < 0.001$. (Redrawn from Srivastava et al. 1991, after Dixson 1998).

have shown that 'a mixture of affinitive and agonistic motivational states' may be associated with homosexual mounting episodes. Grooming often followed mounts, and most often (79% of cases) the dominant female groomed a lower-ranking individual. Sommer *et al.* (2006) calculate that homosexual behaviour accounts for as much as 46% of sexual interactions in adult females. For adult males it is much less prevalent among heterosexually active (high-ranking) males (18% of interactions). However, for lower-ranking males, which rarely have an opportunity to mate, it may represent as much as 95% of their sexual activity. These authors tested a number of theories concerning the basis of homosexual behaviour, including social bonding, alliance formation, dominance, reduction of receptivity in rival females, and as a means of stimulating sexual responsiveness by females of male partners. None was strongly supported, although in general female–female mounts occurred more often in competitive, agonistic contexts, and male–male mounts had 'affinitive motivations'.

Quantitative analyses of the functions of same-sex mounting are also available for bonobos (*Pan paniscus*). Among female bonobos, presentations and mounts, including ventro-ventral mounts with genital contact (GG-rubbing: Figure 9.9) occur in a variety of social contexts, as observed in captivity and in the wild (Kano 1992; Parish 1996). Fruth & Hohmann (2006) have evaluated some possible functions of homosexual behaviour among free-ranging female bonobos. Dominant females were in the top position, during GG-rubbing embraces, more frequently than low-ranking partners. Homosexual interactions were more frequent when females were visiting patches of high-quality food, so that some role in 'tension regulation' is supported by these observations. Although not so pronounced, Fruth and Hohmann also detected some involvement of female homosexual behaviour during reconciliation, after agonistic encounters.

Figure 9.9. Female–female ventro-ventral mounting, with 'G–G rubbing' in the bonobo (*Pan paniscus*). From Dixson (2009), based upon a photograph by Frans DeWaal.

Homosexual behaviour was more likely to involve non-related females rather than kin, but, despite its affiliative aspects, it was not necessarily more frequent among females who associated as grooming partners. The hypothesis that females might engage in GG-rubbing to attract the attention of males, and to solicit copulation, was not supported by these studies.

Information on Hanuman langurs, and bonobos, as well as various macaque and baboon species, thus supports the view that homosexual behaviour may serve widespread functions in primate social communication. Fine-grained analyses are sometimes required to understand the subtle functions of these visual and tactile displays. Not all mounting and presentation postures are necessarily equivalent, for example. Darwin's (1872a) principle of antithesis as expounded in his book on *The Expression of the Emotions in Man and Animals* may be relevant to understanding the form of presentation postures used by female monkeys in submissive and sexual contexts. Hausfater & Takacs (1987) provided an interesting demonstration of this principle, in their studies of hindquarter presentations by yellow baboons. Thus, curvature of the spine and leg-flexure typified the 'cringing' type of presentation made by submissive animals, and this contrasted with the more upright postures used by dominant individuals (Figure 9.10). These variations in the form of presentation postures may be more widespread among Old World monkeys than currently realised; similar variations occur among female mandrills (*Mandrillus sphinx*), for example (Dixson & Frei, unpublished observations). Females may also adopt slightly different positions during homosexual mounts and exhibit thrusting patterns different from those which are used by males. These differences may relate to facilitation of tactile genital stimulation during female–female mounts (Vasey & Duckworth 2006), as will be discussed in the next section.

The widespread occurrence of homosexual mounts and presentations by infants and juveniles in Old World monkeys and apes is also most interesting and relevant to discussions of the functions of homosexual patterns in adulthood. As mentioned above, same-sex mounting is much more frequent among young males than is the case for females. Prenatal exposure of the developing male brain to higher levels of testosterone is important for the emergence of these behavioural sex differences. Mounts and presentations also occur as part of bouts of play, especially the 'rough and tumble' play which is more prevalent among male macaques, baboons, mandrills and other Old World monkeys. These behavioural sexual dimorphisms are not absolute, however, as indicated by the fact that females of many of the species can, and often do, display mounting patterns as adults. Adults thus retain a considerable potential to engage in bisexual behaviour. In Japanese macaques, which have been more extensively studied than other primates to examine this question, pairs of females sometimes form consortships (Vasey 2002a). Mounting between these females is not related to dominance, or to post-conflict reconciliation (Vasey et al. 1998; Vasey 2004). Nor is there any evidence that these females are masculinised neurologically, or that their sexual preferences for other females might be due to prenatal hormonal effects. Thus the sexually dimorphic nucleus of the medial preoptic–anterior hypothalamus (MPO-AH), which is larger in male macaques than in females, is no larger in female Japanese macaques than it is in female rhesus monkeys (Vasey & Pfaus 2005). Vasey et al. (1998) interpret the homosexual consortships and mounting interactions of female Japanese macaques as being the result of 'mutual sexual attraction and gratification' between the participants. Thus it is important to consider the question of whether sexual attraction and pleasurable (hedonic) experiential factors may contribute to homosexual interactions in primates.

Hedonic aspects of sexual behaviour

There is no a priori reason to reject the possibility that non-human primates might engage in sexual

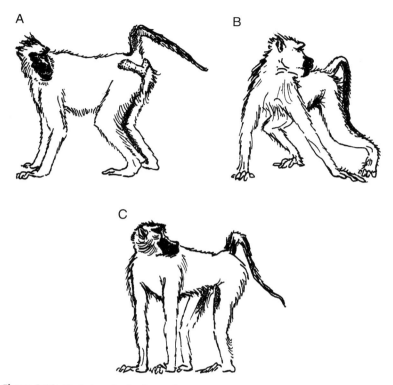

Figure 9.10. Variations in the form of presentation postures in yellow baboons (*Papio cynocephalus*). (A) Submissive presentation by a subadult male (note leg flexure). (B) An adult female presents submissively to a higher-ranking male. Note the flexure of the spine and legs. (C) Sexual presentation posture by an adult female. There is no leg flexure or curvature of the spine. From Dixson (1998), after Hausfater & Takacs (1987).

activity, including homosexual interactions, at least in part because these traits are positively reinforced by pleasurable (hedonic) feedback during mounting and genital stimulation in both sexes. Historically, behavioural scientists may have been reluctant to address these issues because of the risks of anthropomorphism. However, I believe that scientific objectivity can be applied to these questions, just as in other areas of sexology. Further, since human beings are members of the Anthropoidea, it is surely legitimate to seek possible homologues of human sexual response among the monkeys and apes.

It is interesting that both sexes of various anthropoid species have been observed to exhibit orgasm during sexual activities, whether as a result of heterosexual copulation and homosexual mounting, or during auto-erotic (masturbatory) activity. In many primate species, males exhibit responses during copulation which are indicative of orgasm during ejaculation (e.g. cessation of pelvic thrusting, body and leg tremor, muscular spasms, distinctive facial expressions or vocalisations: for a review, see Dixson 1998). Homologues of orgasm also occur in females of some species, although examples are less well documented than is the case for males. As examples, female stump-tails, rhesus macaques, Japanese macaques and chimpanzees all exhibit orgasmic responses, and in many others females show distinctive facial expressions and vocalisations during some copulations. Self-stimulation of the genitalia is also relatively widespread among the primates, and the occurrence of masturbation shows some interesting phylogenetic correlations, which parallel those discussed above, in relation to homosexual mounting in the anthropoids. Thomsen *et al*. (2003) assembled data on occurrences of masturbation by males in 52 primate species. Combining their observations with previously published work (Dixson 1998) and unpublished observations shows that male masturbation occurs most frequently among the Old World anthropoids, less often in the New World monkeys, and is absent in the three prosimian superfamilies. Absence of a behavioural trait is difficult to prove, but whenever further information on prosimian sexual behaviour becomes available, I think it likely that examples of masturbatory patterns will be rare. By contrast, masturbation has been observed in wild or captive males representing 11 genera of the Old World monkeys (i.e. 55% of all cercopithecoid genera) and in 4 out of 5 (80%) of hominoid genera. In the New World primates, species belonging to 6 genera have been reported to engage in male masturbation; this represents 32% of the 19 genera in the superfamily Ceboidea. The occurrence of ejaculation during masturbation is also more prevalent among the Old World anthropoids, having been recorded in 91% of the cercopithecoid genera and 100% of hominoids in which masturbation is known to occur. Among the New World monkeys, ejaculation has only been reported with certainty in one genus (*Brachyteles*: the muriqui) out of 6 (17%) in which males exhibit masturbatory behaviour.

If we allow that masturbation, with or without ejaculation, is perceived by male monkeys and apes as being pleasurable, just as it is in human males, then it is also reasonable to suggest that genital stimulation during male–male mounting might also have hedonic significance. Male Old World monkeys and apes also exhibit homosexual mounts more frequently than other primates; erection, with pelvic thrusting, commonly accompanies these activities. Although anal intromission and ejaculation are rare in this context, it is possible that homosexual mounting might be facilitated by varying degrees of pleasurable tactile feedback, just as in the case of masturbation or heterosexual mounting. In the stump-tail macaque, for example, orogenital contacts may occur between males (Figure 9.4).

Data on the phylogenetic distribution of masturbation in female primates are much less detailed than for males. However, female masturbation is again more common in Old World anthropoids, having been described in the talapoin, various macaques, olive baboons, sooty mangabeys and in all of the great apes, as well as women (reviewed in Dixson 1998). Wolfe (1991) has provided detailed descriptions of such behaviour in female Japanese macaques, which manipulate the clitoris and rub it against the ischial callosities during masturbation (Figure 9.11). Only among the hominoids have females been reported to use objects, such as twigs and leaves, to stimulate their genitalia during masturbation (e.g. in the orang-utan; Rijksen 1978; Nadler 1988).

Given the propensity for females of some anthropoid species to manipulate the clitoris and vulva during masturbation, it is interesting that these areas may be targeted for tactile stimulation during homosexual mounting. Vasey & Duckworth (2006) have made detailed studies of the mechanics of homosexual mounting in female Japanese macaques. Females sometimes mount 'jockey-style' and rub the vulval, perineal and anal (VPA) area against the mountee's back. Mounters also stimulate the VPA region with their tails during same-sex mounts. Their mounting postures and movements are thus different from those that characterise males, and may be specialised to maximise stimulation of the genital and adjacent areas. Similar mounting postures also occur in female stump-tail macaques (see Figure 9.11), in which a distinctive 'climax face' occurs in association with female orgasm during homosexual and heterosexual mounts (Goldfoot *et al*. 1980; Slob *et al*. 1986).The genital rubbing (GG-rubbing) that occurs during the homosexual mounts of female bonobos (Kano 1992) also involves clitoral and vulval stimulation, but in this case tactile stimulation is possible for

Human homosexual behaviour

Figure 9.11. Left: Masturbation in a female Japanese macaque (*Macaca fuscata*). The female rubs her clitoris against an ischial callosity. From Dixson (1998), after Wolfe (1991). Right: Two female stump-tail macaques (*Macaca arctoides*) mounting in a 'piggy-back' or 'jockey-style' position. From Dixson (1998), after Chevalier-Skolnikoff (1974).

both partners, as mounts occur in the ventro–ventral position (see Figure 9.9).

These examples, when viewed in the context of the hedonic significance of sexual behaviour and genital stimulation in primates, serve to emphasise that there is not a rigid dichotomy between the 'sexual', as compared with the 'socio-sexual', functions of homosexual behaviour in the non-human primates. The capacity for monkeys and apes to exhibit bisexual capacities is considerable, and especially so in the Cercopithecoidea and Hominoidea. As *Homo sapiens* is a member of the Hominoidea, and derives from ape-like (australopithecine) ancestors, it is appropriate to conclude this chapter by discussing human homosexual behaviour in comparative, primatological perspective.

Human homosexual behaviour and homosexuality

As explained in Chapters 1 and 2, homosexuality is the term used to denote eroto-sexual preferences for members of the same sex, rather than the opposite sex. In common parlance, men who have a strong orientation to engage in sex with other men are referred to as gay men. Women who find other women to be highly attractive sexually are referred to as lesbians. Men and women whose eroto-sexual attraction is primarily towards members of the opposite sex are referred to as heterosexuals.

Kinsey *et al.* (1948) developed a 7-point scale, which has been widely applied to measure heterosexual–homosexual orientation in human populations. This Kinsey scale, in simplified form, is shown below.

0: Exclusively heterosexual
1: Predominantly heterosexual, only incidentally homosexual
2: Predominantly heterosexual, but more than incidentally homosexual
3: Equally heterosexual and homosexual
4: Predominantly homosexual, but more than incidentally heterosexual

5: Predominantly homosexual, but incidentally heterosexual
6: Exclusively homosexual.

Thus, men scoring as 5 or 6 on the Kinsey scale are regarded as being predominantly or exclusively homosexual, where as those at the opposite end of the scale (scoring 0 or 1) are predominantly or exclusively heterosexual. Individuals with intermediate scores (2, 3, 4) exhibit stronger bisexual preferences to varying degrees. Kinsey *et al.* were thus at pains to point out that 'males do not represent two discrete populations, heterosexual and homosexual. The world is not to be divided into sheep and goats.' From their studies of men in the USA, they concluded that 'any question as to the number of persons in the world who are homosexual and the number who are heterosexual is unanswerable. It is only possible to record the number of those who belong to each of the positions on such a heterosexual-homosexual scale as is given above.'

Figure 9.12 shows graphs by Kinsey *et al.* for development of heterosexuality and homosexuality at various ages in American males. The percentages of males who express 'no socio-sexual response' (as represented by the solid line on the graph) decline rapidly from age 5 years onwards. By contrast, males who rate themselves as predominantly, or exclusively, heterosexual (ratings of 0 or 1: the open line on the graph) increase in number and account for 90% of the adult population. Kinsey *et al.* noted that 'males who are more than incidently homosexual in response or overt activity (ratings 2–6: the dashed line in Figure 9.12) are most abundant in pre-adolescence and through the teens, gradually becoming less abundant with advancing age.' Thus, it appears that matters are less polarised in the younger age groups, indicating a greater spread of sexual orientation ratings and potential for bisexual or homosexual orientation in adolescent males.

Homosexual behaviour is an ancient phenomenon, and has been recorded throughout human history (Boswell 1980). Among the ancient Greeks, mature (and ostensibly heterosexual) men sometimes formed sexual relationships with adolescent males who had yet to develop the secondary sexual traits (facial and body hair) typical of maturity (Dover 1989). Scenes depicting homoerotic interactions are frequently included on pottery from this period (800–200 b.c.e.). An example is

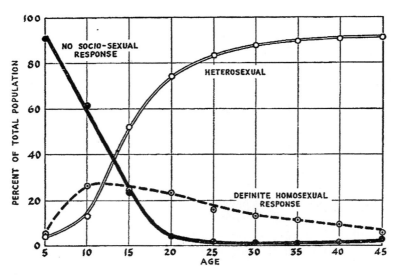

Figure 9.12. Development of heterosexuality and homosexuality by age periods, as measured by Kinsey *et al.* (1948) for males in the USA. Further details are discussed in the text.

shown in Figure 9.13 of a younger male, preparing to engage in non-insertive (crural) sex with a mature, bearded individual. Homosexual behaviour has also been recorded as part of established social and sexual norms in a number of indigenous cultures (see, for example, Herdt 1984a). Among the Sambia of Papua New Guinea that were mentioned in Chapter 5, for example, young boys traditionally left their mothers to join the men's hut when aged between 7 and 10 years. There they were initiated into homosexual activities, and taught to perform fellatio upon older (pubertal and post-pubertal) males. This behaviour was viewed as part of normal development, in which boys were expected to grow big and strong as a result of ingesting semen (Herdt 1981). In Western societies, it is not unusual for homosexual behaviour to occur in all-male environments where men are deprived of opportunities for heterosexual intercourse. This is the case, for example in penal institutions, as Kinsey *et al.* (1948) reported in their classic studies in North America. There is also the phenomenon of the 'macho' male, in which men who rate their orientation as heterosexual may, none the less, play an active (insertor) role during homosexual interactions.

Such examples underscore the fact that the capacity for human males to engage in a spectrum of sexual activities was more common historically, and is still more prevalent than would be expected if a rigid dichotomy exists between heterosexual and homosexual behaviour. Thus, I suggest that human beings retain a significant potential for bisexual behaviour, and share a common evolutionary heritage in this respect with other Old World anthropoids. The degree to which this potential is expressed varies considerably, however, depending upon individual differences, as well as on cultural rules governing sexual expression. If it were possible to apply the equivalent of a Kinsey scale to the Old World monkeys and apes, we might well find that as well as a strong preference to mate with partners of the opposite sex, there would be a willingness to engage in a variety of sexually related behaviours

Figure 9.13. Homosexual behaviour as depicted on ancient Greek pottery. Here an adult (bearded) man and a much younger youth prepare to have non-insertive (crural) sex. From Dover (1989).

with members of the same sex. Humans evolved from ancestral forms (australopithecines) that possessed this type of behavioural flexibility, and the potential for bisexuality and facultative homosexuality in anatomically modern humans is explicable on this basis.

Much more problematic, however, is the existence of that small percentage of men and women who are attracted exclusively, or almost exclusively, to members of the same sex. These are the individuals who consistently rate themselves as 5 or 6 on the Kinsey scale. Cross-cultural studies indicate that perhaps 2%–3% of men rate themselves in this way, whilst the percentage of lesbians is typically lower (Diamond 1993). Berman (2003), for example, refers to the 'puzzle' which male homosexuality presents in terms of explaining the proximate mechanisms underlying the phenomenon, and the ultimate (evolutionary) mechanisms that maintain it in human populations. It is clear that, from early childhood, some males feel and behave differently from the majority, such as the 'sissy boys' studied by Richard Green (1987) that were mentioned in Chapter 3. Elsewhere in this book, Aldo Poiani deals with the possible effects of neuroendocrine, genetic, birth order and other mechanisms in the determination of homosexuality. These are some of the factors that could potentially account for the occurrence of pronounced homosexual (androphilic) preferences in men. However, postnatal environmental factors, as well as genetic and physiological factors, must be entwined in some way. LeVay (1993) states that 'The ultimate challenge will be to establish how the genetic differences among individuals interact with environmental factors to produce the diversity that exists among us.'

Identification of the selective forces that might have shaped human homosexuality has also proven to be somewhat problematic. Wilson's (1978) ideas concerning the possible role of kin selection are only now being tested in humans by using appropriate methodological approaches. One ray of light, among others, concerns the observation that fecundity may be greater in mothers of androphilic males (Camperio-Ciani *et al*. 2004; King *et al*. 2005; Vasey & VanderLaan 2007). If supported by further work, this hypothesis may help to account for the maintenance of homosexuality via pleiotropic effects of genes that enhance maternal fecundity and lifetime reproductive success. Unfortunately no 'model' for studies of this hypothesis is likely to be found among the non-human primates. Human homosexuality, in its extreme form, where men and women are attracted exclusively to members of the same sex, has no strict homologue among the monkeys or apes. In the monkeys and apes, oppositely sexed interactions are actively initiated, or accepted, by males and females that also engage in homosexual mounts, presentations and related patterns of behaviour. Such behaviour is bisexual, rather than being exclusively homosexual.

In conclusion, the evidence reviewed in this chapter indicates that same-sex mounts, and hindquarter presentations, are commonly observed in non-human primates, in free-ranging conditions as well as in captivity. This is especially the case among the anthropoids (monkeys and apes), whereas homosexual behaviour has rarely been documented in the prosimians (lemurs, lorises, galagos and tarsiers). From a phylogenetic perspective, it is also noteworthy that homosexual behaviour occurs much more frequently in species belonging to the Old World anthropoids (superfamilies Cercopithecoidea and Hominoidea), than among the New World monkeys (superfamily Ceboidea).

Same-sex mounts and related patterns are expressed in a variety of social contexts in monkeys and apes. Ritualisation of sexual patterns may have occurred during anthropoid evolution, so that motor patterns normally associated with copulation have been incorporated into socio-sexual behaviour. Examples are discussed in this chapter of the roles played by homosexual mounts and presentations in affiliation, reconciliation, rank-related and other aspects of social communication in monkeys and apes. Sex differences in the expression of mounting and other traits (such as rough and tumble play) typically emerge during infancy and juvenile life, so that males engage in these activities more frequently than females. Exposure of the developing brain in utero to higher levels of testosterone has been shown (e.g. in the rhesus macaque) to masculinise the later expression of mounting and rough and tumble play. These patterns, in turn, serve important functions, underpinning the normal development of socio-sexual behaviour and social relationships within groups of monkeys and apes.

Aside from these functional considerations, there is increasing evidence that pleasurable (hedonic) feedback during same-sex mounting may have facilitated the evolution of this behaviour in the anthropoids. Thus, in Japanese macaques, females position themselves during 'jockey-style' mounts, and use pelvic thrusting movements to maximise tactile stimulation of the vulval, perineal and anal area. Similar patterns of female–female mounting have been observed in some other Old World monkeys, and genital stimulation also occurs during the 'GG-rubbing' displays of female bonobos. Female stumptail macaques sometimes exhibit orgasmic responses during homosexual mounts, as well as during copulations with male partners. Female Japanese macaques may form consortships with other females, and make a series of mounts, similar to the mount series that occur during heterosexual consortships. Further, there is ample evidence that both sexes in many species of Old World monkeys and apes engage in masturbation, so that self-stimulation of the genitalia, as well as stimulation during same-sex partner contacts, provides a source of pleasurable tactile feedback, and reinforces further expression of these patterns.

Observations of the Old World monkeys and apes thus support the conclusion that males and females quite commonly engage in sexual activities with members of the same, as well as the opposite, sex. They exhibit what, in human terms, would be regarded as a considerable degree of bisexual behavioural potential. Kinsey *et al.* (1948) developed a 7-point scale as a means of measuring human sexual preferences. Those individuals who are exclusively or predominantly heterosexual rate themselves as 0 or 1 on the Kinsey Scale.

By contrast, those who rate themselves as 5 or 6 are predominantly, or exclusively, homosexual. Intervening scores (2, 3, and 4) are consistent with various degrees of bisexual preference. Although the majority of adults (>90% in Kinsey's sample) identify themselves as being predominantly heterosexual, there are significant numbers who do not; for males especially, preferences may be much less polarised during the adolescent and teenage years. It is likely that, in common with other Old World anthropoids, humans possess the significant potential to exhibit bisexual preferences. Whether these preferences are actively expressed is affected by individual differences, age-related and cultural factors, including religious taboos concerning same-sex interactions. Examples, such as ritualised homosexuality among the Sambia of Papua New Guinea, sexual liaisons between men and youths in ancient Greece, or the occurrence of facultative homosexuality in all-male institutions (e.g. in prisons) only serve to reinforce the conclusion of Kinsey *et al.* that 'the world is not to be divided into sheep and goats.'

A small percentage of men (approximately 2%–3% in cross-cultural surveys), and a much smaller percentage of women, rate themselves as being exclusively homosexual. The reasons for exclusive homosexuality are likely to involve complex interactions between genetic and experiential factors which are not homologous with the more labile, bisexual preferences seen in some non-human primates. Recent work indicates that genes affecting higher fecundity in women may have pleiotropic influences, and increase a predisposition towards homosexuality in a small proportion of their male offspring.

Summary of main conclusions

- Homosexual patterns of behaviour occur more frequently among the Old World anthropoids (Cercopithecoidea, Hominoidea) than in New World monkeys (Ceboidea) and have rarely been reported in the prosimian primates.

- Although same-sex sexual body contacts are varied in primates, not all of these contacts have a strictly sexual origin or function. Thus caution is required when interpreting such behaviours.
- Homosexual behaviour in primates is at least partly concerned with the ritualised expression of sexual patterns, which serve a variety of social functions.
- Pleasure-seeking is one potential function of same-sex sexual behaviour, especially among the Old World anthropoids.
- Humans retain a significant potential for bisexual behaviour, a trait shared with the rest of the Old World anthropoids.

In the next and final chapter the major findings that have been presented in the previous chapters will be drawn together, with the goal of producing a *Biosocial Model* for the evolution and maintenance of homosexual behaviour in birds and mammals. Some of the practical consequences of the model for various social issues concerning human homosexuality will be also discussed.

A *Biosocial Model* for the evolution and maintenance of homosexual behaviour in birds and mammals

> *Ultimately, though ... it will be necessary to reintegrate the different levels of investigation: the level of genes, synapses, and neurotransmitters, and the level of conscious and unconscious mental processes ... What such a synthesis will ultimately have to say about sexual orientation is difficult to predict.*
>
> Simon LeVay *Queer Science* (1996: 85)

I started this book by introducing animal homosexual behaviour and orientation as an evolutionary paradox. I believe that I am in the position, in this final chapter, to end it with something somewhat more exciting than the Socratic 'one thing only I know and that is that I know nothing', or, as the endocrinologist Louis Gooren put it in a somewhat less dramatic way: 'But it is probably fair to say that we are far away from an understanding of how ... sexual orientation come[s] about' (Gooren 2006: 599). My cautious optimism springs from the results of a wealth of experimental and observational research carried out on diverse taxa by many experts in various fields, from genetics, endocrinology, neurobiology and immunology to behavioural and evolutionary ecology, psychology, anthropology and sociology, that address the causation of same-sex sexuality through various mechanisms and across different levels of analysis. It also springs from the results of the comparative analyses and meta-analytical tests carried out in the various chapters.

In the following discussion of the major findings reviewed in this book, I will highlight those that seem to me to be the most likely contributing factors to the emergence of homosexual behaviour and orientation. In the process I will tend to use a rather optimistic approach. My justification for such optimism is simply that keeping a reasonable hypothesis 'alive' gives researchers the chance of improving the quality of the tests of that hypothesis. If the hypothesis is finally able to withstand the results of a sufficient number of high-quality challenges, then it may be eventually accepted across the board, even if temporarily. Alternatively, the battery of tests may convincingly falsify it and thus consign it to oblivion. If, on the other hand, we knock down the hypothesis after even the flimsiest negative result is obtained (as a naïve interpretation of Popper's falsificationism would suggest we do), then we should not complain if rather soon we are left with nothing but a few 'baby hypotheses' floating in a big waste pool of proverbial bathwater. Much more research is certainly still needed across the board, much variability still remains unexplained, and the fine details of the mechanisms involved are still to a great extent unclear. From the mist, however, a more defined silhouette is starting to emerge.

What is it that we have learned about the biosocial and evolutionary underpinning of homosexuality in vertebrates such as birds and mammals, including humans? From some of the earlier reviews that were mentioned in Chapter 1, we know that same-sex mounting appears early in the ontogeny of many mammals in the context of play and that stages in the oestrous cycle are also correlated with same-sex mounting in female mammals (Dagg 1984), suggesting specific sexual mechanisms modulating the behaviour. Socio-sexual aspects of same-sex mounting such as expression of dominance and coalition formation have also been stressed across a variety of species (Dagg 1984). Similar conclusions of a diversity of functions and contexts for same-sex sexual behaviour have been reached by Vasey & Sommer (2006) in a recent review of the subject. In the case of birds, MacFarlane et al. (2007) have stressed the potential relevance of learning mechanisms, and a sex-specific association of same-sex mounting with mating system.

Primates are at the centre of intense debates regarding homosexual behaviour, with a number of interesting propositions emerging. The distribution of the behaviour across primate taxa is variable, and mainly confined to Old World monkeys, apes and humans, perhaps suggesting a dual contribution of current adaptive value and phylogenetic effects. With the exception of humans, all the other primates seem to express homosexual behaviours only in the context of bisexuality. In addition, social and mating systems may affect the frequency of same-sex sexual behaviour among males and females in primates (Vasey 1995).

Not surprisingly, human homosexuality has gained a prominent position in early and more recent reviews of the subject: not only because it is our species after all, but also because we are one of the few vertebrates known to express exclusive homosexuality. However, the debate about human homosexuality has been, to a certain extent, sidetracked by an excessive, although understandable focus on what are believed to be humans' unique mental faculties (see, for example, Wallen & Parsons 1997; Kirkpatrick 2000), such as those associated with the development of gender identity, for instance. Although we can identify mental states in humans that, in theory at least, are somewhat accessible through language, language can also be used to deceive and make those mental states inaccessible on the one hand (see, for example, Daly & Wilson 1999), and on the other there is no reason to believe that other primates and many other mammals, but also birds, do not possess anything even remotely similar to what we call a 'mental state'. This 'uniqueness' of human homosexuality based on the properties of our mind has been overemphasised, in my opinion, not because human homosexuality is not unique, but because it is not necessarily more unique than homosexuality in other species. One interesting reminder of the perils of taking the uniqueness of human sexuality for granted is the issue of exclusive homosexuality. Exclusive homosexuality has been traditionally described as a unique human trait (see, for example, Gadpaille 1980) but it turns out that we actually share the trait with at least one other mammal species: *Ovis aries* (Chapter 5). Moreover, in both cases homosexual offspring show a trend to be born to a mother that has a high fecundity. By adopting an evolutionary approach, where the focus is on the patterns and processes rather than on any particular species, we gain the clear advantage of bringing together both the uniqueness of each species and also the similarities between the different species in a dynamic and cohesive perspective.

In line with this evolutionary view, recent authors such as Muscarella (2000) have proposed models for the evolution of human homosexuality that involve broader mechanisms also valid for other species, such as sexual segregation of young males, followed by socio-sexual homosexuality associated with affiliation, social bonding and cooperation producing a spread of the trait through interdemic selection. More recently, Muscarella et al. (2001) have suggested female sexual selection for feminised males, who may be more willing to provide parental care, as another possibility. Most of those ideas were first proposed by many earlier authors as indicated in the various chapters and in Table 1.1, and some of the hypotheses are also supported by my own analyses.

The variegated list of hypotheses is reflective of the potential relevance of different specific variables and evolutionary processes. In this book, however, I take on the challenge of integrating those diverse factors into a multicausal evolutionary framework.

Through the list of hypotheses compiled in Table 1.1, the review of published works and the original comparative evidence provided in this book, I have illustrated the many ways in which same-sex mounting across species can be either an adaptation or an exaptation, or it can also be a selectively neutral trait or even, potentially, a maladaptation (e.g. whenever homosexual behaviour is associated with a decrease in inclusive fitness). Behavioural traits that are selectively neutral can be a result of recurrent mutation and can spread in the population via genetic drift, by linkage disequilibrium with selected traits, and by copying. Exaptive traits, in turn, can persist and spread owing to their new adaptive function (for example, mounting that originated in the context of sexual reproduction has been co-opted in the same-sex context for communicative purposes in many social species) and, finally, adaptive traits can spread if they positively affect reproduction of self (e.g. same-sex mounting as expression of dominance) or reproduction of close relatives; for example, same-sex mounting can be one of the few avenues available to engage in sexual behaviours at times of elevated neurohormonal sexual activity for members of a social unit who do not partake in reproduction. Those individuals, by skipping reproduction, may release resources for the reproduction of relatives. Skipping of reproduction in subordinates may be 'voluntary' or it may be 'imposed' by dominants' actions (e.g. aggressiveness) (Creel 2001). In either case, kin selection is expected to facilitate the forfeiting of direct reproduction. Subordinates that do not breed are not always physiologically reproductively suppressed, in which case they are perfectly capable of engaging in same-sex sexual behaviours during the breeding periods of the year (see, for example, Clutton-Brock 1998).

Kin selection, of course is not the only, or even the most important, selective mechanism that could maintain homosexuality in a population. Recent empirical and theoretical work has provided supportive evidence for the role of sexually antagonistic selection in both humans and sheep; i.e. homosexuality in males is associated with increased reproductive success in females within lineages (see Chapter 3).

The hypotheses listed in Table 1.1 also span across various levels of causation. Among the proximate causations of same-sex sexual behaviour, the relative availability of same-sex as compared to other-sex sexual partners during periods of high sexual motivation is a recurrent theme across many taxa.

In the face of such diversity, I have strongly resisted the allure of embracing some simple, all-encompassing concept that, however, explains very little, such as Bruce Bagemihl's *biological exuberance* (Bagemihl 1999). More often than not, a diversity of patterns in evolutionary biology is the characteristic signature of multicausality, and it is a multicausal and integrationist approach that is reflected in this book.

In this final chapter, the major findings will be reviewed in the light of the hypotheses that have been previously suggested to explain homosexual behaviour (see Table 1.1) and I will also endeavour to propose a synthesis in what I call the *Biosocial Model of Homosexuality*. The model will summarise all the mechanisms, at different levels of analysis, that I believe are supported by the evidence available and/or those that are theoretically sound but still require more precise empirical tests. I will then produce a list of novel predictions that can be derived from the model and suggest ways of testing those predictions. I will also mention some areas that should be the focus of more specifically targeted research and will list those potential mechanisms that seem to be of very doubtful importance. The chapter and the book conclude with two sections, the first dealing with the issue of whether homosexuality should be regarded as a pathology or not, and a final section focusing on ethical issues.

The long and winding road to the *Biosocial Model of Homosexuality*

The journey started with an enquiry into the genetic basis of homosexuality. The basic tenet adopted here is that, trivially, all traits, directly or indirectly, are affected by the expression of one or more genes. Homosexual behaviour is very unlikely to be an exception. The issue is, however, to identify and establish the mode of action of those specific genetic mechanisms that determine the expression of homosexual behaviour and the development of a homosexual orientation. It is quite clear from the evidence already available from various taxa (e.g. linkage studies in humans and gene knockout studies in various animals) that there are genes that can affect sexual behaviour and orientation, including human genes. Such genes are spread over various chromosomes, including sex chromosomes and several autosomes. In most cases, it seems reasonable to think that homosexual behaviour is likely to be a polygenic trait, and also that very different mutations could explain the trait in a variety of taxa.

From an evolutionary point of view not one but various mechanisms could plausibly explain the maintenance of genes that in one way or another contribute to the development of homosexual behaviour in any particular species, especially the social ones. Among those highlighted here are:

(1) Linkage disequilibrium between loci affecting homosexuality and loci affecting cooperation: individuals expressing a homosexual phenotype also express a cooperative phenotype.
(2) Pleiotropic effects of single genes, and also
(3) Sexually antagonistic selection for reproductive success (via reproductive rate and/or parental care): homosexuality expressed in one sex is associated with increased reproductive rate, or parental care, in the other sex.
(4) Sexual selection for cooperative individuals in reproduction was also stressed in the case of bisexuality: individuals expressing bisexual behaviours also provide more or better parental care; in the case of exclusive homosexuality, the trait could be also maintained in the population through
(5) Kin selection (homosexuals help close relatives),
(6) Parental manipulation (homosexuals result from ontogenetic effects of parental behaviour, especially during the early periods of development) and
(7) Sibling–sibling conflict (homosexuals result from ontogenetic effects of sibling behaviour, especially during the early periods of development). Finally,
(8) Recurrent mutation can explain those cases where exclusive homosexuality is maintained at low frequency and no direct or indirect benefit of the trait can be proven.

I do not regard all those mechanisms as a set of necessarily alternative hypotheses, we have seen in this book how different cases may be explained by different single mechanisms or a combination of mechanisms. For instance, the case of the homosexual rams reviewed in Chapter 5 could be explained by a mutation or mutations of genes affecting peripheral olfactory tissues but also specific centres in the brain, and from an evolutionary point of view it could be maintained over time by sexually antagonistic selection. By contrast, in humans, pre- and postnatal effects of stress could co-contribute to explaining the ontogeny of at least some cases of homosexuality and also, as was suggested in Chapter 4 for the case of stresses suffered during adrenarche, the transition from homosexuality to transsexuality in males. In addition, same-sex sexual behaviours in humans and other species (e.g. other primates, rodents) may result from behavioural plasticity reflecting plasticity of brain structures and functions in adult individuals. However, as will be shown below, this is not the whole story as far as the causation of human homosexuality is concerned.

One interesting piece of evidence supporting a role for genetic mechanisms underlying homosexuality is the male mutation bias found in both birds and mammals that was mentioned in Chapter 3, a result that nicely fits with the male bias in same-sex sexual behaviour described in both vertebrate classes, and with the higher frequency of exclusive homosexuality found in men compared with women. Other evidence comes from twin studies that strongly suggest a genetic contribution to the inheritance of the trait.

Mutations may produce canalised ontogenies, which eventually could be selected under stabilising selection, or they may produce plastic ontogenies that could be selected under conditions of environmental and social unpredictability, as we have seen in Chapter 4. Both developmental mechanisms will produce very different phenotypes. A highly canalised ontogeny will result in an adult individual that is more likely to have a stable sexual orientation over time; a phenotypically plastic ontogeny will produce individuals with a more labile sexual orientation. In humans, but also other vertebrates, males clearly develop a relatively more stable sexual orientation than females. On the other hand, women are characterised by a considerable degree of plasticity in their preference for sexual partners of a given sex, not only in comparisons between individuals but also within the lifetime of the same individual. This high level of plasticity in sexual orientation is reflected in a higher prevalence of bisexuality among women; this suggests that the specific genetic mechanisms underpinning same-sex sexuality in humans are likely to be somewhat different in males and females.

At a proximate level of analysis, mutations can affect peripheral sensory tissues that in turn may play a crucial role in the expression of same-sex sexual preference in both humans and other animals, without further modifications occurring in the brain (see Chapter 5). That is, changes in the sensory processing of environmental information, e.g. those associated with pheromones, vocalisations, coloration or external body morphology, may be enough to trigger a sexual response towards a same-sex conspecific without any additional modification of the brain centres controlling sexual behaviour. This mechanism may be especially relevant in those cases where the homosexual attraction leads to the expression of sexual behaviours that are more typical of the homosexual's sex (e.g. 'masculinised' gay men, 'feminised' lesbians). Mutations, however, can also directly affect specific brain regions or nuclei that process those peripheral inputs. Here, factors such as apoptosis, neurogenesis, nerve cell architecture (i.e. size of the cell, number of dendrites, number of spikes) and changes in cell membrane receptors (e.g. for steroid hormones) can all help explain sexual orientation variability in both sexes within a species, but also between species, as was argued in Chapters 3, 4 and 5. For instance, there is a very impressive body of evidence that links the well-known ability of some rodents to be plastic in their sexual behaviour with a remarkable neurogenic and neuro-architectural plasticity of their brain (see Chapters 4 and 5). Same-sex sexual patterns that could be described as either 'inversion' (e.g. 'feminised' gay men, 'masculinised' lesbians) or 'third-sex' are more likely to be the result of central nervous system modifications than of changes in the peripheral sensory tissues that simply identify a specific target individual for sexual attraction.

Specific mutations that affect loci controlling developmental programmes may also be responsible for phenomena such as neoteny. Through neoteny juvenile traits are retained into the reproductive adult ages of a derived species. Various juvenile traits that are of importance in the context of the evolution of homosexuality were mentioned in Chapters 4 and 5: (a) the neurogenesis, synaptogenesis and decreased apoptosis that were mentioned above, and (b) bisexual sexual play behaviour. Indeed, under various potential selective mechanisms, canalisation of neotenic traits may favour the expression of a great diversity of sexual behaviours that are characteristic of the juvenile stages of development. That is, although from an *ontogenetic* perspective juvenile homosexual play is a route to efficient heterosexuality in most species, from a *phylogenetic* perspective neotenic processes, by retaining those juvenile traits past the period of sexual maturation throughout evolution, may produce adult homosexuals or bisexuals. In the adult, homosexual behaviours may then undergo further adaptations related to social life. A potential example of this process may be found in the bonobo, where juvenile same-sex sexual behaviours are as common as they are in the congeneric chimpanzee, but in adults they occur much more frequently than in chimpanzees, being expressed in specific socio-sexual contexts. I explained in Chapter 4 how the bonobo also seems to display more neotenic traits than the chimpanzee, with bonobos' sexuality being effectively bisexual.

Sex hormones have been especially targeted in studies of the early development of homosexuality. In fact steroid hormones are significantly involved in the masculinisation/de-masculinisation and feminisation/de-feminisation of brain structures and functions and also behaviour, especially during early ontogeny, although the importance of direct genetic control is also becoming clearer with recent research. With regard to humans, early overexposure to androgens has a much clearer effect on the development of masculine gender role, but also on the development of homosexual behaviour in women, than lack of such exposure (or underexposure) has on feminine gender role and homosexuality in men (for examples, see the various studies reviewed in Chapters 4 and 5). The latter effect is mainly a result of testosterone production in men through various alternative pathways that can buffer testosterone synthesis against specific mutations. Males, on the other hand, may be particularly affected by prenatal exposure to stress, which may exert an effect on brain development through the mediation of adrenal steroids, glucocorticoids, catecholamines, opioids and/or serotonin. Indeed it is prenatal, perinatal but also early postnatal stress that I suggest to be at the core of the proximate mechanisms explaining the *Fraternal Birth Order* and young parent effects described for male homosexuality in humans (see Chapter 5). In addition, I also mentioned in Chapter 4 that stress can even have a role in the development of homosexuality in humans during the prepubertal period when the adrenals start secreting steroids (adrenarche) even before the gonads do (gonadarche). At this point in time all those mechanisms still require further testing in humans in order to account for the great variability observed in the male and female populations.

Early ontogenetic programmes that retain a degree of plasticity, perhaps as a result of neotenic processes, are also subject to the vagaries of changes in the internal and external environments. It is quite clear from the review of early hormonal effects on development that sexual orientation, and even more commonly gender role, could be modified in various vertebrates if the hormonal milieu varies during specific critical periods. Testosterone, in general, tends to masculinise the development of sexual behaviour, either directly, through activating androgen receptors in specific brain tissues, or indirectly through its aromatisation into oestrogens and the effect of the latter via oestrogen receptors (see Chapter 4). This is valid not only for males, but also for females, including women; in the latter, masculinisation may lead to homosexuality or heterosexuality with a touch of masculine gender role, depending on the intensity of the effect. Additional hormonal effects on the early development of a homosexual sexual orientation could be also explained by the endocrine consequences of stress (both in utero and postnatal) as I have already mentioned. The two chief social stressors during postnatal development are represented by interactions with parents (e.g. parental manipulation) and with other members of the social group, siblings in particular (e.g. sibling–sibling conflict). Parent–offspring–offspring interactions are at the very core of the *Synthetic Reproductive Skew Model of Homosexuality*. This model brings together not only *Reproductive Skew* theory, and both sexual and socio-sexual aspects of homosexuality (Chapter 8), but it also represents a synthesis of what in the study of human evolutionary psychology are known as the *Family Dynamics Model* (Sulloway 1996) and the *Confluence Model* (Zajonc *et al.* 1979) (see Chapter 6).

The *Family Dynamics Model* tends to emphasise competitive interactions among siblings, whereas the *Confluence Model* emphasises cooperative interactions. In developmental psychology cooperative sibling–sibling interactions are encapsulated in the concept of *social learning*, whereas competitive interactions are considered in the context of (and are expected to lead to) *sibling de-identification* (see, for example, Whiteman *et al.* 2007). Whiteman *et al.* (2007) suggest that the two mechanisms are complementary in the behavioural development of humans at a proximate level of analysis. I suggest, following the results of the comparative analyses carried out in Chapter 8, that cooperative interactions and competitive conflicts within a social group are also complementary factors in the evolution of same-sex sexual behaviour across a variety of taxa (phylogenetic perspective), but their effects on the development of homosexuality at the individual level (ontogenetic perspective) are not only through learning mechanisms but also, and indeed especially, through other kinds of modifications of the early development of the brain – such as those potentially caused by stress and exposure to gonadal steroids – being also constrained by genetic predispositions.

Whiteman *et al.* (2007: 657) provide a very nice description of this cooperative/competitive synthesis at a proximate level, when they refer to older siblings eliciting 'intense reactions from their younger brothers and sisters, and that these reactions are both positive and negative. Such a pattern corresponds to the characterisation of siblinghood as a love-hate relationship'. Moreover, the *biosocial* approach adopted here also stresses the need to integrate the mechanisms that characterise brain and therefore behavioural plasticity, such as learning, with those that tend to limit such plasticity within an individual but can explain distinct patterns of behaviour across individuals. For instance Whiteman *et al.* (2007) were puzzled by discovering in their study that some of the younger siblings did not fit either the *social learning* or the *de-identification* model, instead they seemed to develop their own way without being significantly influenced by siblings throughout their postnatal life. In order to make sense of this diversity, I suggest that within-family effects on behavioural development should be explicitly expanded to also include the genetic and early pre- and perinatal effects emphasised in this book. Sibling–sibling–parent postnatal interactions do not produce different phenotypes through effects on a plastic brain as if the brain were a piece of play dough. Instead, the adult phenotype is also constrained by the specific genetic make-up of cells, by the degree of canalisation of developmental programmes and by processes occurring during earlier (e.g. prenatal, prepubertal) phases of development.

Nevertheless, social stressors operating on the early development of young vertebrates, including children, could also affect areas of the isocortex involved in learning; here I have also suggested the potential for a mediating role of both dopamine and serotonin (Chapter 5). Dopamine, for instance, could facilitate the learned association between same-sex stimuli and same-sex sexual behaviour via its link with the oxytocin and arginine vasopressin system that affects the conjunct development of both social bonds and sexual behaviour. Why should the ontogeny of a social species, including our own, be so plastic as to produce adult phenotypes that display variable sex roles and a homosexual or bisexual sexual orientation?

It is possible that the original selective context of plasticity in gender role and sexual orientation may have been linked, on the one hand, to the benefits of within-group cooperation that characterises societies with relatively less masculinised individuals, and on the other with the benefits conferred on specific males and females (dominants, leaders) within that society by the skew in control of resources acquired through a more competitive personality (e.g. as expressed through socio-sexual mounting of subordinates, for instance). Subordinates in this case may benefit by enjoying the safety of the group. That is, there are clear evolutionary routes that can stabilise same-sex sexual behaviours via the expression of both 'masculinity' and 'femininity' in both males and females when those sexual behaviours are expressed in a socio-sexual context. Hence the claim that homosexuality should not be exclusively equated with gender inversion. Homosexuals can be masculinised, feminised, hypermasculinised, hyperfeminised, gender-shifted or 'third gender' in behaviour, physiology and/or neuroanatomy. Masculinised males can be and are involved in same-sex interactions, as are feminised females, and the reason for their engaging in same-sex intercourse requires an explanation, as does the involvement of feminised males and masculinised females.

Importantly, homosexual behaviour resulting from selection for behavioural plasticity is more likely to be manifested in a bisexual rather than an obligate homosexual context, as in the former case the socio-sexual functions of same-sex sexual behaviours would not be incompatible with achievement of reproduction. Plasticity in the combination of gender role and sexual orientation would be selected in social, long-living animals if who is going to or not to reproduce is affected by some frequent events such as death or sickness of a dominant. Feminised or masculinised individuals (males or females) may find themselves in the position of dominance after going through a stint as subordinates; if so, homosexual behaviours may be retained throughout via their adaptive value in socio-sexual contexts to both dominants and subordinates. In addition, being a dominant does not automatically confer an exclusive control over reproduction, therefore various strategies that combine diverse sex roles and variable degrees of bisexuality may all be alternative routes to achieving some direct fitness benefits by specific individuals in the group. The Old World primates provide some of the best examples for the above scenario.

This is all very well if same-sex sexual behaviour is expressed in a bisexual individual. The issue, however, is how could any mutation that produces an *exclusive* homosexual be retained in the population, even if at low frequency? I address this issue in the 'Bisexuality vs. exclusive homosexuality' section below.

In adults, same-sex sexual behaviour is controlled by central nervous system mechanisms whose action is triggered by sex-hormonal stimuli (e.g. during breeding periods) or independently from sex-hormonal stimuli (e.g. during non-breeding periods). Although during breeding periods circulating sex hormones may activate same-sex sexual behaviour, the specific short-term expression of the behaviour may be under direct neuronal control, as mentioned above and explained in Chapter 5. The activational role of hormones in adults is uncontroversial. However, hormones may also have an organisational role in the expression of homosexual behaviour in adult individuals, via effects on brain plasticity. This proposition is only now being emphasised in the literature. I provide arguments and evidence from the literature in Chapter 5 that are consistent with such an organisational role in adults; however, I also hope that future studies will throw more light on this aspect of homosexuality. The possibility of an organisational role of sex hormones in adult male-to-female and female-to-male homosexual transsexuals, with regard to some cognitive abilities, has been recently shown experimentally by van Goozen *et al.* (2002).

Ecological, but also social, constraints related to group living and mating system may lead to sexual segregation as shown in Chapter 7. Sexual segregation may be a socioecological factor that is linked to the evolution of homosexual behaviour in mammals, but my comparative analyses indicate that it has not been equally relevant in the evolution of avian same-sex mounting. Care should be taken not to conclude from this result that, therefore, 'sexual segregation at young ages should be conducive to homosexuality in adult mammals'. The comparative result is an evolutionary correlation that has nothing to do with any alleged *ontogenetic effects* of sexual segregation on the development of exclusive homosexuality in adults. In most cases, same-sex mounting occurring during periods of sexual segregation in mammals is a combined result of sexual and socio-sexual functions, and in most circumstances it is finally conducive to heterosexual intercourse when the opportunity arises. In fact, I propose that the evolutionary link between homosexuality and sexual segregation mentioned in Chapter 7 is likely to be the product of segregation selecting, over evolutionary time, for socio-sexual functions of same-sex mounting favoured by availability of same-sex conspecifics during periods of heightened sexual motivation. This is seen, for instance, in the strong tendency for female mammals to establish social relationships with other females and to also engage in same-sex mounting. Incidentally, this also suggests that the ability of women to be more flexible in their sexual orientation than men may be, at least partly, an ancient mammalian trait. In other words, sexual segregation and sociality may have favoured the evolution of affiliative or dominance same-sex mounting, but most adult individuals who engage in such behaviours finally mate heterosexually and reproduce if they have the opportunity to do so. Whenever such experiences contribute to the development of an exclusive homosexual sexual orientation they are likely to do so in the context of a concomitant mutation or specific early (pre-, peri-, postnatal) ontogenetic processes that already contribute to a significant extent to such development.

Once the ability to express same-sex sexual behaviour has evolved, then the actual expression of the behaviour may be also activated by factors such as the current availability of sexual partners, which may explain some aspects of the *intra-specific* variability in the frequency of homosexual mounting observed in various species (see Figure 8.6).

Brain plasticity that can explain cases such as that of political lesbians seems to contradict the above proposition. If exclusive homosexuality could be a result of a conscious decision, as it seems to be the case in political lesbians, why should not it develop from periods of sexual segregation at any stage of development? The fact is that the development of sexual orientation in mammals is obviously buffered, to a considerable degree, against early homosexual experiences, as has been abundantly shown in this book. If this was not the case the frequency of exclusive homosexuals would be much higher than it actually is in all the species experiencing sexual segregation. Many species undergo periods of sexual segregation (homosociality) and yet most individuals develop a heterosexual or at most a bisexual sexual orientation. The proposal that brain plasticity could explain the onset of homosexuality at adult ages points to a state of plasticity of behaviour that in the case of sexual orientation has probably been adaptive in the context of bisexuality and that is consistent with the case of political lesbians. In consequence, as individuals may consciously decide to make the transition from heterosexual to homosexual, never having had any previous indication to do so, for as long as their brain remains plastic and effectively bisexual they will remain free to either reverse that decision or keep on living their new homosexual life. As I have already mentioned, in humans this property is especially developed in women (see also Diamond 2008b).

The above 'freedom of choice' notwithstanding, evolutionary theory predicts that, in the long term, a complete voluntary withdrawal from reproduction will be confined to a minority of individuals in the population; in the same way, effective celibacy among some religious orders is only representative of a small proportion of the overall male and female population. The behaviour, however, could become more widespread if homosexuality and homosociality are also associated with reproductive success: e.g. in the case of two females if a male inseminates one or both of them and then leaves, or in the case of two males if a female leaves a child fathered by one of the males in the care of the homosexual couple. Along the same line of argument, evolutionary theory also predicts that natural selection against exclusive homosexuality should relax at post-reproductive ages, a prediction that is in accord with data available for women (e.g. Diamond 2008a), whereas men, who are still able to sire offspring until old age, do not show a marked increase in the percentage of homosexuals with age (Kinnish *et al.* 2005). On the contrary, as stated by Kertzner (2001: 87): 'Many men, did, however, describe a decrease in the relative importance of homosexual identity over their life course. This may become particularly evident during midlife'.

In some species complete withdrawal from reproduction is not restricted to a small subset of the population, but it is a characteristic of the majority. These are the eusocial species (e.g. honeybees *Apis mellifera*). Eusociality, however, is extremely rare in mammals and unknown in birds and, at this point in our evolution, we humans are definitely not eusocial: exclusive homosexuals are **not** a 'caste' such as the honeybee workers (see also the relatively low percentage of asexuals in humans:

Bogaert 2004, 2006b; Prause & Graham 2007; Brotto *et al.* 2009). At the most, if kin selection has contributed to the evolution of homosexuality, homosexuals could be the equivalent of the 'helpers at the den' described in other social mammals and the 'helpers at the nest' described in some group-living birds (see chapter 8 and Vasey & VanderLaan 2010).

The sexual segregation mentioned above also highlights the issue of the distinction between homosociality and homosexuality. As I have stressed in previous chapters, homosociality and homosexuality are two distinct phenomena that may or may not occur together. I do not regard homosocial individuals as being necessarily homosexuals unless they display same-sex sexual behaviours; conversely, homosexual individuals are not always homosocial. That the two variables are independent is clearly demonstrated by the association, at least in some human populations, of heterosexual male homosociality with homophobia (Britton 1990). Therefore, whenever individuals produce statements such as that of a 25-year-old woman interviewed by Lisa Diamond (2008b: 108): 'I find that it's a lot easier to bond with women, and there's also something really deeply satisfying, there's an understanding that exists with women that doesn't automatically exist with men . . .', what we may be hearing here is the expression of an ancient 'voice' of *homosociality*, so widespread among social mammals (Chapter 7), more than an otherwise 'hidden voice' of *homosexuality*. Once homosociality is established, however, the way is paved, as mentioned above, for the evolution of homosexual behaviours in a socio-sexual adaptive context such as affiliation but also dominance; or for the expression (in the short-term) of same-sex sexual behaviours triggered by specific neuroendocrine states associated with the reproductive cycle.

Female homosociality seen as a potential stepping stone for the further expression of female same-sex sexual behaviours would further benefit from a mating system that is prevalently polygynous. Polygyny, in fact, is common among mammals. Moreover, even among humans, although the institution of marriage across cultures tends to be prevalently monogamous, the actual mating system is mainly polygynous ('general polygyny': 30.6%; 'slight polygyny': 51.6%), with monogamy (16.7%) and polyandry (1.1%) representing a smaller percentage of the human mating systems (Marlowe 2000).

What kind of specific *selective* mechanisms could be responsible for the evolution of homosexuality? As I have already indicated, nine major selective mechanisms are considered in this book: interdemic selection, kin selection, reciprocal altruism, mutualism, parental manipulation, sibling–sibling competition, sexually antagonistic selection, natural selection and sexual selection. Of course, I am not suggesting that they are all always simultaneously active, or have all been simultaneously active in the past for any given species. What I am suggesting instead, is that each one of those selective mechanisms should be evaluated for each species as they may be valid explanations for the evolution of same-sex sexual behaviour in a variety of cases.

Just for illustrative purposes we may consider the following hypothetical evolutionary scenario: neoteny leading to juvenile-like sustained neurogenesis could well produce a derived species that has a larger, more complex and more plastic brain than its ancestor. More complex and plastic brains, needless to say, could be quite adaptive in both the context of complex social environments and the context of unpredictable habitats. *Natural selection* here will be at play and same-sex sexuality may initially arise as a byproduct of the 'juvenilisation' of the brain. Some examples of this process may be found among rodent and primate species, for instance. Following this, however, homosexual behaviours can be subsequently co-opted for competitive and/or cooperative functions. After all, a complex brain can express both aggressive and affiliative behaviours in a social environment. It is also possible, of course, that homosexual socio-sexual behaviour may be a derived state of an initial heterosexual socio-sexuality, an evolutionary process that, at least in some species, might have been also facilitated by neoteny, as young male and female social mammals also play-mount each other. If a complex brain is a trait that enhances acquisition of resources, then it could also be under *sexual selection* (e.g. mutual male and female choice). If by not reproducing, or reproducing less (functional bisexuality), homosexuals release resources for the more successful reproduction of relatives, then *kin selection* would be relevant. Any linkage disequilibrium between loci affecting same-sex sexual behaviour and loci affecting intraspecific cooperation could spread in the population via *interdemic selection*. Cooperative, homosexual dispersers that join groups of non-relatives could enhance their survival via *reciprocal altruism* or *mutualism* (see, for example, the case of dispersing bonobo females). Here, however, a degree of reproductive success is required (i.e. functional bisexuality): after all, survival alone is insufficient for the spread of genes, reproduction of self and/or close relatives is ultimately needed. The above evolutionary mechanisms are likely to be relevant for the evolution of same-sex sexual behaviours among primates, for instance. If heterosexual female relatives of homosexual males have higher reproductive success than heterosexual female relatives of heterosexual males, then *sexually antagonistic selection* may be at play. Exclusive male homosexuality in humans and sheep could be, at least in part, explained by this mechanism. We can see from these hypothetical evolutionary scenarios that interdemic selection, kin selection, reciprocal altruism, mutualism, natural selection, sexually antagonistic selection and sexual selection could quite well be interconnected and play a role, alone or in various combinations, in the evolution of homosexuality and bisexuality. In my view, treating those mechanisms as a set of necessarily alternative, mutually exclusive hypotheses is a mistake.

For instance, recent empirical and theoretical research on the evolution of human (especially male) homosexuality has emphasised *sexually antagonistic selection*, *sexual selection* and *kin selection*. The potential synergistic effects of various kinds of selection mechanisms that I have just mentioned can be clearly illustrated for the case of *sexually antagonistic selection* and *kin selection*.

The operation of *sexually antagonistic selection* results in homosexual offspring being born of mothers with elevated reproductive rates. As reproductive rate increases and the family increases in size, competition will also increase among family members for resources. If dispersal is limited and therefore philopatry is selected in one of the sexes (in humans, males are the philopatric sex), the inclusive fitness of members of the family group may increase if some of those members stop reproducing (e.g. homosexuals), thus making more resources available for the successful reproduction of close relatives (*kin selection*), with inclusive fitness increasing further if, apart from not reproducing, or reproducing less, homosexuals also actively help their kin through the provision of alloparental care. Therefore, sexually antagonistic selection leading to the production of male homosexual offspring in large families may also pave the way for increased cooperative behaviours on the part of those homosexuals under kin selection. In some modern human societies, however, the level of helping received by heterosexual family members from homosexual family members is subject to the caveats already mentioned in Chapter 3, where I discussed the issue of prejudice and discrimination against homosexuals. Homosexuals are more likely to help their kin when they live in a cultural environment where they are not treated as pariahs and their close relatives are not stigmatised by the rest of the community. Incidentally, this tends to be valid for everyone, not just homosexuals. An intriguing corollary of this scenario is that what appears to be evolutionarily maladaptive is not homosexuality as such but the negative discrimination to which homosexuals are subjected.

The *Synthetic Reproductive Skew Model of Homosexuality*

The first partial synthesis produced in this book is the *Synthetic Reproductive Skew Model of Homosexuality*, which was thoroughly tested in Chapter 8 through comparative analyses. With this model three major areas of research in the evolutionary

Table 10.1. Comparative tests of the *Synthetic Reproductive Skew Model of Homosexuality*
The list includes only the statistically significant results obtained in each test

	Birds			Mammals		
Variables	Correlation	Model supported?	Variables	Correlation	Model supported?
Comparative analyses of standardised independent contrasts						
Log-social unit size vs. Same-sex mounting	Negative	Yes	Log-social unit size vs. Same-sex mounting	Negative	Yes
Mating system vs. Plumage sexual dichromatism	Positive	Yes	Ecological constraints vs. Same-sex mounting	Negative	No
Adult sex ratio vs. Same-sex mounting	Positive	Yes	Relatedness vs. Same-sex mounting	Negative	Yes
Sociality vs. Same-sex mounting	Positive	Yes	Mating system vs. Same-sex mounting	Positive	Yes
Dominance mount vs. Sociality	Positive	Yes	Adult sex ratio vs. Sex involved	Negative	Yes
Additional results from the complete correlation matrix of standardised independent contrasts						
Sex involved vs. Same-sex mounting	Increased evolutionary trends towards same-sex mounting are associated with increased female involvement in same-sex mounting		Sex involved vs. Same-sex mounting	Same trend as in birds of greater involvement of females with increased trends to same-sex mounting	
			Ecological constraints vs. Social bonds	As ecological constraints increase (i.e. cooperative breeding increases), social bonds also tend to increase	

Path analyses

(a) The model fits better the bird than the mammal datasets.
(b) But in both cases, the *Synthetic Model* (i.e. *Reproductive Skew* core variables + mainly sexual variables + mainly socio-sexual variables) has a better explanatory capacity than any of the two more restricted models (core variables + mainly sexual variables; core variables + mainly socio-sexual variables).

Direct comparisons between birds and mammals

Variable	Result (M = mammals, B = birds)
Prevalence of same-sex mounting	M > B
Prevalence of polygamous mating systems	M > B
Sociality	M > B
Ecological constraints	M > B
Sexual dimorphism	M > B
Sex ratio	M (female-biased); B (male-biased)
Sex involved	In both M and B, males > females but the bias is smaller in M
Within-group genetic relatedness	M > B

biology of homosexual behaviour are brought together: *Reproductive Skew Theory* and sexual (mate choice, incest avoidance) and socio-sexual (dominance, cooperation) aspects of behaviour, this being a result of what in Chapter 1 I called an *integrationist* approach. Table 10.1 summarises the major statistically significant results of the tests of the model. Although the *Synthetic Model* is generally corroborated by the comparative evidence available from both birds and mammals, it is quite clear that same-sex sexual behaviour in birds and mammals is also the result of taxon-specific evolutionary trends. That is, during the independent evolution of birds and mammals different aspects of the causative mechanisms of homosexuality have been emphasised. Broadly speaking, the model is more successful at explaining same-sex mounting in birds than in mammals, even though same-sex mounting is more prevalent in mammals than in birds. This clearly indicates that same-sex sexual behaviour is a more complex phenomenon in mammals than it is in birds. The difference in the occurrence of same-sex sexual behaviour between the two taxa is in association with a greater degree of sociality and cooperative breeding in mammals, which can favour both sexual and socio-sexual expressions of same-sex mounting. The greater prevalence of polygamy among mammals than birds also favours same-sex mounting, which is especially manifested during periods of segregation of the sexes in monosexual groups. Differences in tertiary sex ratio biases, with sex ratios being more female-biased in mammals, can also contribute to explaining the greater involvement of female mammals than female birds in same-sex mounting, via both sexual and socio-sexual aspects of behaviour.

In spite of its good performance, the *Synthetic Reproductive Skew Model of Homosexuality* is unlikely to provide a full representation of all the major mechanisms involved in the evolution of homosexual behaviour, especially in mammals. I am therefore in the position to introduce the last contribution of this book to the evolutionary understanding of homosexuality: the *Biosocial Model of Homosexuality*.

The *Biosocial Model of Homosexuality*

In Chapter 1 the evolutionary paradox of homosexuality was defined in this manner: *If sexual behaviours such as mounting or genito-genital contact have originally evolved in the context of reproduction, why is it that they occur between members of the same sex where those behaviours cannot obviously lead to immediate fertilisation?* I propose to resolve this paradox by using multiple levels of analysis and a multicausal approach, through a *Biosocial Model of Homosexuality*. The model is summarised in Figure 10.1; as for the *Synthetic Reproductive Skew Model of Homosexuality*, it can be seen that it is *integrationist* in nature. As shown in Figure 10.1 the *Biosocial Model* incorporates the *Synthetic Reproductive Skew Model of Homosexuality* but it also adds novel causative factors such as altriciality, neoteny, social complexity, encephalisation and brain plasticity, early ontogenetic effects and the effect of specific mutations.

Expressed in a brief narrative, the *Biosocial Model of Homosexuality* provides the following resolution to the evolutionary paradox of homosexuality: same-sex sexual behaviours occur across a variety of taxa as a result of a diversity of genetic and neuro-endocrinological mechanisms that act at various levels of causation, from proximate, to ontogenetic, adaptive and evolutionary and that can be affected by aspects of the external environment, the social milieu in particular. More specifically, causation of homosexuality may occur at the different levels in the following manner.

(a) **Proximate causation**: Specific mutations affecting either peripheral sensory tissues, brain areas or both may bias sexual preferences towards members of the same sex. In addition, (i) neuroendocrinological states that characterise the breeding condition of animals, such as when sex hormones reach their highest levels in circulation, along with (ii) socio-sexual functions of sexual behaviours and (iii) the neuroendocrinological link between the reward system (e.g. affected by dopamine), the modulation of the

414 A Biosocial Model for birds and mammals

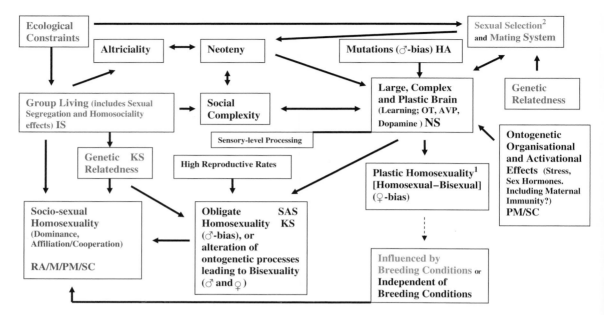

Figure 10.1. A *Biosocial Model* for the evolution and maintenance of homosexual behaviour in birds and mammals. [1]Not highly canalised ontogenetic processes, brain plasticity in adults. [2]Includes all sexual variables that in the *Synthetic Reproductive Skew Model of Homosexuality* were represented by Pair Bond, Sex Ratio, Reproductive Physiology, Mate Choice and EPFs. Variables in colour are components of the *Synthetic Reproductive Skew Model of Homosexuality*: red, core *Reproductive Skew* factors; purple, mainly sexual factors; blue, mainly sociosexual factors. IS, interdemic selection; KS, kin selection; RA/M, reciprocal altruism/mutualism; PM, parental manipulation; SC, sibling–sibling conflicts; NS, natural selection; SAS, sexually antagonistic selection; HA, heterozygote advantage; OT, oxytocin; AVP, arginine vasopressin. Mutations may affect brain centres, peripheral sensory tissues or both, but note also that mutations may also affect other components of the model such as sociality, neoteny, altriciality and reproductive rate. Continuous-line arrows may indicate either (a) proximate causation effects (for example, stress during ontogeny may affect the structure and function of a complex and plastic brain) or (b) ultimate causation effects (for example, sexual selection, evolutionary neotenic processes and a complex social environment may all contribute to the evolution of large and plastic brains). The dotted-line arrow indicates categorisation; for example, plastic homosexuality may be of two kinds, one that is manifested under breeding conditions (e.g. breeding season) and another that is independent of breeding conditions (e.g. non-breeding season). See also colour plate.

activity of brain centres controlling both sexual and bonding behaviours (e.g. via oxytocin and arginine vasopressin action) and associative learning, may co-contribute to the manifestation and mode of operation of same-sex sexuality across individuals and through time (e.g. over the year). More exclusive neurological mechanisms may be in operation in cases of same-sex sexuality occurring during periods of low circulating sex hormones.

(b) **Ontogeny**: Pre-, peri- and early postnatal, including prepubertal, effects of sex hormones and stress hormones seem to be especially important mediators of ontogenetic changes in brain structure and function leading to same-sex sexuality. Some major sources of stresses that could affect the sexual development of young members of a social group are the parents and other members of the group, siblings in particular.

(c) **Adaptation**: In its bisexual manifestation, same-sex sexuality can be a clear adaptation in social organisms in the socio-sexual contexts of dominance (competition) and affiliation (cooperation). In its exclusive homosexual manifestation it could be a result of, for instance, selection for increased indirect fitness, selection for increased reproductive rate in the other sex (e.g. females in the case of male exclusive homosexuality), or a coping strategy in response to interactions with dominant individuals.

(d) **Evolution**: Under the effect of ecological constraints leading to the evolution of sociality, altriciality may have been selected that led to the development of a large and complex brain, with neotenic evolutionary trends enhancing this process in some taxa. Altricial, neotenic species are expected to have the largest, but above all the most plastic brain, which could then sustain a degree of intraspecific variability of sexual behaviours, and even changes in sexual orientation within an individual throughout his/her life. Exclusive homosexuality may be maintained at low frequency by recurrent mutation, sexually antagonistic selection, heterozygote advantage, or kin selection, parental manipulation and sibling–sibling conflict. Bisexuality has a broader spectrum of selectivity compared with exclusive homosexuality, being easily maintained in the population via all the above mechanisms, but also by sexual selection, mutualism and reciprocal altruism.

I emphasise in this book that seeking to point our finger at a single and hopefully simple mechanism that can explain *all* that we can see in terms of homosexual behaviour and orientation across taxa is an illusion. There are various mechanisms that operate at different levels of analysis; they are interconnected at some points but run in parallel at others, and therefore the end homosexual phenotype could be achieved by following different causal routes.

The model suggests that an important evolutionary route that could derive current social functions from initially reproductive behaviours is neoteny, which may be also responsible for elevated encephalisation and eventually increased learning abilities in humans. With increased learning, the possibility of homosexual behaviour spreading through cultural evolution also becomes likely. My suggestion, however, is that such a cultural transmission occurred and is likely to continue to occur more in the context of bisexuality than in that of strict or exclusive homosexuality. Exclusive homosexuality could become more widespread in humans, however, if homosexual couples do reproduce (e.g. through a brief heterosexual sexual encounter, or artificial insemination). Neoteny is also a major evolutionary route to adult endocrine and neurological plasticity that favours the expression of sexual feminine and masculine behaviours in a variety of contexts during interactions between conspecifics. Interestingly, neoteny could perhaps even explain evolutionary patterns towards increased homosexuality at different phylogenetic levels; e.g. from effects within the evolution of anthropoids to more ancient effects in the evolution of mammals (see the higher prevalence of same-sex mounting in mammals as compared with birds).

As I already explained for the case of the more restricted *Synthetic Reproductive Skew Model of Homosexuality*, the *Biosocial Model* is introduced here in a conceptual manner and its main value is heuristic. Still, some qualitative predictions will be drawn from it that can be empirically tested. The *Biosocial Model* clearly includes the *Synthetic Reproductive Skew Model* as a component, as is shown in the coloured sections of Figure 10.1. Admittedly, the *Biosocial Model* is a product of a somewhat optimistic stance, as I mentioned at the beginning of this chapter, in the sense that I included a number of mechanisms that have received some empirical support and others that look plausible although they still need to be tested. I believe that this approach is safer and more productive than the alternative of knocking down every single hypothesis on the basis of even the flimsiest, apparently non-supporting evidence. The empirical process of falsification of a model, as suggested by Karl Popper, requires repetition of tests, control of confounding variables, appropriate methodologies and so forth. It is possible that the *Biosocial Model* may be pruned in the future of some of its complexity. My personal bet, however, is that even more complexity will be added. Time and careful empirical and theoretical work will tell.

I offer the following list of novel predictions derived from the *Biosocial Model* that could be tested through comparative analyses. Clearly, this is just a sample of the specific predictions that can be derived from the model.

(a) Obligate or exclusive homosexuality is more prevalent in species or lineages possessing higher mutation rates per generation. This leads to the expectation that experimentally increased mutation rates (e.g. in invertebrate models) may lead to the production of some exclusive homosexuals.
(b) Across taxa, especially mammals, same-sex sexual behaviour is more prevalent among social, neotenic species.
(c) Higher brain plasticity in adults, as illustrated by the capacity of neurons to undergo neurogenesis and/or synaptogenesis in areas of the brain that control sexual behaviour, is associated with bisexual behaviour across taxa.
(d) Social and/or neotenic species are more likely to respond to prenatal, perinatal or postnatal *stresses* by increasing their level of bisexuality and/or the combination of gender roles (masculine and feminine) into behaviourally androgynous individuals.
(e) Social and/or neotenic species are more likely to respond to prenatal, perinatal or postnatal changes in *exposure to sexual hormones* (e.g. gonadal steroids) by increasing their level of bisexuality and/or combination of gender roles (masculine and feminine) into behaviourally androgynous individuals.
(f) In traditional human societies, exclusive homosexuals are more frequent within large, cooperative families living in habitats where dispersal is less likely. That is, exclusive homosexuality is expected to be associated with high reproductive rates and philopatry.

Bisexuality vs. exclusive homosexuality

In which ways can the *Biosocial Model* contribute to the understanding of the evolutionary appearance and maintenance of, on the one hand, the homosexual behaviour expressed in a bisexual context and, on the other, same-sex sexual behaviour expressed in an exclusive homosexual context? As I have already indicated in this chapter, same-sex sexual behaviours expressed within a bisexual sexual orientation are relatively easy to explain both proximately due to developmental effects, sociosexual functions, and sexual neuroendocrinological mechanisms reviewed in Chapters 4, 5, 8 and 9, and ultimately due to the ability of bisexuals to also exact direct reproductive success when they mate heterosexually. Exclusive homosexuality is, of course, more difficult to explain. The *Biosocial Model* suggests two major proximate routes to exclusive homosexuality: one that is represented by specific mutations that canalise development towards the production of exclusive homosexual individuals, and the other that is the effect of sufficiently intense prenatal or postnatal environmental pressures (e.g. social stress, environmental and maternal steroids) that occur persistently over relevant periods of the development of the individual and that affect those ontogenetic programs that are not strongly canalised. Both mutations and ontogenetic effects can lead to the maintenance of exclusive homosexuality in the population via various evolutionary mechanisms: kin selection, sexually antagonistic selection, heterozygote advantage, parental manipulation and sibling–sibling conflict in an adaptive context, and via recurrence of mutations in a non-adaptive context. On the other hand, I suggest that exclusive homosexuality maintained by brain plasticity during the adult life (e.g. homosexuality adopted as a result of a political choice) is a relatively unstable trait (i.e. a trait that could change at any time during the life of an individual), with such plasticity being more common among women than men, whereas exclusive homosexuality determined by the early expression of specific genes and early developmental effects tends to be more stable over time.

These of course are very broad conclusions derived from a still coarse understanding of the precise mechanisms contributing to the development of exclusive homosexuality. For this reason, giving precise answers to more specific questions – such as why the frequency of male exclusive homosexuality in humans is around 3% and that of male exclusive homosexuality in sheep around 8% – is still beyond our capacity. The various mechanisms included in the *Biosocial Model* can explain why the frequencies are relatively low – because exclusive homosexuality carries high costs in terms of direct fitness – and why the bias is towards males; what is still unclear is the exact nature of that 5% difference. On the other hand, we have also seen in Chapter 5 that the frequency of bullers in domestic cattle is about 1.5%–3.5%. Is this similarity with the percentage of exclusive homosexuals in humans a mere coincidence, or is it a reflection of the operation of similar mechanisms in the two species, such as those described in the *Biosocial Model*?

Homosexuality in men and women

Although in both mammals and birds same-sex mounting is more common among males than among females, in mammals in general the male bias is far less accentuated, with females routinely engaging in same-sex sexual behaviours. The fact that homosexuality in humans is found in both men and women thus reflects, at least in part, a general mammalian pattern. Differences are clear, however, in the specific patterns of sexual orientation described in the two sexes. Two major trends are that (a) women's sexual orientation is more variable than men's, with such variability being expressed both in an increased interindividual spread of sexual orientation distribution among women compared with men (exclusive homosexuality: 3.1% men vs. 2.0 % women, see Chapter 3; bisexuality: 1.8%–33% men vs. 2.8%–65.5% women, see Chapter 1) and in greater sexual orientation variability in women within the same individual over time, these patterns reflect the greater degree of bisexuality in women than men; and (b) the percentage of exclusive homosexuals who manifest their sexual orientation from a very early age is higher among men than among women. I have shown throughout this book how biosocial proximate mechanisms operating in an evolutionary long-term scenario could broadly explain both patterns. Both men's and women's involvement in same-sex sexual behaviour is likely to have evolved in the context of socio-sexual functions such as cooperation and establishment of coalitions, as happens in other mammals (e.g. bonobos, dolphins), but also enforcement of dominance relationships, as is common in many primates and ungulates for instance. Here the hedonistic (pleasurable) aspect of sex is seen as a trait that is linked to both the immediate reproductive (e.g. heterosexual sexual intercourse in both sexes, mate choice and sexual selection but also sperm renewal through masturbation in males) and also the socio-sexual functions of sex (e.g. pair bonding leading to cooperation between heterosexual partners and also between same-sex partners).

Higher variability of sexual orientation in women than men can be expected in a socio-sexual perspective, especially if we consider that women tend to be the dispersing sex (exactly the same scenario holds for our close relative the bonobo). In the hypothetical *Environment of Evolutionary Adaptedness* and throughout their reproductively active period, women were likely to have been under the dual pressure of both establishing coalitions with and exerting their dominance vis-à-vis other reproductive women. Adaptive benefits of such socio-sexual functions of same-sex sexual behaviours are likely to have been valid for women throughout most of their lives as direct benefits of survival and reproduction gave way, at post-menopausal ages, to indirect benefits via enhancing the reproductive output and survival of close relatives (e.g. offspring and grand-offspring). Thus, high levels of same-sex sexuality in women at older ages (see, for example, Francis 2008) is something that could be expected from life-history evolution. Males, on the other hand, even today not only live shorter lives than women (see, for example, Lunenfeld 2001), but they also retain their ability to reproduce until much older ages than women (Kaufman & Vermeulen 1997; Gelfand 2000; Lunenfeld 2001). Thus there is a stronger selective pressure on men to modulate their behaviour over a significant span of their lives so that sooner or later their ability to sire offspring will be realised. Heterosexuality is therefore expected to be prevalent throughout the life of a man. Such differences between the sexes, however, do not preclude men from also engaging in same-sex sexual behaviours for socio-sexual purposes at older ages.

How can then we explain the second major trend in the patterns of homosexuality in humans: that the percentage of **exclusive** homosexuals who manifest their sexual orientation from a young age is higher in men than in women? There are various genetic mechanisms explained in Chapter 3 that can bias mutations towards males and that can explain the male bias in the development of early exclusive homosexuality in a small percentage of the male population: for example, genes found on either Y or X sex chromosomes that will be expressed in males owing to the male sex being heterogametic, and males having higher number of accumulated mutations in their germ cells because those cell lines undergo more mitoses in males than in females. A Y-chromosome carrying a mutation resulting from paternally elevated mitogenesis of germ cells will be expressed in his sons. An X-chromosome of paternal origin also carrying a mutation may be expressed or not in daughters, but it may then be expressed in grandsons. Prenatal and perinatal developmental effects, along with early postnatal social interactions associated with stress (including those occurring around adrenarche) can also help explain the development of an exclusive homosexual orientation in males (Chapter 4). Adaptively, such effects can be understood within the context of *Reproductive Skew* theory – a theory that, it should be remembered, also incorporates kin selection effects – as a result of parent–offspring–offspring conflicts, which, given the ability of males to retain their reproductive capacity for most of their lives and their tendency to be philopatric, would be especially strong among brothers and between father and son (see Chapter 6 and Chapter 8).

Effects occurring in the developing embryo under conditions of environmental or social stress in which the mother may find herself are also expected from Trivers & Willard's (1973) hypothesis of facultative adjustment of sex ratios. The hypothesis predicts a selective advantage for mothers to bias the sex ratio of their offspring towards females under stressful environmental conditions and towards males under more favourable conditions. If the mechanism underpinning the sex-ratio bias (e.g. differential abortion of fetuses according to sex) may also affect specific aspects of brain development in those fetuses that survive (maternal corticosteroids, for example, could achieve both outcomes; see Mulder *et al.* 2002 and the 'Stress Effects' section of Chapter 4), then Trivers and Willard's hypothesis could be subsumed, as is *Reproductive Skew* theory, into the general *Biosocial Model* for the evolution of human homosexuality, with special emphasis in this particular case on male homosexuality. Links between homosexuality and intergenerational mechanisms associated with biases in the sex ratio are also suggested by some preliminary analyses of genealogical data of male homosexuals. George Henry (1941) published some such genealogies that suggest that the mother of male homosexuals tends to come from a family with a female-biased sex ratio among siblings (57.1% families vs. 14.2% male-biased families and 28.5% even sex ratio, $n = 21$), whereas the father of male homosexuals tends to come from a family with a male-biased sex ratio among siblings (60.0% families vs. 25.0% female-biased families and 15.0% even sex ratio, $n = 20$)

($\chi^2_1 = 6.28$, $p = 0.012$; only families of two or more siblings are included). These patterns are consistent with the possibility that maternal effects (perhaps prenatal) may act in synergy with paternal effects (perhaps postnatal) in the development of homosexuality in male offspring.

Sexually antagonistic selection may further contribute to bias exclusive homosexuality towards men given that males are the philopatric sex in humans (see, for example, Towner 1999, 2001; Blum 2004). This is especially so when large numbers of philopatric brothers may affect the development of homosexuality in younger members of the brotherhood not only indirectly via prenatal effects (i.e. effects mediated by the mother during pregnancy), but also directly via postnatal effects.

Historically, the paradox of altruism mentioned in Chapter 1 was initially solved by a relatively simple process of kin selection encapsulated in *Hamilton's Rule*. Very soon, however, the beauty of that simplicity was 'spoiled' by the complexities of real life, and additional processes such as reciprocal altruism, mutualism, parental manipulation, sibling–sibling conflict and interdemic selection were added to kin selection to help resolve the paradox across a large variety of taxa. The resolution of the evolutionary paradox of homosexuality may be achieved in a similar manner: by recruiting the aid of a series of mechanisms that may interact or act independently. The *Biosocial Model* was produced following this general approach.

This conclusion notwithstanding, I should also stress that not all the mechanisms that have been proposed by previous authors are supported by the available empirical evidence. Below I list a series of mechanisms that have been suggested but that I believe are not of great importance, if at all. I also comment on the rescue of immune mechanisms that seem to have gained only lukewarm reception outside the research groups that have been most active in this field; and I will also include some additional comments on kin selection that, in recent times, has been mistakenly consigned to the waste bin of falsified hypotheses.

What does not seem to work

This review does not lend great support to the idea that orgasm is necessarily conducive to exclusive homosexuality. In species that exhibit orgasm, this could be achieved in many different ways, same-sex sexual behaviour being just one of them. This means that although individuals may engage in same-sex sexual behaviours for hedonistic purposes under certain circumstances (see Chapter 9), hedonistic experiences are not sufficient per se to explain the development and maintenance over time of a same-sex sexual orientation. However, they may provide the positive reinforcement to seek outlets for sexual intercourse over an extended period or at high frequency, as in the extreme cases of hypersexuality that were reviewed in Chapter 5. In such circumstances, homosexual intercourse may occur when only same-sex partners are available (e.g. in cases where the sex ratio is biased). Similarly, masturbation is another unlikely route towards the development of a homosexual sexual orientation, as I argued in Chapter 5. Hedonistic behaviours are only correlated with homosexuality; they are not a direct cause of it. Seeking pleasure leads to sexual organ stimulation and intercourse but not necessarily to a *preference*, let alone a *lifetime preference*, for a specific way of achieving that pleasure. The specificity of the sexual partnership requires additional explanations.

Sexual segregation also seems to have been an unlikely factor in the selection of same-sex sexual behaviour in birds during the evolution of the taxa included in this study. This is not the case for mammals, as shown in Chapter 7, although the practice of same-sex mounting during periods of sexual segregation does not necessarily lead to the development of an obligate homosexual orientation in social mammals. In fact, learning processes during juvenile periods of sexual segregation seem to be more adapted to improving heterosexual performance in adulthood and, at most, could be evolutionarily associated with bisexual behaviour in adults; hence the association between same-sex sexual behaviour (usually expressed in individuals that also reproduce; i.e. bisexuals) and sexual segregation in the comparative analyses of mammals.

Aneuploidies associated with Down syndrome, Klinefelter's syndrome, Turner syndrome or XYY trisomy are not associated with the development of a homosexual sexual orientation.

Although specific immune mechanisms are still wanting in detailed empirical support, it is premature to totally dismiss them. The nervous, endocrine and immune systems form a closely integrated network. On that basis one might suspect that in addition to the central nervous system and hormonal mechanisms potentially co-contributing to cause homosexuality, somewhere, somehow, cells and/or molecules produced by tissues of the immune system may also play a non-trivial role (Ellis & Hellberg 2005; Blanchard 2008a). For instance, stress-mediated mechanisms for the development of homosexuality may also involve tissues and molecules of the immune system, as it is well known that stress molecules such as corticosteroids can be immunomodulatory and, conversely, cells of the immune system may affect the activity of the adrenal glands that produce various kinds of steroids in stressful situations (see, for example, Bornstein & Chrousos 1999). We will know whether immune mechanisms play any role in the development of homosexuality as soon as the current research programme shifts from describing patterns (e.g. FBO effect) to actually testing hypotheses of causative mechanisms.

Kin selection has been the focus of a barrage of criticisms with regard to its potential role in the evolution of homosexual behaviour, especially in humans, leading to a popular conclusion that it is another evolutionary mechanism that 'does not seem to work'. However, the criticism that adult homosexuals do not preferentially aid kin has not the strength as an argument against the hypothesis that some suggest. First, there is the obvious counter-argument that the behaviour of homosexuals today is likely to be well removed from the selective context of the *Environment of Evolutionary Adaptedness* where the behaviour may have initially evolved. In this regard, a recent study carried out by Paul Vasey *et al.* (2007) on Samoan *fa'afafine* clearly suggests indirect fitness benefits accruing to male homosexuals in traditional societies where assistance from kin is of significance. Second, I have also argued that forfeiting reproduction could be in itself an aid to kin via the savings in resources that kin can then use for their own reproduction. Why should not individuals who are obligate homosexuals, and who therefore are not reproducing, simply shut down their reproductive physiology and become asexual altogether (Kirkpatrick 2000)? I suggest that although this is a possibility – some individuals are indeed reproductively suppressed in some social species (Wasser & Barash 1983, Creel 2001; see also the cases of asexuality in humans: Bogaert 2004, 2006b; Prause & Graham 2007; Brotto *et al.* 2009) – any small advantage of sexual behaviour (e.g. in the establishment of coalitions) will quickly outweigh any energetic benefit of suppression of that behaviour. In the case of bisexual individuals, they obviously accrue direct genetic benefits from retaining full reproductive capability.

Some interesting areas for future research

In spite of the many advances in our understanding of homosexuality, most evolutionary aspects of homosexuality still require substantial research, as I have suggested throughout this book. Here I highlight some of the most exciting possibilities that have not been stressed yet in this chapter.

Currently, the *organisational* role of sex hormones in adults tends to be dismissed. I provide some arguments and evidence in Chapter 5 to support a re-evaluation of this possibility and suggest the implementation of more specifically targeted research programmes, especially focusing on androgen effects on women's sexual orientation. In addition, future research should also expand the investigation to hormones other than steroids. In this respect the peptides oxytocin and arginine vasopressin are two very interesting targets for research, as suggested in Chapter 5.

Another extremely interesting topic is the potential effect of DNA-methylation on the development of a homosexual orientation. This potential mechanism could not only explain lower levels of heritability of the trait than expected, but also discrepancies in sexual orientation between monozygotic twins. Without a better understanding of DNA-methylation, arguments against the role of specific genes in the development of homosexuality lose a great deal of their power. Genes can exert their phenotypic effects only if they are expressed, if they are silenced through methylation they are still there, but may remain unexpressed for one or more generations. More studies on specific mutations affecting males, both mammals and birds, could also produce a better understanding of the mechanisms that control the expression of homosexuality.

In this book I reviewed the evidence in support of the potential importance of stress on the development of homosexuality. More and even better research should be welcomed on prenatal effects of stress, of course, but I would also suggest that more research be carried out on the interaction between various prenatal and postnatal (e.g. prepubertal) stresses that could lead to the development of a homosexual orientation. One aim could be to determine the interaction between the various prenatal factors that may affect the development of gender role (masculinity, femininity, androgyny) and postnatal stress effects that may lead masculinised girls and feminised boys towards the development of exclusive homosexuality. In circumstances where a less stressful postnatal social environment had pertained, they might have grown up to become heterosexuals or perhaps bisexuals. More research on the potential links between prepubertal stresses and the transition from homosexuality to transsexuality will also be welcome.

From an evolutionary perspective, I have also pointed out the potential relevance of neoteny in the evolution of the expression of homosexual behaviours among adult social vertebrates. The hypothesis still requires proper testing, but I suggest that it could be easily tested through a carefully designed comparative analysis, especially in mammals. In this analysis, however, it is neotenic properties of relevant organs such as the brain, or relevant areas within the brain, that should be studied in correlation with the expression of same-sex sexual behaviours.

One area of research that is reasonably well developed in humans but that has been woefully neglected in many animals is the potential cost of homosexual intercourse in terms of STD transmission. In the model introduced at the end of Chapter 8, I suggest that bisexual behaviour could be the link for the spread of STDs to the whole population, including exclusive male homosexuals, female homosexuals, other bisexuals and heterosexuals (males and females). If so, why is bisexual sexual behaviour so widespread, not only among humans but also among other species, especially primates? Perhaps most current STDs are an evolutionarily recent phenomenon in humans and primates in general? Or perhaps the modalities of same-sex sexual intercourse are such that transmission of STDs is minimised (for example, although anal insertive intercourse between two male non-human mammals has been reported, it is not common)?

Is homosexuality a pathology?

Throughout this book I have argued the case that same-sex sexual behaviour is no indication of an underlying pathology. Homosexual behaviour, especially when expressed in a bisexual context, is common in both young and adults across taxa, especially mammals, Old World primates and anthropoids in particular, including, of course, humans. A more exclusive homosexual sexual orientation can be also expressed in some taxa other than humans. I share the conclusion that homosexuality is not a pathology with other students of animal behaviour such as Bagemihl (1999), Roughgarden (2004) and Sommer & Vasey (2006). I do not deny the possibility that homosexuality may take pathological forms. In fact it does, but then so does heterosexuality. If homosexual behaviour is part of the normal sexual behavioural repertoire of anthropoid primates, especially our closest living relatives the chimpanzee and bonobo, then wholesale treatment of homosexuality as pathology is clearly unwarranted. As such, therefore, homosexuality is no more likely to take pathological forms than any other trait, although in this context defining what 'a pathological form' of a trait means is not always easy (Goldberg 2001). For those fewer cases of life-long exclusive homosexuality, I provide both a review of proximate genetic and developmental mechanisms and a list of ultimate evolutionary mechanisms that could maintain such a trait at low frequency in the human population. In this context exclusive homosexuality again becomes one of the many phenotypes available within the spectrum of sexual orientations that are possible in humans.

In this book I also reject the two classic arguments used by some psychoanalytical schools to still treat homosexuality as pathology: that homosexuality is a result of 'disturbed upbringing' and 'developmental arrest' (see Friedman 1986 for criticisms of these views). I regard the potential development of homosexuality in response to stresses experienced during early development as an adaptive response, not a pathology in need of a cure. My appreciation of Freud's insight that early stresses may contribute to cause homosexuality stops at the point where psychoanalysis concludes that therefore homosexuality is a pathology of development in need of treatment. Adaptive responses have evolved to maintain survival and direct (via bisexuality) or indirect (via helping relatives) reproductive success under challenging environmental (e.g. social) conditions. If so, then homosexuality is adaptive and not a psychopathology. Likewise, the apparent 'developmental arrest' may be associated with more profound neotenic processes experienced by our species and some of our closest relatives during evolution, and it is not a pathology of development. Although more specific studies are still needed, I have argued in this book that neotenic processes have probably contributed to produce one of the most spectacular organs in our species – the brain – and, through brain evolution, adaptive traits such as same-sex sexual behaviours.

Conrad & Angell (2004) provide a very thorough historical review of the issue of homosexuality as pathology, and of the path to de-medicalisation undergone by homosexuality since its initial classification as pathology of sexuality in the 1968 edition of the *Diagnostic and Statistical Manual of Mental Disorders* (DSM-II). The current DSM-IV-TR, in place since the year 2000, does not mention homosexuality as pathology any more, homosexuality having been removed from the DSM in its third edition (DSM-III), in a resolution passed in 1973. However, the DSM still retains the definition of Gender Identity Disorder (GID). I report here the DSM-IV-TR diagnostic criteria for GID as listed in Anon (2003):

A. A strong and persistent cross-gender identification (not merely a desire for any perceived cultural advantages of being the other sex). In children, the disturbance is manifested by four (or more) of the following:
 (1) repeatedly stated desire to be, or insistence that he or she is, the other sex
 (2) in boys, preference for cross-dressing or simulating female attire; in girls, insistence on wearing only stereotypical masculine clothing
 (3) strong and persistent preferences for cross-sex roles in make-believe play or persistent fantasies of being the other sex
 (4) intense desire to participate in the stereotypical games and pastimes of the other sex
 (5) strong preference for playmates of the other sex.
B. Persistent discomfort with his or her sex or sense of inappropriateness in the gender role of that sex.
C. The disturbance is not concurrent with a physical intersex condition.
D. The disturbance causes clinically significant distress or impairment in social, occupational, or other important areas of functioning.

Consequent to the study and review of same-sex sexual behaviour in humans and other vertebrates that we carried out in this book, the inclusion of GID in the DSM on the basis of the above criteria is rather disturbing (see also Bartlett *et al.* 2000 and Vasey & Bartlett 2007, for a very thorough dissection and rejection of GID as a mental disorder). Criteria A–C clearly result from assuming a bipolar concept of masculinity and femininity and from the identification of masculinity with the male sex and femininity with the female sex. What I show in this book, and what other authors have shown in their published works, is that gender roles in humans and other social mammals with biparental care are not static and rigidly organised into clearly masculine traits associated with males and feminine traits associated with females. As far as behaviour is concerned, there are differences but also great regions of overlap between the sexes in both birds and mammals, including humans. This is a profound characteristic of our biology and evolution, not just 'merely a desire for any perceived cultural advantages of being the other sex' (see section A of the diagnostic criteria for GID above). Criterion D should be a cause for some serious reflection. D is a logical consequence of A–C, and the deduction is impeccable, but if A–C are false, then D is false too. If males and females possess a significant region of overlap in gender role then the distress and impairment mentioned in D may well be a consequence of societal misguided expectations based on the belief that gender role is necessarily dichotomous (a similar point has been recently made by Manners 2009). In societies where gender roles are seen in a less dichotomous fashion, it could be predicted that GID would be rather low or absent altogether as a perceived disorder (see also Bartlett *et al.* 2000). In fact, Vasey & Bartlett (2007) report that most Samoan *fa'afafine*, i.e. exclusive homosexual males, do not recall any specific distress associated with their feminised behaviour when they were children. Such significant lack of memories of distress in childhood is likely to be a direct product of the significant acceptance of *fa'afafine* in Samoan society, and if they did experience stresses during some crucial periods in postnatal development, the stress was obviously not chronic. Incidentally, this suggests that *fa'afafine* may be an adaptive phenotype that perhaps develops as a result of prenatal mechanisms. In gender role bipolar societies, on the other hand, it should not surprise us that a feminine boy, who is constantly reminded that feminine behaviours are for girls only, grows up confused about his gender and also likely to experience significant distress. Margaret Mead made exactly the same point more than 70 years ago in *The Deviant* chapter of her book *Sex and Temperament in Three Primitive Societies* (Mead 1935: 297):

> Every time the point of sex-conformity is made, every time the child's sex is invoked as the reason why it should prefer trousers to petticoats, baseball-bats to dolls, fisticuffs to tears, there is planted in the child's mind a fear that indeed, in spite of anatomical evidence to the contrary, it may not really belong to its own sex at all.

Supporters of the idea of keeping GID in the DSM should provide an answer to the full causes of that 'clinically significant distress' mentioned in the DSM. Statements such as this: 'Continuation of GID into adolescence by no means seems to be a rare exception. I believe that treatment should be available for all children, regardless of their eventual sexual orientation, and *should depend on the severity of suffering experienced by the child*' (Cohen-Kettenis 2001; italics mine), are very much in tune with the text of the DSM-IV-TR diagnostic criteria for GID, highlighting the suffering of the child without asking about the full causative origin of that suffering. Treating society as a constant and intervening on the non-conforming individual, especially when she or he is a child, is only one of the possible approaches to try to alleviate the suffering of that child. Changing societal attitudes regarding 'gender non-conforming' individuals is another. If the child's homosexuality is a strongly developmentally canalised trait, psychotherapy to treat GID is likely only to produce unnecessary distress that could cause, in itself, unhappiness or much worse. If the trait is not strongly canalised, heterosexuality or bisexuality may soon develop – if the boy experiences a happy and fulfilling childhood in a welcoming social environment – without any need for professional therapy.

Heino Meyer-Bahlburg (2002) also suggests therapies for the treatment of GID in children, being motivated by the fact that 'childhood GID in boys certainly constitutes a risk factor for exposure to social pressures and adverse emotional consequences. These sequelae of GID are our primary reason for its treatment.' (p. 361), even though he also admits that treatments will only 'speed up the fading of the cross-gender identity which will typically happen in any case' (p. 361). But if cross-gender behaviours and identity are going to fade away with time in most children anyhow, would it not be easier to tell everybody (parents, parents' friends, relatives, school teachers, school mates, etc.), so that the level of social tolerance is elevated and developmental processes simply allowed to run their course without undue interference? If we believe that we can 'teach masculinity' to one individual, why should not we be able to 'teach tolerance' to 50?

It seems to me therefore that the current DSM-IV-TR definition of GID assumes a specific societal perception of a gender role dichotomy that is unquestionable, and that non-conformity to such perception becomes, therefore, a disorder. The onus of change falls upon the individual and the individual alone. Arguments are provided in this book to suggest that this view could indeed be questioned. Gender roles and identities are much more variable than previously thought even within the same society, thus broadening the spectrum of sex/gender role/identity combinations that should be included in our cultural definition of 'conformity'. The recent phenomenon of the 'metrosexuals', i.e. heterosexual males adopting what society defines as feminine behaviours and tastes, stands as clear evidence of the flexibility of gender roles in humans (see, for example, Harrison 2008). This reality is being very quickly exploited by various industries such as cosmetics and fashion. Among male cheerleaders of sport teams in the USA, Anderson (2008) describes a category that he names 'inclusive' cheerleaders, who may self-identify as either heterosexual or homosexual men but who have no problem in adopting what are considered to be feminine behaviours. Although the issue of GID is currently being hotly debated as the new DSM (DSM-V) is being prepared (the DSM-V manual is expected to be available in 2012), I can only hope that this book may contribute an evolutionary biological voice to the de-medicalisation of GID. It should be noted that with this I am not judging parents or researchers from an ethical or moral perspective, mine is a purely technical comment on the definition of disorder as it is currently applied to GID. Ethics, however, will be the focus of the last section of this book.

One last thought

Although the objective in this book has been to explore the evolutionary basis of animal homosexual behaviour, I cannot conclude without mentioning, even if briefly, my own position regarding the ethical and political implications of studies on homosexuality. I am compelled to address this issue not so much as an attempt to prevent the misuse of this book – authors just cannot control the fate of their words – but because I want to state my personal views in writing, so that it will be clear which interpretations reflect my opinions and which do not.

First I address what is generally known as the *Naturalistic Fallacy* (Moore 1903) or the *Is–Ought Problem* (Hume 1740). Both expressions are commonly used to mean that the findings of scientific research aim at being descriptive of nature and not prescriptive of how we should behave in an ethical sense. In other words, through the use of the scientific method we certainly aim to achieve a more accurate description of phenomena, their causative mechanisms and their specific or broader consequences. However, as far as science being normative of our actions, such descriptive knowledge may not be sufficient. Moreover, I strongly oppose, as did John Maynard Smith (1989), the idea of transforming evolutionary biology into an ideology. In the words of an outsider to science, the journalist Chandler Burr, writing in the March 1993 issue of *The Atlantic Monthly*:

> Yet it would be wise to acknowledge that science can be a rickety platform on which to erect an edifice of rights. Science can enlighten, can instruct, can expose the mythologies we sometimes live by. It can make objective distinctions ... But we cannot rely on science to supply full answers to fundamental questions involving human rights, human freedom, and human tolerance
> (Burr 1993: 65)

'Free scientific research' should, however 'provide the type of knowledge that must ultimately weaken the ignorance upon which bias and prejudice rest', as the psychiatrist Richard Friedman put it (Friedman 1992: 2).

Independently of any scientific contribution represented by this book, in my personal ethical stance I subscribe to the General Assembly of the United Nations' *Universal Declaration of Human Rights* in that the

> *... recognition of the inherent dignity and of the equal and inalienable rights of all members of the human family is the foundation of freedom, justice and peace in the world.*

I unreservedly support the fundamental human right of homosexuals, bisexuals, transsexuals and intersex people to express their own self with respect to the laws of the land, and to organise themselves to promote change to those laws and to the stereotypes currently prevalent in that society through the democratic process.

Appendix 1: Glossary

Activational effect Referring to hormonal action: the process by which hormones, by interacting with cells of the nervous system (e.g. in the brain), trigger the display of behaviours.

Affiliative behaviour Behaviour that favours association and cooperation between individuals. The expression *Affinitive behaviour* is also used in this context.

Allozyme Molecular variant of an enzyme.

Altruism Act of helping others at a direct cost to oneself.

Anabolic androgenic steroids Synthetic derivatives of testosterone.

Androgyny From a psychological perspective: the confluence, in the same person, of characteristics that are recognised as masculine and feminine.

Androphilia Sexual attraction towards males. The concept applies to both male and female individuals who are attracted to males.

Aneuploidy A state that results when occasionally, during meiosis, a pair of homologous chromosomes may remain together rather than dividing up and becoming part of haploid spermatozoa or ova. Whenever such chromosomally abnormal reproductive cells are involved in the production of a zygote, the resulting individual will be an aneuploid, that is, his or her cells will have an altered number of a specific chromosome.

Antecedent brother effect Developmental effect suffered especially by male offspring in association with the number of brothers born previously.

Apoptosis Programmed cell death.

Aromatisation Production of a benzene ring in a molecule. It is through aromatisation that, for example, testosterone is converted into oestradiol.

Asexuality Used here to signify sexual unresponsiveness to external sexual stimuli.

Attachment Affective bond between two individuals.

Auditory evoked potentials Weak electric potentials originating from the brain that are recorded from the scalp and are released by brief acoustic stimuli.

Autoimmunity Immune attack against one's own cells and tissues.

Autosexuality Self-sexual stimulation; masturbation.

Birth order effect (fraternal and sororal) Effect of the position within a sibship on the probability of developing a specific trait; in this book the trait of interest is homosexual orientation.

Bisexuality Sexual desire for and/or sexual behaviour performed with individuals of both sexes.

Blastocyst Stage of embryonic development that in mammals is reached by the time of implantation in the uterus.

Bruce effect Blockage of pregnancy in the presence of a new male.

Buller Male bovid or steer that is mounted by other males.

Canalisation A concept first introduced by Conrad Waddington to describe the ability of an organism to achieve full development and functionality in spite of the many perturbations both internal (e.g. mutations) and external that the organism may experience. The more a population is under stabilising selection, the more canalisation the ontogeny of the members of that population is expected to experience.

Chimaera An organism whose cells have different genetic origin and structure; that is, the chimaera is a genetic mosaic. Chimaeras are formed from more than two parent cells fusing together at fertilisation.

Clade A group of taxa that descend from a common ancestor.

Comparative methods Family of techniques used to test evolutionary hypotheses through the analysis of multi-taxon datasets.

Complete control Characteristic of reproductive skew models that assume that reproductive skew is controlled by a single dominant individual.

Conditioning A learning mechanism in which the expression of a behaviour becomes dependent on the occurrence of an associated specific external stimulus, after a period in which the organism has experienced that associated stimulus.

Contagious behaviour A behaviour that is more likely to be exhibited by one individual if a relevant other individual has already expressed that same behaviour.

Coping Set of psychological strategies that allow the individual to withstand stressful situations.

De-identification Maximisation of behavioural differences between adjacent siblings.

Delayed plumage maturation Retention of juvenile-like plumage in sexually mature birds.

Dermatoglyphics Also known as 'fingerprints': are characteristic patterns formed by the dermal ridges in the skin of fingertips, and other parts of the body, that develop in humans between weeks 8 and 16 of foetal life.

Developmental instability Perturbations of developmental processes caused by mutations and/or environmental effects.

Developmental stability The degree of resilience that the developing organism has to withstand the effect of stressors during early ontogeny.

Diencephalon Posterior section of the forebrain that contains the thalamus and hypothalamus.

Dioestrus Period in female mammals between two oestrous cycles.

Disexual Referring to a group or pair of individuals: indicates that the composition of such a group or pair is of individuals of different sexes. Individuals forming disexual groups may or may not engage in sexual behaviours with each other.

Dizygotic twins Twins that are the product of different ova being fertilised by different spermatozoa. Also called Fraternal twins.

DNA-methylation A phenomenon in which cytosine in specific areas of the genome, the CpG islands, acquires a methyl group, a reaction that is catalysed by the enzyme DNA methyltransferase 3-like.

Endocrine-disrupting chemicals Environmental chemicals that have the ability to interfere with the normal action of endogenous hormones on the organism's development and physiology.

Environmental sexual segregation Segregation between the sexes along a spatial axis (it includes Habitat and Spatial sexual segregation).

Environment of evolutionary adaptedness The set of parameters that characterised the environment in which a specific adaptive trait is hypothesised to have evolved. The concept was initially proposed by John Bowlby in the context of his *Attachment Theory*.

Exaptation A trait whose current adaptive function is different from the function that contributed to the initial evolution of the trait.

Familiality effect A pattern of distribution of traits that runs in families.

Family niches A concept proposed by Frank Sulloway whereby the social environment within the family favours the development of specific sets of interindividual interactions and personality features among siblings. Birth order is a particularly powerful variable affecting family niches (e.g. assertive elder vs. creative younger sibling).

Fitness (Darwinian) Contribution of a specific allele at a locus in the genome (or of an individual in the population) to the population of individuals or copies of the genome in the next generation. Relative Darwinian fitness is such a contribution relative to that of the other alleles at the locus or individuals in the population. Darwinian fitness may be direct, when the contribution is made through reproduction, or indirect, when it is made through helping close relatives in their reproductive effort.

Flehmen A behaviour displayed by many mammals, consisting of the curling of the upper lip. In so doing the animals convey chemicals into their vomeronasal organ.

Follicular phase Last phase of the oestrous cycle, in which follicles mature; this phase ends in ovulation.

Gaydar An ability to distinguish between heterosexuals and homosexuals by using indirect cues such as eye gaze and body movements.

Gay man In this book the expression will be used simply to mean a homosexual man.

Gender The behavioural, psychological and cultural aspects defining an individual's sexuality. Cf. **Sex**.

Gender dysphoria A disconformity towards the physical characteristics of one's gender that is also associated with a strong identification with, and a willingness to be a member of, a different gender.

Gender identity Self-definition of an individual's own gender.

Gender inversion Total (or at least very significant) concordance of gender characteristics of an individual with those of a member of the other sex; e.g. feminised males or masculinised females.

Gender role Behavioural characteristics and their associated mental states that are culturally defined as being *masculine* or *feminine* and that are displayed by the individual.

Gender shift Partial concordance of gender characteristics of an individual with those of a member of the other sex; e.g. semi-feminised males or semi-masculinised females.

Gender typing Perception and treatment of an individual in a community according to his/her socially defined gender.

Gene loading The total deleterious genes that are present, but usually hidden, in the genome of individuals in a population and that may be transmitted to offspring.

Genetic accommodation Evolutionary process involving selection for adaptive plasticity.

Genetic assimilation Evolutionary process through which a developmental programme that was initially expressed as a result of the organism's plasticity becomes canalised following the occurrence of mutations.

Genetic drift Random changes in allele frequency distribution in the population that are especially important in small populations.

Genomic imprinting A phenomenon whereby genes that are usually rich in CpG islands undergo parental-specific DNA-methylation. Parental-specific methylation occurs during the germ cells' development into either spermatozoa or ova. The original parental imprint is maintained in the somatic cells of the offspring but erased in its germ cells, which are imprinted again in an individual fashion.

Gonadal anlage Gonadal tissue that is still undifferentiated into male or female gonads; gonadal primordium.

Gynaephilia Sexual attraction towards females. This is valid for both male and female individuals.

Habitat sexual segregation Segregation between the sexes due to females and males using non-overlapping, sex-specific habitats.

Hemizygous Describes individuals that only have one copy of the genes (or most of the genes) on specific chromosomes. For instance, males in mammals (XY) only have one copy of most of the genes found on sex chromosomes.

Heritability In its 'narrow sense', heritability is the proportion of phenotypic variation that one may attribute to additive genetic variation. Realised heritability (h^2), on the other hand, is the ratio between the difference in mean value of the trait between parents and offspring, called *response to selection* (R), and the difference in the mean value for the trait between the population and the parents being selected, called *selection differential* (S).

Heterochrony Differential evolutionary alteration of the timing of developmental processes affecting specific parts of the organism.

Heterosexuality Sexual orientation characterised by sexual attraction to individuals of the other sex.

Heterozygous advantage See **Overdominance**.

Homogametic sex Sex that produces gametes with the same kind of sex chromosomes: X chromosome in female mammals and Z chromosome in male birds.

Homosexual behaviour Behaviour that is unambiguously sexual or of sexual origin and that is performed between members of the same sex.

Homosexuality Sexual orientation characterised by sexual attraction for individuals of the same sex.

Homosociality Involvement in same-sex social consortship (e.g. foraging groups, sports teams) without necessarily engaging in same-sex sexual behaviour. Homosocial animals form monosexual groups.

Hypercarnivory Exclusive reliance on animal tissues as food.

Hypermorphosis An evolutionary process whereby the somatic development remains unchanged throughout speciation but gonadal maturation becomes delayed relative to the rate of development in the ancestor. This process produces species with adult individuals developing their reproductive functions and structures very slowly during their ontogeny.

Hypersexuality Level of sexual activity and interest that is well above the average level and interest displayed by the members of a community or population. The application of the concept is necessarily of a relative nature.

Immune dysregulation Impairment of the mechanisms that regulate activity of the immune system.

Immune privilege Property of specific tissues where immune attack on the cells within those tissues is suppressed.

Immunodeficiency Inability to produce a normal immune response.

Imprinting A learning mechanism in which behavioural patterns are established during a narrow sensitive period in the development of the organism.

Inclusive fitness Darwinian fitness of an individual (or allele) calculated taking into account both the number of offspring (or copies) directly produced and the number of extra offspring-equivalent (or copies) indirectly produced by helping close relatives in their reproductive effort. Offspring that recipients of help would have produced even without the help received are not considered in the calculation of the inclusive fitness of the donor of help.

Incomplete control Model of reproductive skew where a dominant does not completely control the reproductive activity of subordinates.

Interdemic selection Selection of alleles in a population due to their effects on both the individuals that express those alleles and the effects of those individuals on other members of the population.

Isocortex Also known as neocortex; the larger part of the cerebral cortex.

Isoenzymes Slightly different molecular variants of an enzyme.

Isosexuality Usually used to refer to same-sex sexual behaviour in non-human animals. Here isosexuality is regarded as a synonym of homosexuality, the latter term is used by some authors to refer exclusively to human same-sex sexual behaviour, orientation and identity. This choice is dictated by the observation that although same-sex sexual behaviour in humans has characteristics that are unique to the species, so does same-sex sexual behaviour in essentially any other species; both human and non-human homosexualities also share some characteristics in common.

Karyotype Full chromosomal complement of a cell.

Kin selection Selection process that maximises inclusive fitness.

Lee–Boot Effect Variation in the duration of the oestrous cycle of adult rodents occurring when females live in monosexual groups.

Lesbian In this book the term is used simply to mean a homosexual woman.

Lordosis A reflex that results in the arching of the back and the assumption of a mating posture in rats and other rodents.

Luteal phase Stage of the ovulatory cycle starting after ovulation and that is characterised by production of progesterone and oestrogens by the corpus luteum.

Maladaptive trait A trait that is associated with a net decrease in inclusive fitness. An individual possessing a maladaptive trait reproduces less, lives a shorter life and/or has a negative impact on the lifetime reproductive success of its relatives.

Manicheistic From Manicheism, a religion founded by the prophet Mani in the third century c.e. in Persia (Iran). The adjective 'manicheistic' is used in this book to metaphorically characterise dualistic viewpoints, in a manner akin to Manicheism with its emphasis on the light/darkness, good/evil dualisms.

Masturbation Sexual arousal achieved by self-sexual stimulation (autosexuality, self-masturbation) or with the aid of a partner. In the latter case the sexual stimulation is achieved by means other than genito-genital or genito-anal contact (e.g. manual stimulation).

Maternal effects The effects exerted by the maternal genotype (and/or phenotype) on the offspring phenotype.

Meta-analysis Family of analytical tools that use results from various tests of a hypothesis (e.g. from published articles) in order to evaluate the level of empirical corroboration of that hypothesis.

Metaphyseal tissue The cartilage that subsequently becomes bone.

Microsatellite Regions of DNA (loci) that display tandem repeats of short DNA sequences.

Monosexual Referring to a group or pair of individuals: indicates that the composition of such a group or pair is of members of the same sex. Individuals forming monosexual groups may or may not engage in same-sex sexual behaviours.

Monozygotic twins Twins that result from the division into two independent cells of a fertilised ovum. These twins are genetically identical and are also known as identical twins.

Multiallelic trait A phenotypic characteristic in which genetic determination is contributed by a locus that has several alleles in the population.

Mutation Permanent change in nucleic acid sequence that can be transmitted to offspring. Mutations at this level may or may not produce alterations in the phenotype.

Mutualism A kind of interaction between individuals of the same or different species that produces an increase in Darwinian fitness in all parties involved.

Natural selection A process by which beneficial traits increase in relative frequency in the population over successive generations of reproducing organisms, following specific environmental challenges.

Neoteny Evolutionary process involving delayed somatic development in a derived species compared with the ancestor, with gonadal development remaining unchanged. The neotenic derived species will have sexually mature individuals resembling juveniles of the ancestral species.

Neurogenesis The process that results in the production of new nerve cells from cells that act as stem cells.

Ockham's Razor The principle by which, in the production of a hypothesis, one should not multiply entities beyond necessity. It is also variably known as the *Principle of Parsimony*, *Simplicity* or *Economy of Thought*.

Ontogeny Process that encompasses the full development of an organism.

Oophorectomy Removal of one or both ovaries. Female castration.

Operational sex ratio Ratio of sexually active males to sexually active females.

Orchiectomy The removal of one or both testes. Male castration.

Organisational effect Referring to hormones: the effect that they can exert on the development of an individual, usually during its early ontogenetic stages.

Orgasm Physiological, evaluative (sensation) and affective states characterising a specific point that may be reached during the performance of sexual behaviour.

Other-sex mimicry Also known as opposite sex mimicry; a state whereby males or females resemble conspecifics of the other sex in external morphology, behaviour, or both.

Otoacoustic emissions Weak sounds produced in the inner ear of both male and female humans.

Overdominance *Also known as heterozygous advantage:* a situation whereby heterozygotes have higher fitness than homozygotes.

Paedomorphosis A slowing down of the development of individuals in a derived species compared with the ancestor.

Parental manipulation Differences in fitness among siblings, caused by parental activities.

Parity effect Increased likelihood of developmental modifications with birth order.

Parthenogenesis A mode of reproduction in which an ovum that is unfertilised develops into a new individual.

Path analysis A multiple-regression technique used to test causative models.

Peace incentives Benefits conferred by dominants on subordinates in order to achieve complete reproductive control.

Penetrance Proportion of individuals that carry a specific allele and that express it phenotypically.

Peramorphosis An acceleration of development in a derived species compared with its ancestor.

Perinatal Period of development immediately before and after birth.

Periovulatory phase Stage in the ovulatory cycle where an increase in the number of cumulus cells causes an increase in the volume of fluid in the antrum, thus expanding the size of the follicle. This is the stage that immediately precedes ovulation.

Pheromone A chemical that is synthesised by an organism (or obtained and accumulated from the environment) and that is then secreted in order to transmit information to conspecifics.

Phylogenetic trait conservatism The maintenance of a trait within a group of phylogenetically related species.

Phylogeny Branching diagram representing evolutionary relationships between taxa.

Phytoestrogens Plant compounds that have oestrogenic activity in animals.

Plethysmography Technique that is used to measure the changes in blood flow in various organs. In this book it is mentioned in the context of measurements of penile volume.

Polygenic trait A phenotypic characteristic that is determined by the expression of more than one locus or gene.

Polygynandry Mating relationship of more than one male with more than one female.

Polymorphism (genetic) The occurrence in a population of various alleles at the same locus.

Postconception mating Mounting observed outside the fertile period of females.

Proband A specific member of a family who is the focus of a genetic investigation.

Proceptivity An individual's behaviour soliciting mounting activity from another individual.

Proestrus Phase of the oestrous cycle where at least one follicle starts to grow and the endometrium starts its development.

Progenesis An evolutionary process involving accelerated gonadal maturation, with the rate of somatic development remaining unchanged in a derived species compared with its ancestor.

Promoter region A stretch of DNA that is located upstream from a gene and acts as a controlling factor in the expression of that same gene.

Prosocial behaviour Behaviour that facilitates cooperation between individuals.

Prospective studies Studies based on the following up of individuals throughout several stages of their ontogeny.

Pseudopenis Elongated clitoris of female spotted hyaenas that resembles a male's penis.

Punctuational model of evolution Theory of evolutionary change first proposed by Niles Eldredge and Stephen Jay Gould, holding that long periods of stasis in character change are punctuated by shorter periods of rapid evolutionary modifications.

Quantitative genetics Genetic study of traits that vary continuously (e.g. body mass, height).

Queer In the context of sexual orientation: a generalised identity proposed to overcome the narrow distinction between the normal and the alternative. *Queer theory* emphasises the fluidity of sexual orientation.

Receptivity An individual's willingness to accept mounting.

Reciprocal altruism A helping behaviour that is costly for the actor and that is reciprocated by the recipient some time in the future.

Reproductive skew Difference in reproductive success between individuals (e.g. dominant and subordinate) in a group.

Resilient children Children who are able to conduct a normal adult life in spite of having experienced dreadful social conditions during their development.

Response to selection Change in the population mean value of a trait after selection has taken place.

Retrospective studies Studies based on interviews where subjects are asked to recall past events.

Retrotransposon Genetic material that moves to a new location within the DNA by first making an RNA copy of itself, subsequently making a DNA copy using a reverse transcriptase, and finally attaching that copy of DNA into a target DNA region.

Rider Male bovid that mounts another male, the buller.

Ritualised homosexuality As defined by Herdt (1984a): an institutionalised homosexual activity, usually implemented through initiation rites.

Rostrocaudal axis Axis of symmetry running from face (head) to tail (back).

Selection differential The difference between the population mean value of a trait and the mean value of that same trait for the individuals that are selected to be the parents of the next generation.

Series mounting Behavioural pattern whereby interactants repeatedly exchange mounter and mountee position during an encounter.

Sex-typing Development of those attributes and characteristics that are typical of one's own sex.

Sexual aggregation Spatiotemporal distribution of males and females where both sexes overlap.

Sexual orientation Sexual preference for individuals of the same, other, or both sexes.

Sexual segregation Greater distancing in space or time between members of different sexes than between members of the same sex.

Sexual selection Process that leads to the evolution of traits used in the acquisition of sexual mates and the achievement of reproduction.

Sexually antagonistic selection Selection of a trait that is beneficial to one sex but detrimental to the other.

Sexually mutualistic selection Selection of a trait that is beneficial to both sexes.

Social constructionism School of thought that sees homosexuality as the result of the influences of social interactions on the individual and that denies genetic influences on the development of sexual orientation. The French philosopher Michel Foucault was one of the major representative thinkers of this school.

Social sexual segregation Distribution of individuals in the population into monosexual groups that do not overlap in space or time between sexes.

Socionomic sex ratio Ratio of adult males to adult females.

Socio-sexual behaviour In this book the expression will be used to mean the use of sexual behaviours such as mounting in the context of social interactions of dominance or affiliation (see Wickler 1967).

Spatial sexual segregation Microhabitat segregation between the sexes.

Splicing (gene) A process through which one fragment of a DNA molecule is attached to another.

Stay incentives Benefits (e.g. survival, reproduction) offered by dominants to subordinates in order to induce the latter to delay dispersal.

Steer Male bovid, usually a younger one.

Superalliance Unstable type of alliance reported in male *Tursiops* dolphins.

Synaptogenesis Process of formation of new functional connections, known as synapses, between two excitable cells (e.g. two neurons).

Syncretism The practice of reconciling diverse views.

Temporal sexual segregation Activity separation of the sexes along the temporal axis (e.g. males active during the day, females active at night).

Thrombosis Blood clot formation in an artery or a vein.

Transactional models Models that take into account both genetic and environmental contributions to the expression of a phenotypic trait (e.g. a behaviour).

Transcription Process through which a DNA template molecule is used to synthesise a messenger RNA molecule.

Transsexualism Same-sex sexual orientation associated with a desire to fully become a member of the other sex and seek sex reassignment.

Trophoblast The external layer of cells of the blastocyst. The trophoblast attaches the fertilised egg to the uterus and conveys nutrients to the embryo.

Vandenbergh effect Acceleration of puberty in young females associated with the presence of an adult male.

Whitten effect Synchronisation of the oestrous cycle of females in a group following exposure to male pheromones.

Xenobiotic A compound that an organism ingests or somehow absorbs into its body and that can have a direct effect on its physiology, altering normal biological processes.

Appendix 2: Predictions of the *Synthetic Reproductive Skew Model of Homosexuality* and results obtained in the comparative tests of the model carried out in birds and mammals

Results Predicted

Prediction 1

Test a: Same-sex Mounting contrasts and Ecological Constraints contrasts are positively correlated.

Prediction 2

Test a: Same-sex Mounting contrasts are higher at intermediate values of Log-Social Unit Size contrasts.

Prediction 3

Test a: Relatedness contrasts are positively correlated with Affiliative Mount contrasts. (Affiliative Mounts are a specialised kind of same-sex mount.)

Test b: Relatedness contrasts are negatively correlated with Same-sex Mounting contrasts (inbreeding avoidance).

Prediction 4

Test a: Mating System contrasts are positively correlated with Same-sex Mounting contrasts.

Test b: Mating System contrasts are positively correlated with Plumage Sexual Dichromatism (in birds) and Body Mass Sexual Dimorphism (in mammals) contrasts. Note that empirical support for this prediction will also be an indirect test for the quality of our dataset, phylogenies and appropriateness of comparative analysis of independent contrasts. This is so because the association between polygamy and sexual dimorphism/dichromatism is already a well-established pattern from previous independent studies in both birds (see, for example, Dunn *et al.* 2001) and mammals (see, for example, Weckerley 1998).

Test c: Sex Involved contrasts (see Prediction 5, Test a) are positively correlated with Mating System contrasts. That is, F–F mounting would be more common in polygamous species.

Prediction 5

Test a: The greater the bias in sex ratio, the greater the level of homosexual mounting expected in the supernumerary sex. The two variables of importance here are Sex Involved in homosexual mounting and Adult Sex Ratio. The Sex Involved variable was codified in such a way that larger values mean that females are more involved in same-sex mounting than males, whereas the Adult Sex ratio variable was codified so that larger values mean that the sex ratio is more male-biased. Therefore, if Prediction 5 is to be supported, we expect that Adult Sex Ratio contrasts will be negatively correlated with Sex Involved contrasts.

Test b: The greater the bias in the Adult Sex Ratio the greater the level of Same-sex Mounting.

Test c: Adult Sex Ratio contrasts should be positively correlated with Mating System contrasts.

Test d: Pair Bonding contrasts should be positively correlated with Same-sex Mounting contrasts.

Test e: Same-sex Mounting contrasts should be negatively correlated with EPC contrasts.

Prediction 6

Test a: Sociality contrasts should be positively correlated with Same-sex Mounting contrasts.

Test b: Dominance Mount contrasts should be positively correlated with Sociality contrasts.

Test c: Dominance Mount contrasts should be positively correlated with Mating System contrasts.

Prediction 7

Test a: Affiliative Mount contrasts should be positively correlated with Sociality contrasts.

I do not have data to carry out precise comparative tests of independent contrasts of Predictions 8 and 9 on a sufficient number of the species listed in Table 8.2. However, I will carry out some preliminary analyses to determine whether the dataset displays the expected trends.

Prediction 8

Preliminary Test a: Parental manipulation of offspring's early ontogeny often occurs in cooperatively breeding species. If this affects the development of homosexuality then cooperative breeders should be more likely to display same-sex sexual behaviour than non-cooperative breeders. (See also Test a of Prediction 1).

Prediction 9

Preliminary Test a: If short dominance queues increase the probability of attaining breeding status and therefore decrease the probability of developing a homosexual sexual orientation, then among cooperative breeders or species that live in stable social groups with predictable membership, the percentage of all copulations that are same-sex should be positively correlated with group size, at least up to intermediate group sizes.

Results Obtained: Birds

Results in bold are statistically significant (see Table 8.3).

Prediction 1

Test a: Same-sex Mounting contrasts and Ecological Constraints contrasts were not positively correlated, thus contradicting the prediction of the model, although the negative correlation trend was not significant.

Prediction 2

Test a: **Same-sex Mounting contrasts were expected to be higher at intermediate values of Log-Social Unit Size contrasts. The correlation is negative and highly significant, lending at least some partial support to the model (see text).**

Prediction 3

Test a: Relatedness contrasts are not significantly positively correlated with Affiliative Mount contrasts, but the trend is in the direction expected from the model.

Test b: Relatedness contrasts are not significantly and negatively correlated with Same-sex Mounting contrasts and the trend is not in the expected direction either.

Prediction 4

Test a: Mating System contrasts are not significantly correlated with Same-sex Mounting contrasts, but the trend is in the expected positive direction.

Test b: Mating System contrasts are significantly positively correlated with Plumage Sexual Dichromatism contrasts. That is, evolutionary changes towards increased sexual plumage dichromatism tend to be associated with evolutionary shifts towards polygamy. Although this specific result is also shared with other models and it is therefore not exclusive to the *Synthetic Reproductive Skew Model of Homosexuality*, it is nonetheless encouraging that the dataset confirms the well-established association between polygamy and plumage sexual dichromatism.

Test c: Although Sex Involved contrasts are not significantly correlated with Mating System contrasts, the positive sign of the correlation coefficient is in the expected direction.

Prediction 5

Test a: Although Adult Sex Ratio contrasts are not significantly correlated with Sex Involved contrasts, the trend is in the expected negative correlation direction.

Test b: The greater the evolutionary trends towards a male-bias in the Adult Sex Ratio, the significantly greater the evolutionary shifts towards Same-sex Mounting as expected from the model. In birds, the variable Same-sex Mounting is especially reflective of male same-sex mounting.

Test c: Against the prediction of the hypothesis, Adult Sex Ratio contrasts are not positively correlated with Mating System contrasts, although the negative correlation trend is not significant.

Test d: Pair Bonding contrasts are not significantly correlated with Same-sex Mounting contrasts, and the non-significant negative correlation trend runs against the prediction.

Test e: Same-sex Mounting contrasts are not significantly correlated with EPC contrasts, but the negative correlation trend is in the direction expected from the model.

Prediction 6

Test a: As expected from the model, Sociality contrasts are significantly and positively correlated with Same-sex Mounting contrasts.

Test b: Dominance Mount contrasts are also significantly and positively correlated with Sociality contrasts as predicted by the model.

Test c: Dominance Mount contrasts are not correlated with Mating System contrasts and the negative correlation trend obtained also runs against the prediction.

Prediction 7

Test a: Consistent with the prediction of the model, Sociality contrasts tend to be positively correlated with Affiliative Mount contrasts, but the correlation is not statistically significant.

Prediction 8

Preliminary Test a: Although, as expected, 66.6% of all cooperatively breeding species are monotypic for same-sex mounting whereas a smaller 59.2% of all non-cooperatively breeding species are, the difference is not significant: $\chi_1^2 = 0.02$, $p = 0.88$.

Prediction 9

Preliminary Test a: I have information for both percentage of all copulations that are same-sex and social unit size for only 8 of the 12 cooperatively breeding species (Table 8.2). The trend for the correlation between log-social unit size and percentage of all copulations that are same-sex is negative, against the expectation from the model, but not significantly so (Pearson's product-moment correlation: $r = -0.05$, $p = 0.45$, $n = 8$).

Results Obtained: Mammals

Results in bold are statistically significant (see Table 8.6).

Prediction 1

Test a: **Same-sex Mounting contrast residuals and Ecological Constraints contrasts were significantly negatively correlated. This runs against the prediction of the model.**

Prediction 2

Test a: **Same-sex Mounting contrasts were expected to be higher at intermediate values of Log-Social Unit Size contrasts. I obtained a significant negative correla-** tion between the two, which lends partial support to the model (see text).

Prediction 3

Test a: Relatedness contrasts are not significantly correlated with Affiliative Mount contrasts, and the trend is not in the direction expected from the model either.
Test b: **Relatedness contrasts are highly significantly and negatively correlated with Same-sex Mounting contrast residuals, as expected.**

Prediction 4

Test a: **Mating System contrasts are significantly positively correlated with Same-sex Mounting contrast residuals, as expected.**
Test b: Mating System contrasts are not significantly correlated with Sexual Dimorphism contrast residuals. However, the trend goes in the expected direction of a positive association between the two traits. The same result is obtained if the original, i.e. not residual, Sexual Dimorphism contrasts are used.
Test c: Although Sex Involved contrasts are not significantly correlated with Mating System contrasts, the coefficient is in the expected direction of a positive correlation and the probability approaches marginality ($p = 0.11$).

Prediction 5

Test a: **Adult Sex Ratio contrast residuals are significantly negatively correlated with Sex Involved contrasts, as expected from the model.**
Test b: The greater the evolutionary trend towards a male-biased Adult Sex Ratio, the greater the evolutionary shifts towards Same-sex

Mounting, in accordance with the prediction of the model, but the trend is not significant.

Test c: In support of the model, although Adult Sex Ratio contrasts are not statistically correlated with Mating System contrasts, the trend is towards the expected positive association and the non-significance is marginal ($p = 0.10$).

Test d: Social Bonding contrasts are not significantly correlated with Same-sex Mounting contrast residuals, and the non-significant negative correlation trend runs against the prediction.

Test e: Same-sex Mounting contrast residuals are not significantly correlated with EPC contrasts, and the positive correlation trend is against the direction expected from the model.

Prediction 6

Test a: Also against the prediction of the model, Sociality contrast residuals are not significantly correlated with Same-sex Mounting contrast residuals and the trend is in the direction opposite from the expected one.

Test b: Dominance Mount contrasts are not significantly correlated with Sociality contrast residuals, and the sign of the correlation coefficient is the opposite of that expected.

Test c: Dominance Mount contrasts are not correlated with Mating System contrasts but the positive correlation trend is consistent with the prediction of the model and it is also marginally not significant ($p = 0.10$).

Prediction 7

Test a: Against the prediction of the model, Sociality contrast residuals are not correlated with Affiliative Mount contrasts, and the trend is the opposite of that expected.

Prediction 8

Preliminary Test a: Although, as expected, 82.6% of all cooperatively breeding species are monotypic for same-sex mounting and a slightly smaller 79.1% of all non-cooperatively breeding species are, the difference is not significant $\chi^2_1 = 0.001$, $p = 0.98$.

Prediction 9

Preliminary Test a: I have information for both percentage of all copulations that are same-sex and social unit size for 14 species living in more or less stable groups, including 6 cooperative breeders (Table 8.4). The trend for the correlation between the two variables is positive as expected from the model, but not significantly so: Pearson's product-moment correlation: $r = 0.309$, $p = 0.14$, $n = 14$ for all sexes combined, $r = 0.504$, $p = 0.13$, $n = 7$ for males only, and $r = 0.321$, $p = 0.25$, $n = 7$ for females only.

# Appendix 3: Comments on further results of comparative analyses of independent contrasts reported in the full correlation matrices of birds and mammals

In this appendix I briefly comment on the results of comparative analyses of independent contrasts reported in the full correlation matrices of birds and mammals that produced either significant ($p < 0.05$) or marginally not significant ($0.10 < p < 0.05$) correlations, using an α value not corrected with the Dunn–Šidák method.

Birds

Among the results that are significant at the uncorrected $\alpha = 0.05$ (see Table 8.7) it is worth mentioning that:

(a) Increased evolutionary trends towards sociality are associated with evolutionary trends towards polygamy.
(b) As social unit size tends to increase over evolutionary time, affiliative same-sex mounting tends to decrease.

These results are consistent with a higher level of heterosexual competition among members of a social group, as expressed in polygamous mating systems, leading to lower levels of affiliative same-sex mounting, especially in larger groups.

(c) Evolutionary trends towards dominance same-sex mounting are associated with evolutionary shifts towards plumage sexual monochromatism. This trend may result from the marginally non-significant evolutionary association of sexual plumage monochromatism with increased sociality (see below).
(d) Pair bonds tend to decrease as plumage sexual dichromatism increases. This is expected from the strong positive association between plumage sexual dichromatism and polygamy that we already mentioned in the testing of the *Synthetic Reproductive Skew Model of Homosexuality* and that, as I have shown here, can withstand a draconian Dunn–Šidák correction.
(e) Evolutionary trends towards increased ecological constraints on dispersal are positively associated with evolutionary trends towards polygamy in this sample of species.
(f) With regard to relatedness, I detected a significant evolutionary trend in birds for increased relatedness to be associated with an increased level of male-bias in the sex ratio. This can be easily explained by the well-known pattern in birds for males to be the philopatric sex, thus the more males there are in an avian colony, social group or population the more likely it is that the individuals will be more closely related.
(g) As the degree of relatedness between interactants increases, there is a trend for dominance mounting also to increase.

Five tests produced results that were only marginally not significant at the α value of 0.05 (i.e. $0.10 > p \geq 0.05$) and that are also worth mentioning:

(h) As I indicated in (c) (see above), evolutionary trends towards plumage monochromatism are associated with evolutionary trends towards sociality.

(i) The more social the species is, the more females become involved in same-sex mounting.
(j) Increased levels of sociality are also associated with increased degrees of relatedness among members of the group or flock.
(k) From (h) and (j) we can also understand why increased levels of plumage monochromatism are associated with increased levels of intra-group/intra-flock relatedness.
(l) From (i) and (j) we can understand why increased levels of relatedness between interactants are associated with greater involvement of females in same-sex mounting.

Mammals

Ten correlations were significant at the uncorrected $\alpha = 0.05$ (see Table 8.8) that were not mentioned in my previous bivariate tests of the model:

(a) Evolutionary trends towards same-sex mounting are associated with evolutionary trends towards decreased dominance same-sex mounting. This result suggests that in mammals same-sex mounting tends to evolve multiple functions, which, according to the model, are both sexual and socio-sexual: in the latter category, same-sex mounting can have affiliative or competitive functions. That both kinds of socio-sexual function occur in mammals living in complex societies is clearly shown by:
(b) the positive correlation between affiliative mount contrasts and dominance mount contrasts.
(c) The increase in social unit size seems to be linked with an evolutionary trend towards increased ecological constraints.
(d) Ecological constraints to dispersal, in turn, are associated with increased levels of relatedness within social groups.
(e) The relationship between dominance mount contrasts and relatedness contrasts is negative as expected from the action of inbreeding avoidance mechanisms, as I already mentioned above. The trend for affiliative mounting goes in the same negative correlation direction as expected, but the trend is not significant ($p = 0.36$).
(g) As the sex ratio becomes more male-biased, affiliative mounting tends to decrease, suggesting a greater role of females in affiliative mounting interactions in mammals.

A series of correlations that were significant at the uncorrected α also indicate the important role of sexual selection and mating system in the evolution of same-sex mounting in mammals as suggested by the model.

(h) Evolutionary trends towards dominance mounting tend to be associated with evolutionary trends towards sexual body-size monomorphism, whereas
(i) increased levels of ecological constraints are associated with evolutionary trends towards increased sexual body size dimorphism. This result is consistent with the
(j) greater level of relatedness among interactants in polygamous mating systems.

Results (h), (i) and (j) can be better understood when females, the philopatric sex in most mammals, are the sex involved in same-sex mounting. In fact

(k) greater levels of ecological constraints and, to a lesser extent, increased social unit size are associated with a more female-biased sex ratio. Moreover, I also detected a marginally non-significant trend (uncorrected α) for females to be the sex most involved in same-sex mounting as the degree of sociality increases.

REFERENCES

Aastrup, P. 2003. Muskox site fidelity and group cohesion in Jameson Land, East Greenland. *Polar Biology* **27**: 50–5.

Abdi, H. 2003. Partial regression coefficients. In: Lewis-Beck, M., Bryman A. and Futing T. (Eds) *Encyclopedia of Social Sciences Research Methods.* Thousand Oaks, CA: Sage.

Abe, K. and Moran, P. A. P. 1969. Parental age of homosexuals. *British Journal of Psychiatry* **115**: 313–18.

Aboitiz, F. and Montiel, J. 2003. One hundred million years of interhemispheric communication: the history of the corpus callosum. *Brazilian Journal of Medical and Biological Research* **36**: 409–20.

Abraham, K. F., Leafloor, J. O. and Rusch, D. H. 1999. Molt migrant Canada geese in northern Ontario and Western James Bay. *Journal of Wildlife Management* **63**: 649–55.

Adam, B. D. 1985. Age, structure, and sexuality: reflections on the anthropological evidence on homosexual relations. *Journal of Homosexuality* **11**: 19–33.

Adams, E. S. 2005. Bayesian analysis of linear dominance hierarchies. *Animal Behaviour* **69**: 1191–201.

Ader, R., Felten, D. and Cohen, N. 1990. Interactions between the brain and the immune system. *Annual Reviews of Pharmacology and Toxicology* **30**: 561–602.

Adkins-Regan, E. 1988. Sex hormones and sexual orientation in animals. *Psychobiology* **16**: 335–47.

Adkins-Regan, E. and Krakauer, A. 2000. Removal of adult males from the rearing environment increases preference for same-sex partners in the zebra finch. *Animal Behaviour* **60**: 47–53.

Adkins-Regan, E. and Wade, J. 2001. Masculinized sexual partner preference in female zebra finches with sex-reversed gonads. *Hormones and Behavior* **39**: 22–29.

Adriaens, P. R. and A. DeBlock. 2006. The evolution of a social construction: the case of male homosexuality. *Perspectives in Biology and Medicine* **49**: 570–85.

Agate, R. J., Grisham, W., Wade, J. et al. 2003. Neural not gonadal origin of brain sex differences in a gynandromorphic finch. *Proceedings of the National Academy of Sciences, USA* **100**: 4873–8.

Ågmo, A. and Ellingsen, E. 2003. Relevance of non-human animal studies to the understanding of human sexuality. *Scandinavian Journal of Psychology* **44**: 293–301.

Agosta, S. J., Morton, D., Marsh, B. D. and Kuhn, K. M. 2005. Nightly, seasonal, and yearly patterns of bat activity at night roosts in the Central Appalachians. *Journal of Mammalogy* **86**: 1210–19.

Akçakaya, H. R., McCarthy, M. A. and Pearce, J. L. 1995. Linking landscape data with population viability analysis: management options for the helmeted honeyeater *Lichenostomus melanops cassidix*. *Biological Conservation* **73**: 169–76.

Akçay, E. and Roughgarden, J. 2007. Extra-pair paternity in birds: review of the genetic benefits. *Evolutionary Ecology Research* **9**: 855–68.

Akers, J. S. and Conaway, C. H. 1979. Female homosexual behaviour in *Macaca mulatta*. *Archives of Sexual Behavior* **8**: 63–80.

Aketa, K., Asano, S., Wakai, Y. and Kawamura, A. 2001. Apparent digestibility of eelgrass in dugongs (*Dugong dugon*). *Mammalian Science* **82**: 23–34.

Alanko, K., Santtila, P., Harlaar, N. et al. 2009a. Common genetic effects of gender atypical behavior in childhood and sexual orientation in adulthood: a study of Finnish twins. *Archives of Sexual Behavior* **39**: 81–92. DOI 10.1007/S10508-008-9457-3.

Alanko, K., Santtila, P., Witting, K. et al. 2009b. Psychiatric symptoms and same-sex sexual attraction and behavior in light of childhood gender atypical behavior and parental relationships. *Journal of Sex Research* **46**: 1–11.

Alatalo, R. V., Carlson, A., Lundberg, A. and Ulfstrand, S. 1981. The conflict between male polygamy and female monogamy: the case of the pied flycatcher *Ficedula hypoleuca*. *American Naturalist* **117**: 738–53.

Al-Attia, H. M. 1996. Gender identity and role in a pedigree of arabs with intersex due to 5 alpha reductase-2 deficiency. *Psychoneuroendocrinology* **21**: 651–7.

Alberts, S. C., Watts, H. E. and Altmann, J. 2003. Queuing and queue-jumping: long-term patterns of reproductive skew in male savannah baboons, *Papio cynocephalus*. *Animal Behaviour* **65**: 821–40.

Albes, E. 1977. *The Axis Deer in Texas*. Texas Agricultural Experimental Station. Texas A & M University. Caesar Kleberg Research Program in Wildlife Ecology and Department of Wildlife and Fisheries Sciences.

Albonetti, M. E. and Dessì-Fulgheri, F. 1990. Female-female mounting in the European rabbit. *Zeitschrift für Säugetierkunde* **55**: 128–38.

Alcock, J. and Sherman, P. 1994. The utility of the proximate-ultimate dichotomy in ethology. *Ethology* **96**: 58–62.

Alekseyenko, O. V., Waters, P., Zhou, H. and Baum, M. J. 2007. Bilateral damage to the sexually dimorphic medial preoptic area/anterior hypothalamus of male ferrets causes a female-typical preference for and a hypothalamic Fos response to male body odors. *Physiology and Behavior* **90**: 438–49.

Alexander, B. M., Perkins, A., van Kirk, E. A., Moss, G. E. and Fitzgerald, J. A. 1993. Hypothalamic and hypophyseal receptors for estradiol in high and low sexually performing rams. *Hormones and Behavior* **27**: 296–307.

Alexander, B. M., Rose, J. D., Stellflug, J. N., Fitzgerald, J. A. and Moss, G. E. 2001. Low-sexually performing rams but not male-oriented rams can be discriminated by cell size in the amygdala and preoptic area: a morphometric study. *Behavioural Brain Research* **119**: 15–21.

Alexander, B. M., Stellflug, J. N., Rose, J. D., Fitzgerald, J. A. and Moss, G. E. 1999. Behavior and endocrine changes in high-performing, low-performing, and male-oriented domestic rams following exposure to rams and ewes in estrus when copulation is precluded. *Journal of Animal Science* **77**: 1869–74.

Alexander, G. M. 2003. An evolutionary perspective of sex-typed toy preferences: pink, blue, and the brain. *Archives of Sexual Behavior* **32**: 7–14.

Alexander, G. M. and Hines, M. 2002. Sex differences in response to children's toys in nonhuman primates (*Cercopithecus aethiops sabaeus*). *Evolution and Human Behavior* **23**: 467–79.

Alexander, J. E. and Sufka, K. J. 1993. Cerebral lateralization in homosexual males: a preliminary EEG investigation. *International Journal of Psychophysiology* **15**: 269–74.

Alexander, R. D. 1971. The search for an evolutionary philosophy of man. *Proceedings of the Royal Society of Victoria* **84**: 99–120.

Alexander, R. D. 1974. The evolution of social behaviour. *Annual Review of Ecology and Systematics* **5**: 325–83.

Alexander, R. D. 1975. The search for a general theory of behaviour. *Behavioral Science* **20**: 77–100.

Ali, S. 1961. *The Book of Indian Birds*. Mumbai: Bombay Natural History Society.

Allen, C. and Bekoff, M. 1997. *Species of Mind: The Philosophy and Biology of Cognitive Ethology*. Cambridge, MA: MIT Press.

Allen, L. S. and Gorski, R. A. 1991. Sexual dimorphism of the anterior commissure and massa intermedia of the human brain. *Journal of Comparative Neurology* **312**: 97–104.

Allen, L. S. and Gorski, R. A. 1992. Sexual orientation and the size of the anterior commissure in the human brain. *Proceedings of the National Academy of Sciences, USA* **89**: 7199–202.

Allen, L. S., Hines, M., Shryne, J. E. and Gorski, R. A. 1989. Two sexually dimorphic cell groups in the human brain. *Journal of Neuroscience* **9**: 497–506.

Allen, L. S., Richey, M. F., Chai, Y. M. and Gorski, R. A. 1991. Sex differences in the corpus callosum of the living human being. *Journal of Neuroscience* **11**: 933–42.

Allen, M. L. and Lemmon, W. B. 1981. Orgasm in female primates. *American Journal of Primatology* **1**: 15–34.

Allman, D. and Miller, J. P. 2005. B cell development and receptor diversity during aging. *Current Opinion in Immunology* **17**: 463–7.

Altizer, S., Davis, A. K., Cook, K. C. and Cherry, J. J. 2004a. Age, sex, and season affect the risk of mycoplasmal conjunctivitis in a southern house finch population. *Canadian Journal of Zoology* **82**: 755–63.

Altizer, S., Nunn, C. L., Thrall, P. H. *et al.* 2003. Social organization and parasite risk in mammals: integrating theory and empirical studies. *Annual Reviews of Ecology, Evolution and Systematics* **34**: 517–47.

Altizer, S. M., Oberhauser, K. O. and Geurts, K. A. 2004b. Transmission of the protozoan parasite, *Ophryocystis elektroscirrha*, in monarch butterfly populations. In: Oberhauser, K. and Solensky, M. (Eds) *The Monarch Butterfly: Biology and Conservation*. Ithaca, NY: Cornell University Press.

Altmann, J. and Alberts, S. C. 2003. Variability in reproductive success viewed from a life-history perspective in baboons. *American Journal of Human Biology* **15**: 401–9.

Altrichter, M., Drews, C., Carrillo, E. and Sáenz, J. 2001. Sex ratio and breeding of white-lipped peccaries *Tayassu pecari* (Artiodactyla: Tayassuidae) in a Costa Rican rainforest. *Revista de Biología Tropical* **49**: 383–9.

Alvarez, F. and Cónsul, C. 1978. The structure of social behaviour in *Theropithecus gelada*. *Primates* **19**: 45–59.

Alwin, D. F. and Hauser, R. M. 1975. The decomposition of effects in path analysis. *American Sociological Review* **40**: 37–47.

Amateau, S. K. and McCarthy, M. 2004. Induction of PGE_2 by estradiol mediates developmental masculinisation of sex behaviour. *Nature Neuroscience* **7**: 643–50.

Ambady, N., Bernieri, F. J. and Richeson, J. A. 2000. Toward a histology of social behavior: judgmental accuracy from thin slices of the behavioral stream. *Advances in Experimental Social Psychology* **32**: 201–71.

Ambady, N., Hallahan, M. and Conner, B. 1999. Accuracy of judgments of sexual orientation from thin slices of behavior. *Journal of Personality and Social Psychology* **77**: 538–47.

Ambady, N. and Rosenthal, R. 1992. Thin slices of expressive behavior as predictors of interpersonal consequences: a meta-analysis. *Psychological Bulletin* **111**: 256–74.

Ambs, S. M., Boness, D. J., Bowen, W. D., Perry, E. A. and Fleischer, R. C. 1999. Proximate factors associated with high levels of extraconsort fertilization in polygynous grey seals. *Animal Behaviour* **58**: 527–35.

Amidon, K. S. 2008. Sex on the brain: the rise and fall of German sexual science. *Endeavour* **32**: 64–9.

AnAge 2007. Database of the Human Ageing Genomic Resources. www.genomics.senescence.info/species.

Anderson, D. K., Rhees, R. W. and Fleming, D. E. 1985. Effects of prenatal stress on differentiation of the sexually dimorphic nucleus of the preoptic area (SDN-POA) of the rat brain. *Brain Research* **332**: 113–18.

Anderson, E. 2008. "Being masculine is not about who you sleep with...": heterosexual athletes contesting masculinity and the one-time rule of homosexuality. *Sex Roles* **58**: 104–15.

Anderson, P. K. 1982. Studies of dugongs *Dugong dugon* at Shark Bay, Western Australia 1. Analysis of population size, composition, dispersion and habitat use on the basis of aerial survey. *Australian Wildlife Research* **9**: 69–84.

Anderson, P. K. 1997. Shark Bay dugongs in summer I: lek mating. *Behaviour* **134**: 433–62.

Anderson, P. K. 2002. Habitat, niche, and evolution of sirenian mating systems. *Journal of Mammalian Evolution* **9**: 55–98.

Animal Diversity Web 2007. Website of the Museum of Zoology, University of Michigan: http://animaldiversity.ummz.umich.edu

Anon 2003. *DSM-IV-TR* diagnostic criteria for gender identity disorder. *Psychiatric News* **38**: 32.

Apollonio, M., Focardi, S., Toso, S. and Nacci, L. 1998. Habitat selection and group formation pattern of fallow deer *Dama dama* in a subMediterranean environment. *Ecography* **21**: 225–34.

Apollonio, M., Scotti, M. and Gosling, L. M. 2003. Mating success and fidelity to territories in a fallow deer lek: A female removal experiment. *Naturwissenschaften* **90**: 553–7.

Archie, E. A., Morrison, T. A., Foley, C. A. H., Moss, C. J. and Alberts, S. C. 2006a. Dominance rank relationships among wild female African elephants, *Loxodonta africana*. *Animal Behaviour* **71**: 117–27.

Archie, E. A., Moss, C. J. and Alberts, S. C. 2006b. The ties that bind: genetic relatedness predicts the fission and fusion of social groups in wild African elephants. *Proceedings of the Royal Society of London* B **273**: 513–22.

Arletti, R., Bazani, C., Castelli, M. and Bertolini, A. 1985. Oxytocin improves male copulatory performance in rats. *Hormones and Behavior* **19**: 14–20.

Arletti, R., Benelli, A. and Bertolini, A. 1992. Oxytocin involvement in male and female sexual behaviour. *Annals of the New York Academy of Sciences* **652**: 180–93.

Arletti, R. and Bertolini, A. 1985. Oxytocin stimulates lordosis behaviour in female rats. *Neuropeptides* **6**: 247–253.

Armbruster, P. and Lande, R. 1993. A population viability analysis for African elephant (*Loxodonta africana*): how big should reserves be? *Conservation Biology* **7**: 602–10.

Armitage, K. B. 1981. Sociality as life-history tactic of ground squirrels. *Oecologia* **48**: 36–49.

Armitage, K. B. 1998. Reproductive strategies of yellow-bellied marmots: energy conservation and differences between the sexes. *Journal of Mammalogy* **79**: 385–93.

Armitage, K. B., Downhower, J. F. and Svendsen, G. E. 1976. Seasonal changes in weights of marmots. *American Midland Naturalist* **96**: 36–51.

Arnold, A. P. 2004. Sex chromosomes and brain gender. *Nature Reviews* **5**: 1–8.

Arnold, A. P., Xu, J., Grisham, W. et al. 2004. Sex chromosomes and brain sexual differentiation. *Endocrinology* **145**: 1057–62.

Arnold, T. W. and Martin, P. A. 1992. Winter habitat use by male and female American kestrels, *Falco sperverius*, in Southwestern Ontario. *Canadian Field-Naturalist* **106**: 336–41.

Aron, C., Chateau, D., Schaeffer, C. and Roos, J. 1991. Heterotypical sexual behaviour in male mammals: the rat as an experimental model. In: Haug, M., Brain, P. F. and Aron, C. (Eds) *Heterotypical Behaviour in Man and Animals*, pp. 98–126. London: Chapman and Hall.

Aronson, L. R. and Cooper, M. L. 1979. Amygdaloid hypersexuality in male cats re-examined. *Physiology and Behavior* **22**: 257–65.

Arrowood, P. C. 1988. Duetting, pair bonding and agonistic display in parakeet pairs. *Behaviour* **106**: 129–57.

Asa, C. 1999. Male reproductive success in free-ranging feral horses. *Behavioral Ecology and Sociobiology* **47**: 89–93.

Asa, C. S., Goldfoot, D. A. and Ginter, O. J. 1979. Sociosexual behaviour and the ovulatory cycle of ponies (*Equus caballus*) observed in harem groups. *Hormones and Behavior* **13**: 49–65.

Asa, C. S. and Valdespino, C. 1998. Canid reproductive biology: an integration of proximate mechanisms and ultimate causes. *American Zoologist* **38**: 251–9.

Ashworth, A. E. and Walker, W. M. 1972. Social structure and homosexuality: a theoretical appraisal. *The British Journal of Sociology* **23**: 146–58.

Aste, N., Balthazart, J., Absil, P. et al. 1998. Anatomical and neurochemical definition of the nucleus of the stria terminalis in japanese quail (*Coturnix japonica*). *Journal of Comparative Neurology* **396**: 141–57.

Aucoin, M. W. and Wassersug, R. J. 2006. The sexuality and social performance of androgen-deprived (castrated) men throughout history: implications for modern day cancer patients. *Social Science and Medicine* **63**: 3162–73.

Auger, A. P., Moffatt, C. A. and Blaustein, J. D. 1997. Progesterone-independent activation of rat brain progestin receptors by reproductive stimuli. *Endocrinology* **138**: 511–14.

Aujard, F., Heistermann, M., Thierry, B. and Hodges, J. K. 1998. Functional significance of behavioral, morphological, and endocrine correlates across the ovarian cycle in semifree ranging female Yonkean macaques. *American Journal of Primatology* **46**: 285–309.

Baba, T., Endo, T., Honnma, H. et al. 2007. Association between polycystic ovary syndrome and female-to-male transsexuality. *Human Reproduction* **22**: 1011–16.

Bachelot, A., Plu-Bureau, G., Thibaud, E. et al. 2007. Long-term outcome of patients with congenital adrenal hyperplasia due to 21-hydroxylase deficiency. *Hormone Research* **67**: 268–76.

Backhouse, K. M. 1960. The grey seal (*Halichoerus grypus*) outside the breeding season. A preliminary report. *Mammalia* **24**: 307–12.

Bagatell, C. J., Heiman, J. R., Rivier, J. E. and Bremner, W. J. 1994. Effects of endogenous testosterone and estradiol on sexual behavior in normal young men. *Journal of Clinical Endocrinology and Metabolism* **78**: 711–16.

Bagemihl, B. 1999. *Biological Exuberance: Animal Homosexuality and Natural Diversity*. New York: St. Martin's Press.

Bailey, J. M. and Bell, A. P. 1993. Familiality of female and male homosexuality. *Behavior Genetics* **23**: 313–22.

Bailey, J. M. and Benishay, D. S. 1993. Familial aggregation of female sexual orientation. *American Journal of Psychiatry* **150**: 272–7.

Bailey, J. M., Bobrow, D., Wolfe, M. and Mikach, S. 1995. Sexual orientation of adult sons of gay fathers. *Developmental Psychology* **31**: 124–9.

Bailey, J. M., Gaulin, S., Agyei, Y. and Gladue, B. A. 1994. Effects of gender and sexual orientation on evolutionarily relevant aspects of human mating psychology. *Journal of Personality and Social Psychology* **66**: 1081–93.

Bailey, J. M. and Pillard, R. C. 1991. A genetic study of male sexual orientation. *Archives of General Psychiatry* **48**: 1089–96.

Bailey, J. M., Pillard, R. C., Dawood, K. *et al.* 1999. A family history study of male sexual orientation using three independent samples. *Behavior Genetics* **29**: 79–86.

Bailey, J. M., Pillard, R. C., Neale, M. C. and Agyei, Y. 1993. Heritable factors influence sexual orientation in women. *Archives of General Psychiatry* **50**: 217–23.

Bailey, J. M., Willerman, L. and Parks, C. 1991. A test of the maternal stress theory of human homosexuality. *Archives of Sexual Behavior* **20**: 277–93.

Bailey, J. V., Farquhar, C., Owen, C. and Mangtani, P. 2004. Sexually transmitted infections in women who have sex with women. *Sexually Transmitted Infections* **80**: 244–6.

Bailey, N. W. and Zuk, M. 2009. Same-sex sexual behavior and evolution. *Trends in Ecology and Evolution* **24**: 439–46.

Bainbridge, D. R. J. 2000. Evolution of mammalian pregnancy in the presence of the maternal immune system. *Reviews of Reproduction* **5**: 67–74.

Baird, A. D., Wilson, S. J., Bladin, P. F., Saling, M. M. and Reutens, D. C. 2003a. The amygdala and sexual drive: insights from temporal lobe epilepsy surgery. *Annals of Neurology* **55**: 87–96.

Baird, A. D., Wilson, S. J., Bladin, P. F., Saling, M. M. and Reutens, D. C. 2003b. Sexual outcome after epilepsy surgery. *Epilepsy and Behavior* **4**: 268–78.

Baird, J. B. 2006. *The impact of self-esteem, media internalization, sexual orientation, and ethnicity on drive for muscularity in men who workout in gyms.* MA thesis, College of Arts and Sciences, Georgia State University, USA.

Baker, A. J., Pereira, S. L. and Paton, T. A. 2007. Phylogenetic relationships and divergence times of Charadriiformes genera: multigene evidence for the Cretaceous origin of at least 14 clades of shorebirds. *Biological Letters* **3**: 205–9.

Baker, R. R. and Bellis, M. A. 1993a. Human sperm competition: ejaculate adjustment by males and the function of masturbation. *Animal Behaviour* **46**: 861–85.

Baker, R. R. and Bellis, M. A. 1993b. Human sperm competition: ejaculate manipulation by females and a function for the female orgasm. *Animal Behaviour* **46**: 887–909.

Bakker, J., van Ophemert, J. and Slob, A. K. 1993. Organization of partner preference and sexual behaviour and its nocturnal rhythmicity in male rats. *Behavioral Neuroscience* **107**: 1049–58.

Bakker, J., Honda, S.-I., Harada, N. and Balthazart, J. 2002a. Sexual partner preference requires a functional aromatase (*Cyp19*) gene in male mice. *Hormones and Behavior* **42**: 158–71.

Bakker, J., Honda, S.-I., Harada, N. and Balthazart, J. 2002b. The aromatase knock-out mouse provides new evidence that estradiol is required during development in the female for the expression of sociosexual behaviors in adulthood. *Journal of Neuroscience* **22**: 9104–12.

Baldwin, J. D. and Baldwin, J. I. 1989. The socialization of homosexuality and heterosexuality in a non-western society. *Archives of Sexual Behavior* **18**: 13–29.

Baldwin, J. D. and Baldwin, J. I. 1997. Gender differences in sexual interest. *Archives of Sexual Behavior* **26**: 181–210.

Baldwin, J. M. 1896. A new factor in evolution. *American Naturalist* **30**: 441–51.

Baldwin, J. M. 1902. *Development and Evolution*. London: MacMillan Co.

Ball, G. F. and Balthazart, J. 2004. Hormonal regulation of brain circuits mediating male sexual behavior in birds. *Physiology and Behavior* **83**: 329–41.

Ballenberghe, V, van and Miquelle, D. G. 1993. Mating in moose: timing, behaviour, and male access patterns. *Canadian Journal of Zoology* **71**: 1687–90.

Balsam, K. F. and Mohr, J. J. 2007. Adaptation to sexual orientation stigma: a comparison of bisexual and lesbian/gay adults. *Journal of Counseling Psychology* **54**: 305–19.

Balthazart, J., Baillien, M., Charlier, T. D., Cornil, C. A. and Ball, G. F. 2003. The neuroendocrinology of reproductive behaviour in Japanese quail. *Domestic Animal Endocrinology* **25**: 69–82.

Balthazart, J., Castagna, C. and Ball, G. F. 1997. Aromatase inhibition blocks the activation and sexual differentiation of appetitive male sexual behaviour in Japanese quail. *Behavioral Neuroscience* **111**: 381–97.

Bamshad, M., Wooding, S., Salisbury, B. A. and Stephens, J. C. 2004. Deconstructing the relationship between genetics and race. *Nature Genetics* **5**: 598–609.

Bancroft, J. 2005. The endocrinology of sexual arousal. *Journal of Endocrinology* **186**: 411–27.

Bancroft, J., Axworthy, D. and Ratcliffe, S. 1982. The personality and psycho-sexual development of boys with 47 XXY chromosome constitution. *Journal of Child Psychology and Psychiatry* **23**: 169–80.

Bancroft, J., Sanders, D., Davidson, D. and Warner, P. 1983. Mood, sexuality, hormones, and the menstrual cycle. III. Sexuality and the role of androgens. *Psychosomatic Medicine* **45**: 509–16.

Bandeira de Melo, L. F., Lima Sabato, M. A., Vaz Magni, E. M., Young, R. J. and Coelho, C. M. 2007. Secret lives of maned wolves (*Chrysocyon brachyurus* Illiger 1815): as revealed by GPS tracking collars. *Journal of Zoology, London* **271**: 27–36.

Bandura, A. 1977. *Social Learning Theory*. Englewood Cliffs NJ: Prentice-Hall.

Bandura, A. 1989. Social cognitive theory. In: R. Vasta (Ed.) *Annals of Child Development*, vol. 6. pp. 1–60. Six theories of child development. Greenwich, CT: JAI Press.

Banks, A. and Gartrell, N. K. 1995. Hormones and Sexual orientation: a questionable link. *Journal of Homosexuality* **28**: 247–68.

Banks, E. M. 1956. Social organization in red jungle fowl hens (*Gallus gallus* subsp.). *Ecology* **37**: 239–48.

Barash, D. P. 1973. The social biology of the olympic marmot. *Animal Behaviour Monographs* **6**: 171–245.

Barash, D. P. 1974. The social behaviour of the hoary marmot (*Marmota caligata*). *Animal Behaviour* **22**: 256–61.

Barboza, P. S. and Bowyer, R. T. 2000. Sexual segregation in dimorphic deer: a new gastrocentric hypothesis. *Journal of Mammalogy* **81**: 473–89.

Barboza, P. S. and Bowyer, R. T. 2001. Seasonality of sexual segregation in dimorphic deer: extending the gastrocentric model. *Alces* **37**: 275–92.

Barker, F. K., Cibois, A., Schikler, P., Feinstein, J. and Cracraft, J. 2004. Phylogeny and diversification of the largest avian radiation. *Proceedings of the National Academy of Sciences, USA* **101**: 11040–5.

Barnett, S. A. 1958. An analysis of social behaviour in wild rats. *Proceedings of the Zoological Society of London* **130**: 107–52.

Barrette, C. 1977. Social behaviour of captive mutjacs *Muntiacus reevesi* (Ogilby 1839). *Zeitschrift für Tierpsychologie* **43**: 188–213.

Barrette, C. 1991. The size of axis deer fluid groups in Wilpatu National Park, Sri Lanka. *Mammalia* **55**: 207–20.

Bartels, A. and Zeki, S. 2000. The neural basis of romantic love. *NeuroReport* **11**: 3829–34.

Bartlett, N. H. and Vasey, P. L. 2006. A retrospective study of childhood gender-atypical behavior in Samoan *fa'afafine*. *Archives of Sexual Behavior* **35**: 659–66.

Bartlett, N. H., Vasey, P. L. and Bukowski, W. M. 2000. Is gender identity disorder in children a mental disorder? *Sex Roles* **43**: 753–85.

Bartley, A. J., Jones, D. W. and Weinberger, D. R. 1997. Genetic variability of human brain size and cortical gyral patterns. *Brain* **120**: 257–69.

Barton, R. A. 1996. Neocortex size and behavioural ecology in primates. *Proceedings of the Royal Society of London*, B **263**: 173–7.

Barton, R. A. 1998. Visual specialization and brain evolution in primates. *Proceedings of the Royal Society of London*, B **265**: 1933–7.

Barton, R. A. and Harvey, P. H. 2000. Mosaic evolution of brain structure in mammals. *Nature* **405**: 1055–8.

Bartoš, L. and Holečková, J. 2006. Exciting ungulates: male-male mounting in fallow, white-tailed and red deer. In: Sommer, V. and Vasey, P. L. (eds.) *Homosexual Behaviour in Animals: An Evolutionary Perspective*, pp. 154–71. Cambridge: Cambridge University Press.

Bartoš, L., Sustr, P., Janovsky, P. and Bertagnoli, J. 2003. Sika deer (*Cervus nippon*) lekking in a free-ranging population in Northern Austria. *Folia Zoologica* **52**: 1–10.

Bateson, P. and Hinde, R. A. 1987. Developmental changes in sensitivity to experience. In: Bornstein, M. H. (ed.) *Sensitive Periods in Development*, pp. 19–34. New Jersey: Lawrence Erlbaum.

Baucom, D. H. and Danker-Brown, P. 1979. Influence of sex roles on the development of learned helplessness. *Journal of Consulting and Clinical Psychology* **47**: 928–36.

Bauer, G. R. and Welles, S. L. 2001. Beyond assumptions of negligible risk: sexually transmitted diseases and women who have sex with women. *American Journal of Public Health* **91**: 1282–6.

Bauer, M. and Breed, W. G. 2008. Testis mass of the spinifex hopping mouse and its impact on fertility potential. *Journal of Zoology* **274**: 349–56.

Bauers, K. A. and Hearn, J. P. 1994. Patterns of paternity in relation to male social rank in the stumptail macaque, *Macaca arctoides*. *Behaviour* **129**: 149–76.

Baum, M. J. 1976. Effects of testosterone propionate administered perinatally on sexual behaviour of female ferrets. *Journal of Comparative Physiological Psychology* **90**: 399–410.

Baum, M. J. 2006. Mammalian animal models of psychosexual differentiation: when is 'translation' to the human situation possible? *Hormones and Behavior* **50**: 579–88.

Baum, M. J. 2009. Sexual differentiation of pheromone processing: links to male-typical mating behavior and partner preference. *Hormones and Behaviour* **55**: 579–88.

Baum, M. J., Brown, J. J. G., Kica, E. *et al.* 1994. Effect of a null mutation of the c-*fos* proto-oncogene on sexual behaviour of male mice. *Biology of Reproduction* **50**: 1040–8.

Baum, M. J. and Erskine, M. S. 1984. Effect of neonatal gonadectomy and administration of testosterone on coital masculinization in the ferret. *Endocrinology* **115**: 2440–4.

Baum, M. J., Erskine, M. S., Kornberg, E. and Weaver, C. E. 1990. Prenatal and neonatal testosterone exposure interact to affect differentiation of sexual behaviour and partner preferences in female ferrets. *Behavioral Neuroscience* **104**: 183–98.

Baum, M. J., Gallagher, C. A., Martin, J. T. and Damassa, D. A. 1982. Effects of testosterone, dihydrotestosterone, or estradiol administered neonatally on sexual behaviour of female ferrets. *Endocrinology* **111**: 773–80.

Baum, M. J. and Kelliher, K. R. 2009. Complementary roles of the main and accessory olfactory systems in mammalian mate recognition. *Annual Review of Physiology* **71**: 141–60.

Baumeister, R. F. 2000. Gender differences in erotic plasticity: the female sex drive as socially flexible and responsive. *Psychological Bulletin* **126**: 347–74.

Beach, F. A. 1942. Comparison of copulatory behaviour of male rats raised in isolation, cohabitation, and segregation. *Journal of Genetic Psychology* **60**: 121–36.

Beach, F. A. 1968. Factors involved in the control of mounting behaviour by female mammals. In: Diamond, M. (Ed.) *Perspectives in Reproduction and Sexual Behaviour*, pp. 83–131. Bloomington, IN: Indiana University Press.

Beach, F. A. 1987. Alternative interpretations of the development of G-I/R. In: Reinisch, J. M., Rosenblum L. A. and Sanders S. A. (Eds) *Masculinity/Femininity: Basic Perspectives*, pp. 29–34. New York: Oxford University Press.

Bearman, P. S. and Brückner, H. 2002. Opposite-sex twins and adolescent same-sex attraction. *American Journal of Sociology* **107**: 1179–205.

Becker, J. T., Bass, S. M., Dew, M. A. *et al.* 1992. Hand preference, immune system disorder and cognitive function among gay/bisexual men: the Multicenter AIDS/Cohort Study (MACS). *Neuropsychologia* **30**: 229–35.

Beehler, B. and Pruett-Jones, S. G. 1983. Display dispersion and diet of birds of paradise: a comparison of nine species. *Behavioral Ecology and Sociobiology* **13**: 229–38.

Beekman, M., Peeters, C. and O'Riain, M. J. 2006. Developmental divergence: neglected variable in understanding the evolution of reproductive skew in social animals. *Behavioral Ecology* **17**: 622–7.

Beer, J. M. and Horn, J. M. 2000. The influence of rearing order on personality development within two adoption cohorts. *Journal of Personality* **68**: 789–819.

Beers, J. R. 1996. *The desire to parent in gay men*. PhD Thesis, Columbia University, USA.

Bell, A. P. 1975. Research in homosexuality: back to the drawing board. *Archives of Sexual Behavior* **4**: 421–31.

Bell, A. P. and Weinberg, M. S. 1978. *Homosexualities: A Study of Diversity Among Men and Women*. New York: Simon and Schuster.

Bell, A. P., Weinberg M. S. and Hammersmith, S. 1981. *Sexual Preference: its Development in Men and Women*. Bloomington, IN: Indiana University Press.

Bellis, M. A. and Baker, R. R. 1990. Do females promote sperm competition? Data for humans. *Animal Behaviour* **40**: 997–9.

Belsky, J., Steinberg, L. and Draper, P. 1991. Childhood experience, interpersonal development, and reproductive strategy: an evolutionary theory of socialization. *Child Development* **62**: 647–70.

Belthoff, J. R. and Gauthreaux, S. A. Jr. 1991. Partial migration and differential winter distribution of house finches in the Eastern USA. *Condor* **93**: 374–82.

Bem, D. J. 1996. Exotic becomes erotic: a developmental theory of sexual orientation. *Psychological Review* **103**: 320–35.

Bem, D. 2008. Is there a causal link between childhood gender nonconformity and adult homosexuality? *Journal of Gay and Lesbian Mental Health* **12**: 61–79.

Bem, S. L. 1974. The measurement of psychological androgyny. *Journal of Consulting and Clinical Psychology* **42**: 155–62.

Bem, S. L. 1981. Gender schema theory: a cognitive account of sex typing. *Psychological Review* **88**: 354–64.

Bem, S. L. 1987. Masculinity and femininity exist only in the mind of the perceiver. In: Reinisch, J. M., Rosenblum, L. A. and Sanders S. A. (Eds). *Masculinity/Femininity: Basic Perspectives*, pp. 304–11. New York: Oxford University Press.

Bender, L. C. 2000. Relationships between social group size of Columbian black-tailed deer and habitat cover in Washington. *Northwestern Naturalist* **81**: 49–53.

Bene, E. 1965a. On the genesis of male homosexuality: an attempt at clarifying the role of the parents. *British Journal of Psychiatry* **111**: 803–13.

Bene, E. 1965b. On the genesis of female homosexuality. *British Journal of Psychiatry* **111**: 815–21.

Bennett, E. L. and Sebastian, A. C. 1988. Social organization and ecology of proboscis monkeys (*Nasalis larvatus*) in mixed coastal forest in Sarawak. *International Journal of Primatology* **9**: 233–55.

Bennett, P. M. and Owens, I. P. F. 2002. *Evolutionary Ecology of Birds: Life Histories, Mating Systems and Extinction*. Oxford: Oxford University Press.

Bentz, E.-K., Hefler, L. A., Kaufmann, U., Huber, J. C., Kolbus, A. and Tempfer, C. B. 2008. A polymorphism of the *CYP17* gene related to sex steroid metabolism is associated with female-to-male but not male-to-female transexualism. *Fertility and Sterility* **90**: 56–9.

Bercovitch, F. B., Bashaw, M. J. and del Castillo, S. M. 2006. Sociosexual behavior, male mating tactics, and the reproductive cycle of giraffe *Giraffa camelopardalis*. *Hormones and Behavior* **50**: 314–21.

Bergerud, A. T. 1974. Rutting behaviour of Newfoundland caribou. In: Geist, V. and Walther, F. (eds.) *The Behaviour of Ungulates and its Relation to Management*, pp. 395–435. IUCN Publ. New Ser. **24**, Vol. 1.

Berenbaum, S. A. and Bailey, J. M. 2003. Effects on gender identity of prenatal androgens and genital appearance: evidence from girls with congenital adrenal hyperplasia. *Journal of Clinical Endocrinology and Metabolism* **88**: 1102–6.

Berenbaum, S. A., Duck, S. C. and Bryk, K. 2000. Behavioral effects of prenatal versus postnatal androgen excess in children with 21-hydroxylase-deficient congenital adrenal hyperplasia. *Journal of Clinical Endocrinology and Metabolism* **85**: 727–33.

Berenbaum, S. A. and Hines, M. 1992. Early androgens are related to childhood sex-typed toy preferences. *Psychological Science* **3**: 203–6.

Berenbaum, S. A. and Snyder, E. 1995. Early hormonal influences on childhood sex-typed activity and playmate preferences: implications for development of sexual orientation. *Developmental Psychology* **31**: 31–42.

Berge, C. 1998. Heterochronic processes in human evolution: an ontogenetic analysis of the hominid pelvis. *American Journal of Physical Anthropology* **105**: 441–59.

Berger, B. J., Kolton, S., Zenilman, J. M. *et al.* 1995. Bacterial vaginosis in lesbians: a sexually transmitted disease. *Clinical Infectious Diseases* **21**: 1402–5.

Berger, J. 1985. Instances of female-like behaviour in a male ungulate. *Animal Behaviour* **33**: 333–5.

Berger, J. and Gompper, M. E. 1999. Sex ratios in extant ungulates: products of contemporary predation or past life histories? *Journal of Mammalogy* **80**: 1084–1113.

Bergerud, A. T. and Gratson, M. W. 1988. Survival and breeding strategies of grouse. In: Bergerud, A. T. and Gratson, M. W. (Eds) *Adaptive Strategies and Population Ecology of Northern Grouse*, pp. 473–577. Minneapolis, MN: University of Minnesota Press.

Berglund, H., Lindström, P., Dhejne-Helmy, C. and Savic, I. 2008. Male-to-female transsexuals show sex-atypical hypothalamus activation when smelling odorous steroids. *Cerebral Cortex* **18**: 1900–8.

Berglund, H., Lindström, P. and Savic, I. 2006. Brain response to putative pheromones in lesbian women. *Proceedings of the National Academy of Sciences, USA* **103**: 8269–74.

Berman, L. A. 2003. *The Puzzle: Exploring the Evolutionary Puzzle of Male Homosexuality*. Wilmette, IL: Godot Press.

Berndt, R. M. and Berndt, C. H. 1992. *The World of the First Australians: Aboriginal Traditional life: Past and Present*. Canberra: Aboriginal Studies Press. (Ed.) Primate Behavior vol.1, pp. 71–109, New York: Academic Press.

Bernstein, I. S. 1970. Primate status hierarchies. In: Rosenblum, L. A. (Ed.) *Primate Behavior*, vol. 1, pp. 71–109. New York: Academic Press.

Bernstein, I. S. 1975. Activity patterns in a gelada monkey group. *Folia Primatologica* **23**: 50–71.

Bernstein, I. S., Judge, P. G. and Ruehlmann, T. E. 1993. Sex differences in adolescent rhesus monkeys (*Macaca mulatta*) behaviour. *American Journal of Primatology* **31**: 197–210.

Berridge, K. C. 2003. Pleasures of the brain. *Brain and Cognition* **52**: 106–28.

Berry, S. M., Romero, R., Gomez, R. *et al.* 1995. Premature parturition is characterized by in utero activation of the fetal immune system. *American Journal of Obstetrics and Gynecology* **173**: 1315–20.

Berta, A. and Wyss, A. R. 1994. Pinniped phylogeny. In: Berta, A. and Deméré, T. A. (Eds) Contributions in marine mammal paleontology honouring Frank C. Whitmore, Jr. *Proceedings of the San Diego Society of Natural History* **29**: 33–56.

Bertram, B. C. R. 1975. Social factors influencing reproduction in wild lions. *Journal of Zoology, London* **177**: 463–82.

Bertram, B. C. R. 1980. Vigilance and group size in ostriches *Struthio camelus*. *Animal Behaviour* **28**: 278–86.

Bertran, J. and Margalida, A. 1999. Copulatory behavior of the bearded vulture. *Condor* **101**: 164–8.

Bertran, J. and Margalida, A. 2003. Male-male mountings in polyandrous bearded vultures *Gypaetus barbatus*: an unusual behaviour in raptors. *Journal of Avian Biology* **34**: 334–8.

Beyer, C. 1999. Estrogen and the developing mammalian brain. *Anatomical Embryology* **199**: 379–90.

Bielert, C. and van der Walt, L. A. 1982. Male chacma baboon (*Papio ursinus*) sexual arousal: mediation by visual cues from female conspecifics. *Psychoneuroendocrinology* **7**: 31–48.

Bigner, J. J. and Jacobsen, R. B. 1989. The value of children to gay and heterosexual fathers. *Journal of Homosexuality* **18**: 163–72.

Billeter, J.-C., Rideout, E. J., Dornan, A. J. and Goodwin, S. F. 2006. Control of male sexual behaviour in *Drosophila* by the sex determining pathway. *Current Biology* **16**: R766–R776.

Bininda-Emonds, O. R. P., Gittleman, J. L. and Purvis, A. 1999. Building large trees by combining phylogenetic information: a complete phylogeny of the extant Carnivora (Mammalia). *Biological Reviews* **74**: 143–75.

Bird, A. 2002. DNA methylation patterns and epigenetic memory. *Genes and Development* **16**: 6–21.

Birke, L. 2000. Sitting on the fence: biology, feminism and gender-bending environments. *Women's Studies International Forum* **23**: 587–99.

Birkhead, T. R. 1978. Behavioural adaptations to high density nesting in the common guillemot *Uria aalge*. *Animal Behaviour* **26**: 321–31.

Birkhead, T. R., Hatchwell, B. J., Lindner, R. *et al.* 2001. Extra-pair paternity in the common murre. *Condor* **103**: 158–62.

Birkhead, T. R. and Møller, A. P. 1992. *Sperm Competition in Birds: Evolutionary Causes and Consequences.* London: Academic Press.

Birkhead, T. R. and Nettleship, D. N. 1984. Alloparental care in the common murre (*Uria aalge*). *Canadian Journal of Zoology* **62**: 2121–4.

Bissett, C. and Bernard, R. T. F. 2007. Habitat selection and feeding ecology of the cheetah (*Acinonix jubatus*) in thicket vegetation: is the cheetah a savanna specialist? *Journal of Zoology, London* **271**: 310–17.

Black, D., Gates, G., Sanders, S. and Taylor, L. 2000. Demographics of the gay and lesbian population in the United States: evidence from available systematic data sources. *Demography* **37**: 139–54.

Black, I. B. 1986. Trophic molecules and evolution of the nervous system. *Proceedings of the National Academy of Sciences, USA* **83**: 8249–52.

Blackshaw, J. K., Blackshaw, A. W. and McGlone, J. J. 1997. Buller steer syndrome review. *Applied Animal Behaviour Science* **54**: 97–108.

Blaffer Hrdy, S. and Whitten, P. L. 1987. Patterning of sexual activity. In: Smuts, B. B., Cheney, D. L., Seyfarth, R. M., Wrangham, R. W. and Struhsaker, T. T. (Eds). *Primate Societies*, pp. 370–84. Chicago, IL: University of Chicago Press.

Blakey, J. K. 1996. Nest-sharing by female Blue Tits. *British Birds* **89**: 279–80.

Blanchard, R. 1991. Clinical observations and systematic studies of autogynephilia. *Journal of Sex and Marital Therapy* **17**: 235–51.

Blanchard, R. 2001. Fraternal birth order and the maternal immune hypothesis of male homosexuality. *Hormones and Behavior* **40**: 105–14.

Blanchard, R. 2004. Quantitative and theoretical analyses of the relation between older brothers and homosexuality in men. *Journal of Theoretical Biology* **230**: 173–87.

Blanchard, R. 2008a. Review and theory of handedness, birth order, and homosexuality in men. *Laterality* **13**: 51–70.

Blanchard, R. 2008b. Sex ratio of older siblings in heterosexual and homosexual, right-handed and non-right-handed men. *Archives of Sexual Behavior* **37**: 977–81.

Blanchard, R. and Bogaert, A. F. 1996. Homosexuality in men and number of older brothers. *American Journal of Psychiatry* **153**: 27–31.

Blanchard, R. and Bogaert, A. F. 2004. Proportion of homosexual men who owe their sexual orientation to fraternal birth order: an estimate based on two national probability samples. *American Journal of Human Biology* **16**: 151–7.

Blanchard, R., Cantor, J. M., Bogaert, A. F., Breedlove, S. M. and Ellis, L. 2006. Interaction of fraternal birth order and handedness in the development of male homosexuality. *Hormones and Behavior* **49**: 405–14.

Blanchard, R. and Klassen, P. 1997. H-Y antigen and homosexuality in men. *Journal of Theoretical Biology* **185**: 373–8.

Blanchard, R., Leonard, H., Clemmensen, B. A. and Steiner, B. W. 1987. Heterosexual and homosexual gender dysphoria. *Archives of Sexual Behavior* **16**: 139–52.

Blanchard, R. and Lippa, R. A. 2008. The sex ratio of older siblings in non-right-handed homosexual men. *Archives of Sexual Behavior* **37**: 970–6.

Blanchard, R., McConkey, J. G., Roper, V. and Steiner, B. W. 1983. Measuring physical aggressiveness in heterosexual, homosexual, and transsexual males. *Archives of Sexual Behavior* **12**: 511–24.

Blanchard, R., Zucker, K. J., Cohen-Kettens, P. T., Gooren, L. J. G. and Bailey, J. M. 1996. Birth order and sibling sex ratio in two samples of Dutch gender-disphoric homosexual males. *Archives of Sexual Behavior* **25**: 495–514.

Blanchard, R., Zucker, K. J., Siegelman, M., Dickey, R. and Klassen, P. 1998. The relation of birth order to sexual orientation in men and women. *Journal of Biosocial Science* **30**: 511–19.

Blettner, M., Saverbrei, W., Schlehofer, B., Scheuchenpflug, T. and Friedenreich, C. 1999. Traditional reviews, meta-analyses and pooled analyses in epidemiology. *International Journal of Epidemiology* **28**: 1–9.

Block, J. H. 1973. Conceptions of sex role: some cross-cultural and longitudinal perspectives. *American Psychologist* **28**: 512–26.

Blount, B. G. 1990. Issues in bonobo (*Pan paniscus*) sexual behavior. *American Anthropologist* **92**: 702–14.

Blum, E. 2004. *Does natal territory quality predict human dispersal choices? A test of Emlen's model of family formation*. PhD Thesis, University of Pittsburgh, USA.

Blumstein, D., Daniel, J. C. and Sims, R. A. 2003. Group size but not distance to cover influences agile wallaby (*Macropus agilis*) time allocation. *Journal of Mammalogy* **84**: 197–204.

Blumstein, D. T. and Armitage, K. B. 1999. Cooperative breeding in marmots. *Oikos* **84**: 369–82.

Blumstein, P. and Schwartz, P. 1977. Bisexuality: some social psychological issues. *Journal of Social Issues* **33**: 30–45.

Blurton-Jones, M. M., Roberts, J. A. and Tuszynski, M. H. 1999. Estrogen receptor immunoreactivity in the adult primate brain: Neural distribution and association with p75, *trkA*, and choline acetyltransferase. *Journal of Comparative Neurology* **405**: 529–42.

Bobrow, D. and Bailey, J. M. 2001. Is male homosexuality maintained via kin selection? *Evolution and Human Behavior* **22**: 361–8.

Bocklandt, S. and Hamer, D. H. 2003. Beyond hormones: a novel hypothesis for the biological basis of male sexual orientation. *Journal of Endocrinological Investigations* **26**(Suppl 3): 8–12.

Bocklandt, S., Horvath, S., Vilain, E. and Hamer, D. H. 2006. Extreme skewing of X chromosome inactivation in mothers of homosexual men. *Human Genetics* **118**: 671–94.

Bogaert, A. F. 1997. Birth order and sexual orientation in women. *Behavioral Neuroscience* **111**: 1395–7.

Bogaert, A. F. 1998. Physical development and sexual orientation in women: height, weight, and age of puberty comparisons. *Personality and Individual Differences* **24**: 115–21.

Bogaert, A. F. 2000. Birth order and sexual orientation in a national probability sample. *Journal of Sex Research* **37**: 361–8.

Bogaert, A. F. 2003a. The interaction of fraternal birth order and body size in male sexual orientation. *Behavioral Neuroscience* **117**: 381–4.

Bogaert, A. F. 2003b. Number of older brothers and sexual orientation: new tests and the attraction/behaviour distinction in two national probability samples. *Journal of Personality and Social Psychology* **84**: 644–52.

Bogaert, A. F. 2004. Asexuality: prevalence and associated factors in a national probability sample. *Journal of Sex Research* **41**: 279–87.

Bogaert, A. F. 2006a. Biological versus nonbiological older brothers and men's sexual orientation. *Proceedings of the National Academy of Sciences, USA* **103**: 10771–74.

Bogaert, A. F. 2006b. Toward a conceptual understanding of asexuality. *Review of General Psychology* **10**: 241–50.

Bogaert, A. F. 2010. Physical development and sexual orientation in men and women: an analysis of NATSAL-2000. *Archives of Sexual Behavior* **39**: 110–16. DOI 10.007/s10508-008-9398-x.

Bogaert, A. F. and Blanchard, R. 1996. Physical development and sexual orientation in men: height, weight and age of puberty differences. *Personality and Individual Differences* **21**: 77–84.

Bogaert, A. F. and Cairney, J. 2004. The interaction of birth order and parental age on sexual orientation: an examination in two samples. *Journal of Biosocial Science* **36**: 19–37.

Bogaert, A. F. and Friesen, C. 2002. Sexual orientation and height, weight, and age of puberty: new tests from a British national probability sample. *Biological Psychology* **59**: 135–45.

Bogaert, A. F., Friesen, C. and Klentrou, P. 2002. Age of puberty and sexual orientation in a national probability sample. *Archives of Sexual Behavior* **31**: 73–81.

Bogaert, A. F. and Liu, J. 2006. Birth order and sexual orientation in men: evidence for two independent interactions. *Journal of Biosocial Sciences* **38**: 811–19.

Bolhuis, J. J. 1991. Mechanisms of avian imprinting: a review. *Biological Reviews of the Cambridge Philosophical Society* **66**: 303–45.

Bolk, L. 1926. On the problem of anthropogenesis. *Proceedings of the Section of Sciences, Koninklijke Akademie van Wetenschappen te Amsterdam* **29**: 465–75.

Bolwig, N. 1980. Early social development and emancipation of *Macaca nemestrina* and species of *Papio*. *Primates* **21**: 357–75.

Bon, R. and Campan, R. 1989. Social tendencies of the Corsican mouflon *Ovis ammon musimon* in the Caroux-Espinouse Massif (South of France). *Behavioral Processes* **19**: 57–78.

Bon, R. and Campan, R. 1996. Unexplained sexual segregation in polygamous ungulates: a defense of an ontogenetic approach. *Behavioural Processes* **38**: 131–54.

Bonenfant, C., Loe, L. E., Mysterud, A. *et al.* 2004. Multiple causes of sexual segregation in European red deer: enlightenments from varying breeding phenology at high and low latitude. *Proceedings of the Royal Society of London*, B **271**: 883–92.

Boonratana, R. 2002. Social organization of proboscis monkeys (*Nasalis larvatus*) in the lower Kinabatangan, Sabah, Malaysia. *Malayan Nature Journal* **56**: 57–75.

Borges, T., Eisele, G. and Byrd, C. 2001. *Review of androgenic anabolic steroid use*. Report to the Office of Safeguards and Security, US Department of Energy.

Borkowski, J. 2000. Influence of the density of a sika deer population on activity, habitat use, and group size. *Canadian Journal of Zoology* **78**: 1369–74.

Borner, M., Fitzgibbon, C. D., Borner, M. *et al.* 1987. The decline of the Serengeti Thomson's gazelle population. *Oecologia* **83**: 32–40.

Bornstein, S. R. and Chrousos, G. P. 1999. Adrenocorticotropin (ACTH)- and non-ACTH-mediated regulation of the adrenal cortex: neural and immune inputs. *Journal of Clinical Endocrinology and Metabolism* **84**: 1729–36.

Bos, H. M. W., Sandfort, T. G. M., de Bruyn, E. H. and Hakvoort, E. M. 2008. Same-sex attraction, social relationships, psychosocial functioning, and school performance in early adolescence. *Developmental Psychology* **44**: 59–68.

Bos, H. M. W., van Balen, F. and van den Boom, D. C. 2003. Planned lesbian families: their desire and motivation to have children. *Human Reproduction* **18**: 2216–24.

Bosch, P. C. and Svendsen, G. E. 1987. Behavior of male and female vicuna (*Vicugna vicugna Molina* 1782) as it relates to reproductive effort. *Journal of Mammalogy* **68**: 425–9.

Bosinski, H. A. G., Peter, M., Bonatz, G. *et al.* 1997. A higher rate of hyperandrogenic disorders in female-to-male transsexuals. *Psychoneuroendocrinology* **22**: 361–80.

Bossema, I. and Roemers, E. 1985. Mating strategy, including mate choice, in mallards. *Ardea* **73**: 147–57.

Boswell, J. 1980. *Christianity, Social Tolerance and Homosexuality*. Chicago, IL: University of Chicago Press.

Boulden, W. T. 2001. Gay men living in a rural environment. *Journal of Gay and Lesbian Social Services* **12**: 63–75.

Bourc'his, D. and Bestor, T. H. 2004. Meiotic catastrophe and retrotransposon reactivation in male germ cells lacking Dnmt3L. *Nature* **431**: 96–9.

Bourgeois, J. P. 1997. Synaptogenesis, heterochrony and epigenesis in the mammalian neocortex. *Acta Paediatrica Suppl.* **422**: 27–33.

Bouzat, J. L. 2001. The population genetic structure of the Greater Rhea (*Rhea americana*) in an agricultural landscape. *Biological Conservation* **99**: 277–84.

Bowland, A. E., Bishop, K. S., Taylor, P. J. *et al.* 2001. Estimation and management of genetic diversity in small populations of plains zebra (*Equus quagga*) in KwaZulu-Natal, South Africa. *Biochemical Systematics and Ecology* **29**: 563–83.

Bowlby, J. 1969 *Attachment and Loss*, Vol 1: Attachment. New York: Basic Books.

Bowyer, R. T. 2004. Sexual segregation in ruminants: definitions, hypotheses, and implications for conservation and management. *Journal of Mammalogy* **85**: 1039–52.

Bowyer, R. T., Kie, J. G. and van Ballenberghe, V. 1996. Sexual segregation in black-tailed deer: effects of scale. *Journal of Wildlife Management* **60**: 10–7.

Boyd, L. E. 1991. The behaviour of Przewalski's horses and its importance to their management. *Applied Animal Behaviour Science* **29**: 301–18.

Boyd, R. and Richerson, P. J. 1985. *Culture and the Evolutionary Process.* London: University of Chicago Press.

Brackbill, H. 1941. Possible homosexual mating of the Rock Dove. *Auk* **58**: 581.

Bradford, N. J. 1983. Transgenderism and the cult of Yellamma: heat, sex, and sickness in south Indian ritual. *Journal of Anthropological Research* **39**: 307–22.

Braithwaite, L. W. 1981. Ecological studies of the black swan III. Behaviour and social organization. *Australian Wildlife Research* **8**: 135–46.

Brandth, B. 2002. Gender identity in European family farming: a literature review. *Sociologia Ruralis* **42**: 181–200.

Branney, P. and White, A. 2008. Big boys don't cry: depression and men. *Advances in Psychiatric Treatment* **14**: 256–62.

Bredesen, D. E., Rao, R. V. and Mehlen, P. 2006. Cell death in the nervous system. *Nature* **443**: 796–802.

Breed, G. A., Bowen, W. D., McMillan, J. I. and Leonard, M. L. 2006. Sexual segregation of seasonal foraging habitats in a non-migratory marine mammal. *Proceedings of the Royal Society of London,* B **273**: 2319–26.

Breed, W. G. 1983. Sexual dimorphism in the Australian hopping mouse, *Notomys alexis*. *Journal of Mammalogy* **64**: 536–9.

Brennan, T. and Hegarty, P. 2007. Who was Magnus Hirschfeld and why do we need to know? *History and Philosophy of Psychology* **9**: 12–28.

Brereton, A. R. 1995. Coercion-defense hypothesis: the evolution of primate sociality. *Folia Primatologica* **64**: 207–14.

Brett, M. A., Roberts, L. F., Johnson, T. W. and Wassersug, R. J. 2007. Eunuchs in contemporary society: expectations, consequences, and adjustments to castration (Part II). *Journal of Sexual Medicine* **4**: 946–55.

Bribiescas, R. G. 2001. Reproductive ecology and life history of the human male. *Yearbook of Physical Anthropology* **44**: 148–76.

Bried, J., Duriez, O. and Juin, G. 1999. A first case of female-female pairing in the black-faced sheathbill *Chionis minor*. *Emu* **99**: 292–5.

Briskie, J. V. 1998. Avian genitalia. *Auk* **115**: 826–8.

British Broadcasting Corporation 2006. Oslo gay animal show draws crowds. http://news.bbc.co.uk/1/hi/world/europe/6066606.stm. Downloaded on 4 May 2007.

Britton, D. M. 1990. Homophobia and homosociality: an analysis of boundary maintenance. *The Sociological Quarterly* **31**: 423–39.

Broad, K. D., Curley, J. P. and Keverne, E. B. 2006. Mother-infant bonding and the evolution of mammalian social relationships. *Philosophical Transactions of the Royal Society,* B **361**: 2199–214.

Brockelman, W. Y., Reichard, U., Treesucon, U. and Raemaekers, J. J. 1998. Dispersal, pair formation and social structure in gibbons (*Hylobates lar*). *Behavioral Ecology and Sociobiology* **42**: 329–39.

Broders, H. G. and Forbes, G. J. 2004. Interspecific and intersexual variation in roost-site selection of northern long-eared and little brown bats in the Greater Fundy National Park ecosystem. *Journal of Wildlife Management* **68**: 602–10.

Bromhall, C. 2004. *The Eternal Child.* London: Ebury Press.

Brookhart J. M. and Dey F. L. 1941. Reduction of sexual behaviour in male guinea pigs by hypothalamic lesions. *American Journal of Physiology* **133**: 551–4.

Broomhall, L. S., Mills, M. G. L. and DuToit, J. T. 2003. Home range and habitat use by cheetahs (*Acinonyx jubatus*) in the Kruger National Park. *Journal of Zoology* **261**: 119–28.

Brotto, L. A., Knudson, G., Inskip, J., Rhodes, K. and Erskine, Y. 2010. Asexuality: a mixed-methods approach. *Archives of Sexual Behavior* **39**: 599–618. DOI 10.1007/s10508-008-9434-x.

Brown, C. R. and Brown, M. B. 1988. Genetic evidence of multiple parentage in broods of cliff swallows. *Behavioral Ecology and Sociobiology* **23**: 379–87.

Brown, C. R. and Brown, M. B. 1996. *Coloniality in the Cliff Swallow.* Chicago, IL: University of Chicago Press.

Brown, C. R., Covas, R., Anderson, M. D. and Brown, M. B. 2003. Multistate estimates of survival and movement in relation to colony size in the social weaver. *Behavioral Ecology* **14**: 463–71.

Brown, J. L. 1987. *Helping and Communal Breeding in Birds: Ecology and Evolution.* Princeton, NJ: Princeton University Press.

Brown, K. M. 1998. Proximate and ultimate causes of adoption in ring-billed gulls. *Animal Behaviour* **56**: 1520–43.

Brown, M. B. and Brown, C. R. 1988. Access to winter food sources by bright-versus dull colored house finches. *Condor* **90**: 729–31.

Brown, M. F., Gratton, T. P. and Stuart, J. A. 2007. Metabolic rate does not scale with body mass in cultured mammalian cells. *American Journal of Physiology.*

Brown, W. M., Finn, C. J. and Breedlove, S. M. 2002a. Sexual dimorphism in digit-length ratios of laboratory mice. *Anatomical Record* **267**: 231–34.

Brown, W. M., Finn, C. J., Cooke, B. M. and Breedlove, S. M. 2002b. Differences in finger length ratios between self-identified "butch" and "femme" lesbians. *Archive of Sexual Behavior* **31**: 123–7.

Browne, P., Place, N. J., Vidal, J. D. *et al.* 2006. Endocrine differentiation of fetal ovaries and testes of the spotted hyena (*Crocuta crocuta*): timing of androgen-independent versus androgen-driven genital development. *Reproduction* **132**: 649–59.

Browne, S. J. 2004. Some aspects of chaffinch *Fringilla coelebs* biology, based on an analysis of individuals ringed during 1991 to 2003 in Norfolk, England. *Ringing and Migration* **22**: 75–82.

Bruce, L. L. and Butler, A. B. 1984. Telencephalic connections in lizards. I. Projections to cortex. *Journal of Comparative Neurology* **229**: 585–601.

Brugger, C. and Taborsky, M. 1994. Male incubation and its effect on reproductive success in the black swan, *Cygnus atratus*. *Ethology* **96**: 138–46.

Bu, L. and Lephart, E. D. 2007. AVPV neurons containing estrogen receptor-beta in adult male rats are influenced by soy isoflavones. *BMC Neuroscience* **8**: 13–24.

Bubela, T. M. and Happold, D. C. D. 1993. The social organization and mating system of an Australian subalpine rodent, the broad-toothed rat, *Mastacomys fuscus* Thomas. *Wildlife Research* **20**: 405–17.

Bucala, R., Lahita, R. G., Fishman, J. and Cerami, A. 1987. Anti-oestrogen antibodies in users of oral contraceptives, and in patients with systemic lupus erythematosus. *Clinical and Experimental Immunology* **67**:167–75.

Buchanan, O. M. 1966. Homosexual behaviour in wild orange-fronted parakeets. *Condor* **68**: 399–400.

Buechner, H. K. and Schloeth, R. 1965. Ceremonial mating behaviour in Uganda kob (*Adenota kob thomasi* Neumann). *Zeitschrift für Tierpsychologie* **22**: 209–25.

Buhrich, N., Bailey, J. M. and Martin, N. G. 1991. Sexual orientation, sexual identity, and sex-dimorphic behaviors in male twins. *Behavior Genetics* **21**: 75–96.

Bullough, V. L. 2004. Sex will never be the same: the contributions of Alfred C. Kinsey. *Archives of Sexual Behavior* **33**: 277–86.

Burgdorf, J. and Panksepp, J. 2006. The neurobiology of positive emotions. *Neuroscience and Biobehavioral Reviews* **30**: 173–87.

Burger, A. E. 1981. Time budgets, energy needs and kleptoparasitism in breeding Lesser Sheathbills (*Chionis minor*). *Condor* **83**: 106–12.

Burger, J. and Burger, C. G. 1975. Territoriality in the Laughing Gull (*L. atricilla*). *Behaviour* **55**: 301–19.

Burger, J. and Shisler, J. 1978. Nest site selection and competitive interactions of herring and laughing gulls in New Jersey. *Auk* **95**: 252–66.

Burish, M. J., Kueh, H. Y. and Wang, S. S.-H. 2004. Brain architecture and social complexity in modern and ancient birds. *Brain, Behavior and Evolution* **63**: 107–24.

Burke, A. C., Nelson, C. E., Morgan, B. A. and Tabin, C. 1995. Hox genes and the evolution of vertebrate axial morphology. *Development* **121**: 333–46.

Burley, N. T. and Foster, V. S. 2004. Digit ratio varies with sex, egg order and strength of mate preference in zebra finches. *Proceedings of the Royal Society of London, B* **271**: 239–44.

Burley, N. T. and Symanski, R. 1998. "A taste for the beautiful": Latent aesthetic mate preferences for white crests in two species of Australian grassfinches. *American Naturalist* **152**: 792–802.

Burr, C. 1993. Homosexuality and biology. *The Atlantic Monthly* March issue: 47–65.

Buss, D. M. 1989. Sex difference in human mate preferences: evolutionary hypotheses tested in 37 cultures. *Behavioral and Brain Sciences* **12**: 1–49.

Buss, D. M. 1995. Evolutionary psychology: a new paradigm for psychological science. *Psychological Inquiry* **6**: 1–30.

Buss, D. M., Abbott, M., Angleitner, A. *et al.* 1990. International preferences in selecting mates: a study of 37 cultures. *Journal of Cross-Cultural Psychology* **21**: 5–47.

Buss, I. O. and Smith, N. S. 1966. Observations on reproduction and breeding behaviour of the African elephant. *Journal of Wildlife Management* **30**: 375–88.

Buss, R. R., Sun, W. and Oppenheim, R. W. 2006. Adaptive roles of programmed cell death during nervous system development. *Annual Reviews of Neuroscience* **29**: 1–35.

Buston, P. M., Reeve, H. K., Cant, M. A., Vegrencamp, S. L. and Emlen, S. T. 2007. Reproductive skew and the evolution of group dissolution tactics: a synthesis of concession and restraint models. *Animal Behaviour* **74**: 1643–54.

References

Butterstein, G. M., Schadler, M. H., Lysogorski, E., Robin, L. and Sipperly, S. 1985. A naturally occurring plant compound, 6-methoxybenzoxazolinone, stimulates reproductive responses in rats. *Biology of Reproduction* **32**: 1018–23.

Buunk, B. P., Dijkstra, P., Fetchenhauer, D. and Kenrick, D. T. 2002. Age and gender differences in mate selection criteria for various involvement levels. *Personal Relationships* **9**: 271–8.

Byers, J. A. and Bekoff, M. 1981. Social, spacing, and cooperative behavior of the collared peccary, *Tayassu tajacu*. *Journal of Mammalogy* **62**: 767–85.

Byne, W. 1994. The biological evidence challenged. *Scientific American* **270**: 26–31.

Byne, W. 1998. The medial preoptic and anterior hypothalamic regions of the rhesus monkey: cytoarchitectonic comparison with the human and evidence for sexual dimorphism. *Brain Research* **783**: 346–50.

Byne, W., Lasco, M. S., Kemether, E. *et al.* 2000. The interstitial nuclei of the human anterior hypothalamus: an investigation of sexual variation in volume and cell size, number and density. *Brain Research* **856**: 254–8.

Byne, W. and Parsons, B. 1993. Human sexual orientation: the biologic theories reappraised. *Archives of General Psychiatry* **50**: 228–39.

Byne, W., Tobet, S., Mattiace, L. A. *et al.* 2001. The interstitial nuclei of the human anterior hypothalamus: an investigation of variation with sex, sexual orientation, and HIV status. *Hormones and Behavior* **40**: 86–92.

Byrne, R. W. and Whiten, A. 1988. *Machiavellian Intelligence*. Oxford: Oxford University Press.

Cabón-Raczyńska, K., Krasińska, M., Krasiński, Z. A. and Wójcik, J. M. 1987. Rhythm of daily activity and behaviour of European bison in the Białowieża forest in the period without snow cover. *Acta Theriologica* **32**: 335–72.

Cáceres, M., Lachuer, J., Zapala, M. A. *et al.* 2003. Elevated gene expression levels distinguish human from non-human primate brains. *Proceedings of the National Academy of Sciences, USA* **100**: 13030–5.

Caister, L. E., Shields, W. M. and Gosser, A. 2003. Female tannin avoidance: a possible explanation for habitat and dietary segregation of giraffes (*Giraffa camelopardalis peralta*) in Niger. *African Journal of Ecology* **41**: 201–10.

Caldwell, D. K. and Caldwell, M. C. 1977. Cetaceans. In: Sebeok, T. A. (Ed.) *How Animals Communicate*, pp. 794–808. Bloomington, IN: Indiana University Press.

Caldwell, J. D. 1992. Central oxytocin and female sexual behaviour. *Annals of the New York Academy of Sciences* **652**: 166–79.

Caldwell, J. D., Johns, J. M., Faggin, B. M., Senger, M. A. and Pedersen, C. A. 1994. Infusion of an oxytocin antagonist into the medial preoptic area prior to progesterone inhibits sexual receptivity and increases rejection in female rats. *Hormones and Behavior* **28**: 288–302.

Caldwell, M. A. and Peplau, L. A. 1984. The balance of power in lesbian relationships. *Sex Roles* **10**: 27–39.

Calhim, S., Shi, J. and Dunbar, R. I. M. 2006. Sexual segregation among feral goats: testing between alternative hypotheses. *Animal Behaviour* **72**: 31–41.

Campagna, C. and Lewis, M. 1992. Growth and distribution of a Southern elephant seal colony. *Marine Mammal Science* **8**: 387–96.

Campbell, R. A., Gales, N. J., Lento, G. M. and Baker, C. S. 2008. Islands in the sea: extreme female natal site fidelity in the Australian sea lion, *Neophoca cinerea*. *Biology Letters* **4**: 139–42.

Camperio-Ciani, A., Cermelli, P. and Zanzotto, G. 2008a. Sexually antagonistic selection in human male homosexuality. *PloS ONE* **3**: e2282.

Camperio-Ciani, A., Corna, F. and Capiluppi, C. 2004. Evidence for maternally inherited factors favouring male homosexuality and promoting female fecundity. *Proceedings of the Royal Society of London, B* **271**: 2217–21.

Camperio-Ciani, A., Iemmola, F. and Blecher, S. R. 2008b. Genetic factors increase fecundity in female maternal relatives of bisexual men as in homosexuals. *Journal of Sexual Medicine* **6**: 449–55.

Campos-Outcalt, D. and Hurwitz, S. 2002. Female-to-female transmission of syphilis: a case report. *Sexually Transmitted Diseases* **29**: 119–20.

Cant, M. A. 1998. A model for the evolution of reproductive skew without reproductive suppression. *Animal Behaviour* **55**: 163–9.

Cant, M. A. and Johnstone, R. A. 2000. Power struggles, dominance testing, and reproductive skew. *American Naturalist* **155**: 406–17.

Cantor, J. M., Blanchard, R., Paterson, A. D. and Bogaert, A. F. 2002. How many gay men owe their sexual orientation to fraternal birth order? *Archives of Sexual Behavior* **31**: 63–71.

Cao, L., Liu, Z., Wang, X. *et al.* 2005. Winter group size and comparison of blue sheep (*Pseudois nayaur*) in the Helan Mountains, China. *Acta Theriologica Sinica* **25**: 200–4.

Carani, C., Granata, A. R., Rochira, V. *et al.* 2005. Sex steroids and sexual desire in a man with a novel mutation of aromatase gene hypogonadism. *Psychoneuroendocrinology* **30**: 413–17.

Carani, C., Rochira, V., Faustini-Fustini, M., Balestrieri, A. and Granata, A. R. M. 1999. Role of oestrogen in male sexual behaviour: insights from the natural model of aromatase deficiency. *Clinical Endocrinology* **51**: 517–24.

Carbone, D. J. 2008. Treatment of gay men for post-traumatic stress disorder resulting from social ostracism and ridicule. Cognitive behavior therapy and eye movement desensitization and reprocessing approaches. *Archives of Sexual Behavior* **37**: 305–16.

Cardillo, M., Bininda-Emonds, O. R. P., Boakes, E. and Purvis, A. 2004. A species-level phylogenetic supertree of marsupials. *Journal of Zoology, London* **264**: 11–31.

Cardoso, F. L. 2005. Cultural universals and differences in male homosexuality: the case of a Brazilian fishing village. *Archives of Sexual Behavior* **34**: 103–9.

Cardoso, F. L. 2009. Similar faces of same-sex sexual behavior: a comparative ethnographical study in Brazil, Turkey, and Thailand. *Journal of Homosexuality* **56**: 457–84.

Cardoso, F. L. and Werner, D. 2003. Homosexuality. In: Ember, C. R. and Ember, M. (Eds). *Encyclopedia of Sex and Gender: Men and Women in the World's Cultures*, vol. 1: *Topics and Cultures A–K*. pp. 204–15. New York: Kluwer Academic/Plenum Publishers.

Carling, M. D., Wiseman, P. A. and Byers, J. A. 2003. Microsatellite analysis reveals multiple paternity in a population of wild pronghorn antelopes (*Antilocapra americana*). *Journal of Mammalogy* **84**: 1237–43.

Carlson, A. 1989. Courtship feeding and clutch size in red-backed shrikes (*Lanius collurio*). *American Naturalist* **133**: 454–7.

Caro, T. M. 1999. Demography and behaviour of African mammals subject to exploitation. *Biological Conservation* **91**: 91–7.

Caro, T. M. and Collins, D. A. 1987. Male cheetah social organization and territoriality. *Ethology* **74**: 52–64.

Carr, D. 1968. Behaviour of Sand Martins on the ground. *British Birds* **61**: 416–17.

Carranza, J. 1996. Sexual selection for male body mass and the evolution of litter size in mammals. *American Naturalist* **148**: 81–100.

Carrier, J. M. 1985. Mexican male bisexuality. *Journal of Homosexuality* **11**: 75–85.

Carson, A. 1990. Putting her in her place: woman, dirt, and desire. In: Halperin, D. M., Winkler, J. J. and Zeitlin, F. I. (Eds) *Before Sexuality*, pp. 135–69. Princeton, NJ: Princeton University Press.

Carter, C. S. 1992. Oxytocin and sexual behaviour. *Neuroscience and Biobehavioral Reviews* **16**: 131–44.

Carter, C. S. 1998. Neuroendocrine perspectives on social attachment and love. *Psychoneuroendocrinology* **23**: 779–818.

Cassinello, J., Serrano, E., Calabuig, G. and Pérez, J. M. 2004. Range expansion of an exotic ungulate (*Ammotragus lervia*) in southern Spain: ecological and conservation concerns. *Biodiversity and Conservation* **13**: 851–66.

Castle, S. C. 2000. Clinical relevance of age-related immune dysfunction. *Clinical Infectious Diseases* **31**: 578–85.

Castro, I., Minot, E. O., Fordham, R. A. and Birkhead, T. R. 1996. Polygynandry, face-to-face copulation and sperm competition in the Hihi *Notiomystis cincta* (Aves: Meliphagidae). *Ibis* **138**: 765–71.

Castro, L., Toro, M. A. and López-Fanjul, C. 1994. The genetic properties of homosexual copulation behaviour in *Tribolium castaneum*: artificial selection. *Genetics, Selection and Evolution* **26**: 361–7.

Catry, P., Phillips, R. A. and Croxall, J. P. 2005. Sexual segregation in birds: patterns, processes and implications for conservation. In: Ruckstuhl, K. E. and Neuhaus, P. (Eds) *Sexual Segregation in Vertebrates: Ecology of the Two Sexes*, pp. 351–78. Cambridge: Cambridge University Press.

Cavalli-Sforza L. L. 2001. *Genes, Peoples and Languages*. London: Penguin Books.

Cavalli-Sforza, L. L. and Feldman, M. W. 1981. *Cultural Transmission and Evolution: A Quantitative Approach*. Princeton, NJ: Princeton University Press.

Cecim, M. da S. and Hausler, C. L. 1988. Social preferences affect mounting activity in dairy heifers. *Journal of Animal Science* **66**: 231.

Cézilly, F., Johnson, A. R. 1995. Re-mating between and within breeding seasons in the greates flaming. *Phoenicopterus ruber roseus*. *Ibis* **137**: 543–6.

Cézilly, F., Préault, M., Dubois, F., Faivre, B. and Patris, B. 2000. Pair-bonding in birds and the active role of females: a critical review of the empirical evidence. *Behavioral Processes* **51**: 83–92.

Chaline, J., David, B., Magniez-Jannin, F. *et al.* 1998. Quantification de l'évolution morphologique du crâne des hominidés et hétérochronies. *Comptes rendus de L'Academie des Sciences, Series II, Sciences de la Terre et des Planetes* **326**: 291–8.

Chapais, B. and Berman, C. M. (Eds) 2004. *Kinship and Behavior in Primates*. New York: Oxford University Press.

Chapais, B. and Mignault, C. 1991. Homosexual incest avoidance among females in captive Japanese macaques. *American Journal of Primatology* **23**: 171–83.

Chapman, D. I., Chapman, N. G., Horwood, M. T. and Masters, E. H. 1984. Observations on hypogonadism in a perruque sika deer (*Cervus nippon*). *Journal of Zoology, London* **204**: 579–84.

Chapman, N. G., Furlong, M. and Harris, S. 1997. Reproductive strategies and the influence of date of birth on growth and sexual development of an aseasonally-breeding ungulate: Reeve's muntjac (*Muntiacus reevesi*). *Journal of Zoology, London* **241**: 551–70.

Chase, I. D., Tovey, C., Spangler-Martin, D. and Manfredonia, M. 2002. Individual differences versus social dynamics in the formation of animal dominance hierarchies. *Proceedings of the National Academy of Sciences, USA* **99**: 5744–9.

Chatterjee, S. 1999. *Protoavis* and early evolution of birds. *Palaeontographica* **254**: 1–100.

Chavan, S. A. 1981. Observation of homosexual behaviour in Asiatic lion *Panthera leo persica*. *Journal of the Bombay Natural History Society* **78**: 363–4.

Cheke, A. S. 1968. Copulation in the Hammerskop *Scopus umbretta*. *Ibis* **110**: 201–3.

Chen, D.-Q., Wang, K.-X., Gong, W.-M., Wang, D. and Liu, R.-J. 2001. Cycles of sexual masturbation behavior of a male baiji, "Qi-Qi", in captivity. *Acta Hydrobiologica Sinica* **25**: 467–73.

Chen, D.-Q., Zhao, Q.-Z., Wang, K.-X. *et al.* 2002. Relationships between sexual masturbation behavior and serum testosterone of a captive male baiji. *Acta Zoologica Sinica* **48**: 611–7.

Cheney, D. L. and Seyfarth, R. M. 1982. Recognition of individuals within and between groups of free-ranging vervet monkeys. *American Zoologist* **22**: 519–29.

Cherry, J. A. and Baum, M. J. 1990. Effects of lesions of a sexually dimorphic nucleus in the preoptic/anterior hypothalamic area on the expression of androgen- and estrogen-dependent sexual behaviors in male ferrets. *Brain Research* **522**: 191–203.

Chevalier-Skolnikoff, S. 1974. Male-female, female-female, and male-male sexual behaviour in the stumptail monkey, with special attention to the female orgasm. *Archives of Sexual Behavior* **3**: 95–116.

Chevalier-Skolnikoff, S. 1976. Homosexual behaviour in a laboratory group of stumptail monkeys (*Macaca arctoides*): forms, contexts, and possible social functions. *Archives of Sexual Behavior* **5**: 511–27.

Childerhouse, S. and Gales, N. 2001. Fostering behaviour in New Zealand sea lions *Phocarctos hookeri*. *New Zealand Journal of Zoology* **28**: 189–95.

Chiñas, B. 1995. Isthmus Zapotec attitudes toward sex and gender anomalies. In: Murray, S. O. (Ed.) *Latin American Male Homosexualities*, pp. 293–302. Albuquerque, NM: University of New Mexico Press.

Chivers, M. L. 2005. A brief review and discussion of sex differences in the specificity of sexual arousal. *Sexual and Relationship Therapy* **20**: 377–90.

Chivers, M. L. and Bailey, J. M. 2005. A sex difference in features that elicit genital response. *Biological Psychology* **70**: 115–20.

Chivers, M. L., Rieger, G., Latty, E. and Bailey, J. M. 2004. A sex difference in the specificity of sexual arousal. *Psychological Science* **15**: 736–44.

Chivers, M. L., Seto, M. C. and Blanchard, R. 2007. Gender and sexual orientation differences in sexual response to sexual activities versus gender of actors in sexual films. *Journal of Personality and Social Psychology* **93**: 1108–21.

Cho, M. M., Devries, A. C., Williams, J. R. and Carter, C. S. 1999. The effects of oxytocin and vasopressin on partner preferences in male and female prairie voles (*Microtus ochrogaster*). *Behavioral Neuroscience* **113**: 1071–79.

Choi, D. C., Furay, A. R., Evanson, N. K. *et al.* 2007. Bed nucleus of the stria terminalis subregions differentially regulate hypothalamic-pituitary-adrenal axis activity: implications for the integration of limbic inputs. *Journal of Neuroscience* **27**: 2025–34.

Christidis, L. 1987. Phylogeny and systematics of estrildine finches and their relationships to other seed-eating passerines. *Emu* **87**: 119–23.

Chu, P. C. 1994. Historical examination of delayed plumage maturation in the shorebirds (Aves: Charadriiformes). *Evolution* **48**: 327–50.

Chung, W. C. J., de Vries, G. J. and Swaab, D. F. 2002. Sexual differentiation of the bed nucleus of the stria terminalis in humans may extend into adulthood. *Journal of Neuroscience* **22**: 1027–33.

Chunwang, L., Zhigang, J., Yan, Z. and Caie, Y. 2004. Relationship between serum testosterone, dominance and mating success in Père David's deer stags. *Ethology* **110**: 681–91.

Ciesielski, C. A. 2003. Sexually transmitted diseases in men who have sex with men: an epidemiologic review. *Current Infectious Disease Reports* **5**: 145–52.

Činovský, K., Lábady, F. and Koblík, J. 1986. Psychologic follow-up of personality factors in the chromosomal anomaly of Klinefelter's syndrome. *International Urology and Nephrology* **18**: 99–103.

Ciuti, S., Davini, S., Luccarini, S. and Apollonio, M. 2004. Could the predation risk hypothesis explain large-scale spatial sexual segregation in fallow deer (*Dama dama*)? *Behavioral Ecology and Sociobiology* **56**: 552–64.

Clark, M. M. and Galef, B. G. Jr. 1998. Perinatal influences on reproductive behaviour of adult rodents. In: Mousseau, T. A. and Fox, C. W. (Eds) *Maternal Effects as Adaptations*, pp. 261–71. New York: Oxford University Press.

Clark, W. E. LeGros. 1962. *The Antecedents of Man*. Edinburgh: Edinburgh University Press.

Clarke, S., Kraftsik, R., van der Loos, H. and Innocenti, G. M. 1989. Forms and measures of adult and developing human corpus callosum: is there sexual dimorphism? *Journal of Comparative Neurology* **280**: 213–30.

Clayton, D. H. and Moore, H. 1997 *Host–Parasite Evolution, General Principles and Avian Models*. Oxford: Oxford University Press.

Clemens, L. G., Popham, T. V. and Ruppert, P. H. 1979. Neonatal treatment of hamsters with barbiturate alters adult sexual behaviour. *Developmental Psychobiology* **12**: 49–59.

Clutton-Brock, T. H. 1989. Mammalian mating systems. *Proceedings of the Royal Society of London*, B **236**: 339–72.

Clutton-Brock, T. H. 1998. Reproductive skew, concessions and limited control. *Trends in Ecology and Evolution* **13**: 288–92.

Clutton-Brock, T. 2009. Sexual selection in females. *Animal Behaviour* **77**: 3–11.

Clutton-Brock, T. H., Brotherton, P. N. M., Russell, A. F. *et al.* 2001. Cooperation, control, and concession in meerkat groups. *Science* **291**: 478–80.

Clutton-Brock, T. H., Brotherton, P. N. M., Smith, R. *et al.* 1998. Infanticide and expulsion of females in a cooperative mammal. *Proceedings of the Royal Society of London*, B **265**: 2291–95.

Coblentz, B. E. 1980. A unique ungulate breeding pattern. *Journal of Wildlife Management* **44**: 929–33.

Cochran, G. M., Ewald, P. W. and Cochran, K. D. 2002. Infectious causation of disease: an evolutionary perspective. *Perspectives in Biology and Medicine* **43**: 406–48.

Coddington, C. L. and Cockburn, A. 1995. The mating system of free-living emus. *Australian Journal of Zoology* **43**: 365–72.

Codenotti, T. L. and Alvarez, F. 1997. Cooperative breeding between males in the Greater Rhea *Rhea americana*. *Ibis* **139**: 568–71.

Coe, M. J. 1967. "Necking" behaviour in the giraffe. *Journal of Zoology, London* **151**: 313–21.

Cohen-Bendahan, C. C. C., Buitelaar, J. K., van Goozen, S. H. M. *et al.* 2005. Prenatal sex hormone effects on child and adult sex-typed behaviour: methods and findings. *Neuroscience and Biobehavioral Reviews* **29**: 353–84.

Cohen-Kettenis, P. T. 2001. Gender identity disorder in the DSM? *Journal of the American Academy of Child and Adolescent Psychiatry* **40**: 391.

Cohen-Kettenis, P. T. 2005. Is there an effect of prenatal testosterone on aggression and other behavioral traits? A study comparing same-sex and opposite-sex twin girls. *Hormones and Behavior* **47**: 230–7.

Cohen-Kettenis, P. T. and Pfäfflin, F. 2003. *Transgenderism and Intersexuality in Childhood and Adolescence: Making Choices*. Thousand Oaks, CA: Sage.

Colborn, T., vom Saal, F. S. and Soto, A. M. 1993. Developmental effects of endocrine-disrupting chemicals in wildlife and humans. *Environmental Health Perspectives* **101**: 378–84.

Coleman, E., Colgan, P. and Gooren, L. 1992. Male cross-gender behavior in Myanmar (Burma): a description of the acault. *Archives of Sexual Behavior* **21**: 313–21.

Collaer, M. L. and Hines, M. 1995. Human behavioural sex differences: a role for gonadal hormones during early development? *Psychological Bulletin* **118**: 55–107.

Collias, E. C. and Collias, N. E. 1978. Nest building and nesting behaviour of the sociable weaver *Philetairus socius*. *Ibis* **120**: 1–15.

Collias, N. E. and Collias, E. C. 1980. Behavior of the grey-capped social weaver (*Pseudonigrita arnaudi*) in Kenya. *Auk* **97**: 213–26.

Collias, N. E. and Collias, E. C. 1996. Social organization of a red junglefowl, *Gallus gallus*, population related to evolution theory. *Animal Behaviour* **51**: 1337–54.

Coltman, D. W., Bancroft, D. R., Robertson, A. *et al.* 1999. Male reproductive success in a promiscuous mammal: behavioural estimates compared with genetic paternity. *Molecular Ecology* **8**: 1199–209.

Coltman, D. W., Pilkington, J. G. and Pemberton, J. M. 2003. Fine-scale genetic structure in a free-living ungulate population. *Molecular Ecology* **12**: 733–42.

Comings, D. E. 1994. Role of genetic factors in human sexual behaviour based on studies of Tourette syndrome and ADHD probands and their relatives. *American*

Journal of Medical Genetics (Neuropsychiatric Genetics) **54**: 227–41.

Conaway, C. H. and Koford, C. B. 1964. Estrous cycles and mating behaviour in a free-ranging band of rhesus monkeys. *Journal of Mammalogy* **45**: 577–88.

Conley, A. and Hinshelwood, M. 2001. Mammalian aromatases. *Reproduction* **121**: 685–95.

Connell, R. W. and Messerschmidt, J. W. 2005. Hegemonic masculinity: rethinking the concept. *Gender and Society* **19**: 829–59.

Connor, R. C., Heithaus, M. R. and Barre, L. M. 1999. Superalliance of bottlenose dolphins. *Nature* **397**: 571–2.

Connor, R. C. and Smolker, R. A. 1995. Seasonal changes in the stability of male-male bonds in Indian Ocean bottlenose dolphins (*Tursiops* sp.). *Aquatic Mammals* **21**: 213–16.

Conover, M. R. 1984. Consequences of mate loss to incubating Ring-billed and California gulls. *Wilson Bulletin* **96**: 714–16.

Conover, M. R. 1989. What are males good for? *Nature* **342**: 624–5.

Conover, M. R. and Hunt, G. L. Jr. 1984a. Experimental evidence that female-female pairs in gulls result from a shortage of breeding males. *Condor* **86**: 472–6.

Conover, M. R. and Hunt, G. L. Jr. 1984b. Female-female pairing and sex ratios in gulls: an historical perspective. *Wilson Bulletin* **96**: 619–25.

Conover, M. R., Miller, D. E. and Hunt, G. L. Jr. 1979. Female-female pairs and other unusual reproductive associations in ring-billed and California gulls. *Auk* **96**: 6–9.

Conrad, K. F., Johnston, P. V., Crossman, C. *et al.* 2001. High levels of extra-pair paternity in an isolated, low-density, island population of tree swallows (*Tachycineta bicolor*). *Molecular Ecology* **10**: 1301–8.

Conrad, P. and Angell, A. 2004. Homosexuality and remedicalization. *Society* **41**: 32–9.

Conradt, L. 1998. Measuring the degree of sexual segregation in group-living animals. *Journal of Animal Ecology* **67**: 217–26.

Constantinople, A. 2005. Masculinity-femininity: an exception to a famous dictum? *Feminism Psychology* **15**: 385–407.

Cook, J. 2008. *A probabilistic model of the X linked inheritance of human male homosexuality.* Honors Thesis, Marymount University, USA.

Cook, M. 1996. A role for homosexuality in population genetics. *Evolutionary Theory* **11**: 135–51.

Cooke, B., Hegstrom, C. D., Villeneuve, L. S. and Breedlove, S. M. 1998. Sexual differentiation of the vertebrate brain: principles and mechanisms. *Frontiers in Neuroendocrinology* **19**: 323–62.

Cooke, B. M., Breedlove, S. M. and Jordan, C. L. 2003. Both endrogen receptors and androgen receptors contribute to testosterone-induced changes in the morphology of the medial amygdala and sexual arousal in male rats. *Hormones and Behavior* **43**: 336–46.

Cooke B. M., Tabibnia, G. and Breedlove S. M. 1999. A brain sexual dimorphism controlled by adult circulating androgens. *Proceedings of the National Academy of Sciences, USA* **96**: 7538–40.

Cooney, R. and Bennett, N. C. 2000. Inbreeding avoidance and reproductive skew in a cooperative mammal. *Proceedings of the Royal Society of London*, B **267**: 801–6.

Cooper, J. B. 1942. An exploratory study on African lions. *Comparative Psychology Monographs* **17**: 1–48.

Cooper, M. A., Aureli, F. and Singh, M. 2004. Between-group encounters among bonnet macaques (*Macaca radiata*). *Behavioral Ecology and Sociobiology* **56**: 217–27.

Corona, R., Prignano, G., Mele, A. *et al.* 1991. Heterosexual and homosexual transmission of hepatitis C virus: relation with hepatitis B virus and human immunodeficiency virus type 1. *Epidemiology and Infection* **107**: 667–72.

Corradino, C. 1990. Proximity structure in a captive colony of Japanese monkeys (*Macaca fuscata fuscata*): an application of multidimensional scaling. *Primates* **31**: 351–62.

Corti, G., Panning, E., Gordon, S., Hinde, R. J. and Jenkins, R.K.B. 2002. Observations on the puku antelope (*Kobus vardoni* Livingstone, 1857) in the Kilombero Valley, Tanzania. *African Journal of Ecology* **40**: 197–200.

Corti, P. and Shackleton, D. M. 2002. Relationship between predation-risk factors and sexual segregation in Dall's sheep (*Ovis dalli dalli*). *Canadian Journal of Zoology* **80**: 2108–17.

Cosens, S. E. and Blouw, A. 2003. Size-and age-class segregation of bowhead whales summering in northern Foxe Basin: a photogrammetric analysis. *Marine Mammal Science* **19**: 284–96.

Costa, P. T. and McCrae, R. R. 1992. *NEO PI-R. Professional Manual.* Odessa, FL: Psychological Assessment Resources, Inc.

Costa, P. T. Jr., Terracciano, A. and McCrae, R. R. 2001. Gender differences in personality traits across cultures:

robust and surprising findings. *Journal of Personality and Social Psychology* **81**: 322–31.

Côté, K., Earls, C. M. and Lalumière, M. L. 2002. Birth order, birth interval, and deviant sexual preferences among sex offenders. *Sexual Abuse: A Journal of Research and Treatment* **14**: 67–81.

Cote, S. D., Schaeffer, J. A. and Messier, F. 1997. Time budgets and synchrony of activities in muskoxen: the influence of sex, age and season. *Canadian Journal of Zoology* **75**: 1628–35.

Coulson, J. C. and Hickling, G. 1964. The breeding biology of the grey seal, *Halichoerus grypus* (Fab.), on the Faroe Islands, Northumberland. *Journal of Animal Ecology* **33**: 485–512.

Coulson, J. C. and Thomas, C. S. 1985. Changes in the biology of the kittiwake *Rissa tridactyla*: a 31-year study of a breeding colony. *Journal of Animal Ecology* **54**: 9–26.

Counihan K. A., Vertosick F. T. and Kelly R. H. 1991. Anti-estrogen antibodies in systemic lupus erythematosus: a quantitative evaluation of serum levels. *Immunological Investigations* **20**: 317–31.

Covas, R., Dalecky, A., Caizergues, A. and Doutrelant, C. 2006. Kin associations and direct vs. indirect fitness benefits in colonial cooperatively breeding sociable weavers *Philetairus socius*. *Behavioral Ecology and Sociobiology* **60**: 323–31.

Cove, E., Eiseman, M. and Popkin, S. J. 2005. *Resilient Children: Literature Review and Evidence from the HOPE VI Panel Study*. Washington, DC: The Urban Institute.

Cowan, D. P. 1987. Group living in the European rabbit (*Oryctolagus cuniculus*): mutual benefit or resource localization? *Journal of Animal Ecology* **56**: 779–95.

Craig, A. J. F. K. 1974. Reproductive behaviour of the male red bishop bird. *Ostrich* **45**: 149–60.

Craig, A. J. F. K. 1980. Behaviour and evolution in the genus *Euplectes*. *Journal für Ornithologie* **121**: 144–61.

Craig, A. J. F. K. 1982. Mate attraction and breeding success in the Red Bishop. *Ostrich* **53**: 246–8.

Craig, I. W., Harper, E. and Loat, C. S. 2004. The genetic basis for sex differences in human behaviour: role of the sex chromosomes. *Annals of Human Genetics* **68**: 269–84.

Craig, J. L. 1979. Habitat variation in the social organization of a communal gallinule, the pukeko, *Porphyrio porphyrio melanotus*. *Behavioral Ecology and Sociobiology* **5**: 331–58.

Craighead, J. J., Hornocker, M. G. and Craighead, F. C. Jr. 1969. Reproductive biology of young female grizzly bears. *Journal of Reproduction and Fertility*, Suppl. **6**: 447–75.

Craighead, L., Paetkau, D., Reynolds, H. V., Vyse, E. R. and Stroebeck, C. 1995. Microsatellite analysis of paternity and reproduction in Arctic grizzly bears. *Journal of Heredity* **86**: 255–61.

Cramer, R. E., Cupp, R. G. and Kuhn, J. A. 1993. Male attractiveness: masculinity with a feminine touch. *Current Psychology* **12**: 142–50.

Cransac, N. and Hewison, A. J. M. 1997. Seasonal use and selection of habitat by mouflon (*Ovis gmelini*): comparison of the sexes. *Behavioural Processes* **41**: 57–67.

Crapo, R. H. 1995. Factors in the cross-cultural patterning of male homosexuality: a reappraisal of the literature. *Cross-Cultural Research* **29**: 178–202.

Crawford, C. B. 1994. Effects of sex and sex roles on same-sex touch. *Perceptual and Motor Skills* **78**: 391–4.

Creed, G. W. 1984. Sexual subordination: institutionalized homosexuality and social control in Melanesia. *Ethnology* **23**: 157–76.

Creel, S. 2001. Social dominance and stress hormones. *Trends in Ecology and Evolution* **16**: 491–7.

Creel, S., Monfort, S. L., Creel, N. M., Wildt, D. E. and Waser, P. M. 1995. Pregnancy, oestrogens and future reproductive success in Serengeti dwarf mongooses. *Animal Behaviour* **50**: 1132–5.

Crespi, B. J. and Ragsdale, J. E. 2000. A skew model for the evolution of sociality via manipulation: why it is better to be feared than loved. *Proceedings of the Royal Society of London*, B **267**: 821–8.

Crews, D. 1987. Functional associations in behavioural endocrinology. In: Reinisch, J. M., Rosenblum, L. A. and Sanders, S. A.(Eds). *Masculinity/Femininity: Basic Perspectives*, pp. 83–106. New York: Oxford University Press.

Crews, D. and Groothuis, T. 2005. Tinbergen's fourth question, ontogeny: sexual and individual differentiation. *Animal Biology* **55**: 343–70.

Crews, D. and McLachlan, J. A. 2006. Epigenetics, evolution, endocrine disruption, health, and disease. *Endocrinology* **147**: S4–S10.

Crispo, E. 2007. The Baldwin's Effect and genetic assimilation: revisiting two mechanisms of evolutionary change mediated by phenotypic plasticity. *Evolution* **61**: 2469–77.

Crochet, P.-A., Bonhomme, F. and Lebreton, J.-D. 2000. Molecular phylogeny and plumage evolution in gulls (Larini). *Journal of Evolutionary Biology* **13**: 47–57.

Crofoot, M. C. 2007. Mating and feeding competition in white-faced capuchins (*Cebus capucinus*): the importance of short-and long-term strategies. *Behaviour* **144**: 1473–95.

Crosbie-Burnett, M. and Helmbrecht, L. 1993. A descriptive empirical study of gay male stepfamilies. *Family Relations* **42**: 256–62.

Crowley, C. E. and Magrath, R. D. 2004. Shields of offence: signalling competitive ability in the dusky moorhen, *Gallinula tenebrosa*. *Australian Journal of Zoology* **52**: 463–74.

Crozier, I. D. 2000. Taking prisoners: Havelock Ellis, Sigmund Freud, and the construction of homosexuality, 1897–1951. *Social History of Medicine* **13**: 447–66.

Crozier, I. 2008. Nineteenth-century British psychiatric writing about homosexuality before Havelock Ellis: the missing story. *Journal of the History of Medicine and Allied Sciences* **63**: 65–102.

Cubo, J. and Arthur, W. 2001. Patterns of correlated character evolution in flightless birds: a phylogenetic approach. *Evolutionary Ecology* **14**: 693–702.

Cubo, J., Berge, C., Quilhac, A., de Margerie, E. and Castanet, J. 2002. Heterochronic patterns in primate evolution: evidence from endochondral ossification. *European Journal of Morphology* **40**: 81–8.

Cuervo, J. J. 2003. Parental roles and mating system in the black-winged stilt. *Canadian Journal of Zoology* **81**: 947–53.

Cunha, G. R., Place, N. J., Baskin, L. *et al.* 2005. The ontogeny of the urogenital system of the spotted hyena (*Crocuta crocuta Erxleben*). *Biology of Reproduction* **73**: 554–64.

Cunha, G. R., Wang, Y., Place, N. J. *et al.* 2003. Urogenital system of the spotted hyena (*Crocuta crocuta Erxleben*): a functional histological study. *Journal of Morphology* **256**: 205–18.

Dagg, A. I. 1984. Homosexual behavior and female-male mounting in mammals – a first survey. *Mammal Reviews* **14**: 155–85.

Dahlöf, L.-G., Hård, E. and Larsson, K. 1977. Influence of maternal stress on offspring sexual behaviour. *Animal Behaviour* **25**: 958–63.

Dai, H., Chen, Y., Chen, S. *et al.* 2008. The evolution of courtship behaviors through the origination of a new gene in *Drosophila*. *Proceedings of the National Academy of Sciences, USA* **105**: 7478–83.

Daly, J. C. 1981. Effects of social organization and environmental diversity on determining genetic structure of a population of the wild rabbit, *Oryctolagus cuniculus*. *Evolution* **35**: 689–706.

Daly, M. and Wilson, M. I. 1999. Human evolutionary psychology and animal behaviour. *Animal Behaviour* **57**: 509–19.

Dancey, C. P. 1990. Sexual orientation in women: an investigation of hormonal and personality variables. *Biological Psychology* **30**: 251–64.

Daniels, T. J. and Bekoff, M. 1989. Population and social biology of free-ranging dogs, *Canis familiaris*. *Journal of Mammalogy* **70**: 754–62.

Darwin, C. 1859. *On the Origin of Species by Means of Natural Selection, or the Preservation of Favoured Races in the Struggle for Life*. London: John Murray.

Darwin, C. 1871. *The Descent of Man and Selection in Relation to Sex*. London: John Murray.

Darwin, C. R. 1872a. *The Origin of Species by Means of Natural Selection, or the Preservation of Favoured Races in the Struggle for Life*, 6th edn. London: John Murray.

Darwin, C. 1872b. *The Expression of the Emotions in Man and Animals*. London: John Murray.

D'Augelli, A. R. and Hart, M. M. 1987. Gay women, men, and families in rural settings: toward the development of helping communities. *American Journal of Community Psychology* **15**: 79–93.

Davenport, C. W. 1986. A follow-up study of 10 feminine boys. *Archives of Sexual Behavior* **15**: 511–17.

Davidson, J. M., Camargo, C., Smith, E. R. and Kwan, M. 1983. Maintenance of sexual function in a castrated man treated with ovarian steroids. *Archives of Sexual Behavior* **12**: 263–74.

Davis, L. S., Hunter, F. M., Harcourt, R. G. and Heath, S. M. 1998. Reciprocal homosexual mounting in Adélie penguins *Pygoscelis adeliae*. *Emu* **98**: 136–37.

Dawkins, R. 1980. Good strategy or evolutionarily stable strategy? In: G. W. Barlow, and J. Silverberg (Eds) *Sociobiology: Beyond Nature/Nurture*, pp. 331–67. Boulder, CO: Westview.

Dawkins, R. 1982. *The Extended Phenotype*. New York: W. H. Freeman and Co. Ltd.

Dawood, K., Bailey, J. M. and Martin, N. G. 2009. Genetic and environmental influences on sexual orientation. In: Kim, Y.-K. (Ed.) *Handbook of Behavior Genetics*. pp. 269–79. New York: Springer.

Dayaratne, P. and Joseph, L. 1993. *A Study on Dolphin Catches in Sri Lanka*. Madras, India: Bay of Bengal Programme.

Deady, D. K., Law Smith, M. J., Kent, J. P. and Dunbar, R. I. M. 2006. Is priesthood an adaptive strategy? *Human Nature* **17**: 393–404.

Deaux, K. 1987. Psychological constructions of masculinity and femininity. In: Reinisch, J. M., Rosenblum, L. A. and Sanders, S. A. (Eds) *Masculinity/Femininity: Basic Perspectives*. pp. 289–303. New York: Oxford University Press.

De Beer, G. R. 1930. *Embryology and Evolution*. Oxford: Clarendon Press.

De Block, A. and Adriaens, P. 2004. Darwinizing sexual ambivalence: a new evolutionary hypothesis of male homosexuality. *Philosophical Psychology* **17**: 59–76.

De Block, A. and Du Laing, B. 2007. Paving the way for an evolutionary social constructivism. *Biological Theory* **2**: 337–48.

DeBry, R. W. and Sagel, R. M. 2001. Phylogeny of Rodentia (Mammalia) inferred from the nuclear-encoded gene IRBP. *Molecular Phylogenetics and Evolution* **19**: 290–301.

De Kort, S. R., Emery, N. J. and Clayton, N. S. 2006. Food sharing in jackdaws, *Corvus monedula*: what, why and with whom? *Animal Behaviour* **72**: 297–304.

De Lacoste, M. C. and Woodward, D. J. 1988. The corpus callosum in nonhuman primates. *Brain Behavior and Evolution* **31**: 318–23.

De Lacoste-Utamsing, C. and Holloway, R. L. 1982. Sexual dimorphism in the human corpus callosum. *Science* **216**: 1431–32.

DeLamater, J. D. and Hyde, J. S. 1998. Essentialism vs. social constructionism in the study of human sexuality. *Journal of Sex Research* **35**: 10–18.

De Lathouwers, M. and van Elsacker, L. 2006. Comparing infant and juvenile behaviour in bonobos (*Pan paniscus*) and chimpanzees (*Pan troglodytes*): a preliminary study. *Primates* **47**: 287–93.

del Hoyo, J., Elliot, A. and Sargatal, J. 1996. *Handbook of the Birds of the World*. Barcelona: Lynx Ediciones.

Dellovade, T. L., Ottinger, M. A. and Rissman, E. F. 1995. Mating alters gonadotropin-releasing hormone cell number and content. *Endocrinology* **136**: 1648–57.

Delpietro, von H. A. and Russo, R. G. 2002. Observations of the common vampire bat (*Desmodus rotundus*) and the hairy-legged vampire bat (*Diphylla ecaudata*) in captivity. *Mammalian Biology* **67**: 65–78.

de Menocal, P. B. 1995. Plio-Pleistocene African climate. *Science* **270**: 53–59.

de Menocal, P. B. 2004. African climate change and faunal evolution during the Pliocene–Pleistocene. *Earth and Planetary Science Letters* **220**: 3–24.

Demir, E. and Dickson, B. J. 2005. *fruitless* Splicing specifies male courtship behaviour in *Drosophila*. *Cell* **121**: 785–94.

Dempsey, P. J., Townsend, G. C. and Richards, L. C. 1999. Increased tooth crown size in females with twin brothers: evidence for hormonal diffusion between human twins in utero. *American Journal of Human Biology* **11**: 577–86.

Denny, E., Yakolevich, P., Eldridge, M. D. B. and Dickman, C. 2002. Social and genetic analysis of a population of free-living cats (*Felis catus* L.) exploiting a resource-rich habitat. *Wildlife Research* **29**: 405–13.

Derix, R., van Hooff, J., DeVries, H. and Wensing, J. 1993. Male and female mating competition in wolves: female suppression vs. male intervention. *Behaviour* **127**: 141–74.

de Rooij, S. R., Painter, R. C., Swaab, D. F. and Roseboom, T. J. 2009. Sexual orientation and gender identity after prenatal exposure to the Dutch famine. *Archives of Sexual Behavior* **38**: 411–16.

De Silva, M. and de Silva, P. K. 2001. Group composition, sex ratio and seasonality of spotted deer (*Axis axis*) in the Yala Protected Area Complex, Sri Lanka. *Journal of South Asian Natural History* **5**: 135–41.

De Vos, A. and Dowsett, R. J. 1966. The behaviour and population structure of three species of the genus *Kobus*. *Mammalia* **30**: 30–55.

De Vries, G. J., Rissman, E. F., Simerly, R. B. *et al.* 2002. A model system for study of sex chromosome effects on sexually dimorphic neural and behavioural traits. *Journal of Neuroscience* **22**: 9005–14.

De Vries, H. 1998. Finding a dominance order most consistent with a linear hierarchy: a new procedure and review. *Animal Behaviour* **55**: 827–43.

De Waal, F. B. M. 1989. *Peacemaking Among Primates*. Cambridge, MA: Harvard University Press.

De Waal, F. B. M. 1997. *Bonobo: the Forgotten Ape*. Berkeley, CA: University of California Press.

Dewar, C. S. 2003. An association between male homosexuality and reproductive success. *Medical Hypotheses* **60**: 225–32.

Dewing, P., Chiang, C. W. K., Sinchak, K. *et al.* 2006. Direct regulation of adult brain function by the male-specific factor SRY. *Current Biology* **16**: 120–415.

De Witt, T. J., Sih, A. and Wilson, D. S. 1998. Costs and limits of phenotypic plasticity. *Trends in Ecology and Evolution* **13**: 77–81.

Dewsbury, D. A. 1982. Dominance rank, copulatory behavior, and differential reproduction. *Quarterly Review of Biology* **57**: 135–59.

Dewsbury, D. A. and Hodges, A. W. 1987. Copulatory behavior and related phenomena in spiny mice *Acomys cahirinus* and hopping mice *Notomys alexis*. *Journal of Mammalogy* **68**: 49–57.

DeYoung, R. W., Demarais, S., Honeycutt, R. L., *et al.* 2003. Genetic consequences of white-tailed deer (*Odocoileus virginianus*) restoration in Mississippi. *Molecular Ecology* **12**: 3237–52.

Dhondt, A. A., Driscoll, M. J. L. and Swarthout, E. C. H. 2007. House finch *Carpodacus mexicanus* roosting behaviour during the non-breeding season and possible effects of mycoplasmal conjunctivitis. *Ibis* **149**: 1–9.

Diamond, J. and Bond, A. B. 2003. A comparative analysis of social play in birds. *Behaviour* **140**: 1091–115.

Diamond, L. M. 2007. A dynamical system approach to the development and expression of female same-sex sexuality. *Perspectives on Psychological Science* **2**: 142–61.

Diamond, L. M. 2008a. Female bisexuality from adolescence to adulthood: results from a 10-year longitudinal study. *Developmental Psychology* **44**: 5–14.

Diamond, L. M. 2008b. *Sexual Fluidity: Understanding Women's Love and Desire*. Cambridge, MA: Harvard University Press.

Diamond, M. 1965. A critical evaluation of the ontogeny of human sexual behaviour. *Quarterly Review of Biology* **40**: 147–75.

Diamond, M. 1993. Homosexuality and bisexuality in different populations. *Archives of Sexual Behavior* **22**: 291–310.

Diamond, M. 1998. Bisexuality: A biological perspective. In: Haeberle, E. J. and Gindorf, R. (Eds) *Bisexualities – the Ideology and Practice of Sexual Contact with both Men and Women*, pp. 53–80. New York: Continuum.

Diamond, M. 2009. Clinical implications of the organizational and activational effects of hormones. *Hormones and Behavior* **55**: 621–32.

Diamond, M. and Watson, L. A. 2004. Androgen insensitivity syndrome and Klinefelter's syndrome: sex and gender considerations. *Child and Adolescent Psychiatric Clinics of North America* **13**: 623–40.

Díaz-Uriarte, R. and Garland, T. Jr. 1998. Effects of branch length errors on the performance of phylogenetically independent contrasts. *Systematic Biology* **47**: 654–72.

DiBiase, R. and Gunnoe, J. 2004. Gender and culture differences in touching behavior. *Journal of Social Psychology* **144**: 49–62.

Dickemann, M. 1993. Reproductive strategies and gender construction: an evolutionary view of homosexualities. *Journal of Homosexuality* **24**: 55–71.

Dickermann, J. M. 2002. Women who become men: Albanian sworn virgins. *Archives of Sexual Behavior* **31**: 376–9.

Dickins, T. E. and Sergeant, M. J. T. 2008. Social dominance and sexual orientation. *Journal of Evolutionary Psychology* **6**: 57–71.

Dickinson, J., Haydock, J., Koenig, W., Stanback, M. and Pitelka, F. 1995. Genetic monogamy in single-male groups of acorn woodpeckers, *Melanerpes formicivorus*. *Molecular Ecology* **4**: 765–9.

Dickson, N., Paul, C. and Herbison, P. 2003. Same-sex attraction in a birth cohort: prevalence and persistence in early childhood. *Social Science and Medicine* **56**: 1607–15.

Dieckmann, G., Schneider-Jonietz, B. and Schneider, H. 1988. Psychiatric and neuropsychological findings after stereotactic hypothalamotomy, in cases of extreme sexual aggressivity. *Acta Neurochirurgica Suppl.* **44**: 163–6.

Dierendonck, M. C. van, Bandi, N., Batdorj, D., Dügerlham, S. and Munkhtsog, B. 1996. Behavioural observations of reintroduced Takhi or Przewalski horses (*Equus ferus przewalskii*) in Mongolia. *Applied Animal Behaviour Science* **50**: 95–114.

Dietz, J. M. 1984. Ecology and social organization of the maned wolf *Chrysocyon brachyurus*. *Smithsonian Contributions to Zoology* **392**: 1–51.

Dietzman, D. E., Harnisch, J. P., Ray, C. G., Alexander, E. R. and Holmes, K. K. 1977. Hepatitis B surface antigen (HBsAg) and antibody of HBsAg: Prevalence in homosexual and heterosexual men. *Journal of the American Medical Association* **238**: 2625–26.

Di Fiore, A. and Rendall, D. 1994. Evolution of social organization: a reappraisal for primates by using phylogenetic methods. *Proceedings of the National Academy of Sciences, USA* **91**: 9941–5.

Dijkstra, P., Groothof, H. A. K., Poel, G. A. *et al.* 2001. Sex differences in the events that elicit jealousy among homosexuals. *Personal Relationships* **8**: 41–54.

Dilger, W. C. 1960. The comparative ethology of the African parrot genus *Agapornis*. *Zeitschrift für Tierpsychologie* **17**: 649–85.

Dillon, P., Copeland, J. and Peters, R. 1999. Exploring the relationship between male homo/bi-sexuality, body image and steroid use. *Culture, Health and Sexuality* **1**: 317–27.

Dit Durrell, S. E. A. Le van 2000. Individual feeding specialization in shorebirds: population consequences and conservation implications. *Biological Reviews* **75**: 503–18.

Dittami, J., Katina, S., Möstl, E., *et al*. 2008. Urinary androgens and cortisol metabolites in field-sampled bonobos (*Pan paniscus*). *General and Comparative Endocrinology* **155**: 552–7.

Dittmann, R. W., Kappes, M. E. and Kappes, M. H. 1992. Sexual behaviour in adolescent and adult females with congenital adrenal hyperplasia. *Psychoneuroendocrinology* **17**: 153–70.

Dixon, M. M., Reyes, C. J., Leppert, M. F. and Pappas, L. M. 2008. Personality and birth order in large families. *Personality and Individual Differences* **44**: 119–28.

Dixon, R. A. 2004. Phytoestrogens. *Annual Reviews of Plant Biology* **55**: 225–61.

Dixson, A. F. 1977. Observations on the displays, menstrual cycles and sexual behaviour of the "Black ape" of Celebes (*Macaca nigra*). *Journal of Zoology, London* **182**: 63–84.

Dixson, A. F. 1995. Sexual selection and the evolution of copulatory behaviour in nocturnal prosimians. In: Alterman, L., Doyle, G. A. and Izard, M. K. (Eds) *Creatures of the Dark: The Nocturnal Prosimians*, pp. 93–118. New York: Plenum.

Dixson, A. F. 1998. *Primate Sexuality: Comparative Studies of the Prosimians, Monkeys, Apes and Human Beings*. Oxford: Oxford University Press.

Dixson, A. F. 1999. Evolutionary perspectives on primate mating systems and behavior. In: Carter, C. S., Lederhendler, I. I. and Kirkpatrick, B. (Eds) *The Integrative Neurobiology of Affiliation*, pp. 45–64. Cambridge, MA: MIT Press.

Dixson, A. F. 2009. *Sexual Selection and the Origins of Human Mating Systems*. Oxford: Oxford University Press.

Dixson, A. F., Scruton, D. M., and Herbert, J. 1975. Behaviour of the talapoin monkey (*Miopithecus talapoin*) studied in groups in the laboratory. *Journal of Zoology, London* **176**: 177–210.

Dloniak, S. M., French, J. A. and Holekamp, K. E. 2006a. Rank-related maternal effects of androgens on behaviour in wild spotted hyaenas. *Nature* **440**: 1190–3.

Dloniak, S. M., French, J. A. and Holekamp, K. E. 2006b. Faecal androgen concentrations in adult male spotted hyaenas, *Crocuta crocuta*, reflect interactions with socially dominant females. *Animal Behaviour* **71**: 27–37.

Doerr, P., Kockott, G., Vogt, H. J., Pirke, K. M. and Dittmar, F. 1973. Plasma testosterone, estradiol, and semen analysis in male homosexuals. *Archives of General Psychiatry* **29**: 829–33.

Dominguez, J., Riolo, J. V., Xu, Z. and Hull, E. M. 2001. Regulation by the medial amygdala of copulation and medial preoptic dopamine release. *Journal of Neuroscience* **21**: 349–55.

Domning, D. P. 1994. A phylogenetic analysis of the Sirenia. In: Berta, A. and Deméré, T. A. (Eds) Contributions in marine mammal paleontology honouring Frank C. Whitmore Jr. *Proceedings of the San Diego Society of Natural History* **29**: 177–89.

Doran, D. M. 1992. The ontogeny of chimpanzee and pigmy chimpanzee locomotor behaviour: a case study of paedomorphism and its behavioural correlates. *Journal of Human Evolution* **23**: 139–57.

Dörner, G. 1983. Letter to the Editor. *Archives of Sexual Behavior* **12**: 577–82.

Dörner, G., Geier, T., Ahrens, L. *et al*. 1980. Prenatal stress as possible aetiogenetic factor of homosexuality in human males. *Endokrinologie* **75**: 365–8.

Dörner, G., Götz, F., Rohde, W. *et al*. 2001. Genetic and epigenetic effects on sexual brain organization mediated by sex hormones. *Neuroendocrinology Letters* **22**: 403–9.

Dörner, G., Rhode, W., Stahl, F., Krell, L. and Masius, W.-G. 1975. A neuroendocrine predisposition for homosexuality in men. *Archives of Sexual Behavior* **4**: 1–8.

Dörner, G., Schenk, B., Schmiedl, B. and Ahrens, L. 1983. Stress events in prenatal life of bi- and homosexual men. *Experimental and Clinical Endocrinology* **81**: 83–7.

Doutrelant, C., Covas, R., Caizergues, A. and Du Plessis, M. 2004. Unexpected sex ratio adjustment in a colonial cooperative bird: pairs with helpers produce more of the helping sex whereas pairs without helpers do not. *Behavioral Ecology and Sociobiology* **56**: 149–54.

Dover, K. J. 1989. *Greek Homosexuality*. Cambridge, MA: Harvard University Press.

Downey, J., Ehrhardt, A. A., Schiffman, M., Dyrenfurth, I. and Becker, J. 1987. Sex hormones in lesbian and heterosexual women. *Hormones and Behavior* **21**: 347–57.

Drea, C. M., Place, N. J., Weldele, M. L. *et al*. 2002a. Exposure to naturally circulating androgens during foetal life incurs direct reproductive costs in female spotted hyenas, but is prerequisite for male mating. *Proceedings of the Royal Society of London*, B **269**: 1981–87.

Drea, C. M., Vignieri, S. N., Kim, H. S., Weldele, M. L., and Glickman, S. E. 2002b. Responses to olfactory stimuli in spotted hyenas (*Crocuta crocuta*): II. Discrimination of conspecific scent. *Journal of Comparative Psychology* **116**: 342–9.

Drea, C. M., Weldele, M. L., Forger, N. G. et al. 1998. Androgens and masculinisation of genitalia in the spotted hyaena (*Crocuta crocuta*). 2. Effects of prenatal antiandrogens. *Journal of Reproduction and Fertility* **113**: 117–27.

Dressen, W. 1993. On the behaviour and social organization of agile wallabies, *Macropus agilis* (Gould, 1842) in two habitats of northern Australia. *Zeitschrift für Säugetierkunde* **58**: 201–11.

Drori, D. and Folman, Y. 1967. The sexual behavior of male rats unmated to 16 months of age. *Animal Behaviour* **15**: 20–4.

Drummond, K. D., Bradley, S. J., Peterson-Badali, M. and Zucker, K. J. 2008. A follow-up study of girls with gender identity disorder. *Developmental Psychology* **44**: 34–45.

Drüwa, P. 1983. The social behaviour of the bush dog (*Speothos*). *Carnivore* **6**: 46–71.

Dubost, G. 1997. Comportements comparés du pécari à lèvres blanches, *Tayassu pecari*, et du pécari à collier, *T. tajacu* (Artiodactyles, Tayassuidés). *Mammalia* **61**: 313–43.

Dubost, G. and Feer, F. 1981. The behaviour of the male *Antilope cervicapra* L., its development according to age and social rank. *Behaviour* **76**: 62–127.

Dugger, B. N., Morris, J. A., Jordan, C. L. and Breedlove, S. M. 2007. Androgen receptors are required for full masculinisation of the ventromedial hypothalamus (VMH) in rats. *Hormones and Behavior* **51**: 195–201.

Dunbar, R. I. M. 1993. Co-evolution of neocortex size, group size and language in humans. *Behavioral and Brain Sciences* **16**: 681–735.

Dunbar, R. I. M. 1998. The social brain hypothesis. *Evolutionary Anthropology* **6**: 178–90.

Dunlap, J. L., Zadina, J. E. and Gougis, G. 1978. Prenatal stress interacts with prepubertal social isolation to reduce male copulatory behaviour. *Physiology and Behavior* **21**: 873–5.

Dunn, P. O., Afton, A. D., Gloutney, M. L. and Alisauskas, R. T. 1999. Forced copulation results in few extrapair fertilizations in Ross's and lesser snow geese. *Animal Behaviour* **57**: 1071–81.

Dunn P. O., Whittingham, L. A. and Pitcher, T. E. 2001. Mating systems, sperm competition, and the evolution of sexual dimorphism in birds. *Evolution* **55**: 161–75.

Dunne, M. P., Martin, N. G., Bailey, J. M. et al. 1997. Participation bias in a sexuality survey: psychological and behavioural characteristics of responders and non-responders. *International Journal of Epidemiology* **26**: 844–54.

DuPlessis, M. A., Siegfried, W. R. and Armstrong, A. J. 1995. Ecological and life-history correlates of cooperative breeding in South African birds. *Oecologia* **102**: 180–88.

DuPree, M. G., Mustanski, B. S., Bocklandt, S., Nievergelt, C. and Hamer, D. H. 2004. A candidate gene study of CYP19 (Aromatase) and male sexual orientation. *Behavior Genetics* **34**: 243–50.

Du Toit, J. T. 1995. Sexual segregation in kudu: sex differences in competitive ability, predation risk or nutritional needs? *South African Journal of Wildlife Research* **25**: 127–32.

Earle, M. 1987. A flexible body mass in social carnivores. *American Naturalist* **129**: 755–60.

East, M. L., Hofer, H. and Wickler, W. 1993. The erect 'penis' is a flag of submission in a female-dominated society: greetings in Serengeti spotted hyaenas. *Behavioral Ecology and Sociobiology* **33**: 335–70.

Ebensperger, L. A. and Cofré, H. 2001. On the evolution of group-living in the New World cursorial hystricognath rodents. *Behavioral Ecology* **12**: 227–36.

Ebensperger, L. A., Taraborelli, P., Giannoni, S. M. et al. 2006. Nest and space use in a highland population of the southern mountain cavy (*Microcavia australis*). *Journal of Mammalogy* **87**: 834–40.

Eberhardt, L. L. and Knight, R. R. 1996. How many grizzlies in Yellowstone? *Journal of Wildlife Management* **60**: 416–21.

Echteler, S. M. 1984. Connections of the auditory midbrain in a teleost fish, *Cyprinus carpio*. *Journal of Comparative Neurology* **230**: 536–51.

Eckert, E. D., Bouchard, T. J., Bohlen, J. and Heston, L. L. 1986. Homosexuality in monozygotic twins reared apart. *British Journal of Psychiatry* **148**: 421–5.

Edelman, G. M. 1987. *Neural Darwinism: The Theory of Neural Group Selection*. New York: Basic Books.

Edwards, A.-M. A. R. and Todd, J. D. 1991. Homosexual behaviour in wild white-banded gibbons (*Hylobates lar*). *Primates* **32**: 231–6.

Edwards, T. A. 1995. Buller syndrome: what's behind this abnormal sexual behaviour? *Large Animal Veterinarian* **50**: 6–7.

Ehrhardt, A. A. 1987. A transactional perspective on the development of gender differences. In: Reinisch, J. M., Rosenblum, L. A. and Sanders, S. A. (Eds) *Masculinity/Femininity: Basic Perspectives*, pp. 281–5. New York: Oxford University Press.

Ehrhardt, A. A. and Meyer-Bahlburg, H. F. L. 1981. Effects of prenatal sex hormones on gender-related behaviour. *Science* **211**: 1312–18.

Ehrhardt, A. A., Meyer-Bahlburg, H. F. L., Rosen, L. R. *et al.* 1985. Sexual orientation after prenatal exposure to exogenous estrogen. *Archives of Sexual Behavior* **14**: 57–75.

Eibl-Eibesfeldt, I. 1967. *Grundriss der Vergleinchenden Verhaltensforschung. Ethologie.* Münch: Piper, R. and Co. Verlag.

Eisenberg, J. F., McKay, G. M. and Jainudeen, M. R. 1971. Reproductive behaviour of the Asiatic elephant (*Elephas maximus maximus* L.). *Behaviour* **38**: 193–225.

Eising, C. M., Müller, W. and Groothuis, T. G. G. 2006. Avian mothers create different phenotypes by hormone deposition in their eggs. *Biology Letters* **2**: 20–2.

Ellis, H. H. 1928. *Studies in the Psychology of Sex.* Vol. II: Sexual Inversion. Philadelphia, PA: FA Davis.

Ellis, H. H. 1946. *Psychology of Sex.* London: William Heinemann.

Ellis, H. H. 1967. *My Life.* London: Neville Spearman.

Ellis, L. and Ames, M. A. 1987. Neurohormonal functioning and sexual orientation: a theory of homosexuality-heterosexuality. *Psychological Bulletin* **101**: 233–58.

Ellis, L., Ames, M. A., Peckham, W. and Burke, D. 1988. Sexual orientation of human offspring may be altered by severe maternal stress during pregnancy. *Journal of Sex Research* **25**: 152–7.

Ellis, L. and Blanchard, R. 2001. Birth order, sibling sex ratio, and maternal miscarriages in homosexual and heterosexual men and women. *Personality and Individual Differences* **30**: 543–52.

Ellis, L., Ficek, C., Burke, D. and Das, S. 2008. Eye color, hair color, blood type, and the Rhesus factor: exploring possible genetic links to sexual orientation. *Archives of Sexual Behavior* **37**: 145–9.

Ellis, L. and Hellberg, J. 2005. Fetal exposure to prescription drugs and adult sexual orientation. *Personality and Individual Differences* **38**: 225–36.

Ely, J. and Ferrell, R. E. 1990. DNA "fingerprints" and paternity ascertainment in chimpanzees (*Pan troglodytes*). *Zoo Biology* **9**: 91–8.

Emlen, S. T. 1982a. The evolution of helping. I. An ecological constraints model. *American Naturalist* **119**: 29–39.

Emlen, S. T. 1982b. The evolution of helping. II. The role of behavioural conflict. *American Naturalist* **119**: 40–53.

Emlen, S. T. 1995. An evolutionary theory of the family. *Proceedings of the National Academy of Sciences, USA* **92**: 8092–99.

Emlen, S. T. 1996. Reproductive sharing in different types of kin associations. *American Naturalist* **148**: 756–63.

Emlen, S. T., Reeve, H. K. and Keller, L. 1998. Reproductive skew: disentangling concessions from control. *Trends in Ecology and Evolution* **13**: 458–9.

Emmons, S. W. and Lipton, J. 2003. Genetic basis of male sexual behaviour. *Journal of Neurobiology* **54**: 93–110.

Emory, L. E., Williams, D. H., Cole, C. M., Amparo, E. G. and Meyer, W. J. 1991. Anatomic variation of the corpus callosum in persons with gender dysphoria. *Archives of Sexual Behavior* **20**: 409–17.

Endo, A. and Doi, T. 2002. Multiple copulations and post-copulatory guarding in a free-living population of sika deer (*Cervus nippon*). *Ethology* **108**: 739–47.

Engelhardt, A., Hodges, J. K. and Heistermann, M. 2007. Post-conception mating in wild long-tailed macaques (*Macaca fascicularis*): characterization, endocrine correlates and functional significance. *Hormones and Behavior* **51**: 3–10.

Engh, A. L., Funk, S. M., van Horn, R. C. *et al.* 2002. Reproductive skew among males in a female-dominated mammalian society. *Behavioural Ecology* **13**: 193–200.

Engle, M. J., McFalls, J. A., Gallagher, B. J. III and Curtis, K. 2006. The attitudes of American sociologists toward causal theories of male homosexuality. *American Sociologist* **37**: 68–76.

Enomoto, T. 1974. The sexual behaviour of Japanese monkeys. *Journal of Human Evolution* **3**: 351–72.

Enomoto, T. 1990. Social play and sexual behavior of the bonobo (*Pan paniscus*) with special reference to flexibility. *Primates* **31**: 469–80.

Ens, B. J. 1998. Love thine enemy? *Nature* **391**: 635–7.

Entringer, S., Kumsta, R., Hellhammer, R., Wadhwa, D. H. and Wüst, S. 2009. Prenatal exposure to maternal psychosocial stress and HPA axis regulation in young adults. *Hormones and Behavior* **55**: 292–8.

Entwistle, A. C., Racey, P. A. and Speakman, J. R. 2000. Social and population structure of a gleaning bat, *Plecotus auritus*. *Journal of Zoology, London* **252**: 11–17.

Epple, C. 1998. Coming to terms with Navajo "nádleehí": a critique of "berdache", "gay", "alternative gender", and "two-spirit". *American Ethnologist* **25**: 267–90.

Epple, G. 1980. Relationships between aggression, scent-marking and gonadal state in a primate, the tamarin *Saguinus fuscicollis*. In: Müller-Schwarze, D. and Silverstein, R. M. (Eds) *Chemical Signals in Vertebrates*, Vol. 2, *Vertebrates and Aquatic Vertebrates*, pp. 87–105. New York: Plenum.

Epple, G., Belcher, A. M. Kuderling, I. 1993. Make sense out of Scents: species differences in scent glands, scent-marking behaviour, and scent-mark composition in the Callitrichidae. In: Rylands, A. B. (ed) *Marmosets and Tamarins: Systematics, Behaviour and Ecology*, pp. 123–51. Oxford: Oxford University Press.

Ericson, P. G. P. and Johansson, U. S. 2003. Phylogeny of Passerida (Aves: Passeriformes) based on nuclear and mitochondrial sequence data. *Molecular Phylogenetics and Evolution* **29**: 126–38.

Esch, T. and Stefano, G. B. 2004. The neurobiology of pleasure, reward processes, addiction and their health implications. *Neuroendocrinology Letters* **25**: 235–51.

Esch, T. and Stefano, G. B. 2005. The neurobiology of love. *Neuroendocrinology Letters* **26**: 175–92.

Estrada, A. and Sandoval, J. M. 1977. Social relations in a free-ranging troop of stumptail macaques (*Macaca arctoides*): male-care behaviour I. *Primates* **18**: 793–813.

Evans, G. W. and Howard, R. B. 1973. Personal space. *Psychological Bulletin* **80**: 334–44.

Ewald, P. W. 1994. *The Evolution of Infectious Disease*. Oxford: Oxford University Press.

Ewald, P. W. and Rohwer, S. 1980. Age, coloration and dominance in nonbreeding hummingbirds: a test of the asymmetry hypothesis. *Behavioral Ecology and Sociobiology* **7**: 273–9.

Ewen, J. G. and Armstrong, D. P. 2002. Unusual sexual behaviour in the stitchbird (or Hihi) *Notiomystis cincta*. *Ibis* **144**: 530–1.

Ewen, J. G., Armstrong, D. P., Ebert, B. and Hansen, L. H. 2004. Extra-pair copulation and paternity defense in the hihi (stitchbird) *Notiomystis cincta*. *New Zealand Journal of Ecology* **28**: 233–40.

Ewen, J. G., Armstrong, D. P. and Lambert, D. M. 1999. Floater males gain reproductive success through extrapair fertilizations in the stitchbird. *Animal Behaviour* **58**: 321–8.

Exton, M. S., Krüger, T. H. C., Koch, M., *et al*. 2001. Coitus-induced orgasm stimulates prolactin secretion in healthy subjects. *Psychoneuroendocrinology* **26**: 287–94.

Eyre-Walker, A. and Keightley, P. D. 1999. High genomic deleterious mutation rates in hominids. *Nature* **397**: 344–7.

Ezenwa, V. O. 2004. Host social behavior and parasite infection: a multifactorial approach. *Behavioral Ecology* **15**: 446–54.

Fabes, R. A., Martin, C. L. and Hanish, L. D. 2003. Young children's play qualities in same-, other-, and mixed-sex peer groups. *Child Development* **74**: 921–32.

Fain, M. G. and Houde, P. 2007. Multilocus perspectives on the monophyly and phylogeny of the order Charadriiformes (Aves). *BMC Evolutionary Biology* **7**: 35.

Fairbanks, L. A., McGuire, M. T. and Kerber, W. 1977. Sex and aggression during Rhesus monkey group formation. *Aggressive Behavior* **3**: 241–9.

Falconer D. S. and Mackay T. F. C. 1996. *Introduction to Quantitative Genetics*, 4th edn. Essex, UK: Longman.

Falls, J. G., Pulford, D. J., Wylie, A. A. and Jirtle, R. L. 1999. Genomic imprinting: implications for human disease. *American Journal of Pathology* **154**: 635–47.

Fancy, S. G., Farnell, R. F. and Pank, L. F. 1990. Social bonds among adult Barren-Ground caribou. *Journal of Mammalogy* **71**: 458–60.

FAO. 1985. *FAO/UNEP Expert Consultation on Restoration of Przewalski Horse to Mongolia*: Moscow: FAO Corporate Document Repository www.fao.org/ DOCREP/004/AC148E/AC148E01.htm.

Farrell, M. A., Barry, E. and Marples, N. 2000. Breeding behavior in a flock of Chilean flamingos (*Phoenicopterus chilensis*) at Dublin Zoo. *Zoo Biology* **19**: 227–37.

Fauve-Chamoux, A. 1996. Beyond adoption: orphans and family strategies in pre-industrial France. *The History of the Family* **1**: 1–13.

Fedigan, L. 1993. Sex differences and intersexual relations in adult white-faced capuchins (*Cebus capucinus*). *International Journal of Primatology* **14**: 853–77.

Feige, S., Nilsson, K., Phillips, C. J. C. and Johnston, S. D. 2007. Heterosexual and homosexual behaviour and vocalizations in captive female koalas (*Phascolarctos cinereus*). *Applied Animal Behaviour* **103**: 131–45.

Feingold, A. 1994. Gender differences in personality: a meta-analysis. *Psychological Bulletin* **116**: 429–56.

Feist, J. D. and McCullough, D. R. 1976. Behavior patterns and communication in feral horses. *Zeitschrift für Tierpsychologie* **41**: 337–71.

Feldhamer, G. A. 1980. *Cervus nippon*. *Mammalian Species* **128**: 1–7.

Feldman, P. 1984. The homosexual preference. In: Howells, K. (Ed.) *The Psychology of Sexual Diversity*, pp. 5–41. Oxford: Basil Blackwell.

Félix, F. 1997. Organization and social structure of the coastal bottlenose dolphin *Tursiops truncatus* in the Gulf of Guayaquil, Ecuador. *Aquatic Mammals* **23**: 1–16.

Felsenstein, J. 1985. Phylogenies and the comparative method. *American Naturalist* **43**: 300–11.

Fennema, J. S. A., van Ameijden, E. J. C., Coutinho, R. A. *et al*. 1998. HIV surveillance among sexually transmitted disease clinic attenders in Amsterdam, 1991–1996. *AIDS* **12**: 931–8.

Fenstemaker, S. B., Zup, S. L., Frank, L. G., Glickman, S. E. and Forger, N. G. 1999. A sex difference in the hypothalamus of the spotted hyena. *Nature Neuroscience* **2**: 943–5.

Fenton, K. A. and Lowndes, C. M. 2004. Recent trends in epidemiology of sexually transmitted infections in the European Union. *Sexually Transmitted Infections* **80**: 255–63.

Ferguson, T. A. Y. 1995. Alternative sexualities and evolution. *Evolutionary Theory* **11**: 55–64.

Fernández, G. J., Capurro, A. F. and Reboreda, J. C. 2003. Effect of group size on individual and collective vigilance in greater rheas. *Ethology* **109**: 413–25.

Fernández, G. J. and Reboreda, J. C. 2003. Male parental care in greater rheas (*Rhea americana*) in Argentina. *Auk* **120**: 418–28.

Festa-Bianchet, M. 1991. The social system of bighorn sheep: grouping patterns, kinship and female dominance rank. *Animal Behaviour* **42**: 71–82.

Fethers, K., Marks, C., Mindel, A. and Estcourt, C. S. 2000. Sexually transmitted infections and risk behaviours in women who have sex with women. *Sexually Transmitted Infections* **76**: 345–9.

Feuer, L. S. 1957. The principle of simplicity. *Philosophy of Science* **24**: 109–22.

Findlay, C. S. 1992. Phenotypic evolution under gene-culture transmission in structured populations. *Journal of Theoretical Biology* **156**: 387–400.

Fink, B. and Neave, N. 2002. Human body odour, genetic variability, and sexual orientation: a reply to D. Oliva. *Neuroendocrinology Letters* **23**: 289–90.

Finkelhor, D. 1980. Sex among siblings: a survey on prevalence, variety, and effects. *Archives of Sexual Behavior* **9**: 171–94.

Finlay, B. L. and Darlington, R. B. 1995. Linked regularities in the development and evolution of mammalian brains. *Science* **268**: 1578–84.

Finlay, B. L., Wikler, K. C. and Sengelaub, D. R. 1987. Regressive events in brain development and scenarios for vertebrate brain evolution. *Brain Behaviour and Evolution* **30**: 102–17.

Firos Anestis, S. 2004. Female genito-genital rubbing in a group of captive chimpanzees. *International Journal of Primatology* **25**: 477–88.

Fischer, R. B. and Nadler, R. D. 1978. Affiliative, playful and homosexual interactions of adult female lowland gorillas. *Primates* **19**: 657–64.

Fish, F. E. 1993. Power output and propulsive efficiency of swimming bottlenose dolphins (*Tursiops truncatus*). *Journal of Experimental Biology* **185**: 179–93.

Fisher, D. O., Blomberg, S. P. and Owens, I.P.F. 2002. Convergent maternal care strategies in ungulates and macropods. *Evolution* **56**: 167–76.

Fisher, H. E. 1982. Of human bonding. *The Sciences* **22**: 18–23.

Fisher, H. E. 1998. Lust, attraction and attachment in mammalian reproduction. *Human Nature* **9**: 23–52.

Fisher, H. E., Aron, A. and Brown, L. L. 2006. Romantic love: a mammalian brain system for mate choice. *Philosophical Transactions of the Royal Society of London*, B **361**: 2173–86.

Fisher, H. E., Aron, A., Mashek, D., Li, H. and Brown, L. L. 2002. Defining the brain system of lust, romantic attraction, and attachment. *Archives of Sexual Behavior* **31**: 413–19.

Fisher, H. I. 1971. The Laysan Albatross: its incubation, hatching, and associated behaviours. *Living Bird* **10**: 19–78.

Fisher, R. A. 1922. On the dominance ratio. *Proceedings of the Royal Society of Edinburgh* **42**: 321–41.

Fitch, R. H. and Denenberg, V. H. 1998. A role for ovarian hormones in sexual differentiation of the brain. *Behavioral and Brain Sciences* **21**: 311–52.

Fitzgerald, W. A. 2000. Explaining the variety of human sexuality. *Medical Hypotheses* **55**: 435–9.

Flannery, K. A. and Liederman, J. 1994. A test of the immunoreactive theory of the origin of neurodevelopmental disorders: is there an antecedent brother effect? *Developmental Neuropsychology* **10**: 481–92.

Fleming, D., Cinderey, R. N. and Hearn, J. P. 1983. The reproductive biology of Bennett's wallaby (*Macropus rufogriseus rufogriseus*) ranging free at Whipsnade Park. *Journal of Zoology, London* **201**: 283–91.

Fletchall, N. B., Rodden, M. and Taylor, S. (Eds) 1995. Husbandry manual for the maned wolf *Chrysocyon brachyurus*. www.aguaraguazu.or.ar/pdf/husbandry_manual.pdf.

Flux, I., Bradfield, P. and Innes, J. 2006. Breeding biology of North Island kokako (*Callaeas cinerea wilsoni*) at Mapara Wildlife Management Reserve, King Country, New Zealand. *Notornis* **53**: 199–207.

Flynn, J. J. and Nedbal, M. A. 1998. Phylogeny of the Carnivora (Mammalia): congruence vs incompatibility among multiple data sets. *Molecular Phylogenetics and Evolution* **9**: 414–26.

Flynn, J. J., Finarelli, J. A., Zehr, S., Hsu, J. and Nedbal, M. A. 2005. Molecular phylogeny of the Carnivora (Mammalia): assessing the impact of increased sampling on resolving enigmatic relationships. *Systematic Biology* **54**: 317–37.

Focardi, S. and Pecchioli, E. 2005. Social cohesion and foraging decrease with group size in fallow deer (*Dama dama*). *Behavioral Ecology and Sociobiology* **59**: 84–91.

Forastieri, V., Pinto Andrade, C., Lima, A. *et al.* 2002. Evidence against a relationship between dermatoglyphic asymmetry and male sexual orientation. *Human Biology* **74**: 861–70.

Ford, C. S. and Beach, F. A. 1951. *Patterns of Sexual Behaviour*. New York: Harper and Brother.

Ford, F. 2006. A splitting headache: relationships and generic boundaries among Australian murids. *Biological Journal of the Linnean Society* **89**: 117–38.

Forger, N. G. and Breedlove, S. M. 1986. Sexual dimorphism in human and canine spinal cord: Role of early androgen. *Proceedings of the National Academy of Sciences, USA* **83**: 7527–31.

Forger, N. G., Rosen, G. J., Waters, E. M. *et al.* 2004. Deletion of *Bax* eliminates sex differences in the mouse forebrain. *Proceedings of the National Academy of Sciences, USA* **101**: 13666–71.

Fornasari, L., Bottoni, L., Sacchi, N. and Massa, R. 1994. Home range overlapping and socio-sexual relationships in the red-backed shrike *Lanius collurio*. *Ethology, Ecology and Evolution* **6**: 169–77.

Forstmeier, W. 2005. Quantitative genetics and behavioural correlates of digit ratio in the zebra finch. *Proceedings of the Royal Society of London*, B **272**: 2641–49.

Foster, M. 1984. Jewel bird jamboree. *Natural History* **7**: 55–9.

Foster, M. S. 1987. Delayed maturation, neoteny, and social system differences in two manakins of the genus *Chiroxiphia*. *Evolution* **41**: 547–58.

Foucault, M. 1976. *The Will to Knowledge. The History of Sexuality: 1*. London: Penguin Books.

Fox, C. A. and Fox, B. 1971. A comparative study of coital physiology, with special reference to the sexual climax. *Journal of Reproduction and Fertility* **24**: 319–36.

Fox, C. A., Wolff, H. S. and Baker, J. A. 1970. Measurement of intra-vaginal and intra-uterine pressures during human coitus by radio-telemetry. *Journal of Reproduction and Fertility* **22**: 243–51.

Fox, E. A. 2001. Homosexual behaviour in wild Sumatran orangutans (*Pongo pygmaeus abelii*). *American Journal of Primatology* **55**: 177–81.

Fox, E. A. 2002. Female tactics to reduce sexual harassment in the Sumatran orangutan (*Pongo pygmaeus abelii*). *Behavioral Ecology and Sociobiology* **52**: 93–101.

Fox, G. A. and Boersma, D. 1983. Characteristics of supernormal ring-billed gull clutches and their attending adults. *Wilson Bulletin* **95**: 552–9.

Fox, J. R. 1962. Sibling incest. *British Journal of Sociology* **13**: 128–50.

Fraley, R. C., Brumbaugh, C. C. and Marks, M. J. 2005. The evolution and function of adult attachment: a comparative and phylogenetic analysis. *Journal of Personality and Social Psychology* **89**: 731–46.

Francis, A. M. 2008. Family sexual orientation: the family-demographic correlates of homosexuality in men and women. *Journal of Sex Research* **45**: 371–7.

Frank, L. G. 1996. Female masculinisation in the spotted hyena: endocrinology, behavioural ecology, and evolution. In: Gittleman, J. L. (ed.), *Carnivore Behaviour, Ecology and Evolution*, pp. 78–131. Ithaca, NY: Cornell University Press.

Frank, L. G. 1997. Evolution of genital masculinisation: why do female hyaenas have such a large 'penis'? *Trends in Ecology and Evolution* **12**: 58–62.

Frank, L. G., Davidson, J. M. and Smith, E. R. 1985. Androgen levels in the spotted hyaena *Crocuta crocuta*: the influence of social factors. *Journal of Zoology, London* A **206**: 525–31.

Frank, L. G., Glickman, S. E. and Light, P. 1991. Fatal sibling aggression, precocial development, and androgens in neonatal spotted hyenas. *Science* **252**: 702–4.

Franklin, D. C., Smales, I. J., Miller M. A. and Menkhorst, P. W. 1995. The reproductive biology of the helmeted honeyeater *Lichenostomus melanops cassidix*. *Wildlife Research* **22**: 173–91.

Frase, B. A. and Hoffmann, R. S. 1980. *Marmota flaviventris*. *Mammalian Species* **135**: 1–8.

Fratta, W., Biggio, G. and Gessa, G. L. 1977. Homosexual mounting behaviour induced in male rats and rabbits by tryptophan-free diet. *Life Sciences* **21**: 379–84.

Frederick, H. and Johnson, C. N. 1996. Social organisation in the rufous bettong, *Aepyprymnus rufescens*. *Australian Journal of Zoology* **44**: 9–17.

Frederiksen, M. and Bregnballe, T. 2001. Conspecific reproductive success affects age of recruitment in a great cormorant, *Phalacrocorax carbo sinensis*, colony. *Proceedings of the Royal Society of London*, B **268**: 1519–26.

Freud, S. 1905. Three essays on sexual theory. In: Freud, S. 2006, *The Psychology of Love*, pp. 113–220. London: Penguin Books.

Freud, S. 1919. 'A child is being beaten': Contribution to the understanding of the origin of sexual perversions. In: S. Freud, 2006. *The Psychology of Love*, pp. 281–305. London: Penguin Books Ltd.

Freud, S. 1931. On female sexuality. In: Freud, S. 2006. *The Psychology of Love*, pp. 281–305. London: Penguin Books.

Freund, K. and Blanchard, R. 1993. Erotic target location errors in male gender dysphorics, paedophiles, and fetishists. *British Journal of Psychiatry* **162**: 558–63.

Freund, K., Langevin, R., Cibiri, S. and Zajac, Y. 1973. Heterosexual aversion in homosexual males. *British Journal of Psychiatry* **122**: 163–9.

Frid, A. 1999. Huemul (*Hippocamelus bisulcus*) sociality at a periglacial site: sexual aggregation and habitat effects on group size. *Canadian Journal of Zoology* **77**: 1083–91.

Fried, G., Andersson, E., Csöregh, L. et al. 2004. Estrogen receptor beta is expressed in human embryonic brain cells and is regulated by 17β-estradiol. *European Journal of Neuroscience* **20**: 2345–54.

Friedman, R. C. 1986. The psychoanalytic model of male homosexuality: a historical and theoretical critique. *Psychoanalytic Review* **73D**: 79–115.

Friedman, R. C. 1992. Neuropsychiatry and homosexuality: on the need for biopsychological interactionism. *Journal of Neuropsychiatry* **4**: 1–2.

Friedman, R. C. and Downey, J. 1993. Neurobiology and sexual orientation: current relationships. *Journal of Neuropsychiatry and Clinical Neurosciences* **5**: 131–53.

Friedman, R. C. and Downey, J. I. 1998. Psychoanalysis and the model of homosexuality as psychopathology: a historical overview. *American Journal of Psychoanalysis* **58**: 249–70.

Friedman, R. C. and Downey, J. I. 2008. Sexual differentiation of behavior: the foundation of a developmental model of psychosexuality. *Journal of the American Psychoanalytical Association* **56**: 147–75.

Friedrich, W. N. 2000. Cultural differences in sexual behavior in 2–6 year old Dutch and American children. *Journal of Psychology & Human Sexuality* **12**: 117–29.

Friedrich, W. N., Grambsch, P., Broughton, J. K. and Beilke, R. L. 1991. Normative sexual behavior in children. *Pediatrics* **88**: 456–64.

Friesen, V. L., Baker, A. J. and Piatt, J. F. 1996. Phylogenetic relationships within the Alcidae (Charadriiformes: Aves) inferred from total molecular evidence. *Molecular Biology and Evolution* **13**: 359–67.

Friesen, V. L., Montevecchi, W. A., Gaston, A. J., Barrett, R. T. and Davidson, W. S. 1996. Molecular evidence for kin groups in the absence of large-scale genetic differentiation in a migratory bird. *Evolution* **50**: 924–30.

Frigerio, D., Weiss, B. and Kotrschal, K. 2001. Spatial proximity among adult siblings in greylag geese (*Anser anser*): evidence for female bonding? *Acta Ethologica* **3**: 121–5.

Frisch, M. and Hviid, A. 2006. Childhood family correlates of heterosexual and homosexual marriages: a national cohort study of two million Danes. *Archives of Sexual Behavior* **35**: 533–47.

Frith, C. B. 1982. Displays of Count Raggi's bird-of-paradise *Paradisaea raggiana* and congeneric species. *Emu* **81**: 193–201.

Frith, C. B. and Cooper W. T. 1996. Courtship display and mating of Victoria's Riflebird *Ptiloris victoriae* with notes on the courtship displays of congeneric species. *Emu* **96**: 102–13.

Fritsche, S. and Steinhart, H. 1999. Occurrence of hormonally active compounds in food: a review. *European Food Research and Technology* **209**: 153–79.

Frontisi-Ducroux, F. and Lissarrague, F. 1990. From ambiguity to ambivalence: a Dionysiac excursion through the "Anakreontik" vases. In: Halperin, D. M., Winkler, J. J. and Zeitlin, F. I. (Eds). *Before Sexuality: The Construction of Erotic Experience in the Ancient Greek World*, pp. 211–56. Princeton, NJ: Princeton University Press.

Fruth, B. and Hohmann, G. 2006. Social grease for females? Same-sex genital contacts in wild bonobos. In: Sommer, V. and Vasey, P. L. (Eds). *Homosexual Behaviour in Animals: an Evolutionary perspective*, pp. 294–315. Cambridge: Cambridge University Press.

Fry, D. M. and Toone, C. K. 1981. DDT-induced feminization of gull embryos. *Science* **213**: 922–4.

Fry, D. M., Toone, C. K., Speich, S. M. and Peard, R. J. 1987. Sex ratio skew and breeding patterns of gulls: demographic and toxicological considerations. *Studies in Avian Biology* **10**: 26–43.

Fujioka, M. and Yamagishi, S. 1981. Extramarital and pair copulations in the cattle egret. *Auk* **98**: 134–44.

Fuller, W. A. 1960. Behaviour and social organization of the wild bison of Wood Buffalo National Park, Canada. *Arctic* **13**: 3–19.

Furuichi, T. 1989. Social interactions and the life history of female *Pan paniscus* in Wamba, Zaire. *International Journal of Primatology* **10**: 173–97.

Futuyma, D. J. and Risch, S. J. 1984. Sexual orientation, sociobiology and evolution. *Journal of Homosexuality* **9**: 157–68.

Gabriel, W., Lynch, M. and Burger, R. 1993. Muller's ratchet and mutational meltdowns. *Evolution* **47**: 1744–57.

Gadpaille, W. J. 1980. Cross-species and cross-cultural contributions to understanding homosexual activity. *Archives of General Psychiatry* **37**: 349–56.

Gagnon, J. H. 2005. Review of "Solitary sex: a cultural history of masturbation" by Thomas W. Laqueur. *Archives of Sexual Behavior* **34**: 471–3.

Gagnon, J. H. and Simon, W. (Eds.) 1967. *Sexual Deviance*. New York: Harper and Row.

Gahr, M. 2003. Male Japanese quails with female brains do not show male sexual behaviours. *Proceedings of the National Academy of Sciences, USA* **100**: 7959–64.

Gaines, S. O. Jr. and Henderson, M. C. 2002. Impact of attachment style on responses to accommodative dilemmas among same-sex couples. *Personal Relationships* **9**: 89–93.

Gaist, D., Bathum, L., Skytthe, A. *et al.* 2000. Strength and anthropometric measures in identical and fraternal twins: no evidence of masculinization of females with male co-twins. *Epidemiology* **11**: 340–3.

Gales, N., McCauley, R. D., Lanyon, J. and Holley, D. 2004. Change in abundance of dugongs in Shark Bay, Ningaloo and Exmouth Gulf, Western Australia: evidence for large-scale migration. *Wildlife Research* **31**: 283–90.

Gangestad, S. W., Thornhill, R. and Garver-Apgar, C. E. 2005. Women's sexual interests across the ovulatory cycle depend on primary partner developmental instability. *Proceedings of the Royal Society*, B **272**: 2023–7.

Gangestad, S. W., Yeo, R. A., Shaw, P. *et al.* 1996. Human leukocyte antigens and hand preference: preliminary observations. *Neuropsychology* **10**: 423–8.

Garamszegi, L. Z., Eens, M., Hurtrez-Boussès, S. and Møller, A. P. 2005. Testosterone, testes size, and mating success in birds: a comparative study. *Hormones and Behavior* **47**: 389–409.

García-Falgueras, A., Pinos, H., Collado, P. *et al.* 2005. The role of the androgen receptor in CNS masculinization. *Brain Research* **1035**: 13–23.

García-Falgueras, A. and Swaab, D. F. 2008. A sex difference in the hypothalamic uncinate nucleus: relationship to gender identity. *Brain* **131**: 3132–46.

Gariépy, J.-L., Bauer, D. J. and Cairns, R. B. 2001. Selective breeding for differential aggression in mice provides evidence for heterochrony in social behaviours. *Animal Behaviour* **61**: 933–47.

Garland, T. Jr., Harvey, P. H. and Ives, A. R. 1992. Procedures for the analysis of comparative data using phylogenetically independent contrasts. *Systematic Biology* **41**: 18–32.

Garnett, S. T. 1978. The behaviour patterns of the dusky moorhen, *Gallinula tenebrosa* Gould (Aves: Rallidae). *Australian Wildlife Research* **5**: 363–84.

Gartlan, J. S. 1969. Sexual and maternal behaviour of the vervet monkey, *Cercopithecus aethiops*. *Journal of Reproduction and Fertility*, Suppl. **6**: 137–50.

Gastal, M. O., Gastal, E. L., Beg, M. A. and Ginther, O. J. 2007. Elevated plasma testosterone concentrations during stallion-like sexual behaviour in mares (*Equus caballus*). *Hormones and Behavior* **52**: 205–10.

Gates, C. C. and Larter, N. C. 1990. Growth and dispersal of an erupting large herbivore population in Northern Canada: the Mackenzie wood bison (*Bison bison athabascae*). *Arctic* **43**: 231–8.

Gathorne-Hardy, J. 1998. *Sex the Measure of All Things: A Life of Alfred C. Kinsey*. London: Chatto & Windus.

Gavrilets, S. and Rice, W. R. 2006. Genetic models of homosexuality: generating testable predictions. *Proceedings of the Royal Society*, B **273**: 3031–38.

Geary, D. C. 2000. Evolution and proximate expression of human paternal investment. *Psychological Bulletin* **126**: 55–77.

Geary, D. C. 2006. Coevolution of paternal investment and cuckoldry in humans. In: Shackleford, T. K. and Platek, S. (Eds) *Female Infidelity and Paternal Uncertainty*, pp. 14–34. Cambridge: Cambridge University Press.

Geisler, J. H. 2001. New morphological evidence for the phylogeny of Artiodactyla, Cetacea, and Mesonychidae. *American Museum Novitates* **40**: 1–53.

Geist, V. 1968. On the interrelation of external appearance, social behaviour and social structure of mountain sheep. *Zeitschrift für Tierpsychologie* **25**: 199–215.

Geist, V. and Petocz, R. G. 1977. Bighorn sheep in winter: do rams maximize reproductive fitness by spatial and habitat segregation from ewes? *Canadian Journal of Zoology* **55**: 1802–10.

Gelez, H. and Fabre-Nys, C. 2004. The "male effect" in sheep and goats: a review of the respective roles of the two olfactory systems. *Hormones and Behavior* **46**: 257–71.

Gelfand, M. M. 2000. Sexuality among older women. *Journal of Women's Health and Gender-Based Medicine* **9** (Suppl. **1**): 15–20.

Gende, S. M., Womble, J. N., Willson, M. F. and Marston, B. M. 2001. Cooperative foraging by Steller sea lions,

Eumetopias jubatus. *Canadian Field Naturalist* **115**: 355–6.

Gerall, A. A., Hendricks, S. E., Johnson, L. L. and Bounds, T. W. 1967. Effects of early castration in male rats on adult sexual behaviour. *Journal of Comparative and Physiological Psychology* **64**: 206–12.

Gerloff, U., Hartung, B., Fruth, B., Hohmann, G. and Tautz, D. 1999. Intracommunity relationships, dispersal pattern and paternity success in a wild living community of bonobos (*Pan paniscus*) determined from DNA analysis of faecal samples. *Proceedings of the Royal Society of London*, B **266**: 1189–95.

German, R. Z., Hertweck, D. W., Sirianni, J. E. and Swindler, D. R. 1994. Heterochrony and sexual dimorphism in the pigtail macaque (*Macaca nemestrina*). *American Journal of Physical Anthropology* **93**: 373–80.

Geschwind, N. and Galaburda, A. M. 1985. Cerebral lateralization: biological mechanisms, associations, and pathology: II. A hypothesis and a program for research. *Archives of Neurology* **42**: 521–52.

Geschwind, N. and Galaburda, A. M. 1987. *Cerebral Lateralization: Biological Mechanisms, Associations, and Pathology*. Cambridge, MA: MIT Press.

Getz, W. M. 1993. Invasion and maintenance of alleles that influence mating and parental success. *Journal of Theoretical Biology* **162**: 515–37.

Giacoma, C. and Messeri, P. 1992. Attributes and validity of dominance hierarchy in the female pigtail macaque. *Primates* **33**: 181–9.

Gibb, G. C., Kardailsky, O., Kimball, R. T., Braun, E. L. and Penny, D. 2007. Mitochondrial genomes and avian phylogeny: complex characters and resolvability without explosive radiations. *Molecular Biology and Evolution* **24**: 269–80.

Gibson, R. M. and Bradbury, J. W. 1986. Male and female mating strategies on sage grouse leks. In: Rubenstein, D. I. and Wrangham, R. W. (Eds) *Ecological Aspects of Social Evolution*, pp. 379–98. Princeton, NJ: Princeton University Press.

Gibson, R. M. and Guinness, F. E. 1980. Differential reproduction among red deer (*Cervus elaphus*) stags on Rhum. *Journal of Animal Ecology* **49**: 199–208.

Gibson, R. M., Pires, D., Delaney, K. S. and Wayne, R. K. 2005. Microsatellite DNA analysis shows that greater sage grouse leks are not kin groups. *Molecular Ecology* **14**: 4453–59.

Gil, D., Biard, C., Lacroix, A. *et al.* 2007. Evolution of yolk androgens in birds: development, coloniality, and sexual dichromatism. *American Naturalist* **169**: 802–19.

Gilbert, B. K. 1973. Scent marking and territoriality in pronghorn (*Antilocapra americana*) in Yellowstone National Park. *Mammalia* **37**: 25–33.

Gilson, R. J. C., de Ruiter, A., Waite, J. *et al.* 1998. Hepatitis B virus infection in patients attending a genitourinary medicine clinic: risk factors and vaccine coverage. *Sexually Transmitted Infections* **74**: 110–15.

Ginnett, T. F. and Demment, M. W. 1999. Sexual segregation by Masai giraffes at two spatial scales. *African Journal of Ecology* **37**: 93–106.

Ginsberg, J. R. and Rubenstein, D. I. 1990. Sperm competition and variation in zebra mating. *Behavioral Ecology and Sociobiology* **26**: 427–34.

Gittleman, J. L. 1986. Carnivore brain size, behavioural ecology, and phylogeny. *Journal of Mammalogy* **67**: 23–36.

Giuliano, F. and Allard, J. 2001. Dopamine and sexual function. *International Journal of Impotence Research* **13**, Suppl. 3: S18–S28.

Gjershaug, J. O., Folkestad, A. O. and Goksøyr, L. O. 1998. Female-female pairing between a Peregrine Falcon *Falco peregrinus* and a Gyrfalcon *Falco rusticolus* in two successive years. *Fauna Norwegica series C Cinclus* **21**: 87–91.

Gladue, B. A. 1991. Aggressive behavioural characteristics, hormones, and sexual orientation in men and women. *Aggressive Behavior* **17**: 313–26.

Gladue, B. A. 1994. The biopsychology of sexual orientation. *Current Directions in Psychological Science* **3**: 150–54.

Gladue, B. A. and Bailey, J. M. 1995. Aggressiveness, competiveness, and human sexual orientation. *Psychoneuroendocrinology* **20**: 475–85.

Gladue, B. A., Beatty, W. W., Larson, J. and Staton, R. D. 1990. Sexual orientation and spatial ability in men and women. *Psychobiology* **18**: 101–8.

Gleason, M. W. 1990. The semiotics of gender: physiognomy and self-fashioning in the second century c.e. In: Halperin, D. M., Winkler, J. J. and Zeitlin, F. I. (Eds) *Before sexuality: The Construction of Erotic Experience in the Ancient Greek World*, pp. 389–415. Princeton, NJ: Princeton University Press.

Glickman, S. E., Coscia, E. M., Frank, L. G., *et al.* 1998. Androgens and masculinisation of genitalia in spotted hyaena (*Crocuta crocuta*). 3. Effects of juvenile gonadectomy. *Journal of Reproduction and Fertility* **113**: 129–35.

Glickman, S. E., Cunha, G. R., Drea, C. M., Conley, A. J. and Place, N. J. 2006. Mammalian sexual differentiation: lessons from the spotted hyena. *Trends in Endocrinology and Metabolism* **17**: 349–56.

Glickman, S. E., Frank, L. G., Davidson, J. M., Smith, E. R. and Siiteri, P. K. 1987. Androstenedione may organize or activate sex-reversed traits in female spotted hyenas. *Proceedings of the National Academy of Sciences, USA* **84**: 3444–47.

Glickman, S. E., Zabel, C. J., Yoerg, S. I. et al. 1999. Social facilitation, affiliation and dominance in the social life of spotted hyenas. In: Carter, C. S., Lederhendler, I. I. and Kirkpatrick, B. (Eds) *The Integrative Neurobiology of Affiliation*, pp. 131–40. Cambridge, MA: MIT Press.

Głowaciński, Z. and Profus, P. 1997. Potential impact of wolves *Canis lupus* on prey populations in Eastern Poland. *Biological Conservation* **80**: 99–106.

Gobrogge, K. L., Perkins, P. S., Baker, J. H. et al. 2007. Homosexual mating preferences from an evolutionary perspective: sexual selection theory revisited. *Archives of Sexual Behavior* **36**: 717–23.

Goddard K. and Mathis A. 1997. Do opercular flaps of male longear sunfish (*Lepomis megalotis*) serve as sexual ornaments during female mate choice? *Ethology, Ecology and Evolution* **9**: 223–31.

Godoy, J. A., Negro, J. J., Hiraldo, F. and Donázar, J. A. 2004. Phylogeography, genetic structure and diversity in the endangered bearded vulture (*Gypaetus barbatus*, L.) as revealed by mitochondrial DNA. *Molecular Ecology* **13**: 371–90.

Going, J. J. and Dixson, A. 1990. Morphometry of the adult human corpus callosum: lack of sexual dimorphism *Journal of Anatomy* **171**: 163–7.

Goldberg, A. 2001. Depathologizing homosexuality. *Journal of the American Psychoanalytic Association* **49**: 1109–14.

Goldberg, J. 2003. Domestication et comportement. *Bulletin de la Société Zoologique de France* **128**: 175–81.

Goldfoot, D. A., Feder, H. H. and Goy, R. W. 1969. Development of bisexuality in the male rat treated neonatally with androstenedione. *Journal of Comparative and Physiological Psychology* **67**: 41–5.

Goldfoot, D. A. and van der Werff ten Bosch, J. J. 1975. Mounting behaviour of female guinea pigs after prenatal and adult administration of the propionates of testosterone, dihydrotestosterone, and androstanediol. *Hormones and Behaviour* **6**: 139–48.

Goldfoot, D. A., Wallen, K., Neff, D. A., McBrair, M. C. and Goy, R. W. 1984. Social influences on the display of sexually dimorphic behaviour in rhesus monkeys: isosexual rearing. *Archives of Sexual Behavior* **13**: 395–412.

Goldfoot, D. A., Westerborg-Van Loon, H., Groeneveld, W., and Slob, A. K. 1980. Behavioral and physiological evidence of sexual climax in the female stump-tailed macaque (*Macaca arctoides*). *Science* **208**: 1477–9.

Goldfoot, D. A., Wiegand, S. J. and Scheffler, G. 1978. Continued copulation in ovariectomized adrenal-suppressed stumptail macaques (*Macaca arctoides*). *Hormones and Behavior* **11**: 89–99.

Golombok, S. and Tasker, F. 1996. Do parents influence the sexual orientation of their children? Findings from a longitudinal study of lesbian families. *Developmental Psychology* **32**: 3–11.

Gomes, M. N. and Uieda, W. 2004. Diurnal roosts, colony composition, sexual size dimorphism and reproduction of the common vampire bat *Desmodus rotundus* (E. Geoffroy) (Chiroptera, Phyllostomidae) from the state of Sao Paulo, Southeastern Brazil. *Revista Brasileira de Zoologia* **21**: 629–38.

Goncharov, N. P., Tavadian, D. S. and Vorontsov, V. 1981. Diurnal and seasonal rhythms of androgen content of the plasma in *Macaca mulatta*. *Problemy Endokrinologii* **27**: 53–7.

Göncü, A., Mistry, J. and Mosier, C. 2000. Cultural variations in the play of toddlers. *International Journal of Behavioral Development* **24**: 321–9.

Gonzalez, R., Levy, F., Orgeur, P., Poidron, P. and Signoret, J. P. 1991. Female effect in sheep. II. Role of volatile substances from the sexually receptive female. Implication of the sense of smell. *Reproduction, Nutrition, Development* **31**: 103–9.

González-Solís, J., Croxall, J. P. and Wood, A. G. 2000. Sexual dimorphism and sexual segregation in foraging strategies of northern giant petrels, *Macronectes halli*, during incubation. *Oikos* **90**: 390–8.

Goodall, J. 1986. *The Chimpanzees of the Gombe: Patterns of Behavior*. Harvard, MA: Belknap Press.

Goodenough, F. 1934. *Developmental Psychology: An Introduction to the Study of Human Behavior*. New York: Appleton-Century.

Goodman, R. L., Hotchkiss, J., Karsch, F. J. and Knobil, E. 1974. Diurnal variations in serum testosterone concentrations in the adult male rhesus monkey. *Biology of Reproduction* **11**: 624–30.

Goodson, J. L., Saldanha, C. J., Hahn, T. P. and Soma, K. K. 2005. Recent advances in behavioural neuroendocrinology: insights from studies on birds. *Hormones and Behavior* **48**: 461–73.

Goodstrey, A., Carss, D. N., Noble, R. and Piertney, S. B. 1998. Population introgression and differentiation in the

Great Cormorant *Phalacrocorax carbo* in Europe. *Molecular Ecology* **7**: 329–38.

Gooren, L. J. G. 1985. Human male sexual functions do not require aromatization of testosterone: a study using tamoxifen, testolactone, and dihydrotestosterone. *Archives of Sexual Behavior* **14**: 539–48.

Gooren, L. 1986. The neuroendocrine response of luteinizing hormone to estrogen administration in heterosexual, homosexual, and transsexual subjects. *Journal of Clinical Endocrinology and Metabolism* **63**: 583–8.

Gooren, L. J. G. 1993. Concepts and methods of biomedical research into homosexuality and transsexualism. *Journal of Psychology and Human Sexuality* **6**: 5–21.

Gooren, L. 2006. The biology of human psychosexual differentiation. *Hormones and Behavior* **50**: 589–601.

Goossens, B., Setchell, J. M., James, S. S. *et al.* 2006. Philopatry and reproductive success in Bornean orang-utans (*Pongo pygmaeus*). *Molecular Ecology* **15**: 2577–88.

Goostrey, A., Carss, D. N., Noble, L. R. and Piertney, S. B. 1998. Population introgression and differentiation in the great cormorant *Phalacrocorax carbo* in Europe. *Molecular Ecology* **7**: 329–38.

Gordon, T. P. 1981. Reproductive behaviour in the rhesus monkey: social and endocrine variables. *American Zoologist* **21**: 185–95.

Gordon, T. P. and Bernstein, I. S. 1973. Seasonal variation in sexual behaviour of all-male rhesus troops. *American Journal of Physical Anthropology* **38**: 221–6.

Gorman, M. R. 1994. Male homosexual desire: neurological investigations and scientific bias. *Perspectives in Biology and Medicine* **38**: 61–81.

Goronzy, J. J. and Weyand, C. M. 2005. T cell development and receptor diversity during aging. *Current Opinion in Immunology* **17**: 468–75.

Gorski, R. A. 1987. Sex differences in the rodent brain: their nature and origin. In: Reinisch, J. M., Rosenblum, L. A. and Sanders, S. A. (Eds) *Masculinity/Feminity: Basic Perspectives*, pp. 37–67. New York: Oxford University Press.

Gorski, R. A., Gordon, J. H., Shryne, J. E. and Southam, A. M. 1978. Evidence for a morphological sex difference within the medial preoptic area of the rat brain. *Brain Research* **148**: 333–46.

Gorski, R. A., Harlan, R. E., Jacobson, C. D., Shryne, J. E. and Southam, A. M. 1980. Evidence for the existence of a sexually dimorphic nucleus in the preoptic area of the rat. *Journal of Comparative Neurology* **193**: 529–39.

Gorzalka, B. B., Mendelson, S. D. and Watson, N. V. 1990. Serotonin receptor subtypes and sexual behaviour. *Annals of the New York Academy of Sciences* **600**: 435–46.

Gottelli, D., Wang, J., Bashir, S. and Durant, S. M. 2007. Genetic analysis reveals promiscuity among female cheetahs. *Proceedings of the Royal Society of London*, B **274**: 1993–2001.

Gottschalk, L. 2003. Same-sex sexuality and childhood gender non-conformity: a spurious connection. *Journal of Gender Studies* **12**: 35–50.

Götz, F., Rohde, W. and Dörner, G. 1991. Neuroendocrine differentiation of sex-specific gonadotropin secretion, sexual orientation and gender role behaviour. In: Haug, M., Brain, P. F. and Aron, C. (Eds.) *Heterotypical Behaviour in Man and Animals*, pp. 167–94. London: Chapman and Hall.

Gould, S. J. 1977. *Ontogeny and Phylogeny*. Cambridge, MA: Belknap Press.

Gould, S. J. and Lewontin, R. C. 1979. The spandrels of San Marco and the Panglossian paradigm: a critique of the adaptationist programme. *Proceedings of the Royal Society of London*, B **205**: 581–98.

Goulmy, E., Termijtelen, A., Bradley, B. A. and van Rood, J. J. 1977. Y-antigen killing by T cells of women is restricted by HLA. *Nature* **266**: 544–5.

Gouzoules, H. 1974. Harassment of sexual behaviour in the stumptail macaque, *Macaca arctoides*. *Folia Primatologica* **22**: 208–17.

Gouzoules, H. and Goy, R. W. 1983. Physiological and social influences on mounting behavior of troop-living female monkeys (*Macaca fuscata*). *American Journal of Primatology* **5**: 39–49.

Goy, R. W. 1978. Development of play and munting behaviour in female rhesus virilized with esters of testosterone and dihydrotestosterone. In: Chivers, D. J. and Herbert, J. (Eds) *Recent Advances in Primatology*, Vol. 1, *Behaviour*, pp. 449–62. London: Academic Press.

Goy, R. W. and Goldfoot, D. A. 1975. Neuroendocrinology: animal models and problems of human sexuality. *Archives of Sexual Behavior* **4**: 405–20.

Goy, R. W. and McEwen, B. S. 1980. *Sexual Differentiation of the Brain*. Cambridge, MA: MIT Press.

Goy, R. W. and Roy, M. 1991. Heterotypical sexual behaviour in female mammals. In: Haug, M., Brain, P. F. and Aron, C. (Eds) *Heterotypical Behaviour in Man and Animals*. pp. 71–97. London: Chapman and Hall.

Goy, R. W., Wallen, K. and Goldfoot, D. A. 1974. Social factors affecting the development of mounting behaviour in male rhesus monkeys. *Advances in Behavioral Biology* **11**: 223–47.

Goymann, W., Leippert, D. and Hofer, H. 2000. Sexual segregation, roosting, and social behaviour in a free-ranging colony of Indian false vampires (*Megaderma lyra*). *Zeitschrift für Säugetierkunde* **65**: 138–48.

Grady, K. L., Phoenix, C. H. and Young, W. C. 1965. Role of the developing rat testis in differentiation of the neural tissues mediating mating behaviour. *Journal of Comparative and Physiological Psychology* **59**: 176–82.

Grafen, A. 1989. The phylogenetic regression. *Philosophical Transactions of the Royal Society of London*, B **326**: 119–57.

Grafen, A. 1992. The uniqueness of the phylogenetic regression. *Journal of Theoretical Biology* **156**: 405–23.

Grant, J. W. A., Chapman, C. A. and Richardson, K. S. 1992. Defended versus undefended home range size of carnivores, ungulates and primates. *Behavioral Ecology and Sociobiology* **31**: 149–61.

Grant, T. R. 1973. Dominance and association among members of a captive and a free-ranging group of grey kangaroos (*Macropus giganteus*). *Animal Behaviour* **21**: 449–56.

Grau, I. and Doll, J. 2003. Effects of attachment styles on experience of equity in heterosexual couples relationships. *Experimental Psychology* **50**: 298–310.

Gray, G. G. and Simpson, C. D. 1983. Population characteristics of free-ranging barbary sheep in Texas. Journal of Wildlife Management **47**: 954–62.

Green, J. D., Clemente, C. D. and de Groot, J. 1957. Rhinencephalic lesions and behaviour in cats: an analysis of the Klüver-Bucy syndrome with particular reference to normal and abnormal sexual behaviour. *Journal of Comparative Neurology* **108**: 505–45.

Green, R. 1968. Childhood cross-gender identification. *Journal of Nervous and Mental Disease* **147**: 500–9.

Green, R. 1978. Sexual identity of 37 children raised by homosexual or transsexual parents. *American Journal of Psychiatry* **135**: 692–7.

Green, R. 1979. Childhood cross-gender behavior and subsequent sexual preference. *American Journal of Psychiatry* **136**: 106–8.

Green, R. 1985. Gender identity in childhood and later sexual orientation: follow-up of 78 males. *American Journal of Psychiatry* **142**: 339–41.

Green, R. 1987. *The "Sissy Boy Syndrome" and the Development of Homosexuality*. New Haven, CT: Yale University Press.

Green, R. 2000. Birth order and ratio of brothers and sisters in transsexuals. *Psychological Medicine* **30**: 789–95.

Green, R. and Keverne, E. B. 2000. The disparate maternal aunt-uncle ratio in male transsexuals: an explanation invoking genomic imprinting. *Journal of Theoretical Biology* **202**: 55–63.

Green, R., Williams, K. and Goodman, M. 1985. Masculine or feminine gender identity in boys: developmental differences between two diverse family groups. *Sex Roles* **12**: 1155–62.

Greene, E., Lyon, B. E., Muchter, V. R. *et al.* 2000. Disruptive sexual selection for plumage coloration in a passerine bird. *Nature* **407**: 1000–3.

Greenwald, E. and Leitenberg, H. 1989. Long-term effects of sexual experiences with siblings and nonsiblings during childhood. *Archives of Sexual Behavior* **18**: 389–99.

Gregoire, P. E. and Ankney, C. D. 1990. Agonistic behavior and dominance relationships among lesser snow geese during winter and spring migration. *Auk* **107**: 550–60.

Greiling, H. and Buss, D. M. 2000. Women's sexual strategies: the hidden dimension of extra-pair mating. *Personality and Individual Differences* **28**: 929–63.

Greim, H. A. 2004. The endocrine and reproductive system: adverse effects of hormonally active substances? *Pediatrics* **113**: 1070–75.

Griffith, S. C., Owens, I. P. F. and Thuman, K. A. 2002. Extrapair paternity in birds: a review of interspecific variation and adaptive function. *Molecular Ecology* **11**: 2195–451.

Griffiths, A. D. and Brook, B. W. 2005. Body size and growth in tropical small mammals: examining variation using non-linear mixed effects models. *Journal of Zoology, London* **267**: 211–20.

Grosjean, Y., Grillet, M., Augustin, H., Ferveur, J.-F. and Featherstone, D. E. 2008. A glial amino-acid transporter controls synapse strength and courtship in *Drosophila*. *Nature Neuroscience* **11**: 54–61.

Gross, C. G. 2000. Neurogenesis in the adult brain: death of a dogma. *Nature Reviews* **1**: 67–73.

Grossman, L. S., Martis, B. and Fichtner, C. G. 1999. Are sex offenders treatable? A research overview. *Psychiatric Services* **50**: 349–61.

Groves, C. P. 1994. Morphology, habitat and taxonomy. In: Boyd, L. and Houpt, K. A. (Eds) *Przewalski's Horse: the History and Biology of an Endangered Species*, pp. 39–60. New York: State University of New York Press.

Groves, C. P. and Bell, C. H. 2004. New investigations on the taxonomy of the zebras genus *Equus*, subgenus *Hippotigris*. *Mammalian Biology* **69**: 182–96.

Grubb, P. 1973. Distribution, divergence and speciation of the drill and mandrill. *Folia Primatologica* **20**: 161–77.

Gruber, A. J. and Pope, H. G. Jr. 2000. Psychiatric and medical effects of anabolic-androgenic steroid use in women. *Psychotherapy and Psychosomatics* **69**: 19–26.

Gruys, R. C. 1993. Autumn and winter movements and sexual segregation of willow ptarmigan. *Arctic* **46**: 228–39.

Gualtieri, T. and Hicks, R. E. 1985. An immunoreactive theory of selective male affliction. *Behavioural and Brain Sciences* **8**: 427–41.

Guhl, A. M. 1948. Unisexual mating in a flock of white leghorn hens. *Transactions of the Kansas Academy of Sciences* **18**: 107–11.

Guinness, F., Lincoln, G. A. and Short, R. V. 1971. The reproductive cycle of the female red deer, *Cervus elaphus* L. *Journal of Reproduction and Fertility* **27**: 427–38.

Gurevitch, J. and Hedges, L. V. 2001. Meta-analysis: combining the results of independent experiments. In: Scheiner, S. M. and Gurevitch, J. (Eds) *Design and Analysis of Ecological Experiments*, 2nd, edn, pp. 347–69. Oxford: Oxford University Press.

Gurney, M. E. 1981. Hormonal control of cell form and number in the zebra finch song system. *Journal of Neurosciences* **1**: 658–73.

Gust, D. A., Gordon, T. P., Gergits, W. F. *et al.* 1996. Male dominance rank and offspring-initiated affiliative behaviors were not predictors of paternity in a captive group of pigtail macaques (*Macaca nemestrina*). *Primates* **37**: 271–8.

Guyda, H. J. and Friesen, H. G. 1973. Serum prolactin levels in humans from birth to adult life. *Pediatric Research* **7**: 534–40.

Haake, P., Exton, M. S., Haverkamp, J. *et al.* 2002. Absence of orgasm-induced prolactin secretion in a healthy multi-orgasmic male subject. *International Journal of Impotence Research* **14**: 133–5.

Habibi, K. 1987. Overt sexual behaviour among female aoudads. *Southwestern Naturalist* **32**: 148–9.

Habibi, K. 1987. Behavior of aoudad (*Ammotragus lervia*) during the rutting season. *Mammalia* **51**: 497–513.

Habr-Alencar, S. F., Dias, R. G., Teodorov, E. and Bernardi, M. M. 2006. The effect of hetero- and homosexual experience and long-term treatment with fluoxetine on homosexual behaviour in male rats. *Psychopharmacology* **189**: 269–75.

Haddrath, O. and Baker, A. J. 2001. Complete mitochondrial DNA genome sequences of extinct birds: ratite phylogenetics and the vicariance biogeography hypothesis. *Proceedings of the Royal Society of London*, B **268**: 939–45.

Hafner, M. S. and Hafner, J. C. 1984. Brain size, adaptation and heterochrony in geomyoid rodents. *Evolution* **38**: 1088–98.

Hakkarainen, H., Korpimäki, E., Huhta, E. and Palokangas, P. 1993. Delayed maturation in plumage colour: evidence for the female-mimicry hypothesis in the kestrel. *Behavioural Ecology and Sociobiology* **33**: 247–51.

Haldane, J.B.S. 1928. A mathematical theory of natural and artificial selection. Part V: selection and mutation. *Proceedings of the Cambridge Philosophical Society* **23**: 838–44.

Hale, W. G. and Ashcroft, R. P. 1982. Pair formation and pair maintenance in the redshank *Tringa totanus*. *Ibis* **124**: 471–90.

Hale, W. G. and Ashcroft, R. P. 1983. Studies of the courtship behaviour of the Redshank *Tringa totanus*. *Ibis* **125**: 3–23.

Hall, B. K. 2003. Evo-devo: evolutionary developmental mechanisms. *International Journal of Developmental Biology* **47**: 491–95.

Hall, J. A. 1984. *Nonverbal Sex Differences: Communication Accuracy and Expressive Style*. Baltimore, MD: Johns Hopkins University Press.

Hall, J. A. Y. and Kimura, D. 1994. Dermatoglyphic asymmetry and sexual orientation in men. *Behavioral Neuroscience* **108**: 1203–6.

Hall, J. A. Y. and Kimura, D. 1995. Sexual orientation and performance of sexually dimorphic motor tasks. *Archives of Sexual Behavior* **24**: 395–407.

Hall, J. C. 1994. The mating of a fly. *Science* **264**: 1702–14.

Hall, K. R. L. 1962. The sexual, agonistic and derived social behaviour patterns of the wild chacma baboon, *Papio ursinus*. *Proceedings of the Zoological Society of London* **139**: 283–327.

Hall, P. A. and Schaeff, C. M. 2008. Sexual orientation and fluctuating asymmetry in men and women. *Archives of Sexual Behavior* **37**: 158–65.

Halperin, D. M. 1990. Why is Diotima a woman? Platonic erōs and the figuration of gender. In: Halperin, D. M., Winkler, J. J. and Zeitlin, F. I. (Eds) *Before Sexuality: The Construction of Erotic Experience in the Ancient Greek World*, pp. 257–308. Princeton, NJ: Princeton University Press.

Hamer, D. and Copeland, P. 1994. *The Science of Desire: The Search for the Gay Gene and the Biology of Behavior*. New York: Simon and Schuster.

Hamer, D. H., Hu, S., Magnuson, V. L., Hu, N. and Pattatucci, A. M. L. 1993. A linkage between DNA markers and the X chromosome and male sexual orientation. *Science* **261**: 321–7.

Hamilton, S. V. 2001. *Affectional orientation, sex roles, and reasons for living.* MA Thesis, Department of Psychology, East Tennessee State University, USA.

Hamilton, W. D. 1963. The evolution of altruistic behaviour. *American Naturalist* **97**: 354–356.

Hamilton, W. D. 1964a. The genetic evolution of social behaviour I. *Journal of Theoretical Biology* **7**: 17–52.

Hamilton, W. D. 1964b. The genetic evolution of social behaviour II. *Journal of Theoretical Biology* **7**: 1–16.

Hamilton, W. D. 1971. Geometry of the selfish herd. *Journal of Theoretical Biology* **31**: 295–311.

Hammack, P. L. 2005. The life course development of human sexual orientation: an integrative paradigm. *Human Development* **48**: 267–90.

Hammock, E. A. D. and Young, L. J. 2005. Microsatellite instability generates diversity in brain and sociobehavioral traits. *Science* **308**: 1630–34.

Hammond, E. L., Lymbery, A. J., Martin, G. B., Groth, D. and Wetherhall, J. D. 2002. Microsatellite analysis of genetic diversity in wild and farmed emus (*Dromaius novaehollandiae*). *Journal of Heredity* **93**: 376–80.

Hanby, J. P. 1974. Male-male mounting in Japanese monkeys (*Macaca fuscata*). *Animal Behaviour* **22**: 836–49.

Hanby, J. P. and Brown, C. E. 1974. The development of sociosexual behaviours in Japanese macaques *Macaca fuscata*. *Behaviour* **49**: 152–95.

Hand, J. L. 1985. Egalitarian resolution of social conflicts: a study of pair-bonded gulls in nest duty and feeding contexts. *Zeitschrift für Tierpsychologie* **70**: 123–47.

Handen, C. E. and Rodman, P. S. 1980. Social development of bonnet macaques from six months to three years of age: a longitudinal study. *Primates* **21**: 350–6.

Happold, M. 1976. Social behaviour of the conilurine rodents (Muridae) of Australia. *Zeitschrift für Tierpsychologie* **40**: 113–82.

Harcourt, A. H. 1979. Social relationships among adult female mountain gorillas. *Animal Behaviour* **27**: 251–64.

Hård, E. and Larsson, K. 1968. Dependence of adult mating behaviour in male rats on the presence of littermates in infancy. *Brain, Behaviour and Evolution* **1**: 405–19.

Hardy, J. W. 1963. Epigamic and reproductive behaviour of the orange-fronted parakeet. *Condor* **65**: 169–99.

Hare, E. H. and Moran, P. A. P. 1979. Parental age and birth order in homosexual patients: a replication of Slater's study. *British Journal of Psychiatry* **134**: 178–82.

Hare, E. H. and Price, J. S. 1969. Birth order and family size: bias caused by changes in birth rate. *British Journal of Psychiatry* **115**: 647–57.

Hare, L., Bernard, P., Sánchez, F. J. *et al.* 2009. Androgen receptor repeat length polymorphism associated with male-to-female transsexualism. *Biological Psychiatry* **65**: 93–6.

Harlow, H. F. 1971. *Learning to Love.* New York: Ballantine Books.

Harper, J. A., Harn, J. H., Bentley, W. W. and Yocum, C. F. 1967. The status and ecology of Roosevelt elk in California. *Wildlife Monographs* **16**.

Harris, C. R. 2002. Sexual and romantic jealousy in heterosexual and homosexual adults. *Psychological Science* **13**: 7–12.

Harrison, C. 2008. Real men do wear mascara: advertising discourse and masculine identity. *Critical Discourse Studies* **5**: 55–74.

Harrison, G. L., McLenachan, P. A., Phillips, M. J. *et al.* 2004. Four new avian mitochondrial genomes help get to basic evolutionary questions in the Late Cretaceous. *Molecular Biology and Evolution* **21**: 974–83.

Harry, J. 1989. Parental physical abuse and sexual orientation in males. *Archives of Sexual Behavior* **18**: 251–61.

Hart, B. L. and Leedy, M. G. 1983. Female sexual responses in male cats facilitated by olfactory bulbectomy and medial preoptic anterior hypothalamic lesions. *Behavioural Neuroscience* **97**: 608–14.

Hartwich, A. 1951. *Krafft-Ebing's Aberrations of Sexual Life: the Psychopathia Sexualis.* London: A London Panther.

Harvey, P. H. 1999. A bestiary of chaos and biodiversity. *Nature* **397**: 402–3.

Harvey, P. H. and Clutton-Brock, T. H. 1981. Primate home-range size and metabolic needs. *Behavioral Ecology and Sociobiology* **8**: 151–5.

Harvey, P. H. and Pagel, M. 1991. *The Comparative Method in Evolutionary Biology.* Oxford: Oxford University Press.

Hashimoto, C. 1997. Context and development of sexual behavior of wild bonobos (*Pan paniscus*) at Wamba, Zaire. *International Journal of Primatology* **18**: 1–21.

Hass, C. C. and Jenni, D. A. 1991. Structure and ontogeny of dominance relationships among bighorn rams. *Canadian Journal of Zoology* **69**: 471–6.

Hasselquist, D. and Sherman, P. W. 2001. Social mating systems and extrapair fertilizations in passerine birds. *Behavioral Ecology* **12**: 455–66.

Hassold, T. and Hunt, P. 2001. To err (meiotically) is human: the genesis of human aneuploidy. *Nature Genetics* **2**: 280–91.

Hatchwell, B. J. 1988. Intraspecific variation in extra-pair copulations and mate defence in common guillemots *Uria aalge*. *Behaviour* **107**: 157–85.

Haughland, D. L. and Larsen, K. W. 2004. Ecology of North American red squirrels across contrasting habitats: relating natal dispersal to habitat. *Journal of Mammalogy* **85**: 225–36.

Hausfater, G. and Takacs, D. 1987. Structure and functions of hindquarter presentations in yellow baboons (*Papio cynocephalus*). *Ethology* **74**: 297–319.

Hausman, K. 2003. Controversy continues to grow over *DSM*'s GID diagnosis. *Psychiatric News* **38**: 25.

Havelock, J. C., Auchus, R. J. and Rainey, W. E. 2004. The rise in adrenal androgen biosynthesis: adrenarche. *Seminars in Reproductive Medicine* **22**: 337–47.

Hawkins, C. E., Dallas, J. F., Fowler, P. A., Woodroffe, R. and Racey, P. A. 2002. Transient masculinization in the fossa, *Cryptoprocta ferox* (Carnivora, Viverridae). *Biology of Reproduction* **66**: 610–15.

Haydock, J. and Koenig, W. D. 2002. Reproductive skew in the polygynandrous acorn woodpecker. *Proceedings of the National Academy of Sciences, USA* **99**: 7178–83.

Haydock, J. and Koenig, W. D. 2003. Patterns of reproductive skew in the polygynandrous acorn woodpecker. *American Naturalist* **162**: 277–89.

Hayes, A. F. 1995. Age preferences for same- and opposite-sex partners. *Journal of Social Psychology* **135**: 125–33.

Hayes, L. D. 2000. To nest communally or not to nest communally: a review of rodent communal nesting and nursing. *Animal Behaviour* **59**: 677–88.

Hayman, L. A., Rexer, J. L., Pavol, M. A., Strite, D. and Meyers, C. A. 1998. Klüver-Bucy syndrome after bilateral selective damage of amygdala and its cortical connections. *Journal of Neuropsychiatry* **10**: 354–8.

Haynes, L. 2005. The effect of aging on cognate function and development of immune memory. *Current Opinion in Immunology* **17**: 476–9.

Hayward, M. W., Hofmeyr, M., O'Brien, J. and Kerley, G.I.H. 2006. Prey preferences of the cheetah (*Acinonyx jubatus*) (Felidae: Carnivora): morphological limitations of the need to capture rapidly consumable prey before kleptoparasites arrive? *Journal of Zoology* **270**: 615–27.

Hazan, C. and Shaver, P. 1987. Romantic love conceptualized as an attachment process. *Journal of Personality and Social Psychology* **52**: 511–24.

Healey, M. D. and Ellis, B. J. 2007. Birth order, conscientiousness, and openness to experience: tests of the family-niche model of personality using a within-family methodology. *Evolution and Human Behavior* **28**: 55–9.

Heard, D. C. 1992. The effect of wolf predation and snow cover on musk-ox group size. *American Naturalist* **139**: 190–204.

Heard, E. 2004. Recent advances in X-chromosome inactivation. *Current Opinion in Cell Biology* **16**: 247–55.

Hedrick, A. V. and Temeles, E. J. 1989. The evolution of sexual dimorphism in animals: hypotheses and tests. *Trends in Ecology and Evolution* **4**: 136–8.

Heg, D., Ens, B. J., Burke, T., Jenkins, L. and Kruijt, J. P. 1993. Why does the typically monogamous Oystercatcher (*Haematopus ostralegus*) engage in extra-pair copulations? *Behaviour* **126**: 247–89.

Heg, D. and van Treuren, R. 1998. Female-female cooperation in polygynous oystercatchers. *Nature* **391**: 687–91.

Hegelson, V. S. 1994. Prototypes and dimensions of masculinity and femininity. *Sex Role: A Journal of Research* **31**: 653–82.

Hegstrom, C. D., Jordan, C. L. and Breedlove, S. M. 2002. Photoperiod and androgens act independently to induce spinal nucleus of the bulbocavernosus neuromuscular plasticity in the Siberian hamster, *Phodopus sungorus*. *Journal of Neuroendocrinology* **14**: 368–74.

Heimovics, S. A. and Riters, L. V. 2005. Immediate early gene activity in song control nuclei and brain areas regulating motivation relates positively to singing behavior during, but not outside of, a breeding context. *Journal of Neurobiology* **65**: 207–24.

Heinen, J. T. and Singh, G. R. 2001. A census and some management implications for wild buffalo (*Bubalus bubalis*) in Nepal. *Biological Conservation* **101**: 391–4.

Heinsohn, R., Dunn, P., Legge, S. and Double, M. 2000. Coalitions of relatives and reproductive skew in cooperatively breeding White-winged Choughs. *Proceedings of the Royal Society of London*, B **267**: 243–9.

Heitor, F., do Mar Oom, M. and Vicente, L. 2006. Social relationships in a herd of Sorraia horses. Part II. Factors affecting affiliative relationships and sexual behaviours. *Behavioural Processes* **73**: 231–9.

Helfenstein, F., Tirard, C., Danchin, E. and Wagner, R. H. 2004. Low frequency of extra-pair paternity and high frequency of adoption in black-legged kittiwakes. *Condor* **106**: 149–55.

Hemsworth, P. H., Findlay, J. K. and Beilharz, R. G. 1978. The importance of physical contact with other pigs during rearing on the sexual behaviour of the male domestic pig. *Animal Production* **27**: 201–7.

Henderson, A. W., Lehavot, K. and Simoni, J. M. 2008. Ecological models of sexual satisfaction among lesbian/bisexual and heterosexual women. *Archives of Sexual Behavior* **38**: 50–65.

Henderson, B. A. and Berenbaum, S. A. 1998. Sex-typed play in opposite-sex twins. *Developmental Psychology* **31**: 115–23.

Hendricks, S. E., Graber, B. and Rodriguez-Sierra, J. F. 1989. Neuroendocrine responses to exogenous estrogen: no differences between heterosexual and homosexual men. *Psychoneuroendocrinology* **14**: 177–85.

Henley, C. L., Nunez, A. A. and Clemens, L. G. 2009. Estrogen treatment during development alters adult partner preference and reproductive behavior in female laboratory rats. *Hormones and Behavior* **55**: 68–75.

Henley, D. V. and Korach, K. S. 2006. Endocrine-disrupting chemicals use distinct mechanisms of action to modulate endocrine system function. *Endocrinology* (Suppl.) **47**: S25–S32.

Hennessy, A. C., Wallen, K. and Edwards, D. A. 1986. Preoptic lesions increase the display of lordosis by male rats. *Brain Research* **370**: 21–8.

Henningsson, S., Westberg, L., Nilsson, S. *et al.* 2005. Sex steroid-related genes and male-to-female transsexualism. *Psychoneuroendocrinology* **30**: 657–64.

Henry, G. W. 1941. *Sex Variants: a Study of Homosexual Patterns*, vol. I. New York: Paul B. Hoeber.

Henry, J. D. and Herrero, S. M. 1974. Social play in the American black bear: its similarity to canid social play and an examination of its identifying characteristics. *American Zoologist* **14**: 371–89.

Henry, M., Lodi, L. D. and Gastal, M. O. 1998. Sexual behaviour of domesticated donkeys (*Equus asinus*) breeding under controlled or free range management systems. *Applied Animal Behavior Science* **60**: 263–76.

Henry, M., McDonnell, S. M., Lodi, L. D. and Gastal, E. L. 1991. Pasture mating behaviour of donkeys (*Equus asinus*) at natural and induced oestrous. *Journal of Reproduction and Fertility* Suppl. **44**: 77–86.

Hensley, C. 2000. Attitudes toward homosexuality in male and female prison: an exploratory study. *Prison Journal* **80**: 434–41.

Hensley, C. and Tewksbury, R. 2002. Inmate-to-inmate prison sexuality: a review of empirical studies. *Trauma, Violence, and Abuse* **3**: 226–43.

Herdt, G. H. 1981. *Guardians of the Flutes: Idioms of Masculinity*. New York: McGraw-Hill.

Herdt, G. H. 1984a. *Ritualized Homosexuality in Melanesia*. Berkeley, CA: University of California Press.

Herdt, G. H. 1984b. Ritualized homosexual behavior in the male cults of Melanesia, 1862–1983: an introduction. In: Herdt, G. H. (Ed.) *Ritualized Homosexuality in Melanesia*, pp. 1–82. Berkeley, CA: University of California Press.

Herdt, G. 1991. Representations of homosexuality: an essay on cultural ontology and historical comparison Part II. *Journal of the History of Sexuality* **1**: 603–32.

Herdt, G. H. and Davidson, J. 1988. The Sambia "turnim-man": sociocultural and clinical aspects of gender formation in male pseudohermaphrodites with 5-alpha-reductase deficiency in Papua New Guinea. *Archives of Sexual Behavior* **17**: 33–56.

Herdt, G. and McClintock, M. 2000. The magical age of 10. *Archives of Sexual Behavior* **29**: 587–606.

Herremans, M. 1997. Habitat segregation of male and female red-backed shrikes *Lanius collurio* and lesser grey shrikes *Lanius minor* in the Kalahari basin, Botswana. *Journal of Avian Biology* **28**: 240–8.

Hershberger, S. L. 1997. A twin registry study of male and female sexual orientation. *Journal of Sex Research* **34**: 212–22.

Herzer, M. 1985. Kertbeny and the nameless love. *Journal of Homosexuality* **12**: 1–26.

Herzog, D. B., Newman, K. L. and Warshaw, M. 1991. Body image dissatisfaction in homosexual and heterosexual males. *Journal of Nervous and Mental Disorders* **179**: 356–9.

Heston, L. L. and Shields, J. 1968. Homosexuality in twins. *Archives of General Psychiatry* **18**: 149–60.

Het, S., Ramlow, G. and Wolf, O. T. 2005. A meta-analytic review of the effect of acute cortisol administration on human memory. *Psychoneuroendocrinology* **30**: 771–84.

Hetmanski, T. and Wolk, E. 2005. The effect of environmental factors and nesting conditions on clutch overlap in the Feral Pigeon *Columba livia f. urbana* (Gm.). *Polish Journal of Ecology* **53**: 523–34.

Heurgon, J. 1961. *Daily Life of the Etruscans*. London: Phoenix Press.

Heusner, A. A. 1991. Body mass, maintenance and basal metabolism in dogs. *Journal of Nutrition* **121**: S8–S17.

Higgins, P. J., Peter, J. M. and Cowling, S. J. 2006. *Handbook of Australian, New Zealand and Antarctic Birds*. Oxford: Oxford University Press.

Hilborn, R. and Stearns S. C. 1982. On inference in ecology and evolutionary biology: the problem of multiple causes. *Acta Biotheoretica* **32**: 145–64.

Hill, D. A. 1994. Affiliative behaviour between adult males of the genus *Macaca*. *Behaviour* **130**: 293–308.

Hines, M. 2006. Prenatal testosterone and gender-related behaviour. *European Journal of Endocrinology* **155**: S115–S121.

Hines, M., Johnston, K. J., Golombok, S. and the ALSPAC Study Team 2002. Prenatal stress and gender role behaviour in girls and boys: a longitudinal, population study. *Hormones and Behavior* **42**: 126–34.

Hirotani, A. 1990. Social organization of reindeer (*Rangifer tarandus*), with special reference to relationships among females. *Canadian Journal of Zoology* **68**: 743–9.

Hirschenhauser, K., Möstl, E., Wallner, B., Dittami, J. and Kotrschal, K. 2000. Endocrine and behavioural responses of male greylag geese (*Anser anser*) to pairbond challenges during the reproductive season. *Ethology* **106**: 63–77.

Hirschfeld, M. 1914. *The Homosexuality of Men and Women*. Translated by Lombardi-Nash, M. A. and published in 2000. Amherst, NY: Prometheus Books.

Hirth, D. H. 1977. Social behaviour of white-tailed deer in relation to habitat. *Wildlife Monographs* **53**: 1–55.

Hoaken, P. C. S., Clarke, M. and Breslin, M. 1964. Psychopathology in Klinefelter's Syndrome. *Psychosomatic Medicine* **26**: 207–23.

Hogan, J. A. and Bolhuis, J. J. 2005. The development of behaviour: trends since Tinbergen (1963). *Animal Biology* **55**: 371–98.

Hogben, M. and Byrne, D. 1998. Using social learning theory to explain individual differences in human sexuality. *Journal of Sex Research* **35**: 58–71.

Hogg, J. T. 1984. Mating in bighorn sheep: multiple creative male strategies. *Science* **225**: 426–529.

Hogg, J. T. 1987. Intrasexual competition and mate choice in Rocky Mountain bighorn sheep. *Ethology* **75**: 119–44.

Hohmann, G. 2001. Association and social interactions between strangers and residents in bonobos (*Pan paniscus*). *Primates* **42**: 91–9.

Hohmann, G. and Fruth, B. 2000. Use and function of genital contacts among female bonobos. *Animal Behaviour* **60**: 107–20.

Hohmann, G. and Fruth, B. 2003. Intra- and inter-sexual aggression by bonobos in the context of mating. *Behaviour* **140**: 1389–413.

Hohmann, G., Gerloff, U., Tautz, D. and Fruth, B. 1999. Social bonds and genetic ties: kinship, association and affiliation in a community of bonobos (*Pan paniscus*). *Behaviour* **136**: 1219–35.

Hohmann, G., Mundry, R. and Deschner, T. 2009. The relationship between socio-sexual behavior and salivary cortisol in bonobos: tests of the tension regulation hypothesis. *American Journal of Primatology* **71**: 223–232.

Holand, Ø., Askim, K. R., Røed, K. H. et al. 2007. No evidence of inbreeding avoidance in a polygynous ungulate: the reindeer (*Rangifer tarandus*). *Biology Letters* **3**: 36–9.

Holekamp, K. E. and Smale, L. 1998. Behavioral development in the spotted hyena. *Bioscience* **48**: 997–1005.

Holečková, J., Bartoš, L. and Tománek, M. 2000. Inter-male mounting in fallow deer (*Dama dama*) – its seasonal pattern and social meaning. *Folia Zoologica* **49**: 175–81.

Hollister-Smith, J. A., Poole, J. H., Archie, E. A. et al. 2007. Age, musth and paternity success in wild male African elephants, *Loxodonta africana*. *Animal Behaviour* **74**: 287–96.

Holloway, R. L., Anderson, P. J., Defendini, R. and Harper, C. 1993. Sexual dimorphism of the human corpus callosum from three independent samples: relative size of the corpus callosum. *Journal of Physical Anthropology* **92**: 481–98.

Holson, R. R., Gough, B., Sullivan, P., Badger, T. and Sheehan, D. M. 1995. Prenatal dexamethasone or stress but not ACTH or corticosterone alter sexual behaviour in male rats. *Neurotoxicology and Teratology* **17**: 393–401.

Holstege, G., Georgiadis, J. R., Paans, A. M. J. et al. 2003. Brain activation during human male ejaculation. *Journal of Neuroscience* **23**: 9185–93.

Holt, S. B. 1968. *The Genetics of Dermal Ridges*. Springfield, IL: Charles C. Thomas.

Horne, S. G. and Biss, W. J. 2009. Equality discrepancy between women in same-sex relationships: the mediating role of attachment in relationship satisfaction. *Sex Roles* **60**: 721–30.

Horvath, C. D. 2000. Interactionism and innateness in the evolutionary study of human nature. *Biology and Philosophy* **15**: 321–37.

Hough, D., Wong, B., Bennett, G. et al. 1998. Vigilance and group size in emus. *Emu* **98**: 324–7.

Houtsmuller, E. J., Brand, T., Jonge, F. H. de et al. 1994. SDN-POA volume, sexual behaviour, and partner preference of male rats affected by perinatal treatment with ATD. *Physiology and Behavior* **56**: 535–41.

Hrdy, S. B. 1974. Male–male competition and infanticide among the langurs (*Presbytis entellus*) of Abu, Rajasthan. *Folia Primatologica* **22**: 19–58.

Hu, S., Pattatucci, A. M. L., Patterson, C. et al. 1995. Linkage between sexual orientation and chromosome Xq28 in males but not in females. *Nature Genetics* **11**: 248–55.

Hu, S.-H., Wei, N., Wang, Q.-D. et al. 2008. Patterns of brain activation during visually evoked sexual arousal differ between homosexual and heterosexual men. *American Journal of Neuroradiology* **29**: 1890–96.

Huber, R. and Martys, M. 1993. Male-male pairs in greylag geese (*Anser anser*). *Journal für Ornithologie* **134**: S155–S164.

Hudson, Q. J., Wilkins, R. J., Waas, J. R. and Hogg, I. D. 2000. Low genetic variability in small populations of New Zealand kokako *Callaeas cinerea wilsoni*. *Biological Conservation* **96**: 105–12.

Huffman, M. A. 1997. Current evidence for self medication in primates: a multidisciplinary perspective. *Yearbook of Physical Anthropology* **40**: 171–200.

Huffman, M. A. 2003. Animal self-medication and ethno-medicine: exploration and exploitation of the medical properties of plants. *Proceedings of the Nutrition Society* **62**: 371–81.

Hulet, C. V., Blackwell, R. L. and Ercanbrack, S. K. 1964. Observations on sexually inhibited rams. *Journal of Animal Science* **23**: 1095–7.

Hull, E. M. and Dominguez, J. M. 2007. Sexual behavior in male rodents. *Hormones and Behavior* **52**: 45–55.

Hull, E. M., Muschamp, J. W. and Sato, S. 2004. Dopamine and serotonin: influences on male sexual behaviour. *Physiology and Behavior* **83**: 291–307.

Hume D. 1740/1975. *A Treatise on Human Nature*. Oxford: Clarendon Press.

Hundgen, K., Raphael, B. and Sheppard, C. 2000. Egg fertility among vasectomized and non-vasectomized male resident Canada geese at the Wildlife Conservation Park/Bronx Zoo. *Zoo Biology* **19**: 35–40.

Hunt, A. 1998. The great masturbation panic and the discourses of moral regulation in nineteenth- and early twentieth-century Britain. *Journal of the History of Sexuality* **8**: 575–615.

Hunt, G. L. Jr and Hunt, M. W. 1977. Female-female pairing in Western Gulls (*Larus occidentalis*) in Southern California. *Science* **196**: 1466–7.

Hunt, G. L. Jr, Newman, A. L., Warner, M. H., Wingfield, J. C. and Kaiwi, J. 1984. Comparative behaviour of male-female pairs among western gulls prior to egg-laying. *Condor* **86**: 157–62.

Hunt, G. L. Jr, Wingfield, J. C., Newman, A. and Farner, D. S. 1980. Sex ratio of Western Gulls on Santa Barbara Island, California. *Auk* **97**: 473–479.

Hunt, M. 1974. *Sexual Behavior in the 1970's*. Chicago, IL: Playboy Press.

Hunter, F. M., Miller, G. D. and Davis, L. S. 1995. Mate switching and copulation behaviour in the Adelie penguin. *Behaviour* **132**: 691–707.

Hurnik, J. F., King, G. J. and Robertson, H. A. 1975. Estrous and related behaviour in postpartum Holstein cows. *Applied Animal Ethology* **2**: 55–68.

Hutcheon, J. M., Kirsch, J.A.W. and Garland, T. Jr. 2002. A comparative analysis of brain size in relation to foraging ecology and phylogeny in the Chiroptera. *Brain, Behavior and Evolution* **60**: 165–80.

Hutchins, M., Jackson, J. A., Bock, W. J. and Olendorf, D. (Eds) 2002. *Grzimek's Animal Life Encyclopedia*, 2nd edn. Farmington Hills, MI: Gale Group, Inc.

Hutchinson G. E. 1959. Homage to Santa Rosalia, or, why are there so many kinds of animals? *American Naturalist* **93**: 145–56.

Hutchinson, J. B. 1997. Gender-specific steroid metabolism in neural differentiation. *Cellular and Molecular Neurobiology* **17**: 603–26.

Huxley, A. 1963. *Literature and Science*. New Haven, CT: Leete's Island Books.

Huxley, J. (1968) *The Courtship Habits of the Great Crested Grebe*. London: Jonathan Cape. (This is a reprint of Huxley's (1914) paper in the *Proceedings of the Zoological Society of London*.)

Huynen, M. C. 1997. Homosexual interactions in female rhesus monkeys, *Macaca mulatta*. *Advances in Ethology* **32**: 211.

Ibáñez, L., de Zegher, F. and Potau, N. 1999. Anovulation after precocious pubarche: early markers and time course in adolescence. *Journal of Clinical Endocrinology and Metabolism* **84**: 2691–5.

Iemmola, F. and Camperio-Ciani, A. 2009. New evidence of genetic factors influencing orientation in men: female fecundity increase in the maternal line. *Archives of Sexual Behavior* **38**: 393–9.

Imperato-McGinley, J., Miller, M., Wilson, J. D. et al. 1991. A cluster of male pseudohermaprodites with 5α-reductase deficiency in Papua New Guinea. *Clinical Endocrinology* **34**: 293–8.

Innis, A. C. 1958. The behaviour of the giraffe, *Giraffa camelopardalis*, in the Eastern Transvaal. *Proceedings of the Zoological Society of London* **131**: 245–78.

Insel, T. R., Young, L. and Wang, Z. 1997. Central oxytocin and reproductive behaviours. *Reviews of Reproduction* **2**: 28–37.

Irestedt, M., Johansson, U. S., Parsons, T. J. and Ericson, P.G.P. 2001. Phylogeny of major lineages of suboscines (Passeriformes) analysed by nuclear DNA sequence data. *Journal of Avian Biology* **32**: 15–25.

Irvine, A. B., Caffin, J. E. and Kochman, H. E. 1981. *Aerial surveys of manatees and dolphins in western peninsular Florida: with notes on sightings of sea turtles and crocodiles.* Washington, DC: US Fish and Wildlife Service.

Irwin, M. R., Melendy, D. R., Amoss, M. S. and Hutcheson, D. P. 1979. Roles of predisposing factors and gonadal hormones in the buller syndrome of feedlot steers. *Journal of the American Veterinary Medical Association* **174**: 367–70.

Isaac, R. and Shah, A. 2004. Sex roles and marital adjustment in Indian couples. *International Journal of Social Psychiatry* **50**: 129–41.

Isay, R. A. 1990. *Being Homosexual: Gay Men and their Development.* New York: Farrar Straus Giroux.

Ishikawa, A., Sakamoto, H., Katagiri, S. and Takahashi, Y. 2003. Changes in sexual behavior and fecal steroid hormone concentrations during the breeding season in female Hokkaido brown bears (*Ursus arctos yesoensis*) under captive conditions. *Journal of Veterinary Medicine Science* **65**: 99–102.

Isvaran, K. 2005. Variation in male mating behaviour within ungulate populations: patterns and processes. *Current Science* **89**: 1192–9.

Isvaran, K. and Jhala, Y. 1999. Variation in lekking costs in blackbuck (*Antilope cervicapra*): relationship to lek-territory location and female mating patterns. *Behaviour* **137**: 547–63.

IUCN 2005. Bush dog (*Speothos venaticus*). www.canids.org/species/Speothos_venaticus.htm.

Iwaniuk, A. N., Dean, K. M. and Nelson, J. E. 2004. A mosaic pattern characterizes the evolution of the avian brain. *Proceedings of the Royal Society of London*, B **271**: S148–S151.

Iwaniuk, A. N. and Nelson, J. E. 2002. Can endocranial volume be used as an estimate of brain size in birds? *Canadian Journal of Zoology* **80**: 16–23.

Iwaniuk, A. N. and Nelson, J. E. 2003. Developmental differences are correlated with relative brain size in birds: a comparative analysis. *Canadian Journal of Zoology* **81**: 1913–28.

Jabłoński, P. G. and Matyjasiak, P.2002. Male wing-patch asymmetry and aggressive response to intruders in the common chaffinch (*Fringilla coelebs*). *Auk* **119**: 566–72.

Jackson, N. M. 1988. Can individual differences in history of dominance explain the development of linear dominance hierarchies? *Ethology* **79**: 71–7.

Jackson, P. A. 1997. Thai research on male homosexuality and transgenderism and the cultural limits of Foucaultian analysis. *Journal of the History of Sexuality* **8**: 52–85.

Jackson, S. and Scott, S. 2004. The personal is still political: heterosexuality, feminism and monogamy. *Feminism and Psychology* **14**: 151–7.

Jacob, F. 1977. Evolution and tinkering. *Science* **196**: 1161–6.

Jacobson, D. L., Gange, S. J., Rose, N. R. and Graham, N.M.H. 1997. Epidemiology and estimated population burden of selected autoimmune diseases in the United States. *Clinical Immunology and Immunopathology* **84**: 223–43.

Jaczewski, Z. 1958. Reproduction of the European bison, *Bison bonasus* (L.), in reserves. *Acta Theriologica* **1**: 333–76.

Jago, J. G., Cox, N. R., Bass, J. J. and Matthews, L. R. 1997. The effect of prepubertal immunization against gonadotropin-releasing hormone on the development of sexual and social behaviour in bulls. *Journal of Animal Science* **75**: 2609–19.

Jagose, A. 1996. *Queer Theory: an Introduction.* New York: New York University Press.

Jakober, H. and Stauber, W. 1994. Copulation and mate-guarding in the Red-backed Shrike (*Lanius collurio*). *Journal für Ornithologie* **135**: 535–47.

Jakubas, D. and Wojczulanis-Jakubas, K. 2008. Variation in copulatory behavior of the Dovekie. *Waterbirds* **31**: 661–5.

James, E., Chai, J.-G., Dewchand, H. *et al.* 2003. Multiparity induces priming to male-specific minor histocompatibility antigen, HY in mice and humans. *Blood* **102**: 388–93.

James, W. H. 2001. Finger-length ratios, sexual orientation and offspring sex ratios. *Journal of Theoretical Biology* **212**: 273–4.

James, W. H. 2004. A further note on the causes of the fraternal birth order effect in male homosexuality. *Journal of Biosocial Sciences* **36**: 61–2.

James, W. H. 2005. Biological and psychological determinants of male and female human sexual orientation. *Journal of Biosocial Science* **37**: 555–67.

James, W. H. 2006. Two hypotheses on the causes of male homosexuality and paedophilia. *Journal of Biosocial Science* **38**: 745–61.

Jamieson, I. G. 1997. Testing reproductive skew models in a communally breeding bird, the Pukeko *Porphyrio porphyrio*. *Proceedings of the Royal Society of London*, B **264**: 335–40.

Jamieson, I. G. and Craig, J. L. 1987a. Male-male and female-female courtship and copulation behaviour in a communally breeding bird. *Animal Behaviour* **35**: 1251–3.

Jamieson, I. G. and Craig, J. L. 1987b. Dominance and mating in a communally polygynandrous bird: cooperation or indifference towards mating competitors? *Ethology* **75**: 317–27.

Jäncke, L., Staiger, J. F., Schlaug, G., Huang, Y. and Steinmetz, H. 1997. The relationship between corpus callosum size and forebrain volume. *Cerebral Cortex* **7**: 48–56.

Jankowiak, W. R., Hill, E. M. and Donovan, J. M. 1992. The effects of sex and sexual orientation on attractiveness judgments: an evolutionary interpretation. *Ethology and Sociobiology* **13**: 73–85.

Janus, S. and Janus, C. 1993. *The Janus Report on Sexual Behavior*. New York: John Wiley and Sons.

Jarman, P. J. 1991. Social behavior and organization in the Macropodoidea. *Advances in the Study of Behavior* **20**: 1–50.

Jarman, P. J. and Southwell, C. J. 1986. Grouping, associations, and reproductive strategies in eastern grey kangaroos. In: Rubenstein, D. I. and Wrangham, R. W. (Eds) *Ecological Aspects of Social Evolution*, pp. 399–428. Princeton, NJ: Princeton University Press.

Järvi, T., Røskaft, E., Bakken, M. and Zumsteg, B. 1987. Evolution of variation in male secondary sexual characteristics. A test of eight hypotheses applied to pied flycatchers. *Behavioral Ecology and Sociobiology* **20**: 161–9.

Jedrzejewski, W., Branicki, W., Veit, C. *et al.* 2005. Genetic diversity and relatedness within packs in an intensely hunted population of wolves *Canis lupus*. *Acta Theriologica* **50**: 3–22.

Jedrzejewski, W., Schmidt, K., Theverkauf, J., Jedrzejewska, B. and Okarma, H. 2001. Daily movements and territory use by radio-collared wolves (*Canis lupus*) in Bialowieza Promeval Forest in Poland. *Canadian Journal of Zoology* **79**: 1993–2004.

Jefferson, T. Jr., Herbst, J. H. and McCrae, R. R. 1998. Associations between birth order and personality traits: evidence from self-reports and observer ratings. *Journal of Research in Personality* **32**: 498–509.

Jenks, J. A., Leslie, D. M. Jr, Lochmiller, R. L. and Melchiors, M. A. 1994. Variation in gastrointestinal characteristics of male and female white-tailed deer: implications for resource partitioning. *Journal of Mammalogy* **75**: 1045–53.

Jezierski, T. A., Koziorowski, M., Goszczyński, J. and Sieradzka, I. 1989. Homosexual and social behaviours of young bulls of different geno-and phenotypes and plasma concentrations of some hormones. *Applied Animal Behaviour Science* **24**: 101–13.

Jiang, Z.-G., Li, C.-W., Zeng, Y. and Widemo, F. 2004. ''Harem defending'' or ''challenging'': alternative individual mating tactics in Pere David's deer under different fine constraints. *Acta Zoologica Sinica* **50**: 706–13.

Joffe, T. H. 1997. Social pressures have selected for an extended juvenile period in primates. *Journal of Human Evolution* **32**: 593–605.

Joffe, T. H. and Dunbar, R.I.M. 1997. Visual and socio-cognitive information processing in primate brain evolution. *Proceedings of the Royal Society of London*, B **264**: 1303–7.

Johansson, B. B. 2004. Brain plasticity in health and disease. *Keio Journal of Medicine* **53**: 231–46.

Johns, D. W. and Armitage, K. B. 1979. Behavioral ecology of Alpine yellow-bellied marmots. *Behavioral Ecology and Sociobiology* **5**: 133–57.

Johnson, C., Koerner, C., Estrin, M. and Duoos, D. 1980. Alloparental care and kinship in captive social groups of vervet monkeys (*Cercopithecus aethiops sabaeus*). *Primates* **21**: 406–15.

Johnson, C. N. 1989a. Social interactions and reproductive tactics in red-necked wallabies (*Macropus rufogriseus banksianus*). *Journal of Zoology, London* **217**: 267–80.

Johnson, C. N. 1989b. Grouping structure of association in the red-necked wallaby. *Journal of Mammalogy* **70**: 18–26.

Johnson, J. I., Kirsch, J. A. W., Reep, R. L. and Switzer, R. C. III. 1994. Phylogeny through brain traits: more characters for the analysis of mammalian evolution. *Brain Behavior and Evolution* **43**: 319–47.

Johnson, K. L., Gill, S., Reichman, V. and Tassinary, L. G. 2007. Swagger, sway, and sexuality: judging sexual orientation from body motion and morphology. *Journal of Personality and Social Psychology* **93**: 321–34.

Johnson, M. 1995. Transgender men and homosexuality in the Southern Philippines: ethnicity, political violence and the protocols of engendered sexualities amongst the Muslim Tausug and Sama. *South East Asia Research* **3**: 46–66.

Johnson, P. M. 1980. Observations of the behaviour of the rufous rat-kangaroo *Aepyprymnus rufescens* (Gray), in captivity. *Australian Wildlife Research* **7**: 347–57.

Johnson, T. W., Brett, M. A., Roberts, L. F. and Wassersug, R. J. 2007. Eunuchs in contemporary society: characterizing men who are voluntarily castrated (Part I). *Journal of Sexual Medicine* **4**: 930–45.

Johnston, M. W. and Bell, A. P. 1995. Romantic emotional attachment: additional factors in the development of the sexual orientation of men. *Journal of Counseling and Development* **73**: 621–5.

Johnston, V. S. 2006. Mate choice decisions: the role of facial beauty. *Trends in Cognitive Science* **10**: 9–13.

Johnstone, R. A. 2000. Models of reproductive skew: a review and synthesis. *Ethology* **106**: 5–26.

Johnstone, R. A. 2008. Kin selection, local competition, and reproductive skew. *Evolution* **62**: 2592–9.

Johnstone, R. A., Woodroffe, R., Cant, M. A. and Wright, J. 1999. Reproductive skew in multimember groups. *American Naturalist* **153**: 315–31.

Jolicoeur, P., Pirlot, P., Baron, G. and Stephan, H. 1984. Brain structure and correlation patterns in insectivore, chiroptera and primates. *Systematic Zoology* **33**: 14–29.

Jolly, A. 1998. Pair-bonding, female aggression and evolution of lemur societies. *Folia Primatologica* **69**(Suppl. 1): 1–13.

Jones, D. R. and Jones, M. B. 1992. Behavioral contagion in sibships. *Journal of Psychiatric Research* **26**: 149–64.

Jones, H. E., Ruscio, M. A., Keyser, L. A. *et al.* 1997. Prenatal stress alters the size of the rostral anterior commissure in rats. *Brain Research Bulletin* **42**: 341–6.

Jones, M. B. and Blanchard, R. 1998. Birth order and male homosexuality: extension of Slater's Index. *Human Biology* **70**: 775–87.

Jones, M. B. and Jones, D. R. 1994. Testing for behavioral contagion in a case-control design. *Journal of Psychiatric Research* **28**: 35–55.

Jones, M. B. and Jones, D. R. 1995. Preferred pathways of behavioural contagion. *Journal of Psychiatric Research* **29**: 193–209.

Jones. M. E. E., Boon, W. C., McInnes, K. *et al.* 2007. Recognizing rare disorders: aromatase deficiency. *Nature Clinical Practice Endocrinology and Metabolism* **3**: 414–21.

Judson, F. N., Penley, K. A., Robinson, M. E. and Smith, J. K. 1980. Comparative prevalence rates of sexually transmitted diseases in heterosexual and homosexual men. *American Journal of Epidemiology* **112**: 836–43.

Jurke, M. H., Sommovilla, R. H., Harvey, N. C. and Wrangham, R. W. 2001. Behavior and hormonal correlates in bonobos. In: The Apes, *Challenges for the 21st Century, Conference Proceedings, May 10–13, 2000*, pp. 105–6. Brookfield, IL: Chicago Zoological Society.

Jyotika, J., McCutcheon, J., Laroche, J., Blaustein, J. D. and Forger, N. G. 2007. Deletion of the *bax* gene disrupts sexual behavior and modestly impairs motor function in mice. *Developmental Neurobiology* **67**: 1511–19.

Kaeuffer, R., Pontier, D., Devillard, S. and Perrin, N. 2004. Effective size of two feral domestic cat populations (*Felis catus* L.): effect of the mating system. *Molecular Ecology* **13**: 483–90.

Kahlenberg, S. M., Thompson, M. E., Muller, M. M. and Wrangham, R. W. 2008. Immigration costs for female chimpanzees and male protection as an immigrant counterstrategy to intrasexual aggression. *Animal Behaviour* **76**: 1497–509.

Kaiser, S. Brendel, H. and Sachser, N. 2000. Effects of ACTH applications during pregnancy on the female offsprings' endocrine status and behaviour in guinea pigs. *Physiology and Behavior* **70**: 157–62.

Kaiser, S., Kruijver, F. P. M., Swaab, D. F. and Sachser, N. 2003. Early social stress in female guinea pigs induces a masculinization of adult behaviour and corresponding changes in brain and neuroendocrine function. *Behavioural Brain Research* **144**: 199–210.

Kaiser, S. and Sachser, N. 1998. The social environment during pregnancy and lactation affects the female offsprings' endocrine status and behaviour in guinea pigs. *Physiology and Behavior* **63**: 361–6.

Kaiser, S. and Sachser, N. 2001. Social stress during pregnancy and lactation affects in guinea pigs the male offsprings' endocrine status and infantilizes their behaviour. *Psychoneuroendocrinology* **26**: 503–19.

Kallmann F. J. 1952. Comparative twin study on the genetic aspects of male homosexuality. *Journal of Nervous and Mental Diseases* **115**: 283–98.

Kalmbach, E. 2006. Why do goose parents adopt unrelated goslings? A review of hypotheses and empirical evidence, and new research questions. *Ibis* **148**: 66–78.

Kalmijn, M. 2002. Sex segregation of friendship networks. Individual and structural determinants of having cross-sex friends. *European Sociological Review* **18**: 101–17.

Kangasvuo, J. 2003. Bisexuality as a concept in Finland – signs of cultural change? *Lambda Nordica* **2002**: 3–4.

Kano, T. 1992. *The Last Ape: Pygmy Chimpanzee Behavior and Ecology*. Stanford, CA: Stanford University Press.

Kappeler, P. M. and Schäffler, L. 2008. The lemur syndrome unresolved: extreme male reproductive skew in sifakas (*Propithecus verreauxi*), a sexually monomorphic primate with female dominance. *Behavioral Ecology and Sociobiology* **62**: 1007–15.

Kapsalis, E. and Johnson, R. L. 2006. Getting to know you: female-female courtships in free ranging rhesus monkeys. In: Sommer, V. and Vasey, P. L. (Eds). *Homosexual Behaviour in Animals: an Evolutionary Perspective*, pp. 220–37. Cambridge: Cambridge University Press.

Karama, S., Lecours, A. R., Leroux, J.-M. *et al.* 2002. Areas of brain activation in males and females during viewing of erotic film excerpts. *Human Brain Mapping* **16**: 1–13.

Karczmarski, L., Würsig, B., Gailey, G., Larson, K. W. and Vanderlip, C. 2005. Spinner dolphins in a remote Hawaiian atoll: social grouping and population structure. *Behavioral Ecology* **16**: 675–85.

Kardum, I., Hudek-Knežević, J. and Gračanin, A. 2006. Sociosexuality and mate retention in romantic couples. *Psychological Topics* **15**: 277–96.

Katz, I. 1949. Behavioural interactions in a herd of barbary sheep. *Zoologica* **34**: 9–18.

Katz, L. S., Price, E. O., Wallach, S. J. R. and Zenchak, J. J. 1988. Sexual performance of rams reared with or without females after weaning. *Journal of Animal Science* **66**: 1166–73.

Kaufman, J. and Vermeulen, A. 1997. Declining gonadal function in elderly men. *Baillière's Clinical Endocrinology and Metabolism* **11**: 289–309.

Kaufmann, J. H. 1974. Social ethology of the whiptail wallaby, *Macropus parryi*, in northeastern New South Wales. *Animal Behaviour* **22**: 281–369.

Kaufmann, J. H. 1975. Field observations of the social behaviour of the eastern grey kangaroo, *Macropus giganteus*. *Animal Behaviour* **23**: 214–21.

Kauth, M. R. 2000. *True Nature: A Theory of Sexual Attraction*. New York: Kluwer Academics/Plenum.

Kauth, M. R. 2002. Much ado about homosexuality: assumptions underlying current research on sexual orientation. *Journal of Psychology and Human Sexuality* **14**: 1–22.

Kauth, M. R. 2005. Revealing assumptions: explicating sexual orientation and promoting conceptual integrity. *Journal of Bisexuality* **5**: 79–105.

Kawai, M., Dunbar, R., Ohsawa, H. and Mori, U. 1983. Social organization of gelada baboons: social units and definitions. *Primates* **24**: 13–24.

Kelly, S. J., Ostrowski, N. L. and Wilson, M. A. 1999. Gender differences in brain behavior: hormonal and neural bases. *Pharmacology Biochemistry and Behavior* **64**: 655–64.

Kempenaers, B. 1993. A case of polyandry in the Blue Tit: female extra-pair behaviour results in extra male help. *Ornis Scandinavica* **24**: 246–49.

Kempenaers, B. 1994. Polygyny in the blue tit: unbalanced sex ratio and female aggression restrict mate choice. *Animal Behaviour* **47**: 943–57.

Kempner, J. 2008. The chilling effect: how do researchers react to controversy? *PLoS Med* **5**: e222. doi: 10.1371/journal.pmed0050222.

Kendler, K. S., Thornton, L. M., Gilman, S. E. and Kessler, R. C. 2000. Sexual orientation in a U.S. national sample of twin and nontwin sibling pairs. *American Journal of Psychiatry* **157**: 1843–6.

Kennedy, H. C. 1980/81. The "third sex" theory of Karl Heinrich Ulrichs. *Journal of Homosexuality* **6**: 103–11.

Kennedy, H. 1997. Karl Heinrich Ulrichs: first theorist of homosexuality. In: Rosario, V. (Ed.) *Science and Homosexualities*, pp. 26–45. New York: Routledge.

Kenrick, D. T., Keefe, R. C., Bryan, A., Barr, A. and Brown, S. 1995. Age preferences and mate choice among homosexuals and heterosexuals: a case for modular psychological mechanisms. *Journal of Personality and Social Psychology* **69**: 1166–72.

Kenrick, D. T., Sadalla, E. K., Groth, G. and Trost, M. R. 1990. Evolution, traits, and the stages of human courtship: qualifying the parental investment model. *Journal of Personality* **58**: 97–116.

Kenyon, R. F. 1972. Polygyny among Superb Lyrebirds in Sherbrooke Forest Park, Kallista, Victoria. *Emu* **72**: 70–6.

Kerchner, M., Walsbury, C. W., Ward, O. B. and Ward, I. L. 1995. Sexually dimorphic areas in the rat medial amygdala: resistance to the demasculinizing effect of prenatal stress. *Brain Research* **672**: 251–60.

Kerlin, S. P. 2005. Prenatal exposure to diethylstilbestrol (DES) in males and gender-related disorders: results from a 5-year study. Paper presented at the International Behavioral Development Symposium 2005, Minot, North Dakota.

Kerr, W. E. and Freire Maia, N. 1983. Probable inbreeding effect on male homosexuality. *Brazilian Journal of Genetics* **6**: 177–80.

Kertzner, R. M. 2001. The adult life course and homosexual identity in midlife gay men. *Annual Review of Sex Research* **12**: 75–92.

Kessler, C. C. 2002. Eradication of feral goats and pigs and consequences for other biota on Sarigan Island, Commonwealth of the Northern Mariana Islands. In: Veitch, C. R. and Clout, M. N. (Eds) *Turning the Tide: the Eradication of Invasive Species*. IUCN SSC Invasive Species Specialist Group. Gland, Switzerland, and Cambridge, UK: IUCN.

Kessler, S. and Moos, R. H. 1973. Behavioral aspects of chromosomal disorders. *Annual Reviews of Medicine* **24**: 89–102.

Keverne, E. B. and Curley, J. P. 2004. Vasopressin, oxytocin and social behaviour. *Current Opinion in Neurobiology* **14**: 777–83.

Keverne, E. B., Martel, F. L. and Nevison, C. M. 1996. Primate brain evolution: genetic and functional considerations. *Proceedings of the Royal Society of London*, B **262**: 689–96.

Khokhar, A. R. 1991. Population dynamics of house shrew *Suncus murinus* in rice and wheat fields in central Punjab, Pakistan. *Journal of the Bombay Natural History Society* **88**: 384–9.

Kholkute, S. D., Joseph, R., Joshi, U. M. and Munshi, S. R. 1981. Diurnal variations of serum testosterone levels in the male bonnet monkey (*Macaca radiata*). *Primates* **22**: 427–30.

Kie, J. G. and Bowyer, R. T. 1999. Sexual segregation in white-tailed deer: density-dependent changes in use of space, habitat selection, and dietary niche. *Journal of Mammalogy* **80**: 1004–20.

Kihlström, J. E. 1966. Diurnal variation in the spontaneous ejaculations of the male albino rat. *Nature* **209**: 513–14.

Kimball, R. T., Parker, P. G. and Bednarz, J. C. 2003. Occurrence and evolution of cooperative breeding among the diurnal raptors (Accipitridae and Falconidae). *Auk* **120**: 717–29.

Kimura, D. and Carson, M. W. 1995. Dermatoglyphic asymmetry: relation to sex, handedness and cognitive pattern. *Personality and Individual Differences* **19**: 471–8.

Kimwele, C. N. and Graves, J. A. 2003. A molecular genetic analysis of the communal nesting of the ostrich (*Struthio camelus*). *Molecular Ecology* **12**: 229–36.

Kindon, H. A., Baum, M. J. and Paredes, R. J. 1996. Medial preoptic/anterior hypothalamic lesions induce a female-typical profile of sexual partner preference in male ferrets. *Hormones and Behavior* **30**: 514–27.

King, C. E. 2006. Pink flamingos: atypical partnerships and sexual activity in colonially breeding birds. In: Sommer, V. and Vasey, P. L. (Eds) *Homosexual Behaviour in Animals: An Evolutionary Perspective*, pp. 77–106. Cambridge: Cambridge University Press.

King, M., Green, J., Osborn, D. P. J., Arkell, J., Hetherton, J. and Pereira, E. 2005. Family size in white gay and heterosexual men. *Archives of Sexual Behavior* **34**: 117–22.

King, M. and McDonald, E. 1992. Homosexuals who are twins: a study of 46 probands. *British Journal of Psychiatry* **160**: 407–9.

Kinnish, K. K., Strassberg, D. S. and Turner, C. W. 2005. Sex differences in the flexibility of sexual orientation: a multidimensional retrospective assessment. *Archives of Sexual Behavior* **34**: 173–83.

Kinnunen, L. H., Moltz, H., Metz, J. and Cooper, M. 2004. Differential brain activation in exclusively homosexual and heterosexual men produced by the selective serotonin uptake inhibitor fluoxetine. *Brain Research* **1024**: 251–4.

Kinsey, A. C. 1941. Homosexuality: criteria for a hormonal explanation of the homosexual. *Journal of Clinical Endocrinology* **1**: 424–8.

Kinsey, A. C., Pomeroy, W. B. and Martin, C. E. 1948. *Sexual Behavior in the Human Male*. Philadelphia, PA: W. B. Saunders Co.

Kinsey, A. C., Pomeroy, W. B., Martin, C. E. and Gebhard, P. H. 1953. *Sexual Behavior in the Human Female*. Philadelphia, PA: W. B. Saunders Co.

Kirby, J. 2003. A new group selection model for the evolution of homosexuality. *Biology and Philosophy* **18**: 683–94.

Kirk, K. M., Bailey, J. M., Dunne, M. P. and Martin N. G. 2000. Measurement models for sexual orientation in a community twin sample. *Behavior Genetics* **30**: 345–56.

Kirkpatrick, R. C. 2000. The evolution of human homosexual behaviour. *Current Anthropology* **41**: 385–413.

Kitagawa, T. 1988. Ethosociological studies of the black-winged stilt *Himantopus himantopus himantopus* III. Female-female pairing. *Japanese Journal of Ornithology* **37**: 63–7.

Kitchen, D. W. 1974. Social behavior and ecology of the pronghorn. *Wildlife Monographs* **38**: 1–96.

Klar, A. J. 2003. Human handedness and scalp hair-whorl direction develop from a common genetic mechanism. *Genetics* **165**: 269–76.

Klar, A.J.S. 2004. Excess of counterclockwise scalp hair-whorl rotation in homosexual men. *Journal of Genetics* **83**: 251–5.

Kleim, J. A., Lussnig, E., Schwarz, E. R., Comery, T. A. and Greenough, W. T. 1996. Synaptogenesis and FOS expression in the motor cortex of the adult rat after motor skill learning. *Journal of Neuroscience* **16**: 4529–35.

Kleiman, D. G. 1972. Social behaviour of the maned wolf (*Chrysocyon brachyurus*) and bush dog (*Speothos venaticus*): a study in contrast. *Journal of Mammalogy* **53**: 791–806.

Klein, F., Sepekoff, B. and Wolf, T. J. 1985. Sexual orientation: a multi-variate dynamic process. *Journal of Homosexuality* **11**: 35–49.

Klemm, W. R., Sherry, C. J., Schake, L. M. and Sis, R. F. 1983/84. Homosexual behavior in feedlot steers: an aggression hypothesis. *Applied Animal Ethology* **11**: 187–95.

Kling, A. and Dunne, K. 1976. Social-environmental factors affecting behaviour and plasma testosterone in normal and amygdala lesioned *M. speciosa*. *Primates* **17**: 23–42.

Klopman, R. B. 1962. Sexual behavior in the Canada goose. *Living Bird* **1**: 123–9.

Klüver, H. and Bucy, P. C. 1939. Preliminary analysis of functions of the temporal lobes in monkeys. *Archives of Neurology and Psychiatry* **42**: 979–1000.

Knauft, B. 2003. What ever happened to ritualized homosexuality? Modern sexual subjects in Melanesia and elsewhere. *Annual Review of Sex Research* **14**: 137–59.

Knott, G. W., Quairiaux, C., Genoud, C. and Welker, E. 2002. Formation of dendritic spines with GABAergic synapses induced by whisker stimulation in adult mice. *Neuron* **34**: 265–73.

Koch, H. L. 1956. Sissiness and tomboyishness in relation to sibling characteristics. *Journal of Genetic Psychology* **88**: 231–44.

Koch, H. L. 1960. The relation of certain formal attributes of siblings to attitudes held toward each other and toward their parents. *Monographs of the Society for Research in Child Development* **25**: 1–124.

Koehler, N. and Chisholm, J. S. 2007. Early psychosocial stress predicts extra-pair copulations. *Evolutionary Psychology* **5**: 184–201.

Koehler, N., Simmons, L. W. and Rhodes, G. 2004. How well does second-to-fourth-digit ratio in hands correlate with other indications of masculinity in males? *Proceedings of the Royal Society of London*, B **271**: S296–S298.

Koenig, W. D. and Mumme, R. L. 1987. *Population Ecology of the Cooperatively Breeding Acorn Woodpecker*. Princeton, NJ: Princeton University Press.

Koenig, W. D. and Pitelka, F. A. 1981. Ecological factors and kin selection in the evolution of cooperative breeding in birds. In: Alexander, R. D. and Tinkle, D. W. (Eds) *Natural Selection and Social Behavior: Recent Research and New Theory*, pp. 261–80. New York: Chiron Press.

Koepfli, K.-P., Jenks, S. M., Eizirik, E., Zahirpour, T., van Valkenburgh, B. and Wayne, R. K. 2006. Molecular systematics of the Hyaenidae: Relationships of a relictual lineage resolved by a molecular supermatrix. *Molecular Phylogenetics and Evolution* **38**: 603–20.

Koford, C. B. 1957. The vicuña and the puna. *Ecological Monographs* **27**: 153–219.

Koford, R. R. 1982. Mating system of a territorial tree squirrel (*Tamiasciurus douglasii*) in California. *Journal of Mammalogy* **63**: 274–283.

Kohda, M. 1985. Allomothering behaviour of New and Old World monkeys. *Primates* **26**: 28–44.

Kohl, J. V., Atzmueller, M., Fink, B. and Grammer, K. 2001. Human pheromones: integrating neuroendocrinology and ethology. *Neuroendocrinology Letters* **22**: 309–21.

Kohlberg, L. 1966. A cognitive-developmental analysis of children's sex-role concepts and attitudes. In: Maccoby, E. E. (Ed.) *The Development of Sex Differences*, pp. 82–173. Stanford, CA: Stanford University Press.

Kokko, H. 2003. Are reproductive skew models evolutionarily stable? *Proceedings of the Royal Society of London* B **270**: 265–70.

Kokko, H. and Johnstone, R. A. 1999. Social queuing in animal societies: a dynamic model of reproductive skew. *Proceedings of the Royal Society of London* B **266**: 571–8.

Kolodny, R. C., Jacobs, L. S., Masters, W. H., Toro, G. and Daughaday, W. H. 1971. Plasma testosterone and semen analysis in male homosexuals. *New England Journal of Medicine* **285**: 1170–4.

Komers, P. E., Messier, F. and Gates, C. C. 1994. Plasticity of reproductive behaviour in wood bison bulls: when subadults are given a chance. *Ethology, Ecology & Evolution* **6**: 313–30.

Komisaruk, B. R. and Whipple, B. 2005. Functional MRI of the brain during orgasm in women. *Annual Review of Sex Research* **16**: 62–86.

Komisaruk, B. R., Whipple, B., Crawford, A. *et al*. 2004. Brain activation during vaginocervical self-stimulation and orgasm in women with complete spinal cord injury:

fMRI evidence of mediation by the vagus nerves. *Brain Research* **1024**: 77–88.

Koplam, R. B. 1962. Sexual behaviour in the Canada goose. *Living Bird* **1**: 123–9.

Koprowski, J. L. 1992. Do estrous female gray squirrels, *Sciurus carolinensis*, advertise their receptivity? *Canadian Field Naturalist* **106**: 392–4.

Koprowski, J. L. 1993. Alternative reproductive tactics in male eastern gray squirrels: "making the best of a bad job". *Behavioral Ecology* **4**: 165–71.

Koprowski, J. L. 1996. Natal philopatry, communal nesting, and kinship in fox squirrels and gray squirrels. *Journal of Mammalogy* **77**: 1006–16.

Korpimäki, E. 1988. Factors promoting polygyny in European birds of prey – a hypothesis. *Oecologia* **77**: 278–85.

Korpimäki, E., Lahti, K., May, C. A. et al. 1996. Copulatory behaviour and paternity determined by DNA fingerprinting in kestrels: effects of cyclic food abundance. *Animal Behaviour* **51**: 945–55.

Kortlandt, A. 1995. Patterns of pair-formation and nest-building in the European cormorant *Phalacrocorax carbo sinensis*. *Ardea* **83**: 11–25.

Kotrschal, K., Hemetsberger, J. and Weiss, B. M. 2006. Making the best of a bad situation: homosociality in male greylag geese. In: Sommer, V. and Vasey, P. L. (Eds) *Homosexual Behaviour in Animals: An Evolutionary Perspective*, pp. 45–76. Cambridge: Cambridge University Press

Kovacs, K. M., Jonas, K. M. and Welke, S. E. 1990. Sex and age segregation by *Phoca vitulina concolor* at haul-out sites during the breeding season in the Passamaquoddy Bay region, New Brunswick. *Marine Mammal Science* **6**: 204–14.

Kovacs, K. M. and Ryder, J. P. 1981. Nest-site tenacity and mate fidelity in female-female pairs of ring-billed gulls. *Auk* **98**: 625–7.

Kovacs, K. M. and Ryder, J. P. 1985. Morphology and physiology of female-female pair members. *Auk* **102**: 874–8.

Kowal, A. K. and Blinn-Pike, L. 2004. Sibling influences on adolescents' attitudes toward safe sex practices. *Family Relations* **53**: 377–84.

Kozak, D. 1996. Sex role identity and testosterone level among women. *Homo* **46**: 211–26.

Kraaijeveld, K., Carew, P. J., Billing, T., Adcock, G. J. and Moulder, R. A. 2004. Extra-pair paternity does not result in differential sexual selection in the mutually ornamented black swan (*Cygnus atratus*). *Molecular Ecology* **13**: 1625–33.

Kraaijeveld, K. and Mulder, R. A. 2002. The function of triumph ceremonies in the black swan. *Behaviour* **139**: 45–54.

Kraemer, B., Noll, T., Delsignore, A. et al. 2006. Finger length ratio (2D:4D) and dimensions of sexual orientation. *Neuropsychobiology* **53**: 210–14.

Krafft-Ebing, R. von. 1886. *Psychopathia Sexualis*. New York: Rebman Co.

Král, M., Järvi, T. and Bičík, V. 1988. Inter-specific aggression between the Collared Flycatcher and the Pied Flycatcher: the selective agent for the evolution of light-coloured male Pied Flycatcher populations? *Ornis Scandinavica* **19**: 287–9.

Krasiński, Z. and Raczyński, J. 1967. The reproduction biology of European bison living in reserves and in freedom. *Acta Theriologica* **12**: 407–44.

Kristensen, P. and Bjerkedal, T. 2007. Explaining the relation between birth order and intelligence. *Science* **316**: 1717.

Krüger, T. H. C., Haake, P., Cheretah, D. et al. 2003a. Specificity of the neuroendocrine response to orgasm during sexual arousal in men. *Journal of Endocrinology* **177**: 57–64.

Krüger, T. H. C., Haake, P., Haverkamp, J. et al. 2003b. Effects of acute prolactin manipulation on sexual drive and function in males. *Journal of Endocrinology* **179**: 357–65.

Krüger, T. H. C., Hartmann, U. and Schedlowski, M. 2005. Prolactinergic and dopaminergic mechanisms underlying sexual arousal and orgasm in humans. *World Journal of Urology* **23**: 130–8.

Kruijver, F. P. M., Balesar, R., Espila, A. M., Unmehopa, U. A. and Swaab, D. F. 2003. Estrogen-receptor-β distribution in the human hypothalamus: similarities and differences with ERα distribution. *Journal of Comparative Neurology* **466**: 251–77.

Kruijver, F. P. M., Zhou, J.-N., Pool, C. W. et al. 2000. Male-to-female transsexuals have female neuron numbers in a limbic nucleus. *Journal of Clinical Endocrinology and Metabolism* **85**: 2034–41.

Krützen, M., Barré, L. M., Connor, R. C., Mann, J. and Sherwin, W. B. 2004. 'O father: where art thou?' – Paternity assessment in an open fission-fusion society of wild bottlenose dolphins (*Tursiops* sp.) in Shark Bay, Western Australia. *Molecular Ecology* **13**: 1975–90.

Krützen, M., Sherwin, W. B., Connor, R. C. et al. 2003. Contrasting relatedness patterns in bottlenose dolphins (*Tursiops* sp.) with different alliance strategies. *Proceedings of the Royal Society of London* B **270**: 497–502.

Kruuk, H. 1972. *The Spotted Hyena*. Chicago, IL: University of Chicago Press.

Kuhn, T. S. 1962. *The Structure of Scientific Revolutions*. Chicago, IL: University of Chicago Press.

Kummer, H. 1995. In *Quest of the Sacred Baboon*. Princeton, NJ: Princeton University Press.

Kurup, R. K. and Kurup, P. A. 2002. Hypothalamic digoxin, cerebral dominance, and sexual orientation. *Archives of Andrology* **48**: 359–67.

Kuwahara, A., Kamada, M., Irahara, M. *et al.* 1998. Autoantibody against testosterone in a woman with hypergonadotropic hypogonadism. *Journal of Clinical Endocrinology and Metabolism* **83**: 14–16.

Kyle, C. J., Karels, T. J., Davis, C. S. *et al.* 2007. Social structure and facultative mating systems of hoary marmots. *Molecular Ecology* **16**: 1245–55.

La Follette, R. M. 1971. Agonistic behaviour and dominance in confined wallabies *Wallabia rufogrisea frutica*. *Animal Behaviour* **19**: 93–101.

La Freniere, P., Strayer, F. F. and Gauthier, R. 1984. The emergence of same-sex affiliative preferences among preschool peers: a developmental ethological perspective. *Child Development* **55**: 1958–65.

Lagrenade, M.-C. and Mousseau, P. 1983. Female-female pairs and polygynous associations in a Québec ring-billed gull colony. *Auk* **100**: 210–12.

Laland, K. N. 1994. On the evolutionary consequences of sexual imprinting. *Evolution* **48**: 477–89.

Lalumière, M. L., Blanchard, R. and Zucker, K. J. 2000. Sexual orientation and handedness in men and women: a meta-analysis. *Psychological Bulletin* **126**: 575–92.

Lalumière, M. L., Harris, G. T. and Rice, M. E. 1999. Birth order and fluctuating asymmetry: a first look. *Proceedings of the Royal Society of London B* **266**: 2351–4.

Lambert, D. M., Millar, C. D., Jack, K., Anderson, S. and Ctarig, J. L. 1994. Single-and multilocus DNA fingerprinting of communally breeding pukeko: do copulations or dominance ensure reproductive success? *Proceedings of the National Academy of Sciences, USA* **91**: 9641–5.

Lanctot, R. B., Scribner, K. T., Kempenaers, B. and Weatherhead, P. J. 1997. Lekking without a paradox in the buff-breasted sandpiper. *American Naturalist* **149**: 1051–70.

Lanctot, R. B. and Weatherhead, P. J. 1997. Ephemeral lekking behavior in the buff-breasted sandpiper, *Tryngites subruficollis*. *Behavioral Ecology* **8**: 268–78.

Lanctot, R. B., Weatherhead, P. J., Kempenaers, B. and Scriber, K. T. 1998. Male traits, mating tactics and reproductive success in the buff-breasted sandpiper, *Tryngites subruficollis*. *Animal Behaviour* **56**: 419–32.

Landmann, A. and Kollinsky, C. 1995. Age and plumage related territory differences in male black redstarts: the (non)-adaptive significance of delayed plumage maturation. *Ethology Ecology & Evolution* **7**: 147–67.

Landolt, M. A., Bartholomew, K., Saffrey, C., Oram, D. and Perlman, D. 2004. Gender nonconformity, childhood rejection, and adult attachment: a study of gay men. *Archives of Sexual Behavior* **33**: 117–28.

Lane, J. E., Boutin, S., Gunn, M. R., Slate, J. and Coltman, D. W. 2007. Genetic relatedness of mates does not predict patterns of parentage in North American red squirrels. *Animal Behaviour* **74**: 611–19.

Lang, C. and Kuhnle, U. 2008. Intersexuality and alternative gender categories in non-Western cultures. *Hormone Research* **69**: 240–50.

Lang, T. 1940. Studies on the genetic determination of homosexuality. *Journal of Nervous and Mental Diseases* **92**: 55–64.

Langen, T. 2000. Prolonged offspring dependence and cooperative breeding in birds. *Behavioral Ecology* **11**: 367–77.

Langlois, J. H. and Roggman, L. A. 1990. Attractive faces are only average. *Psychological Science* **1**: 115–21.

Langmore, N. E. and Bennett, A.T.D. 1999. Strategic concealment of sexual identity in an estrildid finch. *Proceedings of the Royal Society of London B* **266**: 543–50.

Långström, N., Rahman, Q., Carlström, E. and Lichtenstein, P. 2010. Genetic and environmental effects on same-sex sexual behavior: a population study of twins in Sweden. *Archives of Sexual Behavior* **39**: 75–80. DOI 10.1007/s10508-008-9386-1

Långström, N. and Zucker, K. J. 2005. Transvestic fetishism in the general population. *Journal of Sex and Marital Therapy* **31**: 87–95.

Lank, D., Mineau, P., Rockwell, R. F. and Cooke, F. 1989. Intraspecific nest parasitism and extra-pair copulation in lesser snow geese. *Animal Behaviour* **37**: 74–89.

Lank, D. B., Smith, C. M., Hanotte, O. *et al.* 2002. High frequency of polyandry in a lek mating system. *Behavioral Ecology* **13**: 209–15.

Laqueur, T. W. 2003. *Solitary Sex: A Cultural History of Masturbation*. New York: Zone Books.

Larsson, I. and Svedin, C.-G. 2001. Sexual behaviour in Swedish preschool children, as observed by their parents. *Acta Paediatrica* **90**: 436–44.

Larsson, I. and Svedin, C.-G. 2002. Sexual experiences in childhood: young adults' recollections. *Archives of Sexual Behaviour* **31**: 263–73.

Larsson, I., Svedin, C.-G. and Friedrich, W. N. 2000. Differences and similarities in sexual behaviour among pre-schoolers in Sweden and USA. *Nordic Journal of Psychiatry* **54**: 251–7.

Lasco, M. S., Jordan, T. J., Edgar, M. A., Petito, C. K. and Byne, W. 2002. A lack of dimorphism of sex or sexual orientation in the human anterior commissure. *Brain Research* **936**: 95–8.

Lathe, R. 2001. Hormones and the hippocampus. *Journal of Endocrinology* **169**: 205–31.

LaTorre, R. A. 1980. Devaluation of the human love object: heterosexual rejection as a possible antecedent to fetishism. *Journal of Abnormal Psychology* **89**: 295–8.

Laumann, E., Gagnon, J. H., Michael, R. T. and Michaels, S. 1994. *The Social Organization of Sexuality: Sexual Practices in the United States*. Chicago, IL: University of Chicago Press.

Lawler, R. R., Richard, A. F. and Riley, M. A. 2003. Genetic population structure of the white sifaka (*Propithecus verreauxi verreauxi*) at beza Mahafaly Special Reserve, southwest Madagascar (1992–2001). *Molecular Ecology* **12**: 2307–17.

Lawrence, A. A. 2009. Erotic target location errors: an unappreciated paraphilic dimension. *Journal of Sex Research* **46**: 194–215.

Laws, R. M. 1969. Aspects of reproduction in the African elephant, *Loxodonta africana*. *Journal of Reproduction and Fertility* **6**: 193–217.

Lawson Handley, L. J. and Perrin, N. 2007. Advances in our understanding of mammalian sex-biased dispersal. *Molecular Ecology* **16**: 1559–78.

Lawton, M. F. and Lawton, R. O. 1985. Heterochrony, deferred breeding, and avian sociality. *Current Ornithology* **3**: 187–222.

Layne, J. N. 1954. The biology of the red squirrel, *Tamiasciurus hudsonicus loquax* (Bangs), in central New York. *Ecological Monographs* **24**: 227–67.

Lazo, A. and Soriguer, R. C. 1993. Size-biased foraging behaviour in feral cattle. *Applied Animal Behaviour Science* **36**: 99–110.

Leamy, L. J. and Klingenberg, C. P. 2005. The genetics and evolution of fluctuating asymmetry. *Annual Review of Ecology, Evolution, and Systematics* **36**: 1–21.

Le Boeuf, B. J. 1974. Male-male competition and reproductive success in elephant seals. *American Zoologist* **14**: 163–76.

Le Boeuf, B. J., Crocker, D. E., Costa, D. P. *et al*. 2000. Foraging ecology of northern elephant seals. *Ecological Monographs* **70**: 353–82.

Lebowitz, P. S. 1972. Feminine behavior in boys: aspects of its outcome. *American Journal of Psychiatry* **128**: 102–9.

Lebret, T. 1961. The pair formation in the annual cycle of the mallard, *Anas platyrhynchos* L. *Ardea* **49**: 7–158.

Ledger, H. P. and Smith, N. S. 1964. The carcass and body composition of the Uganda cob. *Journal of Wildlife Management* **28**: 827–39.

Lefebvre, L., Reader, S. M. and Sol, D. 2004. Brains, innovations and evolution in birds and primates. *Brain, Behavior and Evolution* **63**: 233–46.

Legge, S. and Cockburn, A. 2000. Social and mating system of cooperatively breeding laughing kookaburras (*Dacelo novaguineae*). *Behavioral Ecology and Sociobiology* **47**: 220–9.

Lehman, N., Clarkson, P., Mech, L. D., Meier, T. J. and Wayne, R. K. 1992. A study of the genetic relationships within and among wolf packs using DNA fingerprinting and mitochondrial DNA. *Behavioral Ecology and Sociobiology* **30**: 83–94.

Leitenberg, H., Greenwald, E. and Tarran, M. J. 1989. The relation between sexual activity among children during preadolescence and/or early adolescence and sexual behavior and sexual adjustment in young adulthood. *Archives of Sexual Behavior* **18**: 299–313.

Lekuona, J. M. 2002. Kleptoparasitism in wintering grey heron *Ardea cinerea*. *Folia Zoologica* **51**: 215–20.

Lekuona, J. M. and Campos, F. 1998. Extra-pair copulations of grey heron *Ardea cinerea* in Arcachon Bay (SW France). *Miscellania Zoologica* **21**: 73–80.

LeMaoult, J., Szabo, P. and Weksler, M. E. 1997. Effect of age on humoral immunity selection of the B-cell repertoire and B-cell development. *Immunological Reviews* **160**: 115–26.

Leonard, M. L., Zanette, L. and Fairfull, R. W. 1993. Early exposure to females affects interactions between male White Leghorn chickens. *Applied Animal Behaviour Science* **36**: 29–38.

Le Pendu, Y., Briedermann, L., Gerard, J. F. and Maublanc, M. L. 1995. Inter-individual associations and social structure of a mouflon population (*Ovis orientalis musimon*). *Behavioural Processes* **34**: 67–80.

Le Pendu, Y., Ciofolo, I. and Gosser, A. 2000a. The social organization of giraffes in Niger. *African Journal of Ecology* **38**: 78–85.

Le Pendu, Y., Guilhelm, C., Briedermann, L., Maublanc, M.-L. and Gerard, J.-F. 2000b. Interactions and associations between age and sex classes in mouflon sheep (*Ovis gmelini*) during winter. *Behavioural Processes* **52**: 97–107.

Leszczynski, J. P. and Strough, J. 2008. The contextual specificity of masculinity and femininity in early adolescence. *Social Development* **17**: 719–36.

Lettvin, J. Y., Maturana, H. R., McCulloch, W. S. and Pitts, W. H. 1959. What the frog's eye tells the frog's brain. *Proceedings of the Institute of Radio Engineering* **47**: 1940–51.

Leung, A. K. C. and Robson, L. M. 1993. Childhood masturbation. *Clinical Pediatrics* **32**: 238–41.

Leuthold, B. M. 1979. Social organization and behaviour of giraffe in Tsavo East national Park. *African Journal of Ecology* **17**: 19–34.

Leuthold, W. 1966. Variations in territorial behaviour of Uganda kob *Adenota kob thomasi* (Neumann 1869). *Behaviour* **27**: 214–51.

Levan, K. E., Fedina, T. Y. and Lewis, S. M. 2009. Testing multiple hypotheses for the maintenance of male homosexual copulatory behaviour in flour beetles. *Journal of Evolutionary Biology* **22**: 60–70.

LeVay, S. 1991. A difference in hypothalamic structure between heterosexual and homosexual men. *Science* **253**: 1034–37.

LeVay, S. 1993. *The Sexual Brain*. Cambridge, MA: Massachusetts Institute of Technology Press.

LeVay, S. 1996. *Queer Science: The Use and Abuse of Research into Homosexuality*. Cambridge, MA: Massachusetts Institute of Technology Press.

LeVay, S. and Hamer, D. H. 1994. Evidence for a biological influence in male homosexuality. *Scientific American* **270**: 20–5.

Levin, R. J. 2007. Single case studies in human sexuality – important or idiosyncratic? *Sexual and Relationship Therapy* **22**: 457–69.

Levinson, H. Z. and Mori, K. 1983. Chirality determines pheromone activity in flour beetles. *Naturwissenschaften* **70**: 190–2.

Levtin, N. L., Eaton, K. A., Aldrich, W. R. *et al.* 1983. Acquired immunodeficiency syndrome in a colony of macaque monkeys. *Proceedings of the National Academy of Sciences, USA* **80**: 2718–22.

Lewis, K. P. and Barton, R. A. 2006. Amygdala size and hypothalamus size predict social play frequency in nonhuman primates: a comparative analysis using independent contrasts. *Journal of Comparative Psychology* **120**: 31–7.

Lewis, M. 1987. Early sex role behaviour and school age adjustment. In: Reinisch, J. M., Rosenblum, L. A. and Sanders, S. A. (Eds) *Masculinity/Femininity: Basic Perspectives*, pp. 202–26. New York: Oxford University Press.

Leypold, B. G., Yu, C. R., Leinders-Zufall, T. *et al.* 2002. Altered sexual and social behaviour in trp2 mutant mice. *Proceedings of the National Academy of Sciences, USA* **99**: 6376–81.

Li, X., Liu, Z., Wang, X. *et al.* 2007. Group characteristics of blue sheep (*Pseudois nayaur*) during rutting season in the Helan Mountains, China. *Acta Theriologica Sinica* **16**: 39–44.

Licht, P., Frank, L. G., Pavgi, S. *et al.* 1992. Hormonal correlates of masculinisation in female spotted hyaenas (*Crocuta crocuta*). 2. Maternal and fetal steroids. *Journal of Reproduction and Fertility* **95**: 463–74.

Licht, P., Hayes, T., Tsai, P. *et al.* 1998. Androgens and masculinisation of genitalia in the spotted hyaena (*Crocuta crocuta*). 1. Urogenital morphology and placental androgen production during fetal life. *Journal of Reproduction and Fertility* **113**: 105–16.

Liddicoat, R. 1956. *Homosexuality: Results of a Survey*. University of Witwatersrand.

Lieberman, D. E., Carlo, J., Ponce de León, M. and Zollikofer, C.P.E. 2007. A geometric morphometric analysis of heterochrony in the cranium of chimpanzees and bonobos. *Journal of Human Evolution* **52**: 647–62.

Lill, A. 1979. An assessment of male parental investment and pair bonding in the polygamous superb lyrebird. *Auk* **96**: 489–98.

Lillie, F. R. 1916. Theory of the free-martin. *Science* **43**: 611.

Lim, L. C. C. 1995. Present controversies in the genetics of male homosexuality. *Annals of the Academy of Medicine of Singapore* **24**: 759–62.

Lincoln, G. A., Youngson, R. W. and Short, R. V. 1970. The social and sexual behaviour of the red deer stag. *Journal of Reproduction and Fertility*, Suppl. **11**: 71–103.

Lindefors, P. 2002. Sexually antagonistic selection on primate size. *Journal of Evolutionary Biology* **15**: 595–607.

Lindsay, D. R. 1965. The importance of olfactory stimuli in the mating behaviour of the ram. *Animal Behaviour* **13**: 75–8.

Lindstedt, E. R., Oh, K. P. and Badyaev, A. V. 2007. Ecological, social, and genetic contingency of extrapair behavior in a socially monogamous bird. *Journal of Avian Biology* **38**: 214–23.

Ling, W. H. and Jones, P.J.H. 1995. Dietary phytosterols: a review of metabolism, benefits and side effects. *Life Sciences* **57**: 195–206.

Linnankovski, I., Grönroos, M. and Pertovaara, A. 1993. Eye contact as a trigger of male sexual arousal in stump-tailed macaques (*Macaca arctoides*). *Folia Primatologica* **60**: 181–4.

Linstroem, A. 1989. Finch flock size and risk of hawk predation at a migratory stopover site. *Auk* **106**: 225–32.

Lipman-Blumen, J. 1976. Toward a homosocial theory of sex roles: an explanation of the sex segregation of social institutions. *Signs* **1**: 15–31.

Lippa, R. A. 2003a. Handedness, sexual orientation, and gender-related personality traits in men and women. *Archives of Sexual Behavior* **32**: 103–14.

Lippa, R. A. 2003b. Are 2D:4D finger-length ratios related to sexual orientation? Yes for men, no for women. *Journal of Personality and Social Psychology* **85**: 179–88.

Lippa, R. A. 2008a. Sex differences and sexual orientation differences in personality: findings from the BBC Internet Survey. *Archives of Sexual Behavior* **37**: 173–87.

Lippa, R. A. 2008b. Sex differences in personality traits and gender-related occupational preferences across 53 nations: testing evolutionary and social-environmental theories. *Archives of Sexual Behavior* **39**: 619–36. DOI 10.1007/s10508-008-9380-7

Lippa, R. A. and Tan, F. D. 2001. Does culture moderate the relationship between sexual orientation and gender-related personality traits? *Cross-Cultural Research* **35**: 65–87.

Little, J. 2002. Rural geography: rural gender identity and performance of masculinity and femininity in the countryside. *Progress in Human Geography* **26**: 665–70.

Little, J. and Panelli, R. 2003. Gender research in rural geography. *Gender, Place and Culture* **10**: 281–9.

Liu, Y.-C., Salamone, J. D. and Sachs, B. D. 1997. Lesions in medial preoptic area and bed nucleus of stria terminalis: differential effects on copulatory behaviour and noncontact erection in male rats. *Journal of Neuroscience* **17**: 5245–53.

Livezey, B. C. 1997. A phylogenetic classification of waterfowl (Aves: Anseriformes), including selected fossil species. *Annals of the Carnegie Museum* **66**: 457–96.

Livezey, B. C. and Zusi, R. L. 2007. Higher-order phylogeny of modern birds (Theropoda, Aves: Neornithes) based on comparative anatomy. II Analysis and discussion. *Zoological Journal of the Linnean Society* **149**: 1–95.

Lloyd, P. H. and Rasa, O.A.E. 1989. Status, reproductive success and fitness in Cape mountain zebra (*Equus zebra zebra*). *Behavioural Ecology and Sociobiology* **25**: 411–20.

Loehlin, J. C. and McFadden, D. 2003. Otoacoustic emissions, auditory evoked potentials, and traits related to sex and sexual orientation. *Archives of Sexual Behavior* **32**: 115–27.

Loehlin, J. C., McFadden, D., Medland, S. E. and Martin, N. G. 2006. Population differences in finger-length ratios: ethnicity or latitude? *Archives of Sexual Behavior* **35**: 739–42.

Loesch, C. R., Twedt, D. J., Tripp, K., Hunter, W. C. and Woodrey, M. S. 1999. Development and management objectives for waterfowl and shorebirds in the Mississippi Alluvial Valley. In: Bonney, R., Pashley, D. N., Cooper, R. J. and Niles, L. (Eds) *Strategies for Bird Conservation: the Partners in Flight Planning Process*. Ithaca, NY: Cornell Laboratory of Ornithology.

Loison, A., Gaillard, J.-M., Pélabon, C. and Yoccoz, N. G. 1999a. What factors shape sexual size dimorphism in ungulates? *Evolutionary Ecology Research* **1**: 611–33.

Loison, A., Jullien, J.-M. and Menaut, P. 1999b. Subpopulation structure and dispersal in two populations of chamois. *Journal of Mammalogy* **80**: 620–32.

Lombardo, M. P., Bosman, R. M., Faro, C. A., Houtteman, S. G. and Kluisza, T. S. 1994. Homosexual copulations by male Tree Swallows. *Wilson Bulletin* **106**: 555–7.

Lopez Ornat, A. and Greenberg, R. 1990. Sexual segregation by habitat in migratory warblers in Quintana Roo, Mexico. *Auk* **107**: 539–43.

Lorenz, K. 1935. Der Kumpar in der Umvelt des Vögels. *Journal für Ornithologie* **83**: 137–213.

Lorenz, K. 1970 *Studies in Animal and Human Behaviour*. vol. 1. London: Methuen & Co. Ltd.

Lösel, R. and Wehling, M. 2003. Nongenomic actions of steroid hormones. *Nature Reviews Molecular Cell Biology* **4**: 46–56.

Loseto, L. L., Richard, P., Stern, G. A., Orr, J. and Ferguson, S. H. 2006. Segregation of Beaufort Sea beluga whales during open-water season. *Canadian Journal of Zoology* **84**: 1743–51.

Louch, C. D., Ghosh, A. K. and Pal, B. C. 1966. Seasonal changes in weight and reproductive activity of *Suncus murinus* in West Bengal, India. *Journal of Mammalogy* **47**: 73–8.

Lovari, S. and Ale, S. B. 2001. Are there multiple mating strategies in blue sheep? *Behavioural Processes* **53**: 131–5.

Lovari, S. and Locati, M. 1993. Intrasexual social behaviour of female Apennine chamois *Rupicapra pyrenaica ornata* (Neumann 1899). *Ethology, Ecology and Evolution* **5**: 347–56.

Loy, J. and Loy, K. 1974. Behaviour of an all-juvenile group of rhesus monkeys. *American Journal of Physical Anthropology* **40**: 83–96.

Lozano, G. A. and Handford, P. T. 1995. A test of an assumption of delayed plumage maturation hypotheses using female tree swallows. *Wilson Bulletin* **107**: 153–64.

Luenser, K., Fickel, J., Lehmen, A., Speck, S. and Ludwig, A. 2005. Low level of genetic variability in European bisons (*Bison bonasus*) from the Bialowieza National Park in Poland. *European Journal of Wildlife Research* **51**: 84–7.

Lummaa, V., Merilä, J. and Kause, A. 1998. Adaptive sex ratio variation in pre-industrial human (*Homo sapiens*) populations? *Proceedings of the Royal Society of London B* **265**: 563–8.

Lunardini, A. 1989. Social organization in a confined group of Japanese macaques (*Macaca fuscata*): an application of correspondence analysis. *Primates* **30**: 175–85.

Lundy, K. J., Parker, P. G. and Zahavi, A. 1998. Reproduction by subordinates in cooperatively breeding Arabian babblers is uncommon but predictable. *Behavioral Ecology and Sociobiology* **43**: 173–80.

Lunenfeld, B. 2001. The ageing male. In: Fischl, F. H. (Ed.) *Menopause Andropause*, pp. 279–91. Gablitz: Krause & Pachernegg.

Lung, M. A. and Childress, M. J. 2007. The influence of conspecifics and predation risk on the vigilance of elk (*Cervus elaphus*) in Yellowstone National Park. *Behavioral Ecology* **18**: 12–20.

Luppi, P. 2003. How immune mechanisms are affected by pregnancy. *Vaccine* **21**: 3352–57.

Lusseau, D., Schneider, K., Boisseau, O. J. *et al.* 2003. The bottlenose dolphin community of Doubtful Sound features a large proportion of long-lasting associations. Can geographic isolation explain this unique trait? *Behavioral Ecology and Sociobiology* **54**: 396–405.

Lutchmaya, S., Baron-Cohen, S., Raggatt, P., Knickmeyer, R. and Manning, J. T. 2004. 2nd to 4th digit ratios, fetal testosterone and estradiol. *Early Human Development* **77**: 23–8.

Lykken, D. T., Bouchard, T. J. Jr., McGue, M. and Tellegen, A. 2007. The Minnesota twin family registry. www.psych.umn.edu/psylabs/mtfs/mtrf3.htm. Accessed on 22 January 2007.

Lyon, B. E. and Montgomerie, R. D. 1986. Delayed plumage maturation in passerine birds: reliable signalling by subordinate males? *Evolution* **40**: 605–15.

Lyons, D. M., Mendoza, S. P. and Mason, W. A. 1992. Sexual segregation in squirrel monkeys (*Saimiri sciureus*): a transactional analysis of adult social dynamics. *Journal of Comparative Psychology* **106**: 323–30.

McCabe, J. 1989. Psychological androgyny in later life: a psychocultural examination. *Ethos* **17**: 3–31.

McClellan, K. M., Parker, K. L. and Tobet, S. 2006. Development of the ventromedial nucleus of the hypothalamus. *Frontiers in Neuroendocrinology* **27**: 193–209.

McClelland, B. E. 1991. Courtship and agonistic behaviour in mouflon sheep. *Applied Animal Behaviour Science* **29**: 67–85.

Maccoby, E. E. 1987. The varied meanings of "masculine" and "feminine". In: Reinsch, J. M., Rosenblum, L. A. and Sanders, S. A. (Eds) *Masculinity/Femininity: Basic Perspectives*, pp. 227–39. New York: Oxford University Press.

Maccoby, E. E. 1990. Gender and relationships: a developmental account. *American Psychologist* **45**: 513–20.

Maccoby, E. E. 1994. Commentary: gender segregation in childhood. *New Directions for Child Development* **65**: 87–97.

Maccoby, E. E. 1998. *The Two Sexes: Growing Up Apart, Coming Together*. Cambridge MA: Harvard University Press.

Maccoby, E. E. 2000. Perspectives on gender development. *International Journal of Behavioral Development* **24**: 398–406.

Maccoby, E. E. 2002. Gender and group process: a developmental perspective. *Current Directions in Psychological Science* **11**: 54–8.

Maccoby, E. E. and Jacklin, C. N. 1974. *The Psychology of Sex Differences*. Stanford, CA: Stanford University Press.

Maccoby, E. E. and Jacklin, C. N. 1987. Gender segregation in childhood. In: Reese, H. W. (Ed.) *Advances in Child Development and Behavior*, vol. 20, pp. 239–87. Orlando, FL: Academic Press.

McClintock, M. K. and Herdt, G. 1996. Rethinking puberty: the development of sexual attraction. *Current Directions in Psychological Science* **5**: 178–83.

McConaghy, N. and Silove, D. 1992. Does sex-linked behavior in children influence relationships with their parents? *Archives of Sexual Behavior* **21**: 469–79.

McConaghy, N. 1987. Heterosexuality/homosexuality: dichotomy or continuum. *Archives of Sexual Behavior* **16**: 411–24.

McConaghy, N., Hadzi-Pavlovic, D., Stevens, C. *et al.* 2006. Fraternal birth order and ratio of heterosexual/homosexual feelings in women and men. *Journal of Homosexuality* **51**: 161–74.

McCormick, C. M. and Witelson, S. F. 1991. A cognitive profile of homosexual men compared to heterosexual men and women. *Psychoneuroendocrinology* **16**: 457–73.

McCrae, R. R. and John, O. P. 1992. An introduction to the five-factor model and its applications. *Journal of Personality* **60**: 175–215.

MacCulloch, M. J. and Waddington, J. L. 1981. Neuroendocrine mechanisms and aetiology of male and female homosexuality. *British Journal of Psychiatry* **139**: 341–5.

McDonald, B. J. 2005. *Population genetics of dugongs around Australia: implications of gene flow and migration*. PhD Thesis, James Cook University, Australia.

McDonnell, S. M. and Haviland, J.C.S. 1995. Agonistic ethogram of the equid bachelor band. *Applied Animal Behaviour Science* **43**: 147–88.

Macedo, R. H. F., Cariello, M. O., Graves, J. and Schwabl, H. 2004. Reproductive partitioning in communally breeding guira cuckoos, *Guira guira*. *Behavioral Ecology and Sociobiology* **55**: 213–22.

McFadden, D. 1993. A masculinization effect on the auditory systems of human females having male co-twins. *Proceedings of the National Academy of Sciences, USA* **90**: 11900–4.

McFadden, D. and Champlin, C. A. 2000. Comparison of auditory evoked potentials in heterosexual, homosexual, and bisexual males and females. *Journal of the Association for Research in Otolaryngology* **1**: 89–99.

McFadden, D., Loehlin, J. C., Breedlove, S. M. *et al.* 2005. A reanalysis of five studies of sexual orientation and the relative length of the 2nd and 4th fingers (the 2D:4D ratio). *Archives of Sexual Behavior* **34**: 341–56.

McFadden, D. and Pasanen, E. G. 1998. Comparison of the auditory systems of heterosexuals and homosexuals: click-evoked otoacoustic emissions. *Proceedings of the National Academy of Sciences, USA* **95**: 2709–13.

McFadden, D., Pasanen, E. G., Valero, M. D., Roberts, E. K. and Lee, T. M. 2009. Effect of prenatal androgens on click-evoked otoacoustic emissions in male and female sheep (*Ovis aries*). *Hormones and Behavior* **55**: 98–105.

McFadden, D. and Shubel, E. 2002. Relative lengths of fingers and toes in human males and females. *Hormones and Behavior* **42**: 492–500.

MacFarlane, A. M. and Coulson, G. 2005. Synchrony and timing of breeding influences sexual segregation in western grey and red kangaroos (*Macropus fuliginosus* and *M. rufus*). *Journal of Zoology, London* **267**: 419–29.

MacFarlane, G. R., Blomberg, S. P., Kaplan, G. and Rogers, L. J. 2007. Same-sex sexual behaviour in birds: expression is related to social mating system and state of development at hatching. *Behavioural Ecology* **18**: 21–33.

McGary, S., Henry, P. F. P. and Ottinger, M. A. 2001. Impact of vinclozolin on reproductive behaviour and endocrinology in Japanese quail (*Coturnix coturnix japonica*). *Environmental Toxicology and Chemistry* **20**: 2487–93.

McGraw, K. J. and Hill, G. E. 1999. Induced homosexual behaviour in male house finches (*Carpodacus mexicanus*): the "Prisoner Effect". *Ethology, Ecology and Evolution* **11**: 197–201.

McGuire, T. R. 1995. Is homosexuality genetic? A critical review and some suggestions. *Journal of Homosexuality* **28**: 115–45.

Maciejok, J., Saur, B. and Bergman, H.-H. 1995. Behaviour of chaffinches (*Fringilla coelebs*) inside and outside their territories during the breeding period. *Journal für Ornithologie* **136**: 37–45.

MacIntyre, F. and Estep, K. W. 1993. Sperm competition and the persistence of genes for male homosexuality. *Biosystems* **31**: 223–33.

McIntyre, M. H. 2006. The use of digit ratios as markers for perinatal androgen action. *Reproductive Biology and Endocrinology* **4**: 10–18.

McIntyre, M. H., Herrmann, E., Wobber, V. *et al.* 2009. Bonobos have a more human-like second-to-fourth finger length ratio (2D:4D) than chimpanzees: a hypothesized indication of lower prenatal androgens. *Journal of Human Evolution* **56**: 361–5.

McIntyre, M. H., Lipson, S. F. and Ellison, P. T. 2003. Effects of developmental and adult androgens on male abdominal adiposity. *American Journal of Human Biology* **15**: 662–6.

Macke, J. P., Hu, N., Bailey, M. *et al.* 1993. Sequence variation in the androgen receptor gene is not a common determinant of male sexual orientation. *American Journal of Human Genetics* **53**: 844–52.

McKenna, J. J. 1979. The evolution of allomothering behavior among colobine monkeys: function and opportunism in evolution. *American Anthropologist* **81**: 818–40.

McKenna, J. J. 2005. Biosocial functions of grooming behavior among the common Indian langur monkey (*Presbytis entellus*). *American Journal of Physical Anthropology* **48**: 503–9.

Mackie J. L. 1965 Causes and conditions. *American Philosophical Quarterly* **2**: 245–64.

MacKinnon, J. 1974. The behaviour and ecology of wild orang-utans (*Pongo pygmaeus*). *Animal Behaviour* **22**: 3–74.

McLachlan, J. A. 2001. Environmental signalling: what embryos and evolution teach us about endocrine disrupting chemicals. *Endocrine Reviews* **22**: 319–41.

McLoughlin, P. D., Case, R. L., Gau, R. J. *et al.* 2002. Hierarchical habitat selection by barren-ground grizzly bears in the central Canadian Arctic. *Oecologia* **132**: 102–8.

MacLusky, N. J. and Naftolin, F. 1981. Sexual differentiation of the central nervous system. *Science* **211**: 1294–303.

MacLusky, N. J., Naftolin, F. and Goldman-Rakic, P. S. 1986. Estrogen formation and binding in the cerebral cortex of the developing rhesus monkey. *Proceedings of the National Academy of Sciences, USA* **83**: 513–16.

McNab, B. K. 2005. Food habits and the evolution of energetics in birds of paradise (Paradisaeidae). *Journal of Comparative Physiology* B **175**: 117–32.

McNamara, K. J. and McKinney, M. L. 2005. Heterochrony, disparity, and macroevolution. *Paleobiology* **31**: 17–26.

McOrist, S. and Williamson, M. 2007. Ulceration of the anus in groups of pubertal male pigs. *Journal of Swine Health and Production* **15**: 96–8.

McRae, S. B., Emlen, S. T., Rubenstein, D. R. and Bogdanowicz, S. M. 2005. Polymorphic microsatellite loci in a plural breeder, the grey-capped social weaver (*Pseudonigrita arnaudi*), isolated with an improved enrichment protocol using fragment size-selection. *Molecular Ecology Notes* **5**: 16–20.

MacRoberts, M. H. and MacRoberts, B. R. 1976. Social organization and behavior of the Acorn Woodpecker in central coastal California. *Ornithological Monographs* **21**: 1–115.

McShea, W. J., Aung, M., Poszig, D., Wemmer, C. and Monfort, S. 2001. Forage, habitat use, and sexual segregation by a tropical deer (*Cervus eldi thamin*) in a dipterocarp forest. *Journal of Mammalogy* **82**: 848–57.

Maddison, W. P. and Maddison, D. R. 2006. Mesquite: a modular system for evolutionary analysis. Version 1.1. http://mesquiteproject.org.

Madeira, M. D., Ferreira-Silva, L. and Paula-Barbosa, M. M. 2001. Influence of sex and estrus cycle in the sexual dimorphisms of the hypothalamic ventromedial nucleus: stereological evaluation and Golgi study. *Journal of Comparative Neurology* **432**: 329–45.

Magrath, R. D. and Heinsohn, R. G. 2000. Reproductive skew in birds: models, problems and prospects. *Journal of Avian Biology* **31**: 247–58.

Magrath, R. D., Heinsohn, R. G and Johnstone, R. A. 2004. Reproductive skew. In: Koenig, W. D. and Dickinson, J. L. (Eds) *Ecology and Evolution of Cooperative Breeding in Birds*, pp. 157–76. Cambridge: Cambridge University Press.

Mah, K. and Binik, Y. M. 2001. The nature of human orgasm: a critical review of major trends. *Clinical Psychology Review* **21**: 823–56.

Maher, C. R. 1997. Group stability, activity budgets, and movements of pronghorn females. *Southwestern Naturalist* **42**: 25–32.

Main, M. B. and Coblentz, B. E. 1990. Sexual segregation among ungulates: a critique. *Wildlife Society Bulletin* **18**: 204–10.

Main, M. B. and Coblentz, B. E. 1996. Sexual segregation in Rocky Mountain mule deer. *Journal of Wildlife Management* **60**: 497–507.

Main, M. B., Weckerly, F. W. and Bleich, V. C. 1996. Sexual segregation in ungulates: new directions for research. *Journal of Mammalogy* **77**: 449–61.

Makova, K. D. and Li, W.-H. 2002. Strong male-driven evolution of DNA sequences in humans and apes. *Nature* **416**: 624–626.

Malhira A. K., Kestler L. J., Mazzanti C. *et al.* 2002. A functional polymorphism in the COMT gene and performance on a test of prefrontal cognition. *American Journal of Psychiatry* **159**: 652–4.

Malinowski, B. 1932. *The Sexual Life of Savages in North-Western Melanesia*. London: Routledge & Kegan Paul Ltd.

Mani, S. 2003. Emerging concepts in the regulation of female sexual behavior. *Scandinavian Journal of Psychology* **44**: 231–9.

Mani, S. K., Reyna, A. M., Alejandro, M. A., Crowley, J. and Markaverich, B. M. 2005. Disruption of male sexual behavior in rats by tetrahydrofurandiols (THF-diols). *Steroids* **70**: 750–4.

Mann, D. R. and Fraser, H. M. 1996. The neonatal period: a critical interval in male primate development. *Journal of Endocrinology* **149**: 191–7.

Mann, J. 2006. Establishing trust: sociosexual behaviour and the development of male-male bonds among Indian Ocean bottlenose dolphins. In: Sommer, V. and Vasey, P. L. (Eds) *Homosexual Behaviour in Animals: An Evolutionary Perspective*, pp. 107–130. Cambridge: Cambridge University Press.

Manners, P. J. 2009. Gender identity disorder in adolescence: a review of the literature. *Child and Adolescent Mental Health* **14**: 62–8.

Manning, J. T., Barley, L., Walton, J. et al. 2000. The 2nd:4th digit ratio, sexual dimorphism, population differences, and reproductive success: evidence for sexually antagonistic genes? *Evolution and Human Behavior* **21**: 163–83.

Manning, J. T., Bundred, P. E. and Flanagan, B. F. 2002. The ratio of 2nd to 4th digit length: a proxy for transactivation activity of the androgen receptor gene? *Medical Hypotheses* **59**: 334–6.

Manning, J. T., Scutt, D., Wilson, J. and Lewis-Jones, D. I. 1998. The ratio of 2nd to 4th digit length: a predictor of sperm numbers and concentrations of testosterone, luteinizing hormone and oestrogen. *Human Reproduction* **13**: 3000–4.

Manosevitz, M. 1970. Early sexual behaviour in adult homosexual and heterosexual males. *Journal of Abnormal Psychology* **76**: 396–402.

Manson, J. H. 1999. Infant handling in wild *Cebus capucinus*: testing bonds between females. *Animal Behaviour* **57**: 911–21.

Manson, J. H., Perry, S. and Parish, A. R. 1997. Nonconceptive sexual behavior in bonobos and capuchins. *International Journal of Primatology* **18**: 767–86.

Manson, J. H., Rose, L. M., Perry, S. and Gros-Louis, J. 1999. Dynamics of female-female relationships in wild *Cebus capucinus*: data from two Costa Rican sites. *International Journal of Primatology* **20**: 679–706.

Mansukhani, V., Adkins-Regan, E. and Yang, S. 1996. Sexual partner preference in female zebra finches: the role of early hormones and social environment. *Hormones and Behavior* **30**: 506–13.

Marcelli, M., Fusillo, R. and Boitani, L. 2003. Sexual segregation in the activity patterns of European polecats (*Mustela putorius*). *Journal of Zoology, London* **261**: 249–55.

Marchant-Haycox, S. E., McManus, I. C. and Wilson, G. D. 1991. Left-handedness, homosexuality, HIV infection and AIDS. *Cortex* **27**: 49–56.

Marjakangas, A. 1981. A singing Chaffinch *Fringilla coelebs* in female plumage paired with another female-plumaged Chaffinch. *Ornis Fennica* **58**: 90–1.

Marker, L. L. and Dickman, A. J. 2003. Morphology, physical condition, and growth of the cheetah (*Acinonyx jubatus jubatus*). *Journal of Mammalogy* **84**: 840–50.

Marker, L. L., Dickman, A. J., Jeo, R. M., Mills, M. G. L. and MacDonald, D. W. 2003. Demography of the Namibian cheetah, *Acinonyx jubatus jubatus*. *Biological Conservation* **114**: 413–25.

Marker, L. L., Wilkerson, A. J. P., Sarno, R. J. et al. 2008. Molecular genetic insights on cheetah (*Acinonyx jubatus*) ecology and conservation in Namibia. *Journal of Heredity* **99**: 2–13.

Markham, J. A. and Greenough, W. T. 2004. Experience-driven brain plasticity: beyond the synapse. *Neuron Glia Biology* **1**: 351–63.

Marler, P. 1955. Studies of fighting in chaffinches. (2) The effect on dominance relations of disguising females as males. *British Journal of Animal Behaviour* **3**: 137–46.

Marlow, B. J. 1975. The comparative behaviour of the Australasian sea lions *Neophoca cinerea* and *Phocarctos hookeri* (Pinnipedia: Otariidae). *Mammalia* **39**: 159–230.

Marlowe, F. 2000. Paternal investment and the human mating system. *Behavioural Processes* **51**: 45–61.

Marrazzo, J. M. 2000. Sexually transmitted infections in women who have sex with women: who cares? *Sexually Transmitted Infections* **76**: 330–2.

Marrazzo, J. M., Koutsky, L. A., Eschenbach, D. A. et al. 2002. Characterization of vaginal flora and bacterial vaginosis in women who have sex with women. *Journal of Infectious Diseases* **185**: 1307–13.

Marrazzo, J. M., Koutsky, L. A., Stine, K. L. et al. 1998. Genital human papillomavirus infection in women who have sex with women. *Journal of Infectious Diseases* **178**: 1604–9.

Marrazzo, J. M., Stine, K. and Wald, A. 2003. Prevalence and risk factors for infection with herpes simplex virus type-1 and -2 among lesbians. *Sexually Transmitted Diseases* **30**: 890–5.

Marsh, H. 1995. The life history, pattern of breeding, and population dynamics of the dugong. In: O'Shea, T. J., Ackerman, B. B. and Percival, H. F. (Eds) *Population Biology of the Florida Manatee*, pp. 75–83. *Information and Technology Report I*. Washington, DC: USDI National Biological Service.

Marshall, A. J. and Hohmann, G. 2005. Urinary testosterone levels of wiold male bonobos (*Pan paniscus*) in the Lomako forest, Democratic Republic of Congo. *American Journal of Primatology* **65**: 87–92.

Martin, A. R. and da Silva, V.M.F. 2004. River dolphins and flooded forest: seasonal habitat use and sexual segregation of botos (*Inia geoffrensis*) in an extreme cetacean environment. *Journal of Zoology, London* **263**: 295–305.

Martin, C. L. 1990. Attitudes and expectations about children with nontraditional and traditional gender roles. *Sex Roles* **22**: 151–65.

Martin, C. L. 1995. Stereotypes about children with traditional and nontraditional gender roles. *Sex Roles* **33**: 727–51.

Martin, C. L. and Fabes, R. A. 2001. The stability and consequences of young children's same-sex peer interactions. *Developmental Psychology* **37**: 431–46.

Martin, C. L., Fabes, R. A., Evans, S. M. and Wyman, H. 1999. Social cognition on the playground: children's beliefs about playing with girls versus boys and their relations to sex segregated play. *Journal of Social and Personal Relationships* **16**: 751–71.

Martin, J. T. and Nguyen, D. H. 2004. Anthropometric analysis of homosexuals and heterosexuals: implications for early hormone exposure. *Hormones and Behavior* **45**: 31–9.

Martin, J. T., Puts, D. A. and Breedlove, S. M. 2008. Hand asymmetry in heterosexual and homosexual men and women: relationship to 2D:4D digit ratios and other sexually dimorphic anatomical traits. *Archives of Sexual Behavior* **37**: 119–32.

Martin, K., Cooch, F. G., Rockwell, R. F. and Cooke, F. 1985. Reproductive performance in lesser snow geese: are two parents essential? *Behavioral Ecology and Sociobiology* **17**: 257–63.

Martin, K. A. 1993. Gender and sexuality: medical opinion on homosexuality, 1900–1950. *Gender and Society* **7**: 246–60.

Martin, R. D. 1972. Adaptive radiation and behaviour of the Malagasy lemurs. *Philosophical Transactions of the Royal Society, London* B **264**: 295–352.

Martin, R. D. 1990. *Primate Origins and Evolution: A Phylogenetic Reconstruction.* London/ New Jersey: Chapman Hall/Princeton University Press.

Martin-Alguacil, N., Schober, J., Kow, L.-M. and Pfaff, D. 2006. Arousing properties of the vulvar epithelium. *Journal of Urology* **176**: 456–62.

Martins, Y., Preti, G., Crabtree, C. R. *et al.* 2005. Preference for human body odors is influenced by gender and sexual orientation. *Psychological Science* **16**: 694–701.

Mathews, G. A., Fane, B. A., Conway, G. S., Brook, C. G. D. and Hines, M. 2009. Personality and congenital adrenal hyperplasia: possible effects of prenatal androgen exposure. *Hormones and Behavior* **55**: 285–91.

Mathews, N. E. and Porter, W. F. 1993. Effect of social structure on genetic structure of free-ranging white-tailed deer in the Adirondack mountains. *Journal of Mammalogy* **74**: 33–43.

Matsumoto, A. 1991. Synaptogenic action of sex steroids in developing and adult neuroendocrine brain. *Psychoneuroendocrinology* **16**: 25–40.

Matsumoto, A. and Arai, Y. 1983. Sex differences in the volume of the ventromedial nucleus of the hypothalamus in the rat. *Endocrinologia Japonica* **30**: 277–80.

Matsumoto, A. and Arai, Y. 1986. Development of sexual dimorphism in synaptic organization in the ventromedial nucleus of the hypothalamus in rats. *Neuroscience Letters* **68**: 165–8.

Matsumoto, T., Takeyama, K., Sato, T. and Kato, S. 2005. Study of androgen receptor functions by genetic models. *Journal of Biochemistry* **138**: 105–10.

Matsuzaki, O. 2004. Inter-male mating-like behaviour in the domesticated house musk shrew, *Suncus murinus*. *Zoological Sciences* **21**: 43–51.

Matteo S. and Rissman E. F. 1984. Increased sexual activity during the midcycle portion of the human menstrual cycle. *Hormones and Behavior* **18**: 249–55.

Matteson, D. R. 1997. Bisexual and homosexual behavior and HIV risk among Chinese-, Filipino-, and Korean-American men. *Journal of Sex Research* **34**: 93–104.

Mavissakalian, M., Blanchard, E. B., Abel, G. C. and Barlow, D. H. 1975. Responses to complex erotic stimuli in homosexual and heterosexual males. *British Journal of Psychiatry* **126**: 252–7.

Mayer, J. J. and Wetzel, R. M. 1987. *Tayassu pecari*. *Mammalian Species* **293**: 1–7.

Maynard-Smith J. 1964. Group selection and kin selection. *Nature* **201**:1145–7.

Maynard-Smith J. 1971. The origin and maintenance of sex. In Williams, G. C. (Ed.) *Group Selection*, pp. 163–75. Chicago, IL: Aldine-Atherton.

Maynard-Smith J. 1982. *Evolution and the Theory of Games.* Cambridge: Cambridge University Press.

Maynard-Smith, J. 1989. *Did Darwin get it Right? Essays on Games, Sex and Evolution.* New York: Chapman and Hall.

Mayr, E. 1961. Cause and effect in biology. *Science* **134**: 1501–6.

Mayr, E. 1982 *The Growth of Biological Thought: Diversity, Evolution, and Inheritance.* Cambridge, MA: The Belknap Press of Harvard University Press.

Mead, M. 1935. *Sex and Temperament in Three Primitive Societies.* London: George Routledge & Sons, Limited.

Meagher, M. 1986. *Bison bison. Mammalian Species* **266**: 1–8.

Meanley, B. 1955. A nesting study of the little blue heron in eastern Arkansas. *Wilson Bulletin* **67**: 84–99.

Medland, S. E., Loehlin, J. C. and Martin, N. G. 2008. No effects of prenatal hormone transfer on digit ratio in a large sample of same- and opposite-sex dizygotic twins. *Personality and Individual Differences* **44**: 1225–34.

Medrano, E. A., Hernandez, O., Lamothe, C. and Galina, C. S. 1996. Evidence of asynchrony in the onset of signs of oestrus in zebu cattle treated with a progestogen ear implant. *Research in Veterinary Science* **60**: 51–4.

Meek, L. R., Schulz, K. M. and Keith, C. A. 2006. Effects of prenatal stress on sexual partner preference in mice. *Physiology and Behavior* **89**: 133–8.

Meek, S. B. and Barclay, R. M. R. 1996. Settlement patterns and nest-site selection of cliff swallows, *Hirundo pyrrhonota*: males prefer to clump but females settle randomly. *Canadian Journal of Zoology* **74**: 1394–401.

Meenken, D., Fechney, T. and Innes, J. 1994. Population size and breeding success of North Island kokako in the Waipapa Ecological Area, Pureora Forest park. *Notornis* **41**: 109–15.

Meier, P., Finch, A. and Evan, G. 2000. Apoptosis in development. *Nature* **407**: 796–801.

Mele, A., Franco, E., Caprilli, F. *et al.* 1988. Hepatitis B and Delta virus infection among heterosexuals, homosexuals and bisexual men. *European Journal of Epidemiology* **4**: 488–91.

Melnick, D. J. 1987. The genetic consequences of primate social organization: a review of macaques, baboons and vervet monkeys. *Genetica* **73**: 117–35.

Melo, K. F. S., Mendonca, B. B., Billerbeck, A. E. C. *et al.* 2003. Clinical, hormonal, behavioral, and genetic characteristics of androgen insensitivity syndrome in a Brazilian cohort: five novel mutations in the androgen receptor gene. *Journal of Clinical Endocrinology and Metabolism* **88**: 3241–50.

Mercival, R. F., Gibbs, H. L., Galetti, M., Lunardi, V. O. and Galetti, P. M. Jr. 2007. Genetic structure in a tropical lek-breeding bird, the blue manakin (*Chiroxiphia caudata*) in the Brazilian Atlantic forest. *Molecular Ecology* **16**: 4908–18.

Meredith, M. 2001. Human vomeronasal organ function: a critical review of best and worst cases. *Chemical Senses* **26**: 433–45.

Mertens, P. A. 2006. Reproductive and sexual behavioral problems in dogs. *Theriogenology* **66**: 606–9.

Mesoudi, A., Whiten, A. and Laland, K. N. 2004. Is human cultural evolution darwinian? Evidence reviewed from the perspective of *The Origin of Species. Evolution* **58**: 1–11.

Meston, C. M. and Frohlich, P. F. 2000. The neurobiology of sexual function. *Archives of General Psychiatry* **57**: 1012–30.

Meyer, I. H. and Wilson, P. A. 2009. Sampling lesbian, gay, and bisexual populations. *Journal of Counseling Psychology* **56**: 23–31.

Meyer-Bahlburg, H. F. L. 1977. Sex hormones and male homosexuality in comparative perspective. *Archives of Sexual Behavior* **6**: 297–325.

Meyer-Bahlburg, H. F. L. 1984. Psychoendocrine research on sexual orientation: current status and future options. *Progress in Brain Research* **61**: 375–98.

Meyer-Bahlburg, H. F. L. 2002. Gender identity disorder in young boys: a parent-and peer-based treatment protocol. *Clinical Child Psychology and Psychiatry* **7**: 360–6.

Meyer-Bahlburg, H. F. L., Dolezal, C., Baker, S. W. and New, M. I. 2008. Sexual orientation in women with classical or non-classical congenital adrenal hyperplasia as a function of degree of prenatal androgen excess. *Archives of Sexual Behaviour* **37**: 85–99.

Meyer-Bahlburg, H. F. L., Ehrhardt, A. A., Bell, J. J. *et al.* 1985. Idiopathic precocious puberty in girls: psychosexual development. *Journal of Youth and Adolescence* **14**: 339–53.

Michael, R. P. 1961. "Hypersexuality" in male cats without brain damage. *Science* **134**: 553–4.

Midford, P. E., Garland, T. Jr and Maddison, W. P. 2005. PDAP package of Mesquite. Version 1.07.

Migeon, B. R., Brown, T. R., Axelman, J. and Migeon, C. J. 1981. Studies of the locus for androgen receptor: localization on the human X chromosome and evidence for homology with the Tfm locus in the mouse. *Proceedings of the National Academy of Sciences, USA* **78**: 6339–43.

Mikach, S. M. and Bailey, J. M. 1999. What distinguishes women with unusually high numbers of sex partners? *Evolution and Human Behavior* **20**: 141–50.

Milinkovitch, M. C., Orti, G. and Meyer, A. 2000. Revised phylogeny of whales suggested by mitochondrial ribosomal DNA sequences. *Nature* **361**: 346–8.

Miller, E. H. 1975. Walruses ethology. I. The social role of tusks and applications of multidimensional scaling. *Canadian Journal of Zoology* **53**: 590–613.

Miller, E. H. 1976. Walruses ethology. II. Herd structure and activity budgets of summering males. *Canadian Journal of Zoology* **54**: 704–15.

Miller, E. H. and Boness, D. J. 1983. Summer behaviour of Atlantic walruses *Odobenus rosmarus rosmarus (L.)* at Coats Islands, N.W.T. (Canada). *Zeitschrift für Säugetierkunde* **48**: 298–313.

Miller, E. M. 2000. Homosexuality, birth order, and evolution: toward an equilibrium reproductive economics of homosexuality. *Archives of Sexual Behavior* **29**: 1–34.

Miller, K. E., Barnes, G. M., Sabo, D. F., Melnick, M. J. and Farrell, M. P. 2002. Anabolic-androgenic steroid use and other adolescent problem behaviors: rethinking the male athlete assumption. *Sociological Perspectives* **45**: 467–89.

Miller, S. S., Hoffmann, H. L. and Mustanski, B. S. 2008. Fluctuating asymmetry and sexual orientation in men and women. *Archives of Sexual Behavior* **37**: 150–7.

Millet, G., Malebranche, D., Mason, B. and Spikes, P. 2005. Focusing "down low": bisexual black men, HIV risk and heterosexual transmission. *Journal of the National Medical Association* **97**: 52S–59S.

Mills, A. M. 2005. Protogyny in autumn migration: do male birds "play chicken"? *Auk* **122**: 71–81.

Mills, G. S. 1976. American Kestrel sex ratios and habitat separation. *Auk* **93**: 740–8.

Mills, J. A. 1994. Extra-pair copulations in the red-billed gull: females with high quality, attentive males resist. *Behaviour* **128**: 41–64.

Mindell, D. P., Sorenson, M. D., Huddleston, C. J. *et al.* 1997. Phylogenetic relationships among and within selected avian orders based on mitochondrial DNA. In: Mindell, D. P. (Ed.) *Avian Molecular Evolution and Systematics*, pp. 213–47. San Diego, CA: Academic Press.

Ming, G. and Song, H. 2005. Adult neurogenesis in the mammalian central nervous system. *Annual Reviews of Neuroscience* **28**: 223–50.

Miquelle, D. G., Peek, J. M. and van Ballenberghe, V. 1992. Sexual segregation in Alaskan moose. *Wildlife Monographs* **122**: 1–57.

Mischel, W. 1973. Toward a cognitive social learning reconceptualization of personality. *Psychological Review* **80**: 252–83.

Mischel, W., Ebbesen, E. B. and Zeiss, A. R. 1973. Selective attention to the self: situational and dispositional determinants. *Journal of Personality and Social Psychology* **27**: 129–42.

Mitani, J. C., Gros-Louis, J. and Richards, A. F. 1996. Sexual dimorphism, the operational sex ratio, and the intensity of male competition in polygynous primates. *American Naturalist* **147**: 966–80.

Mitani, J. C. and Watts, D. 1997. The evolution of non-maternal caretaking among anthropoid primates: do helpers help? *Behavioral Ecology and Sociobiology* **40**: 213–20.

Mitchell, C. L. 1994. Migration alliances and coalitions among adult male South American squirrel monkeys (*Saimiri sciureus*). *Behaviour* **130**: 169–90.

Mitchell, S. D. and Dietrich, M. R. 2006. Integration without unification: an argument for pluralism in the biological sciences. *American Naturalist* **186**(Suppl.): S73–S79.

Mitteroecker, P., Gunz, P., Bernhard, M., Schaefer, K. and Bookstein, F. L. 2004. Comparison of cranial ontogenetic trajectories among great apes and humans. *Journal of Human Evolution* **46**: 679–98.

Miyamoto, M. M., Porter, C. A. and Goodman, M. 2000. c-Myc gene sequences and the phylogeny of bats and other eutherian mammals. *Systematic Biology* **49**: 501–14.

Moczek, A. P. 2007. Developmental capacitance, genetic accommodation, and adaptive evolution. *Evolution and Development* **9**: 299–305.

Modha, K. L. and Eltringham, S. K. 1976. Population ecology of the Uganda kob (*Adenota kob thomasi* (Neumann)) in relation to the territorial system in the Rwenzori National park, Uganda. *Journal of Applied Ecology* **13**: 453–73.

Moe, S. R. and Wegge, R. 1994. Spacing behaviour and habitat use of axis deer (*Axis axis*) in lowland Nepal. *Canadian Journal of Zoology* **72**: 1735–44.

Mohan Raj, A. B., Moss, B. W., McCaughey, W. J. *et al.* 1991. Behavioural response to mixing of entire bulls, vasectomised bulls and steers. *Applied Animal Behaviour Science* **31**: 157–68.

Moll, A. 1897. *Untersuchungen über die Libido Sexualis*. Berlin: Fischer's Medicinische Buchhandlung, H. Kornfeld.

Møller, A. P. 1987 Behavioural aspects of sperm competition in swallows (*Hirundo rustica*). *Behaviour* **100**: 92–104.

Møller, A. P. 1993. A fungus infecting domestic flies manipulates sexual behaviour of its host. *Behavioural Ecology and Sociobiology* **33**: 403–7.

Møller, A. P. 1997. Immune defence, extra-pair paternity, and sexual selection in birds. *Proceedings of the Royal Society of London* B **264**: 561–6.

Møller, A. P. and Birkhead, T. R. 1993. Cuckoldry and sociality: a comparative study of birds. *American Naturalist* **142**: 118–40.

Molloy, P. L. 1986. Effects of DNA methylation on specific transcription by RNA polymerase II in vitro. *Molecular Biology and Reproduction* **11**: 13–17.

Molsher, R. L. 2006. *The ecology of feral cats,* Felis catus, *in open forest in New South Wales: interactions with food resources and foxes.* PhD Thesis, University of Sydney, Australia.

Mombaerts, P. 2006. Axonal wiring in the mouse olfactory system. *Annual Review of Cell Developmental Biology* **22**: 713–37.

Money, J. 1987. Propaedeutics of dioecious G-I/R: theoretical foundations for understanding dimorphic gender-identity/role. In: Reinisch, J. M., Rosenblum, L. A. and Sanders S. A. (Eds) *Masculinity/Femininity: Basic Perspectives,* pp. 13–28. New York: Oxford University Press.

Money, J. 1988. *Gay, Straight and In-between.* New York: Oxford University Press.

Money, J. 1990. Androgyne becomes bisexual in sexological theory: Plato to Freud and neuroscience. *Journal of the American Academy of Psychoanalysis* **18**: 392–413.

Money, J. and Ehrhardt, A. A. 1971. Fetal hormones and brain: effect on sexual dimorphism of behaviour – A review. *Archives of Sexual Behavior* **1**: 241–62.

Money, J. and Ehrhardt, A. A. 1996. *Man and Woman, Boy and Girl: Gender Identity from Conception to Maturity.* Northvale, NJ: Jason Aronson.

Money, J. and Ogunro, C. 1974. Behavioural sexology: ten cases of genetic male intersexuality with impaired prenatal and pubertal androgenisation. *Archives of Sexual Behavior* **3**: 181–205.

Money, J. and Russo, A. J. 1979. Homosexual outcome of discordant gender identity/role in childhood: longitudinal follow-up. *Journal of Pediatric Psychology* **4**: 29–41.

Money, J., Schwartz, M. and Lewis, V. G. 1984. Adult erotosexual status and fetal hormonal masculinisation and demasculinization: 46, XX congenital virilising adrenal hyperplasia and 46, XY androgen-insensitivity syndrome compared. *Psychoneuroendocrinology* **9**: 405–14.

Montell, H., Fridolfsson, A.-K. and Ellegren, H. 2001. Contrasting levels of nucleotide diversity on the avian Z and W sex chromosomes. *Molecular Biology and Evolution* **18**: 2010–16.

Moore, G. E. 1903. *Principia Ethica.* Cambridge: Cambridge University Press.

Moore, M. C. 1991. Application of organization-activation theory to alternative male reproductive strategies: a review. *Hormones and Behavior* **25**: 154–79.

Moore, N. P., Kelly, P. F., Cahill, J. P. and Hayden, T. J. 1995. Mating strategies and mating success of fallow (*Dama dama*) bucks in a non-lekking population. *Behavioral Ecology and Sociobiology* **36**: 91–100.

Mooring, M. S., Fitzpatrick, T. A., Benjamin, J. E. *et al.* 2003. Sexual segregation in desert bighorn sheep (*Ovis canadensis mexicana*). *Behaviour* **140**: 183–207.

Mooring, M. S., Reisig, D. D., Niemeyer, J. M. and Osborne, E. R. 2002. Sexually and developmentally dimorphic grooming: a comparative survey of the Ungulata. *Ethology* **108**: 911–34.

Mootnick, A. R. and Baker, E. 1994. Masturbation in captive *Hylobates* (Gibbons). *Zoo Biology* **13**: 345–53.

Moran, P.A.P. 1972. Familial effects in schizophrenia and homosexuality. *Australian and New Zealand Journal of Psychiatry* **6**: 116–19.

Morand, S. and Poulin, R. 1998. Density, body mass and parasite species richness of terrestrial mammals. *Evolutionary Ecology* **12**: 717–27.

Moreno, N. and González, A. 2004. Localization and connectivity of the lateral amygdala in anuran amphibians. *Journal of Comparative Neurology* **479**: 130–48.

Morgane, P. J., Galler, J. R. and Mokler, D. J. 2005. A review of systems and networks of the limbic forebrain/limbic midbrain. *Progress in Neurobiology* **75**: 143–60.

Morris, J. A., Gobrogge, K. L., Jordan, C. L. and Breedlove, S. M. 2004a. Brain aromatase: dyed-in-the-wool homosexuality. *Endocrinology* **145**: 475–7.

Morris, J. A., Jordan, C. L., and Breedlove, S. M. 2003. Medial amygdala volume is sexually dimorphic in mice. *Hormones and Behavior* **44**: 65.

Morris, J. A., Jordan, C. L., and Breedlove, S. M. 2004b. Sexual differentiation of the vertebrate nervous system. *Nature Neuroscience* **7**: 1034–9.

Morris, J. A., Jordan, C. L., Dugger, B. N. and Breedlove, S. M. 2005. Partial demasculinization of several brain regions in adult male (XY) rats with a dysfunctional androgen receptor gene. *Journal of Comparative Neurology* **487**: 217–26.

Morris, M. C., Rogers, P. A. and Kinghorn, G. R. 2001. Is bacterial vaginosis a sexually transmitted infection? *Sexually Transmitted Infections* **77**: 63–8.

Morris, R. J. 1990. Aikane: accounts of Hawaiian same-sex relationships in the journals of Captain Cook's Third Voyage (1776–80). *Journal of Homosexuality* **19**: 21–54.

Morrow, G. D. 1989. Bisexuality: an exploratory review. *Annals of Sex Research* **2**: 283–306.

Morton, S. R. 1978a. An ecological study of *Sminthopsis crassicaudata* (Marsupialia: Dasyuridae) II. Behaviour and social organization. *Australian Wildlife Research* **5**: 163–82.

Morton, S. R. 1978b. An ecological study of *Sminthopsis crassicaudata* (Marsupialia: Dasyuridae) III. Reproduction and life history. *Australian Wildlife Research* **5**: 183–211.

Mosher, W. D., Chandrs, A. and Jones, J. 2005. *Sexual behavior and selected health measures: men and women 15–44 years of age, United States, 2002. Advance Data From Vital and Health Statistic, no. 362*. Hyattsville, MD: National Center for Health Statistics.

Moynihan, M. 1990. Social, sexual and pseudosexual behaviour of the blue-bellied roller, *Coracias cyanogaster*: the consequences of crowding or concentration. *Smithsonian Contributions in Zoology* **491**: 1–21.

Moysey, E. D. 1997. A study of resource partitioning within the Helmeted Honeyeater *Lichenostomus melanops cassidix* during the non-breeding season. *Emu* **97**: 207–19.

Muether, V. R., Greene, E. and Ratcliffe, L. 1997. Delayed plumage maturation in Lazuli buntings: tests of the female mimicry and status signalling hypotheses. *Behavioural Ecology and Sociobiology* **41**: 281–90.

Mukandavire, Z., Chiyaka, C., Magombedze, G., Musuka, G. and Malunguza, N. J. 2009. Assessing the effects of homosexuals and bisexuals on the intrinsic dynamics of HIV/AIDS in heterosexual settings. *Mathematical and Computer Modelling* **49**: 1869–82.

Mulder, E. J. H., Robles de Medina, P. G., Huizink, A. C. *et al.* 2002. Prenatal maternal stress: effects on pregnancy and the (unborn) child. *Early Human Development* **70**: 3–14.

Müller, D., Roeder, F. and Orthner, H. 1973. Further results of stereotaxis in the human hypothalamus in sexual deviations. First use of this operation in addition to drugs. *Neurochirurgia* **16**: 113–26.

Muller, M. N. and Wrangham, R. 2002. Sexual mimicry in hyenas. *Quarterly Review of Biology* **77**: 3–16.

Muller, M. N. and Wrangham, R. W. 2004. Dominance, cortisol and stress in wild chimpanzees (*Pan troglodytes schweinfurthii*). *Behavioral Ecology and Sociobiology* **55**: 332–40.

Müller, U. 1996. H-Y antigens. *Human Genetics* **97**: 701–4.

Murai, T. 2004. Social behaviors of all-male proboscis monkeys when joined by females. *Ecological Research* **19**: 451–4.

Murkies, A. L., Wilcox, G. and Davis, S. R. 1998. Phytoestrogens. *Journal of Clinical Endocrinology and Metabolism* **83**: 297–303.

Murphey, R. M., Paranhos da Costa, M. J. R., de Souza Lima, L. O. and de Moura Duarte, F. A. 1991. Communal suckling in water buffalo (*Bubalus bubalis*). *Applied Animal Behaviour Science* **28**: 341–52.

Murphy, W. J., Eizirik, E., Johnson, W. E. *et al.* 2001. Molecular phylogenetics and the origins of placental mammals. *Nature* **409**: 614–18.

Murray, S. B. 1996. "We all love Charles": men in child care and the social construction of gender. *Gender and Society* **10**: 368–85.

Murton, R. K., Coombs, C. F. B. and Thearle, R.J.P. 1972. Ecological studies of the feral pigeon *Columba livia* var. II. Flock behaviour and social organization. *Journal of Applied Ecology* **9**: 875–89.

Muscarella, F. 2000. The evolution of homoerotic behavior in humans. *Journal of Homosexuality* **40**: 51–77.

Muscarella, F., Cevallos, A. M., Silver-Knogl, A. and Peterson, L. M. 2005. The alliance theory of homosexual behavior and the perception of social status and reproduction opportunities. *Neuroendocrinology Letters* **26**: 771–4.

Muscarella, F., Fink, B., Grammer, K. and Kirk-Smith, M. 2001. Homosexual orientation in males: evolutionary and ethological aspects. *Neuroendocrinology Letters* **22**: 393–400.

Mustanski, B. S., Bailey, J. M. and Kaspar, S. 2002. Dermatoglyphics, handedness, sex, and sexual orientation. *Archives of Sexual Behavior* **31**: 113–22.

Mustanski, B. S., Chivers, M. L. and Bailey, J. M. 2002. A critical review of recent biological research on human sexual orientation. *Annual Review of Sex Research* **13**: 89–140.

Mustanski, B. S., DuPree, M. G., Nievergelt, C. M. *et al.* 2005. A genomewide scan of male sexual orientation. *Human Genetics* **116**: 272–8.

Myers, J. P. 1979. Leks, sex, and buff-breasted sandpipers. *American Birds* **33**: 823–5.

Mysterud, A. 2000. The relationship between ecological segregation and sexual body size dimorphism in large herbivores. *Oecologia* **124**: 40–54.

Mysterud, A., Pérez-Barbería, F. J. and Gordon, I. J. 2001. The effect of season, sex and feeding style on home range areas versus body mass scaling in temperate ruminants. *Oecologia* **127**: 30–9.

Nachman, M. W. and Crowell, S. L. 2000. Estimate of the mutation rate per nucleotide in humans. *Genetics* **156**: 297–304.

Nacimiento, W., Töpper, R., Fischer, A. *et al.* 1993. B-50 (GAP-43) in Onuf's nucleus of the adult cat. *Brain Research* **613**: 80–7.

Nadler, R. D. 1986. Sex-related behaviour of immature wild mountain gorillas. *Developmental Psychobiology* **19**: 125–37.

Nadler, R. D. 1988. Sexual and reproductive behavior. In: Schwartz, J. H. (Ed.) *Orang-utan Biology*, pp. 105–16. New York: Oxford University Press.

Naftolin, F., Leranth, C., Horvath, T. L. and Garcia-Segura, L. M. 1996. Potential neuronal mechanisms of estrogen actions in synaptogenesis and synaptic plasticity. *Cellular and Molecular Neurobiology* **16**: 213–23.

Nagy, K. A. 2001. Food requirements of wild animals: predictive equations for free-living mammals, reptiles, and birds. *Nutrition Abstracts and Reviews*, Series B: *Livestock Feeds and Feeding* **71**: 1R–12R.

Najmabadi, A. 2001. Gendered transformations: beauty, love, and sexuality in Qajar Iran. *Iranian Studies* **34**: 89–102.

Nash, S. and Domjan, M. 1991. Learning to discriminate the sex of conspecifics in male Japanese quail (*Coturnix coturnix japonica*) tests of "biological constraints". *Journal of Experimental Psychology* **17**: 342–53.

Natoli, E., Baggio, A. and Pontier, D. 2001. Male and female agonistic and affiliative relationships in a social group of farm cats (*Felis catus L.*). *Behavioural Processes* **53**: 137–43.

Neave, N. and Menaged, M. 1999. Sex differences in cognition: the role of testosterone and sexual orientation. *Brain and Cognition* **41**: 245–62.

Nedergaard, M., Ransom, B. and Goldman, S. A. 2003. New roles for astrocytes: redefining the functional architecture of the brain. *Trends in Neurosciences* **26**: 523–30.

Neelakantan, K. K. 1968. Drumming by, and an instance of homo-sexual behaviour in, the Lesser Golden-backed Woodpecker (*Dinopium benghalense*). *Journal of the Bombay Natural History Society* **59**: 288–90.

Nefdt, R. J. C. 1995. Disruptions of matings, harassment and lek-breeding in Kafue letchwe antelope. *Animal Behaviour* **49**: 419–29.

Nekaris, A. and Bearder, S. K. 2007. The lorisiform primates of Asia and mainland Africa. In: Campbell, C. J., Fuentes, A., Mackinnon, K. C., Panger, M. and Bearder, S. K. (Eds) *Primates in Perspective*, pp. 24–45. Oxford: Oxford University Press.

Nelson, R. J., Demas, G. E. and Klein, S. L. 1998. Photoperiodic mediation of seasonal breeding and immune function in rodents: a multi-factorial approach. *American Zoologist* **38**: 226–37.

Neri, F., Chimini, L., Bonomi, F. *et al.* 2004a. Neuropsychological development of children born to patients with systemic lupus erythematosus. *Lupus* **13**: 805–11.

Neri, F., Chimini, L., Filippini, E., Motta, M., Faden, D. and Tincani, A. 2004b. Pregnancy in patients with rheumatic diseases: psychological implication of a chronic disease and neuropsychological evaluation of the children. *Lupus* **13**: 666–8.

Nesse, R. 2000. www.personal.umich.edu/~nesse/fourquestions.pdf.

Neuchterlein, G. L. and Storer, R. W. 1989. Reverse mounting in grebes. *Condor* **91**: 341–6.

Neuhaus, P. and Ruckstuhl, K. E. 2002a. Foraging behaviour in Alpine ibex (*Capra ibex*): consequences of reproductive status, body size, age and sex. *Ethology, Ecology and Evolution* **14**: 373–81.

Neuhaus, P. and Ruckstuhl, K. E. 2002b. The link between sexual dimorphism, activity budgets, and group cohesion: the case of the plains zebra (*Equus burchelli*). *Canadian Journal of Zoology* **80**: 1437–41.

Newbold, R. R. 1993. Gender-related behavior in women exposed prenatally to diethylstilbestrol. *Environmental Health Perspectives* **101**: 208–13.

Nicholas, C. L. 2004. Gaydar: eye-gaze as identity recognition among gay men and lesbians. *Sexuality and Culture* **8**: 60–86.

Nicoll, A. and Hamers, F. F. 2002. Are trends in HIV, gonorrhoea, and syphilis worsening in western Europe? *British Medical Journal* **324**: 1324–7.

Nicolosi, J. 1997. *Reparative Therapy of Male Homosexuality: A New Clinical Approach*. Northvale, NJ: Jason Aronson.

Niederkorn, J. Y. 2006. See no evil, hear no evil, do no evil: the lessons of immune privilege. *Nature Immunology* **7**: 354–9.

Nieminen, P., Mustonen, A.-M., Lindtröm-Seppä, P. *et al.* 2003. Phytosterols affect endocrinology and metabolism of the field vole (*Microtus agrestis*). *Experimental Biology and Medicine* **228**: 188–93.

Nieminen, P., Mustonen, A.-M., Päiväläinen, P. and Kukkonen, J. V. K. 2004. Reproduction of the tundra vole (*Microtus oeconomus*) with dietary phytosterol supplement. *Food and Chemical Toxicology* **42**: 945–51.

Nieuwenhuijsen, K., de Neef, K. J., van der Werff ten Bosch, J. J. and Slob, A. K. 1987. Testosterone, testis size, seasonality, and behaviour in group-living stumptail macaques (*Macaca arctoides*). *Hormones and Behavior* **21**: 153–69.

Nikaido, M., Kawai, K., Cao, Y., Harada, M., Tomita, S., Okada, N. and Hasegawa, M. 2001. Maximum likelihood analysis of the complete mitochondrial genomes of eutherians and a reevaluation of the phylogeny of bats and insectivores. *Journal of Molecular Evolution* **53**: 508–16.

Nisbet, I. C. T. and Medway, L. 1972. Dispersion, population ecology, and migration of Eastern Great Reed Warbler *Acrocephalus orientalis* wintering in Malaysia. *Ibis* **114**: 451–94.

Nishida, T. 1997. Sexual behavior of adult male chimpanzees of the Mahale Mountains National Park, Tanzania. *Primates* **38**: 379–98.

Nishikawa, K. C. 1997. Emergence of novel functions during brain evolution. *Bioscience* **47**: 341–54.

Niven, D. K. 1993. Male-male nesting behavior in Hooded Warblers. *Wilson Bulletin* **105**: 190–3.

Noble, G. K. and Wurm, M. 1943. The social behavior of the laughing gull. *Annals of the New York Academy of Sciences* **45**: 179–220.

Nonacs, P. 2007. Tug-of-war has no borders: it is the missing model in reproductive skew theory. *Evolution* **61**: 1244–50.

Nordeen, E. J., Nordeen, K. W., Sengelaub, D. R. and Arnold, A. P. 1985. Androgens prevent normally occurring cell death in a sexually dimorphic spinal nucleus. *Science* **229**: 671–3.

Nordgeen, J., Janczak, A. M. and Bakken, M. 2006. Effects of prenatal exposure to corticosterone on filial imprinting in the domestic chick, *Gallus gallus domesticus*. *Animal Behaviour* **72**: 1217–28.

Norman, J. E. and Ashley, M. V. 2000. Phylogenetics of Perissodactyla and tests of the molecular clock. *Journal of Molecular Evolution* **50**: 11–21.

Norris, K. S. and Dohl, T. P. 1980. Behavior of the Hawaiian spinner dolphin, *Stenella longirostris*. *Fishery Bulletin* **77**: 821–49.

Northcutt, R. G. and Kaas, J. H. 1995. The emergence and evolution of mammalian neocortex. *Trends in Neuroscience* **18**: 373–9.

Nottebohm, F. 1980. Testosterone triggers growth of brain vocal control nuclei in adult female canaries. *Brain Research* **189**: 429–36.

Nottebohm, F. 1985. Neuronal replacement in adulthood. *Annals of the New York Academy of Science* **457**: 143–61.

Novacek, M. J. 1992. Fossils, topologies, missing data, and the higher level phylogeny of eutherian mammals. *Systematic Biology* **41**: 58–73.

Novotny, M. 1987. The importance of chemical messengers in mammalian reproduction. In: Reinisch, J. M., Rosenblum, L. A. and Sanders, S. A. (Eds) *Masculinity/Femininity: Basic Perspectives*, pp. 107–28. New York: Oxford University Press.

Nowak, R. 1999. *Walker's Mammals of the World*. Baltimore and London: The Johns Hopkins University Press.

Nowak, R. and Paradiso, J. 1983. *Walker's Mammals of the World*. Baltimore, MD: The John Hopkins University Press.

Nudds, R. L. and Bryant, D. M. 2000. The energetic cost of short flights in birds. *Journal of Experimental Biology* **203**: 1561–72.

Nunn, C. L., Altizer, S., Jones, K. E. and Sechrest, W. 2003. Comparative tests of parasite species richness in primates. *American Naturalist* **162**: 597–614.

O, W.-S., Short, R. V., Renfree, M. B. and Shaw, G. 1988. Primary genetic control of somatic sexual differentiation in a mammal. *Nature* **331**: 716–17.

Oakwood, M. 2002. Spatial and social organization of a carnivorous marsupial *Dasyurus hallucatus* (Marsupialia: Dasyuridae). *Journal of Zoology, London* **257**: 237–48.

Oberhauser, K. S. and Frey, D. 1999. Coerced mating in monarch butterflies. In: Pisanty, I., Oberhauser, K., Merino, L. and Orice, S. (Eds) *Proceedings of the North American Conference on the Monarch Butterfly*, pp. 67–68. Montreal: Commission for Environmental Cooperation.

O'Dell, K. M., Armstrong, J. D., Yang, M. Y. and Kaiser, K. 1995. Functional dissection of the *Drosophila* mushroom bodies by selective feminization of genetically defined subcompartments. *Neuron* **15**: 55–61.

O'Grady, J. J., Brook, B. W., Reed, D. H. *et al.* 2006. Realistic levels of inbreeding depression strongly affect extinction risk in wild populations. *Biological Conservation* **133**: 42–51.

Ohno, S., Geller, L. N. and Young Lai, E. V. 1974. *Tfm* mutation and masculinization versus feminization of the mouse central nervous system. *Cell* **3**: 235–42.

Oi, T. 1990. Patterns of dominance and affiliation in wild pig-tailed macaques (*Macaca nemestrina nemestrina*) in West Sumatra. *International Journal of Primatology* **11**: 339–56.

Okada, A., Tamate, H. B., Minami, M., Ohnishi, N. and Takatsuki, S. 2005. Use of microsatellite markers to assess the spatial genetic structure of a population of sika deer *Cervus nippon* on Kinkazan Island, Japan. *Acta Theriologioca* **50**: 227–40.

Oktenli, C., Yesilova, Z., Kocar, I. H. *et al.* 2002. Study of autoimmunity in Klinefelter's syndrome and idiopathic hypogonadotropic hypogonadism. *Journal of Clinical Immunology* **22**: 137–43.

Oli, M. K. and Armitage, K. B. 2003. Sociality and individual fitness in yellow-bellied marmots: insights from a long-term study (1962–2001). *Oecologia* **136**: 543–50.

Oliva, D. 2002. Mating types in yeast, vomeronasal organ in rodents, homosexuality in humans: does a guiding thread exist? *Neuroendocrinology Letters* **23**: 287–8.

Olmos, F. and Silva e Silva, R. 2002. Breeding biology of the little blue heron (*Egretta caerulea*) in Southern Brazil. *Ornitologia Tropical* **13**: 17–30.

Olsen, K. M. 1985. Pair of apparently adult male kestrels. *British Birds* **78**: 452–3.

O'Neil, J. M. and Egan, J. 1992. Men's and women's gender role journeys: a metaphor for healing, transition, and transformation. In: Rubin Wainrib, B. (Ed.) *Gender Issues across the Life Cycle*, pp. 107–23. New York: Springer.

O'Neill, A. C., Fedigan, L. M. and Ziegler, T. E. 2004a. Ovarian cycle phase and same-sex mating behavior in Japanese macaque females. *American Journal of Primatology* **63**: 25–31.

O'Neill, A. C., Fedigan, L. M. and Ziegler, T. E. 2004b. Relationship between ovarian cycle phase and sexual behavior in female Japanese macaques (*Macaca fuscata*). *American Journal of Physical Anthropology* **125**: 852–62.

Orgeur, P. 1995. sexual play behavior in lambs androgenized in utero. *Physiology and Behavior* **57**: 185–7.

Orgeur, P., Mimouni, P. and Signoret, J. P. 1990. The influence of rearing conditions on the social relationships of young male goats (*Capra hircus*). *Applied Animal Behaviour Science* **27**: 105–13.

Orgeur, P. and Signoret, J. P. 1984. Sexual play and its functional significance in the domestic sheep (*Ovis aries* L.). *Physiology and Behavior* **33**: 111–18.

Orihuela, A. and Galina, C. S. 1997. Social order measured in pasture and pen conditions and its relationship to sexual behavior in Brahman (*Bos indicus*) cows. *Applied Animal Behaviour Science* **52**: 3–11.

Orwin, A., James, S. R. N. and Turner, R. K. 1974. Sex chromosome abnormalities, homosexuality and psychological treatment. *British Journal of Psychiatry* **124**: 293–5.

Östman, J. 1991. Changes in aggressive and sexual behavior between two male bottlenose dolphins (*Tursiops truncatus*) in a captive colony. In: Pryor, K. and Norris, K. S. (Eds) *Dolphin Societies: Discoveries and Puzzles*, pp. 305–17. Berkeley, CA: University of California Press.

Ostovich, J. M. and Sabini, J. 2005. Timing of puberty and sexuality in men and women. *Archives of Sexual Behavior* **34**: 197–206.

Ottinger, M. A., Abdelnabi, M. A., Henry, P. *et al.* 2001. Neuroendocrine and behavioral implications of endocrine disrupting chemicals in quail. *Hormones and Behavior* **40**: 234–47.

Ottinger, M. A., Quinn, M. J. Jr, Lavoie, E. *et al.* 2005. Neuroendocrine and behavioral consequences of embryonic exposure to endocrine disrupting chemicals. In: Dawson, A. and Sharp, P. J. (Eds) *Functional Avian Endocrinology*, pp. 1–14. New Delhi: Narosa Publishing House.

Owens, I. P. F. and Hartley, I. R. 1998. Sexual dimorphism in birds: why are there so many different forms of dimorphism? *Proceedings of the Royal Society of London*, B **265**: 397–407.

Owens, N. W. 1976. The development of sociosexual behaviour in free-living baboons, *Papio anubis*. *Behaviour* **57**: 241–59.

Packer, C. and Pusey, A. E. 1982. Cooperation and competition within coalitions of male lions: kin selection or game theory? *Nature* **296**: 740–2.

Packer, C. and Ruttan, L. 1988. The evolution of cooperative hunting. *American Naturalist* **132**: 159–98.

Pagel, M. D. and Harvey, P. H. 1988. Recent developments in the analysis of comparative data. *Quarterly Review of Biology* **63**: 413–40.

Pal, S. K., Ghosh, B. and Roy, S. 1998. Agonistic behaviour of free-ranging dogs (*Canis familiaris*) in relation to season, sex and age. *Applied Animal Behaviour Science* **59**: 331–48.

Pal, S. K., Ghosh, B. and Roy, S. 1999. Inter- and intra-sexual behaviour of free-ranging dogs (*Canis familiaris*). *Applied Animal Behaviour Science* **62**: 267–78.

Palombit, R. A. 1994. Extra-pair copulations in a monogamous ape. *Animal Behaviour* **47**: 721–3.

Panksepp, J., Nelson, E. and Bekkedal, M. 1997. Brain systems for the mediation of social separation-distress and social-reward. *Annals of the New York Academy of Sciences* **807**: 78–100.

Paoli, T., Palagi, E., Tacconi, G. and Borgognini Tarli, S. 2006. Perineal swelling, intermenstrual cycle, and female sexual behavior in bonobos (*Pan paniscus*). *American Journal of Primatology* **68**: 333–47.

Pardoll, D. M. 1999. Inducing autoimmune disease to treat cancer. *Proceedings of the National Academy of Sciences, USA* **96**: 5340–2.

Paredes, R. G. and Ågmo, A. 2004. Has dopamine a physiological role in the control of sexual behaviour? A critical review of the evidence. *Progress in Neurobiology* **73**: 179–226.

Paredes, R. G. and Baum, M. J. 1995. Altered sexual partner preference in male ferrets given excitotoxic lesions of the preoptic area/anterior hypothalamus. *Journal of Neuroscience* **15**: 6619–30.

Paredes, R. G., Tzschentke, T. and Nakach, N. 1998. Lesions of the medial preoptic area/anterior hypothalamus (MPOA/AH) modify partner preference in male rats. *Brain Research* **813**: 1–8.

Parish, A. R. 1994. Sex and food control in the "uncommon chimpanzee": How bonobo females overcome a phylogenetic legacy of male dominance. *Ethology and Sociobiology* **15**: 157–79.

Parish, A. 1996. Female relationships in bonobos (*Pan paniscus*): evidence for bonding, cooperation, and female dominance in a male-philopatric species. *Human Nature* **7**: 61–96.

Park, S.-R. 1991. Development of social structure in a captive colony of the common vampire bat *Desmodus rotundus*. *Ethology* **89**: 335–41.

Parker, G. A. and Pearson, R. G. 1976. A possible origin and adaptive significance of the mounting behaviour shown by some female mammals in oestrus. *Journal of Natural History* **10**: 241–5.

Parr, B. A., Shea, M. J., Vassileva, G. and McMahon, A. P. 1993. Mouse Wnt genes exhibit discrete domains of expression in the early embryonic CNS and limb buds. *Development* **119**: 247–61.

Parr, D. and Swyer, G.I.M. 1960. Seminal analysis in 22 homosexuals. *British Medical Journal* **2**: 1359–61.

Parr, R. 1980. Population study of Golden Plover *Pluvialis apricaria*, using marked birds. *Ornis Scandinavica* **11**: 179–89.

Parr, R. 1992. Sequential polyandry by Golden Plovers. *British Birds* **85**: 309.

Pati, B. P. 2001. Homosexuality in Asiatic lion: a case study from Gir National Park and Sanctuary. *Journal of the Bombay Natural History Society* **98**: 266.

Paton, T., Haddrath, O. and Baker, A. J. 2002. Complete mitochondrial DNA genome sequences show that modern birds are not descended from transitional shorebirds. *Proceedings of the Royal Society of London* B **269**: 839–46.

Pattatucci, A. M. L. and Hamer, D. H. 1995. Development and familiality of sexual orientation in females. *Behavior Genetics* **25**: 407–20.

Patterson, C. J. 2008. Sexual orientation across the life span: introduction to the special section. *Developmental Psychology* **44**: 1–4.

Paul, S. N., Kato, B. S., Cherkas, L. F., Andrew, T. and Spector, T. D. 2006. Heritability of the second to fourth digit ratio (2d:4d): a twin study. *Twin Research and Human Genetics* **9**: 215–19.

Paul, T., Schiffer, B., Zwarg, T. *et al*. 2008. Brain response to visual sexual stimuli in heterosexual and homosexual males. *Human Brain Mapping* **29**: 726–35.

Pawłowski, B., Lowen, C. B. and Dunbar, R. I. M. 1998. Neocortex size, social skills and mating success in primates. *Behaviour* **135**: 357–68.

Pearcey, S. M., Dochert, K. J. and Dabbs, J. M. 1996. Testosterone and sex role identification in lesbian couples. *Physiology and Behavior* **60**: 1033–5.

Pedersen, J. M., Glickman, S. E., Frank, L. G. and Beach, F. A. 1990. Sex differences in play behaviour of immature spotted hyenas *Crocuta crocuta*. *Hormones and Behavior* **24**: 403–20.

Pellegrini, A. D. 2004. Sexual segregation in childhood: a review of evidence for two hypotheses. *Animal Behaviour* **68**: 435–43.

Pellegrini, A. D., Long, J. D. and Mizerek, E. A. 2005. Sexual segregation in humans. In: Ruckstuhl, K. E. and Neuhaus, P. (Eds) *Sexual Segregation in Vertebrates: Ecology of the Two Sexes*, pp. 200–17. Cambridge: Cambridge University Press.

Pelletier, F. and Festa-Bianchet, M. 2006. Sexual selection and social rank in bighorn rams. *Animal Behaviour* **71**: 649–55.

Pellis, S. M. and Iwaniuk, A. N. 2000. Comparative analysis of the roles of postnatal development in the expression of play fighting in juveniles and adults. *Developmental Psychobiology* **36**: 136–47.

Pemberton, J. M., Coltman, D. W., Smith, J. A. and Pilkington, J. G. 1999. Molecular analysis of a promiscuous, fluctuating mating system. *Biological Journal of the Linnean Society* **68**: 289–301.

Penzhorn, B. L. 1984. A long-term study of social organization and behaviour of Cape Mountain zebras *Equus zebra zebra*. *Zeitschrift für Tierpsychologie* **64**: 97–146.

Penzhorn, B. L. 1988. *Equus zebra*. *Mammalian Species* **314**: 1–7.

Pepin, D., Lamerenx, F. and Chadelaud, H. 1996. Diurnal grouping and activity patterns of the Pyrenean chamois in winter. *Ethology, Ecology and Evolution* **8**: 135–45.

Peplau, L. A. 2001. Rethinking women's sexual orientation: an interdisciplinary relationship-focused approach. *Personal Relationships* **8**: 1–19.

Peplau, L. A., Cochran, S., Rook, K. and Padesky, C. 1978. Loving women: attachment and autonomy in lesbian relationships. *Journal of Social Issues* **34**: 7–27.

Peplau, L. A. and Garnets, L. D. 2000. A new paradigm for understanding women's sexuality and sexual orientation. *Journal of Social Issues* **56**: 329–50.

Peplau, L. A., Garnets, L. D., Spalding, L. R., Conley, T. D. and Veniegas, R. C. 1998. A critique of Bem's "Exotic becomes erotic" theory of sexual orientation. *Psychological Review* **105**: 387–94.

Peplau, L. A., Spalding, L. R., Conley, T. D. and Veniegas, R. C. 1999. The development of sexual orientation in women. *Annual Review of Sex Research* **10**: 70–99.

Pereira, S. L. and Baker, A. J. 2006. A molecular timescale for galliform birds accounting for uncertainty in time estimates and heterogeneity of rates of DNA substitutions across lineages and sites. *Molecular Phylogenetics and Evolution* **38**: 499–509.

Pérez-Barbería, F. J. and Gordon, I. J. 1999. Body size dimorphism and sexual segregation in polygynous ungulates: an experimental test with Soay sheep. *Oecologia* **120**: 258–67.

Pérez-Barbería, F. J. and Gordon, I. J. 2000. Differences in body mass and oral morphology between the sexes in the Artiodactyla: evolutionary relationships with sexual segregation. *Evolutionary Ecology Research* **2**: 667–84.

Pérez-Barbería, F. J. and Gordon, I. J. 2005. Gregariousness increases brain size in ungulates. *Oecologia* **145**: 41–52.

Pérez-Barbería, F. J., Gordon, I. J. and Pagel, M. 2002. The origins of sexual dimorphism in body size in ungulates. *Evolution* **56**: 1276–85.

Pérez-Barbería, F. J., Robertson, E. and Gordon, I. J. 2004. Are social factors sufficient to explain sexual segregation in ungulates? *Animal Behaviour* **69**: 827–34.

Pérez-Barbería, F. J., Robertson, E., Soriguer, R. *et al.* 2007a. Why do polygynous ungulates segregate in space? Testing the activity budget hypothesis in Soay sheep. *Ecological Monographs* **77**: 631–47.

Pérez-Barbería, F. J., Shulktz, S. and Dunbar, R.I.M. 2007b. Evidence for coevolution of sociality and relative brain size in three orders of mammals. *Evolution* **61**: 2811–21.

Perkins, A. and Fitzgerald, J. A. 1992. Luteinizing hormone, testosterone, and behavioural response of male-oriented rams to estrous ewes and rams. *Journal of Animal Science* **70**: 1787–94.

Perkins, A., Fitzgerald, J. A. and Moss, G. E. 1995. A comparison of LH secretion and brain estradiol receptors in heterosexual and homosexual rams and female sheep. *Hormones and Behavior* **29**: 31–41.

Perkins, A., Fitzgerald, J. A. and Price, E. O. 1992. Luteinizing hormone and testosterone response of sexually active and inactive rams. *Journal of Animal Science* **70**: 2086–93.

Perrett, D. I., Lee, K. J., Penton-Voak, I. *et al.* 1998. Effects of sexual dimorphism on facial attractiveness. *Nature* **394**: 884–7.

Perrin, W. F., Dolar, M. L. L., Chan, C. M. and Chivers, S. J. 2005. Length-weight relationships in the spinner dolphin (*Stenella longirostris*). *Marine Mammal Science* **21**: 765–78.

Perrin, W. F. and Mesnick, S. L. 2003. Sexual ecology of the spinner dolphin, *Stenella longirostris*: geographic variation in mating system. *Marine Mammal Science* **19**: 462–83.

Perrins, C. 2003. *The New Encyclopedia of Birds.* Oxford: Oxford University Press.

Perry, S. 1998. Male-male social relationships in wild white-faced capuchins, *Cebus capucinus*. *Behaviour* **135**: 139–72.

Peters, D. K. and Cantrell, P. J. 1993. Gender roles and role conflict in feminist lesbian and heterosexual women. *Sex Roles: A Journal of Research* **28**: 379–92.

Peterson, A. J. 1932. The breeding cycle in Bank Swallow. *Wilson Bulletin* **67**: 235–86.

Petronius. Satyricon. Translated by P. Dinnage. 1998. Hertfordshire, UK: Wordsworth Editions.

Pfaff, D. W. and Sakuma, Y. 1979. Deficit in the lordosis reflex of female rats caused by lesions in the ventromedial nucleus of the hypothalamus. *Journal of Physiology* **288**: 203–10.

Pfaus, J. G. 1999. Neurobiology of sexual behavior. *Current Opinion in Neurobiology* **9**: 751–8.

Pfaus, J. G., Kippin, T. E. and Centeno, S. 2001. Conditioning and sexual behaviour: a review. *Hormones and Behaviour* **40**: 291–321.

Pfaus, J. G., Kippin, T. E. and Coria-Avila, G. 2003. What can animal models tell us about human sexual response? *Annual Review of Sex Research* **14**: 1–63.

Phillips, G. and Over, R. 1995. Differences between heterosexual, bisexual, and lesbian women in recalled childhood experiences. *Archives of Sexual Behavior* **24**: 1–20.

Phillips, R. A., Silk, J. R. D., Phalan, B., Catry, P. and Croxall, J. P. 2004. Seasonal sexual segregation in two *Thalassarche* albatross species: competitive exclusion, reproductive role specialization or foraging niche divergence? *Proceedings of the Royal Society of London*, B **271**: 1283–91.

Phoenix, C. H., Goy, R. W., Gerall, A. A. and Young, W. C. 1959. Organizing action of prenatally administered testosterone propionate on the tissues mediating mating behaviour in the female guinea pig. *Endocrinology* **65**: 369–82.

Piaggio, A. J. and Spicer, G. S. 2001. Molecular phylogeny of the chipmunks inferred from mitochondrial cytochrome *b* and cytochrome oxidase II gene sequences. *Molecular Phylogenetics and Evolution* **20**: 335–50.

Pierotti, R. 1981. Male and female parental roles in the western gull under different environmental conditions. *Auk* **98**: 523–49.

Pigliucci, M. 2007. Do we need an extended evolutionary synthesis? *Evolution* **61**: 2743–9.

Pilastro, A., Pezzo, F., Olmastroni, S. *et al.* 2001. Extrapair paternity in the Adélie Penguin *Pygoscelis adeliae*. *Ibis* **143**: 681–4.

Pillard, R. C. 1990. The Kinsey Scale: is it Familial? In: McWhirter, D. P., Sanders, S. A. and Reisch, J. M. (Eds) *Homosexuality/Heterosexuality: Concepts of Sexual Orientation*, pp. 88–100. Oxford: Oxford University Press.

Pillard, R. C. and Bailey, J. M. 1998. Human sexual orientation has a heritable component. *Human Biology* **70**: 347–65.

Pillard, R. C. and Weinrich, J. D. 1986. Evidence of the familial nature of male homosexuality. *Archives of General Psychiatry* **43**: 808–12.

Pillsworth, E. G. and Haselton, M. G. 2006. Women's sexual strategies: the evolution of long-term bonds and extrapair sex. *Annual Review of Sex Research* **17**: 59–100.

Pilo Boyl, P., Signore M., Annino, A. *et al.* 2001. Otx genes in the development and evolution of the vertebrate brain. *International Journal of Developmental Neuroscience* **19**: 353–63.

Pinckard, K. L., Stellflug, J., Resko, J. A., Roselli, C. E. and Stormshak, F. 2000a. Brain aromatization and other factors affecting male reproductive behaviour with emphasis on the sexual orientation of rams. *Domestic Animal Endocrinology* **18**: 83–96.

Pinckard, K. L., Stellflug, J. and Stormshak, F. 2000b. Influence of castration and estrogen replacement on sexual behaviour of female-oriented, male-oriented, and asexual rams. *Journal of Animal Science* **78**: 1947–53.

Pinkerton, S. D., Bogart, L. M., Cecil, H. and Abramson, P. R. 2002. Factors associated with masturbation in a collegiate sample. *Journal of Psychology and Human Sexuality* **14**: 103–21.

Pinto, V. M., Tancredi, M. V., Neto, A. T. and Buchalla, C. M. 2005. Sexually transmitted disease/HIV risk behaviour among women who have sex with women. *AIDS* **19**(Suppl. 4): S64–S69.

Piper, K. P., McLarnon, A., Arrazi, J. *et al.* 2007. Functional HY-specific CD8$^+$ T cells are found in a high proportion of women following pregnancy with a male fetus. *Biology of Reproduction* **76**: 96–101.

Pitkow, L. J., Sharer, C. A., Ren, X. *et al.* 2001. Facilitation of affiliation and pair-bond formation by vasopressin receptor gene transfer into the ventral forebrain of a monogamous vole. *Journal of Neuroscience* **21**: 7392–6.

Pitman, R. L., Walker, W. A., Everett, W. T. and Gallo-Reynosa, J. P. 2004. Population status, food and foraging of Laysan albatross *Phoebastria immutabilis* nesting on Guadalupe Island, Mexico. *Marine Ornithology* **32**: 159–65.

Pitra, C., Fickel, J., Meijaard, E. and Groves, P. C. 2004. Evolution and phylogeny of old world deer. *Molecular Phylogenetics and Evolution* **33**: 880–95.

Place, N. J., Holekamp, K. E., Sisk, C. L. *et al.* 2002. Effects of prenatal treatment with androgens on luteinizing hormone secretion and sex steroid concentrations in adult spotted hyenas *Crocuta crocuta*. *Biology of Reproduction* **67**: 1405–13.

Platek, S. M. and Shackelford, T. K. 2006. *Female Infidelity and Paternal Uncertainty: Evolutionary Perspectives on Male Anti-cuckoldry tactics*. Cambridge: Cambridge University Press.

Platt, J. R. 1964. Strong inference. *Science* **146**: 347–53.

Ploog, D. W., Blitz, J. and Ploog, F. 1963. Studies on social and sexual behaviour of the squirrel monkey (*Saimiri sciureus*). *Folia Primatologica* **1**: 29–66.

Poasa, K. H., Blanchard, R. and Zucker, K. J. 2004. Birth order in transgendered males from Polynesia: a quantitative study of Samoan *Fa'afāfine*. *Journal of Sex and Marital Therapy* **30**: 13–23.

Poiani, A. 1994. Inter-generational competition and selection for helping behaviour. *Journal of Evolutionary Biology* **7**: 419–34.

Poiani, A. 1995. On bell miners, farmers and the role of metaphors in science: a reply to Loyn. *Emu* **95**: 147–8.

Poiani, A. 2006. Complexity of seminal fluid: a review. *Behavioral Ecology and Sociobiology* **60**: 289–310.

Poiani, A. 2008. Same-sex mounting in birds: comparative test of a Synthetic Reproductive Skew Model of homosexuality. *The Open Ornithology Journal* **1**: 36–45.

Polak, M. 2003. *Developmental Instability: Causes and Consequences.* Oxford: Oxford University Press.

Pollard, I. 1996. Preconceptual programming and sexual orientation: a hypothesis. *Journal of Theoretical Biology* **179**: 269–73.

Polziehn, R. O., Hamr, J., Mallory, F. F. and Strobeck, C. 2000. Microsatellite analysis of North American wapiti (*Cervus elaphus*) populations. *Molecular Ecology* **9**: 1561–76.

Pomerantz, S. M. 1990. Apomorphine facilitates male sexual behaviour of rhesus monkeys. *Pharmacology, Biochemistry and Behavior* **35**: 659–64.

Pomery, E. A., Gibbons, F. X., Gerrard, M. *et al.* 2005. Families and risk: prospective analysis of familial and social influences on adolescent substance use. *Journal of Family Psychology* **19**: 560–70.

Ponseti, J., Bosinski, H. A., Wolff, S. *et al.* 2006. A functional endophenotype for sexual orientation in humans. *NeuroImage* **33**: 825–33.

Ponseti, J., Siebner, H. R., Klöppel, S. *et al.* 2007. Homosexual women have less grey matter in perirhinal cortex than heterosexual women. *PLoS ONE* **2**: e762.

Ponton, L. E. and Judice, S. 2004. Typical adolescent sexual development. *Child and Adolescent Psychiatric Clinics of North America* **13**: 497–511.

Poole, W. E. 1973. A study of breeding grey kangaroos, *Macropus giganteus* Shaw and *M. fuliginosus* (Desmarest), in central New South Wales. *Australian Journal of Zoology* **21**: 183–212.

Poole, W. E. and Catling, P. C. 1974. Reproduction in the two species of grey kangaroos, *Macropus giganteus* Shaw and *M. fuliginosus* (Desmarest) I. Sexual maturity and oestrus. *Australian Journal of Zoology* **22**: 277–302.

Pope, L. C., Blair, D. and Johnson, C. N. 2005. Dispersal and population structure of the rufous bettong, *Aepyprymnus rufescens* (Marsupialia: Potoroidea). *Austral Ecology* **30**: 572–80.

Pope, N. S., Wilson, M. E. and Gordon, T. P. 1987. The effect of season on the induction of sexual behavior by estradiol in female rhesus monkeys. *Biology of Reproduction* **36**: 1047–54.

Pople, A., Cairns, S. C. and Grigg, G. C. 1991. Distribution and abundance of emus *Dromaius novaehollandiae* in relation to environment in the South Australian pastoral zone. *Emu* **91**: 222–9.

Popper, K. R. 1959. *The Logic of Scientific Discovery.* New York: Basic Books.

Popper, K. R. 1969. *Conjectures and Refutations.* London: Routledge and Kegan Paul.

Post, D. M., Armbrust, T. S., Horne, E. A. and Goheen, J. R. 2001. Sexual segregation results in differences in content and quality of bison (*Bos bison*) diets. *Journal of Mammalogy* **82**: 407–13.

Posthuma, D., De Geus, E. J. C., Neale, M. C. *et al.* 2000. Multivariate genetic analysis of brain structure in an extended twin design. *Behaviour Genetics* **30**: 311–19.

Prange, H. D., Anderson, J. F. and Rahn, H. 1979. Scaling of skeletal mass in birds and mammals. *American Naturalist* **113**: 103–22.

Pratt, D. M. and Anderson, V. H. 1985. Giraffe social behaviour. *Journal of Natural History* **19**: 771–81.

Prause, N. and Graham, C. A. 2007. Asexuality: classification and characterization. *Archives of Sexual Behavior* **36**: 341–56.

Preston, B. T., Stevenson, I. R., Pemberton, J. M. and Wilson, K. 2001. Dominant rams lose out by sperm depletion. *Nature* **409**: 681–2.

Preves, S. E. 2003. *Intersex Identity: The Contested Self.* New Brunswick, NJ: Rutgers University Press.

Prevett, J. P. and Barr, J. F. 1976. Lek behaviour of the Buff-breasted Sandpiper. *Wilson Bulletin* **88**: 500–3.

Prevett, J. P. and MacInness, C. D. 1980. Family and other social groups in snow geese. *Wildlife Monographs* **71**: 1–46.

Prévot-Julliard, A.-C., Pradel, R., Lebreton, J.-D. and Cézilly, F. 1998. Evidence for birth-site tenacity in breeding common Black-headed Gulls, *Larus ridibundus. Canadian Journal of Zoology* **76**: 2295–8.

Price, E. O., Borgwardt, R., Blackshaw, J. K. *et al.* 1994. Effect of early experience on the sexual performance of yearling rams. *Applied Animal Behaviour Science* **42**: 41–8.

Price, E. O., Katz, L. S., Wallach, S. J. R. and Zenchak, J. J. 1988. The relationship of male-male mounting to the sexual preference of young rams. *Applied Animal Behaviour Science* **21**: 347–55.

Price, J. S. and Hare, E. H. 1969. Birth order studies: some sources of bias. *British Journal of Psychiatry* **115**: 633–46.

Price, S. A., Bininda-Emonds, O. R. P. and Gittleman, J. L. 2005. A complete phylogeny of the whales, dolphins and even-toed hoofed mammals (Cetartiodactyla). *Biological Reviews* **80**: 445–73.

Priddy, K. D. 1997. Immunologic adaptations during pregnancy. *Journal of Obstetrics, Gynecology and Neonatal Nursing* **26**: 388–94.

Prior, H. M. and Walter, M. A. 1996. Sox genes: architects of development. *Molecular Medicine* **2**: 405–12.

Pryor, D. W., Weinberg, M. S. and Williams, C. J. 1995. *Dual Attraction: Understanding Bisexuality*. New York: Oxford University Press.

Psyllakis, J. M. and Brigham, R. M. 2006. Characteristics of diurnal roosts used by female *Myotis* bats in sub-boreal forests. *Forest Ecology and Management* **223**: 93–102.

Pugesek, B. H. 1983. The relationship between parental age and reproductive effort in the California gull (*Larus californicus*). *Behavioral Ecology and Sociobiology* **13**: 161–71.

Pullen, A. H., Tucker, D. and Martin, J. E. 1997. Morphological and morphometric characterisation of Onuf's nucleus in the spinal cord in man. *Journal of Anatomy* **191**: 201–13.

Pulliam, H. R. and Dunford, C. 1980. *Programmed to Learn: An Essay on the Evolution of Culture*. New York: Columbia University Press.

Purba, J. S., Hofman, M. A., Portegies, P., Troost, D. and Swaab, D. F. 1993. Decreased number of oxytocin neurons in the paraventricular nucleus of the human hypothalamus in AIDS. *Brain* **116**: 795–809.

Purcell, D. W., Blanchard, R. and Zucker, K. J. 2000. Birth order in a contemporary sample of gay men. *Archives of Sexual Behavior* **29**: 349–56.

Puris, K., Landgren, B. M., Cekan, Z. and Diczfalusy, E. 1976. Endocrine effects of masturbation in men. *Journal of Endocrinology* **70**: 439–44.

Purvis, A. 1995. A composite estimate of primate phylogeny. *Philosophical Transactions of the Royal Society of London*, B **348**: 405–21.

Putland, D. A. and Goldizen, A. W. 2001. Juvenile helping behaviour in the Dusky Moorhen, *Gallinula tenebrosa*. *Emu* **101**: 265–7.

Puts, D. A. and Dawood, K. 2006. The evolution of female orgasm: adaptation or byproduct? *Twin Research and Human Genetics* **9**: 467–72.

Puts, D. A., Jordan, C. L. and Breedlove, S. M. 2006. O brother, where art thou? The fraternal birth-order effect on male sexual orientation. *Proceedings of the National Academy of Sciences, USA* **103**: 10531–2.

Putz, D. A., Gaulin, S. J. C., Sporter, R. J. and McBurney, D. H. 2004. Sex hormones and finger length: what does 2D:4D indicate? *Evolution and Human Behavior* **25**: 182–99.

Quigley, C. A. 2002. The postnatal gonadotropin and sex steroid surge – insights from the androgen insensitivity syndrome. *Journal of Clinical Endocrinology and Metabolism* **87**: 24–8.

Quinn, T. W., Davies, J. C., Cooke, F. and White, B. N. 1989. Genetic analysis of offspring of a female-female pair in the lesser snow goose (*Chen c. caerulescens*). *Auk* **106**: 177–84.

Råberg, L. and Stjernman, M. 2003. Natural selection on immune responsiveness in blue tits *Parus caeruleus*. *Evolution* **57**: 1670–8.

Rado, S. 1940. A critical examination of the concept of bisexuality. *Psychosomatic Medicine* **2**: 459–67.

Raff, R. A. and Wray, G. A. 1989. Heterochrony: developmental mechanisms and evolutionary results. *Journal of Evolutionary Biology* **2**: 409–34.

Ragsdale, J. E. 1999. Reproductive skew theory extended: the effect of resource inheritance on social organization. *Evolutionary Ecology Research* **1**: 859–74.

Rahman, Q. 2005. The neurodevelopment of human sexual orientation. *Neuroscience and Biobehavioral Reviews* **29**: 1057–66.

Rahman, Q. 2005. The association between the fraternal birth order effect in male homosexuality and other markers of human sexual orientation. *Biological Letters* **1**: 393–5.

Rahman, Q., Andersson, D. and Govier, E. 2005. A specific sexual orientation-related difference in navigation strategy. *Behavioral Neuroscience* **119**: 311–16.

Rahman, Q., Clarke, K. and Morera, T. 2009. Hair whorl direction and sexual orientation in human males. *Behavioral Neuroscience* **123**: 252–6.

Rahman, Q., Collins, A., Morrison, M. *et al*. 2008. Maternal inheritance and familial fecundity factors in male homosexuality. *Archives of Sexual Behavior* **37**: 962–9.

Rahman, Q. and Hull, M. S. 2005. An empirical test of the kin selection hypothesis for male homosexuality. *Archives of Sexual Behavior* **34**: 461–7.

Rahman, Q., Kumari, V. and Wilson, G. D. 2003. Sexual orientation-related differences in prepulse inhibition of the human startle response. *Behavioural Neuroscience* **117**: 1096–102.

Rahman, Q. and Wilson, G. D. 2003a. Born gay? The psychobiology of human sexual orientation. *Personality and Individual Differences* **34**: 1337–82.

Rahman, Q. and Wilson, G. D. 2003b. Large sexual-orientation-related differences in performance on mental rotation and judgment of line orientation tasks. *Neuropsychology* **17**: 25–31.

Rahn, H., Sotherland, P. R. and Paganelli, C. V. 1985. Interrelationship between egg mass and adult body mass and metabolism among passerine birds. *Journal für Ornithologie* **126**: S263–S271.

Raikow, R. J. 1969. Sexual and agonistic behavior of the common rhea. *Wilson Bulletin* **81**: 196–206.

Raman, T. R. S. 1997. Factors influencing seasonal and monthly changes in the group size of chital or axis deer in southern India. *Journal of Biosciences* **22**: 203–18.

Raman, T. R. S. 1998. Antler cycles and breeding seasonality of the chital (*Axis axis*) in Southern India. *Journal of the Bombay Natural History Society* **95**: 377–91.

Ramo, C. 1993. Extra-pair copulations of grey herons nesting at high densities. *Ardea* **81**: 115–20.

Rampin, O., Bernabé, J. and Giuliano, F. 1997. Spinal control of penile erection. *World Journal of Urology* **15**: 2–13.

Rasa, O. A. E. 1977. The ethology and sociology of the dwarf mongoose (*Helogale undulata rufula*). *Zeitschrift für Tierpsychologie* **43**: 337–406.

Rasa, O. A. E. 1979. The effects of crowding on the social relationships and behaviour of the dwarf mongoose (*Helogale undulata rufula*). *Zeitschrift für Tierpsychologie* **49**: 317–29.

Rasa, O. A. E. and Lloyd, P. H. 1994. Incest avoidance and attainment of dominance by females in a Cape mountain zebra (*Equus zebra zebra*) population. *Behaviour* **128**: 169–88.

Rashid, H., Ormerod, S. and Day, E. 2007. Anabolic androgenic steroids: what the psychiatrist needs to know. *Advances in Psychiatric Treatment* **13**: 203–11.

Ravi, C. and Johnsingh, A. J. T. 1993. Management of Asiatic lions in the Gir Forest, India. *Symposia of the Zoological Society of London* **65**: 409–24.

Reader, S. M. 2003. Innovation and social learning: individual variation and brain evolution. *Animal Biology* **53**: 147–58.

Reader, S. M. and Laland, K. N. 2002. Social intelligence, innovation, and enhanced brain size in primates. *Proceedings of the National Academy of Sciences, USA* **99**: 4436–41.

Redouté, J., Stoléru, S., Grégoire, M.-C. *et al.* 2000. Brain processing of visual sexual stimuli in human males. *Human Brain Mapping* **11**: 162–77.

Reed, C., O'Brien, T. G. and Kinnaird, M. F. 1997. Male social behaviour and dominance hierarchy in the Sulawesi crested black macaque (*Macaca nigra*). *International Journal of Primatology* **18**: 247–60.

Reed-Eckert, M., Meaney, C. and Beauvais, G. P. 2004. *Species assessment for grizzly (brown) bear (Ursus arctos) in Wyoming*. Cheyenne, Wyoming: US Department of the Interior, Bureau of Land Management.

Reeve, H. K. and Emlen, S. T. 2000. Reproductive skew and group size: an N-person staying incentive model. *Behavioral Ecology* **11**: 640–7.

Reeve, H. K., Emlen, S. T. and Keller, L. 1998. Reproductive sharing in animal societies: reproductive incentives or incomplete control by dominant breeders? *Behavioral Ecology* **9**: 267–78.

Reeve, H. K. and Keller, L. 1995. Partitioning of reproduction in mother-daughter versus sibling associations: a test of optimal skew theory. *American Naturalist* **145**: 119–32.

Reeve, H. K. and Keller, L. 1996. Relatedness asymmetry and reproductive sharing in animal societies. *American Naturalist* **148**: 764–9.

Reeve, H. K. and Keller, L. 2001. Tests of reproductive-skew models in social insects. *Annual Reviews of Entomology* **46**: 347–85.

Reeve H. K. and Ratnieks F. L. W. 1993 Queen-queen conflicts in polygynous societies: mutual tolerance and reproductive skew. In: Keller, L. (Ed.) *Queen Number and Sociality in Insects*, pp. 45–85. Oxford: Oxford University Press.

Reeve, H. K. and Shen, S.-F. 2006. A mission model in reproductive skew theory: The bordered tug-of-war. *Proceedings of the National Academy of Sciences, USA* **103**: 8430–4.

Regan, P. C., Medina, R. and Joshi, A. 2001. Partner preferences among homosexual men and women: what is desirable in a sex partner is not necessarily desirable in a romantic partner. *Social Behavior and Personality* **29**: 625–34.

Reik, W. and Walter, J. 2001. Genomic imprinting: parental influence on the genome. *Nature Reviews* **2**: 21–32.

Reinhardt, V. 1983. Flehmen, mounting and copulation among members of a semi-wild cattle herd. *Animal Behaviour* **31**: 641–50.

Reinhardt, V. 1985. Social behaviour in a confined bison herd. *Behaviour* **92**: 209–26.

Reinhardt, V. and Flood, P. F. 1983. Behavioural assessment in muskox calves. *Behaviour* **87**: 1–21.

Reinhardt, V. and Reinhardt, A. 1981. Cohesive relationships in a cattle herd (*Bos indicus*). *Behaviour* **77**: 121–50.

Reinhardt, V., Reinhardt, A., Bercovitch, F. B. and Goy, R. W. 1986. Does intermale mounting function as a dominance demonstration in Rhesus monkeys? *Folia Primatologica* **47**: 55–60.

Reinisch, J. M., Mortensen, E. L. and Sanders, S. A. 2008. Prenatal exposure to progesterone and sexual orientation in humans. *Sexologies* **17**: S157.

Reiss, I. L. 1980. Sexual customs and gender roles in Sweden and America: an analysis and interpretation. In: Lopata, H. (Ed.) *Research in the Interwave of Social Roles: Women and Men*, pp. 191–220. Greenwich, CT: JIA Press.

Reite, M., Sheeder, J., Richardson, D. and Teale, P. 1995. Cerebral laterality in homosexual males: preliminary communication using magnetoencephalography. *Archives of Sexual Behavior* **24**: 585–93.

Reiter, E. O. and Grumbach, M. M. 1982. Neuroendocrine control mechanisms and the onset of puberty. *Annual Review of Physiology* **44**: 595–613.

Rendall, D. and Taylor, L. L. 1991. Female sexual behaviour in the absence of male-male competition in captive Japanese macaques (*Macaca fuscata*). *Zoo Biology* **10**: 319–28.

Rendall, D., Vasey, P. L. and McKenzie, J. 2008. The Queen's English: an alternative biosocial hypothesis for the distinctive features of "gay speech". *Archives of Sexual Behavior* **37**: 188–204.

Resko, J. A., Perkins, A., Roselli, C. E. *et al.* 1996. Endocrine correlates of partner preference behaviour in rams. *Biology of Reproduction* **55**: 120–6.

Reynolds, P. E. 1993. Dynamics of muskox groups in northeastern Alaska. *Rangifer* **13**: 83–9.

Reznikov, A. G., Nosenko, N. D., Tarasenko, L. V., Sinitsyn, P. V. and Polyakova, L. I. 2001. Early and long-term neuroendocrine effects of prenatal stress in male and female rats. *Neuroscience and Behavioral Physiology* **31**: 1–5.

Rhees, R. W. and Fleming, D. E. 1981. Effects of malnutrition, maternal stress, or ACTH injections during pregnancy on sexual behaviour of male offspring. *Physiology and Behavior* **27**: 879–82.

Rhees, R. W., Shryne, J. E. and Gorski, R. A. 1990. Termination of the hormone-sensitive period for differentiation of the sexually dimorphic nucleus of the preoptic area in male and female rats. *Developmental Brain Research* **52**: 17–23.

Rhodes G. 2006. The evolutionary psychology of facial beauty. *Annual Review of Psychology* **57**: 199–226.

Rice, G., Anderson, C., Risch, N. and Ebers, G. 1999. Male homosexuality: absence of linkage to microsatellite markers at Xq28. *Science* **284**: 665–7.

Rice, W. R. 1992. Sexually antagonistic genes: experimental evidence. *Science* **256**:1436–9.

Rice, W. R., Gavrilets, S. and Friberg, U. 2008. Sexually antagonistic "zygotic drive" of the sex chromosomes. *PLoS Genetics* **4**: e1000313.

Richard, A. 1974. Intra-specific variation in the social organization and ecology of *Propithecus verreauxi*. *Folia Primatologica* **22**: 178–207.

Richard, A. F. and Dewar, R. E. 1991. Lemur ecology. *Annual Review of Ecology and Systematics* **22**: 145–75.

Richardson, B., Hayes, R., Wheeler, S. and Yardin, M. 2002. Social structures, genetic structures and dispersal strategies in the Australian rabbit (*Oryctolagus cuniculus*) populations. *Behavioral Ecology and Sociobiology* **51**: 113–21.

Richardson, W. J., Miller, G. W., Davis, R. A. and Koski, W. R. 1995. Feeding, social and migration behavior of bowhead whales, *Balaena mysticetus*, in Baffin Bay vs. the Beaufort Sea – regions with different amounts of human activity. *Marine Mammal Science* **11**: 1–45.

Ridley, J. and Sutherland, W. J. 2002. Kin competition within groups: the offspring depreciation hypothesis. *Proceedings of the Royal Society of London, B* **169**: 1559–64.

Riedman, M. L. 1982. The evolution of alloparental care and adoption in mammals and birds. *Quarterly Review of Biology* **57**: 405–35.

Rieger, G., Chivers, M. L. and Bailey, J. M. 2005. Sexual arousal patterns of bisexual men. *Psychological Science* **16**: 579–84.

Rieger, G., Gygax, L., Linsenmeier, J. A. W. *et al.* 2009. Sex typicality and attractiveness in childhood and adulthood: assessing their relationships from videos. *Archives of Sexual Behavior*. DOI 10.1007/s10508-009-9512-8

Rieger, G., Linsenmeier, J. A. W., Gygax, L. and Bailey, J. M. 2008. Sexual orientation and childhood gender nonconformity: evidence from home videos. *Developmental Psychology* **44**: 46–58.

Rieger, G., Linsenmeier, J. A. W., Gygax, L., García, S. and Bailey, J. M. 2010. Dissecting "Gaydar": accuracy and the role of masculinity-femininity. *Archives of Sexual Behavior* **39**: 124–40.

Rijksen, H. 1978. *A Field Study on Sumatran Orang-utans (Pongo pygmaeus abelii, Lesson, 1827): Ecology, Behaviour and Conservation*. Wageningen, The Netherlands: Zeenman and Zonen.

Rilling, J. K. and Insel, T. R. 1999. Differential expansion of neural projection systems in primate brain evolution. *NeuroReport* **10**: 1453–59.

Ringrose, K. M. 2007. Eunuchs in historical perspective. *History Compass* **5**: 495–506.

Robarts, D. W. and Baum, M. J. 2007. Ventromedial hypothalamic nucleus lesions disrupt olfactory mate recognition and receptivity in female ferrets. *Hormones and Behavior* **51**: 104–13.

Robbins, M. M. 1996. Male-male interactions in heterosexual and all-male wild mammalian gorilla groups. *Ethology* **102**: 942–65.

Robbins, M. M., Sicotte, P. and Stewart, K. J. 2005. *Mountain Gorillas: Three Decades of Research at Karisoke*. Cambridge: Cambridge University Press.

Robbins, T. W. and Everitt, B. J. 1996. Neurobehavioural mechanisms of reward and motivation. *Current Opinion in Neurobiology* **6**: 228–36.

Roberts, M. R., Xie, S. and Mathialagan, N. 1996. Maternal recognition of pregnancy. *Biology of Reproduction* **54**: 214–302.

Robichaud, D., Lefebvre, L. and Robidoux, L. 1995. Dominance affects reseource partitioning in pigeons, but pair bonds do not. *Canadian Journal of Zoology* **74**: 833–40.

Robinson, J. G. and Janson, C. H. 1987. Capuchins, squirrel monkeys, and Atelines: socioecological convergence with old world primates. In: Smuts, B. B., Cheney, D. L., Seyfarth, R. M., Wrangham, R. W. and Struhsaker, T. T. (Eds) *Primate Societies*, pp. 69–82. Chicago, IL: Chicago University Press.

Robinson, S. J. and Manning, J. T. 2000. The ratio of 2nd to 4th digit length and male homosexuality. *Evolution and Human Behavior* **21**: 333–45.

Rochira, V., Balestrieri, A., Madeo, B. et al. 2001. Congenital estrogen deficiency: in search of the estrogen role in human male reproduction. *Molecular and Cellular Endocrinology* **178**: 107–15.

Rochira, V., Scaltriti, S., Zirilli, L. and Carani, C. 2008. Il ruolo degli estrogeni nel maschio. *Caleidoscopio Italiano* **216**: 1–95.

Rochira, V., Zirilli, L., Madeo, B. et al. 2004. Role of estrogens on human male reproductive system. *Journal of Reproduction and Contraception* **15**: 65–80.

Rockwell, R. F., Findlay, C. S. and Cooke, F. 1983. Life history studies of the lesser snow goose (*Anser caerulescens caerulescens*) I. The influence of age and time on fecundity. *Oecologia* **56**: 318–22.

Ródenas, J. M., Herranz, M. T. and Tercedor, J. 1998. Autoimmune progesterone dermatitis: treatment with oophorectomy. *British Journal of Dermatology* **139**: 508–11.

Rodriguez, I., Feinstein, P. and Mombaerts, P. 1999. Variable patterns of axinal projections of sensory neurons in the mouse vomeronasal system. *Cell* **97**: 199–208.

Rodriguez-Sierra, J. F. and Terasawa, E. 1979. Lesions of the preoptic area facilitate lordosis behaviour in male and female guinea pigs. *Brain Research Bulletin* **4**: 513–17.

Rodtian, P., King, G., Subrod, S. and Pongpiachan, P. 1996. Oestrus behaviour of Holstein cows during cooler and hotter tropical seasons. *Animal Reproduction Science* **45**: 47–58.

Røed, K. H., Holand, Ø., Gjostein, H. and Hansen, H. 2005. Variation in male reproductive success in a wild population of reindeer. *Journal of Wildlife Management* **69**: 1163–70.

Røed, K. H., Holand, Ø., Smith, M. E. et al. 2002. reproductive success in reindeer males in a herd with varying sex ratio. *Molecular Ecology* **11**: 1239–43.

Röell, A. 1979. Bigamy in jackdaws. *Ardea* **67**: 123–29.

Roese, N. J., Olson, J. M., Borenstein, M. N., Martin, A. and Sfores, A. L. 1992. Same-sex touching behaviour: the moderating role of homophobic attitudes. *Journal of Nonverbal Behavior* **16**: 249–59.

Rogers, L. J. and McCullock H. 1981. Pair-bonding in the Galah *Cacatua roseicapilla*. *Bird Behaviour* **3**: 80–92.

Rohde, P. A., Atzwanger, K., Butovskaya, M. et al. 2003. Perceived parental favouritism, closeness to kin, and the rebel of the family: the effects of birth order and sex. *Evolution and Human Behavior* **24**: 261–76.

Rohdes, G. 2006. The evolutionary psychology of facial beauty. *Annual Review of Psychology* **57**: 199–226.

Rohland, N., Pollack, J. L., Nagel, D. et al. 2005. The population history of extant and extinct hyenas. *Molecular Biology and Evolution* **22**: 2435–43.

Rohwer, S. 1978. Passerine subadult plumages and the deceptive acquisition of resources: test of a critical assumption. *Condor* **80**: 173–9.

Rohwer, S. 1986. A previously unknown plumage of first-year indigo buntings and theories of delayed plumage maturation. *Auk* **103**: 281–92.

Romano, J. E. and Benech, A. 1996. Effect of service and vaginal-cervical anesthesia on estrus duration in dairy goats. *Theriogenology* **45**: 691–6.

Roney, J. R., Whitham, J. C., Leoni, M. et al. 2004. Relative digit lengths and testosterone levels in Guinea baboons. *Hormones and Behavior* **45**: 285–90.

Rood, J. P. 1970. Ecology and social behaviour of the desert cavy (*Microcavia australis*). *American Midland Naturalist* **83**: 415–54.

Rood, J. P. 1972. Ecologicial and behavioural comparisons of three genera of Argentine cavies. *Animal Behaviour Monographs* **5**: 1–83.

Rood, J. P. 1980. Mating relationships and breeding suppression in the dwarf mongoose *Helogale parvula*. *Animal Behaviour* **28**: 143–50.

Roos, P. E. and Cohen, L. H. 1987. Sex roles and social support as moderators of life stress adjustments. *Journal of Personality and Social Psychology* **52**: 576–85.

Roper, W. G. 1996. The etiology of male homosexuality. *Medical Hypotheses* **46**: 85–8.

Rose, R. J., Kaprio, J., Winter, T. *et al.* 2002. Femininity and fertility in sisters with twin brothers: prenatal androgenization? Cross-sex socialization? *Psychological Science* **13**: 263–7.

Roselli, C. E. 2007. Brain aromatase: roles in reproduction and neuroprotection. *Journal of Steroid Biochemistry and Molecular Biology* **106**: 143–50.

Roselli, C. E., Klosterman, S. and Resko, J. A. 2001. Anatomic relationship between aromatase and androgen receptor mRNA expression in the hypothalamus and amygdala of adult male cynomolgus monkeys. *Journal of Comparative Neurology* **439**: 208–23.

Roselli, C. E., Larkin, K., Resko, J. A., Stellflug, J. N. and Stormshak, F. 2004a. The volume of a sexually dimorphic nucleus in the ovine medial preoptic area/anterior hypothalamus varies with sexual partner preference. *Endocrinology* **145**: 478–83.

Roselli, C. E., Larkin, K., Schrunk, J. M. and Stormshak, F. 2004b. Sexual partner preference, hypothalamic morphology and aromatase in rams. *Physiology and Behavior* **83**: 233–45.

Roselli, C. E., Resko, J. A. and Stormshak, F. 2002a. Hormonal influences on sexual partner preference in rams. *Archives of Sexual Behaviour* **31**: 43–9.

Roselli, C. E., Resko. J. A. and Stormshak, F. 2003. Estrogen synthesis in fetal sheep brain: effect of maternal treatment with an aromatase inhibitor. *Biology of Reproduction* **68**: 370–4.

Roselli, C. E., Resko, J. A. and Stormshak, F. 2006a. Expression of steroid hormone receptors in the fetal sheep brain during the critical period for sexual differentiation. *Brain Research* **1110**: 76–80.

Roselli, C. E., Schrunk, J. M., Stadelman, H. L., Resko, J. A. and Stormshak, F. 2006b. The effect of aromatase inhibition on the sexual differentiation of the sheep brain. *Endocrine* **29**: 501–2.

Roselli, C. E., Stadelman, H., Reeve, R., Bishop, C. V. and Stormshak, F. 2007. The ovine sexually dimorphic nucleus of the medial preoptic area is organized prenatally by testosterone. *Endocrinology* **148**: 4450–7.

Roselli, C. E. and Stormshak, F. 2009a. Prenatal programming of sexual partner preference: the ram model. *Journal of Neuroendocrinology* **21**: 359–364.

Roselli, C. E. and Stormshak, F. 2009b. The neurobiology of sexual partner preferences in rams. *Hormones and Behavior* **55**: 611–20.

Roselli, C. E., Stormshak, F., Stellflug, J. N. and Resko, J. A. 2002b. Relationship of serum testosterone concentrations to mate preferences in rams. *Biology of Reproduction* **67**: 263–8.

Røskaft, E., Järvi, T., Nyholm, N. E. I., Virolainen, M., Winkel, W. and Zang, H. 1986. Geographic variation in secondary sexual plumage colour characteristics of the male Pied Flycatcher. *Ornis Scandinavica* **17**: 293–8.

Ross, G., Sammaritano, L., Nass, R. and Lockshin, M. 2003. Effects of mothers' autoimmune disease during pregnancy on learning disabilities and hand preference in their children. *Archives of Pediatric Adolescent Medicine* **157**: 397–402.

Ross, M. W. 1981. Femininity, masculinity, and sexual orientation: some cross-cultural comparisons. *Journal of Homosexuality* **9**: 27–36.

Ross, M. W., Månsson, S.-A., Daneback, K., Cooper, A. and Tikkanen, R. 2005. Biases in internet sexual health samples: comparison of an internet sexuality survey and a national sexual health survey in Sweden. *Social Science & Medicine* **61**: 245–52.

Ross, M. W. and Wells A. L. 2000. The modernist fallacy in homosexual selection theory: homosexual and homosocial exaptation in South Asian society. *Psychology, Evolution and Gender* **2**: 253–62.

Rosselli, M., Reinhart, K., Imyhurn, B., Keller, P. J. and Dubey, R. K. 2000. Cellular and biochemical mechanisms by which environmental androgens influence reproductive function. *Human Reproduction Update* **6**: 332–50.

Rosser, A. M. 1992. Resource distribution, density, and determinants of mate access in puku. *Behavioral Ecology* **3**: 13–24.

Rothschild, B. M. and Rühli, F. J. 2005. Etiology of reactive arthritis in *Pan paniscus, P. troglodytes troglodytes*, and *P. troglodytes schweinfurthii*. *American Journal of Primatology* **66**: 219–31.

Rothstein, A. 1992. Linearity in dominance hierarchies: a third look at the individual attributes model. *Animal Behaviour* **43**: 684–6.

Rothstein, A. and Griswold, J. G. 1991. Age and sex preferences for social partners by juvenile bison bulls, *Bison bison*. *Animal Behaviour* **41**: 227–37.

Røttingen, J.-A., Cameron, W. and Garnett, G. P. 2001. A systematic review of the epidemiologic interactions between classic sexually transmitted diseases and HIV: how much really is known? *Sexually Transmitted Diseases* **28**: 579–97.

Roughgarden, J. 2004. *Evolution's Rainbow: Diversity, Gender, and Sexuality in Nature and People*. Berkeley, CA: University of California Press.

Rousselot, P., Heintz, N. and Nottenbohm F. 1997. Expression of brain lipid binding protein in the brain of the adult canary and its implications for adult neurogenesis. *Journal of Comparative Neurology* **385**: 415–26.

Rowell, T. E. 1972. Female reproductive cycles and social behaviour in primates. *Advances in the Study of Behavior* **4**: 69–105.

Rowell, T. E. 1973. Social organization of wild talapoin monkeys. *American Journal of Physical Anthropology* **38**: 593–8.

Rowley, I. 1990. *Behavioural Ecology of the Galah Eolophus roseicapillus in the Wheatbelt of Western Australia*. Chipping Norton: Surrey Beatty & Sons Pty. Limited, CSIRO and RAOU.

Ruckstuhl, K. E. 1998. Foraging behaviour and sexual segregation in bighorn sheep. *Animal Behaviour* **56**: 99–106.

Ruckstuhl, K. E. 2007. Sexual segregation in vertebrates: proximate and ultimate causes. *Integrative and Comparative Biology* **47**: 245–57.

Ruckstuhl, K. E. and Kokko, H. 2002. Modelling sexual segregation in ungulates: effects of group size, activity budgets and synchrony. *Animal Behaviour* **64**: 909–14.

Ruckstuhl, K. E., Manica, A., MacColl, A. D. C., Pilkington, J. G. and Clutton-Brock, T. H. 2006. The effects of castration, sex ratio and population density on social segregation and habitat use in Soay sheep. *Behavioral Ecology and Sociobiology* **59**: 694–703.

Ruckstuhl, K. E. and Neuhaus, P. 2000. Sexual segregation in ungulates: a new approach. *Behaviour* **137**: 361–77.

Ruckstuhl, K. E. and Neuhaus, P. 2002. Sexual segregation in ungulates: a comparative test of three hypotheses. *Biological Reviews* **77**: 77–96.

Ruckstuhl, K. E. and Neuhaus, P. (Eds) 2005. *Sexual Segregation in Vertebrates: Ecology of the Two Sexes*. Cambridge: Cambridge University Press.

Rue, L. L. III 1989. *The Deer of North America*. Danbury, CN: Outdoor Life Books.

Ruse, M. 1981. Are there gay genes? Sociobiology and homosexuality. *Journal of Homosexuality* **6**: 5–34.

Ruse, M. 1988. *Homosexuality*. Oxford: Basil Blackwell Ltd.

Russo, D. 2002. Elevation affects the distribution of the two sexes in Daubenton's bats *Myotis daubentonii* (Chiroptera: Vespertilionidae) from Italy. *Mammalia* **66**: 543–51.

Ryan, B. C. and Vandenbergh, J. G. 2002. Intrauterine position effects. *Neuroscience and Biobehavioral Reviews* **26**: 665–78.

Ryder, J. P. 1975. Egg-laying, egg size, and success in relation to immature-mature plumage of Ring-billed gulls. *Wilson Bulletin* **84**: 534–42.

Ryner, L. C., Goodwin, S. F., Castrillon, D. H. *et al*. 1996. Control of male sexual behaviour and sexual orientation in *Drosophila* by the *fruitless* gene. *Cell* **87**: 1079–89.

Saayman, G. S., 1975. The influence of hormonal and ecological factors upon sexual behavior and social organization in Old World primates. In: Tuttle, R. H. (Ed.) *Socioecology and Psychology of Primates*, pp. 181–204. New York: Mouton Publishers.

Sachser, N. and Kaiser, S. 1996. Prenatal social stress masculinizes the females' behaviour in guinea pigs. *Physiology and Behavior* **60**: 589–94.

Sadighi Akha, A. A. and Miller, R. A. 2005. Signal transduction in the aging immune system. *Current Opinion in Immunology* **17**: 486–91.

Saeki, J. 1997. From *nanshoku* to homosexuality: a comparative study of Mishima Yukio's *Confessions of a mask*. *Japan Review* **8**: 127–42.

Saetre, G. P. and Slagsvold, T. 1996. The significance of female mimicry in male contests. *American Naturalist* **147**: 981–95.

Safir, M. P., Sosenmann, A. and Kloner, O. 2003. Tomboyism, sexual orientation, and adult gender roles among Israeli women. *Sex Roles* **48**: 401–10.

Safron, A., Barch, B., Bailey, J. M. *et al*. 2007. Neural correlates of sexual arousal in homosexual and heterosexual men. *Behavioral Neuroscience* **121**: 237–48.

Sagarin, E. 1976. Prison homosexuality and its effect on post-prison sexual behavior. *Psychiatry* **39**: 245–57.

Saino, N., Romano, M. and Innocenti, P. 2006. Length of index and ring fingers differentially influence sexual attractiveness of men's and women's hands. *Behavioral Ecology and Sociobiology* **60**: 447–54.

Sakheim, D. K., Barlow, D. H., Beck, J. G. and Abrahamson, D. J. 1985. A comparison of male heterosexual and male homosexual patterns of sexual arousal. *Journal of Sex Research* **21**: 183–98.

Salais, D. A. and Fischer, R. B. 1995. Sexual preference and altruism. *Journal of Homosexuality* **28**: 185–96.

Salamon, E., Esch, T. and Stefano, G. B. 2005. Role of amygdala in mediating sexual and emotional behaviour via coupled nitric oxide release. *Acta Pharmacologica Sinica* **26**: 389–95.

Salter, R. E. 1979. Observations on social behaviour of Atlantic walruses (*Odobenus rosmarus* (L.)) during terrestrial haul-out. *Canadian Journal of Zoology* **58**: 461–3.

Sameroff, A. 1975. Transactional models in early social relations. *Human Development* **18**: 65–79.

Samuels, A. and Gifford, T. 1997. A quantitative assessment of dominance relations among bottlenose dolphins. *Marine Mammal Science* **13**: 70–99.

Sandberg, D. E., Meyer-Bahlburg, H. F. L., Ehrhardt, A. A. and Yager, T. J. 1993a. The prevalence of gender-atypical behavior in elementary school children. *Journal of the American Academy of Child and Adolescent Psychiatry* **32**: 306–14.

Sandberg, D. E., Meyer-Bahlburg, H. F. L. and Yager, T. J. 1993b. Feminine gender role behavior and academic achievement: Their relation in a community sample of middle childhood boys. *Sex Roles* **29**: 125–40.

Sandberg, D. E., Vena, J. E., Weiner, J. *et al.* 2003. Hormonally active agents in the environment and children's behavior: assessing effects on children's gender-dimorphic outcomes. *Epidemiology* **14**: 148–55.

Sanders, G. and Ross-Field, L. 1986. Sexual orientation and visuo-spatial ability. *Brain and Cognition* **5**: 280–90.

Sanders, S. A., Graham, C. A. and Milhausen, R. R. 2008. Bisexual women differ from lesbian and heterosexual women on several sexuality measures. *Sexologies* **17** (Suppl. 1): S157–S158.

Sanderson, J. T. 2006. The steroid hormone biosynthesis pathway as a target for endocrine-disrupting chemicals. *Toxicological Sciences* **94**: 3–21.

Sandfort, T. G. M. and Cohen-Kettenis, P. T. 2000. Sexual behavior in Dutch and Belgian children as observed by their mothers. *Journal of Psychology & Human Sexuality* **12**: 105–15.

Sandvik, H., Erikstad, K., Barrett, R. T. and Yoccoz, N. G. 2005. The effect of climate and adult survival in five species of North Atlantic seabirds. *Journal of Animal Ecology* **74**: 817–31.

Sannen, A. 2003. *Testosterone and behaviour in bonobos: a male hormone in a female-centred society*. PhD Thesis, Antwerp University, Antwerp.

Sannen, A., van Elsacker, L., Heistermann, M. and Eens, M. 2005. Certain aspects of bonobo female sexual repertoire are related to urinary testosterone metabolic levels. *Folia Primatologica* **76**: 21–32.

Santtila, P., Högbacka, A.-L., Jern, P. *et al.* 2009. Testing Miller's theory of alleles preventing androgenization as an evolutionary explanation for the genetic predisposition for male homosexuality. *Evolution and Human Behavior* **30**: 58–65.

Santtila, P., Sandnabba, N. K., Harlaar, N. *et al.* 2008. Potential for homosexual response is prevalent and genetic. *Biological Psychology* **77**: 102–5.

Sarangthem, K. and Singh, T. N. 2002. Biological activities of phytosterols: a review. *Journal of Phytological Research* **15**: 243–50.

Sastry, P. S. and Subba Rao, K. 2000. Apoptosis and the nervous system. *Journal of Neurochemistry* **74**: 1–20.

Sato, T., Matsumoto, T., Kawano, H. *et al.* 2004. Brain masculinization requires androgen receptor function. *Proceedings of the National Academy of Sciences, USA* **101**: 1673–78.

Sauer, E.G.F. 1972. Aberrant sexual behaviour in the South African Ostrich. *Auk* **89**: 717–37.

Saunders, F. C., McElligott, A. G., Safi, K. and Hayden, T. J. 2005. Mating tactics of male feral goats (*Capra hircus*): risks and benefits. *Acta Ethologica* **8**: 103–10.

Saunders, M. B. and Barclay, R. M. R. 1992. Ecomorphology of insectivorous bats: a test of predictions using two morphologically similar species. *Ecology* **73**: 1335–45.

Savic, I. 2002. Sex differentiated hypothalamic activation by putative pheromones. *Molecular Psychiatry* **7**: 335–6.

Savic, I., Berglund, H. and Linström, P. 2005. Brain response to putative pheromones in homosexual men. *Proceedings of the National Academy of Sciences, USA* **102**: 7356–61.

Savic, I. and Lindström, P. 2008. PET and MRI show differences in cerebral asymmetry and functional connectivity between homo- and heterosexual subjects. *Proceedings of the National Academy of Sciences, USA* **105**: 9403–8.

Savin-Williams, R. C. and Ream, G. L. 2006. Pubertal onset and sexual orientation in an adolescent national probability sample. *Archive of Sexual Behavior* **35**: 279–86.

Savolainen, V. and Lehman, L. 2007. Genetics and bisexuality. *Nature* **445**: 158–9.

Say, L., Naulty, F. and Hayden, T. J. 2003. Genetic and behavioural estimates of reproductive skew in male fallow deer. *Molecular Ecology* **12**: 2793–800.

Sayag, N., Snapir, N., Robinzon, B. *et al.* 1989. Embryonic sex steroids affect mating behaviour and plasma LH in adult chickens. *Physiology and Behaviour* **45**: 1107–12.

Scandura, M. 2004. *The use of microsatellites in the study of social structure in large mammals: Italian wolf and fallow deer as case studies*. PhD Thesis, University of Bielefeld, Germany.

Schaller, G. B. 1967. *The Deer and the Tiger: A Study of Wildlife in India*. Chicago, IL: University of Chicago Press.

Schaller, G. B. 1972. *The Serengeti Lion: A Study of Predator-Prey Relations*. Chicago, IL: University of Chicago Press.

Schaller, G. B. and Hamer, A. 1978. Rutting behaviour of Pere Davids deer, *Elaphurus davidianus*. *Der Zoologische Garten* **48**: 1–15.

Schell, D. M., Saupe, S. M. and Haubenstock, N. 1989. Bowhead whale (*Balaena mysticetus*) growth and feeding as estimated by $\delta^{13}C$ techniques. *Marine Biology* **103**: 433–43.

Schilder, M. B. H. 1988. Dominance relationships between adult plains zebra stallions in semi-captivity. *Behaviour* **104**: 300–19.

Schilder, M. B. H. and Boer, P. L. 1987. Ethological investigations on a herd of plains zebra in a safari park: time-budgets, reproduction and food competition. *Applied Animal Behaviour Science* **18**: 45–56.

Schluter, D. and Repasky, R. R. 1991. Worldwide limitation of finch densities by food and other factors. *Ecology* **72**: 1763–74.

Schmidt, G. and Clement, U. 1995. Does peace prevent homosexuality? *Journal of Homosexuality* **28**: 269–75.

Schmitt, L. H., Bradley, A. J., Kemper, C. M. *et al.* 1989. Ecology and physiology of the northern quoll *Dasyurus hallucatus* (Marsupialia, Dasyuridae), at Mitchell Plateau, Kimberley, Western Australia. *Journal of Zoology, London* **217**: 539–58.

Schreiber, R. W. 1970. Breeding biology of Western gulls (*Larus occidentalis*) on San Nicolas Island, California, 1968. *Condor* **72**: 133–40.

Schreiner, C. and Kling, A. 1953. Behavioral changes following rhinencephalic injury in cat. *Journal of Neurophysiology* **16**: 643–59.

Schreiner, C. and Kling, A. 1956. Rhinencephalon and behaviour. *American Journal of Physiology* **184**: 268–78.

Schuiling, G. A. 2003. The benefit and the doubt: why monogamy? *Journal of Psychosomatic Obstetrics and Gynecology* **24**: 55–61.

Schuiling, G. A. 2004. Death in Venice: the homosexuality enigma. *Journal of Psychosomatic Obstetrics and Gynecology* **25**: 67–76.

Schuiling, G. A. 2005. On sexual behavior and sex-role reversal. *Journal of Psychosomatic Obstetrics and Gynecology* **26**: 217–23.

Schwagmeyer, P. L., St. Clair, R. C., Moodie, J. D. *et al.* 1999. Species differences in male parental care in birds: a re-examination of correlates with paternity. *Auk* **116**: 487–503.

Schwartz, G., Kim, R. M., Kolundzija, A. B., Rieger, G. and Sanders, A. R. 2010. Biodemographic and physical correlates of sexual orientation in men. *Archives of Sexual Behavior* **39**: 90–109.

Schwartz, J. H. 1987. *The Red Ape: Orang-utans and Human Origins*. London: Elm Tree Books.

Schwartz, O. A., Armitage, K. B. and van Vuren, D. 1998. A 32-year demography of yellow-bellied marmots (*Marmota flaviventris*). *Journal of Zoology, London* **246**: 337–46.

Scollay, P. A. and Judge, P. 1981. The dynamics of social organization in a population of squirrel monkeys (*Saimiri sciureus*) in a seminatural environment. *Primates* **22**: 60–9.

Scorolli, C., Ghirlanda, S., Enquist, M., Zattoni, S. and Jannini, E. A. 2007. Relative prevalence of different fetishes. *International Journal of Impotence Research* **19**: 432–7.

Scott, C. J., Clarke, I. J., Rao, A. and Tilbrook, A. J. 2004. Sex differences in the distribution and abundance of androgen receptor mRNA-containing cells in the preoptic area and hypothalamus of the ram and ewe. *Journal of Neuroendocrinology* **16**: 956–63.

Seaborg, D. M. 1984. Sexual orientation, behavioral plasticity, and evolution. *Journal of Homosexuality* **10**: 153–8.

Seavey, M. M. and Mosmann, T. R. 2006. Paternal antigen-bearing cells transferred during insemination do not stimulate anti-paternal $CD8^+$ T cells: role of estradiol in locally inhibiting $CD8^+$ T cell responses. *Journal of Immunology* **177**: 7567–78.

Segarra, A. C. and Strand, F. L. 1989. Perinatal administration of nicotine alters subsequent sexual behavior and testosterone levels of male rats. *Brain Research* **480**: 151–9.

Segovia, S. and Guillamón, A. 1993. Sexual dimorphism in the vomeronasal pathway and sex differences in reproductive behaviours. *Brain Research Reviews* **18**: 51–74.

Seguin, P., Zheng, W. and Souleimanov, A. 2004. Alfalfa phytoestrogen content: impact of plant maturity and herbage components. *Journal of Agronomy and Crop Science* **190**: 211–17.

Sell, R. 1996. The Sell assessment of sexual orientation: background and scoring. *Journal of Gay, Lesbian, and Bisexual Identity*. **1**: 295–310.

Sell, R. L. 1997. Defining and measuring sexual orientation: a review. *Archives of Sexual Behavior* **26**: 643–58.

Seney, M., Goldman, B. D. and Forger, N. G. 2006. Breeding status affects motoneuron number and muscle size in naked mole-rats: recruitment of perineal motoneurons? *Journal of Neurobiology* **66**: 1354–64.

Sergeant, M. J. T., Dickins, T. E., Davies, M. N. O. and Griffiths, M. D. 2006. Aggression, empathy and sexual orientation in males. *Personality and Individual Differences* **40**: 475–86.

Sergeant, M. J. T., Dickins, T. E., Davies, M. N. O. and Griffiths, M. D. 2007. Women's hedonic ratings of body odor of heterosexual and homosexual men. *Archives of Sexual Behavior* **36**: 395–401.

Sergeant, M. J. T., Louie, J. and Wysocki, C. J. 2008. The influence of sexual orientation on human olfactory function. In: Hurst, J. L., Beynon, R. J., Roberts, S. C. and Wyatts, T. D. (Eds) *Chemical Signals in Vertebrates*, Vol. 2, pp. 121–8. New York: Springer.

Sergios, P. and Cody, J. 1986. Importance of physical attractiveness and social assertiveness skills in male homosexual dating behavior and partner selection. *Journal of Homosexuality* **12**: 71–84.

Serrano, J. M., Castro, L., Toro, M. A. and López-Fanjul, C. 1991. The genetic properties of homosexual copulation behavior in *Tribolium castaneum*: diallele analysis. *Behavior Genetics* **21**: 547–58.

Shackleton, D. M. 1991. Social maturation and productivity in bighorn sheep: are young males incompetent? *Applied Animal Behaviour Science* **29**: 173–84.

Shackleton, D. M. and Shank, C. C. 1984. A review of the social behavior of feral and wild sheep and goats. *Journal of Animal Science* **58**: 500–9.

Shane, S. H., Wells, R. S. and Würsig, B. 1986. Ecology, behaviour and social organization of the bottlenose dolphin: a review. *Marine Mammal Science* **2**: 34–63.

Shaner, D. E. and Hutchinson, R. D. 1990. Neuroplasticity and temporal retardation of development (paedomorphic morphology) in human evolution: a consideration of biological requirements for the plasticity of human cognition and the potential acquisition of culturally dependent ethical world views. *Human Evolution* **5**: 175–91.

Shannon, P. W. 2000. Social and reproductive relationships of captive Caribbean flamingos. *Waterbirds* **23**: 173–8.

Shaw, J. S. 1982. Psychological androgyny and stressful life events. *Journal of Personality and Social Psychology* **43**: 145–53.

Shearer, M. K. and Katz, L. S. 2006. Female-female mounting among goats stimulates sexual performance in males. *Hormones and Behavior* **50**: 33–7.

Sheldon, B. C. 1994. Sperm competition in the chaffinch: the role of the female. *Animal Behaviour* **47**: 163–73.

Sheldon, B. C. and Burke, T. 1994. Copulation behavior and paternity in the chaffinch. *Behavioral Ecology and Sociobiology* **34**: 149–56.

Sheldon, F. H. and Slikas, B. 1997. Advances in ciconiiform systematics 1976–1996. *Colonial Waterbirds* **20**: 106–14.

Sheldon, F. H., Whittingham, L. A., Moyle, R. G., Slikas, B. and Winkler, D. W. 2005. Phylogeny of swallows (Aves: Hirundinidae) estimated from nuclear and mitochondrial DNA sequences. *Molecular Phylogenetics and Evolution* **35**: 254–70.

Shen, L., Kondo, Y., Guo, Y. *et al.* 2007. Genome-wide profiling of DNA methylation reveals a class of normally methylated CpG island promoters. *PLoS Genetics* **3**: e181.

Shepherd, L. D., Millar, C. D., Ballard, G. *et al.* 2005. Microevolution and mega-icebergs in the Antarctic. *Proceedings of the National Academy of Sciences, USA* **102**: 16717–22.

Shepperdson, B. 1995. The control of sexuality in young people with Down's syndrome. *Child: Care, Health and Development* **21**: 333–49.

Shernoff, M. 1996. Gay men choosing to be fathers. *Journal of Gay and Lesbian Social Services* **4**: 41–54.

Shifren, J. L. 2002. Androgen deficiency in the oophorectomized woman. *Fertility and Sterility*, Suppl. **4**, 77: S60–S62.

Shimizu, K., Udono, T., Tanaka, C. *et al.* 2003. Comparative study of urinary reproductive hormones in great apes. *Primates* **44**: 183–90.

Shireman, J. F. 1996. Single parent adoptive homes. *Children and Youth Services Review* **18**: 23–36.

Shively, M. G. and De Cecco, J. P. 1977 Components of sexual identity. *Journal of Homosexuality* **3**: 41–8.

Sholl, S. A. and Kim, K. L. 1989. Estrogen receptors in the rhesus monkey brain during fetal development. *Developmental Brain Research* **50**: 189–96.

Shugart, G. W. 1980. Frequency and distribution of polygyny in Great Lakes herring gulls in 1978. *Condor* **82**: 426–9.

Shugart, G. W., Fitch, M. A. and Fox, G. A. 1987. Female floaters and nonbreeding secondary females in herring gulls. *Condor* **89**: 902–6.

Shugart, G. W., Fitch, M. A. and Fox, G. A. 1988. Female pairing: a reproductive strategy for herring gulls? *Condor* **90**: 933–5.

Sibley, C. G. and Ahlquist, J. E. 1990. *Phylogeny and Classification of Birds. A Study in Molecular Evolution.* New Haven, CT: Yale University Press.

Sick, H. 1967. Courtship behavior in the manakins (Pipridae): a review. *Living Bird* **6**: 5–22.

Siegal, M. L. and Bergman, A. 2002. Waddington's canalization revisited: developmental stability and evolution. *Proceedings of the National Academy of Sciences, USA* **99**: 10528–32.

Siegenthaler, A. L. and Bigner, J. J. 2000. The value of children to lesbian and non-lesbian mothers. *Journal of Homosexuality* **39**: 73–91.

Sigusch, V., Schorsch, E., Dannecker, M. and Schmidt, G. 1982. Official statement by the German Society for Sex Research (Deutsche Gesellschaft für Sexualforschung e.V.) on the research of Prof. Dr. Günter Dörner on the subject of homosexuality. *Archives of Sexual Behavior* **11**: 445–9.

Silk, J. B. 1993. Does participation in coalitions influence dominance relationships among male bonnet macaques? *Behaviour* **126**: 171–89.

Silk, J. B. 1994. Social relationships of male bonnet macaques: male bonding in a matrilineal society. *Behaviour* **130**: 271–91.

Silk, J. B., Altmann, J. and Alberts, S. C. 2006a Social relationships among adult female baboons (*Papio cynocephalus*) I. Variation in the strength of social bonds. *Behavioral Ecology and Sociobiology* **61**: 183–95.

Silk, J. B., Alberts, S. C. and Altmann, J. 2006b. Social relationships among adult female baboons (*Papio cynocephalus*) II. Variation in the quality and stability of social bonds. *Behavioral Ecology and Sociobiology* **61**: 197–204.

Silva, J. M. Jr, Silva, F. J. L. and Sazima, I. 2005. Rest, nurture, sex, release and play: diurnal underwater behaviour of the spinner dolphin at Fernando de Noronha Archipelago, SW Atlantic. *AQUA, Journal of Ichthyology and Aquatic Biology* **9**: 161–76.

Silva, M. and Downing, J. A. 1995. *CRC Handbook of Mammalian Body Masses*. Boca Ratón, FL: CRC Press.

Silverthorne, Z. A. and Quinsey, V. L. 2000. Sexual partner age preferences of homosexual and heterosexual men and women. *Archives of Sexual Behavior* **29**: 67–76.

Simeone, A. 1998. *Otx1* and *Otx2* in the development and evolution of the mammalian brain. *EMBO Journal* **17**: 6790–8.

Simerly, R. B. 2002. Wired for reproduction: organization and development of sexually dimorphic circuits in the mammalian forebrain. *Annual Review of Neuroscience* **25**: 507–36.

Simmons, L. W., Firman, R. C., Rhodes, G. and Peters, M. 2004. Human sperm competition: testis size, sperm production and rates of extrapair copulations. *Animal Behaviour* **68**: 297–302.

Simoncini, T. and Genazzani, A. R. 2003. Non-genomic actions of sex steroid hormones. *European Journal of Endocrinology* **148**: 281–92.

Simonds, P. E. 1974. Sex differences in bonnet macaque networks and social structure. *Archives of Sexual Behavior* **3**: 151–66.

Simpson, G. G. 1945. The principles of classification and a classification of mammals. *Bulletin of the American Museum of Natural History* **85**: 1–350.

Singal, R. and Ginder, G. D. 1999. DNA methylation. *Blood* **93**: 4059–70.

Singh, D., Vidaurri, M., Zambarano, R. J. and Dabbs, J. M. Jr. 1999. Lesbian erotic role identification: behavioral, morphological, and hormonal correlates. *Journal of Personality and Social Psychology* **76**: 1035–49.

Singh, J. and Verma, I. C. 1987. Influence of major histo(in)compatibility complex on reproduction. *American Journal of Reproductive Immunology and Microbiology* **15**: 150–2.

Sjare, B. and Stirling, I. 1996. The breeding behaviour of Atlantic walruses, *Odobenus rosmarus rosmarus*, in the Canadian High Arctic. *Canadian Journal of Zoology* **74**: 897–911.

Skaug, H. J. and Givens, G. H. 2007. Relatedness among individuals in BCB bowhead microsatellite samples. Paper SC/59/BRG20 presented to the Scientific Committee of the International Whaling Commission, May 2007.

Slack, K. E., Jones, C. M., Ando, T., Harrison, G. L., Fordyce, R. E., Arnason, U. and Penny, D. 2006. Early penguin fossils, plus mitochondrial genomes, calibrate avian evolution. *Molecular Biology and Evolution* **23**: 1144–55.

Slagsvold, T. and Saetre, G.-P. 1991. Evolution of plumage color in male pied flycatchers (*Ficedula hypoleuca*): evidence for female mimicry. *Evolution* **45**: 910–17.

Slate, J., Kruuk, L. E. B., Marshall, T. C., Pemberton, J. M. and Clutton-Brock, T. H. 2000. Inbreeding depression influences lifetime breeding success in a wild population of red deer (*Cervus elaphus*). *Proceedings of the Royal Society of London*, B **267**: 1657–62.

Slater, E. 1958. The sibs and children of homosexuals. In: Smith, D. R. and Davidson, W. M. (Eds) *Symposium on Nuclear Sex*, pp. 79–83. London: Heinemann.

Slater, E. 1962. Birth order and maternal age of homosexuals. *The Lancet* **1**: 69–71.

Slezak, M. and Pfrieger, F. W. 2003. New roles for astrocytes: regulation of CNS synaptogenesis. *Trends in Neurosciences* **26**: 531–5.

Slob, A. K., de Klerk, L. W. L. and Brand, T. 1987. Homosexual and heterosexual partner preference in ovariectomized female rats: effects of testosterone, estradiol and mating experience. *Physiology and Behavior* **41**: 571–6.

Slob, A. K., Groeneveld, W. H. and van der Werff Ten Bosch, J. J. 1986. Physiological changes during copulation in male and female stumptail macaques (*Macaca arctoides*). *Physiology and Behavior* **38**: 890–5.

Slomkowski, C., Rende, R., Conger, K. J., Simons, R. L. and Conger, R. 2001. Sisters, brothers, and delinquency: evaluating social influence during early and middle adolescence. *Child Development* **72**: 271–83.

Smallwood, J. A. 1987. Sexual segregation by habitat in American kestrels wintering in southcentral Florida: vegetative structure and responses to differential prey availability. *Condor* **89**: 842–9.

Smith, C. C. 1968. The adaptive nature of social organization in the genus of tree squirrels *Tamiasciurus*. *Ecological Monographs* **38**: 31–63.

Smith, C. C. 1981. The indivisible niche of *Tamiasciurus*: an example of nonpartitioning of resources. *Ecological Monographs* **51**: 343–63.

Smith, D. 2004. Central Park Zoo's gay penguins ignite debate. *New York Times*, Saturday 7 February.

Smith, K. K. 2003. Time's arrow: heterochrony and the evolution of development. *International Journal of Developmental Biology* **47**: 613–21.

Smith, L. H. 1988. *The Life of the Lyrebird*. Richmond: William Heinemann Australia.

Smith, R. J. and Jungers, W. L. 1997. Body mass in comparative primatology. *Journal of Human Evolution* **32**: 523–59.

Smith, T. W. 1991. Adult sexual behavior in 1989: number of partners, frequency of intercourse and risk of AIDS. *Family Planning Perspectives* **23**: 102–7.

Smolker, R. A., Richards, A. F., Connor, R. C. and Pepper, J. W. 1992. Sex differences in patterns of association among Indian Ocean bottlenose dolphins. *Behaviour* **123**: 38–67.

Smuts, B. 1987. What are friends for? *Natural History* **96**: 36–45.

Smuts, B. B. and Watanabe, J. M. 1990. Social relationships and ritualized greetings in adult male baboons (*Papio cynocephalus anubis*). *International Journal of Primatology* **11**: 147–72.

Snow, B. K. 1972. A field study of the calfbird *Perissocephalus tricolor*. *Ibis* **114**: 139–62.

Snyder, J., Bank, L. and Burraston, B. 2005. The consequences of antisocial behavior in older male siblings for younger brothers and sisters. *Journal of Family Psychology* **19**: 643–53.

Soares Leal, W., Kuwahara, S., Shi, X. *et al.* 1998. Male-released sex pheromone of the stink bug *Piezodorus hybneri*. *Journal of Chemical Ecology* **24**: 1817–29.

Socarides, C. W. 1978. *Homosexuality*. New York: Jason Aronson.

Solomon, G. F. 1987. Psychoneuroimmunology: interactions between central nervous system and immune system. *Journal of Neuroscience Research* **18**: 1–9.

Solomon, R. 1980. The opponent-process theory of acquired motivation: the costs of pleasure and the benefits of pain. *American Psychologist* **8**: 691–712.

Soman, A., Hedrick, T. L. and Biewener, A. A. 2005. Regional patterns of pectoralis fascicle strain in the pigeon *Columba livia* during level flight. *Journal of Experimental Biology* **208**: 771–86.

Somers, M. J., Rasa, O. A. E. and Penzhorn, B. L. 1995. Group structure and social behaviour of warthogs *Phacochoerus aethiopicus*. *Acta Theriologica* **40**: 257–81.

Sommer, V. 1988. Female-female mounting in langurs (*Presbytis entellus*). *International Journal of Primatology* **8**: 478.

Sommer, V., Schauer, P. and Kyriazis, D. 2006. A wild mixture of motivation: same-sex mounting in Indian langur monkeys. In: Sommer, V. and Vasey, P. L. (Eds)

Sommer, V. and Vasey, P. L. (Eds) 2006. *Homosexual Behaviour in Animals: an Evolutionary Perspective*, pp. 238–72. Cambridge: Cambridge University Press.

Sommer, V. and Vasey, P. L. (Eds) 2006. *Homosexual Behaviour in Animals: an Evolutionary Perspective*. Cambridge: Cambridge University Press.

Sorenson Jamison, C., Meier, R. J. and Campbell, B. C. 1993. Dermatoglyphic asymmetry and testosterone levels in normal males. *American Journal of Physical Anthropology* **90**: 185–98.

Sorin, A. B. 2004. Paternity assignment for white-tailed deer (*Odocoileus virginianus*): mating across age classes and multiple paternity. *Journal of Mammalogy* **85**: 356–62.

Soulé, J., Messaoudi, E. and Bramham, C. R. 2006. Brain-derived neurotrophic factor and control of synaptic consolidation in the adult brain. *Biochemistry Society Transactions* **34**: 600–4.

Spencer H. 1890. The origin of music. *Mind* **15**: 449–68.

Spicer, G. S. and Dunipace, L. 2004. Molecular phylogeny of songbirds (Passeriformes) inferred from mitochondrial 16S ribosomal RNA gene sequences. *Molecular Phylogenetics and Evolution* **30**: 325–35.

Spinage, C. A. 1982. *The Uganda Waterbuck*. London: Academic Press.

Spottiswoode, C. and Møller, A. P. 2004. Extrapair paternity, migration, and breeding synchrony in birds. *Behavioral Ecology* **15**: 41–57.

Sprecher, S., Sullivan Q. and Hatfield E. 1994. Mate selection preferences: gender differences examined in a natural sample. *Journal of Personality and Social Psychology* **66**: 1074–80.

Srivastava, A., Borries, C. and Sommer, V. 1991. Homosexual mounting in free-ranging female hanuman langurs (*Presbytis entellus*). *Archives of Sexual Behavior* **20**: 487–512.

Stacey, J. 2006. Gay parenthood and the decline of paternity as we knew it. *Sexualities* **9**: 27–55.

Stacey, J. and Biblarz, T. J. 2001. (How) does the sexual orientation of parents matter? *American Sociological Review* **66**: 159–83.

Starin, E. D. 2004. Masturbation observations in Temminck's red colobus. *Folia Primatologica* **75**: 114–17.

Stearns, S. C. 2007. Are we stalled part way through a major evolutionary transition from individual to group? *Evolution* **61**: 2275–80.

Steele, M. A. 1998. *Tamiasciurus hudsonicus*. *Mammalian Species* **586**: 1–9.

Stein, D. J., Black, D. W., Shapira, N. A. and Spitzer, R. L. 2001. Hypersexual disorder and preoccupation with internet pornography. *American Journal of Psychiatry* **158**: 1590–4.

Stein, E. 1999. *The Mismeasure of Desire: the Science, Theory, and Ethics of Sexual Orientation*. New York: Oxford University Press.

Stellflug, J. N. 2006. Comparison of cortisol, luteinizing hormone, and testosterone responses to a defined stressor in sexually inactive rams and sexually active female-oriented and male-oriented rams. *Journal of Animal Science* **84**: 1520–5.

Stellflug, J. N. and Berardinelli, J. G. 2002. Ram mating behavior after long-term selection for reproductive rate in Rambouillet ewes. *Journal of Animal Science* **80**: 2588–93.

Stellflug, J. N., Perkins, A. and Lavoie, V. A. 2004. Testosterone and luteinising hormone responses to naloxone help predict sexual performance in rams. *Journal of Animal Science* **82**: 3380–7.

Stepanyants, A., Hof, P. R. and Chklovskii, D. B. 2002. Geometry and structural plasticity of synaptic connectivity. *Neuron* **34**: 275–88.

Stephen, C. L., Devos, J. C. Jr, Lee, T. E. Jr *et al.* 2005. Population genetic analysis of Sonoran pronghorn (*Antilocapra americana sonorensis*). *Journal of Mammalogy* **86**: 782–92.

Stevens, K., Perry, G. C. and Long, S. E. 1982. Effect of ewe urine and vaginal secretions on ram investigative behaviour. *Journal of Chemical Ecology* **8**: 23–9.

Stevick, P. T., Allen, J., Bérubé, M. *et al.* 2003. Segregation of migration by feeding ground origin in North Atlantic whales (*Megaptera novaeangliae*). *Journal of Zoology, London* **259**: 231–7.

Stewart, B. S. 1997. Ontogeny of differential migration and sexual segregation in northern elephant seals. *Journal of Mammalogy* **78**: 1101–16.

Stewart, K. M., Fulbright, T. E., Drawe, D. L. and Bowyer, R. T. 2003. Sexual segregation in white-tailed deer: responses to habitat manipulations. *Wildlife Society Bulletin* **31**: 1210–17.

Stewart, S. A. and German, R. Z. 1999. Sexual dimorphism and ontogenetic allometry of soft tissues in *Rattus norvegicus*. *Journal of Morphology* **242**: 57–66.

Stiles, F. G. 1982. Aggressive and courtship displays of the male Anna's Hummingbird. *Condor* **84**: 208–25.

Stirrat, S. C. and Fuller, M. 1997. The repertoire of social behaviours of agile wallabies, *Macropus agilis*. *Australian Mammalogy* **20**: 71–8.

Stockley, P. 2003. Female multiple mating behaviour, early reproductive failure and litter size variation in mammals. *Proceedings of the Royal Society of London*, B **270**: 271–8.

Stokes, J. P., Damon, W. and McKirnan, D. J. 1997. Predictors of movement toward homosexuality: a longitudinal study of bisexual men. *Journal of Sex Research* **34**: 304–12.

Stokke, S. and du Toit, J. T. 2002. Sexual segregation in habitat use by elephants in Chobe National Park, Botswana. *African Journal of Ecology* **40**: 360–71.

Stoléru, S., Grégoire, M.-C., Gérard, D. *et al.* 1999. Neuroanatomical correlates of visually evoked sexual arousal in human males. *Archives of Sexual Behavior* **28**: 1–21.

Stolte, I. G. and Coutinho, R. A. 2002. Risk behaviour and sexually transmitted diseases are on the rise in gay men, but what is happening with HIV? *Current Opinion in Infectious Diseases* **15**: 37–41.

Stoner, C. J., Caro, T. M. and Graham, C. M. 2003. Ecological and behavioral correlates of coloration in artiodactyls: systematic analyses of conventional hypotheses. *Behavioral Ecology* **14**: 823–40.

Storms, M. D. 1981. A theory of erotic orientation development. *Psychological Review* **88**: 340–53.

Stowers, L., Holy, T. E., Meister, M., Dulac, C. and Koentges, G. 2002. Loss of sex discrimination and male-male aggression in mice deficient for TRP2. *Science* **295**: 1493–500.

Strahan, R. 1983. *The Australian Museum Complete Book of Australian Mammals*. London: Angus and Robertson.

Strasser, R. and Schwabl, H. 2004. Yolk testosterone organizes behaviour and male plumage coloration in house sparrows (*Passer domesticus*). *Behavioral Ecology and Sociobiology* **56**: 491–7.

Stroud, D. C. 1982. Population dynamics of *Rattus rattus* and *R. norvegicus* in a riparian habitat. *Journal of Mammalogy* **63**: 151–4.

Struhsaker, T. T. 1967. Ecology of vervet monkeys (*Cercopithecus aethiops*) in the Masai-Amboseli Game Reserve, Kenya. *Ecology* **48**: 891–904.

Stutchbury, B. J. 1994. Competition for winter territories in a neotropical migrant: the role of age, sex and color. *Auk* **111**: 63–9.

Stutchbury, B. J. and Evans Ogden, L. J. 1996. Fledgling adoption in hooded warblers (*Wilsonia citrina*): does extrapair paternity play a role? *Auk* **113**: 218–20.

Sulloway, F. J. 1979. *Freud, Biologist of the Mind: Beyond the Psychoanalytic Legend*. London: Burneett Books Ltd., in association with André Deutsch Ltd.

Sulloway, F. J. 1995. Birth order and evolutionary psychology: a meta-analytic overview. *Psychological Inquiry* **6**: 75–80.

Sulloway, F. J. 1996. *Born to Rebel: Birth Order, Family Dynamics, and Creative Lives*. New York: Pantheon Books.

Sulloway, F. J. 2001. Birth order, sibling competition, and human behavior. In: Fetzer, J. H. and Holcomb, H. R., III (Eds) *Studies in Cognitive Systems*, Vol. 27. *Conceptual Challenges in Evolutionary Psychology: Innovative Research Strategies*, pp. 39–83. Dordrecht: Kluwer Academic Press.

Sulloway, F. J. 2007a. Birth order. In: Salmon, C. and Shackelford, T. (Eds) *Evolutionary Family Psychology*, pp. 162–82. New York: Oxford University Press.

Sulloway, F. J. 2007b. Birth order and intelligence. *Science* **317**: 1711–12.

Sulloway, F. J. 2007c. Birth order and sibling competition. In: Dunbar, R. and Barrett, L. (Eds) *The Oxford Handbook of Evolutionary Psychology*, pp. 297–311. Oxford: Oxford University Press.

Surani, M. A., Barton, S. C. and Norris, M. L. 1984. Development of reconstituted mouse eggs suggests imprinting of the genome during gametogenesis. *Nature* **308**: 548–50.

Suresha, R. 2008. Commemorating the sixtieth anniversary of the 1948 publication of *Sexual Behavior in the Human Male*, by Alfred C. Kinsey, *et al. Journal of Bisexuality* **8**: 169–72.

Surridge, A. K., Bell, D. J., Ibrahim, K. M. and Hewitt, G. M. 1999. Population structure and genetic variation of European wild rabbits (*Oryctolagus cuniculus*) in East Anglia. *Heredity* **82**: 479–87.

Swaab, D. F. 1995. Development of the human hypothalamus. *Neurochemical Research* **20**: 509–19.

Swaab, D. F. 2003. The human hypothalamus. Basic and clinical aspects. Part I: Nuclei of the hypothalamus. In: Aminoff M. J., Boller F. and Swaab D. F. (Eds) *Handbook of Clinical Neurology*. Amsterdam: Elsevier.

Swaab, D. F. 2007. Sexual differentiation of the brain and behavior. *Best Practice and Research Clinical Endocrinology and Metabolism* **21**: 431–44.

Swaab, D. F., Chung, W. C. J., Kruijver, F. P. M., Hofman, M. A. and Hestiantoro, A. 2003. Sex differences in the hypothalamus in the different stages of human life. *Neurobiology of Aging* **24**: S1–S16.

Swaab, D. F. and Fliers, E. 1985. A sexually dimorphic nucleus in the human brain. *Science* **228**: 1112–15.

Swaab, D. F., Gooren, L. J. G. and Hofman, M. A. 1995. Brain research, gender, and sexual orientation. *Journal of Homosexuality* **28**: 283–301.

Swaab, D. F. and Hofman, M. A. 1990. An enlarged suprachiasmatic nucleus in homosexual men. *Brain Research* **537**: 141–8.

Swaab, D. F. and Hofman, M. A. 1995. Sexual differentiation of the human hypothalamus in relation to gender and sexual orientation. *Trends in Neuroscience* **18**: 264–70.

Swanson D. W. and Stipes, A. H. 1969. Psychiatric aspects of Klinefelter's Syndrome. *American Journal of Psychiatry* **126**: 814–22.

Swedell, L. 2002. Affiliation among females in wild hamadryas baboons (*Papio hamadryas hamadryas*). *International Journal of Primatology* **23**: 1205–26.

Swihart, R. K. 1984. Body size, breeding season length, and life history tactics in lagomorphs. *Oikos* **43**: 282–90.

Sylva, D., Rieger, G., Linsenmeier, J. A. W. and Bailey, J. M. 2010. Concealment of sexual orientation. *Archives of Sexual Behavior* **39**: 141–52.

Symonds, M. R. E. 1999. Life histories of the Insectivora: the role of phylogeny, metabolism and sex differences. *Journal of Zoology, London* **249**: 315–37.

Symonds, M. R. E. 2002. The effect of topological inaccuracy in evolutionary trees of the phylogenetic comparative method of independent contrasts. *Systematic Biology* **51**: 541–53.

Symons, D. 1979. *The Evolution of Human Sexuality.* New York: Oxford University Press.

Szekeres-Bartho, J. 2002. Immunological relationship between the mother and the fetus. *International Review of Immunology* **21**: 471–95.

Szep, T. 1995. Survival rates of Hungarian sand martins and their relationship with Sahel rainfall. *Journal of Applied Statistics* **22**: 891–904.

Tabb, M. M. and Blumberg, B. 2006. New modes of action for endocrine-disrupting chemicals. *Molecular Endocrinology* **20**: 475–82.

Takahata, Y. 1982. The socio-sexual behaviour of Japanese monkeys. *Zeitschrift für Tierpsychologie* **59**: 89–108.

Takenoshita, Y. 1998. Male homosexual behaviour accompanied by ejaculation in a free-ranging troop of Japanese macaques (*Macaca fuscata*). *Folia Primatologica* **69**: 364–7.

Talmage-Riggs, G. and Anschel, S. 1973. Homosexual behaviour and dominance hierarchy in a group of captive female squirrel monkeys (*Saimiri sciureus*). *Folia Primatologica* **19**: 61–72.

Tanner, J. B., Smale, L. and Holekamp, K. E. 2007. Ontogenetic variation in the play behavior of spotted hyenas. *Journal of Developmental Processes* **2**: 5–30.

Tao, G. 2008. Sexual orientation and related viral sexually transmitted disease rates among US women aged 15 to 44 years. *American Journal of Public Health* **98**: 1007–9.

Taub, D. M. 1980. Female choice and mating strategies among wild Barbary macaques (*Macaca sylvanus*). In: Lindburgh, D. G. (Ed.) *The Macaques: Studies in Ecology, Behavior and Evolution*, pp. 287–344. New York: Van Nostrand Reinhold Company.

Tauber, E. S. 1940. Effects of castration upon sexuality of the adult male. *Psychosomatic Medicine* **2**: 74–87.

Tavares, M. B., Costa Fuchs, F., Diligenti, F. *et al.* 2004. Behavioral characteristics of the only child vs first-born and children with siblings. *Revista Brasileira de Psiquiatria* **26**: 16–22.

Taylor, L. F., Booker, C. W., Jim, G. K. and Guichon, P. T. 1997. Epidemiological investigation of the buller steer syndrome (riding behaviour) in a western Canadian feedlot. *Australian Veterinary Journal* **75**: 45–51.

Taylor, V. and Rupp, L. J. 1993. Women's culture and lesbian feminist activism: a reconsideration of cultural feminism. *Signs* **19**: 32–61.

Tchernitchin, A. N. and Tchernitchin, N. 1992. Imprinting of paths of heterodifferentiation by prenatal or neonatal exposure to hormones, pharmaceuticals, pollutants and other agents or conditions. *Medical Science Research* **20**: 391–7.

Tellería, J. L., Virgós, E., Carbonell, R., Pérez-Trís, J. and Santos, T. 2001. Behavioural responses to changing landscapes: flock structure and anti-predator strategies of tits wintering in fragmented forests. *Oikos* **95**: 253–64.

Tenhula, W. N. and Bailey, J. M. 1998. Female sexual orientation and pubertal onset. *Developmental Neuropsychology* **14**: 369–83.

Terman, L. and Miles, C. C. 1936. *Sex and Personality.* New York: McGraw-Hill.

Terranova, M. L. and Laviola, G. 1995. Individual differences in mouse behavioural development: effects of precocious weaning and ongoing sexual segregation. *Animal Behaviour* **50**: 1261–71.

Terranova, M. L., Loggi, G., Chiarotti, F. and Laviola, G. 2000. Attractivity and social preferences in mice (*Mus musculus domesticus*): the role of prepubertal sexual segregation and precocious weaning. *Journal of Comparative Psychology* **114**: 325–34.

Thellin, O. and Heinen, E. 2003. Pregnancy and the immune system: between tolerance and rejection. *Toxicology* **185**: 179–84.

Thierry, B. 1986. Affiliative interference in mounts in a group of Tonkean Macaques (*Macaca tonkeana*). *American Journal of Primatology* **11**: 89–97.

Thierry, B., Gauthier, C. and Peignot, P. 1990. Social grooming in Tonkean macaques (*Macaca tonkeana*). *International Journal of Primatology* **11**: 357–75.

Thierry, B., Heistermann, M., Aujard, F. and Hodges, J. K. 1996. Long-term data on basic reproductive parameters and evaluation of endocrine, morphological, and behavioral measures for monitoring reproductive status in a group of semifree-ranging Tonkean macaques (*Macaca tonkeana*). *American Journal of Primatology* **39**: 47–62.

Thirgood, S. J. 1996. Ecological factors influencing sexual segregation and group size in fallow deer (*Dama dama*). *Journal of Zoology, London* **239**: 783–97.

Thomas, D. W., Fenton, M. B. and Barclay, R. M. R. 1979. Social behaviour of the little brown bat, *Myotis lucifugus*. I. Mating behaviour. *Behavioural Ecology and Sociobiology* **6**: 129–136.

Thomas, G. H., Willis, M. A. and Székely, T. 2004. A supertree approach to shorebird phylogeny. *BMC Evolutionary Biology* **4**: 28.

Thompson, C. W. and Leu, M. 1995. Molts and plumages of orange-breasted buntings (*Passerina leclancherii*): implications for theories of delayed plumage maturation. *Auk* **112**: 1–19.

Thompson, D. C. 1977. Reproductive behaviour of the grey squirrel. *Canadian Journal of Zoology* **55**: 1176–84.

Thompson, D. C. 1978. The social system of the grey squirrel. *Behaviour* **64**: 305–28.

Thompson, P., Cannon, T. D., Narr, K. L. *et al.* 2001. Genetic influences on brain structure. *Nature Neuroscience* **4**: 1253–58.

Thompson, P., Cannon, T. D. and Toga, A. W. 2002. Mapping genetic influences on human brain structure. *Annals of Medicine* **34**: 523–36.

Thomsen, R. and Soltis, J. 2004. Male masturbation in free-ranging Japanese macaques. *International Journal of Primatology* **25**: 1033–41.

Thomsen, R., Soltis, J. and Teltscher, C. 2003. Sperm competition and the function of male masturbation in nonhuman primates. In: Jones, C.B. (Ed.) *Sexual Selection and Reproductive Competition in Primates: New Perspectives and Directions*. ASP-Book Series, *Special Topics in Primatology* **3**: 436–53.

Thornhill, R. and Alcock, J. 1983. *The Evolution of Insect Mating Systems*. Cambridge, MA: Harvard University Press.

Thornton, J., Zehr, J. L. and Loose, M. D. 2009. Effects of prenatal androgens on rhesus monkeys: a model system to explore the organizational hypothesis in primates. *Hormones and Behavior* **55**: 633–44.

Thuman, K. A. and Griffith, S. C. 2004. Genetic similarity and the nonrandom distribution of paternity in a genetically highly polyandrous shorebird. *Animal Behaviour* **69**: 765–70.

Tiersch, T. R., Mitchell, M. J. and Wachtel, S. S. 1991. Studies on the phylogenetic conservation of the SRY gene. *Human Genetics* **87**: 571–3.

Tinbergen, N. 1963. On the aims and methods of Ethology. *Zeitschrift für Tierpsychologie* **20**: 410–33.

Tincani, A., Danieli, E., Nuzzo, M. *et al.* 2006. Impact of *in utero* environment on the offspring of lupus patients. *Lupus* **15**: 801–7.

Tinker, M. T., Kovacs, K. M. and Hamill, M. O. 1995. The reproductive behavior and energetics of male grey seals (*Halichoerus grypus*) breeding on a land-fast ice substrate. *Behavioral Ecology and Sociobiology* **36**: 159–70.

Titman, R. D. and Lowther, J. K. 1975. The breeding behaviour of a crowded population of Mallards. *Canadian Jouranal of Zoology* **53**: 1270–83.

Titus-Ernstoff, L., Perez, K., Hatch, E. E. *et al.* 2003. Psychosexual characteristics of men and women exposed prenatally to diethylstilbestrol. *Epidemiology* **14**: 155–60.

Tobet, S. A., Zahniser, D. J. and Baum, M. J. 1986. Differentiation in male ferrets of a sexually dimorphic nucleus of the preoptic/anterior hypothalamic area requires prenatal estrogen. *Neuroendocrinology* **44**: 299–308.

Toga, A. W. and Thompson, P. M. 2005. Genetics of brain structure and intelligence. *Annual Reviews of Neuroscience* **28**: 1–23.

Torchilin, V. P., Iakoubov, L. Z. and Estrov, Z. 2003. Therapeutic potential of antinuclear autoantibodies in cancer. *Cancer Therapy* **1**: 179–90.

Towner, M. C. 1999. A dynamic model of human dispersal in a land-based economy. *Behavioral Ecology and Sociobiology* **46**: 82–94.

Towner, M. C. 2001. Linking dispersal and resources in humans: life history data from Oakham, Massachusetts (1750–1850). *Human Nature* **12**: 321–49.

Trail, P. W. 1985. Courtship disruption modifies mate choice in a lek-breeding bird. *Science* **227**: 778–80.

Trail, P. W. 1990. Why should lek-breeders be monomorphic? *Evolution* **44**: 1837–2.

Trail, P. W. and Adams, E. S. 1989. Active mate choice at cock-of-the-rock leks: tactics of sampling and comparison. *Behavioral Ecology and Sociobiology* **25**: 283–92.

Trail, P. W. and Koutnik, D. L. 1986. Courtship disruption of the lek in the Guianan cock-of-the-rock. *Ethology* **73**: 197–218.

Travis, J. C. and Holmes, W. N. 1974. Some physiological and behavioural changes associated with oestrus and pregnancy in the squirrel monkey (*Saimiri sciureus*). *Journal of Zoology, London* **174**: 41–66.

Treplin, S. 2006. *Inference of phylogenetic relationships in passerine birds (Aves: Passeriformes) using new molecular markers*. PhD Dissertation, University of Postdam.

Trevathan, W. R., Burleson, M. H. and Gregory, W. L. 1993. No evidence for menstrual synchrony in lesbian couples. *Psychoneuroendocrinology* **18**: 425–35.

Trinh, K. and Storm, D. R. 2003. Vomeronasal organ detects odorants in absence of signaling through main olfactory epithelium. *Nature Neuroscience* **6**: 519–25.

Tripp, C. A. 1975. *The Homosexual Matrix*. New York: Signet.

Trivers R. L. 1971 The evolution of reciprocal altruism. *Quarterly Review of Biology* **46**: 35–57.

Trivers, R. L. 1972 Parental investment and sexual selection. In: B. Campbell (Ed.) *Sexual Selection and the Descent of Man*, pp. 139–79. Chicago, IL: Aldine.

Trivers, R. L. 1974. Parent-offspring conflict. *American Zoologist* **14**: 249–64.

Trivers, R. 1985. *Social Evolution*. Menlo Park, CA: The Benjamin/Cummins Publishing Co.

Trivers, R. L. and Willard, D. E. 1973. Natural selection of parental ability to vary the sex ratio of offspring. *Science* **179**: 90–2.

Troisi, A. 2003. Sexual disorders in the context of Darwinian psychiatry. *Journal of Endocrinological Investigation* **26** (Suppl.): 54–7.

Troisi, A. and Carosi, M. 1998. Female orgasm rate increases with male dominance in Japanese macaques. *Animal Behaviour* **56**: 1261–6.

Tryjanowski, P. and Reuven, Y. 2002. Differences between the spring and autumn migration of the Red-backed Shrike *Lanius collurio*: record from the Eilat stopover (Israel). *Acta Ornithologica* **37**: 85–90.

Tsoi, W. F. 1990. Developmental profile of 200 male and 100 female transsexuals in Singapore. *Archives of Sexual Behavior* **19**: 595–605.

Tuana, N. 1983. Re-fusing nature/nurture. *Women's Studies International Forum* **6**: 621–32.

Tulloch, D. G. 1979. The water buffalo, *Bubalus bubalis*, in Australia: reproductive and parent-offspring behaviour. *Australian Wildlife Research* **6**: 265–87.

Turner, W. C., Jolles, A. E. and Owen-Smith, N. 2005. Alternative sexual segregation during the mating season by male African buffalo (*Syncerus caffer*). *Journal of Zoology, London* **267**: 291–9.

Turner, W. J. 1994. Comments on discordant monozygotic twinning in homosexuality. *Archives of Sexual Behavior* **23**: 115–19.

Turner, W. J. 1995. Homosexuality, type 1: an Xq28 phenomenon. *Archives of Sexual Behavior* **24**: 109–34.

Tuttle, G. E. and Pillard, R. C. 1991. Sexual orientation and cognitive abilities. *Archives of Sexual Behavior* **20**: 307–18.

Tyler, P. A. 1984. Homosexual behaviour in animals. In: K. Howells (Ed.) *The Psychology of Sexual Diversity*, pp. 42–62. Oxford: Basil Blackwell Ltd.

Udry, J. R. 2000. Biological limits of gender construction. *American Sociological Review* **65**: 443–57.

Udry, J. R. 2003. Putting prenatal effects on sex-dimorphic behaviour in perspective: an absolutely complete theory. *Epidemiology* **14**: 135–6.

Valencia, A., Collado, P., Calés, J. M. et al. 1992. Postnatal administration of dihydrotestosterone to the male rat abolishes sexual dimorphism in the accessory olfactory bulb: a volumetric study. *Developmental Brain Research* **68**: 132–5.

van Anders, S. M., Vernon, P. A. and Wilbur, C. J. 2006. Finger-length ratios show evidence of prenatal hormone-transfer between opposite-sex twins. *Hormones and Behavior* **49**: 315–19.

van de Beek, C., van Goozen, S. H. M., Buitelaar, J. K. and Cohen-Kettenis, P. T. 2009. Prenatal sex hormones (maternal and amniotic fluid) and gender-related play behavior in 13-month-old infants. *Archives of Sexual Behavior* **38**: 6–15.

van de Poll, N. E., van Dis, H. 1979. The effect of medial preoptic-anterior hypothalamic lesions on bisexual behaviour of the male rat. *Brain Research Bulletin* **4**: 505–11.

van Goozen, S. H. M., Slabbekoorn, D., Gooren, L. J. G., Sanders, G. and Cohen-Kettenis, P. T. 2002. Organizing and activating effects of sex hormones in homosexual transsexuals. *Behavioral Neuroscience* **116**: 982–8.

van Hoof, J. A. R. A. M. and van Schaik, C. P. 1994. Male bonds: affiliative relationships among nonhuman primate males. *Behaviour* **130**: 310–37.

Van Rhijn, J. 1985. Black-headed gull or black-headed girl? On the advantage of concealing sex by gulls and other colonial birds. *Netherlands Journal of Zoology* **35**: 87–102.

Van Rhijn, J. and Groothuis, T. 1985. Biparental care and the basis for alternative bond-types among gulls, with special reference to black-headed gulls. *Ardea* **73**: 159–74.

Van Rhijn, J. and Groothuis, T. 1987. On the mechanisms of mate selection in black-headed gulls. *Behaviour* **100**: 134–69.

van Solinge, H., Walhout, E. and van Poppel, F. 2000. Determinants of institutionalization of orphans in a nineteenth-century Dutch town. *Continuity and Change* **15**: 139–66.

van Tuinen, M., Sibley, C. G. and Hedges, S. B. 2000. The early history of modern birds inferred from DNA sequences of nuclear and mitochondrial ribosomal genes. *Molecular Biology and Evolution* **17**: 451–7.

van Valkenburgh, B., Wang, X. and Damuth, J. 2004. Cope's rule, hypercarnivory, and extinction in North American canids. *Science* **306**: 101–4.

van Wyk, P. H. and Geist, C. S. 1995. Biology of bisexuality: critique and observations. *Journal of Homosexuality* **28**: 357–73.

Vasey, P. L. 1995. Homosexual behaviour in primates: A review of evidence and theory. *International Journal of Primatology* **16**: 173–204.

Vasey, P. L. 1996. Interventions and alliance formation between female Japanese macaques, *Macaca fuscata*, during homosexual courtships. *Animal Behaviour* **52**: 539–551.

Vasey, P. L. 1998. Female choice and inter-sexual competition for female sexual partners in Japanese macaques. *Behaviour* **135**: 579–597.

Vasey, P. L. 2002a. Sexual partner preference in female Japanese macaques. *Archives of Sexual Behavior* **31**: 51–62.

Vasey, P. L. 2002b. Same-sex sexual partner preference in hormonally and neurologically unmanipulated animals. *Annual Review of Sex Research* **13**: 141–79.

Vasey, P. L. 2004. Sex differences in sexual partner acquisition, retention, and harassment during female homosexual consortships in Japanese macaques. *American Journal of Primatology* **64**: 397–409.

Vasey, P. L. 2006a. The pursuit of pleasure: an evolutionary history of female homosexual behaviour in Japanese macaques. In: Sommer, V. and Vasey, P. L. (Eds) *Homosexual Behaviour in Animals: An Evolutionary Perspective*, pp. 191–219. Cambridge: Cambridge University Press.

Vasey, P. L. 2006b. Where do we go from here? Research on the evolution of homosexual behaviour in animals. In: Sommer, V. and Vasey, P. L. (Eds) *Homosexual Behaviour in Animals: An Evolutionary Perspective*, pp. 349–64. Cambridge: Cambridge University Press.

Vasey, P. L. 2006c. Function and phylogeny: the evolution of same-sex sexual behaviour in primates. *Journal of Psychology and Human Sexuality* **18**: 215–44.

Vasey, P. L. and Bartlett, N. H. 2007. What can the Samoan "fa'afafine" teach us about the Western concept of gender identity disorder in childhood? *Perspectives in Biology and Medicine* **50**: 481–90.

Vasey, P. L., Chapais, B. and Gauthier, C. 1998. Mounting interactions between female Japanese macaques: testing the influence of dominance and aggression. *Ethology* **104**: 387–98.

Vasey, P. L. and Duckworth, N. 2006. Sexual reward via vulvar, perineal, and anal stimulation: a proximate mechanism for female homosexual mounting in Japanese macaques. *Archives of Sexual Behavior* **35**: 523–32.

Vasey, P. L. and Duckworth, N. 2008. Female-male mounting in Japanese macaques: the proximate role of sexual reward. *Behavioural Processes* **77**: 405–7.

Vasey, P. L., Foroud, A., Duckworth, N. and Kovacovsky, S. D. 2006. Male-female and female-female mounting in Japanese macaques: a comparative study of posture and movement. *Archives of Sexual Behavior* **35**: 117–29.

Vasey, P. L. and Gauthier, C. 2000. Skewed sex ratios and female homosexual activity in Japanese macaques: an experimental analysis. *Primates* **41**: 17–25.

Vasey, P. L. and Jiskoot, H. 2009. The biogeography and evolution of female homosexual behavior in Japanese macaques. *Archives of Sexual Behavior*. DOI 10.1007/S10508-009-9578-2

Vasey, P. L. and Pfaus, J. G. 2005. A sexual dimorphic hypothalamic nucleus in a macaque species with frequent female-female mounting and same-sex sexual partner preference. *Behavioural Brain Research* **157**: 265–272.

Vasey, P. L., Pocock, D. S. and VanderLaan, D. P. 2007. Kin selection and male androphilia in Samoan fa'afafine. *Evolution and Human Behavior* **28**: 159–67.

Vasey, P. L., Rains, D., VanderLaan, D. P., Duckworth, N. and Kovacovsky, S. D. 2008a. Courtship behaviour in Japanese macaques during heterosexual and homosexual consortships. *Behavioural Processes* **78**: 401–7.

Vasey, P. L. and Reinhart, C. 2009. Female homosexual behavior in a new group of Japanese macaques: evolutionary implications. *Laboratory Primate Newsletter* **48**: 8–10.

Vasey, P. L. and Sommer, V. 2006. Homosexual behaviour in animals: topics, hypotheses and research trajectories. In: Sommer, V. and Vasey, P. L. (Eds) *Homosexual Behaviour in Animals: an Evolutionary Perspective*, pp. 3–38. Cambridge: Cambridge University Press.

Vasey, P. L. and VanderLaan, D. P. 2007. Birth order and male androphilia in Samoan fa'afafine. *Proceedings of the Royal Society, London* B **274**: 1437–42.

Vasey, P. L. and VanderLaan, D. P. 2009a. Materteral and avuncular tendencies in Samoa: a comparative study of women, men and fa'afafine. *Human Nature* **20**: 269–81.

Vasey, P. L. and VanderLaan, D. P. 2009b. Avuncular tendencies and the evolution of male androphilia in Samoan fa'afafine. *Archives of Sexual Selection*. DOI 10.1007/s10508-008-9404-3

Vasey, P. L., and VanderLaan, D. P. 2010. An adaptive cognitive dissociation between willingness to help kin and nonkin in samoan *fa'afafine*. *Psychological Science*. Published online 14 January 2010. DOI 10.1177/0956797609359623.

Vasey, P. L., VanderLaan, D. P., Rains, D., Duckworth, N. and Kovacovsky, S. D. 2008b. Inter-mount social interactions during heterosexual and homosexual consortships in Japanese macaques. *Ethology* **114**: 564–74.

Veening, J. G. and Coolen, L. M. 1998. Neural activation following sexual behavior in the male and female rat brain. *Behavioural Brain Research* **92**: 181–93.

Vega Matuszczyk, J., Appa, R. S. and Larsson, K. 1994. Age-dependent variations in the sexual preference of male rats. *Physiology and Behavior* **55**: 827–30.

Vega Matuszczyk, J. and Larsson, K. 1995. Sexual preference and feminine and masculine sexual behaviour of male rats prenatally exposed to antiandrogen or antiestrogen. *Hormones and Behavior* **29**: 191–206.

Vehrencamp, S. L. 1983a. Optimal degree of skew in cooperative societies. *American Zoologist* **23**: 327–35.

Vehrencamp, S. L. 1983b. A model for the evolution of despotic versus egalitarian societies. *Animal Behaviour* **31**: 667–82.

Vehrencamp, S. L. 2000. Evolutionary routes to joint-female nesting in birds. *Behavioral Ecology* **11**: 334–44.

Veniegas, R. C. and Conley, D. 2000. Biological research on women's sexual orientations: evaluating the scientific evidence. *Journal of Social Issues* **56**: 267–82.

Vervaecke, H. and Roden, C. 2006. Going with the herd: same-sex interaction and competition in American bison. In: Sommer, V. and Vasey, P. (Eds) *Homosexual Behaviour in Animals: an Evolutionary Perspective*, pp: 131–53. Cambridge: Cambridge University Press.

Vigilant, L., Hofreiter, M., Siedel, H. and Boesch, C. 2001. Paternity and relatedness in wild chimpanzee communities. *Proceedings of the National Academy of Sciences, USA* **98**: 12890–5.

Viglietti-Panzica, C., Montoncello, B., Mura, E., Pessatti, M. and Panzica, G. 2005. Organizational effects of diethylstilbestrol on brain vasotocin and sexual behavior in male quail. *Brain Research Bulletin* **65**: 225–33.

vom Saal, F. S. 1981. Variation in phenotype due to random intrauterine positioning of male and female fetuses in rodents. *Journal of Reproduction and Fertility* **62**: 633–50.

von Hardenberg, A., Bassano, B., Peracino, A. and Lovari, S. 2000. Male alpine chamois occupy territories at hotspots before the mating season. *Ethology* **106**: 617–30.

Voracek, M. and Dressler, S. G. 2007. Digit ratio (2D:4D) in twins: heritability estimates and evidence for a masculinized trait expression in women from opposite-sex pairs. *Psychological Reports* **100**: 115–26.

Waddington, C. H. 1942. Canalization of development and the inheritance of acquired characters. *Nature* **150**: 563–5.

Waddington, C. H. 1953. The "Baldwin effect," "genetic assimilation" and "homeostasis". *Evolution* **7**: 386–7.

Waddington, C. H. 1957. *The Strategy of the Genes*. London: George Allen & Unwin.

Waddington, C. H. 1961. Genetic assimilation. *Advances in Genetics* **10**: 257–90.

Wade, M. J. 1985. Soft selection, hard selection, kin selection, and group selection. *American Naturalist* **125**: 61–73.

Wade, M. J. and McCauley, D. E. 1984. Group selection: the interaction of local deme size and migration in the differentiation of small populations. *Evolution* **38**: 1047–58.

Wagner, A. P., Frank, L. G., Creel, S. and Coscia, E. M. 2007. Transient genital abnormalities in striped hyenas (*Hyaena hyaena*). *Hormones and Behavior* **51**: 626–32.

Wagner, R. H. 1991. The use of extrapair copulations for mate appraisal by razorbills, *Alca torda*. *Behavioral Ecology* **2**: 198–203.

Wagner, R. H. 1992. Extra-pair copulations in a lek: the secondary mating system of monogamous razorbills. *Behavioral Ecology and Sociobiology* **31**: 63–71.

Wagner, R. H. 1996. Male-male mountings by a sexually monomorphic bird: mistaken identity or fighting tactic? *Journal of Avian Biology* **27**: 209–14.

Wahaj, S. A., Guse, K. R., and Holekamp, K. E. 2001. Reconciliation in the spotted hyena (*Crocuta crocuta*). *Ethology* **107**: 1057–74.

Waites, C. L., Craig, A. M. and Garner, C. C. 2005. Mechanisms of vertebrate synaptogenesis. *Annual Reviews of Neuroscience* **28**: 251–74.

Wallen, K. 2001. Sex and context: hormones and primate sexual motivation. *Hormones and Behavior* **40**: 339–57.

Wallen, K. 2005. Hormonal influences on sexually differentiated behavior in nonhuman primates. *Frontiers in Neuroendocrinology* **26**: 7–26.

Wallen, K. and Baum, M. J. 2002. Masculinization and defeminisation in altricial and precocial mammals: comparative aspects of steroid hormone action. In: Pfaff, D. W., Arnold, A. P., Etgen, A. M., Fahrbach, S. E. and Rubin, R. T. (Eds) *Hormones, Brain and Behaviour*, Vol. 4, pp. 385–423. San Diego, CA: Academic Press.

Wallen, K., Goldfoot, D. A. and Goy, R. W. 1981. Peer and maternal influences on the expression of foot-clasp mounting by juvenile male rhesus monkeys. *Developmental Psychobiology* **14**: 299–309.

Wallen, K. and Parsons, W. A. 1997. Sexual behavior in same-sexed nonhuman primates: is it relevant to understanding human homosexuality? *Annual Review of Sex Research* **8**: 195–223.

Wallen, K. and Tannenbaum, P. L. 1999. Hormonal modulation of sexual behaviour and affiliation in rhesus monkeys. In: Carter, C. S., Lederhendler, I. I. and Kirkpatrick, B. (Eds) *The Integrative Neurobiology of Affiliation*, pp. 101–30. Cambridge, MA: MIT Press.

Wallien, M. S. C. and Cohen-Kettenis, P. T. 2008. Psychosexual outcome of gender-dysphoric children. *Journal of the American Academy of Child and Adolescent Psychiatry* **47**: 1413–23.

Wallis, J. 1982. Sexual behaviour of captive chimpanzees (*Pan troglodytes*): pregnant versus cycling females. *American Journal of Primatology* **3**: 77–88.

Walther, F. R. 1978a. Quantitative and functional variations of certain behaviour patterns in male Thomson's gazelle of different social status. *Behaviour* **65**: 212–40.

Walther, F. R. 1978b. Forms of aggression in Thomson's gazelle; their situational motivation and their relative frequency in different sex, age, and social classes. *Zeitschrift für Tierpsychologie* **47**: 113–71.

Wang, C., Makela, T., Hase, T., Adlercreutz, H. and Kurzer, M. S. 1994. Lignans and flavonoids inhibit aromatase enzyme in human preadipocytes. *Journal of Steroid Biochemistry and Molecular Biology* **50**: 205–12.

Wang, X. T., Kruger, D. J. and Wilke, A. 2009. Life history variables and risk-taking propensity. *Evolution and Human Behavior* **30**: 77–84.

Ward, I. L. 1972. Prenatal stress feminizes and demasculinizes the behaviour of males. *Science* **175**: 82–4.

Ward, I. L. 1984. The prenatal stress syndrome: current status. *Psychoneuroendocrinology* **9**: 3–11.

Ward, I. L. and Reed, J. 1985. Prenatal stress and prepubertal social rearing conditions interact to determine sexual behaviour in male rats. *Behavioral Neuroscience* **99**: 301–9.

Ward, I. L. and Stehm, K. E. 1991. Prenatal stress feminizes juvenile play patterns in male rats. *Physiology and Behavior* **50**: 601–5.

Ward, I. L., Ward, O. B., Winn, R. J. and Bielawski, D. 1994. Male and female sexual behaviour potential of male rats prenatally exposed to the influence of alcohol, stress, or both factors. *Behavioral Neuroscience* **108**: 1188–95.

Wasser, S. K. and Barash, D. P. 1983. Reproductive suppression among female mammals: implications for biomedicne and sexual selection theory. *Quarterly Review of Biology* **58**: 513–30.

Wassersug, R. J. and Johnson, T. W. 2007. Modern-day eunuchs: motivations for and consequences of contemporary castration. *Perspectives in Biology and Medicine* **50**: 544–56.

Watson, D. M., Croft, D. B. and Crozier, R. H. 1992. Paternity exclusion and dominance in captive red-necked wallabies, *Macropus rufogriseus* (Marsupialia: Macropodidae). *Australian Mammalogy* **15**: 31–6.

Weatherhead, P. J. and Montgomery, R. 1995. Local resource competetion and sex ratio variation in birds. *Journal of Avian Biology* **26**: 168–71.

Webster, M. and Zelditch, M. L. 2005. Evolutionary modifications of ontogeny: heterochrony and beyond. *Paleobiology* **31**: 354–72.

Weckerly, F. W. 1993. Intersexual resource partitioning in black-tailed deer: a test of the body size hypothesis. *Journal of Wildlife Management* **57**: 475–94.

Weckerly, F. W. 1998. Sexual-size dimorphism: influence of mass and mating systems in the most dimorphic mammals. *Journal of Mammalogy* **79**: 33–52.

Weckerly, F. W. 1999. Social bonding and aggression in female Roosevelt elk. *Canadian Journal of Zoology* **77**: 1379–84.

Weckerly, F. W., Ricca, M. A. and Meyer, K. P. 2001. Sexual segregation in Roosevelt elk: cropping rates and aggression in mixed-sex groups. *Journal of Mammalogy* **82**: 825–35.

Weeks, J. 1985. *Sexualities and its Discontents: Meanings, Myths and Modern Sexualities*. London: Routledge and Kegan Paul.

Wegesin, D. J. 1998a. Event-related potentials in homosexual and heterosexual men and women: sex-dimorphic patterns in verbal asymmetries and mental rotation. *Brain and Cognition* **36**: 73–92.

Wegesin, D. J. 1998b. A neuropsychological profile of homosexual and heterosexual men and women. *Archives of Sexual Behaviour* **27**: 91–108.

Weinberg, M. S., Lottes, I. and Shaver, F. M. 2000. Sociocultural correlates of permissive sexual attitudes: a test of Reiss's Hypotheses about Sweden and the United States. *Journal of Sex Research* **37**: 44–52.

Weinberg, M. S., Williams, C. J. and Calhan, C. 1994. Homosexual foot fetishism. *Archives of Sexual Behavior* **23**: 611–26.

Weinberg, M. S., Williams, C. J. and Calhan, C. 1995. "If the shoe fits": exploring male homosexual foot fetishism. *Journal of Sex Research* **32**: 17–27.

Weinberg, M. S., Williams, C. J. and Pryor, D. W. 1994. *Dual Attraction: Understanding Bisexuality*. New York: Oxford University Press.

Weinrich, J. D. 1985. Transsexuals, homosexuals, and sissy boys: on the mathematics of follow-up studies. *Journal of Sex Research* **21**: 322–8.

Weinrich, J. D. 1987. A new sociobiological theory of homosexuality applicable to societies with universal marriage. *Ethology and Sociobiology* **8**: 37–47.

Weinrich, J. D. and Klein, F. 2003. Bi-gay, bi-straight, and bi-bi: three bisexual subgroups identified using cluster analysis of the Klein Sexual Orientation Grid. *Journal of Bisexuality* **2**: 111–39.

Weinstock, M. 2001. Alterations induced by gestational stress in brain morphology and behaviour of the offspring. *Progress in Neurobiology* **65**: 427–51.

Weis, S., Weber, G., Wegner, E. and Kimbacher, K. 1989. The controversy of the human corpus callosum. *International Journal of Neuroscience* **47**: 169–73.

Weksberg, R., Shuman, C., Caluseriu, O. et al. 2002. Discordant *KCNQ1OT1* imprinting in sets of monozygotic twins discordant for Beckwith-Wiedemann syndrome. *Human Molecular Genetics* **11**: 1317–25.

Weller, A. and Weller, L. 1992. Menstrual synchrony in female couples. *Psychoneuroendocrinology* **17**: 171–7.

Wells, R. S. and Norris, K. S. 1994. Patterns of reproduction. In: Norris, K. S., Würsig, B., Wells, R. S. and Würsig, M. (Eds) *The Hawaiian Spinner Dolphin*, pp. 186–200. Berkeley, CA: University of California Press.

Wemmer, C., Collins, L. R., Beck, B. B. and Rettberg, B. 1983. The ethogram. In: Beck, B. B. and Wemmer, C. (Eds) *The Biology and Management of an Extinct Species, Pere David's Deer*, pp. 91–124. Park Ridge, NJ: Nozes.

Werner, D. 2006. The evolution of male homosexuality and its implications for human psychological and cultural variations. In: Sommer, V. and Vasey, P. L. (Eds) 2006. *Homosexual Behaviour in Animals: An Evolutionary Perspective*, pp. 316–46. Cambridge: Cambridge University Press.

Werschkul, D. F. 1982. Nesting ecology of the little blue heron: promiscuous behaviour. *Condor* **84**: 381–4.

Wesley-Hunt, G. D. and Flynn, J. J. 2005. Phylogeny of the Carnivora: Basal relationships among the carnivoramorphans, and assessment of the position of 'Miacoidea' relative to Carnivora. *Journal of Systematic Palaeontology* **3**: 1–28.

West, D. J. 1959. Parental figures in the genesis of male homosexuality. *International Journal of Social Psychiatry* **5**: 85–97.

Westcott, D. 1992. Inter-and intra-sexual selection: the role of song in a lek mating system. *Animal Behaviour* **44**: 695–703.

Westcott, D. A. 1997a. Neighbours, strangers and male-male aggression as a determinant of lek size. *Behavioral Ecology and Sociobiology* **40**: 235–42.

Westcott, D. A. 1997b. Lek locations and patterns of female movement and distribution in a Neotropical frugivorous bird. *Animal Behaviour* **53**: 235–47.

Westcott, D. A. and Smith, J. N. M. 1994. Behavior and social organization during the breeding season in *Mionectes oleagineus*, a lekking flycatcher. *Condor* **96**: 672–83.

West-Eberhard, M. J. 1975. The evolution of social behaviour by kin selection. *Quarterly Review of Biology* **50**: 1–33.

West-Eberhard, M. J. 2003. *Developmental Plasticity and Evolution.* New York: Oxford University Press.

Weston, K. 1993. Lesbian/gay studies in the house of anthropology. *Annual Review of Anthropology* **22**: 339–367.

Westphal, C. 1869. Die conträre Sexualempfindung. In: Meyer, L. and Westphal, C. (Eds) *Archiv für Psychiatrie und Nervenkrankheiten*, Series II, Vol. 1. Heft 1. Berlin: Verlag von August Hirschwald.

Westwood, G. 1960. *A Minority.* Edinburgh: R. & R. Clark.

Wethington, S. M., Russell, S. M. and West, G. C. 2005. Timing of hummingbird migration in Southeastern Arizona: implications for conservation. *USDA Forest Service General Technical Report.* PSW-GTR-191.

Whalen, R. E. 1974. Sexual differentiation: Models, methods and mechanisms. In: Friedman, R. C., Richart, R. M. and van de Wiele, R. L. (Eds) *Sex Differences in Behaviour*, pp. 467–81. New York: Wiley.

Wheeler, J. C. 2006. Historia natural de la vicuña. In: Vilé, B. (Ed.) *Investigación, Conservación y Manejo de Vicuñas.* Proyecto MACS.

Whitam, F. L. 1977. Childhood indicators of male homosexuality. *Archives of Sexual Behavior* **6**: 89–96.

Whitam, F. L. 1983. Culturally invariable properties of male homosexuality: tentative conclusions from cross-cultural research. *Archives of Sexual Behavior* **12**: 207–26.

Whitam, F. L., Diamond, M. and Martin, J. 1993. Homosexual orientation in twins: a report on 61 pairs and three triplet sets. *Archives of Sexual Behavior* **22**: 187–206.

White, T. D. 1990. Gait selection in the Brush-tail possum (*Trichosurus vulpecula*), the Northern quoll (*Dasyurus hallucatus*), and the Virginia possum (*Didelphis virginiana*). *Journal of Mammalogy* **71**: 79–84.

Whitehead, N. E. 2007. An antibody antibody? Re-examination of the maternal immune hypothesis. *Journal of Biosocial Science* **39**: 905–21.

Whiteman, S. D., McHale, S. M. and Crouter, A. C. 2007. Competing processes of sibling influence: observational learning and sibling deidentification. *Social Development* **16**: 642–61.

Whitney, J. B. and Herrenkohl, L. R. 1977. Effects of anterior hypothalamic lesions on the sexual behaviour of prenatally-stressed male rats. *Physiology and Behavior* **19**: 167–9.

Whitten, P. and Turner, T. R. 2004. Male residence and the patterning of serum testosterone in vervet monkeys (*Cercopithecus aethiops*). *Behavioral Ecology and Sociobiology* **56**: 565–78.

Whitten, P. L., Lewis, C., Russell, E. and Naftolin, F. 1995. Potential adverse effects of phytoestrogens. *Journal of Nutrition* **125**: 771S–776S.

Whittingham, L. A., Dunn, P. O. and Magrath, R. D. 1997. Relatedness, polyandry and extra-group paternity in the cooperatively-breeding white-browed scrubwren (*Sericornis frontalis*). *Behavioral Ecology and Sociobiology* **40**: 261–70.

Wichman, A. L., Rodgers, J. L. and MacCallum, R. C. 2006. A multilevel approach to the relationship between birth order and intelligence. *Personality and Social Psychology Bulletin* **32**: 117–27.

Wickelgren, I. 1999. Discovery of 'gay gene' questioned. *Science* **284**: 571.

Wickler, W. 1967. Socio-sexual signals and their intra-specific imitation among primates. In: Morris, D. (Ed.) *Primate Ethology*, pp. 69–79. Chicago, IL: Aldine.

Widdig, A., Bercovitch, F. B., Streich, W. J. *et al.* 2004. A longitudinal analysis of reproductive skew in male rhesus macaques. *Proceedings of the Royal Society of London*, B **271**: 819–26.

Wielgus, R. B. and Bunnell, F. L. 1995. Tests of hypotheses for sexual segregation in grizzly bears. *Journal of Wildlife Management* **59**: 552–60.

Wiese, J. H. 1976. Courtship and pair formation in the great egret. *Auk* **93**: 709–24.

Wikan, U. 1977. Man becomes woman: transsexualism in Oman as a key to gender roles. *Man* **12**: 304–19.

Wilkins, J. F. 2005. Genomic imprinting and methylation: epigenetic canalization and conflict. *Trends in Genetics* **21**: 356–65.

Wilkins, J. F. and Haig, D. 2003. What good is genomic imprinting: the function of parent-specific gene expression. *Nature Reviews* **4**: 1–10.

Wilkinson, G. S. 1984. Reciprocal food sharing in the vampire bat. *Nature* **308**: 181–4.

Wilkinson, G. S. 1985. The social organization of the common vampire bat. I. Pattern and cause of association. *Behavioral Ecology and Sociobiology* **17**: 111–21.

Wilkinson, G. S. 1986. Social grooming in the common vampire bat, *Desmodus rotundus*. *Animal Behaviour* **34**: 1880–9.

Williams, C. L., Serfass, T. L., Cogan, R. and Rhodes, O. E. Jr. 2002. Microsatellite variation in the reintroduced Pennsylvania elk herd. *Molecular Ecology* **11**: 1299–310.

Williams, D. A. 2004. Female control of reproductive skew in cooperatively breeding brown jays (*Cyanocorax morio*). *Behavioral Ecology and Sociobiology* **55**: 370–80.

Williams, G. C. 1966 *Adaptation and Natural Selection: A Critique of Some Current Evolutionary Thought*, Princeton, NJ: Princeton University Press.

Williams, L. E. and Abee, C. R. 1988. Aggression within mixed age-sex groups of Bolivian squirrel monkeys following single animal introductions and new group formations. *Zoo Biology* **7**: 139–45.

Williams, J. R., Insel, T. R., Harbaugh, C. R. and Carter, C. S. 1994. Oxytocin administered centrally facilitates formation of a partner preference in female prairie voles (*Microtus ochrogaster*). *Journal of Neuroendocrinology* **6**: 247–50.

Williams, T. J., Pepitone, M. E., Christensen, S. E. *et al.* 2000. Finger-length ratios and sexual orientation. *Nature* **404**: 455–6.

Williams, W. L. 1993. Being gay and doing research on homosexuality in non-Western cultures. *Journal of Sex Research* **30**: 115–20.

Williamson, N. B., Morris, R. S., Blood, D. C., Cannon, C. M. and Wright, P. J. 1972. A study of oestrus behaviour and oestrus detection methods in a large commercial dairy herd. *Veterinary Record* **91**: 58–62.

Willmott, M. and Brierley, H. 1984. Cognitive characteristics and homosexuality. *Archives of Sexual Behavior* **13**: 311–19.

Wilmer, J. W., Allen, P. J., Pomeroy, P. P., Twiss, S. D. and Amos, W. 1999. Where have all the fathers gone? An extensive microsatellite analysis of paternity in the grey seal (*Halichoerus grypus*). *Molecular Ecology* **8**: 1417–29.

Wilson D. S. 1975. A theory of group selection. *Proceedings of the National Academy of Sciences, USA* **72**: 143–6.

Wilson, D. S. 2005. Evolutionary social constructivism. In: Gotschall, J. and Wilson, D. S. (Eds) *The Literary Animal: Evolution and the Nature of Narrative*, pp. 20–37. Evanston, IL: Northwestern University Press.

Wilson E. O. 1975. *Sociobiology: The New Synthesis*. Cambridge, MA: Harvard University Press.

Wilson, E. O. 1978. *On Human Nature*. Cambridge, MA: Harvard University Press.

Wilson, G. 1987. An ethological approach to sexual deviation. In: Wilson, G. D. (Ed.) *Variant Sexuality: Research and Theory*, pp. 84–115. Baltimore, MD: Johns Hopkins University Press.

Wilson, G. A. and Strobeck, C. 1999. Genetic variation within and relatedness among wood and plains bison populations. *Genome* **42**: 483–96.

Wilson, J. D. 1999. The role of androgens in male gender role behaviour. *Endocrine Reviews* **20**: 726–37.

Wilson, J. D. and Roehrborn, C. 1999. Long-term consequences of castration in men: lessons from the skoptzy and the eunuchs of the Chinese and Ottoman courts. *Journal of Clinical Endocrinology and Metabolism* **84**: 4324–31.

Wilson, P. 1984. Aspects of reproductive behaviour of Bharal (*Pseudois nayaur*) in Nepal. *Zeitschrift für Säugetierkunde* **49**: 36–42.

Wilson, P. J., Grewal, S., Rodgers, A. *et al.* 2003. Genetic variation and population structure of moose (*Alces alces*) at neutral and functional DNA loci. *Canadian Journal of Zoology* **81**: 670–83.

Winget, C. and Farrell, R. A. 1972. A comparsion of the dreams of homosexual and non-homosexual males. *Psychophysiology* **9**: 119.

Wingfield, J. C., Newman, A. L., Hunt, G. L. and Farner, D. S. 1982. Endocrine aspects of female-female pairing in the Western gull (*Larus occidentalis wymani*). *Animal Behaviour* **30**: 9–22.

Winkler, H., Leisler, B. and Bernroider, G. 2004. Ecological constraints on the evolution of avian brains. *Journal of Ornithology* **145**: 238–44.

Winkler, J. J. 1990 Laying down the law: the oversight of men's sexual behavior in classic Athens. In: Halperin, D. M., Winkler, J. J. and Zeitlin, F. I. (Eds) *Before Sexuality: The Construction of Erotic Experience in the Ancient Greek World*, pp. 171–209. Princeton, NJ: Princeton University Press.

Winnie, J. Jr and Creel, S. 2007. Sex-specific behavioural responses of elk to spatial and temporal variation in the threat of wolf predation. *Animal Behaviour* **83**: 215–25.

Winterbottom, M., Burke, T. and Birkhead, T. R. 2001. The phalloid organ, orgasm and sperm competition in a polygynandrous bird: the red-billed buffalo weaver (*Bubalornis niger*). *Behavioral Ecology and Sociobiology* **50**: 474–82.

Wirtz, P. 1983. Multiple copulations in the waterbuck (*Kobus ellipsiprymnus*). *Zeitschrift für Tierpsychologie* **61**: 78–82.

Wise, R. A. 1996. Neurobiology of addiction. *Current Opinion in Neurobiology* **6**: 243–51.

Witelson, S. F., Kigar, D. L., Scamvougeras, A. *et al.* 2008. Corpus callosum anatomy in right-handed homosexual and heterosexual men. *Archives of Sexual Behavior* **37**: 857–63.

Witt, D. M. 1999. Regulatory mechanisms of oxytocin-mediated sociosexual behaviour. In: Carter,

C. S., Lederhendler, I. I. and Kirkpatrick, B. (Eds) *The Integrative Neurobiology of Affiliation*, pp. 343–357. Cambridge, MA: MIT Press.

Wolf, J. B. W., Kauermann, G. and Trillmich, F. 2005. Males in the shade: habitat use and sexual segregation in the Galápagos sea lion (*Zalophus californianus wollebaeki*). *Behavioral Ecology and Sociobiology* **59**: 293–302.

Wolf, L. L. and Stiles, F. G. 1970. Evolution of pair cooperation in a tropical hummingbird. *Evolution* **24**: 759–73.

Wolfe, L. D. 1986. Sexual strategies of female Japanese macaques (*Macaca fuscata*). *Human Evolution* **1**: 267–75.

Wolfe, L. D. 1991. Human evolution and the sexual behavior of female primates. In: Loy, J. and Peters, C. B. (Eds) *Understanding Behavior: What Primate Studies Tell Us about Human Behavior*, pp. 121–51. New York: Oxford University Press.

Wolitski, R. J., Valdiserri, R. O., Denning, P. H. and Levine, W. C. 2001. Are we headed for a resurgence of the HIV epidemic among men who have sex with men? *American Journal of Public Health* **91**: 883–8.

Wong, B. and Parker, K. L. 1988. Estrus in black-tailed deer. *Journal of Mammalogy* **69**: 168–71.

Woodroffe, R. and MacDonald, D. W. 1995. Female/female competition in European badgers *Meles meles*: effects on breeding success. *Journal of Animal Ecology* **64**: 12–20.

Woodson, J. C. 2002. Including 'learned sexuality' in the organization of sexual behaviour. *Neuroscience and Biobehavioral Reviews* **26**: 69–80.

Woodson, J. C., Balleine, B. W. and Gorski, R. A. 2002. Sexual experience interacts with steroid exposure to shape the partner preferences of rats. *Hormones and Behavior* **42**: 148–57.

Worthington, R. L. and Reynolds, A. L. 2009. Within-group differences in sexual orientation and identity. *Journal of Counseling Psychology* **56**: 44–55.

Wrangham, R. 1993. The evolution of sexuality in chimpanzees and bonobos. *Human Nature* **4**: 47–79.

Wrench, J. S. 2002. The impact of sexual orientation and temperament on physical and verbal aggression. *Journal of Intercultural Communication Research* **31**: 85–106.

Wright, I. C., Sham, P., Murray, R. M., Weinberger, D. R. and Bullmore, E. T. 2002. Genetic contributions to regional variability in human brain structure: methods and preliminary results. *Neuroimage* **17**: 256–71.

Wright, S. 1921. Correlation and causation. *Journal of Agricultural Research* **20**: 557–85.

Wright, S. 1934. The method of path coefficients. *Annals of Mathematical Statistics* **5**: 161–215.

Wrynn, A. S., Sebens, J. B., Koch, T., Leonard, B. E. and Korf, J. 2000. Prolonged c-Jun expression in the basolateral amygdala following bulbectomy: possible implications for antidepressant activity and time of onset. *Molecular Brain Research* **76**: 7–17.

Würsig, B., Guerrero, J. and Silber, G. K. 1993. Social and sexual behaviour of bowhead whales in fall in the Western Arctic: a re-examination of seasonal trends. *Marine Mammal Science* **9**: 103–10.

Würsig, B. and Würsig, M. 1979. Behavior and ecology of the bottlenose dolphin, *Tursiops truncatus*, in the South Atlantic. *Fisheries Bulletin* **77**: 399–412.

Wynne-Edwards, V. C. 1962. *Animal Dispersion in Relation to Social Behavior*. London: Oliver & Boyd.

Yahner, R. H. 1979. Temporal patterns in male mating behavior of captive Reeve's muntjac (*Muntiacus reevesi*). *Journal of Mammalogy* **60**: 560–7.

Yalcinkaya, T. M., Siiteri, P. K., Vigne, J.-L. *et al.* 1993. A mechanism for virilisation of female spotted hyenas in utero. *Science* **260**: 1929–1931.

Yamada, Y. 1999. Quantitative and qualitative differences between adult and juvenile agonistic behavior. *Journal of Ethology* **17**: 63–71.

Yamagiwa, J. 1987. Intra- and inter-group interactions of an all-male group of Virunga mountain gorillas. *Primates* **28**: 1–30.

Yamagiwa, J. 2006. Playful encounters: the development of homosexual behaviour in male mountain gorillas. In: Sommer, V. and Vasey, P. L. (Eds) *Homosexual Behaviour in Animals: An Evolutionary Perspective*, pp. 273–93. Cambridge: Cambridge University Press.

Yamamoto, D. and Nakano, Y. 1999. Sexual behaviour mutants revisited: molecular and cellular basis of *Drosophila* mating. *Cellular and Molecular Life Sciences* **56**: 634–46.

Yamane, A. 1998. Male reproductive tactics and reproductive success of the group-living feral cat (*Felis catus*). *Behavioural Processes* **43**: 239–49.

Yamane, A. 1999. Male homosexual mounting in the group-living feral cat (*Felis catus*). *Ethology, Ecology and Evolution* **11**: 399–406.

Yamane, A. 2006. Frustrated felines: male-male mounting in feral cats. In: Sommer, V, and Vasey, P. L. (Eds) *Homosexual Behaviour in Animals: An Evolutionary Perspective*, pp. 172–88. Cambridge: Cambridge University Press.

Yamane, A., Shotake, T., Mori, A., Boug, A. I. and Iwamoto, T. 2003. Extra-unit paternity of hamadryas baboons (*Papio hamadryas*) in Saudi Arabia. *Ethology, Ecology and Evolution* **15**: 379–87.

Yeager, C. P. 1990. Notes on sexual behaviour of the proboscis monkey (*Nasalis larvatus*). *American Journal of Primatology* **21**: 223–7.

Yearsley, J. M. and Pérez-Barbería, F. J. 2005. Does the activity budget hypothesis explain sexual segregation in ungulates? *Animal Behaviour* **69**: 257–67.

Yerkes, R. M. 1939. Social dominance and sexual status in the chimpanzee. *Quarterly Review of Biology* **14**: 115–36.

Young, A. 2000. *Women who Become Men: Albanian Sworn Virgins*. Oxford: Berg.

Young, L. C., Zaun, B. J. and VanderWerf, E. A. 2008. Successful same-sex pairing in Laysan albatross. *Biological Letters* **4**: 323–5.

Young, L. J. and Wang, Z. 2004. The neurobiology of pair bonding. *Nature Neuroscience* **7**: 1048–54.

Young, L. J., Wang, Z. and Insel, T. R. 1998. Neuroendocrine bases of monogamy. *Trends in Neurosciences* **21**: 71–5.

Zahavi, A. 1974. Communal nesting by the Arabian babbler: a case of individual selection. *Ibis* **116**: 84–7.

Zahavi, A. 1975. Mate selection – a selection for a handicap. *Journal of Theoretical Biology* **53**: 205–14.

Zajonc, R. B., Markus, H. and Markus, G. B. 1979. The birth order puzzle. *Journal of Personality and Social Psychology* **37**: 1325–41.

Zajonc, R. B. and Sulloway, F. J. 2007. The confluence model: birth order as within-family or between-family dynamic? *Personality and Social Psychology Bulletin* **33**: 1187–94.

Zala, S. M. and Penn, D. J. 2004. Abnormal behaviours induced by chemical pollution: a review of the evidence and new challenges. *Animal Behaviour* **68**: 649–64.

Zastowny, T. R., Lehman, A. F. and Dickerson, F. 1987. Klinefelter's syndrome and psychopathology: a case study of the combined effects of nature and nurture. *International Journal of Psychiatry in Medicine* **17**: 155–62.

Zehr, J. L., Maestripieri, D. and Wallen, K. 1998. Estradiol increases female sexual initiation independent of male responsiveness in rhesus monkeys. *Hormones and Behaviour* **33**: 95–103.

Zeier, H. J. and Karten, H. J. 1973. Connections of the anterior commissure in the pigeon (*Columba livia*). *Journal of Comparative Neurology* **150**: 201–16.

Zenchak, J. J. and Anderson, G. C. 1980. Sexual performance levels of rams (*Ovis aries*) as affected by social experiences during rearing. *Journal of Animal Sciences* **50**: 167–74.

Zenchak, J. J., Anderson, G. C. and Schein, 1981. Sexual partner preference of adult rams (*Ovis aries*) as affected by social experiences during rearing. *Applied Animal Ethology* **7**: 157–67.

Zenger, K. R., Eldridge, M. D. B. and Cooper, D. W. 2003. Intraspecific variation, sex-biased dispersal and phylogeography of the eastern grey kangaroo (*Macropus giganteus*). *Heredity* **91**: 153–62.

Zenilman, J. 1988. Sexually transmitted diseases in homosexual adolescents. *Journal of Adolescent Health Care* **9**: 129–38.

Zhang, X.-S. and Hill, W. G. 2005. Genetic variability under mutation selection balance. *Trends in Ecology and Evolution* **20**: 468–70.

Zhou, J.-N., Hofman, M. A., Gooren, L. J. G. and Swaab, D. F. 1995. A sex difference in the human brain and its relation to transsexuality. *Nature* **378**: 68–70.

Zhou, W. and Chen, D. 2008. Encoding human sexual chemosensory cues in the orbitofrontal and fusiform cortices. *Journal of Neuroscience* **28**: 14416–21.

Ziegler, T. E. 2007. Female sexual motivation during non-fertile periods: a primate phenomenon. *Hormones and Behavior* **51**: 1–2.

Zietsch, B. P., Morley, K. I., Shekar, S. N. *et al*. 2008. Genetic factors predisposing to homosexuality may increase mating success in heterosexuals. *Evolution and Human Behavior* **29**: 424–33.

Zimmerman, S. J., Maude, M. B. and Moldawer, M. 1965. Frequent ejaculation and total sperm count, motility, and form in humans. *Fertility and Sterility* **16**: 342–5.

Zink, A. G. and Reeve, H. K. 2005. Predicting the temporal dynamics of reproductive skew and group membership in communal breeders. *Behavioral Ecology* **16**: 880–8.

Zinner, D., Peláez, F. and Torkler, F. 2001. Group composition and adult sex-ratio of Hamadryas Baboons (*Papio hamadryas hamadryas*) in Central Eritrea. *International Journal of Primatology* **22**: 415–30.

Zito, K. and Svoboda, K. 2002. Activity-dependent synaptogenesis in the adult mammalian cortex. *Neuron* **35**: 1015–17.

Zoccali, R., Muscatello, M. R., Bruno, A. *et al*. 2008. Gender role identity in a sample of Italian male homosexuals. *Journal of Homosexuality* **55**: 265–73.

Zucker, K. J. 2008a. Special issue: Biological research on sex-dimorphic behaviour and sexual orientation. *Archives of Sexual Behavior* **37**: 1.

Zucker, K. J. 2008b. On the "natural history" of gender identity disorder in children. *Journal of the American Academy of Child and Adolescent Psychiatry* **47**: 1361–3.

Zucker, K. J., Beaulieu, N., Bradley, S. J., Grimshaw, G. M. and Wilcox, A. 2001. Handedness in boys with gender identity disorder. *Journal of Child Psychology and Psychiatry* **42**: 767–76.

Zucker, K. J., Bradley, S. J. and Sanikhan, M. 1997. Sex differences in referral rates of children with gender identity disorder: some hypotheses. *Journal of Abnormal Child Psychology* **25**: 217–27.

Zucker, K. J., Lightbody, S., Pecore, K., Bradley, S. J. and Blanchard, R. 1998. Birth order in girls with gender identity disorder. *European Child and Adolescent Psychiatry* **7**: 30–5.

Zucker, K. J., Mitchell, J. N., Bradley, S. J. *et al.* 2006. The recalled childhood gender identity/gender role questionnaire: psychometric properties. *Sex Roles* **54**: 469–83.

Zucker, K. J., Wild, J., Bradley, S. J. and Lowry, C. B. 1993. Physical attractiveness of boys with gender identity disorder. *Archives of Sexual Behavior* **22**: 23–36.

Zuckerman, S. 1932. *The Social Life of Monkeys and Apes.* London: Kegan Paul.

Zuger, B. 1989. Homosexuality in families of boys with early effeminate behavior: an epidemiological study. *Archives of Sexual Behavior* **18**: 155–66.

Zuloaga, D. G., Puts, D. A., Jordan, C. L. and Breedlove, S. M. 2008. The role of androgen receptors in the masculinization of brain and behavior: what we've learned from the testicular feminization mutation. *Hormones and Behavior* **53**: 613–26.

Index

abdominal adiposity, 113–114
acquired immunodeficiency syndrome, of apes, 376
activity budget, 299, 301, 302, 303
adaptive evolution, 89–91, 280–415
adaptive mechanism, in stress, 127–130
adaptive neutrality, 34–35
adaptive traits, 402–403
adoption, birth order and, 116–117
adrenal glands, 128
adrenal hyperplasia, congenital, 116, 164, 165, 166
adrenal steroids
    sexual attraction, 130
    stress and, 119
adrenarche, 128
adrenocorticotropic hormone (ACTH), 120, 121, 122
adult brain, sexual dimorphisms, 206–207
adult intervention
    juvenile sexual segregation, 313, 315
    negativity in gender non-conformity, 144
    permissiveness in sex play, 140
adult sex ratio, 358–359, 369–371
adulthood, immunity and homosexuality, 274
age
    age-stratified homosexuality, 8, 129
    and bisexuality, 6, 8
    changes in sexual orientation, 8, 418
    development of sexuality, 129, 397
    in dominance hierarchies, 330
    of chosen partner, 258
    relationship stability, 6
aggregation signal, 89–90
aggressiveness
    empathy and, 249
    female twins, 118
    hyenas
        female, 185–191

new-born, inter-sibling, 186, 187
    masculinity, 240
albatrosses, 311, 347, 347–348
Alces alces gigas (Alaskan moose), 306–307
alcohol, prenatal stress, 120
Alle alle (dovekie), 232
alleles
    feminising, 61–62
    heterozygote advantage hypothesis, 61
    influencing homosexuality, 60, 62
    interdemic selection, 59–60
    mating success and parental care, 59
    phenotypic variation, 60
altered testosterone biosynthesis (ATB), 160–162
altriciality, 415
altruism, 300, 419
amygdala
    medial, 70, 200
    primate, 137
    roles of, 221, 222
anabolic androgenic steroids (AAS), 191–193
anal trauma, 376
anal warts, 378
ancient Greece, homo-erotic interactions, 396–397, 400
androgen deprivation therapy (ADT), 196–197
androgen insensitivity syndrome (AIS), 68–69, 163–164
androgen receptor gene, 69
androgen receptor locus, 79–80
androgen receptors (AR), 104, 123
androgens
    evolution of homosexuality, 406
    postnatal, 119
    prenatal, 119
        intersibling aggression, 187
        masculinised female offspring, 389
        masculinised genitalia, 191
        sexual behaviour differentiation, 106
        sexual play behaviour in lambs, 151
androgyny, 249–250
androstenedione
    pregnant hyenas, 187
    sexual orientation and, 193
    stress and, 119
aneuploidies, 81–82, 420
animal homosexual behaviour, overview, 32
animal same-sex sexual behaviour, early reviews, 11–26
Anseriformes, phylogeny, 42
antecedent brother effect 266, 270
anterior commissure (AC), 209, 213–214

anterior hypothalamus (AH), 208–210
anterior pituitary, 131
anteroventral periventricular nucleus (AVPV), 198
Anthropoidea (anthropoids), 384–385
    classification, 381, 384, 388
    same-sex mounting, 387, 401
    self-medication, 376
antiandrogenic effects, of EDCs, 131–132
antibody production, maternal, 267
antioestrogen, maternal reactivity, 272
antiparasitic hypothesis, 80
antiprogesterone, 267, 272
antitestosterone, 267, 272
Apis mellifera (honeybee), 409–410
apoptosis
    and *Bax* gene, 70
    neuronal loss, 155, 205
Arapesh people, Papua New Guinea, 313
arginine vasopressin, 172
armpit odours, 217–218
aromatase activity
    and brain development, 104
    prenatal, and digit length, 107
    prenatal stress and, 122
aromatase deficiency, congenital, 162–163
Aromatisation Hypothesis 104
Artiodactyla (even-toed ungulates), 43, 301
asexual heterosexuals, 162–163
asexual homosexuals, 37
assertiveness, 240
ATB (altered testosterone biosynthesis), 162
attachment
    mother–infant, 248
    relationship stability, 248
auditory evoked potentials (AEPs), 113
auditory response, 219
autoantibodies, age and, 274–275
autoimmune disorders, handedness and sexuality, 76
autoimmunity, 159–160, 266
autosomal inheritance, 61–62
autosomal locus, reproductive success, 58–59
AVPV (anteroventral periventricular nucleus), 198
avuncular tendencies, *fa'afafine* 94

Bagemihl, Bruce, 11–26
Balaena mysticetus (bowhead whale), 357
Bandura, Albert, 26
bats, 309
Bax gene, 69–70

Bax knock-out mice, 69–70
Bcl-2 protein family, 69
bed nucleus of the stria terminalis (BNST)
   Bax knockout mice, 69
   c-fos proto-oncogene expression, 70
   in sexual behaviour, 200
   principal nucleus, 69–70
   sexual dimorphism, 213, 209
Bedborough, George, 9
behaviour patterns, learning, 134
behaviour, gender non-conforming, 140–145
behavioural development, delayed, 128
behavioural incompatability, 299
behavioural influences, siblings, 72, 74
behavioural learning, 275–276
behavioural masculinisation, in females, 123
behavioural plasticity, 243–246, 408
behavioural variables, 337
   avian data, 346, 368–369
   mammalian data, 349–356, 359, 370–371
Behavioural Contagion hypothesis, 276
Benkert, Karl-Maria (Kertbény), 4
bias potential, 83–84, 85
   datasets used, 38–39
   in twin studies, 86
bidirectional plasticity, 99
bimodal distribution of sexuality, 5, 6
biological essentialism, 26
Biology vs. Culture dichotomy, 48–52
biosocial effects
   family dynamics, 276–281, 280–281
   in sex-type toy preference, 115
   on sexual orientation, 89
      parental age, 271–272
biosocial integration, 26–28
Biosocial Model of Homosexuality 401–403, 413–419
   alternative hypotheses, 419–420
   bisexuality *vs.* exclusive homosexuality, 416–417
   development, 403–413
   diagrammatic summary, 414
   future research, 420–421
   hypotheses, 402–403
   novel evolutionary synthesis, 55–56
birds
   behavioural variables, 368–369
   Biosocial Model of Homosexuality 401–403
   brain centres, 211
   brain evolution, 261
   brain nuclei, 71

   brain steroid receptors, 201–202
   candidate genes, 71
   digit measurements, 109
   Fos protein in sexual behavior, 71
   independent contrasts, 439–440
   learning during early ontogeny, 135
   mammals compared, 367–372, 401–403
   masturbation, 232
   mating systems, 319–320
   migratory, 311, 312
   other-sex mimicry, 150–152
   phylogenetic-based studies, 41–43, 317–318
   polygyny in, 317
   prenatal hormones, 103
   redirected mounting, 232
   same-sex mounting
      body mass in, 319–320
      female–female
         evolution of, 109
         hormones and, 171
      in captivity
         sampling issues, 46–54
         table of species, 47–49
   sexual dimorphism, 211
   sexual segregation, 311–312, 319–320, 321
      same-sex mounting, 285–289, 321
   social complexity, 261
   social play, 134–135
   sociality hypothesis, 317–318
   Synthetic Reproductive Skew Model of Homosexuality 336–348, 367–372
      behavioural variables, 369
      evolution of female same-sex mounting, 109
      life histories and behaviourable variables, 346
      mammals compared, 372
      path analysis, 362–363
      predictions and results, 434–435
      simplified, 364
         core and sexual variables, 364
         core and socio-sexual variables, 365
      social bonding, 357
   Z chromosome, genetic mutations, 78
birth order
   changes, sibling mortality, 279–280
   immunological mechanisms, 274
   IQ score, 303
   parental age effects, 275–281
   personality dimensions, 277
Birth Order Effect 116–117

bisexuality
  age for, 129
  attraction and arousal, 7–8
  changes over time, 8
  defined, 2, 36
  distribution, 6
  evolution of homosexuality, 408, 415
  fru (Drosophila) mutants, 66
  heterozygote advantage, 60–61
  human potential for, 397–398
  otoacoustic emissions, 113
  parental care, 59
  political lesbians, 409
  postnatal stress and, 281
  reproductive success, 323–324
  SDN-POA volume, 210
  sexually antagonistic selection, 91
  sibling birth order, 271
  stability of, 6–7
  STD transmission, 378–379, 421
  vs. exclusive homosexuality, 416–417
blood types, homosexuality and, 79
body mass
  birds, 319–320, 357, 368–369
  mammals, 319–320, 358, 359, 370–371
body mass sexual dimorphism (BMSD), dataset in mammals, 348–357
body size dimorphism, 301
  Bovidae, 304
  Caprinae, 304
  foraging hypothesis, 299
bonding
  mother–infant, 248
  positive affective bonds, 247
Bos bison (wild bison), 304
Bos taurus (domestic cattle),
  buller steer syndrome, 178–179
  hormones and mounting behaviour, 170–171
  sexual segregation, 304
Bovidae, 302–305
boy-inseminating rites, 9
boys *see also* Fraternal Birth Order (FBO)
  gender non-conformity, 140–145
  sexual identity and, 143
brain
  size (encephalisation), 415
  structural complexity, 259–262
brain centres *see also* individual areas
  activation

    cognitive tests, 253
    orgasm, 230
  alterations in *tfm* rats, 69
  emotions and, 246–248
  in sexual arousal, 227–228
  in sexual behaviour, 199–201
  olfactory receptors and sexuality, 184
  reshaping, isoflavones and, 198
  sexual dimorphism, 206–207
  stress effects, 129
brain development *see also* brain plasticity
  and androgen insensitivity, 69
  embryonic steroid hormones, 101
  genetic influences, 69–70
    Bax gene in, 70
    Sry gene, 75–76
  in sex-type toy preference, 115
  lateralisation, 111
  maternal stress hormones, 125
  neoteny in evolution, 154–155
  prenatal steroid hormones, 58
brain nuclei
  Fos protein, 71
  sexual dimorphism, 207
brain plasticity
  biosocial integration, 26
  evolution of homosexuality, 409, 415, 417
    predictions in, 416
  gender role ontogeny, 404, 405
  in learning, 237
  personality traits, 280
  sexual behaviour and, 206
  synaptogenesis, 204
brain steroid receptors, 202
brain tissue, gene expression, 64
brain-derived neurotrophic factor (BDNF) gene, 259
branch lengths, calculation, 41
Brazil, Santa Catarina, 7
breeding success, 255, 328–330
Bubalornis niger (red-billed buffalo weaver), 226–227
buller steer syndrome, 178–179
bullying, sissy boys, 146–147
butch lesbians, 242–243

CAH (congenital adrenal hyperplasia), 116, 166
CAIS (complete androgen insensitivity syndrome), 164
canalisation, 98–100
Candida albicans infection, 376
candidate genes, 66, 72

in birds, 71
in *Drosophila* 68
in humans, 72, 74, 76
in rodents, 71
novel possibilities, 77
testable hypotheses, 58
Capra hircus (goat), 302
Capra ibex (Alpine ibex), 302
Caprinae, 302–303
captivity *see also* prison inmates
    all-male environment, 397
    lock-in model, 173–174
    masturbation, 232–233
    same-sex mounting in, 54
    social partnerships, 367
Carnivora, 43
castration, 196–197
catecholamines, 121–122
catechol-o-methyltransferase (COMT) gene, 259
cation channel, trp, 2, 71
cats, 210, 221
causal relationships
    Biosocial Model of Homosexuality 413–415
    natural phenomena, 34
    path analysis, 44–45
Cavia aperea (guinea pig), 122–123
CD8 immune response, 274
Cebus capucinus (white-faced capuchin), 221
Cebus spp. (capuchins), 385
central nervous system (CNS), 32, 203, 208
    foetal, 265–266
    fru gene expression, 66
    functional and morphological studies, 220
    hypersexuality and homosexuality, 222
central tegmental field (CTF), 70
Cercopithecus aethiops sabacus (vervet monkey), 115
cerebellum, size heritability, 259
cerebral asymmetry, 219
cerebral cortex, 231, 279
Cervidae, 307
Cervus elaphus (European red deer), 306
Cervus elaphus roosevelti (Roosevelt elk), 306
Cervus eldi thamin (thamin), 306
Cetaceae, 42–43
c-fos proto-oncogene, 70
Charadriiformes, 42
CHD1 gene, 78
chemical castration, 197
childhood stress effects, 127

children
    gender identity disorder, 142–143
    gender non-conforming, 145
    interest in having, *Synthetic Reproductive Skew Model of Homosexuality* 195–199
    numbers of, homo-/heterosexual partnerships, 374–381
    same-sex peer preference, 314
    sexual segregation, 313, 314
Chiroptera, 43, 309
Chlamydia infection, 376
cholesterol levels, birds, 347
chromosomal, men, genital anomalies, 163–164
chromosomal, women, genital anomalies, 163
chromosome 7, q36 region, 78–79
click-evoked OAEs (COAEs), 113
clitoris, masculinised, 185–191
Clostridium perfringens infection, 376
coalition formation, 335
coefficient of regression, 44–45
coercion, sexual segregation and, 299, 315–316
cognitive abilities, *265*
cognitive processes, 250–255
Cognitive-Developmental Theory 148–149
collaboration, and subordinate role, 331
collaborative discourse games, 312
community mores, and learning, 250
comparative analyses
    datasets used in, 39
    of independent contrasts
        birds, 439–440
        mammalian sample, 440
        schematic diagram, 40
competition avoidance, 300
competitive games, 312
complete androgen insensitivity syndrome (CAIS), 160–164
Compromise Model 327–328
Concession Model 325–326
Confluence Model 406–407
congenital adrenal hyperplasia (CAH), 116
congenital aromatase deficiency (CAD), 162–163
constructionist paradigm, 10
contagious behaviour theory, 276
continuous distribution of sexuality, 5
continuum theory, of bisexuality, 8
cooperation/competition, family dynamics, 280
cooperative behaviors *see also* kin selection, selection for, 94–95

cooperative breeding, 336–346, 348, 360
copulation rate, 70–71, 166–167
Corcorax melanoramphos (white-winged chough), 331
corpus callosum (CC), 214–215
cortical regions, 231
  grey matter quantity, 260
corticosteroids, 116–119
corticosterone, 121–122
cortisol, 193
Coturnix japonica (Japanese quail)
  learning in sexual orientation, 135
  prenatal hormones, 102–103
coumestrol, 198
courtship ritual
  in birds, 388, 390
  in insects, 66
Crocuta crocuta (spotted hyena)
  group genetic relatedness, 331
  masculinised females, 191
  mate choice by females, 185–191, 331
  mounting behaviour, 189
  MPOA/AH and sexual dimorphism, 207
  pseudopenis, 186
cross-cultural studies, 239–243, 8–9
crowding stress, prenatal, 119
Cryptoprocta ferox (fossa), 190
culture
  gender roles, 243
  homosexual behavior, 238
  juvenile sex play, 138–140
  physical contact, 239
  views on homosexuality, 235–273
Cyanocorax morio (brown jay), 331
CYP17 gene 76–77, 163

Dagg, Anne Innis, 11
Dama dama (fallow deer), 266–306, 43, 360
Danaus plexippus (monarch butterfly), 376
Darwin, Charles, sterile castes paradox, 1
datasets, in comparative analyses, 38–39
defeminised females, mice *Bax* mutations, 70
definitions, in studies of homosexual behavior, 35–38
de-identification, sibling interactions, 277, 406–407
delayed plumage maturation
  neoteny, 152
  other-sex mimicry, 152
  same-sex mounting, 152
Delphinapterus leucas (beluga whale), 283
de-masculinised males, 119–120

dental crown size, 113
derivation, homosexual behaviour in primates, 388–392
dermatoglyphics, 112–113
development
  canalisation in, 98
  environmental EDCs, 131–132
  instability, 98–99
  juvenile sex play, 134–138
  ontogenetic patterns, 32
  plasticity, 98–99
  stability in, 98
Developmental Constraints theory, 259–260
developmental disorders, 266
developmental syndromes, 132–134
dexametasone (DEX), 121–122
diethylstilboestrol (DES), 132–134
digoxin levels, 222
dihydrotestosterone (DHT), deficiency, 162
directional selection, 62–64
dispersal constraints, 331
distribution of homosexuality, family effects, 64
diversity, and evolution, 381–384
dizygotic (DZ) twins, 85–89
DNA-methylation
  antiparasitic hypothesis, 80
  future research, 421
  genomic imprinting, 80–81
  X-chromosome inactivation, 79–80
Dnmt3L gene, 79
dominance
  female sexual behavior, 168
  homosexual mounting in birds, 336
  in masculinity, 239–240
  in prison, 316
  lock-in model, 172, 175, 178, 190
  queuing to breed, 330
  ram homosexuality, 181
  social, 189–191, 299
dominance hierarchies
  age in, 330
  establishment
    in Bovidae, 305
    in Cervidae, 306
  Old World monkeys, 385–386
  Pinnipedia, 310
  reproductive skew theory, 323
  synthetic, 335–336
  reproductive success, 331
  same-sex mounting in birds, 346

dopamine, 223, 224
  metabolism, 259
  prenatal stress and, 122
  reward system, 172
  social stress and, 407
dormancy, evolution of, 39–40
dorso-ventral mounting, female–female, 173
Dörner's model, 206–207, 254
Drosophila melanogaster (fruit fly) *see also fru* gene
  candidate genes, 57
  courtship, 173
  genderblind (gb) gene, 68
  pheromones and, 67

ear breadth, fluctuating asymmetry, 131
early-development effects
  hormonal effects, 103–106
  reproductive skew theory, 330
  same-sex mounting, 402
ecological constraints
  and philopatry, 333
  birds, same-sex mounting in, 345
  cooperative breeding, 360
  evolution of homosexuality, 415
  family groups, 325
  inbreeding avoidance, 334–335
Ecological Theory 10
Ecologically-driven Model 318
EDCs (endocrine-disrupting chemicals), 198–199, 132–141
effeminate men, 141–142
egalitarian homosexuality, 9
ejaculation, brain centres in, 201, 230–231
elephants, 309
Ellis, Henry Havelock, 9–10
embracing behaviour, 388
emotional attachment, 247–248
emotional coping, 249–250
emotions, brain centres, 246–248
empathy, and aggressiveness, 249
endocrine disorders, 159–166
endocrine syndromes, human, 165
endocrine system, 32, 100
endocrine-disrupting chemicals (EDCs), 198–199, 132, 141
Entomophthora muscae (fungus), 375–376
Environment of Evolutionary Adaptedness (EAA), 91–93
environmental aridity, 40, 155
environmental chemicals *see also* endocrine-disrupting chemicals (EDCs),

insect communication, 67
environmental conditions
  brain plasticity and, 205
  endocrine system and, 100
  evolution of homosexuality, 409, 417
  genetic make-up and, 65
  in family studies, 82
  in sexual selection, 259
  pre- and postnatal, 89, 417
epistemological caveats, 33–35
Equidae
  hormones, 170
  masturbation, 232
  sexual segregation, 307
Equus asinus (donkey), 170
Equus caballus (horse), 170
Equus quagga (plains zebra), 307
Erotic Orientation Model 145
erotic stimuli, brain centres in, 200–201
Erotic Target Location Error (ETLE), 148
estra-1,3,5(10),16-tetraen-3-ol (EST), 216–217
estradiol benzoate (EB), 134
estradiol levels
  female sexual behaviour and, 167
  in CAH, 164
  oophorectomy and, 197
  pregnant hyenas, 187
  sexual orientation and, 193
ethical issues
  human genetics and sexuality, 72
  studies in homosexuality, 424–425
ethnicity
  hand digit measurements, 106–111, 108–122
  sociocultural variables, 84–85
ethology, aims of, 34
eunuchs, in history, 196
European countries, juvenile sex play, 138–139, 140
eusociality, 409–410
evolution
  biosocial integration, 28
  brain development, 259–263
  diversity in primates, 381–384
  diversity of mechanisms, 35
  ontogeny and, 152–157
  purpose of orgasm, 225–227
  theories, 10
evolution of homosexuality
  and maintenance, 326–327, 402–403
  Biosocial Model of Homosexuality, 401–403

genetic models, 62, 64, 97, 406
theories, 62–64, 402–403
evolutionary hypotheses
 same-sex mounting, 40–41
 sociality hypothesis, 318, 321
 Synthetic Reproductive Skew Model of Homosexuality, 361–367
evolutionary mechanisms, integration, 64–65
evolutionary paradox of homosexuality, 1–3
 evolution, 413–415
 exclusive, 398
 formulated, 3
evolutionary psychology, 27
exaptations, 35
exaptive hypotheses, 23
exaptive traits, 403
exclusive homosexuality
 at young age in men, 418–419
 evolution of homosexuality
  brain plasticity, 409
  predictions in, 416
  vs. bisexuality, 416–417
 phenotypic trait, 422
 uniqueness, 402
Exotic Becomes Erotic (EBE) model, 145–146
extended family (human), 268
extra-pair copulations (EPC), in birds, 345, 357
extra-pair paternity (EPP), in birds, 336, 357

fa'afafine (feminine homosexual men), 423
facial features, sexual selection, 256
Falco sparverius (American kestrel), 311
Falconiformes, 41
familial effects, 64, 82–85, 83–84
Family Dynamics Model, 325–326, 406–407, 279
family multipoint linkage analysis, 78–79
family relationships
 dominance in, 330
 dynamics, 116, 277–281, 280–281
 stress in, 127, 140, 406–407
family size, 92–93, 116–117, 268
fatherhood
 at older age, 272
 wish for, 373–375
fecundity *see also* maternal fecundity
 and digit measurements, 109–110
feeding precedence, 185
female gender rearing, ATB boys, 160–162
female genitalia in CAIS, 163–164

Female Mimicry hypothesis, 150–152
female partnerships, in birds, 171, 347–348
female sexual aggregation, 315
female sexual orientation
 elevated steroids, 77
 neurophysiological plasticity, 27
female twins
 concordance for sexual orientation, 88–89
 different-sex pairs
  left hand 2D: 4D ratio, 109
  masculinisation, 117–118
  tooth crown measurement, 113
female-associated aneuploidies, 81
female–female embracing, 388
female–female mounting *see also* genito-genital rubbing (GG-rubbing); ventro-ventral mounting
 Crocuta crocuta (spotted hyena), 186
  hormone levels, 166–170
 in birds, evolution of, 109
 in primates, 156–157
 socio-sexual functions, 174–177
females
 homosociality, 312–316
 masculinised, 123, 190
 masturbation, 232
 prenatal stress and, 123
 sexual identity in CAH, 164–165
 sexual orientation changes, 417–418
 subordinate, 123
feminine homosexual men
 cultural perspective, 242–243
 fa'afafine, 423
femininity, and homosexuality, 239–243
feminisation
 in mammals, 103–106
 pheromones and, 67
 prenatal androgens, 111
feminised males (rats), 119–120
feminised men, cultural practices, 6–7
feminising alleles, 61–62
ferrets, 103
fetishism, 147–148
finger prints, dermatoglyphics, 112–113
Finland, 8
fitness consequences of homosexuality, 91
fitness decreasing trait, 89–91
fitness of homosexuality, 91
five factor model, 277
fluctuating asymmetry (FA), 130–131

fluoxetine (FLX), 223–224, 225, 230
foetal antigens, 265–266
foetal CNS. *immune privilege of*, 265
foetal testosterone, 111–112
follicle stimulating hormone (FSH), 193
foodstuffs, EDCs in, 198–199
foraging hypothesis, 299–300
   albatrosses, 311
   Bovidae, 305
   Delphinapteros leucas, 310–311
   in mammals, 319
   Pinnipedia, 43
foraging strategy, 301, 306
forebrain activation, 230
Fos protein, 71
Foucault, Michel, 10
Fragile X mental retardation (*FMR1*) locus, 79–80
Fraternal Birth Order (FBO), 116–117
   and immunity, 266, 270
   effect on sisters, 270–271
   evolution of homosexuality, 327
   gender-typed toy preference, 115
Freemartin Effect, 117–118
Freud, Sigmund, 10–11
friendship networks, same-sex, 314–315
frontal cortex, 260
fru (Drosophila) mutants
   bisexual behaviour, 66
   courtship behaviour, 67
fru gene, 66–67
   expression, 66
   transcript, 66
fru-mAI neurons, 66
functional connectivity, cerebral, 219–220
functions, homosexual behaviour, 388–392

gastrocentric hypothesis, 300, 305
gay fathers
   desire for fatherhood, 374–375
   orientation of children, 149
gaydar concept, 256–258
Gebusi people, 9
gender dysphoria
   in childhood, 142–143
   Klinefelter's syndrome (KS), 273
   prenatal DES exposure, 133
gender identity
   and behaviour, 2
   in CAIS, 163–164

Gender Identity Disorder (GID)
   psychoanalytical theories, 10–11, 422, 423, 424
   tomboy girls, 142–143
Gender inversion model, 143–254
gender inversion, age of partner, 257–258
gender non-conforming children, 140–144
gender prevalence, in human homosexuality, 56–57
gender roles
   cross-cultural studies, 239–243
   development
      5α-reductase deficiency, 162
      ATB boys, 160–162
      plasticity in, 407–408
   human endocrine syndromes, 165
   one locus two allele model, 61–62
gender transitions, plasticity in, 245–246
gender, hand digit measurements, 107
genderblind (gb) gene, 67–68
gene expression
   brain steroid receptors, 202
   brain tissue, 64, 259
   DNA-methylation, 80
gene transmission, 323–324
gene–culture evolution, 27–28, 99–100
genetic accommodation, 99–100
genetic assimilation, 99
genetic disorders, 77
genetic distance, *Synthetic Reproductive Skew Model of Homosexuality*, 358
genetic hypotheses
   study difficulties, 57–58
genetic mechanisms 97
   exclusive homosexuality, 418
   heritability and, 82
   questions or unknowns, 56
   support from twin studies, 86–89
   theoretical work, 62–64
genetic models, 63
genetic mutations
   accumulation, 75
   evolution of homosexuality, 417
   rate variability, 55–56
   recurrent, 56, 64–65
   sex-biased rates, 78
genetic relatedness
   heterosexual groups, 331
   mammals and birds compared, 371
   same-sex groups, 331
   socio-sexual behaviour, 335

544   Index

genetic traits
   and environmental conditions, 65
   Biosocial Model of Homosexuality, 403
genital contact/investigation, primates, 387, 388
genital displays, primates, 387, 388
genital stimulation, 394
genitalia
   feminised, 163–164
   masculinised, 185–191
genito-genital rubbing (GG-rubbing), 38, 168, 391, 394
Geschwind–Galaburda model, 111
gestation, immune mechanisms during, 76
Gestational Neurohormonal Theory, 58
gibbons, 233, 384
Giraffa camelopardalis (giraffe), 357
Giraffa camelopardalis peralta (Niger giraffe), 307
Giraffa camelopardalis tippelskirchi (Masai giraffe), 307
Giraffidae, 307
girls
   gender identity disorder, 142–143
   gender non-conformity, 140–144
   sexual identity, 143
   sororal birth order, 270–271
glial amino acid transporter, 68
glial factors, in synaptogenesis, 204
glossary, 426–436
glucocorticoids, 122
gonad removal, 195–196
gonadal anlage, brain development and, 101–102
gonadal steroids, brain centre activation, 201–202
gonorrhoea infection, 378
Gorilla gorilla (gorilla), 383–384
gregariousness, 240
grey matter
   distribution, 220, 259
   quantity, 279
group size, reproductive skew theory, 331
guinea pigs, 123
Guira guira (Guira cuckoo), 331

habitat segregation, 311, 317
Habitat Social Segregation, 284
hair-whorl direction, 77
Hamilton, William D, 324
Hamilton's Rule, 324, 326–327
hand digit measurements, 107, 111, 122, 131
handedness, 112, 268–269
   left, 76, 112, 133
Hardy–Weinberg equilibrium, 58

hedonic feedback, 392–394, 399
hepatitis B virus (HBV) infection, 378
hepatitis C virus (HVC) infection, 378
heritability
   estimates, 58
   realised (h^2) 82
heterochrony, 152–153
heterosexuality, Kinsey Scale, 395–396
heterosexuals
   age for sexual activity, 129, 396
   asexual, 162–163
   desire for parenthood, 373–375
   fecundity, 374
   sexual plasticity, 243–246
   STD transmission, 375–377
heterozygote advantage, 62–64
   bisexuality and, 60–61
   genetic inheritance, 55, 61–62
Hippocamelus bisulcus (huemul), 306
Hirschfeld, Magnus, 5, 9–10
historical overview, 3–6
   early researchers, 29–31
HIV-AIDS infection
   INAH 1 size, 211
   STD transmission, 377, 378, 379
   susceptibility to, 274
Hominoidea, classification, 381, 383–384
homosexual behavior, 395–399
   defined, 1–2, 35–37
   mounting in birds, 336
   non-human, 11–26
   phylogenetic patterns, 384–388
   potential influences, 323
   predicted influences, 333–336
   with or without attraction, 1–2
homosexual men
   abdominal adiposity, 113–114
   hand digit measurements, 107–108
   orientation among male relatives, 73
homosexual partnerships
   children, numbers of, 374
   wish for fatherhood, 373–375
homosexual rams, 179–185, 211, 402
homosexuality
   defined, 1–2, 35–36
   maintenance of, 62–64, 401–403
   pathology hypothesis, 9–10, 11, 421–424
homosexuals, asexual, 37
homosocial groups, *Capra ibex* 303

homosociality, 37, 312–316
  and homosexuality, 315–316, 408–409, 410
  in child-rearing, 348
hormone levels, 166
  elevated, 76–77
  in primates, 166–170
  orgasm and, 229–230
  same-sex nesting, 347
  sexual orientation and, 191, 193–195
hormones
  activational effects, 100–106, 201–202
  and 2D : 4D digit ratio, 106–111
  antibodies against, 267, 272
  biosynthesis, 160, 161
  brain development and, 57, 58
  buller steer syndrome, 178–179
  EDCs, impact of, 131–132
  effects in mammals, 170–171
  evolution of homosexuality, 406
  exposure to, 103–106, 114–440, 406, 416
  intracellular mechanisms, 203
  non-steroid, 171–191
  organisational effects, 106, 408
  stress and, 119
  tooth crown size, 113
Hox gene, 106
human genetics, and sexuality, 72, 77–78
human genome, study across, 78–79
human homosexuality
  comparative study, 3
  genetic modelling studies, 62
  stress effects, 128
  study sampling difficulties, 58, 57–58
  theoretical mechanisms, 62–64
  uniqueness, 402
human leukocyte antigen (HLA), 76, 265
human populations, fitness studies, 91
human sex chromosome aneuploidies, 81–82, 420
humans
  intrauterine positioning effect, 117–118
  juvenile sex play, 138–140
  male-biased adult sex ratio, 359–360
  prenatal hormones, 103, 104
  sexual pheromones, 215–218
  sexual segregation, 315, 316
  sociality, 298
humoral immunity, 274–275
H-Y antigen
  Klinefelter's syndrome (KS), 273

maternal reactivity, 272–273
parity effects, 266, 274
Hyaena hyaena, (striped hyena) 190
Hyaenidae, phylogeny, 189–200
Hylobates spp. (gibbons), 384
hypercarnivory, 189–191
hypersexuality, 221–222
hypothalamic preoptic area (POA), 101, 102, 122, 137
hypothalamic suprachiasmatic nucleus (SCN), 212
hypothalamus *see also* interstitial nuclei of the anterior hypothalamus (INAH3); medial preoptic area (MPOA);
  activation by pheromones, 217
  anterior (AH), 208
  in sexual behaviour, 200
  mediobasal region, 122
  primate juvenile play, 137
  sexual dimorphism, 208–210
hypothalamus–pituitary axis (HPA), 125–126
hypotheses
  organisation of, 28
  table of, 26

immune dysregulation, 275
immune mechanisms
  during gestation, 76
  evolution of homosexuality, 327
  FBO effect, 270
  parental age effect, 271
immune privilege of fetal CNS, 265
immunology, and homosexuality, 266, 274
Immunoreactive theory, 266
impotence, after orchiectomy, 195
inbreeding avoidance
  ecological constraints, 334–335
  in reduced natal dispersal, 334
  juvenile sex play, 138
incentive, in learning, 236–237
inbreeding avoidance, 147
incest avoidance, 331
inclusive fitness, 127, 323–324, 323–324
Incomplete Control model, 328
independent contrasts
  avian taxa phylogeny, 43
  comparative analyses, 440
  mammalian taxa phylogeny, 44
indigenous cultures, 94
indirect fitness, 93
indirect parental care, 300

individual disadvantage, genetic mutation, 56
industrialised societies, 93
infanticide prevention, 308
infants (human), sexual segregation, 312
infection, susceptibility to, 274
inheritance, of homosexual behaviour, 60
Inia geoffrensis (river dolphins, botos), 310
insect communities, sterile castes paradox, 1
insects *see also Drosophila melanogaster* (fruit fly); *Musca domestica* (domestic fly)
    candidate genes, 68
    communication, 67
integrationist biosocial approach, 237
integrationist theories, 28
integrative pluralism, 34
intellectual development, intersibling learning, 280
interdemic selection, 55, 60, 327–328, 411
interhemispheric connectivity, 219–220
intersexuality, 37
interstitial nuclei of the anterior hypothalamus (INAH3)
    castration and, 196
    sexual dimorphism, 208, 211, 212, 217
Intrauterine Positioning Effect, 118
inversion model of homosexuality, 4
invertebrates, 65–66
IQ scores, and birth order, 279, 280
isocortical regions
    evolution of, 260–261
    grey matter distribution, 220, 259
    in ejaculation/orgasm, 231
isoflavones, 198
Israeli kibbutzniks, juvenile sex play, 138

jealousy, 248
Jewish physicians, early sex research, 4
juvenile development (human)
    same-sex mounting, 37
    sexual segregation
        adult intervention, 283–298
juvenile development (primate)
    same-sex mounting, 399
    social development, 389
juvenile sex play, 134–138, 138–140
juvenile traits, in sensory mechanisms, 405–406

kin selection, 95
    Biosocial Model of Homosexuality, 411
    fraternal birth order, 116
    genes controlling homosexuality, 55
    reproduction hypothesis, 300
    reproductive skew theory, 323
    selection for parental care, 59
Kinsey, Alfred Charles, 5–11
Kinsey Institute for Research in Sex, Gender and Reproduction, 5
Kinsey Reports, 5
Kinsey Scale, 5
Klinefelter's syndrome (KS), 81
Krafft-Ebing, Richard von, 5, 9–10

labour-intensive societies, 240–241
Lagopus lagopus alexandrae (willow ptarmigans), 311
Larus argentatus (Herring gull), 152
Larus californicus (California gull), 347
Larus delawarensis (Ring-billed gull), 347
Larus occidentalis (Western gull), 347
Larus spp., joint nesting behaviour, 348
learning
    early, 134
    early hormonal exposure, 103
    evolution of homosexuality, 415
    ontogeny of homosexuality, 145–147
    plasticity, 28
    processes, 236–239
    sex-typed toy preference, 114, 126
    social context, 134, 148–149
left-handedness, 76, 112, 133
Legitimation League, 9
Lemuroidea (Madagascar lemurs), 385
lesbian mothers
    desire for motherhood, 373–374
    orientation of children, 149
lesbians
    childhood gender atypical behaviour, 142
    evolution of homosexuality, 417–419
    family dynamics, 281
    hand digit measurements, 107–108
    left handedness, 112
    masculine (butch), 242–243
    older fathers, 272, 275
    otoacoustic emissions, 113
    political bisexuals, 409
    political homosexuals, 149
    sibling birth order, 270–271
    sibling homosexuals, 83
    STD transmission, 377–378, 379
    testosterone levels, 194

libido
    loss of after oophorectomy, 197
    loss of after orchiectomy, 195
life histories hypotheses, 22
life-history traits, mammals and birds compared, 369
life-history variables
    avian data, 345, 369
    mammalian data, 349, 358, 370
lignans, 198
limbic–hypothalamic regions, 70
linkage disequilibrium, 55
Lipotes vexillifer (baiji, Chinese river dolphin), 232
lock-in model, 168, 173–178, 190
log-body mass contrasts, in birds, 357
log-body mass standardised
    avian data, 368
    mammalian data, 358
longevity, 92–93
Loxodonta africana (African elephant), 309
luteinising hormone (LH), 182, 193

Macaca arctoides (stumptail macaques), 221
    amygdala lesions, 221
    dominance hierarchies, 328
    female–female mounting style, 395
    lock-in model, 178
    masturbation, 233
Macaca cyclopis (Formosan rock macaque), 376
Macaca fuscata (Japanese macaque)
    lock-in model, 173–177
    masturbation, 234
    MPOA/AH and sexual dimorphism, 211
    same-sex mounting, 167–168
        adult sex ratio, 358–359, 361
        female–female style, 157
Macaca mulatta (rhesus monkey)
    amygdala lesions, 221
    dataset, 348
    dominance hierarchies, 331
    hormonal levels, 110
    and mating activity, 166
    and mounting activity, 167
    learning mounting technique, 137
    lock-in model, 177–178
    mate choice by females, 331
    MPOA/AH and sexual dimorphism, 211
    organisational role of steroids, 105
Macaca nigra (Celebes Sulawesian macaques), 388
Macaca radiata (bonnet macaques), 312, 385

Macaca spp., lock-in model, 178
Macronectes halli (northern giant petrel), 311
Macropus fuliginosus (western grey kangaroo), 308
Macropus rufus (red kangaroo), 308
male birds, same-sex nesting, 347
male foetuses, maternal immune reaction, 266
male gender prevalence
    in human homosexuality, 56–57
    sex-biased mutation rates, 77–78
male homosexuality
    evolution of, 417–419
    interdemic selection, 59–60
    rates in USA, 82–83
male intervention hypotheses, 300, 312
Male Mutation Bias, 77–78
male relatives of homosexuals, sexual orientation of, 73
male twins, concordance for sexual orientation, 88–89
male-associated aneuploidies, 81
male–male bonding, in indigenous societies, 237–239
male–male mounting
    Dama dama (fallow deer), 54, 360
    Porphyrio porphyrio, 38
    primates, 386, 385
mammalian taxa, phylogenetic tree, 43
mammals
    Biosocial Model of Homosexuality, 401–403
    digit measurements, 108
    EDCs and sexual behaviour, 198–199
    female sexual aggregation, 315
    hormones and mounting behaviour, 166–171
    independent contrasts, study, 440
    juvenile sexual play, 135–138
    life-history and behavioural variables, 359, 370–371
    life-history variables, 358
    masturbation, 232–235
    neoteny in evolution, 153–154
    phylogenies in comparative analyses, 41–43
    polygyny, 301, 307–308
    prenatal hormones, 103
    same-sex mounting in captivity, 46–54 349–386
    sexual segregation, 302–311
    social play, 134–135
    sociality hypothesis, 319–321, 349–356
    Synthetic Reproductive Skew Model of Homosexuality,
        367–372, 348–373, 372
    path analysis, 362–363
    predictions and results, 434–435, 437–438
Marsupialia, 43, 308
masculinity, and homosexuality, 239–243

masculinisation
  androgenic steroid use, 192
  behavioural, 123
  Crocuta crocuta (spotted hyena), 185–191
  cultural perspective, 7, 242–243
  female twins, 117–118
  pheromones and, 67
  prenatal androgens, 103–106, 111, 389
  testosterone levels, 194
masturbation
  causal link theory, 5
  homosexuality and, 235–236
  in birds, 232
  in primates, 393–394, 395, 399
  orgasm and forebrain activation, 230
  same-sex sexual behaviour, 419
  segregated living, 284
mate choice
  by females, 331
  incest avoidance, 330–331
maternal effects, 62–64
  learning mounting technique, 137
  mother–infant bonding, 248
  of education level, 126
maternal fecundity
  homosexual offspring, 62, 91, 398
  relatives of homosexuals, 62, 74
  sexually antagonistic genes, 55
Maternal Immune Hypothesis (MIH)
  contradictions, 273–274
  FBO effect and, 266–270
  foetal antigens, 265–266
  mental disorders in offspring, 275
maternal mortality, 185–186, 190–191
maternal stress, 118–124
Maternal Stress Hypothesis, 125
mating success, 58–59, 255–259
mating systems
  as qualitative trait, 41
  mammals and birds compared, 367–369
  phylogenetic-based studies, 317
  polygynandrous or promiscuous, 185
  reproduction hypothesis, 300
  reproductive biology, 335
  sociality hypothesis, 317–318, 319–321
  Synthetic Reproductive Skew Model of Homosexuality 358
Mayer–Rokitansky–Küster–Hauser syndrome, 163
medial amygdala (MeA), 200

medial preoptic area (MPOA)
  c-fos proto-oncogene expression, 70
  hormone receptors, 123
  in sexual behaviour, 196–198, 200
  oxytocin and, 172
  sexual dimorphism, 210, 208–211
medial preoptic nucleus (MPN), 71
mediobasal region, hypothalamus, 122
Megaptera novaeangliae (North Atlantic whale), 310
Megaderma lyra (Indian false vampires), 309
Melanerpes formicivorus (acorn woodpecker), 331
Melanesia, 313–315
men
  brain centres, 231
  distribution of sexuality over time, 6–7
  effeminate, 141–143, 242–243
  oestrogen feedback response, 211
menstrual cycle, 383
  synchronisation, 218
mental state (human), 402
Meriones unguiculatus (Mongolian gerbil), 118
mesodiencephalic transition zone, 231
meta-analytical methods, 45–46
methodology, comparative studies, 28–32
metrosexuals, 424
Mexico, 7
Microtus agrestis (field vole), 199
Microtus oeconomus (tunda vole), 199
migratory birds, 311–312
mimicry, other-sex, 149–152
Miopithecus talapoin (tolapoin monkeys), 312
Mirounga argustirostris (northern elephant seals), 309
modernist fallacy, 240
Modified Socially Driven Model, 318
monkeys *see also* New World monkeys; Old World monkeys
  sexual segregation, 312
  social strees, 315
monosexual groups, *Capra ibex* 302
monosocial groups, *Social Sexual Segregation* 283–284
monozygotic (MZ) twins, 85–89
Mosaic theory brain evolution, 259–260
mother–infant bonding, 248
moult migration, 300
mounting behaviour
  c-fos proto-oncogene in, 71
  reason for focus on, 37
  same-sex, 37
  social development, 389

Sry gene deletion studies, 76
style in primates, 385, 386
Muller's ratchet 74–75
multicausality principle, 33–34
Mus musculus (mouse)
    Bax knock-out, 69–70
    brain plasticity, 204
    candidate genes, 57, 71
    females
        intrauterine positioning, 118
        lacking androgen receptors, 69
    male
        c-fos proto-oncogene, 71
        lacking androgen receptors, 69
        sexual segregation studies, 308
    Sry gene deletions, 75–76
    testicular feminisation mutation (*tfm*), 68–69
    trp 2, olfactory receptors, 71
    vomeronasal organ, 215
    Y^{129} chromosome, 75–76
Mus musculus (Swiss Webster mouse), 123
Musca domestica (domestic fly), 376
Mustela putoricus (European polecate), 308
mutations, genetic variability, 56, 404–406
mutation-selection equilibrium 56, 64
mutualism, 55, 411
Myotis daubentonii (vespertilionid bat), 309
Myotis lucifugus (little brown bat), 309
Myotis septentrionalis (northern long eared bat), 309
Mysticetae, 310

natal dispersal, inbreeding avoidance, 334
natural phenomena, causal complexity, 33–34
natural selection
    brain evolution, 261
    evolution of homosexuality, 410
    genetic mutations, 56
Naturalistic Fallacy, 424–425
navigation, and gender, 253–254
neotenic species, 416
neotenic trait, 152–157
neoteny hypothesis, 9
    childhood stress effects, 128
    delayed plumage maturation, 152
    evolution of homosexuality, 45
        future research, 421
        olfactory sensory mechanisms, 406
    evolutionary process, 100
    homosexuality as a trait, 152–157

nerve cell activation, 202
nesting behaviour, joint, 346–348
nesting success, plumage colour, 151–154
neurobiological hypotheses, 36
neurochemistry, in sexual orientation, 222
neuroendocrine system, 199–201
    evolution of homosexuality, 415
    stress effects, 121–123, 130
neurogenesis, 205
neuronal development
    alcohol and prenatal stress, 120
    Bax knock-out mice, 69–70
    oestrogen and, 104
neuronal mechanisms, evolution of homosexuality, 408
neuronal pathways, plasticity, 202–203
neurons
    apoptosis, 205, 154–155
    INAH3 region of the hypothalamus, 196
neurophysiological plasticity, 26, 27
neurotransmitters, 78–79, 202, 222–224
New Guinea, 359–360
New World monkeys, 382–383, 385
New Zealand, 139
nitric oxide (NO) levels, 222
nocturnal prosimians, 381–382
nomadism, 39–40
non-steroid hormones, 171–191
noradrenaline, 122
Notiomystis cinta (stitchbird), 232
nuclear transcription factor, 70–71
nutritional requirements, Caprinae, 304
nutritional stress, prenatal, 119–120

OAEs (otoacoustic emissions), 113
Ockham's Razor, on causal complexity, 33–34
Odocoileus hemionus columbianus (black-tailed deer), 305
Odocoileus hemionus hemionus (Rocky mountain mule deer), 305
Odocoileus virginanus (white-tailed deer), 305
Odontoceti, 310
oestrogen
    embryonic production, 101–102
    feedback response, 195
    non-steroidal, DES, 132–134
    prenatal, 103–106
    sexual skin swellings, 383
    sexuality in rams, 182–183
oestrogen receptors (ER), 163, 104, 123

oestrogenic effects
    buller steer syndrome, 179
    of EDCs, 132
offspring dispersal, 325
Old World monkeys, 382–383, 385–386
older fathers, 272, 274–275
olfactory bulb, 184
olfactory epithelium, 215–218
olfactory neurocentres, 261
olfactory receptors
    evolution of homosexuality, 404
    sexuality in rams, 181–184
    trp 2 and, 71
one locus two allele model of homosexuality, 61–62
ontogeny of homosexuality
    developmental processes, 32
    early learning, 145–147
    in evolution, 152–157, 414, 416–417
Onuf's nucleus, 218
oophorectomy, 195–198
Ophryocystis elektroscirrha (protozoan), 376
Opponent Process theory 146
orang-utan, 384
orchiectomy, 195–198
organisational sex hormones, 103–106, 408, 420
orgasm
    brain centres in, 231
    forebrain activation, 230
    in primates, 393
    same-sex sexual behaviour, 224–232, 419
other-sex mimicry, 149–152
otoacoustic emissions (OAEs), 113
Ovibos muschatus (muskox) 136
ovine sexually dimorphic nucleus (oSDN), 183–184
Ovis aries (domestic sheep),
    homosexual, exclusive, 402
    juvenile sexual play, 135–136
    olfactory sensory mechanisms, 183–184
    prenatal androgens, 113, 137
    rams
        bisexual, 179–180
        homosexual, 179–185, 211
        dominance in, 181
        same-sex sexual behaviour, 179–185
        sexual preference phenotypes, 180
    Ovis aries (Soay sheep), 302
    Ovis canadensis (bighorn sheep), 303, 328–330, 179
    Ovis canadensis mexicana (desert bighorn sheep), 303–304

Ovis dalli dalli (Dall's sheep), 302–303
Ovis gmelini (mouflon), 304
oxytocin (OT), 171–172

paedomorphosis, 153–154, 156
Pan paniscus (bonobo) *see also* genito-genital rubbing (GG-rubbing)
    play-mounting behaviour, 137–138
    sexual behaviour, 156–157, 168–170
    STD transmission, 376
Pan spp. 376
Pan troglodytes (chimpanzee)
    play-mounting behaviour, 137–138
    sexual behaviour, 156
    sociality, 383–384
    STD transmission, 376
Pan troglodytes verus (chimpanzee) 376
Papio cynocephalus (yellow baboon)
    dominance hierarchies, 331
    mate choice by females, 331
    presentation postures, 393
Papio hamadryas (hamadryas baboons), 388
Papio papio (guinea baboons), 384
Papio spp. (baboons), 384
Papio ursinus (Chamba baboon), 233
Papua New Guinea
    Arapesh people, 313
    Gebusi people, 9
    Sambia people, 397
paradox of altruism, 419
parasite transmission, 375–379
paraventricular nucleus (PVN), 70, 209
parental age effects, 271–272, 274–275, 275–281
parental care, 59, 255–256
parental educational level, 115, 126
parental investment, in offspring, 277, 278–279
parental manipulation, 55, 64, 335
Parental Manipulation theory, 330
parental-specific DNA methylation, 80
parent–offspring interaction, 406
parity effect
    maternal immunity, 266
    parental age and, 271–272
    sibling birth order, 270–271
partner choice, same-sex, 256–258
Passeriformes, 311
Passerina amoena (Lazuli buntings), 151–152
paternal antigens, maternal immune reaction, 265
paternal major histocompatability (MHC) genes, 265

paternity, extra-pair (EPP), in birds, 336–345
path analyses, 44, 45, 54, 361–367
path coefficient, in path analysis, 44–45
pathogen transmission, 375–379
pathology hypothesis, 9–10, 11, 421–424
p-chlorophenylalanine (PCPA), 199
Pearson's product-moment correlations, 84, 317, 318
    avian data, 346, 368–369
    mammalian data, 358, 359, 370–371
peer pressure, 314, 315
peer rejection
    gender non-conformity, 143–144
    sissy boys, 146, 147
peramorphosis, 153
perception, in sex-typed toy preference, 115–116
Perissodactyla, 43, 307
personal wellbeing, 92–93
personality dimensions, birth order, 277
personality structure, 277
personality traits, brain plasticity, 279–280
phalloid organ, in birds, 226–227
phenotypic differentiation, of siblings, 277, 278
phenotypic distribution, inheritance of homosexual behaviour, 60
phenotypic plasticity, 99
phenotypic variability, brain plasticity in, 203–204
pheromones, 216–217
    buller steer syndrome, 179
    genderblind gene, 67–68
    hypothalamic areas and, 217
    insects
        aggregation signal, 89–90
        sexual behaviour, 67
    olfactory reception, 215–218
philopatry
    evolution of homosexuality, 411, 419
    exclusive homosexuality and, 416
    reproductive skew theory, 331
    synthetic reproductive skew theory, 333, 334
Phoca vitulina concolor (harbor seals), 309
Phoebastria ostralegus (Laysan albatross), 347, 348
phonetics of language, 254
phylogenetic relationships, primates, 384–388
phylogenetic trait conservatism, 40
phylogenetic tree
    avian taxa, 42
    mammalian taxa, 43
phylogenetic-based studies
    comparative analyses, 41–43
    independent contrasts calculation, 41
    mating system, 317, 317–318
    Synthetic Reproductive Skew Model of Homosexuality, 345, 367–369
phylogenetic-based studies, 26
phylogeny, Hyaenidae, 189
physical traits, sexual selection, 256
phytoestrogens, 198
phytosterols, 198–199
pied flycatcher, 150–151
'piggy-back' style mounting, 395
Pinnipedia, 43
pituitary gland, anterior, 131
plasticity
    behavioural, 243–246
    bidirectional, 99
    development processes, 98–99
    gene–culture evolution, 27–28, 99–100
    in gender role ontogeny, 148–149, 404, 405, 407
    in learning, 236–237
    in women, 9
    neuronal pathways, 202–203
    Onuf's nucleus, 218
    phenotypic, 99
play, in same-sex groups, 315
pleasure
    primate sexual behaviour, 392–395
    same-sex mounting, 175–176
    social constructionism, 10
Plecotus auritus (brown long-eared bat), 309
pleiotropic effects, 55
plumage
    colour and nesting success, 151
    maturation delayed, 152
    sexual dichromatism, 336, 345
    sexual dimorphism, 109
Podiceps cristatus (great crested grebe), 388, 391
Poephila guttata (zebra finch), 101–103
political homosexuals, studies in homosexuality, 149, 409
polygamy
    dominance mounting, 358
    evolution of, 317
    harems in mammals, 284
    reproduction hypotheses, 300
    Synthetic Reproductive Skew Model of Homosexuality, 358
polygyny
    female homosociality, 410

in mammals, 301, 308
  sexual segregation, 300, 317
population dynamics, interdemic selection, 60
Porphyrio porphyrio (pukeko, purple swamphen)
  dispersal constraints, 331
  dominance hierarchies, 331, 367
  genetic relatedness in same-sex groups, 331
positive affective bonds, 247
post-conception mating, 166–167
posterodorsal preoptic nucleus (PDPN), 70
postnatal development, 117, 134, 147
postnatal environment, 89, 407, 416, 417
postnatal stress, 126, 140, 280–281
power, in social constructionism, 10
predation hypothesis
  Bovidae, 304
  Caprinae, 302, 304
  Cervidae, 306, 307
  Delphinapterus leucas, 310–311
  Giraffidae, 307
predations, Biosocial Model of Homosexuality, 415–416
prednisone, prenatal exposure, 133
preference, in same-sex mounting, 176–177
prejudice, against bisexuals, 6
prenatal environment
  evolution of homosexuality, 406, 417
  exclusive homosexuality, 418
  in twin studies, 89
  stress, 416
prenatal steroid hormones, 116, 132–134
prenatal stress
  fluctuating asymmetry (FA), 131
  future research, 421
Prenatal Stress Syndrome
  feminised males, 120
  neuroendocrine effects, 123
preoptic area (POA), 122
prepulse inhibition (PPI), 220
Pre-school Activities Inventory (PSAI), 126
presentation postures, 392, 393
primates 385
  brain evolution, 262
  brain organisation, 104–105
  brain tissue, gene expression, 64
  digit measurements, 108–109
  diversity, 381–384
  female–female mounting, 156, 157, 180, 393, 401
  homosexual behaviour, 32, 381, 388–392, 402
  hormone levels, 166

juvenile sex play, 137–138
male–male mounting, 386
masturbation, 233–234
MPOA/AH and sexual dimorphism, 211
neoteny in evolution, 156–157
phylogenetic relationships, 43, *383, 388*
prosimian, 383
sexual segregation, 312–316
sociality, 298
STD transmission, 376
prison inmates
  dominance 316
Procolobus badius temminckii (Temminck's red colobus), 234
product-moment correlations
  avian data, 346, *369*
  mammalian data 359, 371
progesterone 134, 164, 347
prolactin (PRL)
  hand digits, 106
  orgasm and, 229, 230
  sexual orientation and, 193
  stress response, 131
promiscuity, STDs and, 375–379
Propithecus verreauxi (sifaka), 385
Prosimii (prosimians), 384–388
proximate causation, 413–414
proximate mechanistic hypotheses, 16
pseudohermaphrodites, 162
pseudopenis, 185–191
pseudopregnancy, 171
psychoanalytical theories, 11, 422
psychological androgyny, 249–250
psychosocial effects, DES in, 132–134
psychosocial neutrality-at-birth (PNAB) theory, 97
psychosocial sexuality-at-birth (PSAB) theory, 97
puberty, ages at, 128–130

q28 region, X-chromosome, 72–76
Qualified Investment Model, 255–256
qualitative traits, calculation, 41
queuing to breed, 328–330

rabbits, 199
Rangifer tarandus (reindeer), 306, 328–330
rank, and socio-sexual mounting, 391
Rattus norvegicus (rat)
  6-MBOA effects, 199

AVPV size, 198
brain centres in sexual activity, 200
candidate genes, 68–71
fluoxetine, 225, 226
industrial pollutants, 198
juvenile sex play, 137
MPOA and sexual dimorphism, 208–210
prenatal hormones, 103
phytoestrogens, 198
preoptic area, 122
social stress, 315
spinal nucleus of bulbocavernosus, 219
synaptogenesis, brain plasticity, 203
testicular feminisation mutation (*tfm*), 68–69
tryptophane deficiency, 199
Rattus spp., masturbation, 232
realized heritability (h^2) 82
receptors, steroid hormones, 203
reciprocal altruism, 55
recurrent mutation, genetic mechanism, 64–65
redirected mounting, 232
reinforcement, in learning, 236–237
rejection, of foetus, 265
relationship stability, 6, 248
reproduction
    paradox of sex, 1
    withdrawal from, 410
reproduction hypotheses, 300, 307, 319
reproductive biology, in synthetic reproductive skew theory, 335
reproductive costs, 186, 191
reproductive rates, exclusive homosexuality and, 416
Reproductive Skew Model of Homosexuality,
    core variables, 332, 333, 334, 360
    exclusive homosexuality, 418
    historical overview, 325–330
    tests of, 331
reproductive strategy
    Cervidae, 305, 306
    Elephantidae, 309
    Lagopus lagopus alexandrae, 311
reproductive success, dominance hierarchies, 331
research studies, ethical and political implications, 424–425
researchers
    past and current list, 29
    potential bias, 39, 83
resilience, childhood stress effects, 127
response to selection (R), 82

Restraint model, 326
restraint stress, 119, 122
Rhesus factor, blood types, 79
Rissa tridactyla (black-legged kittiwake), 337
ritualised behaviour, 388
ritualised homosexuality, 9, 238, 239, 360
Rodentia, 43
rodents *see also Mus musculus* (mouse); *Rattus norvegicus* (rat)
    hormones and mounting behaviour, 170, 171
    juvenile sex play, 136–137
    prenatal stress effects, 124
    VMN and sexual dimorphim, 212

Saimiri boliviensis boliviensis (Bolivian squirrel monkey), 312, 388
Saimiri sciureus (squirrel monkey), 312
Sambia people of Papua New Guinea, 237–239, 397
same-sex embracing, New World monkeys, 388
same-sex female mounting, by hyenas, 189, 190
same-sex female partnerships, birds, 171
same-sex groups, genetic relatedness in, 331
same-sex guardianship, in child-rearing, 348
same-sex mounting
    birds, 153, 319
    Caprinae, 302
    evolutionary models, 40
    full correlation matrix, 363–367
    in captivity, 46
    invertebrate, 89–91
    lock-in model, 178
    mammals, 298, 321, 348
    phylogenetic-based studies, 312, 320
    potential influences, 323
    potential observer bias, 39
    primates, 385, 386, 389, 392
    sociality, 321, 318
    Synthetic Reproductive Skew Model of Homosexuality, 336, 367
same-sex nesting behaviour, 345
same-sex partner choice, 256
same-sex sexual behaviour, segregated living, 284, 316
same-sex sexual behaviour 3, 26
same-sex sexual experience, 226
same-sex sexual play, in normal development, 144–145
Samoa, 423
sample bias, in twin studies, 86
sample size, potential bias, 84
SBO (sororal birth order), 271

scent marking behaviour, 383, 387, 388
schools, sexual segregation in, 313–314, 315
SDN-POA (sexually dimorphic nucleus of the preoptic area of the hypothalamus), 120
selection differential (S), realised heritability (h^2) 82
selection studies, 91
Selective mechanisms, *Biosocial Model of Homosexuality*, 410–411
self-identification, cultural practices, 7
selfish herd effect, 307
self-medication, by apes, 376
Semnopithecus entellus (Hanuman langurs), 390
sensorimotor cortex, 260
sensory mechanisms *see also* olfactory receptors; visual mechanisms,
  evolution of homosexuality, 404, 405
  in sexual selection, 107, 259
  sequential bisexuality, 2
  Sericornis frontalis (white-browed scrubwren), 331
  serotonin levels, 131, 222, 223–224
serotonin reuptake inhibitor, 224, 225, 226, 230
sex chromosome aneuploides, 81–82
sex hormone inhibition, 198–199
  lock-in model, 172–178
  organisational roles, 421
  POM volume, 137
sex offenders, chemical castration, 197
sex play, juvenile, 134–138
sex ratio (M/F), same-sex mounting, 359, 363
sex, paradox in, 1
sex-biased mutation rates, 77–78
sex-chromosomal involvement, 56–57
sex-hormone binding globulin (SHBG), 104
sex-specific androgen production, 68–69
sex-specific familial effects, 83–84
sex-specific splicing, of *fru*, 66
sex-typed toy preference, 114–116, 126
sex-typing, in infant play, 312
sexual abuse, 127
sexual activity
  age and sexuality, 129, 130
  anabolic androgenic steroid use, 192
  hormone levels and orientation, 194
sexual aggregation, 283, 315
sexual arousal *see also*, orgasm brain centres in 7–8, 200–201
sexual attachment, 247
sexual attraction
  adrenal steroids, 130

adrenarche and, 128
between twins, 88
hand digit measurements, 111
to youthful-looking individuals, 250
sexual behaviours
  and androgen production, 68–69
  and CNS centres, 203
  Bax mutations, 70
  differentiation, 101–103
  effects of EDCs, 198–199, 132
  hedonic aspects of, 393–395
  MPOA in, 209
  observations of, 39
  path analysis, 361–363
  prenatal stress effects, 118–124
  SDN-POA in, 210
  social bonding and oxytocin, 171–172
  with or without attraction, 2
sexual development, hormonal effects, 106–111
sexual deviance, FBO and, 268
sexual dimorphism
  body size, 299
  brain and spinal cord nuclei, 207
  brain centres, 206, 211–212
  Delphinapterus leucas, 310–311
  hypothalmus, structures, 217
  Loxodonta africana, 309
  mammals and birds compared, 369
  Mirounga angustirostris, 309
  ventromedial nucleus, 211–212
sexual fantasy scores, 125
sexual identity, 143, 153, 164–165
sexual imprinting, 147
sexual maturation, 128–130
sexual orientation
  and cognitive abilities, 261–265
  Bax mutations, 70
  changes over time, 417–418
  distribution patterns, 5
  gender non-conforming children, 143
  Human Rights and, 425
  juvenile sex play, 134–138
  Kinsey Scale, 395–396
  mental state in, 35, 36
  neurophysiological plasticity, 27
  of researchers, 39
  one locus two allele model, 61–62
  population distribution, 5–6
  sexual identity and, 153

twin studies data, 85–89
sexual patterns, ritualised behaviour, 389–391
sexual plasticity, 228–246
sexual play behaviour, prenatal androgens, 151
sexual preference phenotypes, 180
sexual segregation
    birds, 285, 311–319, 323
    evolution of homosexuality, 408–409, 419–420
    humans, 146, 316, 367, 397
    hypotheses, 298–301
    mammals, 180–181, 298, 302–311
    primates, 312–316
    same-sex mounting, 316–317
    sexual behavioural pattern, 32
    sociality, 285–285, 289–298
    spatio-temporal distribution, 298
    studies, 301–316
sexual selection
    evolution of homosexuality, 411
    in fetishism, 147–148
    influences, 255–259
sexual skin swellings, 383
sexual variables
    bird data, 365
    mammal data, 366
sexuality
    bimodal distribution, 5, 6
    brain development and *Sry* gene, 75–76
    steroid hormones and, 57
sexually antagonistic selection, 62–64
    autosomal inheritance, 61–62
    bisexuality, 91
    evolution of homosexuality, 411
    in insects, 90
    in sheep, 184–185
    maternal line fecundity, 62
    parental care and mating success, 58–59
    parental care by females, 59
    sibling gender, 62
sexually dimorphic nucleus (SDN), 211, 217, 208–211
sexually dimorphic nucleus of the hyena (hSDN), 188
sexually dimorphic nucleus of the preoptic area of the hypothalamus (SDN-POA), 120
sexually mutualistic selection, 58–59
sexually transmitted diseases (STDs), 375–379
siblings
    concordance for sexuality, 83–84
        rates of, 82–85
        twin studies, 87–89
    family dynamics, 276–281
    gender non-conformity in childhood, 143–144
    homosexual
        family multipoint linkage analysis, 78–79
        Xq28 region genetic markers, 72–75
    interactions, intersibling aggression, 186, 187
    interactions, 406–407
    mortality, birth order changes, 279–280
    play and learning in sexuality, 275–276
    sexually antagonistic selection, 62
simultaneous bisexuality, 2
single nucleotide polymorphisms (SNP), 76–77
sissy boys, 140–145
sisters
    birth order, 270–271
    concordance for sexuality, 83–84
    Slater's Index, 266–267
social behaviours, 36–37
    and reproductive skew theory, 325–326
    bonding, 171–172, 357, 371–372
    complex, 259–263
    cooperation, 59–60
    neoteny in evolution, 155–156
Social Cognitive Theory, 148–149
social competition, 123
social conditions
    constraints, 408–409
    hormone levels and, 166
    primate same-sex mounting, 389–392, 398
    ritualised behaviour, 381–391
social constructionist theories, 10–11, 26
social groups
    adult sex ratio, 358–359
    attachment in, 246–247
    birds, 336–345, 346
    Hominoidea, 383–384
    in philopatry, 333–334
    size of, 358, 371
social influences
    birth order and, 116–117
    complex brain evolution, 262–263
    distribution of sexuality data, 6–7
    family studies, 82
    homosexual siblings, 74
Social Intelligence hypothesis, 261
social isolation
    fetishism, 148
    masturbation, 232–233

social learning, 148–149
    and sexual orientation, 275–276
    community mores, 250
    early, 134, 389
    sibling interactions, 406–407
social network, maternal, 126
social play, 134–135, 156–157
social rearing conditions, 120–121
Social Sexual Segregation, 283–284
social species, 416
social stress
    adaptive mechanisms, 130
    adrenarche and, 128–129
    childhood gender non-conformity, 143–144
    GG-rubbing and, 169
    laboratory animals, 315
    prenatal, 122–123
    sibling interactions, 406–407
social traits, 372
social views, on homosexuality, 235–236
sociality
    birds, 285–289, 319
    evolution of homosexuality, 415
    mammals, 289–298
sociality hypothesis
    phylogenetic-based studies, 317, 318, 319–320
    sexual segregation hypothesis, 299
Socially Driven Model, 316–318, 319–320
sociobiology theories, 10
sociocultural variables, 84–85
socio-sexual behaviour, 335, 401–402
    adaptation, 415
    dominance hierarchies, 389–391
    plasticity in, 407–408
    ritualised, 388
socio-sexual communication, 392
socio-sexual functions
    behaviour in prison, 315–316
    changes over time, 417–418
    lock-in model, 173–177
socio-sexual mimicry, 389
socio-sexual variables, 361–363, 365, 366
somatostatin expressing neurons, 213
sororal birth order (SBO), 270–271
spatial abilities, and gender, 252–254
spatial sexual segregation
    bat roosting behaviour, 309
    confinement, 46–54
    ecologically driven model, 317

spatial social segregation, 283
spatio-temporal distribution, 298
Sperm Age hypothesis, 90, 234–235
sphinx gene, 68
spinal cord nuclei, 207, 218
spinal nucleus of bulbocavernosus (SNB), 218–219
splenium, of the CC, 214
spontaneous OAEs (SOAEs), 113
Sry gene, 75–76
stability of attachment, 248
stabilising selection, 98
startle response, 220
sterile castes paradox, 1
stochasticity hypothesis, 298–299
Streptococcus dysgalactiae infection, 376
stress
    attachment and, 247
    family dynamics, 280–281, 406
    future research, 421
    intersibling, 278
    laboratory animals, 315
    pre-/postnatal environment, 406, 416
    stress effects, 118–124, 124–128
    adrenarche and, 128–129
    bullying of sissy boys, 146, 146–147
    exclusive homosexuality, 418–419
    neuroendocrine events, 130
    stress-mediated mechanisms, 420
Struthioniformes, 42
studies of homosexual behaviour
    definitions in, 35–37
    ethical and political implications, 424–425
    methodology, 28–32
    study sampling, difficulties, 57–58
Sturnus vulgaris (European starling), 71
submissiveness, 239–240, 293, 395
subordinate role, 327, 328–330, 331
suprachiasmatic nucleus (SCN), 78–79, 212, 209
Suricata suricatta (meerkat), 331
Sus scrofa (pig), 136, 376
Symons Model, 256–258
synaptic plasticity, 67–68
synaptogenesis, 203–204
Syncerus caffer (African buffalo) 304–305
Synthetic Reproductive Skew Model of Homosexuality, 44, 332–336, 411–413
    children, numbers of, 374
    comparative tests, 336–375, 412

life-history data, 337–344, 346, 349–356, 358, 370–371
parent–offspring interaction, 406
path analysis of core variables, 45
predications and results, 435–436
social and life-history traits, 372
variables, 334
Synthetic Reproductive Skew Model of Homosexuality (simplified), 362–362, 363, 364
bird data, 364
mammal data 365, 366
syphilis infection, 378

Taeniopygia guttata (zebra finch), 109
Taeniopygia guttata castanotis (zebra finch), 109
Tamiasciurus douglasii (Douglas's squirrel), 283
temporal lobe, 220, 221, 222
temporal sexual segregation, 308
terminology, 3
territorial issues, plumage maturation, 152
testable hypotheses, candidate genes, 57–58
testicular feminisation mutation (*tfm*), 68–69
testosterone
female to male transsexuality, 194
fetal, 112, 113
functional metabolism, 122
high levels of in CAH, 164
oophorectomy and, 197
pregnant hyenas, 187
prenatal, 106, 106–107
sexual arousal, 201
sexual orientation, 193, 194
stress and, 119
testosterone biosynthesis
altered (ATB), 160–162
embryonic, 101
evolution of homosexuality, 406
prenatal environment, 406
testosterone levels
buller steer syndrome, 178
in bonobos, 169
sexuality in rams, 182
testosterone propionate (TP), 210
Thalassarche melanophrys (black-browed albatross),
thyroid medication, prenatal exposure, 133
tomboy girls, 145
tooth crown size, 113
touching, sexual or social behaviour, 37
Tourette syndrome (TS), 77

toy preference, sex-type, 116
Tragelaphus strepsiceros (kudu), 304
trait
persistence, 65
selection studies, 91
Transactional model, 326, 328
transsexuals
age for sexual activity, 129
aromatase deficiency, 163
BNST volume, 213
CYP17 SNP allelle, 76
INAH 3 volume, 211
orchiectomy, 196
prenatal DES exposure, 133
response to pheromones, 216
sibling birth order, 270
testosterone levels, 194
Tribolium castaneum (red flour beetle), 89–91
trisomy X, 81
Trobriander people, 313
trp 2, 71
tryptophan deficiency, 199
Turdoides squamiceps (Arabian babbler), 331
Turner syndrome, 420
Tursiops spp. (bottle nose dolphin), 135
twin studies, 85–89
and non-twin siblings, 87
digit measurements, 109
female masculinisation, 118
gender atypical behaviour, 141
grey matter quantity, 279
tooth crown size, 113
tyrosine hydroxylase (TH) activity, 123

Ulrichs, Karl Heinrich, 4
ungulates *see also* individual species
body size dimorphism, 301
juvenile sexual play, 136
phylogeny, 43
sexual segregation, 307
unidirectional plasticity, 99
United States of America (USA), juvenile sex play, 139, 140
Ursus americanus (American black bear), 135
Ursus arctos (grizzly bear), 308
Ursus arctos yesoensis (Hokkaido brown bear), 232

vasoactive intestinal peptide receptor type 2 (*VIPR2*) 79
vasopressin immunoreactive fibres, 75–76

ventromedial nucleus (VMN), 212
ventro-ventral mounting, 157, 393
verbal abilities, and gender, 255
verbal aggression, 249
VIPR2 (vasoactive intestinal peptide receptor type 2) 79
visual cortex, 115
visual mechanisms, 259
VMN (ventromedial nucleus), 212
voice pitch and formant, 254
vomeronasal organ
sexual dimorphism, 218
sexuality in rams, 182, 184
trp 2 cation channel, 71

W chromosome, 78
waist-to-hip ratio, 114
wartime, prenatal stress and, 124–125
Wernicke's cortex, 279
white matter distribution, 220
Wilsonia citrina (hooded warbler), 347
women
brain activation during orgasm, 230, 231
distribution of sexuality over time, 6
homosexuality rates in USA, 82–83
hormone levels and sexual orientation, 194
masculinised, 194
plasticity in sexual orientation, 9
unavailability of, 7
workplace networks, 314–315

X chromosome
homosexuality locus, 59–60
homosexuality trait involvement, 56–57
inactivation, DNA methylation, 79–80
q28 region, 73, 72–74, 76
xenobiotics, environmental, 131–132
X-linked allele, maternal fecundity, 91
X-linked genes, 61–62, 72–74
XYY trisomy, 81, 273, 420

Y chromosome, 56–57, 74–76
Y^{129} chromosome, 75–76
younger parents, 271–272
youthful-looking individuals, 250

Z chromosome, 78
Zalophus californianus wollebaeki (Galápagos sea lion) 309–310